T5-BSH-482

Multivariate Statistical Analysis

Multivariate Statistical Analysis

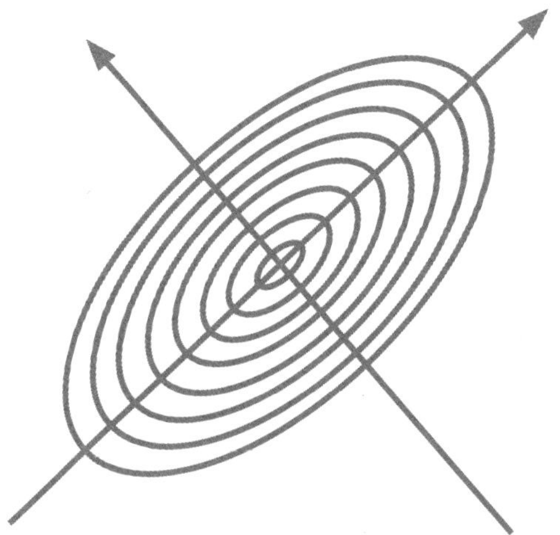

Parimal Mukhopadhyay

Indian Statistical Institute, India

NEW JERSEY • LONDON • SINGAPORE • BEIJING • SHANGHAI • HONG KONG • TAIPEI • CHENNAI

Published by

World Scientific Publishing Co. Pte. Ltd.
5 Toh Tuck Link, Singapore 596224
USA office: 27 Warren Street, Suite 401-402, Hackensack, NJ 07601
UK office: 57 Shelton Street, Covent Garden, London WC2H 9HE

British Library Cataloguing-in-Publication Data
A catalogue record for this book is available from the British Library.

MULTIVARIATE STATISTICAL ANALYSIS

ISBN-13 978-981-279-175-7
ISBN-10 981-279-175-2

Printed in Singapore by World Scientific Printers

To the memory of my parents,

Jnan Sagar Mukhopadhyay
and
Nihar Bala Mukhopadhyay

Preface

Scientific enquiries aimed at explaining different social and physical phenomena frequently require a large body of data for assessing the tenability (or otherwise) of different presumptions and hypotheses. Generally several measurements are done on each individual or object in one or more samples. Also, variables are often added or deleted from the study depending on their importance in explaining the phenomena of interest. In this treatise, we will be concerned with statistical methods designed to elicit information from such kinds of data sets. Because, measurements are taken simultaneously on a multitude of variables, the body of statistical methodology of analyzing such variables is called *multivariate analysis.*

Generally, the variables measured simultaneously on each sampling unit are correlated. A univariate analysis of each variable ignoring the other variables provides only partial information about an interwoven complex system. Clearly, the urge to understand the relationship between a large number of correlated variables makes the multivariate analysis a more difficult subject than the univariate analysis.

Many multivariate techniques are based on a probability model, known as *multivariate normal distribution* or a combination of these distributions. Although there are statistical problems of multiple measurements that cannot be based on the normal distribution and statistical techniques are available that are applicable to samples from other types of distributions, such as multinomial, we shall consider in this treatise statistical analysis based on normal distribution only.

Chapter 1 introduces the subject and gives an outline of various multivariate methods. The next chapter addresses techniques of organizing the multivariate data. This chapter also considers generalized measures of distance and the methods for dealing with the incomplete observations - the so-called missing data techniques.

Chapters 3 and 4 steer around establishing the properties of multivariate normal (MN) distribution and distributions of statistics arising out of sampling from MN distribution, like Wishart, T^2, Wilk's Λ. The next chapter deals with different tests of hypotheses relating to the parameters of the MN distribution.

Chapter 6 veers the discussion around the classical multivariate regression model and studies the impact of explanatory variables on prediction of response variables. Different types of simultaneous confidence intervals have been emphasized. The next chapter applies the theory to the multivariate analysis of variance and covariance models.

Chapters 8 and 9 cover two data-reduction techniques, Principal Component analysis and Factor analysis. The linear association between a set of predictor variables and a set of criteria measures, the so-called *Canonical Correlation* is the subject matter of the next chapter. The problem of classifying an observation into one of several groups or finding functions which sharply discriminate between several groups has been addressed in the concluding chapter.

As noted in Chapter 1, only a limited number of topics has been included in this text. The concepts have been presented in a lucid manner and elucidated through explanatory notes, remarks, examples and exercises. In presentation, I have tried to strike a balance between theory and practice, making it not unduly burdened with data, but showing their use by stepwise calculations.

The foundation of the mathematical exposition in this treatise lies in the theory of statistical inference and matrix algebra. It is expected that the reader will have gone through a founding course of statistical inference and distribution and vector and matrix algebra. For ready reference, a full chapter on matrix algebra has been added as Appendix A.

Different Statistical tables with explanatory notes at the beginning (and also within the text) have been included in Appendix B. In the *subject index*, items which occur frequently in the text have been shown with the page number, where it occurs first or where it is defined.

Theorems, Lemmas, Equations, Tables, Notes, Remarks are numbered sequentially within a section, Figures are numbered sequentially within each chapter. Corollaries are numbered sequentially to the respective theorem.

Standard format has been used in citing references in the text. For a journal article, the year alone suffices, for example, Hoerl and Kennard (1970). For books, in most cases, a page number has been included.

I am indebted to my readers of my other books in Statistics (nine in number) who have given me encouragement to attempt this new project. I will be thankful if this book is also well accepted by them.

My wife Manju, sons, Pabak and Pralay, daughters-in-law, Jayita and Shilpi have encouraged me throughout this project.

Parimal Mukhopadhyay
Kolkata
2nd March, 2008

Contents

Preface vii

Introduction 1 - 2

1: Preliminaries 3 - 8

1.1 Introduction 3
1.2 Outline of Multivariate Methods 5
1.2.1 Dependence methods 6
1.2.2 Interdependence analysis 7

2: Organization of Multivariate Data 9 - 41

2.1 Introduction 9
2.2 The Data Matrix 9
2.3 Summary Statistics 11
2.4 Linear Combination of Variables 18
2.5 Some Important Linear Transformations 20
2.6 Expectation and Covariance of Random Vectors 22
2.7 Expectation and Variance of Quadratic Forms of Random Variables 25
2.8 Measures of Distance 27
2.9 Missing Observations 31
2.10 Exercises and Complements 34
2.11 Appendix 41

3: Multivariate Normal Distributions and Related Distributions **43 - 100**

3.1 Introduction 43
3.2 Multivariate Normal Distribution 43
3.3 Transformation of Normal Data Matrix 58
3.4 Some Results on Quadratic Forms 60
3.5 Multivariate Central Limit Theorem 64
3.6 Maximum Likelihood Estimation of Parameters 68

3.6.1 Multivariate normal likelihood 69
3.6.2 Maximum likelihood estimation 70

3.7 Matrix Normal Distribution 77
3.8 The Multivariate t Distribution 77
3.9 The Dirichlet Distribution 78
3.10 Multivariate Skewness and Kurtosis 80
3.11 Examining the Assumption of Normality 81

3.11.1 Assessing normality for univariate marginal distribution 81
3.11.2 Evaluating bivariate normality 85
3.11.3 Evaluating multivariate normality 86

3.12 Transformations Making the Data Near Normal 88
3.13 Robust Estimation of Location and Scale Parameters 91
3.14 Exercises and Complements 94

4: Distributions Arising Out of the Multivariate Normal Distribution **101 - 127**

4.1 Introduction 101
4.2 Wishart Distribution 101

4.2.1 Properties of the Wishart distribution 103
4.2.2 Distribution of $\mathbf{X}'\mathbf{C}\mathbf{X}$ 108
4.2.3 Non-central Wishart distribution 111
4.2.4 Eigenvalues of a Wishart matrix 111

4.3 Hotelling's T^2 Distribution 111

4.3.1 Non-central Hotelling's T^2 distribution 117

4.4 Wilks' Statistic 118

4.5 Some Statistics Based on Eigenvalues of Wishart Matrices 121

4.5.1 Equivalence of statistics when $m_H = 1$ 124

4.6 Exercises and Complements 125

5: Testing of Hypotheses **129 - 181**

5.1 Introduction 129

5.2 Likelihood Ratio Test 130

5.3 Testing for a Single Population Mean 131

5.3.1 Union-intersection method 132

5.4 Equivalence Between Hotelling's T^2 and Likelihood Ratio Test 134

5.5 Confidence Region and Simultaneous Confidence Intervals 135

5.5.1 T^2-Simultaneous confidence intervals 137

5.5.2 Bonferroni's simultaneous confidence intervals 140

5.6 Large Sample Inference About μ 142

5.7 Comparing Mean Vectors of Two Populations 143

5.7.1 Union-intersection method 147

5.7.2 Behrens-Fisher problem 148

5.7.3 A large-sample test 149

5.7.4 Paired comparison 150

5.7.5 Testing that all components of μ are equal 152

5.7.6 Testing that two subvectors have equal means 155

5.8 Testing for the Variance of a Single Population 156

5.8.1 $H_0(\mathbf{\Sigma} = k\mathbf{\Sigma}_0)$, $\mathbf{\Sigma}_0$ is known, but k unknown 158

5.8.2 $H_0(\mathbf{\Sigma}_{12} = \mathbf{0})$ 158

5.8.3 Blockwise independence 160

5.8.4 Hypothesis of equicorrelation matrix 161

5.9 Test for Equality of Two Dispersion Matrices 162

5.10 Profile Analysis 163

5.11 Multi-Sample Hypotheses 168

5.11.1 Hypothesis $H_a(\mu_1 = \ldots = \mu_k = \mu)$ (unknown), given $\mathbf{\Sigma}_1 = \ldots = \mathbf{\Sigma}_k = \mathbf{\Sigma}$ (unknown) 169

5.11.2 Hypothesis $H_b(\mathbf{\Sigma}_1 = \ldots = \mathbf{\Sigma}_k)$ (test of homogeneity of covariances) 173

5.11.3 Hypothesis that k multinormal populations are equal 175
5.11.4 Union-intersection method 176

5.12 Some Further Tests for $H_0(\mu_1 = \ldots = \mu_k)$, Assuming Unequal Dispersion Matrices 177

5.12.1 Eaton's test 178
5.12.2 James's test 179

5.13 Exercises and Complements 179

6: Multivariate Regression Analysis 183 - 252

6.1 Introduction 183
6.2 Maximum Likelihood Estimators 185

6.2.1 Properties of maximum likelihood estimators 189

6.3 Least Square Estimators 190

6.3.1 Geometrical interpretation of $\hat{\mathbf{B}}$ 193
6.3.2 Properties of the least square estimators 193
6.3.3 Ordinary and generalized least square estimator of $\mathbf{B}$ 196

6.4 Forecasting an Observation 198
6.5 Likelihood Ratio Tests for Regression Parameters 201

6.5.1 The matrix $\mathbf{X}$ is of full rank 202
6.5.2 The matrix $\mathbf{X}$ not of full rank 207

6.6 Restricted Least Square Estimator of $\mathbf{B}$ 209

6.6.1 SSP matrices and their distributions 212
6.6.2 A generalized linear hypothesis 213

6.7 Some Examples 214
6.8 Testing H_0 by Union-Intersection Principle 221

6.8.1 Simultaneous confidence intervals 222

6.9 Mean Centered Model 224
6.10 Classical Linear Regression Model 226

6.10.1 Forecasting a new observation 228
6.10.2 Tests of hypotheses regarding regression parameters 229
6.10.3 Multiple correlation coefficient 230
6.10.4 Selection of variables 233

6.11 Proportion of Variation in $\mathbf{Y}$ Explained by the Multivariate Model 238

6.12 Subset Selection for the Multivariate Regression Model 239
6.12.1 Stepwise procedures 239
6.12.2 All possible subsets 241
6.13 Growth Curve Models 244
6.13.1 Examples 244
6.13.2 A general solution 249
6.14 Exercises and Complements 251

7: Multivariate Analysis of Variance and Covariance 253 - 296

7.1 Introduction 253
7.2 Multivariate One-Way Analysis of Variance 253
7.2.1 Univariate model 253
7.2.2 Multivariate one-way fixed effects model 256
7.2.3 Comparison among MANOVA tests 262
7.2.4 Testing a contrast 263
7.3 Multivariate Two-Way Fixed Effects Model 266
7.3.1 Univariate case: One observation per cell 266
7.3.2 Multivariate two-way fixed effects model (one observation per cell) 269
7.3.3 Univariate case: r observations per cell 270
7.3.4 Multivariate two-way fixed effects model (r observations per cell) 274
7.4 Analysis of Covariance 279
7.4.1 A univariate general linear model 279
7.4.2 Univariate analysis of covariance: Two-way model with one covariate 281
7.4.3 Multivariate analysis of covariance 285
7.5 A Conditional Hypothesis 293
7.6 Exercises and Complements 295

8: Principal Component Analysis 297 - 329

8.1 Introduction 297
8.2 Population Principal Components 298
8.3 Principal Components of a Multivariate Normal Distribution 306

8.4 Sample Principal Components 306
8.5 Principal Components of Covariance Matrices with Special Structures 312
8.6 Geometrical Interpretation of Sample Principal Components 316
8.7 Large Sample Properties of Sample Principal Components 318
8.7.1 Tests of hypotheses 321
8.8 Last Few Principal Components 325
8.8.1 Number of PC's to retain 327
8.9 Exercises and Complements 327

9: Factor Analysis **331 - 361**

9.1 Introduction 331
9.2 The Orthogonal Factor Model 332
9.2.1 Scale-invariance of factor model 335
9.2.2 Non-uniqueness of factor-loadings 336
9.2.3 Interpretation of factors 337
9.3 Estimation of Model-Parameters 338
9.3.1 Principal component method 339
9.3.2 Principal factor solution 343
9.3.3 The maximum likelihood solution 347
9.3.4 Other extraction procedures 347
9.3.5 Different types of rotation of factors 348
9.4 Factor Scores 351
9.4.1 Weighted least squares method 352
9.4.2 The regression method 353
9.5 Determining the Number of Factors 354
9.6 Comparison between Factor Analysis and Principal Component Analysis 357
9.7 Exercises and Complements 357

10: Canonical Correlation **363 - 388**

10.1 Introduction 363
10.2 Canonical Variables and Canonical Correlations 364

10.2.1 Correlation between canonical variables and original variables 372
10.2.2 Relation between canonical correlation and multiple correlation 374

10.3 The Sample Canonical Variables and the Sample Canonical Correlations 375

10.3.1 Sample correlation between original variables and sample canonical variables 378
10.3.2 Sample covariance in terms of canonical coefficients and canonical correlation 380
10.3.3 Approximating sample covariances by first r canonical correlations 380
10.3.4 The proportion of total sample variance explained by the canonical variables 383

10.4 Tests of Independence 384
10.5 Exercises and Complements 386

11: Classification and Discrimination 389 - 451

11.1 Introduction 389
11.2 Classification in Two Groups with Known Distributions and Known Parameters 390

11.2.1 Minimizing the total probability of misclassification (TPM) 391
11.2.2 The likelihood ratio method 395
11.2.3 Minimizing the expected cost of misclassification (ECM) 395
11.2.4 Maximizing the posterior probability 396
11.2.5 Minimax classification 397

11.3 Classification in Two Groups with Known Distributions but Unknown Parameters 398

11.3.1 General Methods 398
11.3.2 Normal Populations 399
11.3.3 Evaluating classification functions: Error rates 402
11.3.4 Some examples 410

11.4 Fisher's Discriminating Function for Separating Two Groups 413

11.4.1 Standardized discriminating functions 415
11.4.2 Tests of significance 417
11.4.3 Using Fisher's discriminating function for classification 417

11.5 Logistic Classification: $g = 2$ 418

11.5.1 Sampling designs 420

11.6 Classification in More than Two Groups 422

11.6.1 Minimum TPM rule 423
11.6.2 Minimum ECM rule 426
11.6.3 Logistic classification 426

11.7 Fisher's Method of Discrimination among $g \geq 2$ Populations 428

11.7.1 Fisher's discriminant procedure for classification 437
11.7.2 Relation between normal theory method and Fisher's discrimination method 437
11.7.3 Tests of significance 439
11.7.4 Contribution of variables in separation of groups 442

11.8 Selection of Variables 444
11.9 Exercises and Complements 446

Appendix A: Matrix Algebra **453 - 491**

Appendix B: Statistical Tables **493 - 524**

Bibliography **525 - 534**

Author Index **535 - 538**

Subject Index **539 - 549**

Introduction

The book germinated out of my lecture notes for an 'Advanced Statistical Methods' course offered by the Indian Statistical Institute, Kolkata. *Multivariate Statistical Analysis* deals with statistical methods for describing and analyzing multivariate data. Researchers in social, biological and physical sciences frequently collect data on a number of variables to seek answers to a number of queries. Analysis of data is an unavoidable path to find reply to these questions. Though the modern computer packages readily yield numerical answers to such complex statistical problems, the researcher should correctly formulate his hypotheses with desired sensitivity and correctly interpret the results. For this, a clear understanding of the reasoning on which the data analysis has been based is required.

Derivation of statistical results underlying the methods is often not an easy task and require sophisticated mathematical skill. On the other hand, being aware only of the techniques without really understanding the supporting results leave one nervous and insecure. The approach of this book is on the middle path. While it deals with the essential statistical results, leaving many details outside its scope, it tries to explain their applications through a number of worked out examples showing stepwise calculations. Also, it does not want the book to be unduly burdened with display of too many data sets.

The methodological tools of multivariate analysis are depicted in Chapters 5 through 11, but they cannot be apprehended vividly without much help from the materials in the introductory chapters 2 through 4. The line of exposition is same (similar) throughout, we have explained the results for the population models first and then the parallel results for the sample have been showcased. Results in Section 6.6, developed following Seber

(1984), depend heavily on the projection operator algebra, considered in Subsections A.17.2 and A.17.3 and may be skipped at the first reading.

It is expected that a student who has taken one or two courses on statistical distribution and inference and a middle level course on matrix algebra will be able to follow the contents of the book. The mathematics of the book is based on matrix algebra for which a full chapter has been devoted as an Appendix.

In writing this book, I have unhesitatingly taken the help of many publications, which have been cited at the proper places and all of which are gratefully acknowledged. I also recall my association with my students and colleagues in universities in different countries around the globe who have encouraged me to write this treatise.

Chapter 1

Preliminaries

1.1 Introduction

Scientific enquiries aimed at explaining different social and physical phenomena frequently require a large body of data for assessing the tenability (or otherwise) of different presumptions and hypotheses. Thus in an experiment to study the air-pollution we may want to examine the effect of speed of wind and solar radiation (predictor variables) on content of $N0_2$ and 0_3 in the air (response variables) simultaneously. For this, a large body of data on these variables must be gathered. Generally several measurements are made on each individual or object in one or more samples. Also variables are often added or deleted from the study depending on their importance in explaining the phenomena of interest. In this treatise we will be concerned with statistical methods designed to elicit information from such kinds of data sets. Because, measurements are taken simultaneously on a multitude of variables, the body of statistical methodology of analyzing such variables is called *multivariate analysis.*

Historically, the subject was initiated from applications in the behavioral and biological sciences. One of the pioneering scientists to treat such statistical problems was the geneticist Francis Galton in the second half of the nineteenth century. Another pioneering scientist was Karl Pearson who developed the theory for studying problems in anthropology, genetics and biology. Analysis of scores on mental tests led to the theory of *Factor Analysis.* Interest in multivariate methods have now spread to the numerous other fields of investigation, e.g., education, chemistry, physics, mining, linguistics, psychology, engineering, medicine, etc. With the present availability of (almost) infinite computing power, the earlier impediment of calculations with large body of data has also been eliminated.

It will be seen that in some cases all the variables are measured in the same scale. Ordinarily, however, the scales differ, as, for example, in measuring height, weight, blood-pressure, heart-rate, etc. In a few techniques such as *Profile Analysis* (Section 5.10), the variables must be commensurable, i.e., measured in the same scale. However, most multivariate techniques do not require this restriction.

Generally the variables measured simultaneously on each sampling unit are correlated. If these were not so, there would be little use in many multivariate techniques and univariate techniques would be sufficient for analysis and hypotheses-testing. Clearly the urge to understand the relationship between a large number of correlated variables makes multivariate analysis a more difficult subject than univariate analysis.

Again, for better understanding the structure underlying a large body of data we may want to reduce the number of variables without sacrificing any essential feature, i.e. to reduce the dimensionality of the problem. Thus one of the goals of multivariate analysis is simplification. In this sense, some multivariate techniques are exploratory in nature. They can often be treated as input to a larger analysis, and serve as means to an end rather than an end by themselves (e.g. *Principal Component Analysis*, Chapter 8).

On the other hand, we may be concerned with the testing of multivariate hypotheses. The problems in this case differ from those in testing a univariate hypothesis in many ways. For example, in multivariate case, the number of possible hypotheses are very large and also in many cases, there are several alternative test-statistics to choose from.

There are several advantages in testing p variables in a single multivariate test rather than in p separate univariate tests. These advantages include preserving the α-level, testing with greater power and determining the contribution of each variable in the presence of other variables. The multivariate tests make allowance for the intercorrelation among other variables, which account in part for the increase in power.

The above discussion imply that multivariate analysis is concerned generally with two aspects - descriptive statistics and inferential statistics. In the descriptive arena we often obtain variables which are optimal linear combinations of original variables. These linear functions may also be useful in further investigation of results from inferential procedures. In the inferential area, we test different multivariate hypotheses, obtain simultaneous confidence intervals for linear combinations of parameters, etc. Many multivariate techniques in this area are direct extensions of corresponding univariate procedures.

It will be subsequently clear that many multivariate techniques are based on a probability model, known as *multivariate normal distribution* or a combination of these distributions. This is primarily due to the fact that mathematical models based on normal distribution are suitable for a large number of cases where multiple measurements are addressed.

The univariate normal distribution arises frequently because the effect studied can be represented as the sum of a large number of independent small random effects (variables). The central limit theorem makes the distribution of such sums approximately univariate normal. Similarly the effects studied through multiple measurements are generally the sum of a large number of small independent effects (vectors of random variables). The multivariate central limit theorem makes the distributions of such sums approximately multivariate normal. Moreover the theory based on multivariate normal distribution is extremely well-developed.

Although there are statistical problems of multiple measurements that cannot be based on the normal distribution and statistical techniques are available that are applicable to samples from other types of distributions, such as multinomial, we shall consider in this treatise statistical analysis based on normal distribution only.

1.2 Outline of Multivariate Methods

The following is a brief outline of some of the multivariate methods.

If the interest focusses on the association between a set of response variables or *criteria measures* or dependent variables and a set of explanatory variables or predictor variables, the appropriate techniques can be designated as *dependence methods.* If the interest centers on mutual association among a set of variables (without any classification of variables), the methods may be designated as *interdependence methods.* The dependence methods seek to predict the values of the response variables or criteria measures on the basis of explanatory variables. The interdependence methods are less explanatory in nature and try to explain the phenomena or structure underlying the data, often through data-reduction.

1.2.1 Dependence methods

The dependence methods can be summarized as follows:

(1) *Multiple Regression*: When we are interested in studying the dependence of a set of response variables on a set of predictor variables with a view towards estimating or predicting the mean values of the dependent variables on the basis of the known values of the predictor variables, the appropriate technique is multiple regression (Chapter 5).

(2) *Discriminant analysis*: This is the problem of discrimination between two or several groups. Suppose we have a vector of p observed measurements, denoted as $\mathbf{X} = (X_1, \ldots, X_p)'$, which is known to belong to either of the two groups (have come from either of the two populations), G_1 and G_2. We want to find an optimum linear combination of $X_1, \ldots, X_p$ such that $\mathbf{X}$ will be placed in either of the groups in a justifiable manner. The problem is similar in case of more than two groups (Chapter 11).

(3) *Logit Analysis*: Logit analysis is appropriate when the single criterion variable is discrete and all predictor variables are also categorical.

(4) *Multivariate Analysis of Variance (MANOVA) and Covariance (MANCOV)*: When multiple response variables are available and the interest centers on studying the impact of different levels of design variables (explanatory variables) on the criteria measures, multivariate analysis of variance is the appropriate technique. When one of the experimental variables is a concomitant variable, measured in a ratio scale, the technique is called multivariate analysis of covariance (Chapter 7).

(5) *Canonical Correlation Analysis*: Here we seek to determine the linear association between a set of predictor variables and a set of criteria measures. We try to find two linear functions, one of the set of dependent variables and the other of the set of predictor variables, such that the product-moment correlation between these two linear functions is maximum. Similarly, we find other pairs of linear functions, which are orthogonal to the earlier functions, such that their product moment correlations are maximum (Chapter 10).

1.2.2 *Interdependence analysis*

If at least one of the variables is measured in ratio scale, then the following multivariate techniques can be applied.

(6) *Principal Component Analysis*: This is a data-reduction technique where the primary objective is to find a linear combination of the variables, which account for as much of the total variance as possible. This is the first principal component. The second principal component is that linear combination of the variables which is orthogonal to the first and has the maximum variance among all such linear combinations. We stop after a few steps when most (80 – 85% or more) of the total variance has been accounted for (Chapter 8).

(7) *Common Factor Analysis*: This is also a data reduction technique which seeks to identify the common factors that account for the total variance as much as possible. However, unlike principal components, these factors are unobservable random variables (Chapter 9).

(8) *Cluster Analysis*: This is also a data-reduction technique which seeks to identify a small number of groups such that the elements placed in the same group are more similar to one another than the elements belonging to different groups.

(9) *Metric Multidimensional Scaling*: The technique deals with the following problem. For a set of observed similarities (or distances) between each of $N(N-1)/2$ pairs of N objects, we want to draw a map of the objects in a reduced space such that the position of an object in the map reflects its observed (dis)similarity with the other objects as closely as possible.

If the variables have only nominal or ordinal scale properties, the following techniques are appropriate.

(10) *Loglinear Models*: This technique investigates the interrelationships between categorical variables that form a contingency table. The cell-probabilities are expressed in terms of main effects and interactions among the categorical variables.

(11) *Nonmetric Multidimensional Scaling*: This technique also allows the researcher to transform the perceived (dis) similarities between a set of objects into distances by placing the objects in a

multidimensional space. However, there is one difference. It is possible to arrange the N objects in a low-dimensional coordinate system using only the rank order of $N(N-1)/2$ original similarities (distances) and not their magnitude. When this ordinal information only is used to map the objects in a reduced space with the objective as in (9) above, the technique is known as nonmetric multidimensional scaling.

The above list is not exhaustive and some of the procedures (items (3), (8), (9), (10), (11)) have not been discussed in this treatise.

The foundation of the mathematical exposition in this treatise lies in the theory of statistical inference and matrix algebra. It is expected that the reader will have gone through a founding course of statistical inference and distribution and vector and matrix theory. For ready reference a full chapter on matrix algebra has been added as the Appendix A. The presentation throughout has been tried to be made very lucid. Explanatory notes, remarks, examples, exercises have been given at various stages to elucidate the concepts.

Chapter 2

Organization of Multivariate Data

2.1 Introduction

This chapter addresses techniques for organizing the multivariate data. Section 2.2 introduces the data matrix, Section 2.3 different summary statistics, like mean vector, covariance matrix, correlation matrix, generalized variance, total variance, measures of intercorrelation. The following section addresses linear combination of variables and the basic statistics of the transformed variables. Section 2.5 considers some important transformations like Mahalanobis transformation, Principal Component transformation. The next two sections deal with properties of random vectors like mean, variance, covariance and the quadratic forms in random variables. Generalized measures of distance are covered in Section 2.8. Lastly we consider methods for dealing with the incomplete observations, - the so-called missing data techniques.

2.2 The Data Matrix

Suppose we have a vector of p random variables $\mathbf{X} = (X_1, \ldots, X_p)'$. Observations n in number, $\mathbf{x}_1, \mathbf{x}_2, \ldots, \mathbf{x}_n$, are taken on the random vector $\mathbf{X}$ with x_{ij} being the value of the random variable X_j in the ith observation $(i = 1, \ldots, n; j = 1, \ldots, p)$. For concreteness, assume that there are n units and $\mathbf{X}$ is measured on each of these units. Then we have a $n \times p$ matrix $\mathbf{X}$ of observations

$$\mathbf{X} = \begin{bmatrix} x_{11} & x_{12} & \cdots & x_{1p} \\ x_{21} & x_{22} & \cdots & x_{2p} \\ \cdot & \cdot & \cdots & \cdot \\ x_{n1} & x_{n2} & \cdots & x_{np} \end{bmatrix} = \left[\mathbf{x}_{(1)}\ \mathbf{x}_{(2)} \cdots \mathbf{x}_{(p)}\right] = \begin{bmatrix} \mathbf{x}'_1 \\ \mathbf{x}'_2 \\ \cdot \\ \cdot \\ \mathbf{x}'_n \end{bmatrix}. \quad (2.2.1)$$

Here $\mathbf{x}_{(j)}$ denotes the vector of all observations on the variable X_j and $\mathbf{x}_i$ is the $p \times 1$ vector of all observations on the unit $i (i = 1, \ldots, n)$. The matrix $\mathbf{X}$ may be called the *Data Matrix*. Generally $n > p$. Note that we have used the same symbol $\mathbf{X}$ to denote a p-dimensional random vector as well as the data matrix. The actual meaning of the symbol will be clear from the context.

Geometrically, an observation $\mathbf{x}_i$ is a point in a p-dimensional space. Thus rows of $\mathbf{X}$ represent n points in a p-dimensional space.

The p variables $X_1, X_2, \ldots, X_p$ represent measurements on a single subject or unit. Since the variables arise from the same sampling unit, they are typically intercorrelated.

In some cases, the variables are measured in the same scale. Ordinarily, however, the scales differ, as, for example, in height, weight, blood pressure, heart rate. In most cases multivariate techniques do not require that the variables be measured in the same scale and many techniques are not affected by change of scales.

EXAMPLE 2.2.1: Suppose there are 5 factories and X_1 = value of output, X_2 = number of workers, X_3 = value of raw materials consumed. We have observations on each of these 5 factories, giving the data matrix

$$\mathbf{X} = \begin{bmatrix} 120 & 12 & 35 \\ 87 & 7 & 19 \\ 76 & 11 & 12 \\ 147 & 15 & 42 \\ 98 & 18 & 22 \end{bmatrix}.$$

Therefore, $\mathbf{x}_3 = (76\ 11\ 12)'$ is the array of values of these variables on the establishment labeled 3. Similarly, $\mathbf{x}_{(3)} = (35\ 19\ 12\ 42\ 22)'$ is the vector of values of raw materials consumed by these factories.

2.3 Summary Statistics

We now consider some statistics which measure some basic characteristics of the variables $X_1, \ldots, X_p$. We have

$$\bar{x}_j = \frac{1}{n}\sum_{i=1}^{n} x_{ij}, \text{ the sample mean of the variable } X_j (j = 1, \ldots, p);$$

$$s_{jj} = \frac{1}{n-1}\sum_{i=1}^{n}(x_{ij}-\bar{x}_j)^2 = s_j^2, \text{ the sample variance of } X_j;$$

$$s_{jk} = \frac{1}{n-1}\sum_{i=1}^{n}(x_{ij}-\bar{x}_j)(x_{ik}-\bar{x}_k), \text{ the sample covariance between } X_j$$

and X_k. Thus we have the sample mean vector

$$\bar{\mathbf{x}} = (\bar{x}_1, \ldots, \bar{x}_p)' = \frac{1}{n}(\mathbf{1}_n'\mathbf{X})' = \frac{1}{n}\mathbf{X}'\mathbf{1}_n, \tag{2.3.1}$$

where $\mathbf{1}_q = (1, \ldots, 1)'_{q\times 1}$; the sample covariance matrix

$$Cov(\mathbf{x}) = \mathbf{S} = ((s_{ij})) = \frac{1}{n-1}\sum_{i=1}^{n}(\mathbf{x}_i - \bar{\mathbf{x}})(\mathbf{x}_i - \bar{\mathbf{x}})' \tag{2.3.2}$$

$$= \frac{1}{n-1}[\sum_{i=1}^{n}\mathbf{x}_i\mathbf{x}_i' - n\bar{\mathbf{x}}\bar{\mathbf{x}}'] \tag{2.3.3}$$

$$= \frac{1}{n-1}[\mathbf{X}'\mathbf{X} - n\bar{\mathbf{x}}\bar{\mathbf{x}}'] \tag{2.3.4}$$

$$\begin{aligned} &= \frac{1}{n-1}[\mathbf{X}'\mathbf{X} - \frac{1}{n}\mathbf{X}'\mathbf{1}_n\mathbf{1}_n'\mathbf{X}] \\ &= \frac{1}{n-1}\mathbf{X}'(\mathbf{I} - \frac{1}{n}\mathbf{J})\mathbf{X} \end{aligned} \tag{2.3.5}$$

where $\mathbf{I}$ is an identity matrix of order n and $\mathbf{J} = \mathbf{1}_n\mathbf{1}_n'$ is a $n \times n$ matrix of 1's. Let

$$\mathbf{H} = \mathbf{I} - \frac{1}{n}\mathbf{J}. \tag{2.3.6}$$

Then

$$\mathbf{S} = \frac{1}{n-1}\mathbf{X}'\mathbf{H}\mathbf{X}. \tag{2.3.7}$$

We can use (2.3.7) to show that $\mathbf{S}$ is at least positive semi-definite. It can be shown by direct multiplication that $\mathbf{H} = \mathbf{H}'\mathbf{H} = \mathbf{H}^2$ so that $\mathbf{H}$ is a symmetric idempotent matrix. Hence

$$\mathbf{S} = \frac{1}{n-1}(\mathbf{H}\mathbf{X})'(\mathbf{H}\mathbf{X}) = \frac{1}{n-1}\mathbf{X}_c'\mathbf{X}_c \tag{2.3.8}$$

where $\mathbf{X}_c = \mathbf{H}\mathbf{X} = (\mathbf{I} - n^{-1}\mathbf{J})\mathbf{X}$ is the *centered form* of the data matrix $\mathbf{X}$. The matrix $\mathbf{H}$ may be called the centering matrix. Since $\mathbf{S}$ is proportional to $\mathbf{X}_c'\mathbf{X}_c$ it is at least positive semidefinite. If the variables are continuous and not linearly related and if $n-1>p$, then the probability is 1 that $\mathbf{S}$ is positive definite (Siotani, Hayakawa and Fujikoshi 1985, p 60).

It is easy to see that

$$\begin{aligned}\mathbf{X}_c &= \mathbf{X} - \tfrac{1}{n}\mathbf{J}\mathbf{X} \\ &= \mathbf{X} - \tfrac{1}{n}\mathbf{1}_n\mathbf{1}_n'\mathbf{X} \\ &= \mathbf{X} - \mathbf{1}_n\bar{\mathbf{x}}' \\ &= \begin{bmatrix} x_{11}-\bar{x}_1 & x_{12}-\bar{x}_2 & \dots & x_{1p}-\bar{x}_p \\ x_{21}-\bar{x}_1 & x_{22}-\bar{x}_2 & \dots & x_{2p}-\bar{x}_p \\ \cdot & \cdot & \dots & \cdot \\ x_{n1}-\bar{x}_1 & x_{n2}-\bar{x}_2 & \dots & x_{np}-\bar{x}_p \end{bmatrix}.\end{aligned} \tag{2.3.9}$$

We shall often write $\mathbf{d}_{(j)} = (x_{1j}-\bar{x}_j, x_{2j}-\bar{x}_j, \dots, x_{nj}-\bar{x}_j)'$, the deviation vector for the variable x_j. Thus

$$\mathbf{X}_c = (\mathbf{d}_{(1)}, \mathbf{d}_{(2)}, \dots, \mathbf{d}_{(p)}). \tag{2.3.10}$$

Sometimes we shall use

$$\mathbf{S}_n = \frac{1}{n}\mathbf{X}'\mathbf{H}\mathbf{X} = \frac{n-1}{n}\mathbf{S} \tag{2.3.11}$$

as the sample covariance matrix.

The matrix

$$\mathbf{M} = \sum_{i=1}^{n} \mathbf{x}_i\mathbf{x}_i' = \mathbf{X}'\mathbf{X} \tag{2.3.12}$$

is called the matrix of sum of squares (SS) and sum of products (SP), SSP matrix. The matrix

$$(n-1)\mathbf{S} = n\mathbf{S}_n = \mathbf{X}'\mathbf{H}\mathbf{X} \tag{2.3.13}$$

may be called the matrix of corrected SS and SP.

The quantity

$$r_{jk} = \frac{s_{jk}}{s_j s_k}$$

is the sample correlation between X_j and X_k. As is well-known, $|r_{jk}| \le 1$. The matrix $\mathbf{R} = ((r_{jk}))$ is the matrix of sample correlation coefficients.

If we write

$$\mathbf{D} = \text{ Diag }(s_1, \dots, s_p),$$

then it can be shown

$$\mathbf{R} = \mathbf{D}^{-1}\mathbf{S}\mathbf{D}^{-1} \tag{2.3.14}$$

or

$$\mathbf{S} = \mathbf{DRD}.$$

Therefore $\mathbf{R}$ is positive semi-definite, since $\mathbf{S}$ is so.

If the data matrix $\mathbf{X} = ((x_{ij}))$ is standardized to $\mathbf{Z} = ((z_{ij}))$ where $z_{ij} = (x_{ij} - \bar{x}_j)/s_j$, then the covariance matrix for the z's is equal to the correlation matrix for the x's:

$$\mathbf{S}_z = \frac{1}{n-1}\mathbf{Z}'\mathbf{Z} = \mathbf{R}. \tag{2.3.15}$$

The sample covariance matrix $\mathbf{S}$ contains p variances and $p(p-1)/2$ covariance terms. Sometimes, it is desirable to obtain a single numerical value for the variations expressed by $\mathbf{S}$. Two such measures are:

(i) Generalized Variance $|\mathbf{S}|$;
(ii) Total Variance, trace $\mathbf{S}$.

For both measures, large values indicate a high degree of scatter around $\bar{\mathbf{x}}$ and low values represent high concentration around $\bar{\mathbf{x}}$. The generalized variance plays an important role in maximum likelihood estimation and the total variance is useful in Principal Component Analysis (Chapter 8).

Likewise it is often advisable to obtain a single numerical measure of overall amount of intercorrelation among variables $X_1, \ldots, X_p$. If one variable in the set is of major interest, we can use the multiple correlation of this variable with the other $(p-1)$ variables. Similarly, we can define the canonical correlations between a set of variables and the set of remaining variables (Chapter 10).

Various authors have suggested different indices of intercorrelation among a set of variables. Hoerl and Kennard (1970) used $\sum_{j=1}^{p}(1/l_j)$ where $l_1, \ldots, l_p (l_1 \geq l_2 \geq \ldots \geq l_p)$ are the eigenvalues of $\mathbf{R}$. Mason et al. (1975) suggested the cndition number l_1/l_p as a measure of intercorrelation. Rencher and Pun (1980) used $\{\sum_{j=1}^{p}(1/l_j)\}/p$. Ferrer and Gleuber (1967) and Haitovsky (1969) suggested $|\mathbf{R}|$ as an index. Gleason and Staelin (1975) used

$$g = \sqrt{\frac{\sum_{j=1}^{p} l_j^2 - p}{p(p-1)}}$$

which ranges from 0 when the variables are independent ($\mathbf{R} = \mathbf{I}$) to 1 when the variables are perfectly correlated ($\mathbf{R} = \mathbf{1}_p\mathbf{1}_p'$). Some other indices of intercorrelation are due to Chatterjee and Price (1977). Heo (1987) considered six different measures of intercorrelation:

(a) $i_1 = (1 - \frac{l_p}{l_1})^{p+2}$,
(2) $i_2 = 1 - \frac{p}{\sum_{j=1}^{p} \frac{1}{l_j}}$,
(c) $i_3 = 1 - \sqrt{|\mathbf{R}|}$,
(d) $i_4 = (\frac{l_1}{p})^{3/2}$,
(e) $i_5 = (1 - \frac{l_p}{p})^5$,
(f) $i_6 = \sum_{j=1}^{p} \frac{1-1/r^{jj}}{p} = \sum_{j=1}^{p} \frac{R_j^2}{p}$ where r^{jj} is the jth diagonal element of $\mathbf{R}^{-1}$ and R_j^2 is the squared multiple correlation of X_j on the remaining variables.

All these indices vary between 0 and 1.

Geometrical Interpretation

The data matrix $\mathbf{X}$ consists of p vectors $\mathbf{x}_{(1)}, \ldots, \mathbf{x}_{(p)}$, each of order $n \times 1$. The vector $\mathbf{x}_{(j)}$ is a point in n-dimensional space R^n. The projection of $\mathbf{x}_{(j)}$ on $\mathbf{1}_n$ is (vide Section A.3),

$$\frac{\mathbf{x}_{(j)}'\mathbf{1}_n}{\mathbf{1}_n'\mathbf{1}_n}\mathbf{1}_n = \bar{x}_j\mathbf{1}_n.$$

The vector $\mathbf{d}_{(j)}$ is the orthogonal projection of $\mathbf{x}_{(j)}$ on $\mathbf{1}_n$. We have

$$(\bar{x}_j\mathbf{1}_n)'\mathbf{d}_{(j)} = 0.$$

Also,

$$||\mathbf{x}_{(j)}||^2 = ||\bar{x}_j\mathbf{1}||^2 + ||\mathbf{d}_{(j)}||^2.$$

The square of the length of the deviation vector $\mathbf{d}_{(j)}$ is $\mathbf{d}_{(j)}'\mathbf{d}_{(j)} = \sum_{i=1}^{n}(x_{ij} - \bar{x}_j)^2 = (n-1)s_{jj}$.

The inner product between deviation vectors $\mathbf{d}_{(j)}$ and $\mathbf{d}_{(k)}$ is $\mathbf{d}_{(j)}'\mathbf{d}_{(k)} = \sum_{i=1}^{n}(x_{ij} - \bar{x}_j)(x_{ik} - \bar{x}_k) = (n-1)s_{jk}$.

The cosine of the angle between $\mathbf{d}_{(j)}$ and $\mathbf{d}_{(k)}$ is

$$\frac{\mathbf{d}_{(j)}'\mathbf{d}_{(k)}}{||\mathbf{d}_{(j)}||||\mathbf{d}_{(k)}||} = \frac{s_{jk}}{\sqrt{s_{jj}s_{kk}}} = r_{jk}.$$

Note 2.3.1: The generalized variance can be geometrically interpreted as follows. If $\mathbf{x}'\mathbf{A}\mathbf{x}$ denotes the squared distance of a point $\mathbf{x} = (x_1, \ldots, x_p)'$ from the origin $\mathbf{0}$ in a p-dim. space, where $\mathbf{A}$ is a positive definite matrix (vide Section 2.8, for discussion on measures of statistical distance), then on substitution $\mathbf{A} = \mathbf{S}^{-1}$, the equation

$$(\mathbf{x} - \bar{\mathbf{x}})'\mathbf{S}^{-1}(\mathbf{x} - \bar{\mathbf{x}}) = c^2 \tag{2.3.16}$$

represents the locus of the points $\mathbf{x}$ which are a constant distance c away from the point $\bar{\mathbf{x}}$. Equation (2.3.15) represents a hyper-ellipsoid (an ellipse for $p = 2$). It can be shown that volume of this hyper-ellipsoid is proportional to $\sqrt{|\mathbf{S}|}$ i.e.

$$\text{Volume of } \{\mathbf{x} : (\mathbf{x} - \bar{\mathbf{x}})'\mathbf{S}^{-1}(\mathbf{x} - \bar{\mathbf{x}}) \leq c^2\} \propto \sqrt{|\mathbf{S}|}$$

or

$$(\text{Volume of ellipsoid})^{\,2} \propto |\mathbf{S}|. \tag{2.3.17}$$

Thus a large value of $|\mathbf{S}|$ denotes a high degree of spread of $\mathbf{x}$ values around $\bar{\mathbf{x}}$. However, as is obvious, the orientation of the ellipsoid is not reflected in the value of its volume and hence in the value of the generalized variance.

Let $(\lambda_i, \mathbf{e}_i)(i = 1, \ldots, p)$ be the eigenvalue, eigenvector pairs of $\mathbf{S}$. The axes of the ellipsoid (2.3.15) are in the direction of the eigenvectors $\mathbf{e}_i$ and their lengths are given by $c\sqrt{\lambda_i}(i = 1, \ldots, p)$. The generalized variance $|\mathbf{S}|$ does not convey this information.

Lemma 2.3.1 : The generalized variance $|\mathbf{S}| = 0$ if and only if the columns of the matrix of deviations $\mathbf{X} - \mathbf{1}\bar{\mathbf{x}}'$ are linearly dependent.

Proof. If the columns of $\mathbf{X} - \mathbf{1}\bar{\mathbf{x}}'$ are linearly dependent, there exist constants $a_1, \ldots, a_p$ not all zeroes, such that

$$a_1\mathbf{d}_{(1)} + \ldots + a_p\mathbf{d}_{(p)} = \mathbf{0}$$

where $\mathbf{d}_{(j)}$ denotes the jth column of the deviation matrix $\mathbf{X}_c = \mathbf{X} - \mathbf{1}\bar{\mathbf{x}}'$. Hence,

$$(\mathbf{X} - \mathbf{1}\bar{\mathbf{x}}')\mathbf{a} = \mathbf{0} \quad \text{for some } \mathbf{a} \neq \mathbf{0}.$$

Then

$$(n-1)\mathbf{S}\mathbf{a} = (\mathbf{X} - \mathbf{1}\bar{\mathbf{x}}')'(\mathbf{X} - \mathbf{1}\bar{\mathbf{x}}')\mathbf{a} = \mathbf{0}.$$

Therefore, $\mathbf{S}\mathbf{a} = \mathbf{0}$. Since $\mathbf{a} \neq \mathbf{0}$ this means $|\mathbf{S}| = 0$.

Alternatively, if $|\mathbf{S}| = 0$, then there exists some linear combination $\mathbf{Sa}$ of columns of $\mathbf{S}$ such that $\mathbf{Sa} = \mathbf{0}$. therefore,

$$\mathbf{a}'(n-1)\mathbf{Sa} = \mathbf{a}'(\mathbf{X} - \mathbf{1}\bar{\mathbf{x}}')'(\mathbf{X} - \mathbf{a}\bar{\mathbf{x}}')\mathbf{a} = \mathbf{0},$$

i.e.

$$[(\mathbf{X} - \mathbf{1}\bar{\mathbf{x}}')\mathbf{a}]'[(\mathbf{X} - \mathbf{1}\bar{\mathbf{x}}')\mathbf{a}] = \mathbf{0}$$

which implies

$$(\mathbf{X} - \mathbf{1}\bar{\mathbf{x}}')\mathbf{a} = \mathbf{0},$$

i.e. the columns of $(\mathbf{X} - \mathbf{1}\bar{\mathbf{x}}')$ are linearly dependent. □

Note that when the generalized variance $|\mathbf{S}| = 0$, the columns of the data matrix may not be linearly dependent. Only the columns of $\mathbf{X}_c$ should be necessarily dependent.

EXAMPLE 2.3.1:

For the data in Example 2.2.1, we have the following results.

$$\bar{x}_1 = 105.6, \quad \bar{x}_2 = 12.6, \quad \bar{x}_3 = 26.0,$$

$$s_{11} = 800.30, \quad s_{12} = 50.30, \quad s_{13} = 341.75,$$

$$s_{22} = 17.30, \quad s_{23} = 18.25, \quad s_{33} = 149.50,$$

$$\mathbf{M} = \mathbf{X}'\mathbf{X} = \begin{bmatrix} 58958 & 6854 & 15095 \\ 6854 & 863 & 1711 \\ 15095 & 1711 & 3978 \end{bmatrix},$$

$$\mathbf{S}^{-1} = \begin{bmatrix} .070474 & -.040124 & -.156201 \\ -.040124 & .089192 & .080834 \\ -.156201 & .080834 & .353890 \end{bmatrix}.$$

Eigenvalues of $\mathbf{S}$ are

$$\lambda_1 = 949.848, \ \lambda_2 = 15.009, \ \lambda_3 = 2.243.$$

The eigenvectors of $\mathbf{S}$ are

$$\begin{aligned} \mathbf{e}_1 &= (-.917690 \ -.057193 \ -.393159)' \\ \mathbf{e}_2 &= (-.045710 \ -.967814 \ .247482)' \\ \mathbf{e}_3 &= (.394659 \ -.245083 \ -.885539)'. \end{aligned}$$

The generalized variance is

$$|\mathbf{S}| = \lambda_1 \times \lambda_2 \times \lambda_3 = 31976.81054.$$

The total variance is

$$\text{trace } \mathbf{S} = \lambda_1 + \lambda_2 + \lambda_3 = s_{11} + s_{22} + s_{33} = 967.1.$$

The axes of the ellipsoid

$$(\mathbf{x} - \bar{\mathbf{x}})'\mathbf{S}^{-1}(\mathbf{x} - \bar{\mathbf{x}}) = c^2$$

are in the directions of $\mathbf{e}_1, \mathbf{e}_2, \mathbf{e}_3$ and their lengths are proportional to $\sqrt{949.848} = 30.82$, $\sqrt{15.009} = 3.87$, $\sqrt{2.243} = 1.5$ units respectively. However, the generalized variance $|\mathbf{S}|$ does not convey this information.

Now we consider different measures of intercorrelation. Here

$$\mathbf{R} = \begin{bmatrix} 1 & .427 & .988 \\ .427 & 1 & .359 \\ .988 & .359 & 1 \end{bmatrix}.$$

Its eigenvalues are

$$l_1 = 2.23777, \ l_2 = .753, \ l_3 = .00923.$$

Hence

$$|\mathbf{R}| = .01555, \ \frac{l_1}{l_3} = 242.445, \ \sum_j \frac{1}{l_j} = 110.117,$$

$$\frac{1}{p}\sum_j \frac{1}{l_j} = 36.706, \ g = .65507,$$

$$i_1 = .979546, \ i_2 = ..97276, \ i_3 = .875289,$$

$$i_4 = .64423, \ i_5 = .99471, \ i_6 = .77034.$$

EXAMPLE 2.3.2:

Let

$$\mathbf{S}_n(\mathbf{a}) = \frac{1}{n}\sum_{i=1}^{n}(\mathbf{x}_i - \mathbf{a})(\mathbf{x}_i - \mathbf{a})'$$

where $\mathbf{a} = (a_1, \ldots, a_p)'$ is a vector of constants. Then

$$n\mathbf{S}_n(\mathbf{a}) = \textstyle\sum_{i=1}^n \{(\mathbf{x}_i - \bar{\mathbf{x}}) + (\bar{\mathbf{x}} - \mathbf{a})\}\{(\mathbf{x}_i - \bar{\mathbf{x}}) + (\bar{\mathbf{x}} - \mathbf{a})\}'$$

$$= \textstyle\sum_{i=1}^n (\mathbf{x}_i - \bar{\mathbf{x}})(\mathbf{x}_i - \bar{\mathbf{x}})' + n(\bar{\mathbf{x}} - \mathbf{a})(\bar{\mathbf{x}} - \mathbf{a})'.$$

Hence,

$$\mathbf{S}_n(\mathbf{a}) = \mathbf{S}_n + (\bar{\mathbf{x}} - \mathbf{a})(\bar{\mathbf{x}} - \mathbf{a})'.$$

Again,

$$|\mathbf{S}_n(\mathbf{a})| = \begin{vmatrix} 1 & (\bar{\mathbf{x}} - \mathbf{a})' \\ -(\bar{\mathbf{x}} - \mathbf{a}) & \mathbf{S}_n \end{vmatrix}. \tag{2.3.18}$$

We now recall a result from the theory of determinants,

$$\begin{vmatrix} \mathbf{A}_{11} & \mathbf{A}_{12} \\ \mathbf{A}_{21} & \mathbf{A}_{22} \end{vmatrix} = |\mathbf{A}_{11}||\mathbf{A}_{22} - \mathbf{A}_{21}\mathbf{A}_{11}^{-1}\mathbf{A}_{12}| = |\mathbf{A}_{22}||\mathbf{A}_{11} - \mathbf{A}_{12}\mathbf{A}_{22}^{-1}\mathbf{A}_{21}|. \tag{2.3.19}$$

Putting in (2.3.19), $\mathbf{A}_{22} = \mathbf{S}_n$ and similarly the other quantities, we have, from (2.3.18),

$$|\mathbf{S}_n(\mathbf{a})| = |\mathbf{S}_n|[1 + (\bar{\mathbf{x}} - \mathbf{a})'\mathbf{S}_n^{-1}(\bar{\mathbf{x}} - \mathbf{a})]. \tag{2.3.20}$$

Therefore,

$$\text{Min.}_{\mathbf{a}}|\mathbf{S}_n(\mathbf{a})| = |\mathbf{S}_n|.$$

Again,

$$|\mathbf{S}_n| = l_1 \times l_2 \times \ldots \times l_p,$$

where $l_1, \ldots, l_p$ are the eigenvalues of $\mathbf{S}_n$. Also,

$$\text{trace } (\mathbf{S}_n) = \sum_{i=1}^{p} l_i.$$

Since,

$$\text{Arithmetic Mean } \geq \text{ Geometric Mean },$$

$$|\mathbf{S}_n| \leq (\frac{1}{n} \text{ trace } (\mathbf{S}_n))^n. \ \square$$

In the next section we shall consider linear functions of the elements of $\mathbf{X}$ and find their variances and covariances in terms of those of the original variables.

2.4 Linear Combination of Variables

Let

$$Y = a_1X_1 + \ldots + a_pX_p \tag{2.4.1}$$

be a linear combination of variables $X_1, \ldots, X_p$, where $a_1, a_2, \ldots, a_p$ are constants. Corresponding to the observation $\mathbf{x}_i$ of $\mathbf{X}(p \times 1)$, the value of Y is

$$y_i = a_1x_{i1} + \ldots + a_px_{ip} \quad (i = 1, \ldots, n).$$

The sample mean of Y is

$$\bar{y} = \frac{1}{n}\sum_{i=1}^{n} y_i = \frac{1}{n}\sum_{i=1}^{n} \mathbf{a}'\mathbf{x}_i = \mathbf{a}'\bar{\mathbf{x}} \tag{2.4.2}$$

where $\mathbf{a} = (a_1, \ldots, a_p)'$. The sample variance of y is

$$\begin{aligned} S_y^2 &= \tfrac{1}{n-1}\textstyle\sum_{i=1}^{n}(y_i - \bar{y}_i)^2 = \tfrac{1}{n-1}\sum_{i=1}^{n} \mathbf{a}'(\mathbf{x}_i - \bar{\mathbf{x}})(\mathbf{x}_i - \bar{\mathbf{x}})\mathbf{a}' \\ &= \mathbf{a}'\mathbf{S}\mathbf{a}. \end{aligned} \tag{2.4.3}$$

In general, let

$$\mathbf{Y}_{q\times 1} = \mathbf{A}_{q\times p}\mathbf{X}_{p\times 1} + \mathbf{B}_{q\times 1} \tag{2.4.4}$$

be a vector of q variables $(Y_1, \ldots, Y_q)$, obtained by a linear combination of the variables in $\mathbf{X} = (X_1, \ldots, X_p)$. Here $\mathbf{A}(q \times p)$ and $\mathbf{B}(q \times 1)$ are matrix and vector of constants. Corresponding to the value $\mathbf{x}_i$ of $\mathbf{X}$ we have the value $\mathbf{y}_i = \mathbf{A}\mathbf{x}_i + \mathbf{B}, i = 1, \ldots, n$. The sample mean vector of $\mathbf{y}_i$'s is

$$\bar{\mathbf{y}} = \mathbf{A}\bar{\mathbf{x}} + \mathbf{B}. \tag{2.4.5}$$

The sample covariance matrix of $\mathbf{y}_i$;s is

$$\begin{aligned} \mathbf{S_y} &= \tfrac{1}{n-1}\textstyle\sum_{i=1}^{n}(\mathbf{y}_i - \bar{\mathbf{y}})(\mathbf{y}_i - \bar{\mathbf{y}})' \\ &= \tfrac{1}{n-1}\mathbf{A}\textstyle\sum_{i=1}^{n}(\mathbf{x}_i - \bar{\mathbf{x}})(\mathbf{x}_i - \bar{\mathbf{x}})\mathbf{A}' \\ &= \mathbf{A}\mathbf{S}\mathbf{A}'. \end{aligned} \tag{2.4.6}$$

If $\mathbf{A}$ is non-singular, $\mathbf{S_y}$ is so and

$$\mathbf{S} = \mathbf{A}^{-1}\mathbf{S_y}(\mathbf{A}')^{-1}. \tag{2.4.7}$$

Let $\mathbf{Z}$ be a vector of q variables,

$$\mathbf{Z} = \mathbf{C}_{r\times p}\mathbf{X}_{p\times 1} + \mathbf{D}_{r\times 1} \tag{2.4.8}$$

where $\mathbf{C}$ is a matrix of constants and $\mathbf{D}$ a vector of constants. The sample covariance matrix between $\mathbf{y}_1, \ldots, \mathbf{y}_n$ and $\mathbf{z}_1, \ldots, \mathbf{z}_n$ is

$$\text{cov}\,(\mathbf{Y}, \mathbf{Z}) = \text{Cov}\,(\mathbf{AX}, \mathbf{CX}) = \mathbf{ASC}'. \tag{2.4.9}$$

Example 2.4.1:

Let

$$\begin{aligned} Y_1 &= X_1 + X_2 + X_3 \\ Y_2 &= X_1 + 2X_2 - X_3 \\ Y_3 &= X_1 + 7X_3. \end{aligned}$$

Here

$$\mathbf{A} = \begin{bmatrix} 1 & 1 & 1 \\ 1 & 2 & -3 \\ 1 & 0 & 7 \end{bmatrix}.$$

Suppose we have 5 observations $\mathbf{x}_1, \ldots, \mathbf{x}_5$ on $\mathbf{X}$ and the observations are given in the data matrix of Example 2.2.1. Hence,

$$\bar{\mathbf{y}} = \mathbf{A}\bar{\mathbf{x}} = (144.2\ 52.8\ 287.6)'.$$

Also

$$\mathbf{S_y} = \mathbf{ASA}' = \begin{bmatrix} 1787.7 & -164.5 & 4758.9 \\ -164.5 & 146.7 & -616.1 \\ 4758.9 & -616.1 & 12910.3 \end{bmatrix}.$$

2.5 Some Important Linear Transformations

We consider in this section some important linear transformations of the variables in $\mathbf{X}$.

(a) *Scaling Transformation* : A vector of variables $\mathbf{Y} = (Y_1, \ldots, Y_p)'$ is said to be obtained by scaling transformation of variables in $\mathbf{X} = (X_1, \ldots, X_p)'$ if

$$\mathbf{Y} = \mathbf{D}^{-1}(\mathbf{X} - \bar{\mathbf{x}}), \tag{2.5.1}$$

where $\mathbf{D} =$ Diag $(s_1, \ldots, s_p)$, defined in (2.3.14). Thus, sample mean of $\mathbf{y}_1, \ldots, \mathbf{y}_n$ is

$$\bar{\mathbf{y}} = \mathbf{0}. \tag{2.5.2}$$

Also, the sample covariance matrix of $\mathbf{y}_i$'s is

$$\text{Cov}\ (\mathbf{y}) = \mathbf{S_y} = \mathbf{D}^{-1}\mathbf{S}\mathbf{D}^{-\mathbf{1}} = \mathbf{R} \quad \text{(by (2.3.14))}. \tag{2.5.3}$$

Therefore, scaling transformation transforms the variables into variables whose sample mean is $\mathbf{0}$ and sample variance-covariance matrix equal to the observed correlation matrix of the original variables.

(b) *Mahalanobis Transformation*: A vector of variables $\mathbf{Y} = (Y_1, \ldots, Y_p)'$ is said to be obtained by *Mahalanobis transformation* of variables in $\mathbf{X} = (X_1, \ldots, X_p)'$ if

$$\mathbf{Y} = \mathbf{S}^{-1/2}(\mathbf{X} - \bar{\mathbf{x}}), \tag{2.5.4}$$

the matrix $\mathbf{S}^{-1/2}$ being the symmetric square root of $\mathbf{S}^{-1}$ (vide A.12.6). Then

$$\bar{\mathbf{y}} = \mathbf{0}, \qquad \mathbf{S_y} = \mathbf{S^{-1/2}SS^{-1/2}} = \mathbf{I}. \tag{2.5.5}$$

Thus the covariance matrix of the transformed observations is an identity matrix.

(c) *Principal Component Transformation:* By Spectral Decomposition Theorem (Theorem A.12.1) we have,

$$\mathbf{S} = \mathbf{GLG}'$$

where $\mathbf{L} = \text{Diag}\,(l_1, l_2, \ldots, l_p)$ is the diagonal matrix of eigenvalues of $\mathbf{S}, (l_1 \geq l_2 \geq \ldots \geq l_p \geq 0), \mathbf{G}$ is an orthogonal matrix whose columns are the eigenvectors corresponding to these eigenvalues. Consider the transformation

$$\mathbf{Y} = \mathbf{G}'(\mathbf{X} - \bar{\mathbf{x}}). \tag{2.5.6}$$

Then

$$\bar{\mathbf{y}} = \mathbf{0}, \quad \mathbf{S_y} = \mathbf{G'SG} = \mathbf{L}. \tag{2.5.7}$$

Thus the sample variances of the transformed variables are the eigenvalues of $\mathbf{S}$ and the sample covariances are zeroes. (The principal components analysis is the subject matter of Chapter 8.)

EXAMPLE 2.5.1: For the data in Example 2.2.1,

$$\mathbf{D} = \text{Diag}\,(28.29 \quad 4.16 \quad 12.23).$$

Hence, scaling transformation is

$$Y_1 = \frac{X_1 - 105.6}{28.29}, \; Y_2 = \frac{X_2 - 12.6}{4.16}, \; Y_3 = \frac{X_3 - 26}{12.23}.$$

Again

$$\begin{aligned}\mathbf{S}^{-1/2} &= \mathbf{GL}^{-1/2}\mathbf{G}' \\ &= \begin{bmatrix} .13184 & -.05145 & -.22452 \\ -.05145 & .28198 & .08378 \\ -.22452 & .08378 & .54431 \end{bmatrix},\end{aligned}$$

where the matrices $\mathbf{G}, \mathbf{L}$ are given from Exercise 2.3.1. Mahalanobis transformation is $\mathbf{Y} = \mathbf{S}^{-1/2}(\mathbf{X} - \bar{\mathbf{x}})$.

The principal transformation is given by $\mathbf{Y} = \mathbf{G}'(\mathbf{X} - \bar{\mathbf{x}})$.

2.6 Expectation and Covariance of Random Vectors

We have so far considered the summary statistics, like sample mean vector, sample covariance matrix based on observations $x_{ij}, i = 1, \ldots, n; j = 1, \ldots, p$. We shall now look at the corresponding (theoretical) random variables.

We note that all the elements x_{ij} in the data matrix $\mathbf{X}$ are actually realizations of random variables X_{ij}. Thus we have a random matrix $\mathbf{X} = ((\mathbf{X}_{ij}))$. We are using small case letters for observed values and the corresponding capital case letters for the random variables. We are using the same symbol $\mathbf{X}$ to denote the vector of variables as well as the data matrix. The vector $\mathbf{X}_i(p \times 1) = (X_{i1}, X_{i2}, \ldots, X_{ip})'$ is the vector of random variables corresponding to the ith unit $(i = 1, \ldots, n)$. The elements of $\mathbf{X}_i$ are not generally independent. However, the observations from one unit to the other are often independent. In fact it is generally assumed that $\mathbf{X}_1, \ldots, \mathbf{X}_p$ are independently and identically distributed (*iid*) with the same mean and the same dispersion matrix. However, in some special situations, like when the multivariate data are obtained from a time-series, the observations on consecutive points of time, $\mathbf{X}_1, \mathbf{X}_2, \ldots$ will be correlated.

When the vectors $\mathbf{X}_1, \ldots \mathbf{X}_n$ are independently distributed, the density function of the random matrix $\mathbf{X}$ will be

$$f_{\mathbf{X}}(\mathbf{x}) = \Pi_i f_{\mathbf{X}_i}(\mathbf{x}_i) = \Pi_{i=1}^n f_i(\mathbf{x}_i),$$

where $f_i(\mathbf{x}_i) = f_{\mathbf{X}_i}(\mathbf{x}_i)$ is the density function of $\mathbf{X}_i$, and $\mathbf{x} = (\mathbf{x}_1, \ldots, \mathbf{x}_n)'$.

We shall first consider the concept of expectation and variance of random matrices and vectors.

Let $\mathbf{Z} = ((Z_{ij}))$ be a $m \times n$ matrix of random variables.

DEFINITION 2.6.1 : The expectation of a random matrix is the matrix of expectation of the corresponding random variables. Thus, $E(\mathbf{Z}) = ((EZ_{ij}))$. The following results are stated without proof.

Theorem 2.6.1: If $\mathbf{A}, \mathbf{B}, \mathbf{C}$ are matrices of constants of order $l \times m, n \times p$ and $l \times p$ respectively and $\mathbf{X}$ is a random matrix of order $m \times n$, then

$$E[\mathbf{AXB} + \mathbf{C}] = \mathbf{A}E(\mathbf{X})\mathbf{B} + \mathbf{C}.$$

Theorem 2.6.2 : If $\mathbf{X}(m \times 1), \mathbf{Y}(n \times 1)$ are vectors of random variables and $\mathbf{A}(q \times m), \mathbf{B}(q \times n)$ are matrices of constants, then

$$E[\mathbf{AX} + \mathbf{BY}] = \mathbf{A}E(\mathbf{X}) + \mathbf{B}E(\mathbf{Y}).$$

In particular, for scalers a, b, $E[a\mathbf{X} + b\mathbf{Y}] = aE(\mathbf{X}) + bE(\mathbf{Y})$.

DEFINITION 2.6.2: If $\mathbf{X}, \mathbf{Y}$ are, respectively, $m \times 1$ and $n \times 1$ vectors of random variables, then the matrix of covariance between $\mathbf{X}, \mathbf{Y}$ is

$$\text{Cov}(\mathbf{X}, \mathbf{Y}) = C(\mathbf{X}, \mathbf{Y}) = ((\text{Cov}(X_i, Y_j)))$$

where

$$\text{Cov }(X_i, Y_j) = E(X_i - E(X_i))(EY_j - E(Y_j))$$

is the population covariance of X_i and Y_j.

Theorem 2.6.3 :

$$C(\mathbf{X}, \mathbf{Y}) = E(\mathbf{X} - E(\mathbf{X}))(\mathbf{Y} - E(\mathbf{Y}))'.$$

DEFINITION 2.6.3 : When $\mathbf{X} = \mathbf{Y}, C(\mathbf{X}, \mathbf{X})$ is the dispersion matrix of $\mathbf{X}$, written as Var $(\mathbf{X})$ or $V(\mathbf{X})$ or $D(\mathbf{X})$.

$$\begin{aligned} D(\mathbf{X}) &= ((\text{ Cov}(X_i, X_j))) \\ &= \begin{bmatrix} V(X_1) & \text{Cov}(X_1, X_2) & \dots & \text{Cov}(X_1, X_m) \\ . & . & \dots & . \\ \text{Cov}(X_m, X_1) & \text{Cov}(X_m, X_2) & \dots & V(X_m) \end{bmatrix}. \end{aligned}$$

Here $V(X_i) = \text{Cov }(X_i, X_i)$ is the variance $V(X_i)$. Clearly, $D(\mathbf{X})$ is a symmetric matrix.

Lemma 2.6.1 : If $\mathbf{a}$ is a vector of constants and $\mathbf{X}$ a vector of random variables,

(a) $D(\mathbf{X} - \mathbf{a}) = D(\mathbf{X})$.
(b) $E[(\mathbf{X} - \mathbf{a})(\mathbf{X} - \mathbf{a})'] = D(\mathbf{X}) + (E(\mathbf{X}) - \mathbf{a})(E(\mathbf{X}) - \mathbf{a})'$.

Theorem 2.6.4 : If $\mathbf{X}, \mathbf{Y}$ are random vectors of orders $m \times 1, n \times 1$ respectively and $\mathbf{A}, \mathbf{B}$ are matrices of constants of orders $q \times m, p \times n$, respectively, then

$$C(\mathbf{AX}, \mathbf{BY}) = \mathbf{A}C(\mathbf{X}, \mathbf{Y})\mathbf{B}'.$$

Corollary 2.6.4.1 : $D(\mathbf{AX}) = \mathbf{A}D(\mathbf{X})\mathbf{A}'$.

Theorem 2.6.5 : If $\mathbf{X}, \mathbf{Y}, \mathbf{U}, \mathbf{V}$ are each $n \times 1$ random vectors, then for all real numbers a, b, c, d,

$$C(a\mathbf{X}+b\mathbf{Y}, c\mathbf{U}+d\mathbf{V}) = acC(\mathbf{X}, \mathbf{U})+adC(\mathbf{X}, \mathbf{V})+bcC(\mathbf{Y}, \mathbf{U})+bdC(\mathbf{Y}, \mathbf{V}).$$

Corollary 2.6.5.1 : $D(a\mathbf{X} + b\mathbf{Y}) = a^2D(\mathbf{X}) + b^2D(\mathbf{Y}) + 2abC(\mathbf{X}, \mathbf{Y})$.

Theorem 2.6.6 : If $\mathbf{X}$ is a vector of random variables such that no element of $\mathbf{X}$ is a linear combination of the remaining elements, i.e., there exists no vector of constants $\mathbf{a}$ and a scalar b such that Prob $(\mathbf{a}'\mathbf{X} = b) = 1$, then $D(\mathbf{X})$ is a positive definite matrix.

Proof. For any vector of constants $\mathbf{c}$,

$$V(\mathbf{c}'\mathbf{X}) = \mathbf{c}'D(\mathbf{X})\mathbf{c} \geq 0,$$

equality holds *iff* $P(\mathbf{c}'\mathbf{X} = \text{ constant }) = 1$ or $\mathbf{c} = \mathbf{0}$. The former condition however cannot hold.

EXAMPLE 2.6.1 : Let $\mathbf{X}, \mathbf{Y}$ be $m \times 1$ and $n \times 1$ vectors of random variables and let $D(\mathbf{X})$ be a *p.d.* matrix. Then there exists a $n \times m$ matrix $\mathbf{A}$ for which

$$C(\mathbf{X}, \mathbf{Y} - \mathbf{AX}) = \mathbf{0}.$$

Proof.

$$\begin{aligned} C(\mathbf{X}, \mathbf{Y} - \mathbf{AX}) &= C(\mathbf{X}, \mathbf{Y}) - C(\mathbf{X}, \mathbf{AX}) = C(\mathbf{X}, \mathbf{Y}) - D(\mathbf{X})\mathbf{A}' = \mathbf{0} \\ &\Rightarrow \mathbf{A} = C(\mathbf{Y}, \mathbf{X})(D(\mathbf{X}))^{-1}, \end{aligned}$$

since $D(\mathbf{X})$ is non-singular.

Note 2.6.1 : A typical element of $D(\mathbf{X})$ will often be denoted as $\sigma_{ij} = \rho_{ij}\sigma_i\sigma_j$ where ρ_{ij} is the correlation coefficient between X_i and X_j. In general, $-1 \leq \rho_{ij} \leq 1$, though in certain cases, it is restricted to smaller interval.

For example, suppose $D(\mathbf{X})$ has the *intraclass correlation structure*

$$D(\mathbf{X}) = \sigma^2 \begin{bmatrix} 1 & \rho & \dots & \rho \\ \rho & 1 & \dots & \rho \\ . & . & \dots & . \\ \rho & \rho & \dots & 1 \end{bmatrix} = \sigma^2[(1-\rho)\mathbf{I} + \rho\mathbf{1}_p\mathbf{1}_p']$$

where $\mathbf{I}$ is a $p \times p$ identity matrix and $\mathbf{1}_p = (1, \dots 1)'$. If $D(\mathbf{X})$ is to be a p.d. covariance matrix, one necessary (but not sufficient) condition is that $|D(\mathbf{X})| > 0$.

Now,

$$|D(\mathbf{X})| = \sigma^{2p}(1-\rho)^p[1 + \frac{p\rho}{1-\rho}].$$

Hence, for $|D(\mathbf{X})| > 0$ it is necessary that $\rho > -\frac{1}{p-1}$. $\square$

A correlation matrix is a matrix of correlation coefficients, often denoted as ρ. Thus

$$\rho(\mathbf{X}) = \begin{bmatrix} 1 & \rho_{12} & \dots & \rho_{1p} \\ . & . & \dots & . \\ \rho_{p1} & \rho_{p2} & \dots & 1 \end{bmatrix}.$$

Clearly,

$$D(\mathbf{X}) = \sqrt{\sigma}\rho(\mathbf{X})\sqrt{\sigma},$$

where $\sigma =$ Diag $(\sigma_{11}, \dots, \sigma_{pp})$.

EXAMPLE 2.6.2: Let $\mathbf{X}_1, \dots, \mathbf{X}_n$ be a random sample from an infinite population with mean μ and variance $\mathbf{\Sigma}$. Then $E(\bar{\mathbf{X}}) = \mu$. Also, $V(\bar{\mathbf{X}}) = n^{-2}[V(\mathbf{X}_1) + \dots + V(\mathbf{X}_n)]$, since C $(\mathbf{X}_i, \mathbf{X}_j) = \mathbf{0}$, $(i \neq j)$, because of independence. Hence $V(\bar{\mathbf{X}}) = \mathbf{\Sigma}/n$.

Note 2.6.2: The sample correlation coefficient r_{ij} is a biased estimate of the population correlation coefficient ρ_{ij}, except when $\rho_{ij} = 0$. Olkin and Pratt (1958) proposed an adjustment of r_{ij} to make it nearly unbiased,

$$r^*_{ij} = r_{ij}[1 + \frac{1 - r^2_{ij}}{2(n-3)}] \tag{2.6.1}$$

and suggested that this bias correction is accurate to within 0.1 for $n \geq 8$ and within .001 for $n \geq 18$. Anderson (1984, p.119) gave a series representation of an unbiased estimator of ρ; the first two terms of the series are similar to equation (2.6.1).

2.7 Expectation and Variance of Quadratic Forms of Random Variables

We state below a few results.

Theorem 2.7.1 : Let $\mathbf{X} = (X_1, \dots, X_p)'$ be a random vector and $\mathbf{A} = ((a_{ij}))$ be a $p \times p$ symmetric matrix of constants. Let $E(\mathbf{X}) = \mu$ and $D(\mathbf{X}) = \mathbf{\Sigma} = ((\sigma_{ij}))$. The

$$E(\mathbf{X}'\mathbf{A}\mathbf{X}) = tr(\mathbf{A}\mathbf{\Sigma}) + \mu'\mathbf{A}\mu$$

where $tr(\mathbf{B})$ denotes the trace of the matrix $\mathbf{B}$.

The straight-line distance between two arbitrary points $P(x_1, \ldots x_p)$ and $Q(y_1, \ldots, y_p)$is given by

$$d(P, Q) = \sqrt{(x_1 - y_1)^2 + \ldots + (x_p - y_p)^2}. \tag{2.8.2}$$

Similarly, we can define the Euclidean angle θ between $\mathbf{x}$ and $\mathbf{y}$ subtended at the origin as

$$\cos\theta = \frac{\mathbf{x}'\mathbf{y}}{||\mathbf{x}||||\mathbf{y}||}, \tag{2.8.3}$$

where $||\mathbf{x}|| = \sqrt{\sum_{i=1}^{p} x_i^2}$. However, the Euclidean distance is not suitable for statistical variables. This is so, because statistical variables are subject to sampling fluctuations, which determine their likely positions. It will, therefore, be erroneous to consider Euclidean distance between two observations as measuring distance between two variables, and thereby ignoring some measures of their sampling fluctuations, as for example, variances and covariances.

When the co-ordinates represent independent variables that are subject to random fluctuations of different magnitudes, it is desirable first to standardize the variables and then to take the Euclidean distance between standardized co-ordinates.

Thus, if X_1, X_2 are two independent variables with sample variances s_{11}, s_{22}, respectively (obtained from n observations on both of them), then the standardized variables (co-ordinates) are taken as

$$X_1^* = \frac{X_1}{\sqrt{s_{11}}}, \quad X_2^* = \frac{X_2}{\sqrt{s_{22}}}.$$

The statistical distance of the point $P(x_1, x_2)$ from the origin is taken as the Euclidean distance between the origin and the standardized observations (x_1^*, x_2^*),

$$d_s(O, P) = \sqrt{x_1 *^2 + x_2 *^2} = \sqrt{\frac{x_1^2}{s_{11}} + \frac{x_2^2}{s_{22}}}.$$

For p independent variables $X_1, \ldots, X_p$ with s_{ii} as the variance for X_i, the distance of the point $P(x_1, \ldots, x_p)$ from the origin O$(0, \ldots, 0)$ is

$$\sqrt{\frac{x_1^2}{s_{11}} + \frac{x_2^2}{s_{22}} + \ldots + \frac{x_p^2}{s_{pp}}}.$$

If $(x_1, \ldots, x_p), (y_1, \ldots, y_p)$ are two observations out of a random sample of n observations drawn from a p-variate population, with sample covariance matrix $\mathbf{s} = ((s_{ij}))$, the distance between two points $P(x_1, \ldots, x_p)$ and $Q(y_1, \ldots, y_p)$ is given by

$$d_s(P, Q) = \sqrt{\frac{(x_1 - y_1)^2}{s_{11}} + \frac{(x_2 - y_2)^2}{s_{22}} + \ldots + \frac{(x_p - y_p)^2}{s_{pp}}}. \qquad (2.8.4)$$

If $s_{11} = s_{22} = \ldots = s_{pp}$, then $d_s(P.Q)$ reduces to $d(P, Q)$ given in (2.8.2).

In general, the squared statistical distance between the points $P(\mathbf{x})$ and $Q(\mathbf{y})$, where the variables $X_1, \ldots, X_p$ are not necessarily uncorrelated, is of the form

$$d_s^2(P, Q) = (\mathbf{x} - \mathbf{y})'\mathbf{A}(\mathbf{x} - \mathbf{y}), \qquad (2.8.5)$$

where $\mathbf{A}$ is a symmetric positive definite matrix.

Contours of constant distance computed from (2.8.5) are hyper-ellipsoids. Its volume and orientation are determined by the eigenvalues and eigenvectors of $\mathbf{A}^{-1}$. For example, suppose $p = 2$. Then the points $\mathbf{x} = (x_1, x_2)$ of constant distance c from the origin have the locus

$$\mathbf{x}'\mathbf{A}\mathbf{x} = a_{11}x_1^2 + 2a_{12}x_1x_2 + a_{22}x_2^2 = c^2. \qquad (2.8.6)$$

By the spectral decomposition theorem (Theorem A.12.1), if $\lambda_i(> 0), \mathbf{e}_i(i = 1, 2)$ are the eigenvalue-eigenvector pairs of $\mathbf{A}$,

$$\mathbf{A} = \lambda_1\mathbf{e}_1\mathbf{e}_1' + \lambda_2\mathbf{e}_2\mathbf{e}_2'$$

so that, from (2.8.6),

$$\mathbf{x}'\mathbf{A}\mathbf{x} = \lambda_1(\mathbf{x}'\mathbf{e}_1)^2 + \lambda_2(\mathbf{x}_2'\mathbf{e}_2)^2 = c^2. \qquad (2.8.7)$$

The equation (2.8.7) represents an ellipse in co-ordinates $y_1 = \mathbf{x}'\mathbf{e}_1$ and $\mathbf{y}_2 = \mathbf{x}'\mathbf{e}_2$. We see that the points

$$\mathbf{x} = \frac{c\mathbf{e}_i}{\sqrt{\lambda}}, \quad i = 1, 2$$

satisfy (2.8.7). Thus, the points at distance c from the origin lie on an ellipse whose axes are given by the eigenvectors of $\mathbf{A}$ with lengths proportional to the reciprocal of the square root of its eigenvalues. The constant of proportionality is c.

We note that any distance measure must satisfy the following properties:

(a) $d(P, Q) = d(Q, P)$;

(b) $d(P, Q) > 0$, if $P \neq Q$;
(c) $d(P, Q) = 0$ if $P = Q$;
(d) $d(P, Q) \leq d(P, R) + d(R, Q)$ for any intermediate point R (triangle inequality).

Mahalanobis distance

If $\mathbf{x}_1, \ldots, \mathbf{x}_n$ constitute a random sample from a population having sample covariance matrix $\mathbf{S}$, then Mahalanobis distance between two points $\mathbf{x}_r, \mathbf{x}_s$ is given by

$$D_M = D(\mathbf{x}_r, \mathbf{x}_s) = \{(\mathbf{x}_r - \mathbf{x}_s)'\mathbf{S}^{-1}(\mathbf{x}_r - \mathbf{x}_s)\}^{1/2}. \tag{2.8.8}$$

Similarly, we can define the Mahalanobis angle θ between $\mathbf{x}_r$ and $\mathbf{x}_s$, subtended at the origin by

$$\cos\theta = \frac{\mathbf{x}_r'\mathbf{S}^{-1}\mathbf{x}_s}{D(\mathbf{x}_r, \mathbf{0})D(\mathbf{x}_s, \mathbf{0})}. \tag{2.8.9}$$

The distance between $\mathbf{x}$ and $\bar{\mathbf{x}}$ is defined as

$$D(\mathbf{x}, \bar{\mathbf{x}}) = \{(\mathbf{x} - \bar{\mathbf{x}})'\mathbf{S}^{-1}(\mathbf{x} - \bar{\mathbf{x}})\}^{1/2}. \tag{2.8.10}$$

The population analogue of (2.8.10) may be defined as follows. If $\mathbf{X}$ is a vector of random variables with mean μ and dispersion matrix $\mathbf{\Sigma}$, we define the Mahalanobis distance between $\mathbf{X}$ and μ as

$$\Delta(\mathbf{X}, \mu) = \{(\mathbf{X} - \mu)'\mathbf{\Sigma}^{-1}(\mathbf{X} - \mu)\}^{1/2}. \tag{2.8.11}$$

The population analogue of (2.8.8) is obtained by replacing $\mathbf{S}$ by $\mathbf{\Sigma}$ in the same expresion.

In the case of two populations with means μ_1 and μ_2, and common dispersion matrix $\mathbf{\Sigma}$, we can define distance between μ_1, μ_2 as

$$\Delta(\mu_1, \mu_2) = \{(\mu_1 - \mu_2)'\mathbf{\Sigma}^{-1}(\mu_1 - \mu_2)\}^{1/2}. \tag{2.8.12}$$

Similarly, if we have a sample of size n_i, with sample mean $\mathbf{x}_i$ and sample variance $\mathbf{S}_i$, from a population with mean μ_i and dispersion matrix $\mathbf{\Sigma}$, $(i = 1, 2)$, we define the sample analogue of (2.8.12) as

$$D(\bar{\mathbf{x}}_1, \bar{\mathbf{x}}_2) = \{(\bar{\mathbf{x}}_1 - \bar{\mathbf{x}}_2)'\mathbf{S}_p^{-1}(\bar{\mathbf{x}}_1 - \bar{\mathbf{x}}_2)\}^{1/2} \tag{2.8.13}$$

where $\mathbf{S}_p = \{(n_1 - 1)\mathbf{S}_1 + (n_2 - 1)\mathbf{S}_2\}/(n_1 + n_2 - 2)$ is a pooled unbiased estimate of $\mathbf{\Sigma}$.

We note that the Mahalanobis distance (2.8.8) is in fact the Euclidean distance between the transformed points (obtained through Mahalanobis transformation) $\mathbf{z}_r = \mathbf{S}^{-1/2}(\mathbf{x}_r - \bar{\mathbf{x}})$ and $\mathbf{z}_s = \mathbf{S}^{-1/2}(\mathbf{x}_s - \bar{\mathbf{x}})$,

$$||\mathbf{z}_r - \mathbf{z}_s|| = \mathbf{z}_r'\mathbf{z}_s = D_M^2.$$

Similarly, one can consider Principal Component transformation and define another distance measure.

2.9 Missing Observations

We often find missing values for one or more variables in some of the observation vectors. A small number of missing entries in the data matrix $\mathbf{X}$ can easily be taken care of by deleting the row if it has one or more missing values. However, for a large number of missing values spread over different rows, this approach will lead to deletion of most of the data.

We consider four basic methods that have been proposed for dealing with the problem of missing values. All these methods assume that the missing values are scattered at random throughout the data matrix rather than missing values depend on the variables. Rubin (1976) distinguished between *missing at random* (MAR) and *missing completely at random* (MCAR).

(a) *Listwise deletion* : This approach considers only those observation vectors which are complete. However, this approach is not satisfactory if a large number of missing values are scattered over different rows. Moreover, if the missing values do not occur at random, then exclusion of their data vector will leave a non-random sample which is not representative of the population. A variation of this procedure is to delete either variables or observation vectors, i.e., either columns or rows of the data matrix (Hemel, *et al.*, 1987).

(b) *Pairwise deletion*: This method uses all available observations when calculating $\bar{\mathbf{x}}$ and all available pairs of observations when calculating $\mathbf{S}$ or $\mathbf{R}$. This procedure is not generally suitable, because the covariance matrix $\mathbf{S}$ or the correlation matrix $\mathbf{R}$ calculated in this way is usually not positive definite or even positive semidefinite. Such a result invalidates many multivariate techniques that uses $\mathbf{R}$ or $\mathbf{S}$ (Heilberger, 1977).

(c) *Smoothing procedure*: This method is due to Schwertman and Allen (1973, 1979). In this method we find the positive definite matrix $\mathbf{T}$ that is 'closest' (in the least square sense) to the matrix $\mathbf{S}$ found by method (2) above. The expression

$$\sum_{j=1}^{p}\sum_{k=1}^{p}(t_{jk} - s_{jk})^2 \tag{2.9.1}$$

is minimized where $\mathbf{T} = ((t_{jk})), \mathbf{S} = ((s_{jk}))$. The matrix $\mathbf{T}$ that minimizes (2.9.1) is

$$\mathbf{T} = \sum_{j:\lambda_j>0} \lambda_j \mathbf{a}_j \mathbf{a}_j' \tag{2.9.2}$$

where λ_j is a positive eigenvalue of $\mathbf{S}$ and $\mathbf{a}_j$ the corresponding eigenvector, $\mathbf{a}_j'\mathbf{a}_j = 1$. Since, in most cases, there will be some negative eigenvalues of $\mathbf{S}$ that will be deleted, $\mathbf{T}$ will be singular and cannot be used in applications where $\mathbf{T}^{-1}$ is required. However, this method can be used in applications like Principal Component Analysis (Chapter 8) where inverse is not required.

(d)*Estimation of missing values*: We shall now discuss various schemes for estimation of missing values. The use of these procedures is based on the assumption that the missing values occur at random. If the incidence of missing values depend on the variable-values, the methods may become substantially biased.

Wilks (1932) proposed replacing each missing value by the mean of the corresponding variables, i.e., by the respective column mean. This will reduce the sample variances as well as the absolute values of covariances. As a result, the sample covariance matrix $\mathbf{S}$ will retain the property of positive-definiteness but will be biased.

Bock (1960) suggested a regression approach for estimating a missing value x_{ij}. Consider the submatrix of $\mathbf{X}$ which contains all the available values. We use this submatrix to obtain the regression equation of x_j on $x_1, x_2, \ldots, x_{j-1}, x_{j+1}, \ldots, x_p$, i.e., $\hat{x}_j = b_0 + b_1 x_1 + b_2 x_2 + \ldots + b_{j-1} x_{j-1} + b_{j+1} x_{j+1} + \ldots + b_p x_p$. We then calculate the covariance matrix using the predicted missing values and other available values. Use $\mathbf{S}$ to obtain new regression equation for x_j that will produce new predicted values for x_{ij}. The procedure is continued until the predicted values stabilize.

If most of the observations have missing values, it may not be possible to obtain sufficient data to initiate initial regression estimates of missing values. In that case we may use sample means to obtain initial estimates and then use regression method in subsequent iteration.

In most cases, the regression method is preferable to Wilks' method. If the multiple correlation of x_j on other variables is not high, there is little to gain in using regression method over Wilk's method. The regression technique underestimates variances and covariances but the bias is less than that for Wilk's method.

Several authors have suggested a maximum likelihood approach, based on the assumption of a multivariate normal distribution. Orchard and Woodbury (1977), Dempster, Leird and Rubin (1977) proposed the EM (expectation-maximization) approach. At the expectation step, a missing value is estimated by means of a conditional expectation given the observed values and the current population estimates. At the maximization step, maximum likelihood estimates of the parameters are found using the above estimated values. The new parameter estimates are then used to find estimates for the missing values. This iterative procedure is continued until the estimated parameter values converge. Little (1992) has shown that in case of multivariate normal distribution, the EM algorithm produces essentially the same estimates as the iterative regression approach.

Gleason and Staelin (1978) proposed the following method based on singular value decomposition (SVD). Consider the standardized data matrix $\mathbf{Z}_{n\times p}$ of rank p. It is known from (2.3.15) that $\mathbf{Z}'\mathbf{Z} = (n-1)\mathbf{R}$. By A.8(3), the normalized eigenvectors of $\mathbf{R}$ and $\mathbf{Z}'\mathbf{Z}$ are the same. By SVD (Theorem A.12.2) we can write

$$\mathbf{Z} = \sum_{i=1}^{p} \sqrt{\lambda_i}\mathbf{u}_i\mathbf{v}_i' = \mathbf{ULV}' \tag{2.9.3}$$

where $\mathbf{U} = [\mathbf{u}_1, \mathbf{u}_2, \ldots, \mathbf{u}_p]_{n\times p}, \mathbf{V} = [\mathbf{v}_1, \mathbf{v}_2, \ldots, \mathbf{v}_p]_{p\times p}, \mathbf{L} =$ Diag $(\sqrt{\lambda_1}, \ldots, \sqrt{\lambda_p})$. Here $\mathbf{ZZ}'_{n\times n}$ has (eigenvalue, eigenvector) pairs $(\lambda_i, \mathbf{u}_i)$ so that

$$\mathbf{ZZ}'\mathbf{u}_i = \lambda_i\mathbf{u}_i \tag{2.9.4}$$

with $\lambda_1, \lambda_2, \ldots, \lambda_p > 0 = \lambda_{p+1} = \ldots$. Similarly, $(\lambda_i, \mathbf{v}_i)(i = 1, \ldots, p)$ are the (eigenvalue, eigenvector) pairs of $\mathbf{Z}'\mathbf{Z} = (n-1)\mathbf{R}$. From (2.9.3) we can write

$$\mathbf{U} = \mathbf{Z}(\mathbf{LV}')^{-1} = \mathbf{ZVL}^{-1} \tag{2.9.5}$$

since $\mathbf{V}$ is an orthogonal matrix.

We now consider some approximations to $\mathbf{U}, \mathbf{V}$ and $\mathbf{L}$. Suppose we consider only the r largest eigenvalues $\lambda_1, \lambda_2, \ldots, \lambda_r$ where r is so chosen that $\sum_{j=1}^{r}\lambda_j / \sum_{j=1}^{p}\lambda_j$ is a desired proportion ($r < p$). Denote Diag $(\lambda_1, \ldots, \lambda_r)_{r\times r}$ as $\hat{\mathbf{L}}$, the $n \times r$ submatrix $[\mathbf{u}_1, \ldots, \mathbf{u}_r]$ as $\hat{\mathbf{U}}$, the $p \times r$ submatrix $[\mathbf{v}_1, \ldots, \mathbf{v}_r]$ as $\hat{\mathbf{V}}$.

Hence, from (2.9.3) and (2.9.5) we get the approximation

$$\begin{aligned}\hat{\mathbf{Z}} &= \hat{\mathbf{U}}\hat{\mathbf{L}}\hat{\mathbf{V}}' = \hat{\mathbf{Z}}\hat{\mathbf{V}}\hat{\mathbf{L}}^{-1}\hat{\mathbf{L}}\hat{\mathbf{V}}' \\ &= \mathbf{Z}\hat{\mathbf{V}}\hat{\mathbf{V}}'.\end{aligned} \tag{2.9.6}$$

Suppose now that $\mathbf{Z} = (\mathbf{Z}_1, \mathbf{Z}_2)$where $\mathbf{Z}_1$ contains columns with missing values. Let the corresponding partition of $\hat{\mathbf{V}}$ be

$$\hat{\mathbf{V}}_{p\times r} = \begin{bmatrix} \mathbf{V}_1(p_1 \times r) \\ \mathbf{V}_2(p_2 \times r) \end{bmatrix}.$$

Then from (2.9.6),

$$(\hat{\mathbf{Z}}_1, \hat{\mathbf{Z}}_2) = (\mathbf{Z}_1, \mathbf{Z}_2) \begin{bmatrix} \hat{\mathbf{V}}_1\hat{\mathbf{V}}_1' & \hat{\mathbf{V}}_1\hat{\mathbf{V}}_2' \\ \hat{\mathbf{V}}_2\hat{\mathbf{V}}_1' & \hat{\mathbf{V}}_2\hat{\mathbf{V}}_2' \end{bmatrix}.$$

Hence

$$\hat{\mathbf{Z}}_1 = \mathbf{Z}_1\hat{\mathbf{V}}_1\hat{\mathbf{V}}_1' + \mathbf{Z}_2\hat{\mathbf{V}}_2\hat{\mathbf{V}}_2'. \tag{2.9.7}$$

Replacing $\mathbf{Z}_1$ by $\hat{\mathbf{Z}}_1$ in the right side of (2.9.7) we get an initial estimate of $\mathbf{Z}_1$ as

$$\hat{\mathbf{Z}}^{(1)} = (\mathbf{I}_{p_1} - \hat{\mathbf{V}}_1\hat{\mathbf{V}}_1')^{-1}\mathbf{Z}_2\hat{\mathbf{V}}_2\hat{\mathbf{V}}_2'. \tag{2.9.8}$$

The value of $\hat{\mathbf{Z}}_1^{(1)}$ along with $\mathbf{Z}_2$ can be used to obtain an approximation to $\mathbf{R}$ and hence new approximate value for $\hat{\mathbf{V}}_1$ and hence the second approximation $\hat{\mathbf{Z}}_1^{(2)}$. The process can thus be iterated.

The method requires an initial value of $\mathbf{R}$ which can be obtained by estimating the missing values by Wilk's method.

2.10 Exercises and Complements

2.1 The data matrix $\mathbf{X}$ show the marks of four students on X_1 = first test, X_2 = second test, X_3 = final test.

$$\begin{bmatrix} 17 & 15 & 11 \\ 4 & 19 & 23 \\ 6 & 11 & 18 \\ 5 & 12 & 22 \end{bmatrix}.$$

Calculate

(a) the matrix of deviations, $\mathbf{X}_c = \mathbf{X} - \mathbf{1}\bar{\mathbf{x}}'$;
(b) the corrected sum of squares and sum of products matrix $\mathbf{M}$;
(c) the covariance matrix $\mathbf{S}$ and the correlation matrix $\mathbf{R}$;
(d) the generalized variance $|\mathbf{S}|$;
(e) the total variance;
(f) the equation of the ellipsoid $(\mathbf{x} - \bar{\mathbf{x}})'\mathbf{S}^{-1}(\mathbf{x} - \bar{\mathbf{x}}) \leq c^2$, where c is a non-negative constant.

Verify the relation (2.3.14).

2.2 For the following covariance matrices sketch the solid ellipsoids $(\mathbf{x}-\bar{\mathbf{x}})'\mathbf{S}^{-1}(\mathbf{x}-\bar{\mathbf{x}}) \leq 1$:

$$\mathbf{S} = \begin{bmatrix} 8 & 5 \\ 5 & 6 \end{bmatrix}, \quad \begin{bmatrix} 8 & -5 \\ -5 & 6 \end{bmatrix}.$$

2.3 For the following data matrix

$$\mathbf{X} = \begin{bmatrix} 3 & 1 & 0 \\ 6 & 4 & 6 \\ 4 & 2 & 2 \\ 7 & 0 & 3 \\ 5 & 3 & 4 \end{bmatrix},$$

calculate $\mathbf{S}$. Show that the generalized variance is zero. Verify that the columns of $\mathbf{S}$ are linearly dependent. Are the columns of $\mathbf{X}$ linearly dependent?

2.4 Consider the transformations $\mathbf{Y} = \mathbf{AX} + \mathbf{b}$ and $\mathbf{Z} = \mathbf{CX} + \mathbf{d}$ where

$$\mathbf{A} = \begin{bmatrix} 1 & 1 & 1 \\ -1 & 2 & 0 \\ 0 & 1 & -1 \\ 2 & 1 & 1 \end{bmatrix}, \quad \mathbf{C} = \begin{bmatrix} 2 & 1 & 0 \\ 1 & -1 & 2 \\ 0 & 0 & 1 \\ 1 & 1 & -2 \end{bmatrix},$$

$$\mathbf{b} = [1 \ \ 3 \ \ 0 \ \ 2], \quad \mathbf{d} = [0 \ \ 2 \ \ 1 \ \ 3].$$

Considering the data matrix given in Exercise 2.1, find the mean vectors of $\mathbf{Y}$ and $\mathbf{Z}$, their dispersion matrices and covariance matrix between $\mathbf{Y}$ and $\mathbf{Z}$.

2.5 Replace $\bar{\mathbf{x}}$ by $\mathbf{a}$ in $\mathbf{S}$ to obtain $\mathbf{S}_a = \sum_{i=1}^{n}(\mathbf{x}_i - \mathbf{a})(\mathbf{x} - \mathbf{a})'/(n-1)$.

(1) Show that $\mathbf{S}_a = \mathbf{S} + n(\mathbf{x} - \bar{\mathbf{x}})(\mathbf{x} - \bar{\mathbf{x}})/(n-1)$.
(2) Show that $|\mathbf{S}_a| = |\mathbf{S}|[1 + n(\bar{\mathbf{x}} - \mathbf{a})\mathbf{S}^{-1}(\bar{\mathbf{x}} - \mathbf{a})/(n-1)]$ and that $\min_a |\mathbf{S}_a| = \mathbf{S}$;
(3) Show that $\min_a \operatorname{tr}(\mathbf{S}_a) = \operatorname{tr}(\mathbf{S})$.

2.6 Show that for any vector constants $\mathbf{a}$, $V(\mathbf{a}'\mathbf{X}) = \mathbf{a}\boldsymbol{\Sigma}\mathbf{a}$. Hence, or otherwise, show that $\boldsymbol{\Sigma}$ is at least positive semidefinite.

2.7 Prove the result (2.4.9).

2.8 Show that *Mahalanobis* distance (2.8.8) is in fact the Euclidean distance between two transformed points obtained through *Mahalanobis* transformation.

Table 2.E.1: Data on Body Measurements of Female Sparrows

Bird	X_1	X_2	X_3	X_4	X_5
1	156	245	31.6	18.5	20.5
2	154	240	30.4	17.9	19.6
3	153	240	31.0	18.4	20.6
4	153	236	30.9	17.7	20.2
5	155	243	31.5	18.6	20.3
6	163	247	32.0	19.0	20.9
7	157	238	30.9	18.4	20.2
8	155	239	32.8	18.6	21.2
9	164	248	32.7	19.1	21.1
10	158	238	31.0	18.8	22.0
11	158	240	31.3	18.6	22.0
12	160	244	31.1	18.6	20.5
13	161	246	32.3	19.3	21.8
14	157	245	32.0	19.1	20.0
15	157	235	31.5	18.1	19.8
16	156	237	30.9	18.0	20.3
17	158	244	31.4	18.5	21.6
18	153	238	30.5	18.2	20.9
19	155	236	30.3	18.5	20.1
20	163	246	32.5	18.6	21.9
21	159	236	31.5	18.0	21.5
22	155	240	31.4	18.0	20.7
23	156	240	31.5	18.2	20.6
24	160	242	32.6	18.8	21.7
25	152	232	30.3	17.2	19.8
26	160	250	31.7	18.8	22.5
27	155	237	31.0	18.5	20.0
28	157	245	32.2	18.5	21.4
29	165	245	33.1	19.8	22.7
30	153	231	30.1	17.3	19.8
31	162	239	30.3	18.0	23.1

(Continued)

(Table 2.E.1 continued)

Bird	X_1	X_2	X_3	X_4	X_5
32	162	243	31.6	18.8	21.3
33	159	245	31.8	18.5	21.7
34	159	247	30.9	18.1	19.0
35	155	243	30.9	18.5	21.3
36	162	252	31.9	19.1	22.2
37	152	230	30.4	17.3	18.6
38	159	242	30.8	18.2	20.5
39	155	238	31.	17.9	19.3
40	163	249	33.4	19.5	22.8
41	163	242	31.0	18.1	20.7
42	156	237	31.7	18.2	20.3
43	159	238	31.5	18.4	20.3
44	161	245	32.1	19.1	20.8
45	155	235	30.7	17.7	19.6
46	162	247	31.9	19.1	20.4
47	153	237	30.6	18.6	20.4
48	162	245	32.5	18.5	21.1
49	164	248	32.3	18.8	20.9

Source: Manly (1994, pp 1 - 3)

2.9 Obtain a distance function as the Euclidean distance between two points obtained through Principal Component transformation.

2.10 For the data matrix given in Exercise 2.1, obtain (a) Scaling transformation (b) Mahalanobis transformation (c) Principal component transformation.

2.11 From the data matrix given in Exercise 2.1, obtain Mahalanobis distance between $\mathbf{x}_1$ and $\mathbf{x}_2, \mathbf{x}_1$ and $\mathbf{x}_3$ and $\mathbf{x}_2$ and $\mathbf{x}_4$.

2.12 For the bird data in Table 2.E.1, obtain variance-covariance matrix for the survivor birds and the non-survivor birds.

2.13 (a) If $\mathbf{X}, \mathbf{Y}$ are $m \times 1$ and $n \times 1$ random vectors and $\mathbf{a}, \mathbf{b}$ are $m \times 1, n \times 1$ vectors of constants, prove that

$$C[\mathbf{X} - \mathbf{a}, \mathbf{Y} - \mathbf{b}] = C[\mathbf{X}, \mathbf{Y}].$$

(b) Prove that

$$C(\mathbf{X}, \mathbf{Y}) = E(\mathbf{XY}') - (E(\mathbf{X}))(E(\mathbf{Y}))'.$$

2.14 Let $\mathbf{X} = (X_1, \ldots, X_n)'$ be a vector of random variables and let

$$Y_1 = X_1, Y_i = X_i - X_{i-1},\ i = 2, \ldots, n.$$

If the Y_i's are mutually independent random variables each with unit variance, find $D(\mathbf{X})$.

2.15 If $X_1, \ldots, X_n$ are random variables each with variance σ^2 and $X_{i+1} = \rho X_i + a (i = 1, \ldots, n)$, where a and ρ are constants, find $D(\mathbf{X})$.

2.16 If $\mathbf{A}$ is a symmetric matrix and $\mathbf{X}$ is a random vector, prove that

$$E(\mathbf{X}'\mathbf{AX}) = tr[\mathbf{A}E(\mathbf{XX}')].$$

2.17 If $\mathbf{X}$ is $N_p(\mu, \mathbf{\Sigma})$ prove that

(a) $E(\mathbf{X}'\mathbf{AX}) = \text{ tr } (\mathbf{A\Sigma}) + \mu'\mathbf{A}\mu;$
(b) $V(\mathbf{X}'\mathbf{AX}) = 2[\text{ tr } (\mathbf{A\Sigma A\Sigma}) + 2\mu'\mathbf{A\Sigma A}\mu].$

2.18 If $X_1, \ldots, X_n$ are mutually independent random variables with common mean μ and variance $\sigma_1^2, \ldots, \sigma_n^2$ respectively, prove that

$$E[\frac{1}{n(n-1)} \sum_{i=1}^{n} (X_i - \bar{X})^2] = V(\bar{X}).$$

2.19 The random variables $X_1, \ldots, X_n$ have a common mean μ, a common variance σ^2 and a common correlation coefficient ρ. Find $V(\bar{X})$. Show that

$$E\{\sum_{i=1}^{n} (X_i - \bar{X})^2\} = (n-1)(1-\rho)\sigma^2.$$

2.20 Let $X_1, \ldots, X_n$ be *iid* $N(\mu, \sigma^2)$ variables. Define

$$S^2 = \sum_{i=1}^{n} (X_i - \bar{X})^2/(n-1),\ Q = \sum_{i=1}^{n-1} (X_{i+1} - X_i)^2/2(n-1).$$

(a) Prove that $V(S^2) = 2\sigma^4/(n-1)$.

(b) Show that $E(Q) = \sigma^2$.

2.21 The Table 2.E.1 gives the body measurements of female sparrows ($\mathbf{X}_1$ = total length, X_2 = alar extent, X_3 = length of beak and head, X_4 = length of humerus, X_5 = length of keel and sternum, all in mm.). Birds 1 to 21 survived a certain treatment, while the remaining died.

For these data calculate $\bar{\mathbf{x}}, \mathbf{M}, \mathbf{S}, \mathbf{R}$, eigenvalues and the eigenvectors of $\mathbf{S}$, generalized variance $|\mathbf{S}|$, total variance, constant-distance ellipsoid (2.3.15), Mahalanobis transformation (2.5.4) and Principal Component transformation (2.5.6). Verify the relation (2.3.14). Also obtain different measures of intercorrelation. Comment on the values of the statistics.

Suppose that the following observations are missing: X_3, X_4 for observation 3 5, 11, 20; X_1, X_3, X_5 for observation 7, 9, 23; X_2, X_5 for observations 1, 9, 15, 32, 37, 49; X_1, X_3, X_4 for observations 3, 12, 18, 24, 30, 35 42, 45. Estimate the missing values by the method of means and the regression approach.

2.22 A survey was carried out by the Indian Statistical Institute, Kolkata in 1998 for the estimation of population along with its socio-economic characteristics in the two districts of West Bengal, India. The Table 2.E.2 gives the socio-economic data for some households in two villages. The items are:

(i) Vill. : village,
(ii) Unit: household serial number,
(iii) x_1: household size,
(iv) x_2: area cultivated (in acres),
(v) x_3:crop-income (in Rupees),
(vi) x_4: income from livestock, poultry, forestry and fishing (in Rupees),
(vii) x_5: Employment income (in Rupees),
(viii) x_6: Total income (in Rupees).

Table 2.E.2: Data on Agricultural Land and Household Income for Two Villages in West Bengal

Vill.	Unit	x_1	x_2	x_3	x_4	x_5	x_6
1	1	7	0.82	2279	10000	2635	25114
	2	6	3.00	7518	3810	0	11473
	3	3	3.32	8494	400	490	9504
	4	3	5.00	5909	110	560	6654
	5	9	7.92	20693	2800	0	26653

(Continued)

(Table 2.E.2 continued)

Vill.	Unit	x_1	x_2	x_3	x_4	x_5	x_6
	6	5	1.65	7506	100	0	7606
	7	5	2.66	6350	5600	0	11950
	8	8	4.66	10901	1760	940	13601
	9	9	6.64	24700	3060	180	31196
	10	13	18.00	30707	13100	38400	82207
	11	4	0.00	0	0	4800	11280
	12	6	0.16	1060	300	1680	17480
	13	4	0.50	1073	0	2265	4477
	14	4	4.08	4653	1800	320	6883
	15	6	2.63	5621	11800	0	17486
	16	4	1.0	1355	300	2270	4060
	17	3	2.00	5429	0	1960	7389
	18	7	0.44	23302	600	4260	30242
	19	5	1.98	4108	400	1755	9973
	20	5	3.57	15410	1400	0	16810
	21	7	3.94	18943	3200	420	22873
	22	7	7.32	24270	6800	0	35210
2	1	4	0.99	1975	0	1500	4315
	2	4	3.16	4127	0	0	5027
	3	3	1.83	3440	0	0	3672
	4	5	2.82	4187	2000	0	11827
	5	8	3.14	9240	200	0	11827
	6	2	1.32	7572	2000	2610	12232
	7	11	10.99	22665	9700	0	38365
	8	7	4.82	7720	600	700	9134
	9	4	2.33	3940	0	5000	8970
	10	6	2.99	4424	3100	0	7664
	11	7	2.33	1683	0	6555	8238
	12	5	3.16	3869	300	0	4214
	13	8	7.32	5010	2600	0	7630
	14	6	4.49	3910	0	0	3910
	15	4	0.93	3059	400	17120	20604
	16	5	1.66	2920	400	1600	4940
	17	4	6.93	4955	0	0	5035
	18	9	3.40	19830	4300	0	24194
	19	8	6.41	9145	7700	0	17705
	20	6	4.66	7255	900	0	8215

Source: Mukhopadhyay (1998)

For each village and also for the two villages combined, calculate $\bar{\mathbf{x}}, \mathbf{S}, \mathbf{R}$, eigenvalues and eigenvectors of $\mathbf{S}$, generalized variance $|\mathbf{S}|$, total variance, constant-density ellipsoid (2.3.15), Mahalanobis transformation (2.5.4) and Principal Component transformation (2.5.6). Also obtain different measures of intercorrelation.

2.11 Appendix

Ellipsoids

Let $\mathbf{A}$ be a positive-definite matrix. Then the equation

$$(\mathbf{x}-\alpha)'\mathbf{A}^{-1}(\mathbf{x}-\alpha) = c^2 \tag{2.11.1}$$

represents an ellipsoid in n-dimensional Euclidean space. The center of ellipsoid is at $\mathbf{x} = \alpha$.

DEFINITION 2.11.1: Let $\mathbf{x}$ be a point on the ellipsoid (2.11.1) and let $f(\mathbf{x}) = ||\mathbf{x}-\alpha||^2$, the squared Euclidean distance between α and $\mathbf{x}$. A line through α and $\mathbf{x}$ for which $\mathbf{x}$ is a stationary point of $f(\mathbf{x})$ is called a *principal axis* of the ellipsoid. For this $\mathbf{x}$, the distance $||\mathbf{x}-\alpha||$ is half the length of the principal axis.

Theorem 2.11.1: Let $\lambda_1, \ldots, \lambda_n$ be the eigenvalues of $\mathbf{A}$ satisfying $\lambda_1 > \lambda_2 > \ldots > \lambda_n$ with $\gamma_1, \gamma_2, \ldots, \gamma_n$ the corresponding eigenvectors. Then

(a) the direction cosine of the ith prncipal axis of the ellipsoid (2.11.1) is γ_i;

(b) the length of the ith principal axis of the ellipsoid (2.11.1) is $c\sqrt{\lambda_i}$.

Chapter 3

Multivariate Normal Distribution and Related Distributions

3.1 Introduction

This chapter considers multivariate normal distribution and some related distributions. Section 3.2 addresses in details properties of a multivariate normal distribution. The next section deals with distribution of transformations of a normal data matrix. Distribution of quadratic forms in multivariate normal variables are examined subsequently. Sections 3.5 and 3.6 consider respectively multivariate central limit theorem and maximum likelihood estimation of parameters of a multivariate normal distribution in both unrestricted and restricted cases. Subsequently, matrix normal distribution and multivariate t distribution are examined. (The Dirichlet distribution, though not related to a multivariate normal distribution is also looked into, because of its usefulness.) Measures of multivariate skewness and kurtosis are also recalled. Section 3.11 examines different measures for assessing the assumption of normality. The next section considers some transformations which make the data look near normal. Lastly robust estimation of location and scale parameters are addressed.

3.2 Multivariate Normal Distribution

Let $\mathbf{X} = (X_1, \ldots, X_p)'$ be a random vector of p random variables. Let $\mu = (\mu_1, \ldots, \mu_p)'$ be a vector of real constants and $\mathbf{A}$ be a $p \times p$ real, symmetric and positive definite (p.d.) matrix.

DEFINITION 3.2.1. The random variables $X_1, \ldots, X_p$ have a joint multivariate normal distribution if their joint probability density can be written as

$$f_{\mathbf{X}'}(x_1, \ldots, x_p) = C.\exp[-\frac{1}{2}(\mathbf{x} - \mu)'\mathbf{A}(\mathbf{x} - \mu)], \tag{3.2.1}$$

where $\mathbf{x} = (x_1, \ldots, x_p)'$ and C is a suitable constant. In other words, $\mathbf{X}$ follows a p-variate normal distribution if its density can be written in the form

$$f_{\mathbf{X}'} = C.\exp[-\frac{1}{2} \text{ (positive definite quadratic form in } \mathbf{x})]. \tag{3.2.2}$$

It is obvious that C must be a function of μ and $\mathbf{A}$. In order to determine the constant C we consider the moment generating function (m.g.f.) of $\mathbf{X}$ about $\mathbf{0} = (0, \ldots, 0)'$. The m.g.f. is

$$M_{\mathbf{X}}(\mathbf{t}) = E(e^{\mathbf{t}'\mathbf{X}})$$

where $\mathbf{t} = (t_1, \ldots, t_p)'$, a set of arbitrary real numbers, if the expectation exists in some neighborhood of $\mathbf{0}$.

Theorem 3.2.1: The m.g.f of the distribution in (3.2.1) is given by

$$M(t_1, \ldots, t_p) = \exp\{\mathbf{t}'\mu + \frac{1}{2}\mathbf{t}'\mathbf{A}^{-1}\mathbf{t}\}. \tag{3.2.3}$$

Proof.

$$M(\mathbf{t}) = C\int_{-\infty}^{\infty} \cdots \int_{-\infty}^{\infty} \exp[-\frac{1}{2}(\mathbf{x} - \mu)'\mathbf{A}(\mathbf{x} - \mu) + \mathbf{t}'\mathbf{x}]d\mathbf{x}.$$

Let $x_i - \mu_i = y_i, i = 1, \ldots, p, \mathbf{y} = (y_1, \ldots, y_p)'$. Hence

$$M(\mathbf{t}) = C.e^{\mathbf{t}'\mu}\int_{-\infty}^{\infty} \cdots \int_{-\infty}^{\infty} \exp[-\frac{1}{2}\mathbf{y}'\mathbf{A}\mathbf{y} + \mathbf{t}'\mathbf{y}]d\mathbf{y}. \tag{3.2.4}$$

Since $\mathbf{A}$ is p.d. we can write $\mathbf{A} = \mathbf{H}'\mathbf{H}$ where $\mathbf{H}$ is a non-singular matrix. We now transform $\mathbf{z} = \mathbf{H}\mathbf{y}$, where $\mathbf{z} = (z_1, \ldots z_p)'$. The Jacobian of transformation is $|J| = 1/|\mathbf{H}| = (\sqrt{|\mathbf{A}|})^{-1}$. Hence (3.2.4) reduces to

$$M(\mathbf{t}) = \frac{C}{|\mathbf{H}|}e^{\mathbf{t}'\mu}\int_{-\infty}^{\infty} \cdots \int_{-\infty}^{\infty} \exp[-\frac{1}{2}\mathbf{z}'\mathbf{z} + \mathbf{t}'\mathbf{H}^{-1}\mathbf{z}]d\mathbf{z} \tag{3.2.5}$$

$$= \frac{C}{|\mathbf{H}|}e^{\mathbf{t}'\mu}\int_{-\infty}^{\infty} \cdots \int_{-\infty}^{\infty} \exp[-\frac{1}{2}\sum_{i=1}^{p}(z_i^2 + 2b_i z_i)]|\Pi dz_i, \tag{3.2.6}$$

where $\mathbf{b}' = \mathbf{t}'\mathbf{H}^{-1}$.

$$= \frac{C}{|\mathbf{H}|}\exp(\mathbf{t}'\mu + \frac{1}{2}\mathbf{b}'\mathbf{b})\int_{-\infty}^{\infty} \cdots \int_{-\infty}^{\infty} \exp[-\frac{1}{2}(z_i + b_i)^2]\Pi dz_i$$

$$= \frac{C}{|\mathbf{H}|}(2\pi)^{p/2}\exp\{\mathbf{t}'\mu + \frac{1}{2}\mathbf{b}'\mathbf{b}\}.$$

Now $\mathbf{b}'\mathbf{b} = \mathbf{t}'(\mathbf{H}'\mathbf{H})^{-1}\mathbf{t} = \mathbf{t}'\mathbf{A}^{-1}\mathbf{t}$. Hence

$$M(\mathbf{t}) = \frac{C}{\sqrt{|\mathbf{A}|}}(2\pi)^{p/2}\exp\{\mathbf{t}'\mu + \frac{1}{2}\mathbf{t}'\mathbf{A}^{-1}\mathbf{t}\}. \tag{3.2.7}$$

Since $M(\mathbf{0}) = 1$,

$$C = \sqrt{|\mathbf{A}|}(2\pi)^{-p/2}. \tag{3.2.8}$$

Hence the proof. □

The joint density function (3.2.1) is, therefore,

$$f_{\mathbf{X}'}(x_1, \ldots, x_p) = \frac{\sqrt{|\mathbf{A}|}}{(2\pi)^{p/2}}\exp\{-\frac{1}{2}(\mathbf{x} - \mu)'\mathbf{A}(\mathbf{x} - \mu)\}. \tag{3.2.9}$$

We now find values of μ and $\mathbf{A}$ in terms of moments of the distribution. From (3.2.3),

$$\begin{aligned} E(\mathbf{X}) &= \mu \\ D(\mathbf{X}) &= \mathbf{\Sigma} \text{ (say) } = ((\sigma_{ij})) = \mathbf{A}^{-1}. \end{aligned} \tag{3.2.10}$$

Hence,

$$f_{\mathbf{X}'}(\mathbf{x}') = \frac{1}{(2\pi)^{p/2}\sqrt{|\mathbf{\Sigma}|}}\exp\{-\frac{1}{2}(\mathbf{x} - \mu)'\mathbf{\Sigma}^{-1}(\mathbf{x} - \mu)\}. \tag{3.2.11}$$

We shall denote in this case $\mathbf{X} \cap N_p(\mu, \mathbf{\Sigma})$. The p-variate normal density (3.2.11) has a maximum value when $\mathbf{x} = \mu$. Thus μ is the *mode* as well as the expected value of $\mathbf{X}$. A small value of $|\mathbf{\Sigma}|$ indicates that $\mathbf{x}$ are concentrated closer to μ than in the case when $|\mathbf{\Sigma}|$ is large. A small value of $|\mathbf{\Sigma}|$ may also indicate a high degree of *multicollinearity* among the variables, signifying that $\mathbf{x}$'s tend to occupy a smaller dimension subspace of the p dimensions.

Note 3.2.1: The m.g.f in (3.2.3) shows that all cumulants and cross-cumulants of order higher than two of the distribution are zero.

Note 3.2.2: The density function (3.2.11) is constant whenever the quadratic form in the exponent is so, so that it is constant on the ellipsoid

$$\{\mathbf{x} : (\mathbf{x} - \mu)'\mathbf{\Sigma}^{-1}(\mathbf{x} - \mu) = c^2\} \tag{3.2.12}$$

in $\mathcal{R}^p, c$ being a constant. This ellipsoid has center μ and $\mathbf{\Sigma}$ determines its shape and orientation. The axes of this ellipsoid are

$$\pm c\sqrt{\lambda_i}\mathbf{e}_i \tag{3.2.13}$$

where $(\lambda_i, \mathbf{e}_i)$ are the eigenvalue-eigenvector pairs of $\mathbf{\Sigma}(i = 1, \ldots, p)$. Thus the axes of these ellipsoids are in the direction of eigenvectors of $\mathbf{\Sigma}$ (or $\mathbf{\Sigma}^{-1}$). These ellipsoids are called the contours of the distribution or the *ellipsoids of equal concentration.* For $\mu = \mathbf{0}$, these contours are centered at $\mathbf{0}$ and when $\mathbf{\Sigma} = \mathbf{I}$ the contours are circles or in higher dimensions spheres or hyperspheres.

The principal component transformation (vide Section 2.5) facilitates interpretation of the ellipsoids of equal concentration. By spectral decomposition theorem (Theorem A.12.1), $\mathbf{\Sigma} = \mathbf{\Gamma\Lambda\Gamma}'$, where $\mathbf{\Lambda} =$ Diag. $(\lambda_1, \ldots, \lambda_p)$ is the matrix of eigenvalues of $\mathbf{\Sigma}$ and $\mathbf{\Gamma}$ is an orthogonal matrix whose columns are the corresponding normalized eigenvectors. The principal component transformation is $\mathbf{y} = \mathbf{\Gamma}'(\mathbf{x} - \mu)$. In terms of $\mathbf{y}$, the ellipsoid in (3.2.12) becomes

$$\{\mathbf{y} : \sum_{i=1}^{p} \frac{y_i^2}{\lambda_i} = c^2 \}.$$

This is the equation of an ellipsoid with semimajor axes of length $c\sqrt{\lambda_j}; y_1, \ldots, y_p$ represent the different axes of the ellipsoid. To find the value of c that corresponds to a given portion of observations contained in the ellipsoid, we use the fact that $\sum_{i=1}^{n} y_i^2/\lambda_i$ has a chi-squared distribution (see Theorem 3.4.2(a)); for example, $P(\sum_{i=1}^{n} y_i^2/\lambda_i \leq \chi^2_{p;\alpha}) = 1 - \alpha$. The direction of the axes are given by the eigenvectors of $\mathbf{\Sigma}$, since $\mathbf{y} = \mathbf{\Gamma}'(\mathbf{x} - \mu)$ is a rotation and $\mathbf{\Gamma}$ is orthogonal with eigenvectors of $\mathbf{\Sigma}$ as columns. Figure 3.1 shows ellipses of equal concentration for the bivariate normal distribution, along with the principal components y_1, y_2.

EXAMPLE 3.2.1 : The bivariate normal density is

$$f(x_1, x_2) = \frac{1}{2\pi\sigma_1\sigma_2\sqrt{(1-\rho_{12}^2)}}$$

$$\exp\left\{-\frac{1}{2(1-\rho_{12}^2)}[(\frac{x_1-\mu_1}{\sigma_1})^2 - 2\rho(\frac{x_1-\mu_1}{\sigma_1})(\frac{x_2-\mu_2}{\sigma_2}) + (\frac{x_2-\mu_2}{\sigma_2})^2]\right\}$$

where $\sigma_1 = \sqrt{\sigma_{11}}, \sigma_2 = \sqrt{\sigma_{22}}$ and $\rho_{12} = \frac{\sigma_{12}}{\sigma_1\sigma_2}$. To obtain contours of bivariate normal density we have to find the eigenvalues and eigenvectors of $\mathbf{\Sigma}$. Suppose

$$\mu_1 = 5,\ \mu_2 = 8,\ \mathbf{\Sigma} = \begin{bmatrix} 12 & 9 \\ 9 & 25 \end{bmatrix}.$$

Here the equation $|\mathbf{\Sigma} - \lambda\mathbf{I}| = 0$ becomes

$$0 = \begin{vmatrix} 12-\lambda & 9 \\ 9 & 25-\lambda \end{vmatrix} = (12-\lambda)(25-\lambda) - 81 = 0.$$

The equation gives $\lambda_1 = 29.6018, \lambda_2 = 7.3982$. The corresponding normalized eigenvectors are

$$\mathbf{e}_1 = \begin{bmatrix} .455252 \\ .890362 \end{bmatrix}, \quad \mathbf{e}_2 = \begin{bmatrix} .890362 \\ -.455252 \end{bmatrix}.$$

The ellipse of the constant probability density

$$\{\mathbf{x} : (\mathbf{x} - \mu)'\mathbf{\Sigma}^{-1}(\mathbf{x} - \mu) = c^2\} \tag{i}$$

has center at $\mu = (5, 8)'$ and axes $\underline{+}\, 5.4408c\mathbf{e}_1$ and $\underline{+}\, 2.72c\mathbf{e}_2$. The eigenvector $\mathbf{e}_1$ lies along the line which makes an angle of $\cos^{-1}(.455252) = 62.92^0$ with the X_1 axis and passes through the point (5, 8) in the (X_1, X_2) plane. By principal component transformation $\mathbf{y} = \mathbf{\Gamma}'(\mathbf{x} - \mu)$ where $\mathbf{\Gamma} = [\mathbf{e}_1, \mathbf{e}_2]$, the ellipse (i) has the equation

$$\{\mathbf{y} : \ \frac{y_1^2}{\lambda_1} + \frac{y_2^2}{\lambda_2} = c^2 \ \}.$$

The ellipse has center at $(y_1 = 0, y_2 = 0)$ and axes $\underline{+}\, c.\sqrt{\lambda_i}\tilde{\mathbf{e}}_i$, where $\tilde{e}_i, i = 1, 2$ are the eigenvectors of $\Lambda =$ Diag.(λ_1, λ_2). The eigenvalues of Λ are λ_1, λ_2. In terms of the new co-ordinate axes Y_1, Y_2, the eigenvectors are $\tilde{e_1} = (1, 0)', \tilde{e_2} = (0, 1)'$. The lengths of the axes are $c\sqrt{\tilde{e_i}}(i = 1, 2)$. (The axes Y_1, Y_2 are the Principal Components of $\mathbf{\Sigma}$.)

Note 3.2.3: If $\mathbf{A}$ is only positive semi-definite (p.s.d.) (i.e. $|\mathbf{A}| = 0$), the joint distribution of $(X_1, \ldots, X_p)'$ is called a *singular multinormal distribution.*

It follows from (3.2.3) that $M(0, \ldots, 0, t_i, 0 \ldots, 0) = \exp(t_i\mu_i + \frac{t_i^2}{2}\sigma_{ii})$ is the m.g.f. of $X_i (i = 1, \ldots, p)$. Thus, each $X_i \cap N(\mu_i, \sigma_{ii}), i = 1, \ldots, p$.

For $i \neq j$, the joint m.g.f. of X_i, X_j is

$$M(0, \ldots, 0, t_i, 0, \ldots, 0, t_j, 0, \ldots, 0)$$

$$= \exp\{t_i\mu_i + t_j\mu_j + \frac{1}{2}(\sigma_{ii}t_i^2 + 2\sigma_{ij}t_it_j + t_j^2\sigma_{jj})\}.$$

This is the m.g.f of a bivariate normal distribution with means μ_i, μ_j, variances σ_{ii}, σ_{jj} and covariance $\sigma_{ij} = \rho_{ij}\sigma_i\sigma_j$ where $\sigma_i = \sqrt{\sigma_{ii}}$ and ρ_{ij} is the correlation coefficient between X_i and $X_j (i \neq j)$.

It follows similarly that the marginal distribution of any subset $(X_{i_1}, \ldots, X_{i_r})$ of $(X_1, \ldots, X_p)(r < p)$ is r-variate normal with mean vector $(\mu_{i_1}, \ldots, \mu_{i_r})'$ and dispersion matrix $((\sigma_{i_t i_{t'}}, t, t' = 1, \ldots, r))$ (vide Corollary 3.2.2.2).

If $\mathbf{\Sigma} =$ Diag $(\sigma_{ii}; i = 1, \ldots, p)$, then $X_1, \ldots, X_p$ are all mutually independent. It readily follows that if $X_1, \ldots, X_p$ are mutually independent, then $\mathbf{\Sigma}$ is a Diagonal matrix, Diag $(\sigma_{ii}; i = 1, \ldots, p)$. Thus for a multivariate normal distribution zero-covariance implies independence and conversely (vide also Theorem 3.2.4).

We shall now consider several properties of a multivariate normal distribution.

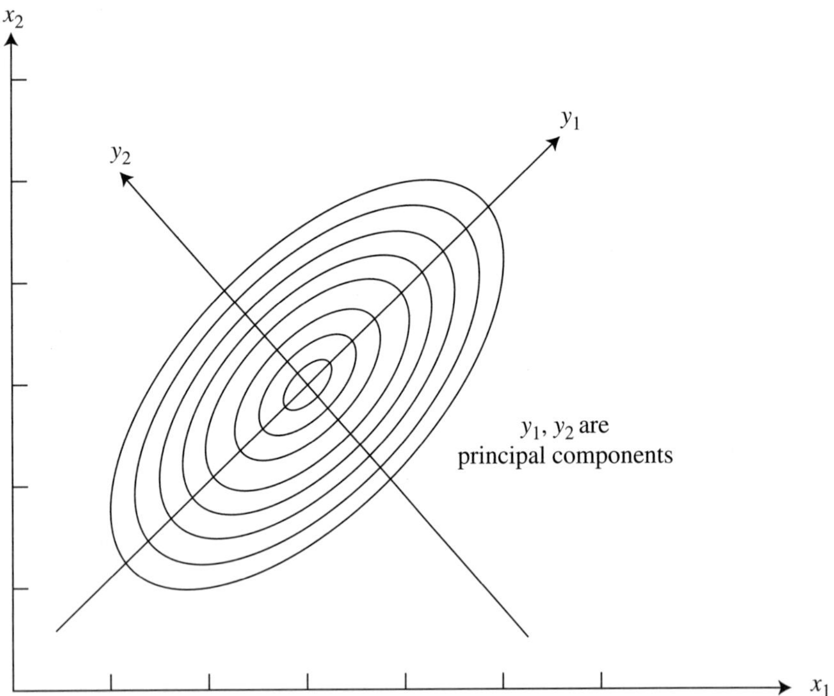

Fig. 3.1: Ellipse of equal concentration for the bivariate normal distribution

For a multivariate normal distribution all central moments of odd order are zero. Fourth central moments of the distribution are given by

$$E(X_i - \mu_i)(X_j - \mu_j)(X_k - \mu_k)(X_l - \mu_l)] = \sigma_{ij}\sigma_{kl} + \sigma_{ik}\sigma_{jl} + \sigma_{il}\sigma_{jk}$$

with special cases as follows:

$$E[(X_i - \mu_i)^2(X_j - \mu_j)(X_k - \mu_k)] = \sigma_{ii}\sigma_{jk} + 2\sigma_{ij}\sigma_{ik},$$

$$E[(X_i - \mu_i)^2(X_j - \mu_j)^2] = \sigma_{ii}\sigma_{jj} + 2\sigma_{ij}^2,$$

$$E(X_i - \mu_i)^4 = 3\sigma_{ii}^2.$$

Since the multivariate normal distribution is fairly well characterized by its first four moments, these moments can be used for checking for normality. The tests for multivariate normality have been considered in Section 3.11.

Theorem 3.2.2: If $\mathbf{X} \cap N_p(\mu, \mathbf{\Sigma})$ and $\mathbf{C}$ is a $m \times p$ matrix with rank m, then

$$\mathbf{CX} \cap N_m(\mathbf{C}\mu, \mathbf{C\Sigma C}').$$

Proof. Let $\mathbf{Y} = \mathbf{CX}$. Then for any real vector $\mathbf{t}$,

$$\begin{aligned} E\{\exp(\mathbf{t}'\mathbf{Y})\} &= E\{\exp(\mathbf{t}'\mathbf{CX})\} = E\{\exp(\mathbf{S}'\mathbf{X})\} \text{ where } \mathbf{S}' = \mathbf{t}'\mathbf{C} \\ &= \exp(\mathbf{S}'\mu + \tfrac{1}{2}\mathbf{S}'\mathbf{\Sigma}\mathbf{S}) = \exp[\mathbf{t}'\mathbf{C}\mu + \tfrac{1}{2}\mathbf{t}'(\mathbf{C\Sigma C}')\mathbf{t}]. \end{aligned} \quad (3.2.14)$$

Since $\mathbf{C\Sigma C}'$ is positive definite, (3.2.14) is the m.g.f. of a $N_m(\mathbf{C}\mu, \mathbf{C\Sigma C}')$ distribution. The result follows from the uniqueness of the m.g.f.

Corollary 3.2.1.1: Any linear combination of $X_1, \ldots, X_p$ is a univariate normal variable. In particular, every marginal distribution is a univariate normal distribution.

Corollary 3.2.2.2: The marginal distribution of any subset of the elements of $\mathbf{X}$ is also multivariate normal.

Proof. Suppose $\mathbf{Y} = (X_1, \ldots, X_m)'(m < p)$. Put $\mathbf{C} = (I_m, \mathbf{0})$.

Corollary 3.2.2.3: Suppose $\mathbf{X} \sim N_p(\mu I_p, \mathbf{\Sigma})$, where $\mathbf{\Sigma} = \sigma^2 I_p$. Consider $\mathbf{Y} = \mathbf{LX}$ where $\mathbf{L}$ is an orthogonal matrix. Then $\mathbf{Y} \sim N_p(\mu\mathbf{L}, \mathbf{\Sigma})$.

The corollary states that mutually independent normal variables with the same variance remains mutually independent normal variables with the same variance under orthogonal transformation. The above invariance property is a characterization of a normal distribution.

Corollary 3.2.2.4: Let $\mathbf{Y} = \mathbf{\Sigma}^{-1/2}(\mathbf{X} - \mu)$, where $\mathbf{\Sigma}^{1/2}$ is the symmetric positive definite square root of $\mathbf{\Sigma}$. Then $Y_1, \ldots, Y_p$ are independent $N(0,1)$ variables.

EXAMPLE 3.2.2: Let $X_1 \ldots, X_n$ be mutually independent $N(\mu, \sigma^2)$ variables. Prove that $\bar{X}$ is statistically independent of

$$Q = \sum_{i=1}^{n} (X_i - \bar{X})^2/\sigma^2 \text{ and } Q \sim \chi^2_{(n-1)}.$$

Proof. Let $Y_i = (X_i - \mu)/\sigma$. Then Y_i follows $N(0,1)$. Consider an orthogonal transformation (Helmert's transformation),

$$\begin{aligned}
Z_1 &= \tfrac{Y_1 + \ldots + Y_n}{\sqrt{n}}, \\
Z_2 &= \tfrac{Y_1 - Y_2}{\sqrt{2.1}}, \\
Z_3 &= \tfrac{Y_1 + Y_2 - 2Y_3}{\sqrt{3.2}}, \\
&\ldots \\
Z_n &= \tfrac{Y_1 + \ldots + Y_{n-1} - (n-1)Y_n}{\sqrt{n(n-1)}}.
\end{aligned}$$

By Corollary 3.2.2.3, $Z_1, \ldots, Z_n$ are independent normal $N(0,1)$ variables. Now

$$\sum_{i=1}^{n} Z_i^2 = \sum_{i=1}^{n} Y_i^2 = n\bar{Y}^2 + \sum_{i=1}^{n} (Y_i - \bar{Y})^2. \tag{3.2.15}$$

Hence, $Q = \sum_{i=1}^{n} (Y_i - \bar{Y})^2 = \sum_{i=2}^{n} Z_i^2$. Since Z_1 is independent of $Z_2, \ldots, Z_n$, it is independent of $\sum_{i=2}^{n} Z_i^2 = Q$. Also, $Z_i^2 \sim \chi^2_{(1)} (i = 2, \ldots, n)$, so that $Q \sim \chi^2_{(n-1)}$.

Note 3.2.4: We have noted that the marginal distribution of each component of $\mathbf{X}$, when $\mathbf{X}$ has a multivariate normal distribution, is univariate normal. The converse is not true in general. (Obviously, the converse is also true if the components of $\mathbf{X}$ are all independent and normal.)

For example, suppose that the random variables X_1, X_2 have joint distribution function

$$F(X_1, X_2) = \Phi(x_1)\Phi(x_2)[1 + \alpha(1 - \Phi(x_1))(1 - \Phi(x_2))],$$

where $|\alpha| < 1, \Phi(x)$ is the distribution function of the standard normal distribution. It can be shown that the marginal distributions of X_1 and X_2 are standard normal (see also Exercises 3.18 and 3.19).

Theorem 3.2.3: The random vector $\mathbf{X} = (X_1, \ldots, X_p)'$ has a p-variate normal distribution *iff* every linear function of components of $\mathbf{X}, \mathbf{l}'\mathbf{X} = l_1X_1 + \ldots + l_pX_p$ follows a univariate normal distribution.

Proof. Suppose $\mathbf{l}'\mathbf{X}$ has a univariate normal distribution for any $\mathbf{l}$. Then the m.g.f of $\mathbf{l}'\mathbf{X}$ is

$$\Phi(t) = \exp\{bt + \frac{1}{2}t^2\sigma^2\},$$

where $b = E(\mathbf{l}'\mathbf{X}) = \mathbf{l}'\mu$ (say), $\sigma^2 = V(\mathbf{l}'\mathbf{X})$ and t is any real number. Thus

$$\Phi(t) = \exp\{\mathbf{l}'\mu t + \frac{1}{2}t^2\mathbf{l}'\boldsymbol{\Sigma}\mathbf{l}\}.$$

Let $t = 1$. Then

$$\Phi(1) = \exp\{\mathbf{l}'\mu + \frac{1}{2}\mathbf{l}'\boldsymbol{\Sigma}\mathbf{l}\} = M(l_1, \ldots, l_p)$$

which is the m.g.f of a $N_p(\mu, \boldsymbol{\Sigma})$ distribution. The converse follows from corollary 3.2.2.1.

Note 3.2.5: Note that for the results of Theorem 3.2.3 it is necessary that every linear combination $\mathbf{l}'\mathbf{X}$ is univariate normal and not just $X_1, \ldots, X_p$. The relevance of the note 3.2.4 should now be clear.

Theorem 3.2.3 also states that the multivariate normal distribution is the only multivariate distribution which has the property that every linear combination $\mathbf{l}'\mathbf{X}$ is a univariate normal variable.

Corollary 3.2.3.1: If $\mathbf{X} = (X_1, X_2)'$ and the marginal distributions of X_1 and X_2 are both univariate normal, then $\mathbf{X}$ has a bivariate normal distribution *iff* $l_1X_1 + l_2X_2$ is univariate normal for all real l_1 and l_2.

Theorem 3.2.4: Let $\mathbf{X} \cap N_p(\mu, \boldsymbol{\Sigma})$ and suppose that $\mathbf{X} = (\mathbf{X}^{(1)'}, \mathbf{X}^{(2)'})'$, where $\mathbf{X}^{(1)}$ has m elements, $m < p$. Then $\mathbf{X}^{(1)}, \mathbf{X}^{(2)}$ are statistically independent *iff* Cov $(\mathbf{X}^{(1)}, \mathbf{X}^{(2)}) = \mathbf{0}$.

Proof. If $\mathbf{X}^{(1)}, \mathbf{X}^{(2)}$ are independent, then all pairs of elements $X_i^{(1)}, X_j^{(2)}$ are so. Hence,

$$\text{Cov }(X_i^{1)}, X_j^{(2)}) = 0 \text{ and Cov }(\mathbf{X}^{(1)}, \mathbf{X}^{(2)}) = \mathbf{0}.$$

Conversely, if Cov $(\mathbf{X}^{(1)}, \mathbf{X}^{(2)}) = \boldsymbol{\Sigma}_{12} = \mathbf{0}$, then

$$\boldsymbol{\Sigma} = \begin{bmatrix} \boldsymbol{\Sigma}_{11} & \boldsymbol{\Sigma}_{12} \\ \boldsymbol{\Sigma}_{21} & \boldsymbol{\Sigma}_{22} \end{bmatrix} = \begin{bmatrix} \boldsymbol{\Sigma}_{11} & \mathbf{0} \\ \mathbf{0} & \boldsymbol{\Sigma}_{22} \end{bmatrix}.$$

Here

$$|\mathbf{\Sigma}| = |\mathbf{\Sigma}_{11}|.|\mathbf{\Sigma}_{22}|; \mathbf{\Sigma}^{-1} = \begin{bmatrix} \mathbf{\Sigma}_{11}^{-1} & \mathbf{0} \\ \mathbf{0} & \mathbf{\Sigma}_{22}^{-1} \end{bmatrix}.$$

Therefore,

$$f_{\mathbf{X}'}(\mathbf{x}') = \{\frac{1}{(2\pi)^{m/2}\sqrt{\mathbf{\Sigma}_{11}}} \exp[-\frac{1}{2}(\mathbf{x}^{(1)} - \mu^{(1)})'\mathbf{\Sigma}_{11}^{-1}(\mathbf{x}^{(1)} - \mu^{(1)})]\}.$$

$$\{\frac{1}{(2\pi)^{(p-m)/2}\sqrt{|\mathbf{\Sigma}_{22}|}} \exp[-\frac{1}{2}(\mathbf{x}^{(2)} - \mu^{(2)})'\mathbf{\Sigma}_{22}^{-1}(\mathbf{x}^{(2)} - \mu^{(2)})]\},$$

so that $\mathbf{X}^{(1)}, \mathbf{X}^{(2)}$ are independent.

Corollary 3.2.4.1: If $\mathbf{X} = (\mathbf{X}^{(1)'}, \mathbf{X}^{(2)'}, \ldots, \mathbf{X}^{(r)'})'$ and Cov $(\mathbf{X}^{(i)}, \mathbf{X}^{(j)}) = \mathbf{0}$ for each $i \neq j (= 1, \ldots, r)$ and if $\mathbf{X} \cap N_p(\mu, \mathbf{\Sigma})$, then $\mathbf{X}^{(i)}$'s are mutually independent and not just pairwise independent.

Corollary 3.2.4.2: If $\mathbf{X} \cap N_p(\mu, \mathbf{\Sigma})$, then $\mathbf{AX}, \mathbf{BX}$ are independent *iff* $\mathbf{A\Sigma B}' = \mathbf{0}$.

Proof. The proof follows from Theorem 3.2.4, observing that

$$\text{Cov }(\mathbf{AX}, \mathbf{BX}) = \mathbf{A\Sigma B}'.$$

Corollary 3.2.4.3: If $\mathbf{X} \sim N_p(\mu, \sigma^2\mathbf{I})$ and $\mathbf{G}$ is any row-orthogonal $(q \times p)$ matrix such that $\mathbf{GG}' = \mathbf{I}$, then $\mathbf{GX}$ is independent of $(\mathbf{I} - \mathbf{G}'\mathbf{G})\mathbf{X}$.

Proof. Follows since $\mathbf{G}(\mathbf{I} - \mathbf{G}'\mathbf{G})' = \mathbf{0}$.

Corollary 3.2.4.4: Let $\mathbf{X}_1, \ldots, \mathbf{X}_n$ be a set of mutually independently and identically distributed p vectors such that

$$\mathbf{X} = \sum_{i=1}^{n} a_i \mathbf{X}_i, \ \mathbf{Y} = \sum_{i=1}^{n} b_i \mathbf{X}_i$$

where $a_1, \ldots, a_n, b_1, \ldots, b_n$ are two sets of real constants. If $\mathbf{X}_1, \ldots, \mathbf{X}_n$ are independently normally distributed p vectors and if $\sum_{i=1}^{n} a_i b_i = 0$, then $\mathbf{X}, \mathbf{Y}$ are independent.

Proof. The vectors $\mathbf{X}, \mathbf{Y}$ have p variate normal distributions. Now,

$$Cov(\mathbf{X}, \mathbf{Y}) = \sum_{i=1}^{n} a_i b_i V(\mathbf{X}_i) = \mathbf{\Sigma}.0 = \mathbf{0}.$$

Theorem 3.2.5: If the $p \times 1$ random vectors $\mathbf{X}$ and $\mathbf{Y}$ are independent and $\mathbf{X} + \mathbf{Y}$ has a p-variate normal distribution, then both $\mathbf{X}$ and $\mathbf{Y}$ are normal.

Proof. See Muirhead (1982), p.14.

We have considered two characterizing properties of normal distributions in Corollary 3.2.2.3 and Theorem 3.2.3. Other characterizations of univariate and multivariate normal distributions are due to Lukacs (1956), Laha (1957), Rao (1969, 1972, 1973), Kingman and Graybill (1970), Patil and Boswell (1970).

Theorem 3.2.6: If $\mathbf{X} = (\mathbf{X}_1', \mathbf{X}_2')' \cap N_p(\mu, \mathbf{\Sigma})$, then $\mathbf{X}_2, \mathbf{X}_{1.2} = \mathbf{X}_1 - \mathbf{\Sigma}_{12}\mathbf{\Sigma}_{22}^{-1}\mathbf{X}_2$ have the following properties and they are statistically independent.

$$\mathbf{X}_2 \cap N_s(\mu_2, \mathbf{\Sigma}_{22}), \quad \mathbf{X}_{1.2} \sim N_r(\mu_{1.2}, \mathbf{\Sigma}_{11.2})$$

where

$$r + s = p, \quad \mu_{1.2} = \mu_1 - \mathbf{\Sigma}_{12}\mathbf{\Sigma}_{22}^{-1}\mu_2, \quad \mathbf{\Sigma}_{11.2} = \mathbf{\Sigma}_{11} - \mathbf{\Sigma}_{12}\mathbf{\Sigma}_{22}^{-1}\mathbf{\Sigma}_{21}. \quad (3.2.16)$$

Proof. Note that

$$\mathbf{X}_2 = [\mathbf{0} \quad \mathbf{I}_s]\,\mathbf{X} = \mathbf{AX},$$

$$\begin{aligned}\mathbf{X}_{1.2} &= \begin{bmatrix}\mathbf{I}_r & -\mathbf{\Sigma}_{12}\mathbf{\Sigma}_{22}^{-1}\end{bmatrix}\begin{bmatrix}\mathbf{X}_1\\ \mathbf{X}_2\end{bmatrix}\\ &= \mathbf{BX}.\end{aligned}$$

Now,

$$\mathbf{A\Sigma B}' = [\mathbf{0} \quad \mathbf{I_s}]\begin{bmatrix}\mathbf{\Sigma}_{11} & \mathbf{\Sigma}_{12}\\ \mathbf{\Sigma}_{21} & \mathbf{\Sigma}_{22}\end{bmatrix}\begin{bmatrix}\mathbf{I}\\ -\mathbf{\Sigma_{22}^{-1}\Sigma_{21}}\end{bmatrix}$$

$$= [\mathbf{\Sigma}_{21} \quad \mathbf{\Sigma}_{22}]\begin{bmatrix}\mathbf{I}\\ -\mathbf{\Sigma}_{22}^{-1}\mathbf{\Sigma}_{21}\end{bmatrix} = \mathbf{\Sigma}_{21} - \mathbf{\Sigma_{21}} = \mathbf{0}.$$

Again, $\mathbf{X}_2 \cap N_s(\mu_2, \mathbf{\Sigma}_{22}), \mathbf{X}_{1.2} \cap N_r(\mu_{1.12}, \mathbf{\Sigma}_{11.2})$. Also, by Theorem 3.2.4, $\mathbf{X}_2, \mathbf{X}_{1.2}$ are independent. Hence the theorem.

We shall now use this theorem to find the conditional distribution of $\mathbf{X}_1$ given $\mathbf{X}_2 = \mathbf{x}_2$.

Theorem 3.2.7: Under the assumptions and notations of Theorem 3.2.6, the conditional distribution of $\mathbf{X}_1$ given $\mathbf{X}_2 = \mathbf{x}_2$ is r-variate normal with mean

$$\mu_1 + \mathbf{\Sigma}_{12}\mathbf{\Sigma}_{22}^{-1}(\mathbf{x}_2 - \mu_2)$$

and variance

$$\mathbf{\Sigma}_{11.2} = \mathbf{\Sigma}_{11} - \mathbf{\Sigma}_{12}\mathbf{\Sigma}_{22}^{-1}\mathbf{\Sigma}_{21}. \tag{3.2.17}$$

Proof. Since $\mathbf{X}_{1.2}$ is independent of $\mathbf{X}_2$, the conditional distribution of $\mathbf{X}_{1.2}$ is the same as its marginal distribution. Now

$$\mathbf{X}_1 = \mathbf{X}_{1.2} + \mathbf{\Sigma}_{12}\mathbf{\Sigma}_{22}^{-1}\mathbf{X}_2.$$

When $\mathbf{X}_2$ is given, $\mathbf{\Sigma}_{12}\mathbf{\Sigma}_{22}^{-1}\mathbf{x}_2$ is a constant. Hence the conditional distribution of $\mathbf{X}_1$ given $\mathbf{X}_2 = \mathbf{x}_2$ is normal with mean

$$\begin{aligned} &\mu_{1.2} + \mathbf{\Sigma}_{12}\mathbf{\Sigma}_{22}^{-1}\mathbf{x}_2 \\ &= \mu_1 + \mathbf{\Sigma}_{12}\mathbf{\Sigma}_{22}^{-1}(\mathbf{x}_2 - \mu_2) \end{aligned}$$

and variance $\mathbf{\Sigma}_{11.2}$.

Alternative Proof. The theorem can be proved directly as follows.

We partition the matrix $\mathbf{A}(= \mathbf{\Sigma}^{-1})$ at the rth row and rth column as

$$\mathbf{A} = \begin{bmatrix} \mathbf{A}_{11} & \mathbf{A}_{12} \\ \mathbf{A}_{21} & \mathbf{A}_{22} \end{bmatrix}.$$

(Note that $\mathbf{A}_{12} = \mathbf{A}'_{21}$). Consider the $p \times p$ matrix

$$\mathbf{C} = \begin{bmatrix} \mathbf{I}_r & \mathbf{0} \\ \mathbf{A}_{21}\mathbf{A}_{11}^{-1} & \mathbf{I}_s \end{bmatrix};$$

$$\mathbf{CAC}' = \begin{bmatrix} \mathbf{A}_{11} & \mathbf{0} \\ \mathbf{0} & \mathbf{A}_{22} - \mathbf{A}_{21}\mathbf{A}_{11}^{-1}\mathbf{A}_{12} \end{bmatrix}. \tag{3.2.18}$$

Transform $(\mathbf{x} - \mu)' = \mathbf{y}'\mathbf{C}$. Now

$$(\mathbf{x} - \mu)'\mathbf{A}(\mathbf{x} - \mu) = \mathbf{y}'\mathbf{CAC}'\mathbf{y} = \mathbf{y}_1'\mathbf{A}_{11}\mathbf{y}_1 + \mathbf{y}_2'\mathbf{D}\mathbf{y}_2,$$

where

$$\mathbf{y}_1 = (y_1, \ldots, y_r)', \mathbf{y}_2 = (y_{r+1}, \ldots, y_p)' \text{ and } \mathbf{D} = \mathbf{A}_{22} - \mathbf{A}_{21}\mathbf{A}_{11}^{-1}\mathbf{A}_{12}.$$

The joint distribution of $\mathbf{Y} = (Y_1, \ldots, Y_p)'$ [where $(\mathbf{X} - \mu)' = \mathbf{Y}'\mathbf{C}$] is, therefore,

$$f_{\mathbf{Y}'}(\mathbf{y}_1', \mathbf{y}_2') = \{\frac{\sqrt{|\mathbf{A}_{11}}}{(2\pi)^{r/2}} \exp(-\frac{1}{2}\mathbf{y}_1'\mathbf{A}_{11}\mathbf{y}_1)\},$$

$$\{\frac{\sqrt{|\mathbf{D}|}}{(2\pi)^{s/2}}\exp(-\frac{1}{2}\mathbf{y}_2'\mathbf{D}\mathbf{y}_2)\}, \tag{3.2.19}$$

since $|\mathbf{A}| = |\mathbf{A}_{11}|.|\mathbf{D}|$ (from (3.2.18), because $|\mathbf{C}| = 1$.)

It follows from (3.2.19) that the random vectors $\mathbf{Y}_1 = (Y_1, \ldots, Y_r)'$ and $\mathbf{Y}_2 = (Y_{r+1}, \ldots, Y_p)'$ are independent of each other and each has a multivariate normal distribution. Again,

$$\mathbf{Y}_2' = (\mathbf{X}_2 - \mu_2)'; \;\; \mathbf{Y}_1' = (\mathbf{X}_1 - \mu_1)' + (\mathbf{X}_2 - \mu_2)'\mathbf{A}_{21}\mathbf{A}_{11}^{-1}.$$

Therefore,

(a) $(X_{r+1}, \ldots X_p)'$ have a multinormal distribution with mean vector $(\mu_{r+1}, \ldots \mu_p)'$ and variance-covariance matrix

$$\mathbf{D}^{-1} = (\mathbf{A}_{22} - \mathbf{A}_{21}\mathbf{A}_{11}^{-1}\mathbf{A}_{12})^{-1} = \mathbf{\Sigma}_{22}.$$

(b) Since the second factor in (3.2.19) is the marginal distribution of $\mathbf{X}_2$, the first factor must be the conditional distribution of $\mathbf{X}_1$ given $\mathbf{X}_2$. The conditional joint distribution of $\mathbf{X}_1 = (X_1, \ldots, X_r)'$ given $(X_{r+1}, \ldots, X_p)' = \mathbf{x}_2$ is r-variate normal with expected value

$$\mu_1 - \mathbf{A}_{11}^{-1}\mathbf{A}_{12}(\mathbf{x}_2 - \mu_2) = \mu_1 + \mathbf{\Sigma}_{12}\mathbf{\Sigma}_{22}^{-1}(\mathbf{x}_2 - \mu_2) \tag{3.2.20}$$

and variance-covariance matrix $\mathbf{A}_{11}^{-1} = \mathbf{\Sigma}_{11.2}$.

(The matrix relations used in (a), (b) follow from results (A.6.9)-(A.6.12)).

Note 3.2.6: The result (b) (and formula (3.2.17)) shows that the regression of each of the variables $X_1, \ldots, X_r$ on the set $\mathbf{X}_2 = (X_{r+1}, \ldots, X_p)'$ is linear and the conditional distributions are homoscedastic (since $\mathbf{A}_{11}^{-1}$ does not depend on $\mathbf{X}_2$). If $\mathbf{X}$ has a p-variate normal distribution, the true regression of each of $X_1, \ldots, X_r$ on the set $\mathbf{X}_2$ is linear.

Special cases:

r = 1: Consider the distribution of X_1 given $X_2, \ldots, X_p$:

$$E(X_1 \mid x_2, \ldots, x_p) = \mu_1 - \sum_{j=2}^{p}(x_j - \mu_j)\frac{\sigma^{1j}}{\sigma^{11}}; \;\; V(X_1 \mid x_2, \ldots, x_p) = \frac{1}{\sigma^{11}}.$$

The quantity

$$1 - \frac{V(X_1|x_2, \ldots, x_p)}{V(X_1)} = \rho_{1.2\ldots p}^2,$$

where $\rho^2_{1.2\ldots,p}$ is the squared multiple population multiple correlation coefficient of X_1 on $X_2, \ldots X_p$. Hence,

$$\rho^2_{1.2\ldots p} = 1 - \frac{1}{\sigma^{11}\sigma_{11}}.$$

r = 2: Here

$$E(X_1|x_3, \ldots, x_p) = \mu_1 - \frac{\sigma^{22}}{H}\sum_{j=3}^{p}(x_j - \mu_j)\sigma^{j1} + \frac{\sigma^{21}}{H}\sum_{j=3}^{p}(x_j - \mu_j)\sigma^{j2},$$

$$E(X_2|x_3, \ldots, x_p) = \mu_2 - \frac{\sigma^{11}}{H}\sum_{j=3}^{p}(x_j - \mu_j)\sigma^{j2} + \frac{\sigma^{12}}{H}\sum_{j=3}^{p}(x_j - \mu_j)\sigma^{j1},$$

where $H = \sigma^{11}\sigma^{22} - (\sigma^{12})^2$.

Also,

$$D\left(\begin{pmatrix} X_1 \\ X_2 \end{pmatrix} \mid x_3, \ldots, x_p\right) = \begin{bmatrix} \sigma^{11} & \sigma^{12} \\ \sigma^{21} & \sigma^{22} \end{bmatrix}^{-1};$$

$$\text{Corr. } (x_1, x_2|X_3, \ldots, X_p) = -\frac{\sigma^{12}}{\sqrt{\sigma^{11}\sigma^{22}}}.$$

The correlation$(X_1, X_2|x_3, \ldots, x_p)$ is called the partial correlation between X_1, X_2 eliminating the effects of $X_3, \ldots, X_p$ and is denoted as $\rho_{12.3\ldots p}$. For a multivariate normal population this is independent of values of $X_3, \ldots, X_p$.

EXAMPLE 3.2.3: Suppose $\mathbf{\Sigma}$ is the equicorrelation matrix

$$\mathbf{\Sigma} = \begin{bmatrix} 1 & \rho & \ldots & \rho \\ \rho & 1 & \ldots & \rho \\ . & . & \ldots & . \\ \rho & \rho & \ldots & 1 \end{bmatrix} = (1-\rho)\mathbf{I} + \rho\mathbf{1}\mathbf{1}'$$

and we want to find the conditional distribution of $\mathbf{X}_2$ given $\mathbf{X}_1 = \mathbf{x}_1$. We have

$$E(\mathbf{X}_2|\mathbf{x}_1) = \mu_2 + \mathbf{\Sigma}_{21}\mathbf{\Sigma}_{11}^{-1}(\mathbf{x}_1 - \mu_1)$$

and variance

$$V(\mathbf{X}_2|\mathbf{x}_1) = \mathbf{\Sigma}_{22.1} = \mathbf{\Sigma}_{22} - \mathbf{\Sigma}_{21}\mathbf{\Sigma}_{11}^{-1}\mathbf{\Sigma}_{12}.$$

Here

$$\mathbf{\Sigma}_{11} = (1-\rho)\mathbf{I}_r + \rho\mathbf{1}_r\mathbf{1}_r',$$

$$\mathbf{\Sigma}_{22} = (1-\rho)\mathbf{I}_s + \rho\mathbf{1}_s\mathbf{1}_s',$$

$$\mathbf{\Sigma}_{12} = \rho\mathbf{1}_r\mathbf{1}_s'.$$

Furthermore,

$$\mathbf{\Sigma}_{11}^{-1} = (1-\rho)^{-1}[\mathbf{I}_r - \alpha\mathbf{1}_r\mathbf{1}_r'], \quad \alpha = \frac{\rho}{1+(r-1)\rho}.$$

Now,

$$\begin{aligned}\mathbf{\Sigma}_{21}\mathbf{\Sigma}_{11}^{-1} &= \rho\mathbf{1}_s\mathbf{1}_r'(1-\rho)^{-1}[\mathbf{I}_r - \alpha\mathbf{1}_r\mathbf{1}_r'] \\ &= \alpha\mathbf{1}_s\mathbf{1}_r'.\end{aligned}$$

Also,

$$\mathbf{\Sigma}_{22.1} = (1-\rho)\mathbf{I}_s + \rho(1-r\alpha)\mathbf{1}_s\mathbf{1}_s'.$$

Therefore, the conditional mean is

$$E[\mathbf{X}_2|\mathbf{x}_1] = \mu_2 + \alpha(X_{10} - \mu_{10})\mathbf{1}_s$$

where $X_{10} = \sum_{j=1}^r x_j, \mu_{10} = \sum_{j=1}^r \mu_j$.

Theorem 3.2.8: Suppose $\mathbf{X} \cap N_p(\mu, \sigma^2\mathbf{I}_p)$ and let $\mathbf{U} = \mathbf{AX}, \mathbf{V} = \mathbf{BX}$. Let $\mathbf{A}_1$ represent the linearly independent row vectors of $\mathbf{A}$ and $\mathbf{U}_1 = \mathbf{A}_1\mathbf{X}$. Suppose Cov $(\mathbf{U}, \mathbf{V}) = \mathbf{0}$. Then

(a) $\mathbf{U}_1$ is independent of $\mathbf{V}'\mathbf{V}$.
(b) $\mathbf{U}'\mathbf{U}$ is independent of $\mathbf{V}'\mathbf{V}$.

Proof. Cov $(\mathbf{U}, \mathbf{V}) = \mathbf{A}D(\mathbf{X})\mathbf{B}' = \sigma^2\mathbf{AB}' = \mathbf{0}$. Hence, every row of $\mathbf{A}$ is orthogonal to every row of $\mathbf{B}$. Let $\mathbf{B}_1$ represent the linearly independent rows of $\mathbf{B}$. Let $\mathbf{C} = \begin{bmatrix}\mathbf{A}_1 \\ \mathbf{B}_1\end{bmatrix}$.

The rows of $\mathbf{C}$ are linearly independent. The vector $\mathbf{CX}$ has a multivariate normal distribution.

Let $\mathbf{V}_1 = \mathbf{B}_1\mathbf{X}$. We have, Cov $(\mathbf{U}_1, \mathbf{V}_1) = \sigma^2\mathbf{A}_1\mathbf{B}_1' = \mathbf{0}$.

Hence, $\mathbf{U}_1, \mathbf{V}_1$ are independent.

Without loss of generality assume $\mathbf{B} = \begin{bmatrix}\mathbf{B}_1 \\ \mathbf{B}_2\end{bmatrix}$.

Since rows of $\mathbf{B}_2$ are linearly dependent on those of $\mathbf{B}_1$, let $\mathbf{B}_2 = \mathbf{HB}_1$. Hence,

$$\mathbf{V} = \mathbf{BX} = \begin{bmatrix}\mathbf{I} \\ \mathbf{H}\end{bmatrix}\mathbf{B}_1\mathbf{X} = \mathbf{MV}_1 \text{ (say) };$$

$$\mathbf{V}'\mathbf{V} = \mathbf{V}_1'\mathbf{M}'\mathbf{M}\mathbf{V}_1.$$

Since $\mathbf{U}_1$ is independent of $\mathbf{V}_1$, it is independent of $\mathbf{V}'\mathbf{V}$. Similarly, $\mathbf{U}$ can be expressed as $\mathbf{U} = \mathbf{L}\mathbf{U}_1$ and since $\mathbf{V}_1$ is independent of $\mathbf{U}_1, \mathbf{V}'\mathbf{V}$ is independent of $\mathbf{U}'\mathbf{U}$.

EXAMPLE 3.2.4: Suppose X_i's $(i = 1, \ldots, n)$ are *iid* $N(0, \sigma^2)$ variables. Show that $\bar{X}$ and $Q = \sum_{i=1}^n (X_i - \bar{X})^2$ are statistically independent ($\bar{X} = \sum_{i=1}^n X_i/n$). .

Proof. We have $\mathbf{X} \cap N_n(\mathbf{0}, \sigma^2\mathbf{I})$. Let $\bar{X} = U, \mathbf{V} = \mathbf{X} - \bar{X}\mathbf{1}_n$. Now, $\text{Cov}(\bar{X}, V_i) = 0, i = 1, \ldots, n$. Hence, $\text{Cov}(\bar{X}, \mathbf{V}) = \mathbf{0}$. Therefore, $\bar{X}$ and $\mathbf{V}'\mathbf{V}$ are statistically independent.

Note 3.2.7: It can be shown that if $X_1, \ldots, X_p$ have a joint p-variate normal distribution, then

$$\text{Prob. } [\cap(|X_j - \mu_j| \leq C_j)] \geq \Pi_{j=1}^p \text{ Prob.} [|X_j - \mu_j| \leq C_j] .$$

The standardized multivariate normal distribution has mean vector $\mathbf{0}$ and dispersion matrix $\mathbf{R} = ((\rho_{ij}))$, the correlation matrix of $\mathbf{X}$. Hence, the density function of a standardized multivariate normal distribution is

$$f_{\mathbf{X}}(\mathbf{x}) = \frac{1}{(2\pi)^{p/2}\sqrt{|\mathbf{R}|}} \exp[-\frac{1}{2}\mathbf{x}'\mathbf{R}^{-1}\mathbf{x}].$$

Generalizing the univariate probability integral $\Phi(z)$, we define

$$\Phi_p(z_1, \ldots, z_p) = \text{ Prob. } [\cap_{j=1}^p (X_j \leq z_j)]$$

$$= \frac{1}{(2\pi)^{p/2}\sqrt{|\mathbf{R}|}} \int_{-\infty}^{z_1} \cdots \int_{-\infty}^{z_p} \exp[-\frac{1}{2}\mathbf{x}'\mathbf{R}^{-1}\mathbf{x}]d\mathbf{x}. \ \square$$

In this section we have considered some properties of a multivariate normal distribution. In the next section we shall consider a random matrix whose rows are independent samples from a $N_p(\mu, \mathbf{\Sigma})$ distribution and variables obtained by transformation of such matrices.

3.3 Transformation of Normal Data Matrix

Let $\mathbf{X}_1, \ldots, \mathbf{X}_n$ be a random sample from $N_p(\mu, \mathbf{\Sigma})$. We call $\mathbf{X} = (\mathbf{X}_1, \ldots, \mathbf{X}_n)'$ a data matrix from $N_p(\mu, \mathbf{\Sigma})$. We shall consider in this section linear functions $\mathbf{AXB}$ where $\mathbf{A}, \mathbf{B}$ are matrices of fixed real numbers.

The most important functions are $\bar{\mathbf{X}}' = \frac{1}{n}\mathbf{1}'\mathbf{X}$, where $\mathbf{A} = n^{-1}\mathbf{1}'$ and $\mathbf{B} = \mathbf{I}_p$.

Theorem 3.3.1: If $\mathbf{X}(n \times p)$ is a data matrix from $N_p(\mu, \mathbf{\Sigma})$, then $\bar{\mathbf{X}} \sim N_p(\mu, \mathbf{\Sigma}/n)$.

Proof. We have $\bar{\mathbf{X}} = \frac{1}{n}\sum_{i=1}^{n}\mathbf{X}_i$; hence, $E(\bar{\mathbf{X}}) = \mu, V(\bar{\mathbf{X}}) = \mathbf{\Sigma}/n$.

We may ask under what conditions $\mathbf{Y} = \mathbf{AXB}$ form a normal data matrix, that is, rows of $\mathbf{Y}$ are independently normally distributed. This is given in the following theorem.

Theorem 3.3.2: If $\mathbf{X}(n \times p)$ is a data matrix from $N_p(\mu, \mathbf{\Sigma})$ and if $\mathbf{Y} = \mathbf{AXB}$ (where $\mathbf{A}$ is $m \times n$ and $\mathbf{B}$ is $p \times q$) then $\mathbf{Y}$ is a normal data matrix *iff*

(a) $\mathbf{A1} = \alpha\mathbf{1}$ for some α or $\mathbf{B}'\mu = \mathbf{0}$ and
(b) $\mathbf{AA}' = \beta\mathbf{I}$ for some β or $\mathbf{B}'\mathbf{\Sigma}\mathbf{B} = \mathbf{0}$.

If both conditions (a) and (b) are satisfied, $\mathbf{Y}(m \times q)$ is a normal data matrix from $N_q(\alpha\mathbf{B}'\mu, \beta\mathbf{B}'\mathbf{\Sigma}\mathbf{B})$.

Proof. Omitted.

Theorem 3.3.3: If $\mathbf{X}$ is a data matrix from $N_p(\mu, \mathbf{\Sigma})$ and $\mathbf{Y} = \mathbf{AXB}, \mathbf{Z} = \mathbf{CXD}$, then the elements of $\mathbf{Y}$ are independent of elements of $\mathbf{Z}$ *iff* either (a) $\mathbf{B}'\mathbf{\Sigma}\mathbf{D} = \mathbf{0}$ or (b) $\mathbf{AC}' = \mathbf{0}$.

The result remains valid even if rows of $\mathbf{X}$ have different means.

Proof. Omitted. (For proof of Theorems 3.3.2 and 3.3.3, interested readers may refer to Mardia *et al.* (1979, p.65)

Corollary 3.3.3.1: Under the conditions of Theorem 3.3.3, if $\mathbf{X} = (\mathbf{X}_1(n\times r), \mathbf{X}_2(n \times s))$, then $\mathbf{X}_1$ and $\mathbf{X}_{2.1} = \mathbf{X}_2 - \mathbf{X}_1\mathbf{\Sigma}_{11}^{-1}\mathbf{\Sigma}_{12}$ are independent. Also, $\mathbf{X}_1$ is a data matrix from $N_r(\mu_1, \mathbf{\Sigma}_{11})$ and $\mathbf{X}_{2.1}$ is a data matrix from $N_s(\mu_{2.1}, \mathbf{\Sigma}_{22.1})$ when $\mu_{2.1} = \mu_2 - \mathbf{\Sigma}_{21}\mathbf{\Sigma}_{11}^{-1}\mu_1, \mathbf{\Sigma}_{22.1} = \mathbf{\Sigma}_{22} - \mathbf{\Sigma}_{21}\mathbf{\Sigma}_{11}^{-1}\mathbf{\Sigma}_{12}$.

Proof. We have

$$\mathbf{I}_n\,[\mathbf{X}_1 \;\; \mathbf{X}_2]\begin{bmatrix}\mathbf{I}_r\\ \mathbf{0}\end{bmatrix} = \mathbf{X}_1.$$

Thus, considering conditions for Theorem 3.3.3,

$$\mathbf{A} = \mathbf{I}_n, \quad \mathbf{B} = \begin{bmatrix}\mathbf{I}_r\\ \mathbf{0}\end{bmatrix}.$$

Also,

$$\mathbf{I}_n \, [\mathbf{X}_1 \;\; \mathbf{X}_2] \begin{bmatrix} -\mathbf{\Sigma}_{11}^{-1}\mathbf{\Sigma}_{12} \\ \mathbf{I}_s \end{bmatrix} = \mathbf{X}_{2.1}.$$

Here

$$\mathbf{C} = \mathbf{I}_n, \quad \mathbf{D} = \begin{bmatrix} -\mathbf{\Sigma}_{11}^{-1}\mathbf{\Sigma}_{12} \\ \mathbf{I}_s \end{bmatrix}.$$

It can be checked that $\mathbf{B}'\mathbf{\Sigma}\mathbf{D} = \mathbf{0}$. Hence, by Theorem 3.3.3, $\mathbf{X}_1$ and $\mathbf{X}_{2.1}$ are independent.
Also, (considering conditions of Theorem 3.3.2) for $\mathbf{X}_1, \mathbf{A} = \mathbf{I}_n, \alpha = 1, \beta = 1$. Hence, $\mathbf{X}_1$ is a data matrix from $N_r(\alpha\mathbf{B}'\mu = \mu_1, \beta\mathbf{B}'\mathbf{\Sigma}\mathbf{B} = \mathbf{\Sigma}_{11})$.
Similarly, for $\mathbf{X}_{2.1}, \mathbf{C} = \mathbf{I}_n, \alpha = 1, \beta = 1$. Hence, $\mathbf{X}_{2.1}$ is a data matrix from $N_s(\alpha\mathbf{D}'\mu = \mu_{2.1}, \beta\mathbf{D}'\mathbf{\Sigma}\mathbf{D} = \mathbf{\Sigma}_{22.1})$.

Corollary 3.3.3.2: Under the conditions of Theorem 3.3.3, $\bar{\mathbf{X}}' = n^{-1}\mathbf{1}_n'\mathbf{X}$ is independent of $\mathbf{HX}$ and hence is independent of $n^{-1}\mathbf{X}'\mathbf{HX} = \mathbf{S}$ when $\mathbf{H} = \mathbf{I}_n - \frac{1}{n}\mathbf{1}_n\mathbf{1}_n'$.

Proof. We have $\bar{\mathbf{X}}' = \mathbf{AXB}$ where $\mathbf{A} = n^{-1}\mathbf{1}_n', \mathbf{B} = \mathbf{I}_p$. For $\mathbf{HX}, \mathbf{C} = \mathbf{H}, \mathbf{D} = \mathbf{I}_p$. Hence, $\mathbf{AC}' = n^{-1}\mathbf{1}_n'\mathbf{H} = n\mathbf{1}_n - n^{-2}\mathbf{1}_n'\mathbf{1}_n\mathbf{1}_n' = \mathbf{0}$. Hence, $\bar{\mathbf{X}}'$ and $\mathbf{HX}$ are independent.

3.4 Some Results on Quadratic Forms

We consider in this section some results on quadratic forms in multivariate normal variables.

Theorem 3.4.1: Let $\mathbf{X} \sim N_p(\mu, \sigma^2\mathbf{I}_p)$ and $\mathbf{P}$ be a non-negative symmetric matrix of rank r. Then $Q = (\mathbf{X} - \mu)'\mathbf{P}(\mathbf{X} - \mu)/\sigma^2$ is distributed as $\chi^2_{(r)}$ *iff* $\mathbf{P}$ is an idempotent matrix.

Proof. Suppose $\mathbf{P}$ is idempotent. Since $\mathbf{P}^2 = \mathbf{P}, \mathbf{P}$ has r eigenvalues, each equal to unity and the rest $(p - r)$ eigenvalues, each equal to zero. Hence, there exists an orthogonal matrix $\mathbf{T}$ such that

$$\mathbf{T}'\mathbf{PT} = \text{ Diag. } (\lambda_1, \ldots, \lambda_p) = \Lambda \text{ (say)},$$

where λ_i's are eigenvalues of $\mathbf{P}$. Let $\mathbf{Y} = \mathbf{T}'(\mathbf{X} - \mu)$. Then, $\mathbf{Y} \cap N_n(\mathbf{0}, \sigma^2\mathbf{I}_n)$. Therefore, $Q = \mathbf{Y}'\Lambda\mathbf{Y}/\sigma^2 \sim \chi^2_{(r)}$.

Conversely, suppose $Q \sim \chi^2_{(r)}$. Then

$$E(e^{tQ}) = (1 - 2t)^{-r/2}, \tag{3.4.1}$$

being the moment generating function of a Gamma distribution. Since $\mathbf{P}$ is symmetric there exists an orthogonal matrix $\mathbf{S}$ such that $\mathbf{S}'\mathbf{P}\mathbf{S} = \Lambda$ (Theorem A.12.1). Let $\mathbf{X} - \mu = \mathbf{S}\mathbf{Y}$. Therefore, $Q = \sum_{i=1}^{n} \lambda_i Y_i^2/\sigma^2$. Now

$$E[\exp(tQ)] = E[\exp(t\sum_{i=1}^{n} \lambda_i Y_i^2/\sigma^2)].$$

But $\mathbf{Y} = \mathbf{S}'(\mathbf{X} - \mu) \cap N_p(\mathbf{0}, \sigma^2 \mathbf{I}_p)$ so that Y_i's are independent $N(0, \sigma^2)$ variables. Hence, $E(\exp(t\lambda_i Y_i^2/\sigma^2)) = (1 - 2\lambda_i t)^{-1/2}$. Therefore, from (3.4.1),

$$(1-2t)^{-r/2} = \Pi_{i=1}^{p}(1-2t\lambda_i)^{-1/2}$$

identically in t for sufficiently small $|t|$. By the uniqueness property of polynomial roots we must have r λ_i's, each equal to unity and the rest λ_i's, each equal to zero. Hence, $\mathbf{P}^2 = \mathbf{P}$ and rank $(\mathbf{P}) = r$.

EXAMPLE 3.4.1: Let $\mathbf{X} \cap N_p(\mathbf{0}, \mathbf{I}_p), \mathbf{X}\mathbf{X}' = \mathbf{X}'\mathbf{A}\mathbf{X} + \mathbf{X}'\mathbf{B}\mathbf{X}$ and $\mathbf{X}'\mathbf{A}\mathbf{X} \cap \chi^2_{(r)}$; then

$$\mathbf{X}'\mathbf{B}\mathbf{X} \cap \chi^2_{(p-r)}.$$

Here $\mathbf{A}$ is symmetric idempotent of rank r. Also, $\mathbf{B} = \mathbf{I}_p - \mathbf{A}$ is symmetric and $\mathbf{B}^2 = \mathbf{B}$ so that $\mathbf{B}$ is idempotent. Hence, $\mathbf{X}'\mathbf{B}\mathbf{X} \cap \chi^2_{(p)}$. Now

$$p = \text{ rank } \mathbf{B} = \text{trace } \mathbf{B} = p - \text{ trace } \mathbf{A} = p - \text{ rank } \mathbf{A} = p - r.$$

EXAMPLE 3.4.2: If $\mathbf{X} \cap N_p(\mathbf{0}, \mathbf{I}_p)$, prove that

$$\sum_{i=1}^{p}(X_i - \bar{X})^2 \cap \chi^2_{(p-1)}.$$

Proof. We have

$$\sum_{i=1}^{p}(X_i - \bar{X})^2 = \mathbf{X}'\mathbf{A}\mathbf{X}$$

where

$$\mathbf{A} = (\delta_{ij} - \frac{1}{p}),\ \delta_{ij} = 1(0)$$

if $i = (\neq) j$. Hence, $\mathbf{A}$ is idempotent. Also, rank $(\mathbf{A}) =$ trace $(\mathbf{A}) = p - 1$.

Theorem 3.4.2: If $\mathbf{X}$ is $N_p(\mu, \mathbf{\Sigma})$ and $\mathbf{\Sigma}$ is non-singular, then

(a) $(\mathbf{X} - \mu)'\mathbf{\Sigma}^{-1}(\mathbf{X} - \mu) \sim \chi^2_{(p)}$;
(b) $\mathbf{X}'\mathbf{\Sigma}^{-1}\mathbf{X} \cap \chi^2(p, \lambda)$ where $\lambda = \mu'\mathbf{\Sigma}^{-1}\mu$.

Proof. (a) Let $\mathbf{\Sigma} = \mathbf{C}\mathbf{C}'$ where $\mathbf{C}$ is non-singular. But $\mathbf{U} = \mathbf{C}^{-1}(\mathbf{X} - \mu)$ so that $\mathbf{U}$ is $N_p(\mathbf{0}, \mathbf{I}_p)$. Also

$$(\mathbf{X} - \mu)'\mathbf{\Sigma}^{-1}(\mathbf{X} - \mu) = \mathbf{U}'\mathbf{U}$$

which is a sum of squares of p-independent $N(0,1)$ variables and hence is $\chi^2_{(p)}$.

(b) Let $\mathbf{V} = \mathbf{C}^{-1}\mathbf{X}$. Hence

$$\mathbf{V} \cap N_p(\mathbf{C}^{-1}\mu, \mathbf{I}_p) \text{ and } \mathbf{X}'\mathbf{\Sigma}^{-1}\mu = \mathbf{V}'\mathbf{V}$$

which follows $\chi^2(p, \lambda)$ where $\lambda = (\mathbf{C}^{-1}\mu)'(\mathbf{C}^{-1}\mu) = \mu'\mathbf{\Sigma}^{-1}\mu$.

Note 3.4.1: The theorem is important in testing the hypothesis about the mean vector of a normal population when $\mathbf{\Sigma}$ is known.

Suppose that $\mathbf{X}_1, \ldots, \mathbf{X}_n$ is a random sample from $N_p(\mu, \mathbf{\Sigma})$. Then $\bar{\mathbf{X}} = (\bar{X}_1, \ldots, \bar{X}_p)'$ where $\bar{X}_j = \sum_{i=1}^n X_{ij}/n (j = 1, \ldots, p)$ follows $N_p(\mu, \mathbf{\Sigma}/n)$ by Theorem 3.3.1.

Suppose we want to test $H_0(\mu = \mu_0)$ against $H(\mu \neq \mu_0), \mathbf{\Sigma}$ being known. If the null hypothesis is true, the statistic

$$W = n(\bar{\mathbf{X}} - \mu_0)'\mathbf{\Sigma}^{-1}(\bar{\mathbf{X}} - \mu_0) \cap \chi^2_{(p)}.$$

The critical region for H_0 against H would be to reject H_0 if

$$W > \chi^2_{p;\alpha}$$

where $\chi^2_{p;\alpha}$ is the upper $100\alpha\%$ point of the $\chi^2_{(p)}$ distribution.

If the null hypothesis is not true, $(\bar{\mathbf{X}} - \mu_0) \cap N_p((\mu - \mu_0), n^{-1}\mathbf{\Sigma})$. Hence, by the second part of Theorem 3.4.2, W follows $\chi^2(p, \lambda)$ where $\lambda = n(\mu - \mu_0)'\mathbf{\Sigma}^{-1}(\mu - \mu_0)$. Here $\chi^2(p, \lambda)$ denotes a noncentral χ^2 with p degrees of freedom and noncentrality parameter (ncp) λ. The power of the test is

$$\gamma(\lambda) = P\{\chi^2(p, \lambda) > \chi^2_{p;\alpha}\}$$

the value of which can be found from the table of $\chi^2(p, \lambda)$ distribution.

EXAMPLE 3.4.3: Considering the variables X_1, X_2, X_3 of the bird data in Table 2.E.1, we have the sample mean vector $\bar{\mathbf{x}}' = (157.98,\ 241.33,\ 31.46)$. Suppose we want to test $\mu = \mu_0 = (150,\ 250,\ 30)'$ and it is known that

$$\mathbf{\Sigma} = \begin{bmatrix} 15 & 15 & 2 \\ 15 & 30 & 3 \\ 2 & 3 & 1 \end{bmatrix}.$$

Here $(\bar{\mathbf{x}} - \mu) = (7.98,\ -8.67,\ 1.46)'$. The statistic W has the value 1280.32. The null hypothesis $H_0(\mu = \mu_0)$ is rejected.

Theorem 3.4.3: If $\mathbf{X}$ follows $N_p(\mu, \mathbf{I}_p)$ and $\mathbf{B}$ is a $p \times p$ symmetric matrix then $\mathbf{X}'\mathbf{B}\mathbf{X}$ follows non-central χ^2 distribution *iff* $\mathbf{B}$ is idempotent. In this case *d.f.* of χ^2 is r =rank $(\mathbf{B})$ and the non-centrality parameter $\lambda = \mu'\mathbf{B}\mu$.

Proof. Omitted.

Corollary 3.4.3.1: If $\mathbf{X} \cap N_p(\mu, \mathbf{\Sigma})$ where $\mathbf{\Sigma}$ is non-singular and $\mathbf{X}, \mu, \mathbf{\Sigma}$ are partitioned as

$$\mathbf{X} = \begin{bmatrix} \mathbf{X}_1 \\ \mathbf{X}_2 \end{bmatrix}, \quad \mu = \begin{bmatrix} \mu_1 \\ \mu_2 \end{bmatrix}, \quad \mathbf{\Sigma} = \begin{bmatrix} \mathbf{\Sigma}_{11} & \mathbf{\Sigma}_{12} \\ \mathbf{\Sigma}_{21} & \mathbf{\Sigma}_{22} \end{bmatrix}$$

where $\mathbf{X}_1$ and μ_1 are $r \times r$ and $\mathbf{\Sigma}_{11}$ is $r \times r$, then

$$Q = (\mathbf{X} - \mu)'\mathbf{\Sigma}^{-1}(\mathbf{X} - \mu) - (\mathbf{X}_1 - \mu_1)'\mathbf{\Sigma}^{-1}(\mathbf{X}_1 - \mu_1)$$

follows $\chi^2_{(p-r)}$.

Proof. Let $\mathbf{\Sigma} = \mathbf{C}\mathbf{C}'$ where $\mathbf{C}$ is non-singular. Partition $\mathbf{C}$ as

$$\mathbf{C} = \begin{bmatrix} \mathbf{C}_1 \\ \mathbf{C}_2 \end{bmatrix}$$

where $\mathbf{C}_1$ is $(r \times p)$ so that $\mathbf{\Sigma}_{11} = \mathbf{C}_1\mathbf{C}_1'$. Let $\mathbf{U} = \mathbf{C}^{-1}(\mathbf{X} - \mu)$. Then

$$\mathbf{U} \cap N_p(\mathbf{0}, \mathbf{I}_p) \text{ and } \mathbf{X}_1 - \mu_1 = \mathbf{C}_1\mathbf{U}.$$

Then

$$\begin{aligned} Q &= \mathbf{U}'\mathbf{U} - \mathbf{U}'\mathbf{C}_1'(\mathbf{C}_1\mathbf{C}_1')^{-1}\mathbf{C}_1\mathbf{U} \\ &= \mathbf{U}'[\mathbf{I}_p - \mathbf{C}_1'(\mathbf{C}_1\mathbf{C}_1')^{-1}\mathbf{C}_1]\mathbf{U} \\ &= \mathbf{U}'(\mathbf{I}_p - \mathbf{P})\mathbf{U} \end{aligned}$$

where $\mathbf{P} = \mathbf{C}_1'(\mathbf{C}_1\mathbf{C}_1')^{-1}\mathbf{C}_1$. Now $\mathbf{P}$ is symmetric and idempotent. Hence, $\mathbf{B} = (\mathbf{I}_p - \mathbf{P})$ is also so. Therefore, the result follows by Theorem 3.4.3.

The Theorem 3.4.3 is generalized into Theorems 3.4.4 and 3.4.5 below.

Theorem 3.4.4: Suppose $\mathbf{X} \cap N_p(\mathbf{0}, \mathbf{\Sigma})$. A necessary and sufficient condition for $\mathbf{X}'\mathbf{B}\mathbf{X}$ to have a χ^2 distribution is $\mathbf{\Sigma}\mathbf{B}\mathbf{\Sigma}\mathbf{B}\mathbf{\Sigma} = \mathbf{\Sigma}\mathbf{B}\mathbf{\Sigma}$, in which case the *d.f.* of χ^2 is $r =$ rank $(\mathbf{B}\mathbf{\Sigma})$.

Proof. Omitted.

Theorem 3.4.5: If $\mathbf{X}$ is $N_p(\mu, \mathbf{\Sigma})$, where $\mathbf{\Sigma}$ is non-singular and $\mathbf{B}$ is $p \times p$ symmetric matrix, then $\mathbf{X}'\mathbf{B}\mathbf{X}$ is $\chi^2(r, \lambda)$ where $r =$ rank $(\mathbf{B}), \lambda = \mu'\mathbf{\Sigma}\mu$, *iff* $\mathbf{B}\mathbf{\Sigma}$ is idempotent.

Proof. Let $\mathbf{C}$ be a non-singular $p \times p$ matrix such that $\mathbf{C\Sigma C'} = \mathbf{I}_p$. Put $\mathbf{Y} = \mathbf{CX}$. Then $\mathbf{X'BX} = \mathbf{Y'C^{-1'}BC^{-1}Y}$, where $\mathbf{Y} \cap N_p(\mathbf{C}\mu, \mathbf{I}_p)$. It follows from Theorem 3.4.3 that $\mathbf{X'BX}$ is non-central χ^2 *iff* $\mathbf{C^{-1'}BC^{-1}}$ is idempotent. It is, therefore, sufficient to show that this is so *iff* $\mathbf{B\Sigma}$ is idempotent. If $\mathbf{B\Sigma}$ is idempotent, $\mathbf{B} = \mathbf{B\Sigma B} = \mathbf{BC^{-1}C^{-1'}B}$; hence

$$\mathbf{C^{-1'}BC^{-1}} = (\mathbf{C^{-1'}BC^{-1}})(\mathbf{C^{-1'}BC^{-1}})$$

so that $\mathbf{C^{-1'}BC^{-1}}$ is idempotent. If $\mathbf{C^{-1'}BC^{-1}}$ is idempotent, then

$$\mathbf{C^{-1'}BC^{-1}} = \mathbf{C^{-1'}BC^{-1}C^{-1'}BC^{-1}} = \mathbf{C^{-1'}B\Sigma BC^{-1}}$$

so that $\mathbf{B} = \mathbf{B\Sigma B}$ and hence $\mathbf{B\Sigma}$ is idempotent.

3.5 Multivariate Central Limit Theorem

We now consider multivariate central limit theorem due to Cramer (1946) and Anderson (1984).

Theorem 3.5.1: Let $\mathbf{X}_1, \mathbf{X}_2, \ldots$ be a sequence of independently and identically distributed p-dimensional random variables each with mean μ and covariance matrix $\mathbf{\Sigma}$. Let

$$\bar{\mathbf{X}}_n = \frac{1}{n}\sum_{i=1}^{n}\mathbf{X}_i, \ n \geq 1.$$

Then as $n \to \infty$, the asymptotic distribution of

$$\sqrt{n}(\bar{\mathbf{X}}_n - \mu) = n^{-1/2}\sum_{i=1}^{n}(\mathbf{X}_i - \mu)$$

is $N(\mathbf{0}, \mathbf{\Sigma})$.

Proof. Let

$$\mathbf{Y}_n = n^{-1/2}\sum_{i=1}^{n}(\mathbf{X}_i - \mu).$$

By the continuity theorem of the characteristic function it suffices to show that $\phi_n(t)$, the characteristic function of $\mathbf{Y}_n$ converges to

$$\exp(-\frac{1}{2}\mathbf{t'\Sigma t}),$$

the characteristic function of $N_p(\mathbf{0}, \mathbf{\Sigma})$ distribution. Now the characteristic function of $\mathbf{t'Y}_n$, when $\mathbf{t} \in \mathcal{R}^p$, is

$$f_n(\alpha, \mathbf{t}) = E[\exp(i\alpha\mathbf{t'Y}_n)]$$

considered as a function of $\alpha \in \mathcal{R}^1$. Again,

$$\mathbf{t}'\mathbf{Y}_n = n^{-1/2}\sum_{j=1}^{n}(\mathbf{t}'\mathbf{X}_j - \mathbf{t}'\mu).$$

Now, $\mathbf{t}'\mathbf{X}_j - \mathbf{t}'\mu$ is a random variable in $\mathcal{R}^1$. It, therefore, follows by the univariate central limit theorem that as $n \to \infty$, the asymptotic distribution of $\mathbf{t}'\mathbf{Y}_n$ is $N(0, \mathbf{t}'\mathbf{\Sigma}\mathbf{t})$ and hence, as $n \to \infty$,

$$f_n(\alpha, \mathbf{t}) \to \exp(-\frac{1}{2}\alpha^2\mathbf{t}'\mathbf{\Sigma}\mathbf{t})$$

for all $\mathbf{t}$ and α. Putting $\alpha = 1$,

$$\phi_n(\mathbf{t}) = f_n(1, \mathbf{t}) \to \exp(-\frac{1}{2}\mathbf{t}'\mathbf{\Sigma}\mathbf{t}),$$

as $n \to \infty$. Hence the proof.

Corollary 3.5.1.1: Under the conditions of Theorem 3.5.1, for large n

$$n(\bar{\mathbf{X}} - \mu)'\mathbf{\Sigma}^{-1}(\bar{\mathbf{X}} - \mu) \tag{3.5.1}$$

is distributed approximately as $\chi^2_{(p)}$ (vide Theorem 3.4.2(a)). Again, since for large n, $\mathbf{S}$ is very close t$\mathbf{\Sigma}$ with high probability (vide note 2.8.1),

$$n(\bar{\mathbf{X}} - \mu)'\mathbf{S}^{-1}(\bar{\mathbf{X}} - \mu) \tag{3.5.2}$$

is distributed approximately as $\chi^2_{(p)}$. The approximation (3.5.2) holds if n is large relative to p.

When the random vector $\mathbf{X}_i$ are independently but not identically distributed, the following central limit theorem holds.

Theorem 3.5.2: Let $\{\mathbf{X}_i\}$ be independent random vectors with means $\{\mu_i\}$ and dispersion matrix $\{\mathbf{\Sigma}_i\}$ and distribution function $\{F_i\}$. Suppose that

$$\frac{\mathbf{\Sigma}_1 + \ldots \mathbf{\Sigma}_n}{n} \to \mathbf{\Sigma}, \quad \text{as } n \to \infty$$

and that

$$\frac{1}{n}\sum_{i=1}^{n}\int_{||\mathbf{x}_i-\mu_i||\geq\epsilon\sqrt{n}} ||\mathbf{x}_i - \mu_i||^2 dP(\mathbf{x}_i) \to \infty, \quad \text{as } n \to \infty, \quad \text{for each } \epsilon > 0.$$

Then $\frac{1}{n}\sum_{i=1}^{n}\mathbf{X}_i$ is asymptotically normal $N(\frac{1}{n}\sum_{i=1}^{n}\mu_i, \frac{1}{n}\mathbf{\Sigma})$. Here $||\mathbf{z}|| = \sqrt{\sum_{i=1}^{n} z_i^2}$ where $\mathbf{z} = (z_1, \ldots, z_p)'$.

The Theorem 3.5.3 below follows as an application of the central limit theorem 3.5.1.

Consider the following notation. If $\mathbf{T}$ is a $p \times q$ matrix

$$\mathbf{T}(\mathbf{t}_1, \mathbf{t}_2, \ldots, \mathbf{t}_q)$$

where $\mathbf{t}_i$ is $p \times 1 (i = 1, \ldots, q)$, then

$$\text{vec }(\mathbf{T}) = \begin{bmatrix} \mathbf{t}_1 \\ \mathbf{t}_2 \\ \cdot \\ \cdot \\ \cdot \\ \cdot \\ \mathbf{t}_q \end{bmatrix}$$

which is $pq \times 1$.

In the following theorem when we talk about the asymptotic normality of a random matrix $\mathbf{T}$ we mean the asymptotic normality of vec($\mathbf{T}$).

Theorem 3.5.3: Let $\mathbf{X}_1, \mathbf{X}_2, \ldots$ be a sequence of independently and identically distributed random vectors with finite fourth order moments and mean μ and covariance matrix $\mathbf{\Sigma}$ and let

$$\mathbf{A}(n) = \sum_{i=1}^{n} (\mathbf{X}_i - \bar{\mathbf{X}}_n)(\mathbf{X}_i - \bar{\mathbf{X}}_n)'$$

where $\bar{\mathbf{X}}_n = \frac{1}{n}\sum_{i=1}^{n} \mathbf{X}_i$. Then the asymptotic distribution of $\mathbf{T}(n) = n^{-1/2}(\mathbf{A}(n) - n\mathbf{\Sigma})$ is normal with mean $\mathbf{0}$ and covariance matrix

$$\mathbf{V} = \text{ Cov } [\text{ vec } ((\mathbf{X}_1 - \mu)(\mathbf{X}_1 - \mu)')].$$

Corollary 3.5.3.1: Let $\mathbf{S} = (n-1)^{-1}\mathbf{A}(n)$. Under the conditions of Theorem 3.5.3, the asymptotic distribution of $\mathbf{U}(n) = \sqrt{n}(\mathbf{S}(n) - \mathbf{\Sigma})$ is normal with mean $\mathbf{0}$ and covariance matrix $\mathbf{V}$.

It may be noted that this asymptotic normal distribution is singular, because $\mathbf{V}$ is singular. This is due to the fact that $\mathbf{V}$ is $p^2 \times p^2$ covariance matrix in the asymptotic distribution of vec($\mathbf{T}(n)$) or vec ($\mathbf{U}(n)$) and, because $\mathbf{T}(n)$ and $\mathbf{U}(n)$ are symmetric, these vectors have repeated elements.

EXAMPLE 3.5.1: Let $\mathbf{X}_1 = (X_{11}, X_{12})', \mathbf{X}_2, \ldots$ be a sequence of *iid* 2×1 random vector, having mean $\mu = (\mu_1, \mu_2)'$, covariance matrix $\mathbf{\Sigma}$ and finite fourth order moments. Let

$$\mathbf{S}(n) = \frac{1}{n-1} \sum_{i=1}^{n} (\mathbf{X}_i - \bar{\mathbf{X}}_n)(\mathbf{X}_i - \bar{\mathbf{X}}_n)'$$

where $\bar{\mathbf{X}}_n = \frac{1}{n}\sum_{i=1}^{n}\mathbf{X}_i$. Then the asymptotic distribution of the sample covariance matrix $\mathbf{S}(n)$ is asymptotically 4-variate normal with mean $\mathbf{\Sigma}$ and covariance matrix $\mathbf{V}/n$ where

$$\mathbf{V} = \text{Cov}\,[\text{ vec }(\mathbf{X}_1 - \bar{\mathbf{X}})(X_1 - \bar{\mathbf{X}}_1)']$$

$$= \text{Cov}\begin{bmatrix} (X_{11} - \bar{X}_1)^2 \\ (X_{11} - \bar{X}_1)(X_{12} - \bar{X}_1) \\ (X_{12} - \bar{X}_1)(X_{11} - \bar{X}_1) \\ (X_{12} - \bar{X}_1)^2 \end{bmatrix},$$

where $\bar{\mathbf{X}}_1 = (\bar{X}_{11},\ \bar{X}_{12})'$. This asymptotic normal distribution is singular, because the elements in the vector, vec $(\mathbf{X}_1 - \bar{\mathbf{X}}_1)(\mathbf{X}_1 - \bar{\mathbf{X}}_1)'$ are not all mutually independent. Again, the normality of the random vector $\mathbf{S}(n)$ means the asymptotic normality of vec $(\mathbf{S}(n))$.

Remark 3.5.1: Given an underlying distribution for the $\mathbf{X}_i$, it is rather tedious to find the elements of asymptotic covariance matrix, since this involves finding all the fourth order mixed moments of the distribution. However, the calculations are straightforward if the sampling is from a $N_p(\mu, \mathbf{\Sigma})$ distribution. In this case, the elements of asymptotic covariance matrix are given by $\text{Cov}(u_{ij}(n), u_{ke}(n)) = \sigma_{ik}\sigma_{je} + \sigma_{ie}\sigma_{jk}$.

We note below an asymptotic property of a multivariate normal distribution.

Theorem 3.5.4: Let $\mathbf{U}(n) = (U_1(n), \ldots, U_m(n))'$ be a random vector (depending on sample size n) and $\mathbf{b} = (b_1, \ldots, b_m)'$ be a fixed vector. Assume $\sqrt{n}(\mathbf{U}(n) - \mathbf{b})$ is asymptotically $N_m(\mathbf{0}, \mathbf{T})$.

Let $\mathbf{V} = f(\mathbf{U})$ be a function of $\mathbf{U}$ with first and second order derivatives in the neighborhood of $\mathbf{U} = \mathbf{b}$. Let $\partial f(\mathbf{U})/\partial U_i]_{\mathbf{U}=\mathbf{b}}$ be the ith component (column vector) of $\mathbf{\Phi_b}$. Then the distribution of

$$\sqrt{n}[f(\mathbf{U}) - f(\mathbf{b})] \text{ is } N(0, \mathbf{\Phi}_{\mathbf{b}}'\mathbf{T}\mathbf{\Phi}_{\mathbf{b}}).$$

The theorem is proved in Cramer (1946, p.366).

EXAMPLE 3.5.2: Let $\mathbf{X}$ be a multinomial random variable with parameters $a_1, \ldots, a_p (a_i > 0, i = 1, \ldots, p; \sum_{i=1}^{p} a_i = 1)$ and n, a positive integer.

$$P(\mathbf{X} = \mathbf{n}) = \begin{cases} \frac{n!}{n_1! \ldots n_p!}\Pi_{i=1}^{p} a_i^{n_i}, \ n_i \geq 0, (i = 1, \ldots p), \ \sum_{i=1}^{p} n_i = n \\ 0, \qquad \text{otherwise.} \end{cases}$$

To find the mean and the variance of $\mathbf{X}$, we proceed as follows.

Let $\mathbf{e}_j = (0, \ldots, 0, 1, 0, \ldots, 0)'$ be a $p \times 1$ vector with 1 in the jth place and zero elsewhere. Let $\mathbf{Y}_1, \mathbf{Y}_2, \ldots$ be *iid* random vectors with $P(\mathbf{Y} = \mathbf{e}_j) = a_j (j = 1, \ldots, p)$. Then

$$E(\mathbf{Y}) = \sum_{j=1}^{p} a_j \mathbf{e}_j = \mathbf{a} = (a_1, \ldots, a_p)',$$

$$V(\mathbf{Y}) = E(\mathbf{Y}\mathbf{Y}') - \mathbf{a}\mathbf{a}' = \text{ Diag. } (\mathbf{a}) - \mathbf{a}\mathbf{a}' = \mathbf{\Sigma} \text{ (say) }.$$

Let $\mathbf{X}(n) = \sum_{i=1}^{n} Y_i$. Clearly, $\mathbf{X}(n)$ has a multinomial distribution as given above.

$$\begin{aligned} E(\mathbf{X}(n)) &= n\mathbf{a}, \\ V(\mathbf{X}(n)) &= n[\text{ Diag. } (\mathbf{a} - \mathbf{a}\mathbf{a}')] = n\mathbf{\Sigma}. \end{aligned}$$

By the Central Limit Theorem 3.5.1,

$$\sqrt{n}(\bar{\mathbf{Y}}(n) - \mathbf{a}) = n^{-1/2}(\mathbf{X}(n) - n) \sim N_p(\mathbf{0}, \mathbf{\Sigma}), \quad \text{asymptotically },$$

where $\bar{\mathbf{Y}}(n) = \sum_{i=1}^{n} \mathbf{Y}_i/n$, that is, $\mathbf{X}(n) \sim N_p(n\mathbf{a}, n\mathbf{\Sigma})$ asymptotically.

Let

$$\mathbf{V}(n) = (\sqrt{\bar{Y}_1(n)}, \ldots, \sqrt{\bar{Y}_p(n)})' = f(\bar{\mathbf{Y}}(n)),$$

$$f(\mathbf{a}) = (\sqrt{a_1}, \ldots, \sqrt{a_p})' = \mathbf{g} \text{ (say) }.$$

Now, $\left.\frac{\partial f(\bar{\mathbf{Y}})}{\partial \bar{Y}_i}\right]_{\bar{\mathbf{Y}}=\mathbf{a}} = (0, \ldots, 0, \frac{1}{2\sqrt{a_i}}, 0, \ldots, 0)'$. By Theorem 3.5.3,

$$\sqrt{n}[\mathbf{V}(n) - f(\mathbf{a})] \sim N_p(\mathbf{0}, \frac{1}{4}\mathbf{A}\mathbf{\Sigma}\mathbf{A} = \frac{1}{4}(\mathbf{I} - \mathbf{g}\mathbf{g}') = \mathbf{W} \text{ (say) })$$

where $\mathbf{A} = \text{ Diag. } (\frac{1}{\sqrt{a_1}}, \ldots, \frac{1}{\sqrt{a_p}})$. Note that since $\sum_{i=1}^{p} g_i^2 = 1, \mathbf{W}\mathbf{g} = \mathbf{0}$ and hence $\mathbf{W}$ is singular.

In the next section we shall consider the estimation of parameters of the distribution.

3.6 Maximum Likelihood Estimators of Parameters

In this section we shall consider maximum likelihood estimation of parameters μ and $\mathbf{\Sigma}$ of a multivariate normal distribution.

3.6.1 *Multivariate normal likelihood*

Let $\mathbf{x}_1, \ldots \mathbf{x}_n$ be a random sample from a $N(\mu, \mathbf{\Sigma})$ population. The joint density function of the observations $f(\mathbf{x}_1, \ldots, \mathbf{x}_n)$ is the product of all marginal densities $f(\mathbf{x}_i)(i = 1, \ldots, n)$ and considered as a function of $(\mu, \mathbf{\Sigma})$ is the likelihood function of $(\mu, \mathbf{\Sigma})$ given the values of the observations $\mathbf{x}_1, \ldots, \mathbf{x}_p$. Hence, from (3.2.11), the likelihood function of $(\mu, \mathbf{\Sigma})$ is

$$L(\mu, \mathbf{\Sigma} : \mathbf{X}) = |2\pi\mathbf{\Sigma}|^{-n/2} \exp\{-\frac{1}{2}\sum_{i=1}^{n}(\mathbf{x}_i - \mu)'\mathbf{\Sigma}^{-1}(\mathbf{x}_i - \mu)\}. \tag{3.6.1}$$

Hence,

$$l(\mu, \mathbf{\Sigma} : \mathbf{X}) = \log L(\mu, \mathbf{\Sigma}; \mathbf{X}) = -\frac{n}{2}\log|2\pi\mathbf{\Sigma}| - \frac{1}{2}\sum_{i=1}^{n}(\mathbf{x}_i - \mu)'\mathbf{\Sigma}^{-1}(\mathbf{x}_i - \mu). \tag{3.6.2}$$

Now,

$$(\mathbf{x}_i - \mu)'\mathbf{\Sigma}^{-1}(\mathbf{x}_i - \mu) = (\mathbf{x}_i - \bar{\mathbf{x}})'\mathbf{\Sigma}^{-1}(\mathbf{x}_i - \bar{\mathbf{x}}) + (\bar{\mathbf{x}} - \mu)'\mathbf{\Sigma}^{-1}(\bar{\mathbf{x}} - \mu) + 2(\bar{\mathbf{x}} - \mu)'\mathbf{\Sigma}^{-1}(\mathbf{x}_i - \bar{\mathbf{x}}). \tag{3.6.3}$$

When summed over $i = 1, \ldots, n$, the final term in the right hand side of (3.6.3) vanishes. Again,

$$(\mathbf{x}_i - \bar{\mathbf{x}})'\mathbf{\Sigma}^{-1}(\mathbf{x}_i - \bar{\mathbf{x}}) = \text{ trace } \{\mathbf{\Sigma}^{-1}(\mathbf{x}_i - \bar{\mathbf{x}})(\mathbf{x}_i - \bar{\mathbf{x}})'\} \text{ by (A.9(7))}. \tag{3.6.4}$$

Hence,

$$\sum_{i=1}^{n}(\mathbf{x}_i - \mu)'\mathbf{\Sigma}^{-1}(\mathbf{x}_i - \mu) = \text{ tr } \{\mathbf{\Sigma}^{-1}\sum_{i=1}^{n}(\mathbf{x}_i - \bar{\mathbf{x}})(\mathbf{x}_i - \bar{\mathbf{x}})'\} + n(\bar{\mathbf{x}} - \mu)'\mathbf{\Sigma}^{-1}(\bar{\mathbf{x}} - \mu). \tag{3.6.5}$$

Writing $\sum_{i=1}^{n}(x_i - \bar{\mathbf{x}})(\mathbf{x}_i - \bar{\mathbf{x}})' = n\mathbf{S}_n$,

$$\begin{aligned} l(\mu, \mathbf{\Sigma} : \mathbf{X}) &= -\tfrac{n}{2}\log|2\pi\mathbf{\Sigma}| - \tfrac{n}{2} \text{ tr } \mathbf{\Sigma}^{-1}\mathbf{S}_n - \tfrac{n}{2}(\bar{\mathbf{x}} - \mu)'\mathbf{\Sigma}^{-1}(\bar{\mathbf{x}} - \mu) \\ &= -\tfrac{n}{2}\log|2\pi\mathbf{\Sigma}| - \tfrac{n}{2} \text{ tr } (\mathbf{\Sigma}^{-1})\{\mathbf{S}_n + (\bar{\mathbf{x}} - \mu)(\bar{\mathbf{x}} - \mu)'\}. \end{aligned} \tag{3.6.6}$$

In particular, when $\mu = \mathbf{0}$ and $\mathbf{\Sigma} = \mathbf{I}$ (3.6.6) becomes

$$l(\mu, \mathbf{\Sigma} : \mathbf{X}) = -\frac{np}{2}\log(2\pi) - \frac{n}{2} \text{ tr } \mathbf{S}_n - \frac{n}{2}\bar{\mathbf{x}}\bar{\mathbf{x}}'. \tag{3.6.7}$$

3.6.2 *Maximum likelihood estimation*

The maximum likelihood estimates (*mle*'s) of parameters are those values of μ and $\mathbf{\Sigma}$ for which the likelihood (3.6.1) or the log-likelihood function (3.6.2) is maximum.

(a) The unrestricted case: We first consider the case when there is no restriction on the parameters, except that $\mu \in \mathcal{R}^p, \mathbf{\Sigma}$ is positive definite.

Assume that $n \geq p+1$. Note that by Corollary 4.2.9.1, $\mathbf{S}_n > 0$ with probability one. Writing $\mathbf{A} = ((a_{ij}))(= \mathbf{\Sigma}^{-1})$, we have, from (3.6.6), the log likelihood function

$$l(\mu, \mathbf{A} : \mathbf{X}) = -\frac{np}{2}\log(2\pi) + \frac{n}{2}\log|\mathbf{A}| - \frac{n}{2}\text{ tr }(\mathbf{A}\mathbf{S}_n) - \frac{n}{2}\text{ tr }\mathbf{A}(\bar{\mathbf{x}}-\mu)(\bar{\mathbf{x}}-\mu)'. \tag{3.6.8}$$

Differentiating with respect to μ,

$$\frac{\partial l}{\partial \mu} = n\mathbf{A}(\bar{\mathbf{x}} - \mu), \quad \text{by (A.15.10)} . \tag{3.6.9}$$

To calculate $\frac{\partial l}{\partial \mathbf{A}}$, consider

$$\frac{\partial \log|\mathbf{A}|}{\partial a_{ij}} = \frac{1}{|\mathbf{A}|}.\frac{\partial|\mathbf{A}|}{\partial a_{ij}} = \begin{cases} A_{ii}/|\mathbf{A}|, & i = j \\ 2A_{ij}/|\mathbf{A}|, & i \neq j, \end{cases}$$

where A_{ij} is the cofactor of a_{ij} in $|\mathbf{A}|$. Thus,

$$\frac{\partial \log|\mathbf{A}|}{\partial \mathbf{A}} = \frac{1}{|\mathbf{A}|}\begin{bmatrix} A_{11} & 2A_{12} & \ldots & 2A_{1p} \\ 2A_{21} & A_{22} & \ldots & 2A_{2p} \\ . & . & \ldots & . \\ . & . & \ldots & . \\ 2A_{p1} & 2A_{p2} & \ldots & 2A_{pp} \end{bmatrix}$$

$$= 2\mathbf{\Sigma} - \text{ Diag. } \mathbf{\Sigma}.$$

Again

$$\frac{\partial}{\partial \mathbf{A}}\text{ tr }(\mathbf{A}\mathbf{S}_n) = 2\mathbf{S}_n - \text{ Diag. }(\mathbf{S}_n),$$

and

$$\frac{\partial \text{ tr } \mathbf{A}(\bar{\mathbf{x}}-\mu)(\bar{\mathbf{x}}-\mu)'}{\partial \mathbf{A}} = 2(\bar{\mathbf{x}}-\mu)(\bar{\mathbf{x}}-\mu)' -$$

$$\text{Diag }\{(\bar{\mathbf{x}}-\mu)(\bar{\mathbf{x}}-\mu)\}, \quad \text{by (A.15.9)}.$$

Combining these equations

$$\frac{\partial l}{\partial \mathbf{A}} = \frac{n}{2}(2\boldsymbol{\Sigma} - \text{ Diag } \boldsymbol{\Sigma}) - \frac{n}{2}(2\mathbf{S}_n - \text{ Diag } \mathbf{S}_n) - \frac{n}{2}\{2(\bar{\mathbf{x}} - \mu)(\bar{\mathbf{x}} - \mu)' - \text{Diag } [(\bar{\mathbf{x}} - \mu)(\bar{\mathbf{x}} - \mu)']\}$$
$$= \frac{n}{2}(2\mathbf{M} - \text{ Diag } \mathbf{M}) \qquad (3.6.10)$$

where $\mathbf{M} = \boldsymbol{\Sigma} - \mathbf{S}_n - (\bar{\mathbf{x}} - \mu)(\bar{\mathbf{x}} - \mu)'$.

To find the m.l.e.'s of $\mu, \boldsymbol{\Sigma}$, we must solve,

$$\frac{\partial l}{\partial \mu} = \mathbf{0} \qquad (3.6.11.1)$$

$$\frac{\partial l}{\partial \mathbf{A}} = \mathbf{0}. \qquad (3.6.11.2)$$

From (3.6.11.1) and (3.6.9), the m.l.e. of μ is

$$\hat{\mu} = \bar{\mathbf{x}}.$$

From (3.6.11.2) and (3.6.10), $2\mathbf{M} -$ Diag. $(\mathbf{M}) = \mathbf{0}$. This implies $\mathbf{M} = \mathbf{0}$, i.e., the m.l.e. of $\boldsymbol{\Sigma}$ is given by

$$\hat{\boldsymbol{\Sigma}} = \mathbf{S}_n + (\bar{\mathbf{x}} - \mu)(\bar{\mathbf{x}} - \mu)'. \qquad (3.6.12)$$

Since $\hat{\mu} = \bar{\mathbf{x}}$, (3.6.12) gives $\hat{\boldsymbol{\Sigma}} = \mathbf{S}_n$.

The above results only prove that $\bar{\mathbf{x}}$ and $\mathbf{S}_n$ are the stationary points of the likelihood function. In order to show that $\bar{\mathbf{x}}, \mathbf{S}_n$ give the maximum value of the likelihood, we consider the Theorem 3.6.1 below. For proving this theorem we require the following lemma.

Lemma 3.6.1: If $u > 0$, then

$$f(u) = u - 1 - \log u$$

has a minimum value 0 when $u = 1$. Hence, if b and h are the arithmetic mean (a.m.) and geometric mean (g.m.) of a set of positive numbers, then

$$b \geq \log h + 1, \qquad (3.6.13)$$

the equality holds when each positive number equals unity.

Proof. We have $f'(u) > 0$ when $u > 1, f'(u) < 0$, when $u < 1$, zero when $u = 1$. Therefore, $u \geq \log u + 1$. Now if $b = (u_1 + \ldots u_m)/m$, where u_i's are positive quantities, then $\log h = \sum_{i=1}^{m} \log u_i/m$. Since, $u_i \geq \log u_i + 1, (i = 1, \ldots m)$, (3.6.13) holds.

Theorem 3.6.1: For any fixed matrix $\mathbf{B} > 0$,

$$f(\mathbf{\Sigma}) = |\mathbf{\Sigma}|^{-n/2} \exp(-\frac{1}{2} \text{ tr } \mathbf{\Sigma}^{-1}\mathbf{B})$$

is maximized over $\mathbf{\Sigma} > 0$ by $\mathbf{\Sigma} = n^{-1}\mathbf{B}$, and

$$f(n^{-1}\mathbf{B}) = |n^{-1}\mathbf{B}|^{-n/2}e^{-np/2}.$$

Proof. It is seen that $f(n^{-1}\mathbf{B})$ takes the form given above. We have

$$\begin{aligned}\log f(n^{-1}\mathbf{B}) - \log f(\mathbf{\Sigma}) &= -\tfrac{n}{2}\log(|n^{-1}\mathbf{\Sigma}^{-1}\mathbf{B}|) - \tfrac{np}{2} + \tfrac{1}{2}anp \\ &= \tfrac{1}{2}np(a - 1 - \log g)\end{aligned}$$

where $a = \text{ tr } (\mathbf{\Sigma}^{-1}\mathbf{B})/np$ and $g = |n^{-1}\mathbf{\Sigma}^{-1}\mathbf{B}|^{1/p}$ are the arithmetic mean and geometric mean of the eigenvalues of $n^{-1}\mathbf{\Sigma}^{-1}\mathbf{B}$ (vide A.8(1)). Note that all these eigenvalues are positive (vide Section A.8 (7)). Hence, the Theorem follows by Lemma 3.6.1. □

Now, when $\hat{\mu} = \bar{\mathbf{x}}$, (3.6.8) reduces to

$$l(\bar{\mathbf{x}}, \mathbf{A} : \mathbf{X}) = -\frac{np}{2}\log(2\pi) + \frac{n}{2}\log|\mathbf{A}| - \frac{n}{2} \text{ tr } (\mathbf{A}\mathbf{S}_n) \tag{3.6.14}$$

which is maximized by Theorem 3.6.1 for $\hat{\mathbf{\Sigma}} = \mathbf{S}_n$.

Note 3.6.1: The maximum value of the likelihood in the unrestricted case is, therefore,

$$\frac{1}{(2\pi)^{np/2}|\mathbf{S}_n|^{n/2}}e^{-np/2}. \tag{3.6.15}$$

Discussion: If we maximize the likelihood (3.6.1) over $\mathbf{\Sigma}$ for fixed μ, we find from Theorem 3.6.1 that the m.l.e. $\hat{\mathbf{\Sigma}}$ is given by (3.6.12). Now,

$$\begin{aligned}l(\mu, \mathbf{S}_n + (\bar{\mathbf{x}} - \mu)(\bar{\mathbf{x}} - \mu)' : \mathbf{X}) &= -\tfrac{n}{2}|\mathbf{S}_n + (\bar{\mathbf{x}} - \mu)(\bar{\mathbf{x}} - \mu)'| \\ &= -\tfrac{n}{2}|\mathbf{S}_n|\{1 + (\bar{\mathbf{x}} - \mu)\mathbf{S}_n^{-1}(\bar{\mathbf{x}} - \mu)'\} \\ &\leq -\frac{n}{2}|\mathbf{S}_n|,\end{aligned} \tag{3.6.16}$$

(by (A.5.6)) and hence the log likelihood is maximized by $\hat{\mu} = \bar{\mathbf{x}}$. Alternatively, we could maximize over μ first.

We note that the m.l.e. of μ can be deduced from (3.6.1) directly whether $\mathbf{\Sigma}$ is known or not. Since $\mathbf{\Sigma}^{-1} > 0, -(\bar{\mathbf{x}} - \mu)'\mathbf{\Sigma}^{-1}(\bar{\mathbf{x}} - \mu) \leq 0$ and is maximized when $\mu = \bar{\mathbf{x}}$. Thus, the m.l.e. of μ is $\bar{\mathbf{x}}$, whether $\mathbf{\Sigma}$ is known or not, constrained or not. However, in case μ is constrained, the m.l.e. of $\mathbf{\Sigma}$ will be affected.

From (3.6.8) it follows that the joint density depends on the whole set of observations $\mathbf{x}_1, \ldots, \mathbf{x}_n$, only through sample mean $\bar{\mathbf{x}}$ and $\sum_{i=1}^{n}(\mathbf{x}_i - \bar{\mathbf{x}})(\mathbf{x}_i - \bar{\mathbf{x}})' = n\mathbf{S}_n$. Hence $(\bar{\mathbf{x}}, \mathbf{S}_n)$ are jointly sufficient for $(\mu, \mathbf{\Sigma})$.

The theoretical justifications for using $\hat{\mu} = \bar{\mathbf{x}}$ and $\hat{\mathbf{\Sigma}} = \mathbf{S}_n$(or the unbiased estimate $\mathbf{S} = n\mathbf{S}_n/(n-1)$), namely, sufficiency, consistency, efficiency and Bayesian characteristics have been proved by several authors (see for Example, Giri, 1977, Chapter 5). The Bayesian estimate of $\mathbf{\Sigma}$ is discussed by Chen (1979).

The use of $\bar{\mathbf{x}}$ as an estimate of μ has been criticized on the grounds that it is inadmissible for particular loss functions when $p \geq 3$ (Stein, 1955, James and Stein 1966). Let $\mathbf{t}$ be an estimate of μ. Using the loss function $(\mathbf{t} - \mu)'\mathbf{\Sigma}^{-1}(\mathbf{t} - \mu)$, James and Stein (1961) showed that for certain positive ϵ,

$$\mathbf{t} = (1 - \frac{\epsilon}{\bar{\mathbf{x}}'\mathbf{S}^{-1}\bar{\mathbf{x}}})\bar{\mathbf{x}}$$

has smaller risk than $\bar{\mathbf{x}}$ for μ.

(b) Constraint on the mean vector μ:

Suppose

$$\mu = k\mu_0, \tag{3.6.17}$$

where μ_0 is known. If $\mathbf{\Sigma}$ is known, the log likelihood function (3.6.2) becomes

$$l(k : \mathbf{\Sigma}, \mathbf{X}) = -\frac{n}{2}\log|2\pi\mathbf{\Sigma}| - \frac{n}{2}\text{ tr } \mathbf{\Sigma}^{-1}\mathbf{S}_n - \frac{n}{2}(\bar{\mathbf{x}} - k\mu_0)'\mathbf{\Sigma}^{-1}(\bar{\mathbf{x}} - k\mu_0).$$

The m.l.e. of k is given by the solution of the equation $\partial l/\partial k = 0$. Now,

$$\frac{\partial l}{\partial k} = n(\bar{\mathbf{x}} - k\mu_0)'\mathbf{\Sigma}^{-1}\mu_0 = \mathbf{0} \tag{3.6.18}$$

$$\Rightarrow \hat{k} = \frac{\bar{\mathbf{x}}'\mathbf{\Sigma}^{-1}\mu_0}{\mu_0'\mathbf{\Sigma}^{-1}\mu_0}. \tag{3.6.19}$$

The estimator $\hat{k}$ is unbiased with

$$V(\hat{k}) = (n\mu_0'\mathbf{\Sigma}^{-1}\mu_0)^{-1}.$$

When $\mathbf{\Sigma}$ is unknown, the likelihood equations are $\frac{\partial l}{\partial k} = 0$ which gives the solution (3.6.19) and $\frac{\partial l}{\partial \mathbf{A}} = \mathbf{0}$ where $\mathbf{A} = \mathbf{\Sigma}^{-1}$, which gives the equation (3.6.12),

$$\hat{\Sigma} = \mathbf{S}_n + (\bar{\mathbf{x}} - \mu)(\bar{\mathbf{x}} - \mu)'. \qquad (3.6.20)$$

Pre- and post-multiplying (3.6.20) by $\mathbf{\Sigma}^{-1}$ and $\mathbf{S}_n^{-1}$, respectively,

$$\mathbf{S}_n^{-1} = \hat{\mathbf{\Sigma}}^{-1} + \hat{\mathbf{\Sigma}}^{-1}(\bar{\mathbf{x}} - \mu)(\bar{\mathbf{x}} - \mu)'\mathbf{S}_n^{-1}. \qquad (3.6.21)$$

Pre-multiplying (3.6.21) by μ_0',

$$\mu_0'\mathbf{S}_n^{-1} = \mu_0'\hat{\mathbf{\Sigma}}^{-1} + \mu_0'\hat{\mathbf{\Sigma}}^{-1}(\bar{\mathbf{x}} - \mu)(\bar{\mathbf{x}} - \mu)'\mathbf{S}_n^{-1}$$

$$= \mu_0'\hat{\mathbf{\Sigma}}^{-1} \quad \text{by (3.6.18)}.$$

From (3.6.19), the m.l.e. is

$$\tilde{k} = \frac{\bar{\mathbf{x}}'\mathbf{S}_n^{-1}\mu_0}{\mu_0'\mathbf{S}_n^{-1}\mu_0}. \qquad (3.6.22)$$

We now consider another type of constraint on μ,

$$\mathbf{H}\mu = \mathbf{h} \qquad (3.6.23)$$

where $\mathbf{H}$ and $\mathbf{h}$ are pre-specified. Suppose $\mathbf{\Sigma}$ is known. To find the m.l.e. of μ we maximize the expression

$$l^* = l - n\lambda'(\mathbf{H}\mu - \mathbf{h}),$$

where λ is a vector of Lagrangian multipliers and l is given by (3.6.2). We are required to find λ for which the solution of the equation

$$\frac{\partial l^*}{\partial \mu} = \mathbf{0} \qquad (3.6.24)$$

satisfies the constraint $\mathbf{H}\mu = \mathbf{h}$. From (3.6.9),

$$\frac{\partial l^*}{\partial \mu} = n\mathbf{\Sigma}^{-1}(\bar{\mathbf{x}} - \mu) - n\mathbf{H}'\lambda.$$

Therefore,

$$\bar{\mathbf{x}} - \mu = \mathbf{\Sigma}\mathbf{H}'\lambda.$$

Pre-multiplying by $\mathbf{H}$ and using (3.6.23), we have $\mathbf{H}\bar{\mathbf{x}} - \mathbf{h} = (\mathbf{H}\mathbf{\Sigma}\mathbf{H}')\lambda$. Thus, we take

$$\lambda = (\mathbf{H}\mathbf{\Sigma}\mathbf{H}')^{-1}(\mathbf{H}\bar{\mathbf{x}} - \mathbf{h}),$$

so that

$$\hat{\mu} = \bar{\mathbf{x}} - \mathbf{\Sigma}\mathbf{H}'\lambda = \bar{\mathbf{x}} - \mathbf{\Sigma}\mathbf{H}'(\mathbf{H}\mathbf{\Sigma}\mathbf{H}')^{-1}(\mathbf{H}\bar{\mathbf{x}} - \mathbf{h}). \qquad (3.6.25)$$

If $\mathbf{\Sigma}$ is unknown, the m.l.e. of μ is

$$\hat{\mu} = \bar{\mathbf{x}} - \mathbf{S}_n \mathbf{H}'(\mathbf{HSH}')^{-1}(\mathbf{H}\bar{\mathbf{x}} - \mathbf{h}). \tag{3.6.26}$$

(c) Constraints on $\mathbf{\Sigma}$:

Suppose

$$\mathbf{\Sigma} = k\mathbf{\Sigma}_0 \tag{3.6.27}$$

where $\mathbf{\Sigma}_0$ is known. From (3.6.2),

$$2n^{-1}l(k, \mu : \mathbf{X}) = -p \log k - \log |2\pi \mathbf{\Sigma}_0| - k^{-1}\theta$$

where $\theta = \text{ tr } \mathbf{\Sigma}_0^{-1}\mathbf{S}_n + (\bar{\mathbf{x}} - \mu)'\mathbf{\Sigma}_0^{-1}(\bar{\mathbf{x}} - \mu_0)$ and is independent of k. If μ is known, to obtain the m.l.e. of k we solve the equation $\partial l/\partial k = 0$. This gives

$$-\frac{p}{k} + \frac{\theta}{k^2} = 0.$$

Hence, the m.l.e. of k is

$$\hat{k} = \theta/p.$$

If μ is unknown and unconstrained, the likelihood equations are $\partial l/\partial \mu = \mathbf{0}$ and $\partial l/\partial k = 0$. These together give the m.l.e. of k,

$$\hat{k} = \text{ tr } (\mathbf{\Sigma}_0^{-1}\mathbf{S}_n/p). \tag{3.6.28}$$

If the constraint on $\mathbf{\Sigma}$ is $\mathbf{\Sigma}_{12} = \mathbf{0}$, the constraint implies that the two corresponding groups of variables are independent. The m.l.e of $\mathbf{\Sigma}$ when μ is unconstrained and unknown is

$$\hat{\mathbf{\Sigma}} = \begin{bmatrix} \mathbf{S}_{11} & \mathbf{0} \\ \mathbf{0} & \mathbf{S}_{22} \end{bmatrix}$$

and is found by considering each group separately. Here, both $\mathbf{S}_{11}$ and $\mathbf{S}_{22}$ have denominator n (and not $n - 1$).

Samples from several populations with constrained parameters

Suppose we have independent data matrices $\mathbf{X}_1, \ldots \mathbf{X}_k$ where each row of $\mathbf{X}_i(n_i \times p)$ is an independent sample from $N_p(\mu_i, \mathbf{\Sigma}_i)(i = 1, \ldots, k)$. We assume that the parameters μ_i and $\mathbf{\Sigma}_i$ are related. The most common constraints are:

(a) $\mathbf{\Sigma}_1 = \ldots \mathbf{\Sigma}_k$.

(b) $\mathbf{\Sigma}_1 = \ldots = \mathbf{\Sigma}_k$ and $\mu_1 = \ldots = \mu_k$.

If (b) holds then we can consider all the $n(= \sum_{i=1}^{k} n_i)$ samples as samples from a $N_p(\mu, \mathbf{\Sigma})$ population and find estimates of μ and $\mathbf{\Sigma}$ as before.

In case of (a), we have, from (3.6.6), the log likelihood function given by

$$l = -\frac{1}{2}\sum_{i=1}^{k}[n_i \log|2\pi\mathbf{\Sigma}| + n_i \text{ tr } [\mathbf{\Sigma}^{-1}(\mathbf{S}_{i(n)} + \mathbf{d}_i\mathbf{d}_i')], \qquad (3.6.29)$$

where $\mathbf{S}_{i(n)}$ is the sample covariance matrix in the ith sample,

$$\mathbf{S}_{i(n)} = \frac{1}{n_i}\sum_{r=1}^{n_i}(\mathbf{x}_{ri} - \bar{\mathbf{x}}_i)(\mathbf{x}_{ri} - \bar{\mathbf{x}}_i)',$$

$\mathbf{x}_{ri}$ is the rth observation in the ith sample $\mathbf{x}_{ri} = (x_{ri1}, x_{ri2}, \ldots, x_{irp})', i = 1, \ldots, k$ and $\mathbf{d}_i = \bar{\mathbf{x}}_i - \mu_i$. In the absence of any restriction on population means, the m.l.e. of μ_i is $\bar{\mathbf{x}}_i$ and setting this value of μ_i in (3.6.29), the likelihood becomes

$$l = -\frac{n}{2}\log|2\pi\mathbf{\Sigma}| - \frac{1}{2} \text{ tr } \mathbf{\Sigma}^{-1}\mathbf{W}, \text{ where } \mathbf{W} = \sum_{i=1}^{n} n_i\mathbf{S}_{i(n)}. \qquad (3.6.30)$$

Differentiating (3.6.30) with respect to $\mathbf{\Sigma}$ and equating the resulting expression to zero gives the m.l.e. of $\mathbf{\Sigma}$ as $\hat{\mathbf{\Sigma}} = n^{-1}\mathbf{W}$.

EXERCISE 3.6.1: For the bird data of Table 2.E.1, considering the variables X_1, X_2, X_3 only, the sample mean vector and covariance matrix are

$$\bar{\mathbf{x}} = (157.98, 241.33, 31.459)', \quad \mathbf{S}_n = \begin{bmatrix} 13.354 & 13.611 & 1.922 \\ 13.611 & 25.683 & 2.714 \\ 1.922 & 2.714 & 0.632 \end{bmatrix}.$$

The $\bar{\mathbf{x}}, \mathbf{S}_n$ are the unconstrained m.l.e.'s of μ and $\mathbf{\Sigma}$, respectively. If from earlier information, it is known that $\mu = \mu_0 = (150,\ 250,\ 30)'$, then m.l.e.of $\mathbf{\Sigma} = \tilde{\mathbf{\Sigma}}$ is given by (3.6.12). Now,

$$(\bar{\mathbf{x}} - \mu) = (7.98,\ -8.67,\ 1.46)',$$

$$\hat{\mathbf{\Sigma}} = \mathbf{S}_n + (\bar{\mathbf{x}} - \mu_0)(\bar{\mathbf{x}} - \mu_0)'$$

$$= \begin{bmatrix} 77.034 & -55.576 & 13.573 \\ -55.576 & 100.852 & -9.944 \\ 13.573 & -9.944 & 2.764 \end{bmatrix}.$$

If we assume $\mu = k(150,\ 250,\ 30)'$, then from (3.6.22), $\hat{k} = \mu_0'\mathbf{S}_n^{-1}\bar{\mathbf{x}}/\mu_0'\mathbf{S}_n^{-1}\mu_0 = .98455$. so the m.l.e. for μ is $\tilde{\mu} = (147.683,\ 246.138,\ 29.536)'$. The m.l.e. of $\mathbf{\Sigma}$ is now

$$\hat{\mathbf{\Sigma}} = \mathbf{S}_n + (\bar{\mathbf{x}} - \tilde{\mu})(\bar{\mathbf{x}} - \tilde{\mu})'$$

$$= \begin{bmatrix} 119.444 & -36.138 & 21.698 \\ -36.138 & 49.012 & -6.560 \\ 21.698 & -6.560 & 4.318 \end{bmatrix}.$$

If the covariance matrix is known to be proportional to

$$\mathbf{\Sigma}_0 = \begin{bmatrix} 7 & 7 & 2 \\ 7 & 14 & 3 \\ 2 & 3 & 1 \end{bmatrix},$$

and μ is unconstrained and unknown, then from (3.6.28), $\hat{k} = 1.9681$ and

$$\hat{\mathbf{\Sigma}} = \begin{bmatrix} 13.776 & 13.776 & 3.936 \\ 13.776 & 27.552 & 5.904 \\ 3.936 & 5.904 & 1.968 \end{bmatrix}. \ \square$$

We shall, in the next two sections, consider some distributions which are related to normal distribution.

3.7 Matrix Normal Distribution

Let X be a matrix whose n rows $\mathbf{x}_1', \ldots \mathbf{x}_n'$ are independently distributed as $N_p(\mu, \mathbf{\Sigma})$, i.e. $\mathbf{X}$ is a data matrix from $N_p(\mu, \mathbf{\Sigma})$. Then $\mathbf{X}$ has the *matrix normal distribution* and represents a normal sample from $N_p(\mu, \mathbf{\Sigma})$. The *pdf* of $\mathbf{X}$ is

$$f(\mathbf{X}) = |2\pi\mathbf{\Sigma}|^{-n/2} \exp\{-\tfrac{1}{2}\textstyle\sum_{r=1}^{n}(\mathbf{x}_r - \mu)'\mathbf{\Sigma}^{-1}(\mathbf{x}_r - \mu)\}$$

$$= |2\pi\mathbf{\Sigma}|^{-n/2} \exp\left\{-\tfrac{1}{2}tr[\mathbf{\Sigma}^{-1}(\mathbf{X} - \mathbf{1}\mu')'(\mathbf{X} - \mathbf{1}\mu')]\right\},$$

where $\mathbf{1}$ is a $n \times 1$ vector of 1's.

3.8 The Multivariate t Distribution

Let $\mathbf{X}$ be a $p \times 1$ random vector; then $\mathbf{X}$ has a t-distribution with d degrees of freedom, mean vector μ and precision matrix $\mathbf{T}$ if the *pdf* of $\mathbf{X}$ is

$$f(\mathbf{x}|d, \mu, \mathbf{T}) \propto [1 + \frac{(\mathbf{x} - \mu)'\mathbf{T}(\mathbf{x} - \mu)}{d}]^{-(d+p)/2}, \mathbf{x} \in \mathcal{R}^p$$

where $d > 0$ and $\mathbf{T}$ is a $p \times p$ symmetric positive definite matrix.

It can be shown that when $p = 1, X$ has a univariate t-distribution and that $\sqrt{T}(x - \mu)$ has Student's t-distribution with d degrees of freedom. An interesting property of the multivariate t distribution is that if $p \geq 2$, the components of $\mathbf{X}$ cannot be independent, but they can be uncorrelated since the dispersion matrix of $\mathbf{X}$ is

$$D(\mathbf{X}) = \frac{d}{d-2}\mathbf{T}^{-1}, d > 2.$$

The mean vector of $\mathbf{X}$ is $E(\mathbf{X}) = \mu$ and the marginal distribution of $\mathbf{X}^{(1)}$ is also a t-distribution with d degrees of freedom, location parameter $\mu^{(1)}$ and precision matrix $\mathbf{T}_{11} - \mathbf{T}_{12}\mathbf{T}_{22}^{-1}\mathbf{T}_{21}$ where

$$\mathbf{T} = \begin{pmatrix} \mathbf{T}_{11} & \mathbf{T}_{12} \\ \mathbf{T}_{21} & \mathbf{T}_{22} \end{pmatrix}$$

and $\mathbf{T}_{11}$ is of order r. Here $\mathbf{X}^{(1)}$ consists of the first r components of $\mathbf{X}$ and μ is partitioned as

$$\mu = \begin{pmatrix} \mu^{(1)} \\ \mu^{(2)} \end{pmatrix}$$

where $\mu^{(1)}$ is $r \times 1$.

The conditional distribution of $\mathbf{X}^{(1)}$ given $\mathbf{X}^{(2)}$ is also a t where the conditional mean vector of $\mathbf{X}^{(1)}$ is

$$E[\mathbf{X}^{(1)}|\mathbf{X}^{(2)} = \mathbf{x}^{(2)}] = \mu^{(1)} - \mathbf{T}_{11}^{-1}\mathbf{T}_{12}[\mathbf{x}^{(2)} - \mu^{(2)}],$$

and the conditional precision matrix is

$$\frac{(d+p-r)\mathbf{T}_{11}}{d + [\mathbf{x}^{(2)} - \mu^{(2)}]'[\mathbf{T}_{22} - \mathbf{T}_{21}\mathbf{T}_{11}^{-1}\mathbf{T}_{12}][\mathbf{x}^{(2)} - \mu^{(2)}]}.$$

Unlike the normal distribution, the precision matrix of the conditional distribution of $\mathbf{X}^{(1)}$ given $\mathbf{X}^{(2)}$ depends on the value of the conditioning variables.

3.9 The Dirichlet Distribution

This distribution is not related to multivariate normal distribution. However, this distribution is noted here due to its usefulness.

A multivariate generalization of the univariate beta distribution is the Dirichlet distribution.

The random vector $\mathbf{X} = (X_1, \ldots, X_p)'$ follows a p-variate Dirichlet distribution $D(\gamma_1, \ldots, \gamma_p; \gamma_{p+1})$ (with parameters $\gamma_1, \ldots, \gamma_{p+1}$) if its *pdf* is given by

$$f_{\mathbf{X}}(\mathbf{x}) = \frac{\Gamma(\gamma_1 + \ldots + \gamma_{p+1})}{\Gamma(\gamma_1) \ldots \Gamma(\gamma_{p+1})} x_1^{\gamma_1 - 1} \ldots x_p^{\gamma_p - 1}(1 - x_1 - \ldots - x_p)^{\gamma_{p+1} - 1}$$

whenever

$$x_k \geq 0 \quad (k = 1, \ldots, p), \quad \sum_{k=1}^{p} x_k \leq 1,$$

$$\gamma_k > 0, \quad k = 1, \ldots, p + 1.$$

The moments about the origin are given by

$$E(X_1^{r_1} \ldots X_p^{r_p}) = \frac{\Gamma(\gamma_1 + r_1) \ldots \Gamma(\gamma_p + r_p)\Gamma(\gamma_1 + \ldots + \gamma_{p+1})}{\Gamma(\gamma_1) \ldots \Gamma(\gamma_p)\Gamma(\gamma_1 + \ldots + \gamma_{p+1} + r_1 + \ldots + r_p)}.$$

Thus

$$E(X_k) = \frac{\gamma_k}{\gamma_1 + \ldots + \gamma_{k+1}}, \quad k = 1, \ldots, p$$

$$V(X_k) = \frac{\gamma_k(\gamma_1 + \ldots + \gamma_{p+1} - \gamma_k)}{(\gamma_1 + \ldots + \gamma_{p+1})^2(\gamma_1 + \ldots + \gamma_{k+1} + 1)}, \quad k = 1, \ldots, p$$

$$\text{Cov}\,(X_j, X_k) = -\frac{\gamma_j \gamma_k}{(\gamma_1 + \ldots + \gamma_{p+1})^2(\gamma_1 + \ldots + \gamma_{p+1} + 1)}, \quad (j \neq k = 1, \ldots, p).$$

Again if $\mathbf{X} \cap D(\gamma_1, \ldots, \gamma_p; \gamma_{p+1})$ then

$$\mathbf{X}^{(1)} = (X_1, \ldots, X_r)'(r < p) \cap D(\gamma_1, \ldots, \gamma_r; \gamma_{r+1} + \ldots + \gamma_{p+1}).$$

It follows that if the random variable Y_k has a univariate gamma density $f_G(\gamma_k) = e^{-y}y^{\gamma_k - 1}/\Gamma(\gamma_k)(y > 0, \gamma_k > 0)(k = 1, \ldots, p + 1)$ and Y_k's are independent then the variable $\mathbf{X} = (X_1, \ldots, X_p)'$ where

$$X_k = \frac{Y_k}{Y_1 + \ldots + Y_{p+1}}, k = 1, \ldots, p$$

follows $D(\gamma_1, \ldots, \gamma_p; \gamma_{p+1})$.

3.10 Multivariate Skewness and Kurtosis

Multivariate measures of skewness and kurtosis are useful, particularly, for normality and robustness studies. Mardia (1974, 1975) introduced the following measures:

$$\beta_{1,p} = E[\{(\mathbf{x}-\mu)\mathbf{\Sigma}^{-1}(\mathbf{y}-\mu)\}^3]$$

and

$$\beta_{2,p} = E[\{(\mathbf{x}-\mu)'\mathbf{\Sigma}^{-1}(\mathbf{y}-\mu)\}^2],$$

where $\mathbf{y}$ is independent of $\mathbf{x}$ but has the same distribution. When $\mathbf{X} \cap N_p(\mu, \mathbf{\Sigma}), \beta_{1,p} = 0$ and $\beta_{2,p} = p(p+2)$.

If $p = 1$,

$$\sqrt{\beta_1} = \sqrt{\beta_{1,1}} = \frac{\mu_3}{\sigma^3},$$

and

$$\beta_2 = \beta_{2,2} = \frac{\mu_4}{\sigma^4},$$

which are the usual measures of skewness and kurtosis. A more common measure of univariate kurtosis is $\gamma_2 = \beta_3 - 3$.

If $\mathbf{x}_1, \ldots, \mathbf{x}_n$ constitute a random sample from the distribution, the sample estimates of skewness and kurtosis are

$$b_{1,p} = \frac{1}{n^2}\sum_{i=1}^{n}\sum_{k=1}^{n}\{(\mathbf{x}_i-\bar{\mathbf{x}})'\hat{\mathbf{\Sigma}}^{-1}(\mathbf{x}_i-\bar{\mathbf{x}})\}^2 \tag{3.10.1}$$

and

$$b_{2,p} = \frac{1}{n}\sum_{i=1}^{n}\{(\mathbf{x}_i-\bar{\mathbf{x}})'\hat{\mathbf{\Sigma}}^{-1}(\mathbf{x}_i-\bar{\mathbf{x}})\}^2, \tag{3.10.2}$$

where $\hat{\mathbf{\Sigma}} = n^{-1}\sum_{i=1}^{n}(\mathbf{x}_i-\bar{\mathbf{x}})(\mathbf{x}_i-\bar{\mathbf{x}})'/n = \mathbf{S}_n$. If the parent distribution is normal,

$$E(b_{1,p}) = 0, E(b_{2,p}) = \frac{p(p+2)(n-1)}{n+1}.$$

3.11 Examining the Assumption of Normality

In many multivariate techniques we assume that each vector observation $\mathbf{X}_i$ comes from a p-variate normal distribution. In situations, where the sample size is large and the technique depends on the distribution of $\bar{\mathbf{X}}$ or the distance-statistic $n(\bar{\mathbf{X}} - \mu)'\mathbf{S}^{-1}(\bar{\mathbf{X}} - \mu)$, the assumption of normality is less crucial. However, we want to examine how far the distribution of observation $\mathbf{X}_i$ depart from normality.

It is known that if $\mathbf{X}$ follows multivariate normal, each component X_i of $\mathbf{X}$ follows univariate normal and each pair of components (X_i, X_j) follows bivariate normal distribution. However, the converse is not necessarily true (Exercise 3.18).

In examining the multivariate normality of $\mathbf{X}$, we shall mostly depend on examining the normality of each component and bivariate normality of each pair of components. It has been found extremely difficult to construct a good overall test of joint normality in more than two dimensions. For most practical purposes one-dimensional and two-dimensional investigations are sufficient.

3.11.1 *Assessing normality for univariate marginal distribution*

(a) *Using dot-diagram or histogram*: A dot diagram for small values of n or a histogram for $n > 25$ will reveal if the distribution of observations on X_i appears symmetric. If the distribution is symmetric, we can compare the proportion of observations in specific intervals with expected proportions in a normal distribution. A univariate normal distribution assigns probability .683 to the interval $(\mu_i - \sqrt{\sigma_{ii}}, \mu_i + \sqrt{\sigma_{ii}})$ and the probability .954 to $(\mu_i - 2\sqrt{\sigma_{ii}}, \mu_i + 2\sqrt{\sigma_{ii}})$. Consequently, if n is large we can expect .683 proportion of observations to lie in the interval $(\bar{x}_i - \sqrt{s_{ii}}, \bar{x}_i + \sqrt{s_{ii}})$ and .954 proportion of observations to lie in the interval $(\bar{x}_i - 2\sqrt{s_{ii}}, \bar{x}_i + 2\sqrt{s_{ii}})$. If we denote these two sample proportions by $\hat{p}_{1i}$ and $\hat{p}_{2i}$ respectively, then it is known by the normal approximation to the distribution of sample proportion that the following inequalities are satisfied with a very small probability:

$$|\hat{p}_{1i} - .683| > 3\sqrt{\frac{.683 \times .317}{n}} = \frac{1.396}{n},$$

$$|\hat{p}_{2i} - .954| > 3\sqrt{\frac{.954 \times .046}{n}} = \frac{.628}{n}.$$

Consequently, if at least one of the above inequalities is satisfied, we suspect the normality of the distribution of X_i.

(b) *Tests based on skewness and kurtosis*: The classical methods of evaluating the normality of univariate observations $x_1, \ldots, x_n$ are based on skewness and kurtosis coefficients,

$$\sqrt{b_1} = \frac{\sqrt{n}\sum_{i=1}^{n}(x_i - \bar{x})^3}{\{\sum_{i=1}^{n}(x_i - \bar{x})^2\}^{3/2}},$$

$$b_2 = \frac{n\sum_{i=1}^{n}(x_i - \bar{x})^4}{\{\sum_{i=1}^{n}(x_i - \bar{x})^2\}^2}.$$

These coefficients are invariant with respect to the changes in origin and scale and hence their distributions are independent of population mean μ and variance σ^2. If the underlying distribution is normal, the corresponding population coefficients $\sqrt{\beta_1}$ and β_2 are 0 and 3 respectively.

Tables of approximate 5% and 1% points of these two statistics may be found in Pearson and Hartley (1966, pp. 207 - 208).

Exact percentage points of the null distribution of $\sqrt{b_1}$ for $4 \leq n \leq 25$ due to Mullholland (1977) are reproduced in Table B.5. For $n \geq 8$ D'Agostino and Pearson (1973) gave values of δ and $1/\lambda$ such that the transformed variable

$$Z(\sqrt{b_1}) = \delta \sinh^{-1}(\frac{\sqrt{b_1}}{\lambda})$$

follows approximately a $N(0,1)$ distribution. These are reproduced in Table B.6.

D'Agostino and Tietjan (1971) simulated the percentiles for the distribution of b_2 for selected values of $7 \leq n \leq 50$ (reproduced in Table B.7).

(c) *Some tests based on order statistic*:

Q-Q Plot: A Q-Q (or quantile versus quantile) plot is a plot of the sample quantiles versus the quantiles one would observe if the observations were normally distributed.

Let $x_1, \ldots, x_n$ denote observations on any single characteristic X_i. Let $x_{(1)} < x_{(2)} < \ldots < x_{(n)}$ represent the observations after they are ordered according to magnitude. Then $x_{(j)}$ is $\{(j-1/2)/n\}$th order sample quantile, after making correction for continuity.

Let the $\{(j-1/2)/n\}$th order quantile of a standard normal distribution be denoted by q_j. Then

$$P[Z \leq q_j] = \frac{1}{\sqrt{2\pi}} \int_{-\infty}^{q_j} e^{-z^2/2} dz = \frac{j-1/2}{n} = p_{(j)}, \tag{3.11.1}$$

say. If the data arise from a $N(\mu_i, \sqrt{\sigma_{ii}})$-population, then $x_{(j)}$ and q_j will be approximately linearly related, since $E(x_{(j)}) \approx \mu_i + q_{(j)}\sqrt{\sigma_{ii}}$.

The plot of $x_{(j)}$ against $q_j (j = 1, \ldots, n)$ is called a Q-Q plot. If the plot is approximately a straight line, assumption of normality is tenable.

The inference from the Q-Q plot is not valid unless the sample size n is at least moderate, for instance, $n \geq 25$.

The straightness of Q-Q plot can be measured by the correlation coefficient r_Q between q_j and $x_{(j)}$,

$$r_Q = \frac{\sum_{j=1}^{n}(x_{(j)} - \bar{x})(q_j - \bar{q})}{\sqrt{\sum_{j=1}^{n}(x_{(j)} - \bar{x})^2}\sqrt{\sum_{j=1}^{n}(q_{(j)} - \bar{q})^2}}. \tag{3.11.2}$$

The critical points of the distribution of r_Q for values of n in $[5, 300]$ have been given in Johnson and Wichern (1998, Table 4.2, p.193). We reject the hypothesis of normality at level of significance α if r_Q falls below the appropriate value.

D'Agostino's test: D'Agostino (1971, 1972) calculated the percentage points of the statistic (reproduced in Table B.9)

$$Y = \frac{\sqrt{n}[D - (2\pi)^{-1}]}{0.2998598} \tag{3.11.3}$$

where

$$D = \frac{\sum_{i=1}^{n}[i - \frac{1}{2}(n+1)]x_{(i)}}{n^{3/2}\tilde{s}}$$

and $\tilde{s}^2 = \sum_{i=1}^{n}(x_{(i)} - \bar{s})^2$. As the range of Y is too small, D should be calculated to 5 decimal places. The hypothesis of normality is rejected if $|Y|$ is too large.

Anderson-Darling test: Percentage points of the statistic

$$A_n^2 = -\frac{1}{n}\{\sum_{i=1}^{n}(2i-1)[\log z_i + \log(1 - z_{n+1-i})]\} - n, \tag{3.11.4}$$

where $z_i = \Phi([x_{(i)} - \bar{x}]/s)$ and Φ is the distribution function for $N(0,1)$ has been calculated (Pettitt, 1977). Given $p = P[A_n^2 \leq a_n], a_n$ is calculated from the expression

$$a_n \approx a_\infty[1 + c_1 n^{-1} + c_2 n^{-2}].$$

The values are reproduced in Table B.8. The hypothesis of normality is rejected if A_n^2 is too large.

EXAMPLE 3.11.1: A sample of 10 observations gives the values in the following table.

Table 3.11.1: Calculations for Q-Q Plot

Ordered observation	Probability levels $(j-1/2)/n$	Standard normal quantile $q_{(j)}$
55.5	.05	-1.645
57.4	.15	-1.036
58.1	.25	-.674
58.4	.35	-.385
59.2	.45	-.125
59.4	.55	.125
60.2	.65	.385
61.3	.75	.674
62.4	.85	1.036
62.5	.95	1.645

Figure 3.2 shows the Q-Q plot for the above data which is a plot of the ordered data $x_{(j)}$ against the normal quantiles $q_{(j)}$. The pairs of points $(q_{(j)}, x_{(j)})$ are located approximately on a straight line and the assumption of normality may be accepted, particularly when the sample size is as small as 10.

The value of r_Q, correlation between $x_{(j)}$ and $q_{(j)}$ is $0.986 > .9198$. Hence, the hypothesis of normality can be accepted at 0.5% level of significance.

(d) *Another test*: Shapiro and Wilk (1965) suggested the statistic

$$W = \frac{(\sum_{i=1}^n a_i x_{(i)})^2}{\sum_{i=1}^n (x_i - \bar{x})^2}$$

where $x_{(1)} \leq x_{(2)} \leq \ldots \leq x_{(n)}$ denote the ordered observations and $\mathbf{a} = (a_1, \ldots, a_n)'$ is defined as

$$\mathbf{a}' = \frac{\mathbf{m}'\mathbf{V}^{-1}}{||\mathbf{m}'\mathbf{V}^{-1}||}$$

where $\mathbf{m}$ and $\mathbf{V}$, respectively are the vector of expected values and covariance matrix of standard normal order statistic. Shapiro and Wilk provided tables of values of coefficients a_i for $n = 2(1)50$ and percentage points of distribution of W for $n = 3(1)50$. Small values of W correspond to departure from normality.

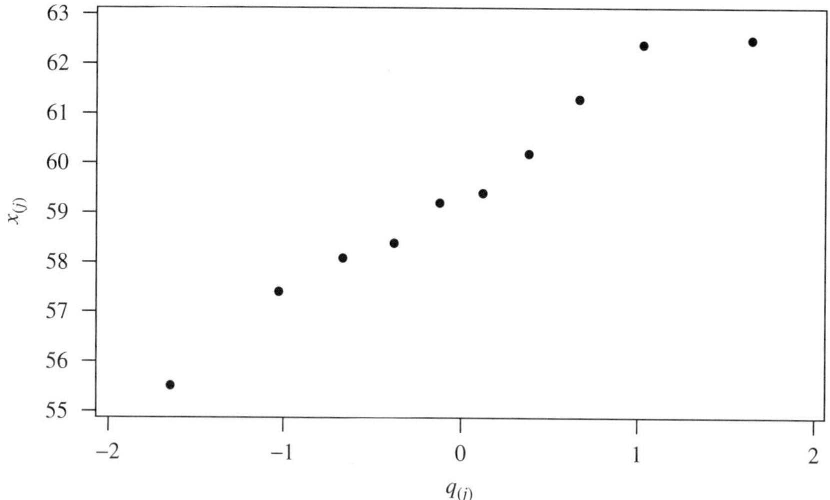

Fig. 3.2: $Q - Q$ plot of the sample data

3.11.2 *Evaluating bivariate normality*

We know that if the observations are generated by a multivariate normal distribution, each pair of variables would have a bivariate normal distribution and the constant density ellipsoid would be an ellipse. The scatter plot of observations should conform to this structure and show an overall structure that is nearly elliptic.

Again it is known that for a bivariate normal distribution $(\mathbf{X}-\mu)'\mathbf{\Sigma}^{-1}(\mathbf{X}-\mu)$ follows a $\chi^2_{(2)}$ distribution and hence

$$P[\mathbf{x} : (\mathbf{x} - \mu)'\mathbf{\Sigma}^{-1}(\mathbf{x} - \mu) \leq \chi^2_{2;0.5}] = 0.5. \tag{3.11.5}$$

Hence, we should expect roughly 50% of the sample observations to lie in the ellipse

$$(\mathbf{x}-\bar{\mathbf{x}})'\mathbf{S}^{-1}(\mathbf{x}-\bar{\mathbf{x}}) \leq \chi^2_{2,0.5}. \tag{3.11.6}$$

If not the normality assumption is under suspicion.

3.11.3 *Evaluating multivariate normality*

The method in (3.11.5) and (3.11.6) can be extended to p variables. However, the method is a rather rough procedure.

Another method of judging the joint normality is based on the squared generalized distance

$$d_j^2 = (\mathbf{x}_j - \bar{\mathbf{x}})'\mathbf{S}^{-1}(\mathbf{x}_j - \bar{\mathbf{x}}) \tag{3.11.7}$$

where $\mathbf{x}_1, \ldots, \mathbf{x}_n$ are sample observations. When the parent population is multivariate normal and both n and $n-p$ are greater than 25, each of the squared distances $d_1^2, \ldots, d_n^2$ should behave like a $\chi^2_{(p)}$ variables. We assume the variables to be independent.

We arrange the squared distances as $d_{(1)}^2 \leq d_{(2)}^2 \leq \cdots \leq d_{(n)}^2$. We plot the $d_{(j)}^2$ values against $\chi^2_{(p)}[(j-1/2)/n], j = 1, \ldots, n$ where $\chi^2_{(p)}[(j-1/2)/n]$ denotes the $\{(j-1/2)/n\}$th order quantile of a $\chi^2_{(p)}$ distribution. If the observations come from a multivariate normal distribution, the χ^2-plot should be a straight line. A systematic curve pattern shows lack of normality. A few points lying away from a straight line indicate outlying observations. For further details in this area, the reader may refer to Krzanowski (1988, p.211 - 215).

Mardia (1970, 1975) proposed a large-sample test for multivariate normality based on measures of multivariate skewness and kurtosis, $b_{1,p}, b_{2,p}$ given in (3.10.1) and (3.10.2). Under joint normality $nb_{1,p}/6$ follows a χ^2 distribution with $p(p+1)(p+2)/6$ d.f. and $b_{2,p}$ follows a normal distribution with mean $p(p+2)$ and variance $8p(p+2)/n$. The tests are valid in large samples only. However, it has been empirically established that many of the classical multivariate techniques remain unaffected by non-normality, provided the data come from some other centrally symmetric distributions.

Mardia (1980) provided a survey of various tests for normality.

EXAMPLE 3.11.2: Consider the values of the variables X_1, X_2 for the first

$n = 10$ observations in the Bird-data of Table 2.E.1. These data give

$$\bar{\mathbf{x}} = \begin{bmatrix} 156.8 \\ 241.4 \end{bmatrix}, \quad \mathbf{S} = \begin{bmatrix} 15.0667 & 11.7556 \\ 11.7556 & 16.9333 \end{bmatrix},$$

$$\mathbf{S}^{-1} = \begin{bmatrix} 0.144810 & -0.100531 \\ -0.100531 & .128847 \end{bmatrix}.$$

Now $\chi^2_{2;.05} = 1.39$. Therefore, any observation $\mathbf{x}_j$ satisfying

$$d_j^2 = (\mathbf{x} - \bar{\mathbf{x}})'\mathbf{S}^{-1}(\mathbf{x} - \bar{\mathbf{x}}) \leq 1.39$$

where $\mathbf{x}_j = (x_{1j}, x_{2j})'$ will lie on or inside the estimated 50% contour of the $\chi^2_{(2)}$ distribution.

For the given data the values of d_j^2 are as follows:

$$d_1^2 = 2.3416,\ d_2^2 = 0.2.5997,\ d_3^2 = 1.2739\ d_4^2 = 1.7224\ d_5^2 = 1.3781,$$

$$d_6^2 = 2.6262,\ d_7^2 = 1.6320,\ d_8^2 = 0.3428, d_9^2 = 3.5650, d_{10}^2 = 2.5183.$$

Thus 4 out of 10 observations (roughly 50% of observations) lie within the estimated contour.

The ordered distances and the corresponding chi-square percentiles for $p = 2, n = 10$ are given below.

Table 3.11.2 showing the Calculations for the Chi-square plot of the Generalized Distance

j	$d^2_{(j)}$	$\chi^2_{(2)}[(j-1/2)/10]$
1	.34	.10
2	.60	.33
3	1.27	.58
4	1.38	.86
5	1.63	1.20
6	1.72	1.60
7	2.34	2.10
8	2.52	2.77
9	2.62	3.79
10	3.57	5.99

The chi-square plot is shown in Figure 3.3. The figure shows that the plot is not far away from a straight line. We do not reject the assumption of bivariate normality.

3.12 Transformations Making the Data Near Normal

When the normality assumption is not tenable, by considering some transformations, the transformed data can sometimes be made normal-looking. Normal theory analysis can then be carried out with the transformed data. Results of the application of the technique can then be transformed back (using the inverse of the original transformation) so that they can be expressed in terms of the original observations. We first consider the univariate case.

The following transformations are often used to achieve near-normality:

Original scale	Transformed scale
Count, x	$\sqrt{x}$
Proportion, $\hat{p}$	logit $(\hat{p}) = \frac{1}{2}\log(\frac{\hat{p}}{1-\hat{p}})$
Correlation, r	Fisher's $z = \frac{1}{2}\log_e \frac{1+r}{1-r}$

The above transformations are suggested from theoretical considerations, such as the objective of stabilizing variance, or inducing additivity or normality in the context of hypothesis testing or analysis of variance.

Sometimes the transformations are suggested by the data themselves. An important family of transformations in this area is the family of power transformations,

$$x^{(\lambda)} = \begin{cases} x^{\lambda}, & \lambda \neq 0, \\ \log x, & \lambda = 0 \text{ and } x > 0. \end{cases}$$

This family studied in detail by Tukey (1957) for $|\lambda| \leq 1$ contains the log, square-root and inverse transformations. Value of appropriate λ is found by trial and error by checking normality by dot-diagram or histogram or Q-Q plot.

Box and Cox (1964) considered the slightly modified family of power transformations for univariate observations:

$$x^{(\lambda)} = \begin{cases} \frac{x^{\lambda}-1}{\lambda}, \ \lambda \neq 0, \\ \log x, \ \lambda = 0, x > 0, \end{cases} \tag{3.12.1}$$

which is continuous in λ for $x > 0$. If the transformed observations $x_i^{(\lambda)}$ are *iid* $N(\mu, \sigma^2)$, then the likelihood for the untransformed data $x_i (i = 1, \ldots, n)$ is

$$(2\pi\sigma^2)^{-n/2}\left[\exp\left\{-\sum_{i=1}^{n}\frac{(x_i^{(\lambda)}-\mu)^2}{2\sigma^2}\right\}\right]\left[\Pi_{i=1}^{n}x_i^{\lambda-1}\right]. \tag{3.12.2}$$

The last term in the square bracket is the Jacobian of transformation and does not contain μ or σ^2. The maximum likelihood estimates of μ and σ^2 for given μ are

$$\begin{aligned}\hat{\mu} &= \bar{x}(\lambda) = \frac{1}{n}\sum_{i=1}^{n} x_i^{(\lambda)} = \frac{1}{n}\sum_{i=1}^{n}\frac{x_i^{\lambda}-1}{\lambda},\\ \hat{\sigma}^2 &= \frac{1}{n}\sum_{i=1}^{n}(x_i^{(\lambda)} - \bar{x}(\lambda))^2.\end{aligned} \tag{3.12.3}$$

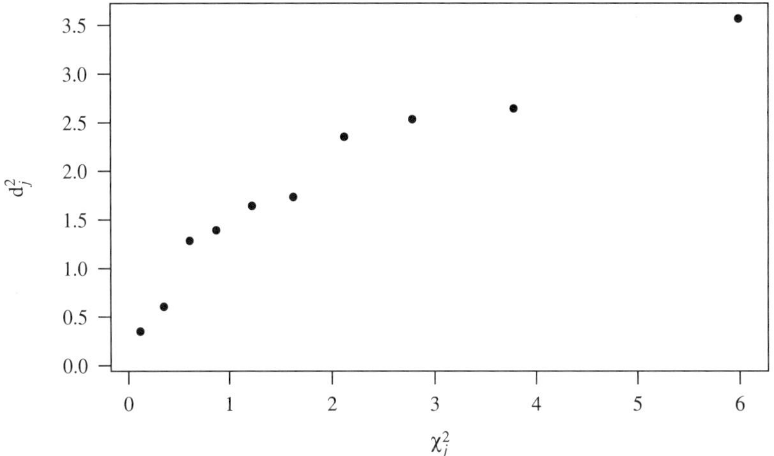

Fig. 3.3: A chi-squared plot of the ordered distances

The maximum value of the log likelihood for given λ is obtained by substituting $\hat{\mu}$ and $\hat{\sigma}^2$ for μ and σ^2 respectively in the logarithm of (3.12.2) and is (apart from a constant)

$$\begin{aligned}L(\lambda) &= -\tfrac{1}{2}n\log\hat{\sigma}^2 + (\lambda-1)\sum_{i=1}^{n}\log x_i\\ &= -\frac{1}{2}n\log\hat{\sigma}^2 + n\log(\tilde{x}^{(\lambda-1)})\\ &= -\frac{1}{2}n\log\hat{\sigma}_z^2,\end{aligned} \tag{3.12.4}$$

where $\tilde{x} = (x_1.x_2.\ldots.x_n)^{1/n}$ is the geometric mean of the x_i's and $z_i = x_i^{(\lambda)}/(\tilde{x})^{\lambda-1}$. The optimum value of $\lambda, \hat{\lambda}$ is found by maximizing $L(\lambda)$ with respect to λ. This value can be found out by solving $dL(\lambda)/d\lambda = 0$ iteratively or by plotting $L(\lambda)$ against λ. To test $H_0(\lambda = \lambda_0)$ we treat $2[L(\hat{\lambda}) - L(\lambda_0)]$ as being approximately $\chi^2_{(1)}$ under H_0. An approximate $100(1-\alpha)\%$ confidence region for the true value of λ is the set of all λ satisfying

$$L(\hat{\lambda}) - L(\lambda) \leq \frac{1}{2}\chi^2_{1;\alpha}.$$

Some other references in this area are Andrews (1971), Atkinson (1973) and Carroll (1980).

If some of the x's are negative one may add a positive constant ψ to each x_i to make them positive. The likelihood function $L(\lambda)$ in (3.12.3) will now be replaced by $L(\lambda, \psi)$ and the optimum value of λ and ψ are found by maximizing the likelihood with respect to λ and ψ.

The power transformation (3.12.1) makes the skewed distributions more symmetrical. For example, the effect of the square root transformation or the logarithmic transformation is to pull in one tail of the distribution.

John and Draper (1980) suggested an alternative family of transformations,

$$x^{[\lambda]} = \begin{cases} (\text{sign of } x)\left[\frac{(|x|+1)^\lambda - 1}{\lambda}\right], \ \lambda \neq 0, \\ (\text{sign of } x)[\log(|x|+1)], \ \lambda = 0. \end{cases} \tag{3.12.5}$$

The modulus transformation is effective on a distribution that already possess approximate symmetry about some central point and makes it more normal. If all the data are positive the power transformation and modulus transformation are equivalent.

Hinkley (1975, 1977) proposed two rough and ready methods of transforming the data to obtain approximate symmetry.

For multivariate observations, a power transformation must be chosen for each variable. Let $\hat{\lambda}_1, \ldots, \hat{\lambda}_p$ be the power transformations for the p variables. This procedure makes each marginal distribution approximately normal. Although normal marginals are not sufficient to make the joint distribution normal, for all practical purposes, the procedure is good enough.

Alternatively, we have the following generalization to Box and Cox's technique. Assuming that the transformed variables $\mathbf{x}_i^{(\lambda)} = (x_{i1}^{\lambda_1}, \ldots, x_{ip}^{\lambda_p})'$ to be *iid* $N_p(\mu, \mathbf{\Sigma})$, the likelihood for $\lambda = (\lambda_1, \ldots, \lambda_p)'$, corresponding to (3.12.4) is

$$L(\lambda) = -\frac{n}{2}\log|\mathbf{S}(\lambda)| + \sum_{j=1}^{p}(\lambda_j - 1)\sum_{i=1}^{n}\log x_{ij}, \tag{3.12.6}$$

where

$$\mathbf{S} = \frac{1}{n}\sum_{i=1}^{n}(\mathbf{x}_i^{(\lambda)} - \bar{\mathbf{x}}^{(\lambda)})(\mathbf{x}_i^{(\lambda)} - \bar{\mathbf{x}}^{(\lambda)})'.$$

We choose $\lambda = \hat{\lambda}$ where $\hat{\lambda}$ maximizes $L(\lambda)$ in (3.12.6). To test $H_0(\lambda = \lambda_0)$ we treat $2[L(\hat{\lambda}) - L(\lambda_0)]$ to be approximately $\chi^2_{(p)}$ under H_0. This statistic can also be used to find a confidence region for λ. Gnanadesikan (1977) gave some interesting plots displaying the transformations stated above.

3.13 Robust Estimation of Location and Scale Parameters

If $\mathbf{X}$ has a multivariate normal distribution, then it is known that usual unbiased estimator of μ is $\bar{\mathbf{X}}$ and of $\mathbf{\Sigma}$ is $\mathbf{S}$. These estimators have a number of desirable properties, such as consistency and efficiency. These properties are however valid under the specific assumption about the underlying distribution. Deviation from the multivariate model may be caused by the occurrence of outliers or gross errors, i.e. observations that follow a distribution different from the assumed model, the difference (sometimes) consisting in the mass of the outlier distribution being concentrated around much larger (or smaller) values than typically encountered in the model.

In this section we shall consider estimators of location and scale parameters that would remain relatively unaffected by deviations from the assumed model. Such estimators are generally termed *robust*. The present section has been developed following Gnanadesikan (1997).

Some robust estimators of multivariate location that are simply vectors of the univariate robust estimators are as follows.

(i) $\mathbf{x}^*_M$, the vector of medians of the observations of the variables (Mood, 1941).

(ii) $\mathbf{x}^*_{HL}$, the vector of Hodges-Lehmann estimators (i.e. the median of the average of pairs of observations) for each variable (Bickel, 1964).

(iii) $\mathbf{x}^*_{T(\alpha)}$, the vector of α-trimmed means (i.e., the mean of the data remaining after deleting a proportion α of the smallest and of the largest observations) for each variable.

Let $x_{(1)} \leq x_{(2)} \leq \ldots \leq x_{(n)}$. Let also $n\alpha = r + \theta$, where r is an integer and $0 \leq \theta < 1$. Then the α-trimmed mean of the variable

x is

$$x^*_{T(\alpha)} = \frac{(1-\theta)x_{(r+1)} + x_{(r+2)} + \ldots + x_{(n-r-1)} + (1-\theta)x_{(n-r)}}{n(1-2\alpha)}.$$

(iv) $\mathbf{x}^*_{W(\alpha)}$, the vector of α-Winsorized mean for each variable.

α-Winsorized mean is defined as follows. Let $0 < \alpha < 1$. For the α-Winsorized mean we replace each of the lower r values $x_{(1)}, \ldots, x_{(r)}$ by $x_{(r+1)}$ and each of the upper r values $x_{(n-r+1)}, x_{(n-r+2)}, \ldots, x_{(n)}$ by $x_{(n-r)}$ and take their average. The α-Winsorized mean is

$$x^*_{W(\alpha)} = \frac{1}{n}[rx_{(r+1)} + x_{(r+2)} + \ldots x_{(n-r)} + rx_{(n-r)}].$$

In general, the trimmed means are preferred to Winsorized means.

For normal data, the $\alpha-$Winsorized mean has an efficiency of approximately $1 - \frac{2}{3}\alpha$ ranging from 1 $(\alpha = 0)$ to $\frac{2}{3}(\alpha = 0.5)$. For non-normal data we are faced with the problem of choosing an appropriate α. A popular choice is $\alpha = 0.1$.

(v) $\mathbf{x}^*_m$, a vector of m-estimators of univariate locations for each variable X_j. An m-estimator for location of the variable X_j, when its scale parameter is unknown, is generally defined as the solution t_j of the equation

$$\sum_{i=1}^{n} \psi(\frac{x_{ij} - t_j}{\tilde{s}_j}) = 0$$

where $\tilde{s}_j$ is a robust estimator of the scale of the variable X_j such as the median of the absolute deviation(MAD) of the observations from the median. The common choice for the function ψ are:

$$\psi(u) = \begin{cases} -1.5 & \text{if } u \leq -1.5 \\ u & \text{if} |u| \leq 1.5 \\ 1.5 & \text{if } u > 1.5, \end{cases}$$

leading to Huber's (1977) m-estimate;

$$\psi(u) = \begin{cases} 0 & \text{if } u < -c \\ (u/c)(1-u/c) & \text{if } |u| \leq c \\ 0 & \text{if } u > c, \end{cases}$$

with values of c in the range $[6, 9]$, leading to *bisquare* or *biweight* estimate proposed by Tukey (1960). Calculation of these

m-estimators involve iterative computations. The m-estimators are written in the form

$$\frac{\sum_{i=1}^{n} w_{(i)} x_{(i)j}}{\sum_{i=1}^{n} w_{(i)}}$$

where $x_{(1)j} \leq x_{(2)j} \leq \ldots \leq x_{(n)j}$ are the ordered observations. For starting the iteration common practice is to take the median of the observations as the initial value.

Some of the robust estimators of σ_{jj} suggested in the literature are the following:

(i) The square of the median absolute deviation, $(MAD)^2$.
(ii) Trimmed variance from an α-trimmed sample.
(iii) Winsorized variance from an α-Winsorized sample.

For robust estimation of covariance and correlation, consider two variables X_1, X_2. We have

$$\text{cov }(X_1, X_2) = \frac{1}{4}[V(X_1 + X_2) - V(X_1 - X_2)]. \tag{3.13.1}$$

One robust estimator s_{12}^* of $\text{Cov}(X_1, X_2)$ may be

$$s_{12}^* = \frac{1}{4}[\hat{\sigma}_1^{*2} - \hat{\sigma}_2^{*2}] \tag{3.13.2}$$

where $\hat{\sigma}_1^{*2}$ and $\hat{\sigma}_2^{*2}$ are, respectively, robust estimators of $V(X_1 + X_2)$ and $V(X_1 - X_2)$ and may be obtained by any of the methods described above.

A robust estimator of the correlation coefficient is

$$r_{12}^* = \frac{s_{12}^*}{\sqrt{s_{11}^*}\sqrt{s_{22}^*}} \tag{3.13.3}$$

where s_{jj}^* is a robust estimator of $\sigma_{jj} = V(X_j), j = 1, 2$. However, r_{12}^* may not lie in the range $[-1, 1]$. A modification is therefore suggested. Let $Z_j = X_j/\sqrt{s_{jj}^*}$. Then define

$$\hat{\rho}_{12}^* = \frac{\hat{\sigma}_3^{*2} - \hat{\sigma}_4^{*2}}{\hat{\sigma}_3^{*2} + \hat{\sigma}_4^{*2}} \tag{3.13.4}$$

where $\hat{\sigma}_3^{*2}$ and $\hat{\sigma}_4^{*2}$ are, respectively, robust estimators of $V(Z_1 + Z_2)$ and $V(Z_1 - Z_2)$. Corresponding to $\hat{\rho}_{12}^*$, which necessarily lies in the range $[-1, 1]$, a covariance estimator may be written as

$$\tilde{\sigma}_{12}^* = \hat{\rho}_{12}^* \sqrt{s_{11}^* s_{22}^*}. \tag{3.13.5}$$

However, $r_{12}^*, \hat{\rho}_{12}^*$ are biased estimators in small samples.

A direct method of obtaining a robust estimator of the covariance matrix and correlation matrix is to put those estimators together in a matrix. Thus, corresponding to each of the two methods for obtaining an estimate of the correlation coefficient, a robust estimate of the covariance matrix is

$$\mathbf{S}_i^* = \mathbf{D}\mathbf{R}_i\mathbf{D},\ i = 1, 2 \tag{3.13.6}$$

where $\mathbf{D}$ is a diagonal matrix with elements $\sqrt{s_{jj}^*}$. However, $\mathbf{S}_1^*, \mathbf{S}_2^*$ are not necessarily positive definite.

Gnanadesikan and Kettenring (1972) proposed some ad hoc methods of obtaining positive-definite estimates of covariance matrices.

3.14 Exercises and Complements

3.1 Suppose $\mathbf{X} = (X_1, X_2, X_3)' \cap N_3(\mu, \mathbf{\Sigma})$ where

$$\mu = \begin{bmatrix} 2 \\ 1 \\ 2 \end{bmatrix}, \quad \mathbf{\Sigma} = \begin{bmatrix} 2 & 1 & 1 \\ 1 & 3 & 0 \\ 1 & 0 & 1 \end{bmatrix}.$$

Find the joint distribution of $Y_1 = X_1 + X_2 + X_3$ and $Y_2 = X_1 - X_2$.

3.2 Let $X_1, \ldots, X_n$ be independent $N(0, 1)$ variables. Find the joint moment generating function of $\bar{X}, X_1 - \bar{X}, X_2 - \bar{X}, \ldots, X_n - \bar{X}$ and hence deduce that $\bar{X}$ and $\sum_{i=1}^n (X_i - \bar{X})^2$ are statistically independent.

(Hogg and Craig, 1970)

3.3 Let $\mathbf{X} \cap N_n(\mu, \mathbf{\Sigma})$ where $\sigma_{ii} = \sigma^2 (\forall\ i)$ and $\sigma_{ij} = (1-\rho)\sigma^2 (i \neq j), \rho < 1$. Using Helmert's transformation, prove that $\bar{X}$ is statistically independent of $\sum_{i=1}^n (X_i - \bar{X})^2$ and $Q \cap \chi^2_{(n-1)}$.

3.4 If $\mathbf{X} \cap N_p(\mu, \mathbf{\Sigma})$, show that $\mathbf{X}$ and $\mathbf{GX}$ have the same distribution for all orthogonal matrices $\mathbf{G}$ *iff* $\mu = \mathbf{0}$ and $\mathbf{\Sigma} = \sigma^2 \mathbf{I}$.

3.5 If $\mathbf{x}_1, \ldots, \mathbf{x}_m$ is a random sample from $N_p(\mathbf{0}, \mathbf{\Sigma})$, show that $\mathbf{x} = (\mathbf{x}_1' \ldots, \mathbf{x}_m)' \cap N_{pm}(\mathbf{0}, \mathbf{I}_m \otimes \mathbf{\Sigma})$.

3.6 Let $\mathbf{Y} \cap (\mathbf{0}, \mathbf{I}_n)$ and let $\mathbf{X} = \mathbf{AY}, \mathbf{U} = \mathbf{BY}, \mathbf{V} = \mathbf{CY}$, where $\mathbf{A}, \mathbf{B}$ and $\mathbf{C}$ are all $r \times n$ matrices of rank $r(< n)$. If $C(\mathbf{X}, \mathbf{U}) = \mathbf{0}$ and $C(\mathbf{X}, \mathbf{V}) = \mathbf{0}$, prove that $\mathbf{X}$ is independent of $\mathbf{U} + \mathbf{V}$.

3.7 Let $\mathbf{X}_1, \mathbf{X}_2$ be two random p-vectors and let $\mathbf{X} = \begin{bmatrix} \mathbf{X}_1 \\ \mathbf{X}_2 \end{bmatrix}$. If $\mathbf{X}$ has a multivariate normal distribution with mean vector $\mathbf{0}$ and dispersion matrix

$$\begin{bmatrix} \mathbf{A} & \mathbf{B} \\ \mathbf{B}' & \mathbf{C} \end{bmatrix}$$

when $\mathbf{A}, \mathbf{B}, \mathbf{C}$ are all $p \times p$ matrices, show that a necessary and sufficient condition for $\mathbf{X}_1 + \mathbf{X}_2$ be independent of $\mathbf{X}_1 - \mathbf{X}_2$ is $\mathbf{A} = \mathbf{C}, \mathbf{B} = \mathbf{B}'$.

3.8 Let $\mathbf{Y}_\alpha \cap N_p(a_\alpha \mu, \mathbf{\Sigma}), \alpha = 1, \ldots, n$ be independent. Show that $Q = \sum_{\alpha=1}^n \mathbf{Y}_\alpha \mathbf{Y}'_\alpha$ and $\mathbf{ZZ}'$ are independently distributed where

$$\mathbf{Z} = \frac{\sum_{\alpha=1}^n a_\alpha \mathbf{Y}_\alpha}{\sqrt{\sum_{\alpha=1}^n a_\alpha^2}}.$$

3.9 If $\mathbf{X}$ is $N_p(\mu, \mathbf{\Sigma})$, prove that

(a) $E(\mathbf{X}'\mathbf{AX}) = tr(\mathbf{A\Sigma}) + \mu'\mathbf{A}\mu$;
(b) $V(\mathbf{X}'\mathbf{AX}) = 2[tr(\mathbf{A\Sigma A\Sigma}) + 2\mu'\mathbf{A\Sigma A}\mu]$.

3.10 Suppose that the vector $\begin{bmatrix} Y \\ \mathbf{X} \end{bmatrix}$ when Y is $1 \times 1, \mathbf{X}$ is $(p-1) \times 1$, has mean vector $\mu\mathbf{1}$, and covariance matrix

$$\begin{bmatrix} \sigma_{11} & \mathbf{\Sigma}_{12} \\ \mathbf{\Sigma}'_{12} & \mathbf{\Sigma}_{22} \end{bmatrix}$$

where $\sigma_{11} = V(Y), \mathbf{\Sigma}_{22} =$ Cov $(\mathbf{X})$. Find the coefficient vector α of a linear function $\alpha'\mathbf{X}$ which minimizes $V(Y - \alpha'\mathbf{X})$ subject to the condition $E(\alpha'\mathbf{X}) = E(Y)$.

3.11 Let $\mathbf{X} = (X_1, \ldots, X_p)'$ be distributed as $N_p(\mu, \mathbf{\Sigma})$. Show that for $\mu = \mathbf{1}_p$,

$$\mathbf{\Sigma} = \begin{bmatrix} 1 & 1 & 1 & \ldots & 1 & 1 \\ 1 & 2 & 2 & \ldots & 2 & 2 \\ 1 & 2 & 3 & \ldots & 3 & 3 \\ . & . & . & \ldots & . & . \\ . & . & . & \ldots & . & . \\ 1 & 2 & 3 & \ldots & p-1 & p-1 \\ 1 & 2 & 3 & \ldots & p-1 & p \end{bmatrix},$$

the random variable

$$R = (X_2 - X_1)^2 + \ldots + (X_p - X_{p-1})^2$$

has a chi-square distribution.

3.12 Show that if $\mathbf{X}_n \to^L N(\mu, \mathbf{\Sigma})$ and if $\mathbf{C}$ be a $m \times k$ matrix then $\mathbf{C}\mathbf{X}_n \to^L N(\mathbf{C}\mu, \mathbf{C}\mathbf{\Sigma}\mathbf{C}')$.

3.13 Derive the characteristic function of the multivariate normal distribution with mean vector $\mu = \mathbf{0}$ and dispersion matrix $\mathbf{\Sigma} = ((\sigma_{ij}))$. Show that

$$E(X_i X_j X_k X_l) = \sigma_{ij}\sigma_{kl} + \sigma_{il}\sigma_{jk} + \sigma_{ik}\sigma_{jl}.$$

3.14 Let $\mathbf{A}, \mathbf{B}$ be real real symmetric matrices of order $n \times n$ and let $X_1, \ldots, X_n$ be normal variates with zero mean and unit variances. Show that necessary and sufficient conditions for $\mathbf{X}'\mathbf{A}\mathbf{X}$ and $\mathbf{X}'\mathbf{B}\mathbf{X}$ to be distributed independently as chi-squared variates are $\mathbf{AB} = \mathbf{0}$, and $\mathbf{A}^2 = \mathbf{A}, \mathbf{B}^2 = \mathbf{B}$.

3.15 Show that if rows of $\mathbf{X}$ are *iid* $N_p(\mu, \mathbf{\Sigma})$ and if $\mathbf{C}_1, \ldots, \mathbf{C}_k$ are symmetric matrices, then $\mathbf{X}'\mathbf{C}_1\mathbf{X}, \ldots, \mathbf{X}'\mathbf{C}_k\mathbf{X}$ are jointly independent *iff* $\mathbf{C}_r\mathbf{C}_s = \mathbf{0}$ for all $r \neq s$.

3.16 Let $\mathbf{X}'\mathbf{A}\mathbf{X}, \mathbf{X}'\mathbf{B}\mathbf{X}$ be two quadratic forms in independent normal variates $X_1, \ldots, X_n$ each of unit variance. Show that their joint moment generating function is given by $|\mathbf{I} - \theta_1\mathbf{A} - \theta_2\mathbf{B}|^{-1/2}$ where $\theta_1 = 2t_1, \theta_2 = 2t_2$. Hence derive Cochran's theorem that a necessary and sufficient condition for the two forms to be independent is that

$$|\mathbf{I} - \theta_1\mathbf{A} - \theta_2\mathbf{B}| = |\mathbf{I} - \theta_1\mathbf{A}||\mathbf{I} - \theta_2\mathbf{B}|$$

identically.

3.17 Suppose that $\mathbf{X}$ is $N_3(\mathbf{0}, \mathbf{\Sigma})$ where

$$\mathbf{\Sigma} = \begin{bmatrix} 1 & \rho & 0 \\ \rho & 1 & \rho \\ 0 & \rho & 1 \end{bmatrix}.$$

Is there any value of ρ for which $X_1 + X_2 + X_3$ and $X_1 - X_2 - X_3$ are independent?

3.18 Show that for the following bivariate distributions the marginal distributions of X_1 anX_2 are normal.

(a) $f(x_1, x_2) = \frac{1}{2\pi}\exp[-\frac{1}{2}(x_1^2 + x_2^2)]\{1 + x_1x_2\exp[-\frac{1}{2}(x_1^2 + x_2^2)]\}$, $-\infty < x_1, x_2 < \infty$,

(b) $f(x_1, x_2) = \frac{1}{2}\exp[-\frac{1}{2}(x_1^2 + x_2^2)](1 - \frac{x_1 x_2}{(1+x_1^2)(1+x_2^2)})$.

3.19 Let X_1 be $N(0,1)$ and

$$X_2 = \begin{cases} -X_1 & \text{if} \quad -1 \leq X_1 \leq 1 \\ X_1 & \quad\quad \text{otherwise.} \end{cases}$$

Show that

(a) X_2 has a $N(0,1)$ distribution;
(b) X_1, X_2 do not have a bivariate normal distribution.

3.20 Suppose that two variables X, Y follow a bivariate normal distribution $N(\mu, \mathbf{\Sigma})$ where

$$\mu = \begin{bmatrix} 6 \\ 2 \end{bmatrix}, \quad \mathbf{\Sigma} = \begin{bmatrix} 9 & 3 \\ 3 & 4 \end{bmatrix}.$$

Find (a) $E(Y|X), E(X|Y), (b) V(Y|X), (c) P(X \leq 5), (d) P(2 \leq X \leq 5, 1 \leq Y \leq 3)$.

3.21 Let $\mathbf{X} \cap N(\mu, \Sigma), |\mathbf{\Sigma}| > 0$. It is found that first q variables form a group, so $\mathbf{X}$ is partitioned as

$$\mathbf{X} = \begin{bmatrix} \mathbf{X}^{(1)} \\ \mathbf{X}^{(2)} \end{bmatrix}, \quad \mathbf{X}^{(1)} : q \times 1.$$

Find the distribution of $\mathbf{AX}^{(1)} + \mathbf{BX}^{(2)}$, where $\mathbf{A} : q \times q$ and $\mathbf{B} : q \times (p - q), 2q < p$.

3.22 Suppose $\mathbf{X} = (X_1, X_2, X_3)'$ follows $N_3(\mu, \mathbf{\Sigma})$, where

$$\mu = \begin{bmatrix} 3 \\ 10 \\ 8 \end{bmatrix}, \quad \mathbf{\Sigma} = \begin{bmatrix} 1 & 3 & 1 \\ 3 & 16 & 2 \\ 1 & 2 & 4 \end{bmatrix}.$$

Find the distribution of (a) X_1, for given $(X_2, X_3), (b)(X_1, X_2)$ for given X_4 and (c) X_3, for given X_1 and X_2.

3.23 Suppose the covariance matrix is

$$\mathbf{\Sigma} = \begin{bmatrix} 1 & \rho & \rho^2 \\ \rho & 1 & 0 \\ \rho^2 & 0 & 1 \end{bmatrix}.$$

Show that the conditional distribution of (X_1, X_2) given X_3 has the mean vector $(\mu_1 + \rho^2(x_3 - \mu_3), \mu_3)'$ and covariance matrix

$$\begin{bmatrix} 1-\rho^4 & \rho \\ \rho & 1 \end{bmatrix}.$$

3.24 If $\mathbf{X}_1, \mathbf{X}_2, \mathbf{X}_3$ are *iid* $N_p(\mu, \mathbf{\Sigma})$ random variables and $\mathbf{Y}_1 = \mathbf{X}_1 + \mathbf{X}_2, \mathbf{Y}_2 = \mathbf{X}_2 + \mathbf{X}_3, \mathbf{Y}_3 = \mathbf{X}_3 + \mathbf{X}_1$, obtain the conditional distribution of $\mathbf{Y}_1$ given $\mathbf{Y}_2$ and of $\mathbf{Y}_1$ given $\mathbf{Y}_2, \mathbf{Y}_3$.

3.25 If $\mathbf{X}$ is $N_m(\mu, \mathbf{\Sigma})$, where $\mathbf{\Sigma}$ is positive definite, $\mathbf{A}$ is an $m \times m$ symmetric matrix and $\mathbf{B}$ is an $r \times m$ matrix, prove that $\mathbf{X}'\mathbf{A}\mathbf{X}$ and $\mathbf{B}\mathbf{X}$ are independent *iff* $\mathbf{B\Sigma A} = \mathbf{0}$.

3.26 *(Fre'chet (1951) Inequality)* Let X, Y be random variables with *c.d.f* $H(x; y)$ and marginal *cdfs*, $F(x)$ and $G(y)$. Show that

$$\max\ (F + G - 1, 0) \leq H \leq\ \min\ (F, G).$$

3.27 Let $\mathbf{X}_j = (X_{j1}, \ldots, X_{jn_j})' \cap N_{n_j}(\mu \mathbf{1}_{n_j}, \tau^2 \mathbf{I}_{n_j} + \sigma^2 \mathbf{1}_{n_j} \mathbf{1}'_{n_j})$ where $\mathbf{I}_p$ is an identity matrix of order p and $\mathbf{1}_q = (1, \ldots, 1)'_{q \times 1}$. Suppose we have m such random vectors $\mathbf{X}_j (j = 1, \ldots, m)$. Let

$$\bar{X}_j = \sum_{i=1}^{n_j} X_{ji}/n_j, \quad \bar{X}_{00} = \sum_{j=1}^{m} n_j \bar{X}_j / n_T,$$

where

$$n_T = \sum_{j=1}^{m} n_j,$$

$$MSB = \sum_{j=1}^{m} n_j (\bar{X}_j - \bar{X}_{00})^2/(m-1), \quad MSW = \sum_{j=1}^{m} \sum_{i=1}^{n_j} (X_{ji} - \bar{X}_j)^2/(n_T - m).$$

Show that

$$E(MSW) = \tau^2, \quad V(MSW) = 2\tau^4/(n_T - m),$$

$$E(MSB) = \tau^2 + \left(n_T - (\sum_{j=1}^{m} n_j^2)/n_T \right) \frac{\sigma^2}{m-1},$$

$$V(MSB) = \frac{2}{(m-1)^2} \left[\sum_{j=1}^{m} \sigma_j^4 + (\sum_{j=1}^{m} n_j \sigma_j^2)^2/n_T^2 - 2 \sum_{j=1}^{m} n_j \sigma_j^4 / n_t \right],$$

where $\sigma_j^2 = \tau^2 + \sigma^2 n_j (1 \le j \le m)$.

[*Hints:* $\sum_{i=1}^{n_j}(X_{ji} - \bar{X}_j)^2 \cap \chi^2_{(n_j-1)}.\tau^2$, since $(X_{ji} - \bar{X}_j) \cap N(0, \tau^2)$. Again, $\sqrt{n_j}(\bar{X}_j - \mu)$ are independent $N(0, \sigma_j^2)$ variables.

Let $Z_j = \sqrt{n_j}(\bar{X}_j - \mu)\sigma_j, \mathbf{Z} = (Z_1, \ldots, Z_m)', \mathbf{A} = \mathbf{D} - \mathbf{UU}'$, where $\mathbf{D} =$ Diag. $(\sigma_1^2, \ldots, \sigma_m^2), \mathbf{U} = \frac{1}{\sqrt{n_T}}(\sqrt{n_1}\sigma_1, \ldots, \sqrt{n_m}\sigma_m)'$. Then MSB $= \mathbf{Z}'\mathbf{AZ}/(m-1)$. Now $E(\mathbf{Z}'\mathbf{AZ}) = tr(\mathbf{A})$. Also, $V(\mathbf{Z}'\mathbf{AZ}) = 2tr(\mathbf{A}^2)$].

(Ghosh and Meeden, 1986)

3.28 Consider the model $Y_i = \theta + \epsilon_i (i = 1, \ldots N)$, where $\theta, \epsilon_1, \ldots, \epsilon_N$ are independent random variables with $\theta \cap N(\mu, \sigma^2)$ and ϵ_i's are *iid* $N(0, \tau^2)$. Show that the joint probability distribution of $\mathbf{Y} = (Y_1, \ldots, Y_N)'$ given μ, σ^2, τ^2 is $N_N(\mu\mathbf{1}_N, \tau^2\mathbf{I}_N + \sigma^2\mathbf{1}_N\mathbf{1}_N')$. Show also that the joint conditional distribution of $Y_{n+1}, \ldots, Y_N$ given that $Y_i = y_i (i = 1, \ldots, n)$ is $(N-n)$-variate normal with mean vector $\frac{M\mu + n\bar{y}}{M+n}\mathbf{1}_{N-n}$ and dispersion matrix $\tau^2[\mathbf{I}_{N-n} + \mathbf{1}_{N-n}\mathbf{1}_{N-n}'/(M+n)]$ where $M = \tau^2/\sigma^2$ and $\bar{y} = \frac{1}{n}\sum_{i=1}^n y_i$. Hence, show that

$$E\left(\sum_{j=1}^N Y_j \mid Y_i = y_i, i = 1, \ldots, n\right) = n\bar{y} + (N-n)\{B\mu + (1-B)\bar{y}\}$$

where $B = \frac{M}{M+n}$.

3.29 *A multivariate Poisson distribution* Let $U_0, U_1, \ldots, U_p$ be independent Poisson variables with parameters $\lambda_0, \lambda_1 - \lambda_0, \ldots, \lambda_p - \lambda_0$ respectively. Write down the joint distribution of $X_i = U_i + U_0 (i = 1, \ldots, p)$ and show that the marginal distribution of $X_1, \ldots, X_p$ are all Poisson. For $p = 2$, show that the *pmf* is given by

$$f(x_1, x_2) = \exp(-\lambda_1 - \lambda_2 + \lambda_0)\frac{a^{x_1}b^{x_2}}{x_1!x_2!}\sum_{r=0}^{s}\frac{x_1^{(r)}x_2^{(r)}\lambda_0^r}{a^r b^r r!}$$

where $s =$ min $(x_1, x_2), a = \lambda_1 - \lambda_0, b = \lambda_2 - \lambda_0, \lambda_1 > \lambda_0 > 0, \lambda_2 > \lambda_0 > 0$, and $x^{(r)} = x(x-1)\ldots(x-r+1)$. Furthermore, show that

$$E(X_2|x_1) = b + (\frac{\lambda_0}{\lambda_1})x_1, \quad \text{Var } (X_2|x_1) = b + \{\frac{a\lambda_0}{\lambda_1^2}\}x_1.$$

(Mardia, et al., 1979)

3.30 Consider the bird-data in Table 2.E.1.

(a) Construct $Q-Q$ plot for each of the variables. Do the marginal distributions seem to be normal? Explain.

(b) Carry out the tests of normality based on correlation coefficient r_Q for each of the variables. Let the significance level be $\alpha = 0.05$. Do the results of these tests corroborate the results in (a)?

(c) For two variables X_1, X_2, calculate the squared generalized distances $(\mathbf{x}_i - \bar{\mathbf{x}})'\mathbf{S}^{-1}(\mathbf{x}_i - \bar{\mathbf{x}}), i = 1, 2, \ldots$ where $\mathbf{x}_i = (x_{i1}, x_{i2})'$.

(d) Using the distances in (c) above, determine the proportion of observations falling within the estimated 50% probability contour of a bivariate normal distribution and draw appropriate conclusion.

(e) For the two variables X_1, X_2 draw a chi-square plot. Are these data approximately bivariate normal? Explain.

(f) Obtain a chi-square plot for all the variables X_1, X_2, X_3. Are these data approximately trivariate normal? Explain.

3.31 For the data of Table 2.E.2, for each village and for both the villages combined, test (a) the univariate normality of the distributions of X_2, X_3 (b) bivariate normality of the joint distribution of X_2, X_3.

Chapter 4

Distributions Arising Out of the Multivariate Normal Distribution

4.1 Introduction

In this chapter we consider distributions of some statistics obtained from sampling from a multivariate normal distribution. Section 4.2 considers Wishart's distribution, Section 4.3 Hotelling's T^2 and Section 4.4 Wilk's (1932) statistic. The last section considers some statistics based on the eigenvalues of Wishart matrices.

4.2 Wishart Distribution

In Section 3.4 we considered the distribution of the quadratic form $\mathbf{X}'\mathbf{AX}$ where $\mathbf{X} \cap N_p(\mu, \mathbf{\Sigma})$. In Section 3.3 we considered the distribution of the function $\mathbf{AXB}$ where $\mathbf{X}$ is a data matrix from $N_p(\mu, \mathbf{\Sigma})$. Here, we shall address the distribution of the function $\mathbf{X}'\mathbf{AX}$ where $\mathbf{X}$ is a data matrix from $N_p(\mathbf{0}, \mathbf{\Sigma})$.

Definition 4.2.1: Suppose that $\mathbf{X}_1, \ldots, \mathbf{X}_m$ are independently and identically distributed as $N_p(\mathbf{0}, \mathbf{\Sigma})$; then

$$\mathbf{W} = \sum_{i=1}^{m} \mathbf{X}_i \mathbf{X}_i' \tag{4.2.1}$$

is said to have a Wishart distribution $W_p(m, \mathbf{\Sigma})$ with scale parameter $\mathbf{\Sigma}$ and degrees of freedom m. When $\mathbf{\Sigma} = \mathbf{I}$, the distribution is said to be in the standard form.

In other words, if $\mathbf{X}(m \times p)$ is a data matrix from $N_p(\mathbf{0}, \mathbf{\Sigma})$ (that is, the row vectors of $\mathbf{X}$ are independent samples from $N_p(\mathbf{0}, \mathbf{\Sigma})$), then

$$\mathbf{W} = \mathbf{X}'\mathbf{X} \tag{4.2.2}$$

is said to have a Wishart distribution $W_p(m, \mathbf{\Sigma})$.

If $\mathbf{\Sigma} > 0$, and $m \geq p$, then it can be shown that $\mathbf{W} > 0$ with probability one (Corollary 4.2.9.1). In this case we have a non-singular Wishart distribution. Otherwise, by the definition above, $\mathbf{W} \geq 0$ and there is a non-zero probability that $|\mathbf{W}| = 0$. In the latter case we have a singular Wishart distribution.

We have

$$E(\mathbf{W}) = E(\sum_{i=1}^{m} \mathbf{X}_i \mathbf{X}_i') = m\mathbf{\Sigma}, \tag{4.2.3}$$

because, $E(\mathbf{X}_i \mathbf{X}_i') = V(\mathbf{X}_i) = \mathbf{\Sigma}$.

Note 4.2.1: When $p = 1, W = \sum_{i=1}^{m} X_i^2$, when $X_i \cap N_1(0, \sigma^2)$. Hence, $W_1(m, \sigma^2)$ follows $\sigma^2 \chi^2_{(m)}$ distribution.

The density function of the Wishart distribution is given in Definition 4.2.2 below.

DEFINITION 4.2.2: Let $\mathbf{W} = ((w_{ij}))$ be a $p \times p$ symmetric matrix of random variables, that is positive definite with probability one and let $\mathbf{\Sigma}$ be a $p \times p$ positive definite matrix. If m is an integer such that $m \geq p$, then $\mathbf{W}$ is said to have a (non-singular) Wishart distribution with m degrees of freedom if the joint density function of the $\frac{p(p+1)}{2}$ distinct elements of $\mathbf{W}$ (in, say, the upper triangle) is

$$f(w_{11}, w_{12}, \ldots, w_{pp}) = c^{-1} |\mathbf{W}|^{(m-p-1)/2} \exp\{ \text{ trace } (-\frac{1}{2}\mathbf{\Sigma}^{-1}\mathbf{W})\} \tag{4.2.4}$$

where

$$c = 2^{mp/2} |\mathbf{\Sigma}|^{m/2} \pi^{\frac{p(p-1)}{4}} \Pi_{j=1}^{p} \Gamma[\frac{1}{2}(m + 1 - j)]. \tag{4.2.5}$$

Note that the Definition 4.2.2 requires $\mathbf{\Sigma}$ to be positive definite, while for Definition 4.2.1, it is enough to assume $\mathbf{\Sigma} \geq 0$.

We shall use the Definition 4.2.1, as it does not require the manipulation of the density function. Also, we shall assume $\mathbf{\Sigma} > 0$.

4.2.1 *Properties of Wishart Distribution*

We shall now study some properties of the distribution.

Theorem 4.2.1: If $\mathbf{W} \cap W_p(m, \mathbf{\Sigma})$ and $\mathbf{B}(q \times p)$ is of rank q then $\mathbf{BWB}'$ is $W_q(m, \mathbf{B\Sigma B}')$.

Proof. Let $\mathbf{W} = \sum_{i=1}^m \mathbf{X}_i\mathbf{X}_i'$ where $\mathbf{X}_i$ is $N_p(\mathbf{0}, \mathbf{\Sigma})$. Then, $\mathbf{Y}_i = \mathbf{BX}_i$ is $N_q(\mathbf{0}, \mathbf{B\Sigma B}')$. Therefore, $\sum_{i=1}^m \mathbf{Y}_i\mathbf{Y}_i' = \mathbf{BWB}' \cap W_q(m, \mathbf{B\Sigma B}')$.

Corollary 4.2.1.1: Consider a transformation $\mathbf{W} = \mathbf{CAC}'$ when $\mathbf{C}$ is non-singular $p \times p$. If $\mathbf{W}$ is distributed as $W_p(m, \mathbf{\Sigma})$ then $\mathbf{A}$ is distributed as $W_p(m, \Phi)$ where $\mathbf{\Phi} = \mathbf{C}^{-1}\mathbf{\Sigma}(\mathbf{C}^{-1})'$.

Proof. Let $\mathbf{W} = \sum_{i=1}^m \mathbf{X}_i\mathbf{X}_i'$ where $\mathbf{X}_i$ is $N_p(\mathbf{0}, \mathbf{\Sigma})$. Let $\mathbf{Y}_i = \mathbf{C}^{-1}\mathbf{X}_i$. Then $\mathbf{Y}_i \cap N_p(\mathbf{0}, \mathbf{C}^{-1}\mathbf{\Sigma}(\mathbf{C}^{-1})')$. Hence, $\sum_{i=1}^m \mathbf{Y}_i\mathbf{Y}_i' = \mathbf{C}^{-1}\mathbf{W}(\mathbf{C}^{-1})' = \mathbf{A} \cap W_p(m, \mathbf{\Phi})$.

Corollary 4.2.1.2: If $\mathbf{W} \cap W_p(m, \mathbf{\Sigma})$ then $\mathbf{\Sigma}^{-1/2}\mathbf{W}\mathbf{\Sigma}^{-1/2} \cap W_p(m, \mathbf{I})$.

Corollary 4.2.1.3: If $\mathbf{W} \cap W_p(m, \mathbf{I})$ and $\mathbf{B}(p \times q)$ of rank q satisfies $\mathbf{B}'\mathbf{B} = \mathbf{I}_q$, then $\mathbf{B}'\mathbf{WB} \cap W_q(m, \mathbf{I})$.

Corollary 4.2.1.4: Each diagonal element of $\mathbf{W}, w_{jj} = \sum_{i=1}^m X_{ij}^2$ follows $W_1(m, \sigma_{jj}) = \sigma_{jj}\chi^2_{(m)}$, where $\mathbf{\Sigma} = ((\sigma_{ij}))$.

Theorem 4.2.2: Let $\mathbf{W} \cap W_p(m, \mathbf{\Sigma})$. Let also $\mathbf{W}, \mathbf{\Sigma}$ be partitioned as

$$\mathbf{W} = \begin{bmatrix} \mathbf{W}_{11} & \mathbf{W}_{12} \\ \mathbf{W}_{21} & \mathbf{W}_{22} \end{bmatrix}, \quad \mathbf{\Sigma} = \begin{bmatrix} \mathbf{\Sigma}_{11} & \mathbf{\Sigma}_{12} \\ \mathbf{\Sigma}_{21} & \mathbf{\Sigma}_{22} \end{bmatrix},$$

where $\mathbf{W}_{11}, \mathbf{\Sigma}_{11}$ are $q \times q$. Then, $\mathbf{W}_{11} \cap W_q(m, \mathbf{\Sigma}_{11}), \mathbf{W}_{22} \cap W_{(p-q)}(m, \mathbf{\Sigma}_{22})$.

Proof. We have, $\mathbf{W} = \sum_{i=1}^m \mathbf{X}_i\mathbf{X}_i'$ where $\mathbf{X}_i$'s are independent $N_p(\mathbf{0}, \mathbf{\Sigma})$. Partition

$$\mathbf{X}_i = \begin{bmatrix} \mathbf{X}_i^{(1)} \\ \mathbf{X}_i^{(2)} \end{bmatrix}$$

where $\mathbf{X}_i^{(1)}$ is $q \times 1$. Thus, $\mathbf{X}_i^{(1)} \cap N_q(\mathbf{0}, \mathbf{\Sigma}_{11})$. Hence, $\mathbf{W}_{11} = \sum_{i=1}^m \mathbf{X}_i^{(1)}\mathbf{X}_i^{(1)'}$ is distributed as $W_q(m, \mathbf{\Sigma}_{11})$. Similar result holds for $\mathbf{W}_{22}$.

Note 4.2.2. : However, $\mathbf{W}_{11}$ and $\mathbf{W}_{22}$ are not independent in general (vide Theorems 4.2.3 and 4.2.7).

Theorem 4.2.3: Let $\mathbf{W}$ and $\mathbf{\Sigma}$ be partitioned into $(p_1, \ldots, p_q)$ rows and

$(p_1, \ldots, p_q)$ columns with $\sum_{j=1}^{q} p_j = p$.

$$\mathbf{W} = \begin{bmatrix} \mathbf{W}_{11} & \ldots & \mathbf{W}_{1q} \\ . & \cdots & . \\ . & \cdots & . \\ \mathbf{W}_{q1} & \ldots & \mathbf{W}_{qq} \end{bmatrix}, \quad \mathbf{\Sigma} = \begin{bmatrix} \mathbf{\Sigma}_{11} & \ldots & \mathbf{\Sigma}_{1q} \\ . & \cdots & . \\ . & \cdots & . \\ \mathbf{\Sigma}_{q1} & \ldots & \mathbf{\Sigma}_{qq} \end{bmatrix},$$

where $\mathbf{W}_{ij}, \mathbf{\Sigma}_{ij}$ are of order $p_i \times p_j (i, j = 1, \ldots, q)$. If $\mathbf{\Sigma}_{ij} = \mathbf{0}, \ \ i \neq j$ and if $\mathbf{W} \cap W_p(m, \mathbf{\Sigma})$, then $\mathbf{W}_{11}, \ldots \mathbf{W}_{qq}$ are independently distributed and $\mathbf{W}_{jj} \cap W_{p_j}(m, \mathbf{\Sigma}_{jj})$.

Proof. $\mathbf{W}$ is distributed as $\sum_{i=1}^{m} \mathbf{X}_i \mathbf{X}_i'$ where $\mathbf{X}_i$ is $N_p(\mathbf{0}, \mathbf{\Sigma})$. Let $\mathbf{X}_i$ be partitioned as

$$\mathbf{X}_i = \begin{bmatrix} \mathbf{X}_i^{(1)} \\ . \\ . \\ . \\ \mathbf{X}_i^{(q)} \end{bmatrix},$$

where $\mathbf{X}_i^{(j)}$ is of order $p_j \times 1$. Since $\mathbf{\Sigma}_{ij} = \mathbf{0}, \mathbf{X}_i^{(j)} (j = 1, \ldots, q)$ are mutually independent. Also, $\mathbf{W}_{jj} = \sum_{i=1}^{m} \mathbf{X}_i^{(j)} \mathbf{X}_i^{(j)'}$ is independent of $\mathbf{W}_{kk} = \sum_{i=1}^{m} \mathbf{X}_i^{(k)} \mathbf{X}_i^{(k)'} (j \neq k)$. The result now follows Theorem 4.2.2.

Theorem 4.2.4: If $\mathbf{W} \cap W_p(m, \mathbf{\Sigma})$ and $\mathbf{a}(p \times 1)$ is any fixed vector, then

$$\frac{\mathbf{a}'\mathbf{W}\mathbf{a}}{\mathbf{a}'\mathbf{\Sigma}\mathbf{a}} \cap \chi^2_{(m)} \ \text{ provided } \mathbf{a}'\mathbf{\Sigma}\mathbf{a} \neq 0.$$

Proof. From Theorem 4.2.1, $\mathbf{a}'\mathbf{W}\mathbf{a} \cap W_1(m, \mathbf{a}'\mathbf{\Sigma}\mathbf{a}) = (\mathbf{a}'\mathbf{\Sigma}\mathbf{a})\chi^2_{(m)}$.

[The converse is not necessarily true. That is, if $\mathbf{a}'\mathbf{W}\mathbf{a}/\mathbf{a}'\mathbf{\Sigma}\mathbf{a} \cap \chi^2((m)$ for all $\mathbf{a}$, then $\mathbf{W}$ does not necessarily follow Wishart (Mitra, 1969)].

(A similar result occurs in Theorem 4.2.8)

Theorem 4.2.5: If $\mathbf{W} \cap W_p(m, \mathbf{\Sigma})$, and $\mathbf{Y}$ is any $p \times 1$ random vector which is independent of $\mathbf{W}$ with $P(\mathbf{Y} = \mathbf{0}) = 0$, then $\mathbf{Y}'\mathbf{W}\mathbf{Y}/(\mathbf{Y}'\mathbf{\Sigma}\mathbf{Y}) \cap \chi^2_{(m)}$ and is independent of $\mathbf{Y}$.

Proof. By Theorem 4.2.4, conditional on $\mathbf{Y}$, $\mathbf{Y}'\mathbf{W}\mathbf{Y}$ follows $W_1(m, \mathbf{Y}'\mathbf{\Sigma}\mathbf{Y})$. Therefore, $\mathbf{Y}'\mathbf{W}\mathbf{Y}/(\mathbf{Y}'\mathbf{\Sigma}\mathbf{Y})$ follows $\chi^2_{(m)}$. Since this distribution does not depend on $\mathbf{Y}$, it is also the unconditional distribution of $\mathbf{Y}'\mathbf{W}\mathbf{Y}/(\mathbf{Y}'\mathbf{\Sigma}\mathbf{Y})$.

Corollary 4.2.5.1: If $\bar{\mathbf{X}}$ and $\mathbf{S} = (n-1)^{-1} \sum_{i=1}^{n} (\mathbf{X}_i - \bar{\mathbf{X}}_i)(\mathbf{X}_i - \bar{\mathbf{X}})'$ are the mean and covariance vector of a random sample $\mathbf{X}_i (i = 1, \ldots, n)$ from

$N_p(\mu, \mathbf{\Sigma})$, then

$$\frac{(n-1)\bar{\mathbf{X}}'\mathbf{S}\bar{\mathbf{X}}}{\bar{\mathbf{X}}'\mathbf{\Sigma}\bar{\mathbf{X}}}$$

follows $\chi^2_{(n-1)}$ and is independent of $\bar{\mathbf{X}}$.

Proof. It is known that $\bar{\mathbf{X}}$ and $\mathbf{S}$ are independently distributed (vide Corollary 3.3.3.2). Also, $(n-1)\mathbf{S}$ is $\mathbf{W}_p(n-1, \mathbf{\Sigma})$ (vide Theorem 4.2.10). Hence, the result follows by Theorem 4.2.5.

Theorem 4.2.6: If $\mathbf{W}_1 \cap W_p(m_1, \mathbf{\Sigma}), \mathbf{W}_2 \cap W_p(m_2, \mathbf{\Sigma})$, and $\mathbf{W}_1, \mathbf{W}_2$ are independent, then $\mathbf{W} = \mathbf{W}_1 + \mathbf{W}_2 \cap W_p(m_1 + m_2, \mathbf{\Sigma})$.

Proof. Let $\mathbf{W}_i = \mathbf{X}i'\mathbf{X}_i (i = 1, 2)$ where $\mathbf{X}_i(m_i \times p)$ is a data matrix from $N_p(\mathbf{0}, \mathbf{\Sigma})$ (that is, all the m_i rows of $\mathbf{X}_i$ are independent samples from $N_p(\mathbf{0}, \mathbf{\Sigma})$). Also, all the rows of $\mathbf{X}_1$ and $\mathbf{X}_2$ are independent. Now

$$\mathbf{W} = \mathbf{W}_1 + \mathbf{W}_2 = \mathbf{X}_1'\mathbf{X}_1 + \mathbf{X}_2'\mathbf{X}_2 = \mathbf{X}'\mathbf{X}$$

where $\mathbf{X} = \begin{bmatrix} \mathbf{X}_1 \\ \mathbf{X}_2 \end{bmatrix}$ is a data matrix of order $(m_1 + m_2) \times p$ from $N_p(\mathbf{0}, \mathbf{\Sigma})$. Hence, $\mathbf{W}$ follows $W_p(m_1 + m_2, \mathbf{\Sigma})$.

Corollary 4.2.6.1: If $\mathbf{X} = (\mathbf{X}_1, \ldots, \mathbf{X}_m)'$ is a data matrix from $N_p(\mathbf{0}, \mathbf{\Sigma})$, then each of $\mathbf{X}_i\mathbf{X}_i'$ follow independent $W_p(1, \mathbf{\Sigma})$ distribution.

Proof. We have $\mathbf{W} = \mathbf{X}'\mathbf{X} = \sum_{i=1}^{m} \mathbf{X}_i\mathbf{X}_i'$ follows $W_p(m, \mathbf{\Sigma})$ distribution. Also, $\mathbf{X}_i$'s are independent. Hence, each of $\mathbf{X}_i\mathbf{X}_i'$ must have an $W_p(1, \mathbf{\Sigma})$ distribution.

We now consider some more theorems on partitioned Wishart matrices. Let

$$\mathbf{W} = \begin{bmatrix} \mathbf{W}_{11} & \mathbf{W}_{12} \\ \mathbf{W}_{21} & \mathbf{W}_{22} \end{bmatrix},$$

where $\mathbf{W}_{11}$ is $q \times q$.

Theorem 4.2.7: Let $\mathbf{W} \cap W_p(m, \mathbf{\Sigma}), m > q$. Then

(a) $\mathbf{W}_{22.1} = \mathbf{W}_{22} - \mathbf{W}_{21}\mathbf{W}_{11}^{-1}\mathbf{W}_{12} \cap \mathbf{W}_{p-q}(m - q, \mathbf{\Sigma}_{22.1})$ where $\mathbf{\Sigma}_{22.1} = \mathbf{\Sigma}_{22} - \mathbf{\Sigma}_{21}\mathbf{\Sigma}_{11}^{-1}\mathbf{\Sigma}_{12}$ and is independent of $(\mathbf{W}_{11}, \mathbf{W}_{12})$.

(b) If $\mathbf{\Sigma}_{12} = \mathbf{0}$, then $\mathbf{W}_{22} - \mathbf{W}_{22.1} = \mathbf{W}_{21}\mathbf{W}_{11}^{-1}\mathbf{W}_{12}$ has the $W_{p-q}(q, \mathbf{\Sigma}_{22})$ distribution and $\mathbf{W}_{21}\mathbf{W}_{11}^{-1}\mathbf{W}_{12}, \mathbf{W}_{22.1}$ and $\mathbf{W}_{11}$ are jointly independent.

Proof. Omitted. Interested readers may refer to Mardia, *et al.* (1979, Section 3.4.3)

Therefore, if $\mathbf{\Sigma}_{12} = \mathbf{0}$, then the Wishart matrix $\mathbf{W}_{22} = \mathbf{W}_{22.1} + \mathbf{W}_{21}\mathbf{W}_{11}^{-1}\mathbf{W}_{12}$ can be decomposed into two independent Wishart matrices, $\mathbf{W}_{22.1} \cap W_{p-q}(m-q, \mathbf{\Sigma}_{22.1} = \mathbf{\Sigma}_{22}), \mathbf{W}_{21}\mathbf{W}_{11}^{-1}\mathbf{W}_{12} \cap W_{p-q}(q, \mathbf{\Sigma}_{22})$ and their sum $\mathbf{W}_{22} \cap W_{p-q}(m, \mathbf{\Sigma}_{22})$.

Note that

$$\mathbf{W}^{22} = \mathbf{W}_{22.1}^{-1} \text{ where } \mathbf{W}^{-1} = ((\mathbf{W}^{ij}, i, j = 1, 2)).$$

Hence we have the following corollary.

Corollary 4.2.7.1: Under the assumptions of Theorem 4.2.7 the following results hold:

(a) $(\mathbf{W}^{22})^{-1} \cap W_{p-q}(m - q, (\mathbf{\Sigma}^{22})^{-1})$ and is independent of $(\mathbf{W}_{11}, \mathbf{W}_{12})$.
(b) If $\mathbf{\Sigma}_{12} = \mathbf{0}$, then $\mathbf{W}_{22} - (\mathbf{W}^{22})^{-1} \cap W_{p-q}(q, \mathbf{\Sigma}_{22})$ and $(\mathbf{W}^{22})^{-1}, \mathbf{W}_{22} - (\mathbf{W}^{22})^{-1}$ and $\mathbf{W}_{11}$ are jointly independent.

We will note in Theorem 4.2.10 that if $\mathbf{S} = ((s_{ij})) = (n-1)^{-1}\mathbf{X}'\mathbf{H}\mathbf{X}$ is the sample covariance matrix, then $(n-1)\mathbf{S} = \sum_{i=1}^{n}(\mathbf{X}_i - \bar{\mathbf{X}})(\mathbf{X}_i - \bar{\mathbf{X}})' \cap W_p(n-1, \mathbf{\Sigma})$. Therefore, for $p = 2$, writing $n - 1 = m$,

$$w_{11} = ms_{11} = ms_1^2 \cap \sigma_1^2\chi_{(m)}^2;$$

$$w_{12} = ms_{12}, \quad w_{22} = ms_{22} = ms_2^2 \cap \sigma_2^2\chi_{(m)}^2;$$

$$w_{22.1} = w_{22} - w_{21}w_{11}^{-1}w_{12} = ms_2^2(1 - r^2) \cap \sigma_2^2(1 - \rho^2)\chi_{(m-1)}^2.$$

If $\sigma_{12} = 0$,

$$w_{21}w_{11}^{-1}w_{12} = w_{22} - w_{22.1} = mr^2s_2^2 \cap \sigma_2^2\chi_{(1)}^2.$$

In this case, the three variables $w_{22} - w_{22.1}, w_{22.1}$ and w_{11} are jointly independent.

Theorem 4.2.8: If $\mathbf{W} \cap W_p(m, \mathbf{\Sigma}), m > p$, then

(a) the ratio $\mathbf{a}'\mathbf{\Sigma}^{-1}\mathbf{a}/\mathbf{a}'\mathbf{W}^{-1}\mathbf{a}$ has a $\chi_{(m-p+1)}^2$ distribution for any fixed vector $\mathbf{a}$ and in particular $\sigma^{ii}/w^{ii} \cap \chi_{(m-p+1)}^2, i = 1, \ldots p$ (here $\mathbf{\Sigma}^{-1} = ((\sigma^{ij}))$).
(b) w^{ii} is independent of all elements of $\mathbf{W}$ except w_{ii}.

Proof. Omitted. Interested readers may refer to Mardia *et al.* (1979, Section 3.4.3). A sketch of the proof can be obtained from proof of Lemma 4.3.3.

(For an extension of the result (a) see Exercise 4.13.)

Theorem 4.2.9: If $\mathbf{W} \cap W_p(m, \mathbf{\Sigma})$, then $|\mathbf{W}| = |\mathbf{\Sigma}|\chi^2_{(m)} \cdot \chi^2_{(m-1)} \cdots \chi^2_{(m-p+1)}$, where all the χ^2's are independent.

Proof. The result is proved by induction. For $p = 1, |W| \cap \sigma^2\chi^2_{(m)}$, so that the result is true. For $p > 1$, let

$$\mathbf{W} = \begin{bmatrix} \mathbf{W}_{11} & \mathbf{W}_{12} \\ \mathbf{W}_{21} & W_{22} \end{bmatrix},$$

where $\mathbf{W}_{11}$ is of order $(p-1) \times (p-1)$ and W_{22} is a scalar. Suppose

$$|\mathbf{W}_{11}| = |\mathbf{\Sigma}_{11}|\chi^2_{(m)}\chi^2_{(m-1)} \cdots \chi^2_{(m-p+2)}.$$

Now,

$$\left.\begin{aligned} |\mathbf{W}| &= |\mathbf{W}_{11}||\mathbf{W}_{22.1}| \\ &= \{|\mathbf{\Sigma}_{11}|\chi^2_{(m)}\chi^2_{(m-1)} \cdots \chi^2_{(m-p+2)}\}\{|\mathbf{\Sigma}_{22.1}|\chi^2_{(m-p+1)}\} \\ &= |\mathbf{\Sigma}_{11}||\mathbf{\Sigma}_{22.1}|\chi^2_{(m)}\chi^2_{(m-1)} \cdots \chi^2_{(m-p+1)} \\ &= |\mathbf{\Sigma}|\chi^2_{(m)}\chi^2_{(m-1)} \cdots \chi^2_{(m-p+1)} \end{aligned}\right\},$$

because, by Theorem 4.2.7, $\mathbf{W}_{11}$ is independent of $W_{22.1}$ and $W_{22.1} \cap \mathbf{\Sigma}_{22.1}\chi^2_{(m-p+1)}$.

Corollary 4.2.9.1: If $\mathbf{W} \cap W_p(m, \mathbf{\Sigma}), \mathbf{\Sigma} > 0$ and $m \geq p$, then $\mathbf{W} > 0$ with probability one.

Proof. We have $|\mathbf{W}| \cap |\mathbf{\Sigma}|\chi^2_{(m)} \cdots \chi^2_{(m-p+1)}$. Since a chi-square variable is strictly positive and $|\mathbf{\Sigma}| > 0$, it follows that $|\mathbf{W}| > 0$. Again, since $\mathbf{W} = \sum_{i=1}^m \mathbf{X}_i\mathbf{X}_i'$ (where $\mathbf{X}_i$ is $N_p(\mathbf{0}, \mathbf{\Sigma})$) is positive semi-definite, all its eigenvalues are strictly positive with probability one and $\mathbf{W}$ is positive definite.

Note 4.2.3: We have seen that if $\mathbf{X}_1, \ldots, \mathbf{X}_n$ are independent samples from $N_p(\mu, \mathbf{\Sigma})$, then the corrected SSP matrix $(n-1)\mathbf{S} \cap W_p(n-1, \mathbf{\Sigma})$. Therefore, the generalized variance

$$|\mathbf{S}| \cap (n-1)^{-p}|\mathbf{\Sigma}|\chi^2_{(n-1)}\chi^2_{(n-2)} \cdots \chi^2_{(n-p)}$$

and these chi-squares are all independent.

4.2.2 *Distribution of* $\mathbf{X}'\mathbf{C}\mathbf{X}$

We shall now consider random matrices of the form $\mathbf{X}'\mathbf{C}\mathbf{X}$ where $\mathbf{X}$ is a normal data matrix and $\mathbf{C}$ is a symmetric matrix.

Theorem 4.2.10: (Cochran, 1934) Let $\mathbf{X}(n \times p)$ be a data matrix from $N_p(\mathbf{0}, \mathbf{\Sigma})$ and $\mathbf{C}$ a symmetric matrix. Then

(a) $\mathbf{X}'\mathbf{C}\mathbf{X}$ has the same distribution as the weighted sum of independent $\mathbf{W}_p(1, \mathbf{\Sigma})$ variables, where the weights are the eigenvalues of $\mathbf{C}$.

(b) $\mathbf{X}'\mathbf{C}\mathbf{X}$ has the Wishart distribution *iff* $\mathbf{C}$ is a symmetric idempotent matrix, in which case $\mathbf{X}'\mathbf{C}\mathbf{X} \cap \mathbf{W}_p(r, \mathbf{\Sigma})$, where r is the rank of $\mathbf{C}$.

(c) The matrix $(n-1)\mathbf{S} = \mathbf{X}'\mathbf{H}\mathbf{X}$ where $\mathbf{H} = \mathbf{I}_n - \frac{1}{n}\mathbf{1}_n\mathbf{1}'_n$, has $\mathbf{W}_p(n-1, \mathbf{\Sigma})$ distribution.

Proof. By spectral decomposition theorem (Theorem A.12.1),

$$\mathbf{C} = \sum_{i=1}^{n} \lambda_i \gamma_i \gamma_i'$$

where $(\lambda_i, \gamma_i), (i = 1, \ldots, p)$ are the (eigenvalue, eigenvector) pairs of $\mathbf{C}$. Then

$$\begin{aligned} \mathbf{X}'\mathbf{C}\mathbf{X} &= \textstyle\sum_{i=1}^{n} \lambda_i \mathbf{X}'\gamma_i\gamma_i'\mathbf{X} \\ &= \textstyle\sum_{i=1}^{n} \lambda_i \mathbf{Y}_i \mathbf{Y}_i', \end{aligned} \tag{4.2.6}$$

where $\mathbf{Y}_i = \mathbf{X}'\gamma_i$. Let

$$\begin{aligned} \mathbf{Y}(n \times p) &= \begin{bmatrix} \mathbf{Y}_1' \\ \cdot \\ \cdot \\ \mathbf{Y}_n' \end{bmatrix} = \begin{bmatrix} \gamma_1' \\ \cdot \\ \cdot \\ \gamma_n' \end{bmatrix} \mathbf{X} \\ &= \mathbf{\Gamma}\mathbf{X} \text{ (say)}. \end{aligned}$$

Now, considering Theorem 3.3.2, $\mathbf{A} = \mathbf{\Gamma}, \mathbf{B} = \mathbf{I}_p, \mathbf{B}'\mu = \mathbf{0}, \mathbf{A}\mathbf{A}' = \mathbf{I}_p$, so that $\mathbf{Y}$ is a data matrix from $N_p(\mathbf{0}, \mathbf{\Sigma})$. Therefore each $\mathbf{Y}_i\mathbf{Y}_i'$ follows $\mathbf{W}_p(1, \mathbf{\Sigma})$ distribution (Corollary 4.2.1.1). Hence, the result follows by (4.2.6).

(b) If $\mathbf{C}$ is symmetric idempotent, $\lambda_i = 1$ or 0 and rank $\mathbf{C} = r =$ number of non-zero eigenvalues of $\mathbf{C}$. Hence $\mathbf{X}'\mathbf{C}\mathbf{X} = \mathbf{W}_p(1, \mathbf{\Sigma}) + \ldots + \mathbf{W}_p(1, \mathbf{\Sigma}) = \mathbf{W}_p(r, \mathbf{\Sigma})$.

(c) $\mathbf{S} = (n-1)^{-1}\mathbf{X}'\mathbf{H}\mathbf{X}$. Here $\mathbf{H}$ is symmetric idempotent of rank $n-1$. Hence, $(n-1)\mathbf{S} \cap \mathbf{W}_p(n-1, \mathbf{\Sigma})$.

Corollary 4.2.10.1 : Let $\mathbf{X}(n \times p)$ be a data matrix from $N_p(\mu, \mathbf{\Sigma})$ and $\mathbf{Y} = \mathbf{X} - \mu\mathbf{1}_n'$. Thus $\mathbf{Y}$ is a data matrix from $N_p(\mathbf{0}, \mathbf{\Sigma})$. Suppose that $\mathbf{C}(n \times n)$ is symmetric and idempotent. Then,

(a) $\mathbf{X}'\mathbf{C}\mathbf{X} = \mathbf{Y}'\mathbf{C}\mathbf{Y} + \mathbf{Y}'\mathbf{C}\mathbf{1}\mu' + \mu\mathbf{1}'\mathbf{C}\mathbf{Y} + (\mathbf{1}'\mathbf{C}\mathbf{1})\mu\mu'$. Thus, $\mathbf{X}'\mathbf{C}\mathbf{X}$ is the sum of a $\mathbf{W}_p(r, \mathbf{\Sigma})$ variable, two normal variables (dependent on each other) and a constant, where $r =$ rank $\mathbf{C}$.

(b) If further, (i) $\mu = \mathbf{0}$ or (ii) $\mathbf{C}\mathbf{1} = \mathbf{0}$, then $\mathbf{X}'\mathbf{C}\mathbf{X} \cap \mathbf{W}_p(r, \mathbf{\Sigma})$ where $r =$ rank of $\mathbf{C}$.

(c) $(n-1)\mathbf{S} = \mathbf{X}'\mathbf{H}\mathbf{X} \cap \mathbf{W}_p(n-1, \mathbf{\Sigma})$.

Proof. (a)

$$\begin{aligned} \mathbf{X}'\mathbf{C}\mathbf{X} &= (\mathbf{Y} + \mathbf{1}\mu')'\mathbf{C}(\mathbf{Y} + \mathbf{1}\mu') \\ &= \mathbf{Y}'\mathbf{C}\mathbf{Y} + \mathbf{Y}'\mathbf{C}\mathbf{1}\mu' + \mu\mathbf{1}'\mathbf{C}\mathbf{Y} + (\mathbf{1}'\mathbf{C}\mathbf{1})\mu\mu'. \end{aligned} \tag{4.2.7}$$

By Theorem 4.2.10, $\mathbf{Y}'\mathbf{C}\mathbf{Y} \cap W_p(r, \mathbf{\Sigma})$.

(b) If $\mu = \mathbf{0}$ or $\mathbf{C}\mathbf{1} = \mathbf{0}$, the last three terms of (4.2.7) vanish.

(c) $\mathbf{H}$ satisfies $\mathbf{H}\mathbf{1} = \mathbf{0}$ and rank $(\mathbf{H}) = n-1$.

Note 4.2.4: We note from Theorem 4.2.10 and Corollary 4.2.10.1 that $(n-1)\mathbf{S} = \mathbf{X}'\mathbf{H}\mathbf{X} \cap W_p(n-1, \mathbf{\Sigma})$, when $\mathbf{X}$ is a data matrix from $N_p(\mu, \mathbf{\Sigma})$, even if $\mu \neq \mathbf{0}$.

Theorem 4.2.11: (Craig, 1943) Let $\mathbf{X}$ be a data matrix from $N_p(\mu, \mathbf{\Sigma})$ and $\mathbf{C}_1, \ldots, \mathbf{C}_k$ be symmetric matrices. Then $\mathbf{X}'\mathbf{C}_1\mathbf{X}, \ldots, \mathbf{X}'\mathbf{C}_k\mathbf{X}$ are jointly independent if $\mathbf{C}_r\mathbf{C}_s = \mathbf{0} \; \forall \; r \neq s$.

Proof. Consider $k = 2$ and let us write $\mathbf{C}_1 = \mathbf{C}, \mathbf{C}_2 = \mathbf{D}$. Let $(\lambda_i, \gamma_i), (\psi_j, \delta_j)$ be the (eigenvalue, eigenvector) pairs for $\mathbf{C}, \mathbf{D}$, respectively. Now, by spectral decomposition (Theorem A.12.1)

$$\mathbf{X}'\mathbf{C}\mathbf{X} = \mathbf{X}'(\sum_i \lambda_i\gamma_i\gamma_i')\mathbf{X} = \sum \lambda_i\mathbf{y}_i\mathbf{y}_i' \text{ where } \mathbf{y}_i = \mathbf{X}'\gamma_i.$$

Similarly,

$$\mathbf{X}'\mathbf{D}\mathbf{X} = \sum \psi_j\mathbf{z}_j\mathbf{z}_j' \quad \text{where } \mathbf{z}_j = \mathbf{X}'\delta_j.$$

Therefore, the np-dimensional normal vectors $(\sqrt{\lambda_1}\mathbf{y}_1', \ldots, \sqrt{\lambda_n}\mathbf{y}_n')'$ and $(\sqrt{\psi_1}\mathbf{z}_1, \ldots, \sqrt{\psi_n}\mathbf{z}_n)'$ will be independent *iff*

$$\lambda_i\psi_j\gamma_i\delta_j = 0 \; \forall \; i, j. \tag{4.2.8}$$

Now

$$\begin{aligned} \mathbf{CD} &= (\sum \lambda_i \gamma_i \gamma_i)(\sum \psi_j \delta_j \delta_j' \\ &= \sum_i \sum_j \lambda_i \psi_j \gamma_i \gamma_i' \delta_j \delta_j'. \end{aligned} \tag{4.2.9}$$

If $\mathbf{CD} = \mathbf{0}$, pre-multiplying (4.2.9) by γ_u' and post-multiplying by δ_v gives $\lambda_u \psi_v \gamma_u' \delta_v = 0$. Since, u, v are arbitrary, this relation should hold for all i, j. Therefore, (4.2.8) holds and since $\mathbf{X'CX}, \mathbf{X'DX}$ are functions of independent normal np-vectors, they are independent. The result holds similarly for $k > 2$.

(The converse of this theorem is also true. (Ogawa (1949)).

Corollary 4.2.11.1: if $\mu = \mathbf{0}$ and $\mathbf{C}_j$'s are symmetric idempotent, then if $\mathbf{C}_r\mathbf{C}_s = \mathbf{0}(\forall\ r \neq s), \mathbf{X'C}_1\mathbf{X}, \ldots, \mathbf{X'C}_k\mathbf{X}$ have independent Wishart distributions.

Theorem 4.2.12: If $\mathbf{X}$ is a data matrix from $N_p(\mu, \mathbf{\Sigma})$ and if $\mathbf{C}$ is a symmetric $n \times n$ matrix, then $\mathbf{X'CX}$ and $\mathbf{AXB}$ are independent if either $\mathbf{B'\Sigma} = \mathbf{0}$ or $\mathbf{AC} = \mathbf{0}$.

Proof. We have

$$\mathbf{X'CX} = \sum_{i=1}^{n} \lambda_i \mathbf{y}_i \mathbf{y}_i' \quad \text{where } \mathbf{y}_i = \mathbf{X}'\gamma_i,$$

$(\lambda_i, \gamma_i), (i = 1, \ldots, p)$ being the (eigenvalue, eigenvector) pairs of $\mathbf{C}$. By Theorem 3.3.3, $\mathbf{AXB}$ is independent of $\sqrt{\lambda_i}\gamma_i'\mathbf{X}$ if either $\mathbf{B'\Sigma} = \mathbf{0}$ or

$$\sqrt{\lambda_i}\mathbf{A}\gamma_i = \mathbf{0}\ \forall\ i. \tag{4.2.10}$$

Now,

$$\mathbf{AC} = \sum_i \lambda_i \mathbf{A} \gamma_i \gamma_i'. \tag{4.2.11}$$

If $\mathbf{AC} = \mathbf{0}$, post-multiplying both sides of (4.2.11) by $\gamma_u, \lambda_u \mathbf{A}\gamma_u = \mathbf{0}$, that is, $\mathbf{A}\gamma_u = \mathbf{0}$, whenever $\lambda_u \neq 0$. Since, u is arbitrary, it means $\mathbf{A}\gamma_i = \mathbf{0}$, whenever $\lambda_i \neq 0 (i = 1, \ldots, n)$. Hence, if $\mathbf{AC} = \mathbf{0}$ (4.2.10) holds and hence $\mathbf{X'CX}$ and $\mathbf{AXB}$ are independent. Hence the proof.

Note 4.2.5: Theorems 4.2.10, 4.2.11 and 4.2.12 can be extended to the case $\mathbf{XCX}'(n \times n)$. This is done by noting that if rows of $\mathbf{X}(n \times p)$ are *iid* $N_p(\mu, \mathbf{\Sigma})$, then in general, rows of $\mathbf{X}'$ are not *iid*. However, if $\mu = \mathbf{0}, \mathbf{\Sigma} = \mathbf{I}$, then the rows of $\mathbf{X}'$ are *iid* $N_n(\mathbf{0}, \mathbf{I})$ since every element of $\mathbf{X}$ is $N_1(0, 1)$. (vide Exercises 4.9, 4.10).

4.2.3 *Non-central Wishart distribution*

Let $\mathbf{X}_1, \ldots, \mathbf{X}_m$ be distributed as $N_p(\mu_i, \mathbf{\Sigma})(i = 1, \ldots, m)$. We can define $\mathbf{X}'\mathbf{X}$ to have a non-central Wishart distribution $W_p(m, \mathbf{\Sigma}, \mathbf{\Delta})$ with non-centrality matrix $\mathbf{\Delta}$ with

$$\begin{aligned}\mathbf{\Delta} &= \textstyle\sum_{i=1}^{m}(\mathbf{\Sigma}^{-1/2}\mu_i)(\mathbf{\Sigma}^{-1/2}\mu_i)' \\ &= \mathbf{\Sigma}^{-1/2}\mathbf{M}'\mathbf{M}\mathbf{\Sigma}^{-1/2},\end{aligned}$$

where $\mathbf{M}(m \times p) = (\mu_1, \ldots, \mu_m)'$. Clearly,

$$E(\mathbf{X}'\mathbf{X}) = m\mathbf{\Sigma} + \mathbf{M}'\mathbf{M}.$$

When $\mathbf{M} = \mathbf{0}, \mathbf{\Delta} = \mathbf{0}$ and we get central Wishart distribution $W_p(m, \mathbf{\Sigma})$.

A particular case is the 'linear non-central distribution' in which rank($\mathbf{\Delta}$) = rank ($\mathbf{M}'$) = 1. Some references in this area are Gleser (1976), Muirhead (1982).

4.2.4 *Eigenvalues of a Wishart matrix*

Suppose that $\mathbf{W} \cap W_p(m, \mathbf{I}_p), m \geq p$ and let $c_1 \geq c_2 \geq \ldots \geq c_p \geq 0$ be the ordered eigenvalues of $\mathbf{W}$. The joint distribution of $\mathbf{c} = (c_1, \ldots, c_p)'$ is given by

$$f(\mathbf{c}) = K[\exp(-\frac{1}{2}\sum_{j=1}^{p} c_j)](\Pi_{j=1}^{p} c_j)^{(m-p-1)/2}$$

$$\Pi_{i<j}^{p}(c_i - c_j), \quad c_1 \geq c_2 \geq \ldots \geq c_p > 0 \tag{4.2.12}$$

where K is a constant (Fisher (1939), Hsu (1939), Roy (1939)). For the general case where $\mathbf{W} \cap W_p(m, \mathbf{\Sigma})$, various exact and asymptotic results for the distribution of c_i's in terms of the eigenvalues $\lambda_1 \geq \lambda_2 \geq \ldots \lambda_p > 0$ of $\mathbf{\Sigma}$ are given in Muirhead (1982).

4.3 Hotelling's T^2 Distribution

We now consider Hotelling's T^2 distribution (Hotelling, 1931).

DEFINITION 4.3.1: Suppose $\mathbf{Y}$ is distributed as $N_p(\mathbf{0}, \mathbf{I}), \mathbf{M}$ is $W_p(m, \mathbf{I})$ and $\mathbf{Y}, \mathbf{M}$ are independently distributed. We say

$$T^2 = m\mathbf{Y}'\mathbf{M}^{-1}\mathbf{Y} \tag{4.3.1}$$

has the Hotelling's T^2 distribution with parameters (degrees of freedom) (p, m). We shall write $T^2 \cap T^2(p, m)$.

Theorem 4.3.1: Let $\mathbf{X} \cap N_p(\mu, \mathbf{\Sigma}), \mathbf{W} \cap W_p(m, \mathbf{\Sigma})$ and $\mathbf{X}, \mathbf{W}$ be independent. Then the statistic

$$T^2 = m(\mathbf{X} - \mu)'\mathbf{W}^{-1}(\mathbf{X} - \mu) \tag{4.3.2}$$

follows Hotelling's T^2 distribution with (p, m) degrees of freedom.

Proof. Let $\mathbf{Y} = \mathbf{\Sigma}^{-1/2}(\mathbf{X} - \mu), \mathbf{M} = \mathbf{\Sigma}^{-1/2}\mathbf{W}\mathbf{\Sigma}^{-1/2}$. Then, $\mathbf{Y} \cap N_p(\mathbf{0}, \mathbf{I}), \mathbf{M} \cap W_p(m, \mathbf{I})$ and

$$\begin{aligned} T^2 &= m\mathbf{Y}'\mathbf{M}^{-1}\mathbf{Y} \\ &= m(\mathbf{X} - \mu)'\mathbf{W}^{-1}(\mathbf{X} - \mu). \end{aligned}$$

Note 4.3.1: For $p = 1, X \cap N_1(\mu, \sigma^2), W \cap \sigma^2\chi^2_{(m)}, X$ and W are independent,

$$t = \frac{(X - \mu)/\sigma}{\sqrt{W/(m\sigma^2)}}$$

follows Student's t distribution with m degrees of freedom. Hence t^2 follows $F_{1,m}$ distribution. We shall show in Theorem 4.3.2, T^2 given in (4.3.2) follows $cF_{p,m-p+1}$ distribution, where c is a suitable constant. We first note the following lemmas.

Lemma 4.3.1: Let $\mathbf{Y} = (Y_1, \ldots, Y_p)' \cap N_p(\theta, \mathbf{\Sigma})$. Then the conditional distribution of Y_p given $Y_1 = y_1, \ldots, Y_{p-1} = y_{p-1}$ is univariate normal with mean $\beta_o + \sum_{j=1}^{p-1} \beta_j y_j$ and variance $1/\sigma^{pp}$ where $\mathbf{\Sigma}^{-1}(= ((\sigma^{ij})))$.

Proof. Let $\mathbf{Y}_1 = (Y_1, \ldots, Y_{p-1})', \theta^{(1)} = (\theta_1, \ldots, \theta_{p-1})'$ and

$$\mathbf{\Sigma} = \begin{bmatrix} \mathbf{\Sigma}_{11} & \mathbf{\Sigma}_{12} \\ \mathbf{\Sigma}_{21} & \sigma_{pp} \end{bmatrix}.$$

It follows from Theorem 3.2.7, that the conditional distribution of Y_p given $y_2, \ldots, y_{p-1}$, is normal with

$$\begin{aligned} E[Y_p|\mathbf{y}_1] &= \theta_p + \mathbf{\Sigma}_{21}\mathbf{\Sigma}_{11}^{-1}(\mathbf{y}_1 - \theta^{(1)}) \\ &= (\theta_p - \mathbf{\Sigma}_{21}\mathbf{\Sigma}_{11}^{-1}\theta^{(1)}) + \mathbf{\Sigma}_{21}\mathbf{\Sigma}_{11}^{-1}\mathbf{y}_1 \\ &= \beta_0 + \beta'\mathbf{y}_1 \quad \text{(say)}. \end{aligned}$$

Since $\mathbf{\Sigma} > 0, \mathbf{\Sigma}_{11} > 0$. Also,

$$|\mathbf{\Sigma}| = |\mathbf{\Sigma}_{11}|(\sigma_{pp} - \mathbf{\Sigma}_{21}\mathbf{\Sigma}_{11}^{-1}\mathbf{\Sigma}_{12}) \quad \text{(vide A.5.7)}.$$

Hence,

$$\begin{aligned} V(Y_p|\mathbf{y}_1) &= \sigma_{pp} - \mathbf{\Sigma}_{21}\mathbf{\Sigma}_{11}^{-1}\mathbf{\Sigma}_{12} \\ &= \frac{|\mathbf{\Sigma}|}{|\mathbf{\Sigma}_{11}|} \\ &= \frac{1}{\sigma^{pp}}. \end{aligned}$$

Lemma 4.3.2: Consider the linear regression model $\mathbf{Z} = \mathbf{A}\beta + \epsilon$, where $\mathbf{A}(m \times p)$ is of rank p and $\epsilon \cap N_m(\mathbf{0}, \sigma^2 \mathbf{I}_m)$. Then the quantity

$$Q = (\mathbf{Z} - \mathbf{A}\hat{\beta})'(\mathbf{Z} - \mathbf{A}\hat{\beta}) = 1/u^{p+1\ p+1} \cap \sigma^2 \chi^2_{(m-p)}$$

where $\hat{\beta} = (\mathbf{A}'\mathbf{A})^{-1}(\mathbf{A}'\mathbf{Z}), \mathbf{U}^{-1} = ((u^{ij}))$ and

$$\mathbf{U} = \begin{bmatrix} \mathbf{A}'\mathbf{A} & \mathbf{A}'\mathbf{Z} \\ \mathbf{Z}'\mathbf{A} & \mathbf{Z}'\mathbf{Z} \end{bmatrix}.$$

Proof. We have

$$\begin{aligned} Q &= \mathbf{Z}'\mathbf{Z} - (\mathbf{A}\hat{\beta})'(\mathbf{A}\hat{\beta}) \\ &= \mathbf{Z}'(\mathbf{I}_m - \mathbf{R})\mathbf{Z}, \end{aligned}$$

where

$$\mathbf{R} = \mathbf{A}(\mathbf{A}'\mathbf{A})^{-1}\mathbf{A}'$$

is a symmetric idempotent matrix of rank p. Clearly, $(\mathbf{I}_m - \mathbf{R})$ is symmetric idempotent of rank $m - p$. Also,

$$\begin{aligned} |\mathbf{U}| &= |\mathbf{A}'\mathbf{A}|(\mathbf{Z}'\mathbf{Z} - \mathbf{Z}'\mathbf{A}(\mathbf{A}'\mathbf{A})^{-1}\mathbf{A}'\mathbf{Z}) \\ &= |\mathbf{A}'\mathbf{A}|(\mathbf{Z}'\mathbf{Z} - (\mathbf{A}\hat{\beta})'(\mathbf{A}\hat{\beta})) \end{aligned}$$

so that

$$Q = \frac{|\mathbf{U}|}{|\mathbf{A}'\mathbf{A}|} = \frac{1}{u^{p+1\ p+1}}.$$

It easily follows

$$\begin{aligned} Q &= \mathbf{Z}'(\mathbf{I}_m - \mathbf{R})\mathbf{Z} \\ &= \epsilon'(\mathbf{I}_m - \mathbf{R})\epsilon. \end{aligned}$$

Hence by Theorem 3.4.1, $Q/\sigma^2 \cap \chi^2_{(m-p)}$.

Lemma 4.3.3: Let $\mathbf{W} \cap W_p(m, \mathbf{\Sigma})$, where $m \geq p$. The following results hold:

(a) $\sigma^{pp}/w^{pp} \cap \chi^2_{(m-p+1)}$ and is independent of $w_{jk} (j, k = 1, \ldots, p-1)$.
(b) $\mathbf{l}'\mathbf{\Sigma}^{-1}\mathbf{l}/(\mathbf{l}'\mathbf{W}^{-1}\mathbf{l}) \cap \chi^2_{(m-p+1)}$ for any fixed $(\mathbf{l} \neq \mathbf{0})$.

Proof. Let $\mathbf{W} = \sum_{i=1}^m \mathbf{X}_i \mathbf{X}_i'$ where $\mathbf{X}_i \cap N_p(\mathbf{0}, \mathbf{\Sigma})$. If $\mathbf{X}_i = (X_{i1}, \ldots, X_{ip})'$, then by Lemma 4.3.1, the conditional distribution of X_{ip} given $(X_{i1}, \ldots X_{ip-1})'$ is $N_1(\sum_{j=1}^{p-1} \beta_j x_{ij}, 1/\sigma^{pp})$. Therefore, by Lemma 4.3.2, considering the regression equation of X_p on $X_1, \ldots, X_{p-1}$, (X_j being the variable whose values are $X_{ij}, i = 1, \ldots, m$), the error sum of squares

$$Q = \sum_{i=1}^{m} (X_{ip} - \sum_{j=1}^{p-1} \hat{\beta}_j x_{ij})^2 = \frac{1}{w^{pp}} \cap \chi^2_{(m-p+1)} \cdot \frac{1}{\sigma^{pp}},$$

since $V(X_p|X_1,\ldots,X_{p-1}) = 1/\sigma^{pp}$, by Lemma 4.3.1. Here $\hat{\beta}_j$is the best linear estimate of β_j under the regression model.

Since the conditional distribution of Q does not depend on the conditioning variables $X_1,\ldots,X_{p-1}$, it is also the unconditional distribution of Q. Thus, Q is independent of $x_{ij}(i=1,\ldots,m; j=1,\ldots,p-1)$ and hence is independent of $w_{jk} = \sum_{i=1}^{m} x_{ij}x_{ik}(j \neq p, k \neq p)$.

(b) Let $\mathbf{L}$ be any $p \times p$ orthogonal matrix with its last row equal to $\mathbf{l}'/||\mathbf{l}||$. Now, by Theorem 4.2.1, $\mathbf{LWL}' \cap W_p(m, \mathbf{L\Sigma L}')$. Again,

$$(\mathbf{LWL}')^{-1} = \mathbf{LW}^{-1}\mathbf{L}'$$

$$(\mathbf{L\Sigma L}')^{-1} = \mathbf{L\Sigma}^{-1}\mathbf{L}'.$$

Applying part (a) to $\mathbf{LWL}'$ we have

$$\frac{p\text{th diagonal element of } (\mathbf{L\Sigma L}')^{-1}}{p\text{th diagonal element of } (\mathbf{LWL}')^{-1}} \cap \chi^2_{(m-p+1)}.$$

Again pth diagonal element of $(\mathbf{L\Sigma L}')^{-1} = p$th diagonal element of $\mathbf{L\Sigma}^{-1}\mathbf{L}' = \mathbf{l}'\mathbf{\Sigma}^{-1}\mathbf{l}$. Similarly, the pth diagonal element of $(\mathbf{LWL}') = \mathbf{l}'\mathbf{W}^{-1}\mathbf{l}$. Hence, $\mathbf{l}'\mathbf{\Sigma}^{-1}\mathbf{l}/\mathbf{l}'\mathbf{W}^{-1}\mathbf{l} \cap \chi^2_{(m-p+1)}$.

We shall now obtain the distribution of T^2 using Lemmas 4.3.1-4.3.3.

Theorem 4.3.2: Let $T^2 = m\mathbf{Y}'\mathbf{W}^{-1}\mathbf{Y}$, where $\mathbf{Y} \cap N_p(\mathbf{0}, \mathbf{\Sigma}), \mathbf{W} \cap W_p(m, \mathbf{\Sigma})$ and $\mathbf{Y}, \mathbf{W}$ are statistically independent. Assume also that $\mathbf{\Sigma} > 0, m \geq p$, so that $\mathbf{W}^{-1}$ exists with probability one and the Wishart distribution is non-singular. Then

$$\frac{m-p+1}{p}.\frac{T^2}{m} \cap F_{p,m-p+1}.$$

Proof. We have

$$\begin{aligned}\tfrac{T^2}{m} &= \mathbf{Y}'\mathbf{W}^{-1}\mathbf{Y}\\ &= \tfrac{\mathbf{Y}'\mathbf{\Sigma}^{-1}\mathbf{Y}}{\mathbf{Y}'\mathbf{\Sigma}^{-1}\mathbf{Y}/(\mathbf{Y}'\mathbf{W}^{-1}\mathbf{Y})}\\ &= \tfrac{A}{B} \text{ (say)}.\end{aligned}$$

It follows by Lemma 4.3.3(b), that for a given value of $\mathbf{Y} = \mathbf{l}$, the conditional distribution of B is $\chi^2_{(m-p+1)}$. Since this conditional distribution does not depend on $\mathbf{Y}$ it is also the unconditional distribution of B. Also, the distribution of B is independent of $\mathbf{Y}$ and hence is independent of A. Again, from Theorem 3.4.1, $A \cap \chi^2_{(p)}$ so that T^2/m is the ratio of two independent chi-square variables. Hence

$$\frac{m-p+1}{p}\frac{T^2}{m} = \frac{A/p}{B/(m-p+1)} \cap F_{p,m-p+1}.$$

Corollary 4.3.2.1: Let $\mathbf{X} \cap N_p(\mu, \alpha^{-1}\mathbf{\Sigma}), \mathbf{W} \cap W_p(m, \mathbf{\Sigma})$ and $\mathbf{X}, \mathbf{W}$ be independent, where α is a constant. Then

$$\begin{aligned} T^2 &= \alpha m(\mathbf{X}-\mu)\mathbf{W}^{-1}(\mathbf{X}-\mu) \\ &= m(\mathbf{X}-\mu)'(\mathbf{W}/\alpha)^{-1}(\mathbf{X}-\mu). \end{aligned}$$

Corollary 4.3.2.2: It is known that if $\mathbf{X}_1, \ldots \mathbf{X}_n$ are independent samples from $N_p(\mu, \mathbf{\Sigma})$, then $\bar{\mathbf{X}} \cap N_p(\mu, \mathbf{\Sigma}/n)$. Also $(n-1)\mathbf{S} = \sum_{i=1}^{n}(\mathbf{X}_i - \bar{\mathbf{X}})(\mathbf{X}_i - \bar{\mathbf{X}})' \cap W_p(n-1, \mathbf{\Sigma})$ and $\bar{\mathbf{X}}, \mathbf{S}$ are independent. Therefore

$$\begin{aligned} T^2 &= (n-1)n(\bar{\mathbf{x}}-\mu)'((n-1)\mathbf{S})^{-1}(\bar{\mathbf{x}}-\mu) \\ &= n(\bar{\mathbf{x}}-\mu)'\mathbf{S}^{-1}(\bar{\mathbf{x}}-\mu) \cap T^2(p, n-1). \end{aligned}$$

Also,

$$\{\frac{n(n-p)}{(n-1)p}\}(\bar{\mathbf{x}}-\mu)'\mathbf{S}^{-1}(\bar{\mathbf{x}}-\mu) \cap F_{p,n-p}.$$

Corollary 4.3.2.3: If $\mathbf{X}_1, \ldots, \mathbf{X}_n$ constitute a random sample from an infinite population with mean μ and variance $\mathbf{\Sigma}$, then, if $n-p$ is large,

$$T^2 = n(\bar{\mathbf{X}}-\mu)'\mathbf{S}^{-1}(\bar{\mathbf{X}}-\mu)$$

follows approximately a $\chi^2_{(p)}$ distribution. Note that the normality of the distribution of $\mathbf{X}$ has not been assumed here.

By the Central Limit Theorem, $\bar{\mathbf{X}} \sim N(\mu, \mathbf{\Sigma}/n)$. Hence, by Theorem 3.4.2(a),

$$n(\bar{\mathbf{X}}-\mu)'\mathbf{\Sigma}^{-1}(\bar{\mathbf{X}}-\mu) \sim \chi^2_p$$

approximately. Again, as has been noted in note 2.8.1, $\mathbf{S}$ is very close to $\mathbf{\Sigma}$ with high probability and hence T^2 follows approximately a $\chi^2_{(p)}$ distribution.

Theorem 4.3.3: The T^2 statistic is invariant under non-singular transformation of the variable from $\mathbf{x}$ to $\mathbf{x}^* = \mathbf{A}\mathbf{x} + \mathbf{b}$, where $\mathbf{A}$ is a non-singular matrix of constants.

Proof. If $\mathbf{x} \cap N_p(\mu, \mathbf{\Sigma}), \mathbf{x}^* \cap N_p(\mathbf{A}\mu + \mathbf{b}, \mathbf{A\Sigma A}')$. Let $\mathbf{W}^* = \mathbf{AW}$; If $\mathbf{W} \cap W_p(m, \mathbf{\Sigma}), \mathbf{W}^* \cap W_p(m, \mathbf{A\Sigma A}')$. Hence,

$$\begin{aligned} T^2(\mathbf{x}^*) &= m((\mathbf{x}^* - \mathbf{A}\mu - \mathbf{b})'(\mathbf{A\Sigma A}')^{-1}(\mathbf{x}^* - \mathbf{A}\mu - \mathbf{b}) \\ &= m(\mathbf{x} - \mu)'\mathbf{\Sigma}^{-1}(\mathbf{x} - \mu) = T^2(\mathbf{x}). \end{aligned}$$

Theorem 4.3.4: Let $\mathbf{Y} \cap N_p(\mathbf{0}, \mathbf{I}), \mathbf{W} \cap W_p(m, \mathbf{I})$. Then

$$\frac{|\mathbf{W}|}{|\mathbf{W} + \mathbf{YY}'|} \cap B(\frac{m-p+1}{2}, \frac{p}{2}).$$

Proof. We have

$$\begin{aligned} \tfrac{|\mathbf{W}|}{|\mathbf{W}+\mathbf{YY}'|} &= \tfrac{|\mathbf{W}|}{|\mathbf{W}|(1+\mathbf{Y}'\mathbf{W}^{-1}\mathbf{Y})} \quad (\text{by } A.5.9) \\ &= \tfrac{m}{m+T^2}. \end{aligned}$$

Now,

$$\frac{T^2}{m}\frac{m-p+1}{p} \cap F_{p,m-p+1}.$$

Hence, the result follows by the result, if $F \cap F_{u,v}$, then $Y = (1+(u/v)F)^{-1} \cap B(v/2, u/2)$.

Theorem 4.3.5: If $\mathbf{Y} \cap N_p(\mathbf{0}, \mathbf{I}), \mathbf{W} \cap W_p(m, \mathbf{I})$, and $\mathbf{Y}, \mathbf{W}$ are independently distributed, then

$$\mathbf{Y}'\mathbf{Y}(1 + \frac{1}{\mathbf{Y}'\mathbf{W}^{-1}\mathbf{Y}}) \cap \chi^2_{(m+1)}$$

and is distributed independently of $\mathbf{Y}'\mathbf{W}^{-1}\mathbf{Y}$.

Proof. By Theorem 4.2.8, the quantity $\beta = \mathbf{Y}'\mathbf{Y}/(\mathbf{Y}'\mathbf{W}^{-1}\mathbf{Y}) \cap \chi^2_{(m-p+1)}$, for a given value $\mathbf{Y}$. Again, since this distribution is independent of value of $\mathbf{Y}$, it is also the unconditional distribution of β. Hence, β and $\mathbf{Y}'\mathbf{Y}$ have independent chi-square distributions and their ratio $\mathbf{Y}'\mathbf{Y}/\beta = \mathbf{Y}'\mathbf{W}^{-1}\mathbf{Y}$ is independent of their sum $\mathbf{Y}'\mathbf{Y} + \beta$ which follows $\chi^2_{(m+1)}$. Hence the result.

Theorem 4.3.6: If $\mathbf{Y}$ and $\mathbf{W}$ are independently distributed as $N_p(\mathbf{0}, \mathbf{I})$ and $W_p(m, \mathbf{I})$respectively, then $\mathbf{Y}'\mathbf{W}^{-1}\mathbf{Y}$ is distributed independently of $\mathbf{W} + \mathbf{YY}'$.

Proof. Omitted.

Note 4.3.1: We have seen that Mahalanobis distance between two populations with means μ_1, μ_2 and common variance $\mathbf{\Sigma}$ is given by

$$\Delta^2 = (\mu_1 - \mu_2)'\mathbf{\Sigma}^{-1}(\mu_1 - \mu_2).$$

If we have two samples of sizes n_1, n_2, then the sample analogue of Δ^2 is given by

$$D^2 = (\bar{\mathbf{x}}_1 - \bar{\mathbf{x}}_2)'\mathbf{S}_p^{-1}(\bar{\mathbf{x}}_1 - \bar{\mathbf{x}}_2), \tag{4.3.3}$$

where $\mathbf{S}_p = \{(n_1 - 1)\mathbf{S}_1 + (n_2 - 1)\mathbf{S}_2)\}/(n_1 + n_2 - 2)$. It can be easily checked that if $\mu_1 = \mu_2$, then

$$\frac{n_1 n_2}{n} D^2 \cap T^2(p, n_1 + n_2 - 2).$$

EXAMPLE 4.3.1: Consider the data matrix

$$\mathbf{X} = \begin{bmatrix} 8 & 12 \\ 16 & 20 \\ 15 & 10 \end{bmatrix}.$$

Evaluate T^2 for $\mu = (12,\ 15)'$. Also, find its sampling distribution.

Here $\bar{\mathbf{x}}' = (13,\ 14)$.

$$s_{11} = \frac{1}{2}[(8 - 13)^2 + (16 - 13)^2 + (15 - 13)^2] = 19.$$

Similarly, $s_{12} = 10, s_{22} = 29.5$. Thus

$$\mathbf{S} = \begin{bmatrix} 19 & 10 \\ 10 & 29.5 \end{bmatrix}, \mathbf{S}^{-1} = \begin{bmatrix} .06406 & -.02172 \\ -.02172 & .04126 \end{bmatrix}.$$

Therefore the value of

$$T^2 = n(\bar{\mathbf{x}} - \mu)'\mathbf{S}^{-1}(\bar{\mathbf{x}} - \mu) = .46628.$$

The sampling distribution of T^2 is a $T^2(2, 2)$ distribution, that is, a $F_{2,1}$ distribution.

4.3.1 *Non-central Hotelling's T^2 distribution*

If $\mathbf{Y} \cap N_p(\theta, \mathbf{\Sigma})$ and $\mathbf{W} \cap W_p(m, \mathbf{\Sigma})$ and $\mathbf{Y}$ is independent of $\mathbf{W}$, then $T^2 = m\mathbf{Y}'\mathbf{W}^{-1}\mathbf{Y}$ is said to have a non-central T^2 distribution with non-centrality parameter $\delta = \theta'\mathbf{\Sigma}^{-1}\theta$.

Theorem 4.3.7: If T^2 has a non-central T^2 distribution with parameters (m, p) and non-centrality parameter δ, then

$$\frac{T^2}{m}\frac{m - p + 1}{p} \cap F_{(p, m-p+1, \delta)}.$$

Proof. Considering the ratio A/B in the proof of Theorem 4.3.1, we proved $B \cap \chi^2_{(m-p+1)}$ and is independent of A. This result remains valid even if $\theta \neq \mathbf{0}$. From Theorem 3.4.2 $\mathbf{Y}'\mathbf{\Sigma}^{-1}\mathbf{Y}$ has a non-central $\chi^2(p, \delta)$ distribution. Hence,

$$\frac{A/p}{B/(m - p + 1)} = \frac{\chi^2(p, \delta)}{\chi^2_{(m-p+1)}}$$

and the result follows.

4.4 Wilks' Statistic

In testing of different hypotheses for a normal population model $N_p(\mu, \mathbf{\Sigma})$ the likelihood ratio test statistic involves

$$\Lambda = \frac{|\hat{\mathbf{\Sigma}}_{H_0 \cup H}|}{|\hat{\mathbf{\Sigma}}_{H_0}|}, \tag{4.4.1}$$

where H_0 is the null hypothesis and H, the alternative hypothesis. Since a sample-variance is an unbiased estimate of $\mathbf{\Sigma}, \Lambda$ would be a ratio of the determinants of two sample variance-covariance matrices and more specifically would be of the form

$$\Lambda(p; m_H, m_E) = \frac{|\mathbf{E}|}{|\mathbf{E}+\mathbf{H}|} = |\mathbf{I}+\mathbf{E}^{-1}\mathbf{H}|^{-1}, \tag{4.4.2}$$

where $\mathbf{E}, \mathbf{H}$ are two symmetric $p \times p$ matrices,

$$\mathbf{E} \cap W_p(m_E, \mathbf{\Sigma}),\ \mathbf{H} \cap W_p(m_H, \mathbf{\Sigma}),$$

$\mathbf{E}, \mathbf{H}$ are independent, $m_E \geq p$. In order for the determinants in (4.4.2) to be positive it is necessary that $m_E \geq p$ (Corollary 4.2.9.1). The statistic (4.4.2) is known as Wilks' (1932) statistic with (m_H, m_E) degrees of freedom. The range of Λ is $0 \leq \Lambda \leq 1$. Like the T^2 statistic, the Λ statistic is invariant under change of the scale parameters of $\mathbf{E}$ and $\mathbf{H}$.

The statistic Λ is also known as U statistic with dimension p and degrees of freedom m_H, m_E and we write $U \sim U_{p,m_H,m_E}$. Wilks' Λ in (4.4.2) can be expressed in terms of eigenvalues ϕ_i of $\mathbf{E}^{-1}\mathbf{H}$ as follows:

$$\Lambda = \Pi_{i=1}^{s} \frac{1}{1+\phi_i} \tag{4.4.3}$$

(by A.8(1) and A.8(3)). The number of nonzero eigenvalues of $\mathbf{E}^{-1}\mathbf{H}$ is $s =$ min (p, m_H). The matrix $\mathbf{E}^{-1}\mathbf{H}$ has the same eigenvalues as $\mathbf{H}\mathbf{E}^{-1}$ (A.8(8)).

Theorem 4.4.1: If $\mathbf{E} \cap W_p(m_E, \mathbf{\Sigma})$ and $\mathbf{H} \cap W_p(m_H, \mathbf{\Sigma})$ are independent and if $m_E \geq p, m_H \geq p$, then

$$\psi = |\mathbf{E}^{-1}\mathbf{H}| = \frac{|\mathbf{H}|}{|\mathbf{E}|} \propto \Pi_{i=1}^{p} \psi_i$$

where $\psi_i \cap F_{(m_H-i+1, m_E-i+1)}$.

Proof. From Theorem 4.2.9, $\mathbf{E}$ and $\mathbf{H}$ are each $\mathbf{\Sigma}$ times the product of p independent χ^2 variables. Therefore, ψ is the product of p ratios of independent χ^2 variables. The ith ratio is $\chi^2_{(m_H-i+1)}/\chi^2_{(m_E-i+1)}$, i.e.$\{(m_H-i+1)/(m_E-i+1)\}F_{(m_H-i+1, m_E-i+1)}$. Hence the result.

Theorem 4.4.2: We have

$$\Lambda(p; m_H, m_E) = \Pi_{i=1}^{m_H} u_i \tag{4.4.4}$$

where $u_1, \ldots, u_{m_H}$ are m_H independent variables and $u_i \cap B(\frac{M_E+i-p}{2}, \frac{p}{2}), i = 1, \ldots, m_H$.

Proof. Let us write $\mathbf{H} = \mathbf{X}'\mathbf{X}$, where m_H rows of $\mathbf{X}_i$ are *iid* $N_p(\mathbf{0}, \mathbf{I})$. Let $\mathbf{X}_i(i \times p)$ be the matrix consisting of the first i rows, $\mathbf{x}'_1, \ldots, \mathbf{x}'_i$ of $\mathbf{X}$. Let

$$\mathbf{W}_i = \mathbf{X}'_i\mathbf{X}_i + \mathbf{E},\ i = 1, \ldots, m_H.$$

Then,

$$\mathbf{W}_0 = \mathbf{E}, \mathbf{W}_{m_H} = \mathbf{X}'\mathbf{X} + \mathbf{E}, \mathbf{W}_i = \mathbf{W}_{i-1} + \mathbf{x}_i\mathbf{x}'_i.$$

Now,

$$\begin{aligned}\Lambda(p; m_H, m_E) &= \tfrac{|\mathbf{E}|}{|\mathbf{E}+\mathbf{H}|} = \tfrac{|\mathbf{W}_0|}{|\mathbf{W}_{m_H}|} = \tfrac{|\mathbf{W}_0|}{|\mathbf{W}_1|}\tfrac{|\mathbf{W}_1|}{|\mathbf{W}_2|}\cdots\tfrac{|\mathbf{W}_{m_H-1}|}{|\mathbf{W}_{m_H}|}\\ &= u_1.u_2.\ldots u_{m_H},\end{aligned}$$

where

$$u_i = \frac{|\mathbf{W}_{i-1}|}{|\mathbf{W}_i|}, i = 1, \ldots, m_H.$$

Now, $u_i = \frac{|\mathbf{W}_{i-1}|}{|\mathbf{W}_{i-1}+\mathbf{x}_i\mathbf{x}'_i|} \cap B(\frac{m_E+i-p}{2}, \frac{p}{2})$ by Theorem 4.3.4.

Again, from Theorem 4.3.6, $\mathbf{W}_i$ is statistically independent of $1 + \mathbf{x}'_i\mathbf{W}_{i-1}^{-1}\mathbf{x}_i = |\mathbf{W}_i|/|\mathbf{W}_{i-1}| = u_i^{-1}$. Again, since u_i is independent of $\mathbf{x}_{i+1}, \ldots, \mathbf{x}_n$ and $W_{i+k} = \mathbf{W}_i + \sum_{j=1}^k \mathbf{x}_{i+j}\mathbf{x}'_{i+j}$, it follows that u_i is independent of $\mathbf{W}_{i+1}, \mathbf{W}_{i+2}, \ldots, \mathbf{W}_{m_H}$ and hence independent of $u_{i+1}, u_{i+2}, \ldots, u_{m_H}$.

Theorem 4.4.3: The $\Lambda(p; m_H, m_E)$ and $\Lambda(m_H; p, m_E + m_H - p)$ distributions are the same.

Proof. Omitted.

It can be shown that

(1) if $m_H = 1$, then

$$\frac{(1-\Lambda)/p}{\Lambda/(m_E - p + 1)} \cap F_{(p, m_E-p+1)}. \tag{4.4.5}$$

(2) If $m_H = 2$, then

$$\frac{(1-\sqrt{\Lambda})/p}{\sqrt{\Lambda}/(m_E - p + 1)} \cap F_{(2p; 2m_E-2p+2)}, p \geq 2. \tag{4.4.6}$$

(3) For $p = 1$,

$$\frac{(1-\Lambda)/m_H}{\Lambda/m_E} \cap F_{m_H, m_E}. \tag{4.4.7}$$

(4) For $p = 2$,

$$\frac{(1-\sqrt{\Lambda})/m_H}{\sqrt{\Lambda}/(m_E - 1)} \sim F_{(2m_H, 2(m_E-1))}, m_H \geq 2. \tag{4.4.8}$$

Bartlett (1938) showed that for large $m_E, W = -f \ln \Lambda$ is approximately distributed as $\chi^2_{(pm_H)}$, where

$$f = m_E - (p - m_H + 1)/2. \tag{4.4.9}$$

These approximations are accurate for the usual critical values, to three decimal places if $p^2 + m_H^2 \leq f/3$.

Rao (1951) showed that

$$\frac{1 - \Lambda^{1/t}(p; m_H, m_E)}{\Lambda^{1/t}(p; m_H, m_E)} \cdot \frac{ft - g}{pm_H} \tag{4.4.10}$$

is approximately $F_{(pm_H, ft-g)}$ where

$$t = \left\{ \frac{p^2 m_H^2 - 4}{p^2 + m_H^2 - 5} \right\}^{1/2}, \quad g = \frac{pm_H - 2}{2}. \tag{4.4.11}$$

Let

$$s = \min\ (p, m_H), \quad \nu_1 = \frac{1}{2}(|m_H - p) - 1),$$

$$\nu_2 = \frac{1}{2}(m_E - p - 1). \tag{4.4.12}$$

If $s = 1$,

$$\frac{1 - \Lambda(p; m_H, m_E)}{\Lambda(p; m_H, m_E)} \cdot \frac{\nu_2 + 1}{\nu_1 + 1} \sim F_{(2\nu_1 + 2, 2\nu_2 + 2)}. \tag{4.4.13}$$

If $s = 2$,

$$\frac{1 - \Lambda^{(1/2}(p; m_H, m_E)}{\Lambda^{1/2}(p; m_H, m_E)} \cdot \frac{2\nu_2 + 2}{2\nu_1 + 3} \sim F_{(4\nu_1 + 6, 4\nu_2 + 4)}. \tag{4.4.14}$$

The statistic $\Lambda(p; m_H, m_E)$ is used for testing of hypotheses by the likelihood ratio statistic. H_0 is rejected if Λ is too small. Exact critical values $\Lambda_{p, m_H, m_E; \alpha}$ for Λ have been given in Table B.10 of the Appendix B for $p = 3$ and $p = 4$ (in the table, $d = p$). For these and other values of p, Tables are available in Schatzoff (1966), Lee (1972), Pillai and Gupta (1969) and Davis (1979).

4.5 Some Statistics Based on Eigenvalues of Wishart Matrices

Suppose that $H \cap \chi^2_{m_H}\sigma^2$ and $E \cap \chi^2_{m_E}\sigma^2$ where H and E are statistically independent. For example, H may be the sum of squares due to hypothesis and E the sum of squares due to error in an analysis of variance. We are often interested in the statistics H/E and $H/(E+H)$, which are known to have type II Beta $(m_H/2, m_E/2)$ and type I Beta $(m_H/2, m_E/2)$ distributions respectively.

We generalize this situation. Let $\mathbf{H}$ and $\mathbf{E}$ be matrices with independent nonsingular Wishart distributions, $\mathbf{H} \cap \mathbf{W}_p(m_H, \mathbf{\Sigma}), \mathbf{E} \cap \mathbf{W}_p(m_E, \mathbf{\Sigma})$, where $m_H, m_E \geq p$. By analogy with the above univariate approach we could consider $\mathbf{HE}^{-1}$ and $\mathbf{H}(\mathbf{E}+\mathbf{H})^{-1}$. However, these matrices are not symmetric and do not lead to useful density functions.

In the sequel we will be concerned with the eigenvalues of $\mathbf{HE}^{-1}$ and $\mathbf{H}(\mathbf{E}+\mathbf{H})^{-1}$.

Since $\mathbf{E}$ and $\mathbf{H}$ are positive definite with probability one (Corollary 4.2.9.1), we can obtain symmetry by defining the positive definite matrices (see A.11.3)

$$\mathbf{T} = \mathbf{E}^{-1/2}\mathbf{H}\mathbf{E}^{-1/2}, \tag{4.5.1}$$

$$\mathbf{V} = (\mathbf{E}+\mathbf{H})^{-1/2}\mathbf{H}(\mathbf{E}+\mathbf{H})^{-1/2} \tag{4.5.2}$$

where $\mathbf{E}^{1/2}$ and $(\mathbf{E}+\mathbf{H})^{1/2}$ are the symmetric square roots of $\mathbf{E}$ and $\mathbf{E}+\mathbf{H}$ respectively (see A.12.4).

Note that the matrix $\mathbf{HE}^{-1}$ has the same eigenvalues as the symmetric matrix $\mathbf{T}$. This follows from the result that the eigenvalues of $\mathbf{AB}$ are the same as those of $\mathbf{BA}$ (A8(8)). Let $\mathbf{A} = \mathbf{E}^{-1/2}\mathbf{H}, \mathbf{B} = \mathbf{E}^{-1/2}$. Then $\mathbf{AB} = \mathbf{T}$. Also, $\mathbf{BA} = \mathbf{E}^{-1}\mathbf{H}$. Also, eigenvalues of $\mathbf{E}^{-1}\mathbf{H}$ are the same as those of $\mathbf{HE}^{-1}$. Hence, $\mathbf{T}$ and $\mathbf{HE}^{-1}$ have the same eigenvalues. Also, the eigenvectors of $\mathbf{AB} = \mathbf{T}$ are of the form $\mathbf{Aa} = \mathbf{E}^{-1/2}\mathbf{Ha}$ where $\mathbf{a}$ is the eigenvector of $\mathbf{BA} = \mathbf{E}^{-1}\mathbf{H}$ (vide A.8(8)).

Similarly, it follows that eigenvalues of $\mathbf{H}(\mathbf{E}+\mathbf{H})^{-1}$ are the same as those of $\mathbf{V}$.

Suppose θ is an eigenvalue of $\mathbf{V}$. Then θ is a root of

$$|\mathbf{V} - \theta\mathbf{I}_p| = 0,$$

or

$$|(\mathbf{E}+\mathbf{H})^{-1/2}\mathbf{H}(\mathbf{E}+\mathbf{H})^{-1/2} - \theta \mathbf{I}_p| = 0$$

or

$$|\mathbf{E}+\mathbf{H}|^{-1}|\mathbf{H} - \theta(\mathbf{E}+\mathbf{H})| = 0$$

or

$$|\mathbf{H} - \theta(\mathbf{E}+\mathbf{H})| = 0. \tag{4.5.3}$$

Since $\mathbf{V} > 0$ with probability 1, $\theta > 0$ with probability 1 (A.11 (9)). If $\theta > 1, -(\theta - 1)\mathbf{H} - \theta\mathbf{E} < 0$ with probability 1 and (4.5.3) holds with probability 0. Hence $0 < \theta < 1$ with probability 1 and we can express (4.5.3) in the form

$$|\mathbf{H}(1-\theta) - \theta\mathbf{E}| = 0$$

or

$$|\mathbf{H} - \phi\mathbf{E}| = 0 \tag{4.5.4}$$

or

$$|\mathbf{H}\mathbf{E}^{-1} - \phi\mathbf{I}_p| = 0, \tag{4.5.5}$$

where $\phi = \theta/(1-\theta)$. We note that both $\mathbf{H} > 0, \mathbf{E} > 0$ with probability 1 and the eigenvalues of both $\mathbf{H}$ and $\mathbf{E}$ are distinct with probability 1 (A.7 (13)). Hence, eigenvalues of $\mathbf{H}\mathbf{E}^{-1}$ are distinct and therefore, eigenvalues of $\mathbf{V}$ are distinct with probability 1. We order them as $1 > \theta_1 > \theta_2 > \ldots > \theta_p > 0$. Here, the number of non-zero eigenvalues of $\mathbf{H}\mathbf{E}^{-1}$ is p, because $m_H \geq p$.

Roy (1939) obtained the distribution of $\theta_{max} = \theta_1$. This statistic is used for testing the hypothesis by Roy's (1957) union-intersection Principle (Section 5.3.1). The tables of percentage points of θ_1 has been constructed by several authors [vide Pearson and Hartley (1972), Pillai (1964, 1965) and Johnson and Koltz, 1972, pp. 18 - 26 for further references]. More extensive tables have been given by Pillai and Flury (1984). In Table B.11 the upper percentage points of θ_{max}, the largest eigenvalue of $\mathbf{H}(\mathbf{E}+\mathbf{H})^{-1}$ are given for $s = \min(p, m_H) = 2(1)7, \nu_1 = (|p-m_H|-1)/2$, and $\nu_2 = (m_E-p-1)/2$ for $\alpha = 0.05$ and 0.01.

Note that

$$\theta_j = \frac{\phi_j}{(1+\phi_j)}, j = 1, \ldots, p; \tag{4.5.6}$$

$$\phi_{max} = \frac{\theta_1}{1-\theta_1}. \tag{4.5.7}$$

Also, if $\lambda_1, \lambda_2, \ldots, \lambda_p$ are the eigenvalues of $(\mathbf{E}+\mathbf{H})^{-1}\mathbf{E}$, then

$$\lambda_j = 1-\theta_j, j=1,\ldots,p. \tag{4.5.8}$$

Here $\lambda_1 < \lambda_2 < \ldots < \lambda_p < 1$. Clearly, $\lambda_{max} = 1-\theta_{min}$.

No satisfactory F-approximation to θ_1 or ϕ_1 is available. Some software programmes calculate

$$\mathcal{F} = \frac{(m_E-d-1)\phi_1}{d} \tag{4.5.9}$$

where $d =$ max (p, m_H). Theoretically,

$$\mathcal{F}_{d,m_E-d-1;\alpha} \geq F_{d,m_E-d-1;\alpha} \tag{4.5.10}$$

where $\mathcal{F}_{d,m_E-d-1;\alpha}$ is obtained by putting $\phi_{1;\alpha}$, the upper $100\alpha\%$ point of ϕ_1 for ϕ_1 in (4.5.9). If the observed $\mathcal{F}$ is, therefore, less than or equal to $F_{d,m_E-d-1;\alpha}$ so that H_0 is accepted, we are sure of our decision; but if H_0 is rejected ($\mathcal{F} > F_{d,m_E-d-1;\alpha}$) we are less sure of our decision.

We now consider two trace statistics which are also used for testing H_0.

Lawley-Hotelling Statistics:

In the context of testing of hypotheses, Lawley (1938) and Hotelling (1951) considered the following statistic when $m_E \geq p$, namely

$$T_g^2 = m_E \text{ tr } [\mathbf{HE}^{-1}] = m_E U^{(s)}, \tag{4.5.11}$$

where

$$U^{(s)} = \sum_{j=1}^{s} \phi_j = \sum_{j=1}^{s} \frac{\theta_j}{1-\theta_j} \tag{4.5.12}$$

and $s =$ min (p, m_H). Exact and approximate 5% and 1% upper tail critical values for various values of p have been given by Davis (1970 a,b; 1980), Pillai and Young (1971), among many others. McKeon (1974) has shown that the distribution of $U^{(s)}$ can be approximated by $kF_{(a,b)}$ where

$$a = pm_H,\ b = 4+(a+2)/(B-1),\ k = a(b-2)/b(m_E-p-1),$$

$$B = \frac{(m_E+m_H-p-1)(m_E-1)}{(m_E-p-3)(m_E-p)}.$$

When $m_E \to \infty, T_g^2 \sim \chi^2_{(pm_H)}$ approximately.

In Table B.12 upper percentage points of T_g^2/m_H are tabulated for selected values of p with $p \leq m_H, m_E$ (in the table, $d = p$). If $m_H < p$, we make the transformation $p \to m_H, m_H \to p, m_E \to m_E + m_H - p$, and enter these tables using the transformed values of the triple (p, m_H, m_E). The statistic T_g^2 is also known as *Hotelling's generalized* T_0^2 *statistic.*

Pillai's Statistic:

One of the several statistics proposed by Pillai (1953) is

$$V^{(s)} = \text{ tr } [\mathbf{H}(\mathbf{E}+\mathbf{H})^{-1}] = \sum_{j=1}^{s} \theta_j \tag{4.5.13}$$

where $s =$ min (p, m_H). Tables of percentage points have been given by Pillai (1960), John (1976, 1977), Schuurmann *et al.* (1975), among others. It has been shown that when $m_E \to \infty, m_E V^{(s)} \sim \chi^2_{(pm_H)}$ approximately.

All the above statistical tables are available in Kres (1983).

4.5.1 *Equivalence of statistics when* $m_H = 1$

When $m_H = 1$, al the five statistics, Hotelling's $T^2, \Lambda(p, m_E, m_H)$, Roy's statistic, θ_{max}, T_g^2 and $V^{(s)}$ are functions of each other. This can be seen as follows.

Since rank $(\mathbf{H}) =$ rank $[\mathbf{H}(\mathbf{H}+\mathbf{E})^{-1}] = 1$, the equation $|\mathbf{H} - \theta(\mathbf{H}+\mathbf{E})| = 0$ has only one root, which is θ_{max}. Thus

$$\begin{gathered} U^{(1)} = \theta_{max}/(1-\theta_{max}), \qquad (4.5.14) \\ T_g^2 = m_E U^{(1)}, \\ V^{(1)} = \theta_{max}, \end{gathered}$$

and

$$\Lambda = 1 - \theta_{max} \text{ (by (4.4.4) and (4.5.6) .}$$

Again, since rank $(\mathbf{H}) = 1$ and $\mathbf{H}$ has a Wishart distribution, $\mathbf{H}$ can be expressed as $\mathbf{H} = \mathbf{xx}'$, where $\mathbf{x} \cap N_p(0, \mathbf{\Sigma})$ and $\mathbf{x}$ is independent of $\mathbf{E}$ (because, by assumption, $\mathbf{H}$ is independent of $\mathbf{E}$). Thus

$$\begin{aligned} T_g^2 &= m_E U^{(s)} \\ &= m_E \text{ tr } [\mathbf{HE}^{-1}] = m_E \text{ tr } [\mathbf{xx}'\mathbf{E}^{-1}] \\ &= m_E[\text{ tr } \mathbf{x}'\mathbf{E}^{-1}\mathbf{x}] \text{ (by A.9(1))} \\ &= m_E \mathbf{x}'\mathbf{E}^{-1}\mathbf{x} \\ &= T^2 \sim T^2_{p,m_E}. \end{aligned}$$

Again, by Theorem 4.3.2,

$$\frac{T^2(p,m)}{m} \cdot \frac{m-p+1}{p} = F_{(p,m-p+1)}. \tag{4.5.15}$$

Combining (4.5.14) and (4.5.15),

$$\frac{\theta_{max}}{1-\theta_{max}} \cdot \frac{m_E - p + 1}{p} \sim F_{(p,m_E-p+1)}.$$

Thus when $m_H = 1$, all the above five statistics are related.

4.6 Exercises and Complements

4.1 Let $\mathbf{W} \cap W_p(m, \mathbf{\Sigma})$. Show that for any vector of constants $\mathbf{a}_{p\times 1}$, $\mathbf{a}'\mathbf{W}\mathbf{a} \cap (\mathbf{a}'\mathbf{\Sigma}\mathbf{a})\chi^2_{(m)}$.

4.2 Let $\mathbf{x}_1$ and $\mathbf{x}_2$ be independently distributed as $N_p(\mathbf{0}, \mathbf{\Sigma})$. For what values of the constants a, b will the quantity $a(\mathbf{x}_1\mathbf{x}_1' + \mathbf{x}_2\mathbf{x}_2') + b(\mathbf{x}_1\mathbf{x}_2' + \mathbf{x}_2\mathbf{x}_1')$ have a Wishart distribution?

4.3 Suppose $\mathbf{W} \cap W_p(m, \mathbf{\Sigma})$. Show that the characteristic function of $\mathbf{W}$ is

$$\begin{aligned}\Phi(\theta) &= E(e^{i\theta \mathbf{W}}) \\ &= \frac{|\mathbf{\Sigma}^{-1}|^{m/2}}{|\mathbf{\Sigma}^{-1} - 2i\theta|^{m/2}},\end{aligned}$$

where θ is a real $p \times p$ symmetric matrix of real elements.

4.4 Show that the following conditions are necessary and sufficient for $\mathbf{W}$ to have the $W_p(m, \mathbf{\Sigma})$ distribution:

(a) $\mathbf{W}$ is symmetric, and if $\mathbf{a}'\mathbf{\Sigma}\mathbf{a} = 0$, then $\mathbf{a}'\mathbf{W}\mathbf{a} = 0$ with probability one.

(b) for every $(q \times p)$ matrix $\mathbf{L}$ which satisfies $\mathbf{L}\mathbf{\Sigma}\mathbf{L}' = \mathbf{I}$, the diagonal elements of $\mathbf{L}\mathbf{W}\mathbf{L}'$ are independent $\chi^2_{(m)}$ variables.

(Mitra, 1969)

4.5 Let $(n-1)\mathbf{S} = \sum_{i=1}^{n}(\mathbf{X}_i - \bar{\mathbf{X}})(\mathbf{X}_i - \bar{\mathbf{X}})'$ be the covariance matrix, where $\mathbf{X}_i (i = 1, \ldots, n)$ is a random sample from a $N_p(\mu, \mathbf{\Sigma})$ distribution and $\mathbf{S} = ((s_{ij}))$. Show that $(n-1)\mathbf{S} \cap W_p(n-1, \mathbf{\Sigma})$ and hence

$$\begin{aligned}E(s_{ij}) &= \sigma_{ij} \\ E(s_{ij}s_{kl}) &= \sigma_{ij}\sigma_{kl} + \tfrac{1}{n-1}\{\sigma_{ik}\sigma_{jl} + \sigma_{il}\sigma_{jk}\}.\end{aligned}$$

4.6 If $\mathbf{W} \cap W_p(m, \mathbf{\Sigma})$, show that $\mathbf{a}'\mathbf{W}\mathbf{a}$ and $\mathbf{b}'\mathbf{W}\mathbf{b}$ are statistically independent if $\mathbf{a}'\mathbf{\Sigma}\mathbf{b} = 0$. Hence, show that w_{ii} and w_{jj} are independent if $\sigma_{ij} = 0$, and that when $\mathbf{\Sigma} = \mathbf{I}$, trace $\mathbf{W}$ has a $\chi_{(mp)}$ distribution.

[*Hints:* Use Theorem 3.3.3.]

4.7 If $\mathbf{W}$ follows a Wishart distribution, show that the distribution of trace($\mathbf{W}$) is the same as the distribution of a linear combination of chi-square variables.

4.8 Let $\mathbf{X}$ be a $(n \times p)$ data matrix from $N_p(\mu, \mathbf{\Sigma})$ and let $\mathbf{C}_1, \ldots \mathbf{C}_k$ be $k(n \times n)$ symmetric idempotent matrices such that $\mathbf{C}_1 + \ldots + \mathbf{C}_k = \mathbf{I}_n$. Show that

(a) $\mathbf{C}_r\mathbf{C}_s = \mathbf{0}(r \neq s)$. Hence, if $\mathbf{W}_i = \mathbf{X}'\mathbf{C}_i\mathbf{X}$ show that $\mathbf{X}'\mathbf{X} = \mathbf{W} = \sum_{i=1}^{k} \mathbf{W}_i$ is the sum of k independent matrices $(\mathbf{W}_i)$.
(b) If $\mu = \mathbf{0}, \mathbf{W}_i \cap W_p(r_i, \mathbf{\Sigma})$ where $r_i =$ trace $(\mathbf{C}_i)$.
(c) For general μ, if $\mathbf{C}_1 = \frac{1}{n}\mathbf{1}_n\mathbf{1}_n'$, then $\mathbf{W}_i \cap W_p(r_i, \mathbf{\Sigma})(i = 2, \ldots, k)$.

[*Hints.* (a)

$$\mathbf{C}_1 + \ldots + \mathbf{C}_k = \mathbf{I}_n$$

Multiplying by $\mathbf{C}_r$, we have $\mathbf{C}_r \sum_{s(\neq r)} \mathbf{C}_s = \mathbf{0}$.

(b) Follows by Theorem 4.2.10.

(c) Since $\mathbf{C}_r\mathbf{C}_1 = \mathbf{0}, \mathbf{C}_r\mathbf{1} = \mathbf{0}\ \forall r = 2, \ldots, n$. The result follows by Corollary 4.2.10.1]

4.9 (*Extensions of Theorems 4.2.10 - 4.2.12*) Show that if $\mathbf{X}$ is a data matrix from $N_p(\mathbf{0}, \mathbf{I})$ and $\mathbf{C}$ and $\mathbf{D}$ are symmetric matrices then the following results hold:

(a) $\mathbf{XCX}' \cap W_r(r, \mathbf{I})$ *iff* $\mathbf{C}$ is idempotent, where $r =$ rank $(\mathbf{C})$.
(b) $\mathbf{XCX}'$ and $\mathbf{XDX}'$ are independent *iff* $\mathbf{CD} = \mathbf{0}$.
(c) $\mathbf{XCX}'$ and $\mathbf{AX}'\mathbf{B}$ are independent *iff* either $\mathbf{B} = \mathbf{0}$ or $\mathbf{AC} = \mathbf{0}$.

4.10 (*Extension of Exercise 4.9*) If $\mathbf{X}$ is a data matrix from $N_p(\mathbf{0}, \mathbf{\Sigma}), \mathbf{\Sigma} > 0$ and $\mathbf{C}, \mathbf{D}$ are symmetric matrices, then the following results hold.

(a) $\mathbf{XCX}' \cap W_n(r, \mathbf{I})$ *iff* $\mathbf{C\Sigma C} = \mathbf{C}$, in which case $r =$ trace $(\mathbf{C\Sigma})$.
(b) $\mathbf{XCX}'$ and $\mathbf{XDX}'$ are independent *iff* $\mathbf{C\Sigma D} = \mathbf{0}$.
(c) $\mathbf{XCX}'$ and $\mathbf{AX}'\mathbf{B}$ are independent *iff* either $\mathbf{A\Sigma C} = \mathbf{0}$ or $\mathbf{B} = \mathbf{0}$.

4.11 If $\mathbf{W} \cap W_p(m, \mathbf{\Sigma})$, and $\mathbf{A}(k \times p)$ has rank k, then

$$(\mathbf{A}\mathbf{W}^{-1}\mathbf{A}')^{-1} \cap W_k(m - p + k, (\mathbf{A}\mathbf{\Sigma}^{-1}\mathbf{A}')^{-1}).$$

4.12 Suppose $\mathbf{W} \cap W_p(m, \mathbf{\Sigma})$, where $\mathbf{\Sigma} > 0$. Show that $|\mathbf{W}|/|\mathbf{\Sigma}|$ is distributed as the product of p independent chi-square variables with degrees of freedom $m, m-1, \ldots, m-p+1$, respectively.

Chapter 5

Tests of Hypotheses

5.1 Introduction

In this chapter we shall be concerned with the testing of multivariate hypotheses. The problems of testing a multivariate hypothesis differ from those in testing a univariate hypothesis in many ways. For example, in multivariate case, the number of possible hypotheses are very large and also in many cases, there are several alternative test-statistics to choose from. For example, the p-variate normal distribution has $p(p+1)/2$ parameters for each of which we could specify a hypothesis, as well as for subsets or functions of parameters. We will consider various approaches to test construction, including the likelihood ratio, the union-intersection method and other techniques.

There are several advantages in testing p variables in a single multivariate test rather than in p separate univariate tests. These advantages include preserving the α-level, testing with greater power and determining the contribution of each variable in the presence of other variables. The multivariate tests make allowance for the intercorrelation among other variables, which account in part for the increase in power.

Section 5.2 considers the likelihood ratio test procedure and its properties. Sections 5.3 and 5.4 address single sample problem $H_0(\mu = \mu_0)$. These sections also prove equivalence between the likelihood ratio test, union-intersection test and T^2 test for this problem. Confidence region for μ, simultaneous confidence intervals for linear functions of μ and large sample test for μ are considered in Sections 5.5 and 5.6. The next section deals with several hypotheses relating to mean vectors of two populations. Hypotheses relating to dispersion matrix, for a single population and two populations are the subject matters of Sections 5.8 and 5.9 respectively. Section 5.10 considers profile analysis for two populations. Lastly, multi-sample hypotheses relating to $k(\geq 2)$ populations are considered. Throughout, the parent populations are assumed to be multivariate normal.

5.2 Likelihood Ratio Test

Let $\mathbf{X} = (\mathbf{X}_1, \ldots, \mathbf{X}_n)'$ have a joint distribution $L(\mathbf{x}_1, \ldots, \mathbf{x}_n \mid \theta)$ depending on an parameter θ. The likelihood function of θ for fixed value of $(\mathbf{x}_1, \ldots, \mathbf{x}_n)'$ is

$$L(\theta \mid \mathbf{x}_1, \ldots, \mathbf{x}_n).$$

Suppose we want to test the simple hypothesis $H_0(\theta = \theta_0)$ against the composite alternative $H_1(\theta \neq \theta_0)$. Let $\Theta_0 = \{\theta = \theta_0\}, \Theta_1 = \{\theta : \theta \neq \theta_0\}, \Theta = \Theta_0 \cup \Theta_1$. A measure of how far the value θ_0 of θ is supported by the observation $\mathbf{x} = (\mathbf{x}_1, \ldots, \mathbf{x}_n)'$ is given by the likelihood ratio (LR),

$$\lambda(\mathbf{x}) = \frac{L(\theta_0 \mid \mathbf{x})}{\sup_{\theta \in \Theta} L(\theta \mid \mathbf{x})}. \tag{5.2.1}$$

A small value of λ will lead to the rejection of H_0. The size of the critical region is given by

$$w : \{\mathbf{x} : \lambda(\mathbf{x}) \leq \lambda_0\}$$

where

$$P\{\lambda(\mathbf{x}) \leq \lambda_0 \mid \theta_0\} = \alpha. \tag{5.2.2}$$

If the null hypothesis is composite, say, $H_0(\theta \in \Theta_0)$, we consider the maximum value that the likelihood can attain when the null hypothesis is true. This is reasonable, because, if the ratio of the maximum to that of the overall maximum is not acceptable, then any smaller value of the likelihood should also be not acceptable. Here the test is based on the LR

$$\lambda(\mathbf{x}) = \frac{\sup_{\theta \in \Theta_0} L(\theta|\mathbf{x})}{\sup_{\theta \in \Theta} L(\theta|\mathbf{x})}.$$

The critical value of the test is

$$w = \{\mathbf{x} : \lambda(\mathbf{x}) \leq \lambda_0\},$$

where λ_0 is such that

$$\sup_{\theta \in \Theta_0} P_\theta\{\mathbf{x} \in w\} = \alpha.$$

The LR-test has the following important asymptotic property.

Theorem 5.2.1 If Θ_0 is a r-dimensional subspace of Θ which is a space in $\mathcal{R}^m$, then under suitable regularity conditions, for each $\theta \in \Theta_0, -2\log_e \lambda$ has an asymptotic $\chi^2_{(m-r)}$ distribution as the sample size $n \to \infty$.

For a proof, see for example Silvey (1970, p.113).

In the next section we consider the LR test for the mean of a Normal population.

5.3 Testing for a Single Population Mean

Let $\mathbf{x}_1, \ldots, \mathbf{x}_n$ be a random sample from $N_p(\mu, \mathbf{\Sigma})$. We want to test $H_0(\mu = \mu_0)$ against $H(\mu \neq \mu_0)$.

If $\mathbf{\Sigma}$ is known, then under H_0,

$$n(\bar{\mathbf{x}} - \mu_0)'\mathbf{\Sigma}^{-1}(\bar{\mathbf{x}} - \mu) \cap \chi^2_{(p)}$$

and hence the critical region for H_0 is

$$w : \{\mathbf{x} : n(\bar{\mathbf{x}} - \mu_0)'\mathbf{\Sigma}^{-1}(\bar{\mathbf{x}} - \mu) \ \geq \chi^2_{p;\alpha}\}.$$

When $\mathbf{\Sigma}$ is unknown, it is known that under H_0

$$T_0^2 = n(\bar{\mathbf{x}} - \mu_0)'\mathbf{S}^{-1}(\bar{\mathbf{x}} - \mu_0) \tag{5.3.1}$$

has a $T^2(p, n-1)$-distribution and hence by Theorem 4.3.2,

$$F_0 = \frac{T_0^2}{n-1}.\frac{n-p}{p} \cap F_{(p,n-p)}. \tag{5.3.2}$$

The null hypothesis H_0 is rejected if $F_0 > F_{p,n-p;\alpha}$ where $P[F_{(p,n-p)} \geq F_{p,n-p;\alpha}] = \alpha$.

When H_0 is false, T_0^2 has a non-central $T^2(p, n-1)$ distribution with non-centrality parameter $\delta = n(\mu - \mu_0)'\mathbf{\Sigma}^{-1}(\mu - \mu_0)$ (Subsection 4.3.1). Thus

$$\frac{T_0^2}{n-1}.\frac{n-p}{p} \cap F_{(p,n-p)}(\delta).$$

The power of the test for a critical region of size α is

$$P[F_{(p,n-p)}(\delta) \geq F_{p,n-p;\alpha}]$$
$$= \gamma(p, n-p; \delta).$$

The function γ has been extensively tabulated by Tiku (1967, 1972) for a range of values of $p, n-p$ and standardized non-centrality parameter $\phi = [\delta/(p+1)]^{1/2}$.

5.3.1 *Union-intersection method*

The union-intersection method of test construction was introduced by Roy (1957). The procedure involves linear functions of parameters.

It is clear that $H_0(\mu = \mu_0)$ is true *iff* $\mathbf{a}'\mu = \mathbf{a}'\mu_0 \ \forall \ \mathbf{a}$ and H_0 is false *iff* $\mathbf{a}'\mu \neq \mathbf{a}'\mu_0$ for at least one $\mathbf{a}$. We then reject H_0 if at least one vector $\mathbf{a}$ can be found such that the univariate hypothesis $H_{0a} : \mathbf{a}'\mu = \mathbf{a}'\mu_0$ is rejected. The rejection region is therefore, the union over $\mathbf{a}$ of all rejection regions for the univariate hypothesis H_{0a}. Similarly, we accept H_0 *iff* every univariate hypothesis H_{0a} is accepted. The acceptance region is, therefore, the intersection over all $\mathbf{a}$ of the acceptance regions of the univariate hypotheses H_{0a}.

Now, $\bar{\mathbf{x}}$ follows $N_p(\mu, \mathbf{\Sigma}/n)$ and hence, $\mathbf{a}'\bar{\mathbf{x}}$ is N $(\mathbf{a}'\mu, \mathbf{a}'\mathbf{\Sigma}\mathbf{a}/n)$. We can, therefore test the univariate hypothesis H_{0a} by the t-statistic

$$t(\mathbf{a}) = \frac{\mathbf{a}'\bar{\mathbf{x}} - \mathbf{a}'\mu_0}{\sqrt{\mathbf{a}'\mathbf{S}\mathbf{a}/n}} \tag{5.3.3}$$

which for a given $\mathbf{a}$ has the acceptance region

$$|t(\mathbf{a})| < c,$$

where c is selected to provide size α to the test. Note that the value of c does not depend on $\mathbf{a}$. The acceptance region for H_0 is, therefore,

$$\cap_{\mathbf{a}}[|t(\mathbf{a})| < c]. \tag{5.3.4}$$

The rejection region for H_0 is, therefore,

$$\cup_{\mathbf{a}}[|t(\mathbf{a})| \geq c]. \tag{5.3.5}$$

Thus, we will accept H_0 if $\max_{\mathbf{a}}|t(\mathbf{a})| \leq c$.

To find $\max_{\mathbf{a}}|t(\mathbf{a})|$, it is more convenient to work with $t^2(\mathbf{a})$ which is

$$t^2(\mathbf{a}) = \frac{n[\mathbf{a}'(\bar{\mathbf{x}} - \mu_0)]^2}{\mathbf{a}'\mathbf{S}\mathbf{a}}. \tag{5.3.6}$$

The value of $\mathbf{a}$ that maximizes (5.3.6) can be found by differentiating $t^2(\mathbf{a})$ with respect to $\mathbf{a}$ and equating the result with $\mathbf{0}$. Now,

$$\frac{\partial t^2(\mathbf{a})}{\partial \mathbf{a}} = \frac{2n\mathbf{a}'\mathbf{S}\mathbf{a}[\mathbf{a}'(\bar{\mathbf{x}} - \mu_0)](\bar{\mathbf{x}} - \mu_0) - 2n[\mathbf{a}'(\bar{\mathbf{x}} - \mu_0)]^2\mathbf{S}\mathbf{a}}{(\mathbf{a}'\mathbf{S}\mathbf{a})^2} = \mathbf{0}$$

which simplifies to

$$\mathbf{a}'\mathbf{S}\mathbf{a}(\bar{\mathbf{x}} - \mu_0) - \mathbf{a}'(\bar{\mathbf{x}} - \mu_0)\mathbf{S}\mathbf{a} = \mathbf{0},$$

from which we get

$$\mathbf{a} \propto \mathbf{S}^{-1}(\bar{\mathbf{x}} - \mu_0). \tag{5.3.7}$$

Substituting (5.3.7) into (5.3.6), we obtain

$$\max{}_{\mathbf{a}} t^2(\mathbf{a}) = n(\bar{\mathbf{x}} - \mu_0)'\mathbf{S}^{-1}(\bar{\mathbf{x}} - \mu_0) = T^2$$

which follows a $T^2(p, n-1)$ distribution under H_0. Hence, acceptance region for H_0 is

$$\{\mathbf{x} : n(\bar{\mathbf{x}} - \mu_0)'\mathbf{S}^{-1}(\bar{\mathbf{x}} - \mu_0) \leq \frac{(n-1)p}{n-p} F_{p,n-p;\alpha}\}. \tag{5.3.8}$$

Thus, the union-intersection method leads to T^2 test.

(For another method of maximizing t^2 in (5.3.6), Section 5.5 (portion just before Theorem 5.5.1) may be seen.)

Note 5.3.1: From (5.3.7) it is clear that any multiple of $\mathbf{a} = \mathbf{S}^{-1}(\bar{\mathbf{x}} - \mu_0)$ maximizes $t^2(\mathbf{a})$, i.e., maximally separates $\mathbf{a}'\bar{\mathbf{x}}$ from $\mathbf{a}'\mu_0$. The linear function $z = \mathbf{a}'\mathbf{x}$ with coefficient vector $\mathbf{a} = \mathbf{S}^{-1}(\bar{\mathbf{x}} - \mu_0)$ is called the *discriminant function.* The discriminant functions have been considered in Chapter 11.

Note 5.3.2: The discriminant function provides a multivariate analysis of the effect of the individual variables on T^2. Suppose that the T^2 test leads to the rejection of $H_0(\mu = \mu_0)$. Then by the result above, $t^2(\mathbf{a})$ with $\mathbf{a} = \mathbf{S}^{-1}(\bar{\mathbf{x}} - \mu_0)$ will lead to the rejection of $H_{0a}(\mathbf{a}'\mu = \mathbf{a}'\mu_0)$. Thus we can examine the values of $a_j (j = 1, \ldots, p)$ to determine which of the x_j's contribute most to the rejection of $H_0(\mu = \mu_o)$. The effect of a variable on T^2 is often altered by the presence of other variables.

EXAMPLE 5.3.1: For the bird data of Table 2.E.1, considering the variables X_1, X_2, X_3 only, test the hypothesis $H_0(\mu_1 = 160, \mu_2 = 240, \mu_3 = 30)$ for the survivor birds (that is, birds with serial numbers 1 to 21).

Here $\bar{\mathbf{x}} = (157.38,\ 241,\ 31.43)'$. Also

$$\mathbf{S} = \begin{bmatrix} 11.0476 & 9.1000 & 1.5567 \\ 9.1000 & 17.5000 & 1.9100 \\ 1.5567 & 1.9100 & 0.533 \end{bmatrix},$$

$$\mathbf{S}^{-1} = \begin{bmatrix} .1876 & -.0618 & -.3274 \\ -.0618 & .1144 & -.2302 \\ -.3274 & -.2302 & 3.6689 \end{bmatrix}.$$

Here

$$T_0^2 = 21(\bar{\mathbf{x}} - \mu_0)'\mathbf{S}^{-1}(\bar{\mathbf{x}} - \mu_0) = 21 \times 11.0227.$$

Hence

$$F_0 = \frac{T_0^2}{20}.\frac{21-3}{3} = 69.443 >> F_{3,18;.05} = 3.16.$$

The test therefore strongly rejects the null hypothesis.

5.4 Equivalence Between Hotelling's T^2 and Likelihood Ratio Test

The likelihood ratio test for the hypothesis $H_0(\mu = \mu_0)$ is given by

$$\lambda = \frac{\sup_{\boldsymbol{\Sigma}} L(\mu_0, \boldsymbol{\Sigma})}{\sup_{(\mu,\boldsymbol{\Sigma})} L(\mu, \boldsymbol{\Sigma})}.$$

From Theorem 3.6.1,

$$\begin{aligned} \sup_{(\mu,\boldsymbol{\Sigma})} L(\mu, \boldsymbol{\Sigma}) &= L(\bar{\mathbf{x}}, \mathbf{S}_n) \\ &= (2\pi)^{-np/2}|\mathbf{S}_n|^{-n/2}e^{-np/2}. \end{aligned} \tag{5.4.1}$$

When $\mu = \mu_0$, the maximum likelihood estimate (m.l.e.) of $\boldsymbol{\Sigma}$ is $\hat{\boldsymbol{\Sigma}}^* = \mathbf{S}_n^* = n^{-1}\sum_i(\mathbf{x}_i - \mu_0)(\mathbf{x}_i - \mu_0)'$. This is obtained by using the similar procedure used to obtain the m.l.e. $\hat{\boldsymbol{\Sigma}}$ in Theorem 3.6.1. Hence,

$$\begin{aligned} \sup_{\boldsymbol{\Sigma}} L(\mu_0, \boldsymbol{\Sigma}) &= L(\mu_0, \hat{\boldsymbol{\Sigma}}^*) \\ &= (2\pi)^{-np/2}|\hat{\boldsymbol{\Sigma}}^*|^{-n/2}e^{-np/2}. \end{aligned} \tag{5.4.2}$$

Hence

$$\lambda = \frac{|\mathbf{S}_n|^{n/2}}{|\mathbf{S}_n^*|^{n/2}} = \frac{|\mathbf{Q}|^{n/2}}{|\mathbf{Q}^*|^{n/2}}$$

where $\mathbf{Q} = \sum_i(\mathbf{x}_i - \bar{\mathbf{x}})(\mathbf{x}_i - \bar{\mathbf{x}})'$, $\mathbf{Q}^* = \sum_i(\mathbf{x}_i - \mu_0)(\mathbf{x}_i - \mu_0)' = \mathbf{Q} + n(\bar{\mathbf{x}} - \mu_0)(\bar{\mathbf{x}} - \mu_0)'$. Now

$$\begin{aligned} |\mathbf{Q}^*| &= |\mathbf{Q} + [\sqrt{n}(\bar{\mathbf{x}} - \mu_0)][\sqrt{n}(\bar{\mathbf{x}} - \mu_0)]'| \\ &= \begin{vmatrix} 1 & \sqrt{n}(\bar{\mathbf{x}} - \mu_0)' \\ -\sqrt{n}(\bar{\mathbf{x}} - \mu_0) & \mathbf{Q} \end{vmatrix} \\ &= |\mathbf{Q}|[1 + n(\bar{\mathbf{x}} - \mu_0)'\mathbf{Q}^{-1}(\bar{\mathbf{x}} - \mu_0)] \end{aligned}$$

(by (A.5.6)). Hence

$$\begin{aligned} \lambda^{2/n} &= \frac{1}{1+n(\bar{\mathbf{x}}-\mu_0)'\mathbf{Q}^{-1}(\bar{\mathbf{x}}-\mu_o)} \\ &= \frac{1}{1+\frac{T_0^2}{n-1}}. \end{aligned} \tag{5.4.3}$$

By the LR-test we reject H_0 if λ is too small, that is, if T_0^2 is too large. By Theorem 5.2.1, $-2\ln\lambda$ follows approximately a $\chi^2_{(m-r)}$ distribution in large samples. Here m = dimension of $\Theta = p(p+1)/2 + p, r$ = dimension of $\Theta_0 = p(p+1)/2$. Hence, $-2\ln\lambda$ follows approximately a $\chi^2_{(p)}$ distribution in large samples.

5.5 Confidence Region and Simultaneous Confidence Intervals

If $\mathbf{X}_1, \ldots, \mathbf{X}_n$ is a random sample from $N_p(\mu, \mathbf{\Sigma})$, then it has been found

$$P\{n(\bar{\mathbf{X}} - \mu)'\mathbf{S}^{-1}(\bar{\mathbf{X}} - \mu) \leq \frac{(n-1)p}{n-p} F_{p,n-p;\alpha}\} = 1 - \alpha.$$

Hence, a $100(1-\alpha)\%$ confidence region for the mean μ is the ellipsoid defined by

$$\{\mu : n(\mu - \bar{\mathbf{x}})'\mathbf{S}^{-1}(\mu - \bar{\mathbf{x}}) \leq \frac{(n-1)p}{n-p} F_{p,n-p;\alpha}\}, \tag{5.5.1}$$

where $\bar{\mathbf{x}}$ is the observed sample mean. As has been noted in (3.2.12), the center of the ellipsoid is $\bar{\mathbf{x}}$ and beginning at the center, the axes of the ellipsoid are

$$\pm \left\{ \sqrt{\lambda_i} \sqrt{\frac{(n-1)p}{n(n-p)} F_{p,n-p;\alpha}} \right\} \mathbf{e}_i,$$

where $(\lambda_i, \mathbf{e}_i)$ $(i = 1, \ldots, p)$ are the (eigenvalue, eigenvector) pairs of $\mathbf{S}$.

If the point μ_0 lies in the ellipsoid (5.5.1) the hypothesis $H_0(\mu = \mu_0)$ is accepted.

EXAMPLE 5.5.1: In the bird-data of Table 2.E.1 consider variables X_1, X_2 only for the survivor birds. Here $\bar{x}_1 = 157.38, \bar{x}_2 = 241.00$,

$$\mathbf{S} = \begin{bmatrix} 11.0476 & 9.1000 \\ 9.1000 & 17.9000 \end{bmatrix}, \quad \mathbf{S}^{-1} = \begin{bmatrix} .1583 & -.0823 \\ -.0823 & .1000 \end{bmatrix}.$$

The eigenvalues and eigenvectors of $\mathbf{S}$ are

$$\lambda_1 = 23.9288, \ \mathbf{e}_1 = (.5770, \ .8167)',$$

$$\lambda_2 = 4.6188, \ \mathbf{e}_2 = (.8167, \ -.5770)'.$$

The 95% confidence ellipse for μ consists of all values (μ_1, μ_2) satisfying

$$21(157.38 - \mu_1, \ 241.00 - \mu_2)\mathbf{S}^{-1}(157.38 - \mu_1, \ 241.00 - \mu_2)'$$

$$\leq \frac{(20)(2)}{19} F_{2,19;.05} = 7.4105. \qquad (i)$$

To see if the point $\mu_0 = (155,\ 240)'$ lies in the confidence region, we compute the right side of (i) using these values. This is found to be 12.7035. Hence μ_0 lies outside the confidence region. Equivalently, the hypothesis $H_0(\mu = \mu_0)$ will be rejected at $\alpha = .05$ level of significance.

The confidence ellipsoid is plotted in Figure 5.1. The center is at $\bar{\mathbf{x}} = (157.38, 241.00)'$. The half-lengths of the major and minor axes are given by

$$\sqrt{\lambda_1}\sqrt{\frac{p(n-1)}{n(n-p)}F_{p,n-p;\alpha}} = 2.9059 \text{ and } 1.2767$$

units respectively. The axes lie along $\mathbf{e}_1' = (.5770,\ .8167)'$ and $e_2' = (.8167,\ -.5770)'$ when these vectors are plotted with $\bar{\mathbf{x}}$ as the origin.

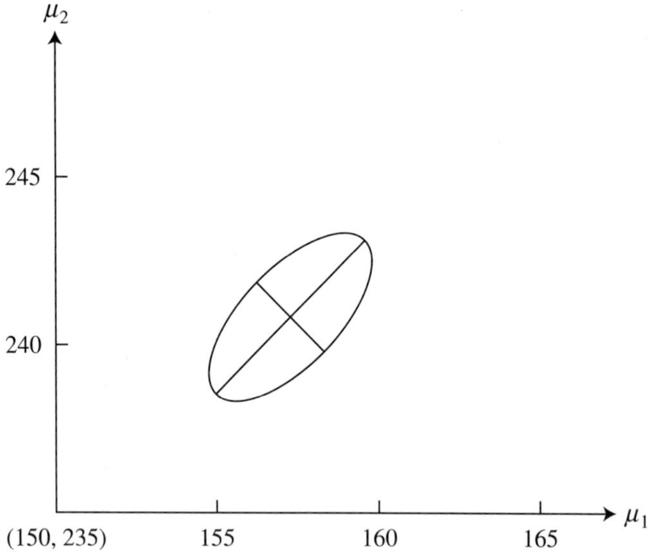

Fig. 5.1: A 95% confidence ellipse for μ based on the bird-data

5.5.1 T^2-*Simultaneous confidence intervals*

The expression (5.5.1) states that μ should lie within the stated ellipsoid with probability $(1-\alpha)$. However, we are more interested in finding confidence intervals regarding individual components or linear combinations of components of μ. In doing so we want to find intervals which will contain such linear functions of μ simultaneously with a very high probability.

Consider a transformation $Z = \mathbf{a}'\mathbf{X}$ where $\mathbf{a} = (a_1, \ldots, a_p)'$, a set of given constants. Clearly, $E(Z) = \mu_z = \mathbf{a}'\mu, V(Z) = \sigma_z^2 = \mathbf{a}'\mathbf{\Sigma}\mathbf{a}$. Corresponding to observations $\mathbf{x}_1 \ldots, \mathbf{x}_n$ of $\mathbf{X}$ we have n observations $z_i = \mathbf{a}'\mathbf{x}_i, i = 1, \ldots, n$. The sample mean of the z-values is $\bar{z} = \sum_{i=1}^n z_i/n$ and the sample variance is $s_z^2 = \sum_{i=1}^n (z_i - \bar{z})^2/(n-1) = \mathbf{a}'\mathbf{S}\mathbf{a}$. It is known that $\sqrt{n}(\bar{z} - \mu_z)/\sigma_z$ follows $N(0,1)$ distribution and when σ_z is unknown, $\sqrt{n}(\bar{z} - \mu_z)/s_z \cap t_{n-1}$, i.e.

$$\frac{\sqrt{n}(\mathbf{a}'\bar{\mathbf{x}} - \mathbf{a}'\mu)}{\sqrt{\mathbf{a}'\mathbf{S}\mathbf{a}}} \cap t_{(n-1)}. \tag{5.5.2}$$

Hence $100(1-\alpha)\%$ confidence interval for $\mathbf{a}'\mu$ is:

$$[\mathbf{a}'\bar{\mathbf{x}} - t_{n-1;\alpha/2}\frac{\sqrt{\mathbf{a}'\mathbf{S}\mathbf{a}}}{\sqrt{n}}, \quad \mathbf{a}'\bar{\mathbf{x}} + t_{n-1;\alpha/2}\frac{\sqrt{\mathbf{a}'\mathbf{S}\mathbf{a}}}{\sqrt{n}}]. \tag{5.5.3}$$

When $\mathbf{a} = (0, \ldots, 0, 1, 0, \ldots, 0)'$ the confidence interval for μ_j is

$$[\bar{x}_j - t_{n-1;\alpha/2}\frac{s_j}{\sqrt{n}}, \ \bar{x}_j + t_{n-1;\alpha/2}\frac{s_j}{\sqrt{n}}], \tag{5.5.4}$$

where $\bar{x}_j = \sum_{i=1}^n x_{ij}/n, s_j^2 = \sum_{i=1}^n (x_{ij} - \bar{x}_j)^2/(n-1)$.

Taking $\mathbf{a} = (0, \ldots, 0, 1, 0, \ldots, 0, -1, 0, \ldots, 0)'$ the confidence interval for $\mu_j - \mu_k$ is

$$[(\bar{x}_j - \bar{x}_k) - t_{n-1;\alpha/2}\sqrt{\frac{s_{jj} - 2s_{jk} + s_{kk}}{n}},$$

$$(\bar{x}_j - \bar{x}_k) + t_{n-1;\alpha/2}\sqrt{\frac{s_{jj} - 2s_{jk} + s_{kk}}{n}}]. \tag{5.5.5}$$

For different choices of $\mathbf{a}$ we shall have different such confidence intervals. The statement (5.5.4) is valid only for μ_j. Similarly, the statement (5.5.5) is valid only for $\mu_j - \mu_k$.

We could make several such confidence statements about the functions involving the components of μ, each with the associated confidence coefficient

$(1-\alpha)$, by choosing different coefficient vectors $\mathbf{a}$. However, the probability that all of these statements taken together is true is not $1-\alpha$. To see this more clearly, suppose $X_1, \ldots, X_p$ are independent normal random variates. Then the joint probability that each of μ_j lies in the interval (5.5.4) for $j = 1, \ldots, p$ is $(1-\alpha)^p$ which is much less than $1-\alpha$. If $\alpha = 0.05$ and $p = 5$, this probability is $(.95)^5 = .77$. To ensure that the joint probability is $1-\alpha$, the individual intervals must be wider than the separate t-intervals in (5.5.4) depending on n and p.

We want to make a statement regarding the confidence interval for $\mathbf{a}'\mu$ which should hold simultaneously for all choices of $\mathbf{a}$.

Now, (5.5.2) implies that $100(1-\alpha)\%$ confidence interval for $\mathbf{a}'\mu$ is comprised of those values of μ for which

$$\left|\frac{\sqrt{n}(\mathbf{a}'\bar{\mathbf{x}} - \mathbf{a}'\mu)}{\sqrt{\mathbf{a}'\mathbf{S}\mathbf{a}}}\right| < t_{n-1;\alpha/2}$$

or

$$\frac{n(\mathbf{a}'\bar{\mathbf{x}} - \mathbf{a}'\mu)^2}{\mathbf{a}'\mathbf{S}\mathbf{a}} = \psi(\mathbf{a}; \bar{\mathbf{x}}, \mu, \mathbf{S}) \text{ (say) } \leq t^2_{n-1;\alpha/2}. \tag{5.5.6}$$

The inequality (5.5.6) should hold for all $\mathbf{a}$ if we can ensure

$$\begin{aligned} \psi(\bar{\mathbf{x}}, \mu, \mathbf{S}) &= \max{}_{\mathbf{a}} \psi(\mathbf{a}; \bar{\mathbf{x}}, \mu, \mathbf{S}) \\ &= \max{}_{\mathbf{a}} \left[\tfrac{n(\mathbf{a}'\bar{\mathbf{x}} - \mathbf{a}'\mu)^2}{\mathbf{a}'\mathbf{S}\mathbf{a}}\right] \leq c^2 \end{aligned}$$

where we require

$$P[\psi(\bar{\mathbf{x}}, \mu, \mathbf{S}) \leq c^2] = 1-\alpha. \tag{5.5.7}$$

This would give $100(1-\alpha)\%$ simultaneous confidence interval for $\mathbf{a}'\mu$ $\forall \mathbf{a}$. However, for those $\mathbf{a}$ for which $\psi(\mathbf{a}; \bar{\mathbf{x}}, \mu, \mathbf{S})$ is less than its maximum value, $P[\psi(\mathbf{a}; \bar{\mathbf{x}}, \mu, \mathbf{S}) \leq c^2]$ will be greater than $1-\alpha$. Hence, for these ψ's, this method provides a wider confidence interval for the corresponding $\mathbf{a}'\mu$ and yet states that the coverage probability is $1-\alpha$. This method is conservative in the sense that to protect the $\max{}_{\mathbf{a}}\psi$-value, it makes conservative statement about the other ψ-values.

Considering Theorem A.14.1, if we put $\mathbf{x} = \sqrt{n}\,\mathbf{a}, \mathbf{b} = (\bar{\mathbf{x}} - \mu), \mathbf{C} = \mathbf{S}/n$, then

$$\begin{aligned} \max{}_{\mathbf{a}}\left\{\tfrac{n(\mathbf{a}'\bar{\mathbf{x}} - \mathbf{a}'\mu)^2}{\mathbf{a}'\mathbf{S}\mathbf{a}}\right\} &= n(\bar{\mathbf{x}} - \mu)'\mathbf{S}^{-1}(\bar{\mathbf{x}} - \mu) \\ &= T^2. \end{aligned}$$

The maximum value is attained if $\mathbf{a} \propto \mathbf{S}^{-1}(\bar{\mathbf{x}} - \mu)$. Therefore, from (5.5.1),

$$c^2 = \frac{(n-1)p}{n-p} F_{p,n-p;\alpha}.$$

Therefore, for all values of $\mathbf{a}$,

$$P\left[\left|\frac{\sqrt{n}(\mathbf{a}'\bar{\mathbf{x}} - \mathbf{a}'\mu)}{\sqrt{\mathbf{a}'\mathbf{S}\mathbf{a}}}\right| < \sqrt{\frac{(n-1)p}{n-p}F_{p,n-p;\alpha}}\right] \geq 1-\alpha.$$

Hence, we have the following theorem.

Theorem 5.5.1 Let $\mathbf{x}_1, \ldots, \mathbf{x}_n$ be a random sample from $N_p(\mu, \mathbf{\Sigma})$. Then $100(1-\alpha)\%$ simultaneous confidence intervals for $\mathbf{a}'\mu$ are

$$\left[\mathbf{a}'\bar{\mathbf{x}} - \sqrt{\frac{(n-1)p}{n-p}F_{p,n-p;\alpha}\frac{\mathbf{a}'\mathbf{S}\mathbf{a}}{n}},\right.$$

$$\left.\mathbf{a}'\bar{\mathbf{x}} + \sqrt{\frac{(n-1)p}{n-p}F_{p,n-p;\alpha}\frac{\mathbf{a}'\mathbf{S}\mathbf{a}}{n}}\right], \qquad (5.5.8)$$

in the sense that the above interval will contain $\mathbf{a}'\mu$ with a probability $(1-\alpha)$, simultaneously for all $\mathbf{a}$. □

As noted before, the interval (5.5.8) is wider than the interval (5.5.3) obtained for the individual function $\mathbf{a}'\mu$. The simultaneous intervals (5.5.8) are also referred to as T^2-intervals.

The simultaneous confidence interval for μ_i is the projection of of the confidence ellipsoid (5.5.1) on the μ_i axis (Figure 5.2).

In addition to (5.5.8) we can obtain $(1-\alpha)$-point simultaneous confidence ellipse for (μ_i, μ_k) as follows.

$$P\{n(\bar{x}_i - \mu_i,\ \bar{x}_k - \mu_k)\begin{bmatrix} s_{ii} & s_{ik} \\ s_{ki} & s_{kk}\end{bmatrix}^{-1}\begin{bmatrix}\bar{x}_i - \mu_i \\ \bar{x}_k - \mu_k\end{bmatrix} \leq$$

$$\frac{(n-1)p}{n-p}F_{p,n-p;\alpha}\} = 1-\alpha.$$

EXAMPLE 5.5.2: For the data in Example 5.5.1, let us compute one-at-a-time confidence interval (5.5.4) and T^2-simultaneous confidence interval (5.5.8) for μ_1. By (5.5.4), $t-$confidence interval for μ_1 is

$$\bar{x}_1 - t_{20;.025}\sqrt{\frac{s_{11}}{21}},\ \bar{x}_1 + t_{20;.025}\sqrt{\frac{s_{11}}{21}} = (155.87,\ 158.89). \qquad (ii)$$

By formula (5.5.8), simultaneous confidence interval for μ_1 is

$$\bar{x}_1 - \sqrt{\frac{(20)(2)}{(19)(21)}s_{11}F_{2,19;.05}}, \bar{x}_1 + \sqrt{\frac{(20)(2)}{(19)(21)}s_{11}F_{2,19;.05}}$$

$$= (155.41,\ 159.35). \qquad (iii)$$

Clearly, the T^2 interval (iii) is wider than the t-interval (ii).

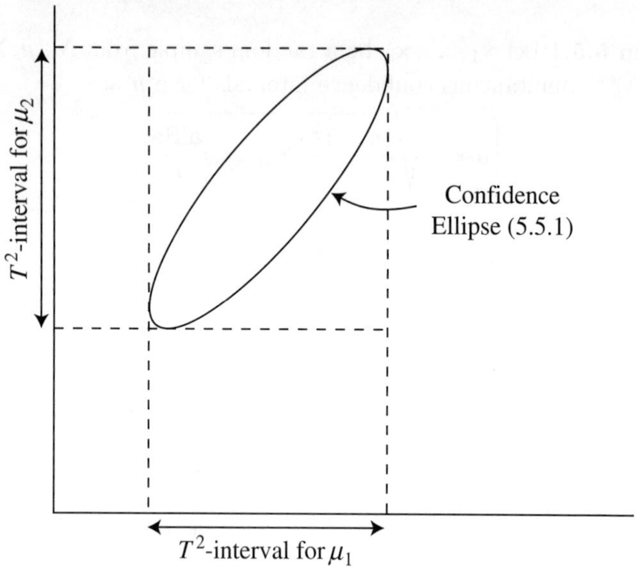

Fig. 5.2: Simultaneous confidence intervals for μ_1, μ_2 as projections of confidence ellipse

5.5.2 *Bonferroni's simultaneous confidence intervals*

If we are interested in finding simultaneous confidence intervals for a small number m of linear combinations $\mathbf{a}_i'\mu(i = 1, \ldots, m)$, then an alternative set of simultaneous confidence intervals can be used. This set was developed by Bonferroni using a probability inequality and is better than T^2-intervals in the sense of shorter length for each interval. We derive the results below.

It is known that if $A_1, \ldots, A_t$ are t events,

$$(\cap_{i=1}^t A_i)^c = \cup_{i=1}^t A_i^c \text{ (by De Morgan's law).}$$

Hence,

$$P[(\cap_{i=1}^t A_i)^c] = P[\cup_{i=1}^t A_i^c],$$

or

$$P(\cap_{i=1}^t A_i) = 1 - P(\cup_{i=1}^t A_i^c). \tag{5.5.9}$$

Now, by Boole's inequality,

$$P(\cup_{i=1}^t A_i) \leq \sum_{i=1}^t P(A_i).$$

Hence,

$$P(\cup_{i=1}^t A_i^c) \leq \sum_{i=1}^t P(A_i^c).$$

Therefore, from (5.5.9),

$$P(\cap_{i=1}^t A_i) \geq 1 - \sum_{i=1}^t P(A_i^c). \tag{5.5.10}$$

Let C_i be a confidence statement about $\mathbf{a}_i'\mu$ (e.g., C_i : the interval $\mathbf{a}_i'\bar{\mathbf{x}} - h_i, \mathbf{a}_i'\bar{\mathbf{x}} + h_i$ contains $\mathbf{a}_i'\mu$, where h_i is a suitable quantity. $P(C_i$ is true) $= 1 - \alpha_i$ if the confidence coefficient of the interval is $1 - \alpha_i$.)

Let A_i be the event that C_i is true. Then (5.5.10) states that

$$P(\text{ all the confidence statements } C_1, \ldots, C_m \text{ are simultaneously true })$$

$$\geq 1 - \sum_{i=1}^m \alpha_i. \tag{5.5.11}$$

One solution to ensure that the right hand side of (5.5.11) is $\geq (1 - \alpha)$ is to choose $\alpha_i = \alpha/m$. Hence, from (5.5.3), Bonferroni's simultaneous confidence intervals for $\mathbf{a}_1'\mu, \ldots, \mathbf{a}_m'\mu$ are

$$I_j : \left[\mathbf{a}_j'\bar{\mathbf{x}} - t_{n-1;\alpha/2m}\sqrt{\frac{\mathbf{a}_j'\mathbf{S}\mathbf{a}_j}{n}}, \right.$$

$$\left. \mathbf{a}_j'\bar{\mathbf{x}} + t_{n-1;\alpha/2m}\sqrt{\frac{\mathbf{a}_j'\mathbf{S}\mathbf{a}_j}{n}}\right], j = 1, \ldots, m. \tag{5.5.12}$$

In particular for $m = p, \mathbf{a}_j'\mu = \mu_j, I_j = [\bar{x}_j - t_{n-1;\alpha/2p}\frac{s_j}{\sqrt{n}}, \bar{x}_j - t_{n-1;\alpha/2p}\frac{s_j}{\sqrt{n}}] (j = 1, \ldots, p)$.

Under the assumptions that $\mathbf{x}_1, \ldots, \mathbf{x}_n$ constitute a random sample from a $N_p(\mu, \boldsymbol{\Sigma})$ population we have, therefore, three types of confidence intervals for linear functions of μ: individual confidence interval (J_1) given by (5.5.3), T^2-simultaneous confidence intervals (J_2) given by (5.5.8) and Bonferroni simultaneous confidence intervals (J_3), given by (5.5.12), which is applicable if the number of such linear functions is small. As has been noted. J_1 gives the shortest intervals, next come J_3-intervals and then J_2 intervals.

In the next section we shall find another set of simultaneous confidence intervals, which are valid only in large samples.

EXAMPLE 5.5.3: For the data in Example 5.5.2, let us compute Bonferroni simultaneous (for μ_1, μ_2) confidence interval for μ_1. Here, $m = 2$. The interval is

$$\bar{x}_1 - t_{20;.0125}\sqrt{\frac{s_{11}}{21}},\ \bar{x}_1 + t_{20;.0125}\sqrt{\frac{s_{11}}{21}} = (155.62,\ 159.14), \qquad (iv)$$

since $t_{20;.0125} = 2.4231$. It is seen from (ii), (iii) and (iv) that interval (ii) is the shortest, than comes the interval (iv) and then (iii) which is the widest.

5.6 Large Sample Inference About μ

We have seen in Corollary 3.5.1.1 that if $\mathbf{x}_1, \ldots, \mathbf{x}_n$ constitute a random sample from a population with mean μ and variance $\boldsymbol{\Sigma}$, then for large value of $(n-p)$, the statistic

$$n(\bar{\mathbf{x}} - \mu)'\mathbf{S}^{-1}(\bar{\mathbf{x}} - \mu)$$

approximately follows a $\chi^2_{(p)}$ distribution. Hence, a large sample test for testing $H_0(\mu = \mu_0)$ against $H(\mu \neq \mu_0)$ is to reject H_0 if

$$n(\bar{\mathbf{x}} - \mu_0)'\mathbf{S}^{-1}(\bar{\mathbf{x}} - \mu_0) > \chi^2_{p;\alpha},$$

where $\chi^2_{p;\alpha}$ is the upper $100(1-\alpha)\%$ point of a $\chi^2_{(p)}$ distribution.

As in the case of simultaneous T^2-confidence intervals, simultaneous large-sample confidence intervals for $\mathbf{a}'\mu$ are

$$\mathbf{a}'\bar{\mathbf{x}} \mp \sqrt{\chi^2_{p;\alpha}}\sqrt{\frac{\mathbf{a}'\mathbf{S}\mathbf{a}}{n}}. \qquad (5.6.1)$$

It will be found that the confidence intervals given by (5.5.8) and (5.6.1) are almost identical, whenever the result (5.6.1) holds. This is because, for large $n-p$, the value of $(n-1)pF_{p,n-p;\alpha}/(n-p)$ is essentially same as that of $\chi^2_{p;\alpha}$.

In addition to (5.6.1) we can obtain $(1-\alpha)$-point large-sample simultaneous confidence ellipse for (μ_i, μ_k) as follows.

$$P\{n(\bar{x}_i - \mu_i,\ \bar{x}_k - \mu_k) \begin{bmatrix} s_{ii} & s_{ik} \\ s_{ki} & s_{kk} \end{bmatrix}^{-1} \begin{bmatrix} \bar{x}_i - \mu_i \\ \bar{x}_k - \mu_k \end{bmatrix} \leq \chi^2_{p;\alpha}\} = 1 - \alpha.$$

EXAMPLE 5.6.1: For the data of Example 5.5.2, simultaneous large-sample confidence interval for μ_1 is

$$(\bar{x}_1 - \sqrt{\chi^2_{2;.05}}\sqrt{\frac{s_{11}}{21}},\ \bar{x}_1 + \sqrt{\chi^2_{2;05}}\sqrt{\frac{s_{11}}{21}}) = (155.60,\ 159.16). \qquad (v)$$

5.7 Comparing Mean Vectors of Two Populations

Let $\mathbf{X}_1 = (\mathbf{X}_{11}, \ldots, \mathbf{X}_{1n_1})'$ be a random sample of size n_1 from a $N_p(\mu_1, \mathbf{\Sigma})$-population. Similarly, let $\mathbf{X}_2 = (\mathbf{X}_{21}, \ldots, \mathbf{X}_{2n_2})'$ be a random sample of size n_2 from a $N_p(\mu_2, \mathbf{\Sigma})$-population. Also, assume that the samples $\mathbf{X}_1, \mathbf{X}_2$ are independent. We want to test $H_0(\mu_1 - \mu_2 = \delta_0)$ against $H_1 : (\mu_1 - \mu_2 \neq \delta)$. Note that we have assumed that both the populations have the same, but unknown, dispersion matrix $\mathbf{\Sigma}$. (The case when this assumption is relaxed will be considered in Subsections 5.7.2 and 5.7.3).

In this case, an unbiased estimate of $\mathbf{\Sigma}$ is

$$\begin{aligned} \mathbf{S}_p &= \tfrac{(n_1-1)\mathbf{S}_1+(n_2-1)\mathbf{S}_2}{n_1+n_2-2} \\ &= \tfrac{\sum_{i=1}^{n_1}(\mathbf{x}_{1i}-\bar{\mathbf{x}}_1)(\mathbf{x}_{1i}-\bar{\mathbf{x}}_1)'+\sum_{i=1}^{n_2}(\mathbf{x}_{2i}-\bar{\mathbf{x}}_2)(\mathbf{x}_{2i}-\bar{\mathbf{x}}_2)'}{n_1+n_2-2}. \end{aligned}$$

Now $(n_1 + n_2 - 2)\mathbf{S}_p \cap W_p(n_1 + n_2 - 2, \mathbf{\Sigma})$. Under $H_0, E(\bar{\mathbf{X}}_1 - \bar{\mathbf{X}}_2) = \delta_0, V(\bar{\mathbf{X}}_1 - \bar{\mathbf{X}}_2) = \mathbf{\Sigma}(1/n_1 + 1/n_2)$. Hence, under H_0,

$$(\bar{\mathbf{X}}_1 - \bar{\mathbf{X}}_2 - \delta_0)\sqrt{\frac{n_1 n_2}{n_1 + n_2}} \cap N_p(\mathbf{0}, \mathbf{\Sigma})$$

and is distributed independently of $\mathbf{S}_p$. Therefore, under H_0,

$$T_0^2 = \frac{n_1 n_2}{n_1 + n_2}(\bar{\mathbf{X}}_1 - \bar{\mathbf{X}}_2 - \delta_0)'\mathbf{S}_p^{-1}(\bar{\mathbf{X}}_1 - \bar{\mathbf{X}}_2 - \delta_0) \cap T^2(p, n_1 + n_2 - 2). \quad (5.7.1)$$

Therefore, H_0 is rejected if

$$T_0^2 \geq \frac{p(n_1 + n_2 - 2)}{n_1 + n_2 - p - 1} F_{p,n_1+n_2-p-1;\alpha} = c_0^2 \ \text{(say)}\ . \qquad (5.7.2)$$

The confidence ellipsoid for $\mu_1 - \mu_2$ is given by those values of $\delta = \mu_1 - \mu_2$ for which $T_0^2 \le c^2$ where we must replace δ_0 by $\delta = \mu_1 - \mu_2$ and the variable $\bar{\mathbf{X}}_1 - \bar{\mathbf{X}}_2$ by its observed value $\bar{\mathbf{x}}_1 - \bar{\mathbf{x}}_2$.

As in equation (5.5.2), for any constant vector $\mathbf{a}$,

$$\frac{\mathbf{a}'(\bar{\mathbf{x}}_1 - \bar{\mathbf{x}}_2) - \mathbf{a}'(\mu_1 - \mu_2)}{\sqrt{\mathbf{a}'\mathbf{S}_p\mathbf{a}(\frac{1}{n_1} + \frac{1}{n_2})}} \cap t_{(n_1+n_2-2)}.$$

Hence, one-at-a-time $100(1-\alpha)\%$ confidence interval for $\mathbf{a}'(\mu_1 - \mu_2)$ is

$$\mathbf{a}'(\bar{\mathbf{x}}_1 - \bar{\mathbf{x}}_2) \underset{+}{-} t_{n_1+n_2-2;\alpha/2}\sqrt{\mathbf{a}'\mathbf{S}_p\mathbf{a}(\frac{1}{n_1} + \frac{1}{n_2})}. \tag{5.7.3}$$

In particular for $\mathbf{a} = (0, \ldots, 0, 1, 0, \ldots, 0)'$, t-confidence interval for $\mu_{1i} - \mu_{2i}$ is, where $\mu_j = (\mu_{j1}, \mu_{j2}, \ldots, \mu_{jp})', (j = 1, 2)$,

$$\bar{x}_{1i} - \bar{x}_{2i} - t_{n_1+n_2-2;\alpha/2}\sqrt{s_{iip}(\frac{1}{n_1} + \frac{1}{n_2})} \tag{5.7.4}$$

where $\mathbf{S}_p = ((s_{ijp}))$. For $\mathbf{a} = (0, \ldots, 0, 1, 0, \ldots, 0, -1, 0, \ldots, 0)'$, t-interval for $(\mu_{1i} - \mu_{2i}) - (\mu_{1j} - \mu_{2j})$ is

$$(\bar{x}_{1i} - \bar{x}_{2i}) - (\bar{x}_{1j} - \bar{x}_{2j}) \underset{+}{-} t_{n_1+n_2-2;\alpha/2}\sqrt{(s_{iip} - 2s_{ijp} + s_{jjp})(\frac{1}{n_1} + \frac{1}{n_2})}. \tag{5.7.5}$$

As in Theorem 5.5.1, the simultaneous confidence intervals for $\mathbf{a}'(\mu_1 - \mu_2)$ is given by

$$[\mathbf{a}'(\bar{\mathbf{x}}_1 - \bar{\mathbf{x}}_2) - c_0\sqrt{\mathbf{a}'\mathbf{S}_p\mathbf{a}(\frac{1}{n_1} + \frac{1}{n_2})},$$

$$\mathbf{a}'(\bar{\mathbf{x}}_1 - \bar{\mathbf{x}}_2) + c_0\sqrt{\mathbf{a}'\mathbf{S}_p\mathbf{a}(\frac{1}{n_1} + \frac{1}{n_2})}] \tag{5.7.6}$$

where c_0 is given by (5.7.2).

If we are interested in finding simultaneous confidence intervals for a small number m of linear combinations $\mathbf{a}'(\mu_1 - \mu_2)$ Bonferroni- $100(1-\alpha)\%$ simultaneous confidence intervals for the population differences $\mathbf{a}_j'(\mu_1 - \mu_2)$ is given by, as in (5.5.12),

$$\mathbf{a}_j'(\bar{\mathbf{x}}_1 - \bar{\mathbf{x}}_2) \underset{+}{-} t_{n_1+n_2-2;\alpha/(2m)}\sqrt{\mathbf{a}_j'\mathbf{S}_p\mathbf{a}_j(\frac{1}{n_1} + \frac{1}{n_2})}, j = 1, \ldots, m. \tag{5.7.7}$$

In particular, Bonferroni's confidence interval for $\mu_{1i} - \mu_{2i}$ is given by,

$$\mu_{1i} - \mu_{2i} \in [(\bar{x}_{1i} - \bar{x}_{2i}) - t_{n_1+n_2-2;\alpha/(2m)}\sqrt{s_{iip}(\frac{1}{n_1} + \frac{1}{n_2})},$$

$$(\bar{x}_{1i} - \bar{x}_{2i}) + t_{n_1+n_2-2;\alpha/(2m)}\sqrt{s_{iip}(\frac{1}{n_1} + \frac{1}{n_2})}]. \tag{5.7.8}$$

Another set of confidence intervals, - large-sample simultaneous intervals, - which do not require the assumption of normality has been considered in Subsection 5.7.3.

EXAMPLE 5.7.1: Considering Table 2.E.1, suppose that the data for the survivor birds and perished birds constitute two independent random samples from populations 1 and 2 respectively and we are interested only in three variables X_1, X_2, X_3. Test $H_0(\mu_1 = \mu_2)$ against $H(\mu_2 \neq \mu_2)$. Find 95% percent confidence ellipsoid for $\mu_1 - \mu_2$. Also, find confidence intervals for (i) $\mu_{11} - \mu_{21}$ and (ii) $(\mu_{11} - \mu_{21}) - (\mu_{12} - \mu_{22})$.

Here,

$$\bar{x}_1 = 157.38,\ \bar{x}_2 = 241.00,\ \bar{x}_3 = 31.43,\ n_1 = 21,$$

$$\bar{y}_1 = 158.43,\ \bar{y}_2 = 241.57,\ \bar{y}_3 = 31.48,\ n_2 = 28,$$

$$\mathbf{S}_1 = \begin{bmatrix} 11.0476 & 9.1000 & 1.5567 \\ 9.1000 & 17.5000 & 1.9100 \\ 1.5567 & 1.9100 & .5313 \end{bmatrix},$$

$$\mathbf{S}_2 = \begin{bmatrix} 15.0689 & 17.1905 & 2.2429 \\ 17.1905 & 32.5503 & 3.3979 \\ 2.2429 & 3.3979 & .7284 \end{bmatrix},$$

$$\mathbf{S}_p = \begin{bmatrix} 13.3577 & 13.7477 & 1.9509 \\ 13.7477 & 26.7204 & 2.7647 \\ 1.9509 & 2.7647 & .6445 \end{bmatrix},$$

$$\mathbf{S}_p^{-1} = \begin{bmatrix} .1816 & -.0658 & -.2678 \\ -.0658 & .0911 & -.1917 \\ -.2678 & -.1917 & 3.1847 \end{bmatrix}.$$

Eigenvalues and eigenvectors of $\mathbf{S}_p$ are

$$\lambda_1 = 35.6506,\ \lambda_2 = 4.7616,\ \lambda_3 = 0.3106,$$

$$\begin{aligned} \mathbf{e}_1' &= (-.5286,\ -.8434,\ -.0961), \\ \mathbf{e}_2' &= (.8444,\ -.5339,\ .0416), \\ \mathbf{e}_3' &= (.0864,\ .0591,\ -.9945). \end{aligned}$$

By formulae (5.7.1) and (5.7.2),

$$c_0^2 = \frac{(3)(47)}{45} F_{3,45;.05} = 8.8987, \ c = 2.98276,$$

$$T_0^2 = \frac{(21)(28)}{49} (\bar{\mathbf{x}} - \bar{\mathbf{y}})' \mathbf{S}_p^{-1} (\bar{\mathbf{x}} - \bar{\mathbf{y}}) = 1.46365 < c_0^2.$$

Hence, $H_0(\mu_1 = \mu_2)$ is accepted at .05-point level of significance.

The confidence ellipsoid for $(\mu_1 - \mu_2)$ is given by

$$\{\mu_1 - \mu_2 : [(\mu_1 - \mu_2) - (\bar{\mathbf{x}} - \bar{\mathbf{y}})]' \mathbf{S}_p^{-1} [(\mu_1 - \mu_2) - (\bar{\mathbf{x}} - \bar{\mathbf{y}})]\}$$

$$\leq c_0^2 \frac{n_1 + n_2}{n_1 n_2} = 0.74156\}. \qquad (vi)$$

The center of this ellipsoid is at $(\bar{\mathbf{x}} - \bar{\mathbf{y}})$ and the lengths of the semiaxes are

$$\sqrt{\lambda}_1 \sqrt{c_0^2 \frac{n_1 + n_2}{n_1 n_2}} = 5.1417, 1.8791 \text{ and } 0.4799$$

units from the center. The direction cosines of the axes are given by $\mathbf{e}_1, \mathbf{e}_2, \mathbf{e}_3$ respectively, when these vectors are plotted with $\bar{\mathbf{x}} - \bar{\mathbf{y}}$ as the origin.

One-at-a-time .05-point confidence interval for $\mu_{11} - \mu_{21}$ is given by, vide formula (5.7.4),

$$\bar{x}_1 - \bar{y}_1 \mp t_{47;.025} \sqrt{s_{11} (\frac{1}{n_1} + \frac{1}{n_2})} = (-3.1823, \ 1.0823). \qquad (vii)$$

Similarly, the t-confidence interval for $(\mu_{11} - \mu_{21}) - (\mu_{12} - \mu_{22})$ is given by formula (5.7.5) as

$$(\bar{x}_1 - \bar{y}_1) - (\bar{x}_2 - \bar{y}_2) \mp t_{47;.05} \sqrt{(s_{11} - 2s_{12} + s_{22})(\frac{1}{n_1} + \frac{1}{n_2})}$$

$$= (-2.5495, \ 1.5894). \qquad (viii)$$

The simultaneous T^2 confidence interval for $\mu_{11} - \mu_{21}$ is given by formula (5.7.6) as

$$\bar{x}_1 - \bar{y}_1 \mp c_0 \sqrt{s_{11} (\frac{1}{n_1} + \frac{1}{n_2})} = (-4.1970, \ 2.0970) \qquad (ix)$$

and for $(\mu_{11} - \mu_{21}) - (\mu_{12} - \mu_{22})$ by

$$(\bar{x}_1 - \bar{y}_1) - (\bar{x}_2 - \bar{y}_2) \mp c_0 \sqrt{(s_{11} - 2s_{12} + s_{22})(\frac{1}{n_1} + \frac{1}{n_2})}$$

$$= (-3.5343,\ 2.5743). \tag{x}$$

Suppose Bonferroni simultaneous confidence intervals cover only 4 linear functions $\mathbf{a}_j'(\mu_1 - \mu_2)$, that is $m = 4$. Under this system, 95% confidence interval for $(\mu_{11} - \mu_{21})$ is

$$\bar{x}_{11} - \bar{x}_{21} \mp t_{47,.00625}\sqrt{s_{11p}(\frac{1}{n_1} + \frac{1}{n_2})}$$

$$= (-3.75,\ 1.69) \tag{xi}$$

and similarly, for $(\mu_{11} - \mu_{21}) - (\mu_{12} - \mu_{22})$ is

$$(-3.34,\ 2.38). \tag{xii}$$

The comments made earlier about the lengths of the three types of confidence intervals in the single population problem also apply here.

5.7.1 *Union-intersection method*

In the union-intersection method we work with the hypothesis $H_{0a}(\mathbf{a}'\mu_1 = \mathbf{a}'\mu_2)$ and reject $H_0(\mu_1 = \mu_2)$ if $\max_{\mathbf{a}} t^2(\mathbf{a}) \geq c$, where

$$t^2(\mathbf{a}) = \frac{[\mathbf{a}'(\bar{\mathbf{x}}_1 - \bar{\mathbf{x}}_2)]^2}{[\frac{n_1+n_2}{n_1 n_2}]\mathbf{a}'\mathbf{S}_p\mathbf{a}}. \tag{5.7.9}$$

The maximum value of $t^2(\mathbf{a})$ is given by

$$\max_{\mathbf{a}} t^2(\mathbf{a}) = T^2 = \frac{n_1 n_2}{n_1 + n_2}(\bar{\mathbf{x}}_1 - \bar{\mathbf{x}}_2)\mathbf{S}_p^{-1}(\bar{\mathbf{x}}_1 - \bar{\mathbf{x}}_2). \tag{5.7.10}$$

The proof follows as in the proof of the hypothesis $H(\mu = \mu_0)$ in (5.3.6) or in Section 5.5. The vector $\mathbf{a}$ that maximizes $t^2(\mathbf{a})$ in (5.7.9) is given by

$$\mathbf{a} \propto \mathbf{S}_p^{-1}(\bar{\mathbf{x}}_1 - \bar{\mathbf{x}}_2). \tag{5.7.11}$$

The linear function $z = \mathbf{a}'\mathbf{x}$ with coefficient vector $\mathbf{a} = \mathbf{S}_p^{-1}(\bar{\mathbf{x}}_1 - \bar{\mathbf{x}}_2)$ which maximally separates $\mathbf{a}'\bar{\mathbf{x}}_1$ from $\mathbf{a}'\bar{\mathbf{x}}_2$ is the discriminant function between two normal populations $N_p(\mu_1, \mathbf{\Sigma}), N_p(\mu_2, \mathbf{\Sigma})$. The discriminant functions have been considered in Chapter 11.

5.7.2 *Behrens-Fisher problem*

We now consider the case, $\mathbf{\Sigma}_1 \neq \mathbf{\Sigma}_2$.

When $\mathbf{\Sigma}_1 \neq \mathbf{\Sigma}_2, \sum_{i=1}^{n_1}(\mathbf{X}_{1i} - \bar{X}_1)(\mathbf{X}_{1i} - \bar{\mathbf{X}}_1)' + \sum_{i=1}^{n_2}(\mathbf{X}_{2i} - \bar{X}_2)(\mathbf{X}_{2i} - \bar{\mathbf{X}}_2)'$ does not have a Wishart distribution with covariance matrix given by a multiple of $(\mathbf{\Sigma}_1/n_1 + \mathbf{\Sigma}_2/n_2)$. Hence the above technique does not apply. When $n_1 \neq n_2$, the corresponding problem in the univariate case of unequal variances is known as the Behrens-Fisher problem and various solutions have been proposed. One solution, proposed by Scheffe' (1943) has been generalized to the multivariate case by Bennett (1951) and extended to more than two populations by Anderson (1963 a). Eaton (1969) gave a more general solution which includes al previous solutions as special cases. His solution has been considered in Section 5.12.

Assume $n_1 < n_2$. Define

$$\mathbf{Y}_i = \mathbf{X}_{1i} - \sqrt{\frac{n_1}{n_2}}\mathbf{X}_{2i} + \frac{1}{\sqrt{n_1 n_2}}\sum_{j=1}^{n_1}\mathbf{X}_{2j} - \frac{1}{n_2}\sum_{k=1}^{n_2}\mathbf{X}_{2k},\ i = 1, \ldots, n_1. \tag{5.7.12}$$

Then, $E(\mathbf{Y}_i) = \mu_1 - \mu_2$. Also

$$E(\mathbf{Y}_i - E(\mathbf{Y}_i))(\mathbf{Y}_t - E(\mathbf{Y}_t))'$$

$$= E[(\mathbf{X}_{1i} - \mu_i) - \sqrt{\frac{n_1}{n_2}}(\mathbf{X}_{2i} - \mu_2) + \frac{1}{\sqrt{n_1 n_2}}\sum_{j=1}^{n_1}(\mathbf{X}_{2j} - \mu_2) - \frac{1}{n_2}\sum_{k=1}^{n_2}(\mathbf{X}_{2k} - \mu_2)]$$

$$[(\mathbf{X}_{1t} - \mu_1) - \sqrt{\frac{n_1}{n_2}}(\mathbf{X}_{2t} - \mu_2) + \frac{1}{\sqrt{n_1 n_2}}\sum_{j=1}^{n_1}(\mathbf{X}_{2j} - \mu_2) - \frac{1}{n_2}\sum_{k=1}^{n_2}(\mathbf{X}_{2k} - \mu_2)]'$$

$$= \delta_{it}\mathbf{\Sigma}_1 + \delta_{it}\frac{n_1}{n_2}\mathbf{\Sigma}_2 + \mathbf{\Sigma}_2(-\frac{2}{n_2} + \frac{2}{n_2}\sqrt{\frac{n_1}{n_2}} + \frac{n_1}{n_1 n_2} - 2\frac{n_1}{\sqrt{n_1 n_2} n_2} + \frac{n_2}{n_2^2})$$

$$= \delta_{it}(\mathbf{\Sigma}_1/n_1 + \mathbf{\Sigma}_2/n_2)n_1 \tag{5.7.13}$$

where $\delta_{it} = 1(0)$ if $i = (\neq)t$. Therefore, $\bar{\mathbf{Y}}$'s is normal with mean $\mu_1 - \mu_2$ and covariance matrix $(\mathbf{\Sigma}/n_1 + \mathbf{\Sigma}/n_2)$. Hence, under H_0,

$$T_0^2 = n_1\bar{\mathbf{Y}}'\mathbf{S}_y^{-1}\bar{\mathbf{Y}} \tag{5.7.14}$$

where $(n_1 - 1)\mathbf{S}_y = \sum_{i=1}^{n_1}(\mathbf{Y}_i - \bar{\mathbf{Y}})(\mathbf{Y}_i - \bar{\mathbf{Y}})'$, follows T^2 with $(n_1 - 1)$ d.f.

Note that if $n_1 = n_2$, then T^2 can be used. In this case, under $H_0, \bar{\mathbf{Y}} \cap N_p(\mathbf{0}, (\mathbf{\Sigma}_1 + \mathbf{\Sigma}_2)/n)$. Also,

$$(n-1)\mathbf{S}_y = \sum_{i=1}^{n}(\mathbf{Y}_i - \bar{\mathbf{Y}})(\mathbf{Y}_i - \bar{\mathbf{Y}})' \cap W_p(n-1, \mathbf{\Sigma}_1 + \mathbf{\Sigma}_2). \qquad (5.7.15)$$

Therefore, $n\bar{\mathbf{Y}}'\mathbf{S}_y^{-1}\bar{\mathbf{Y}}$ follows T^2 with $(n-1)$ d.f. under H_0.

Bartlett's (1938) test (vide Subsection 5.11.3) is used to test the equality of $\mathbf{\Sigma}_1$ and $\mathbf{\Sigma}_2$ in terms of the generalized variances. This test is, however, very sensitive to the assumption of normality. Tiku and Balakrishnan (1985) have proposed a test which is less sensitive to the assumption of normality.

5.7.3 *A large-sample test*

If sample sizes are large such that $n_1 - p, n_2 - p$ are large, then a large sample test exists for $H_0(\mu_1 - \mu_2 = \delta_0)$. We assume that $\mathbf{X}_{i1}, \ldots, \mathbf{X}_{in_i} (i = 1, 2)$ constitute a random sample from an infinite population with mean μ_i and variance $\mathbf{\Sigma}_i (i = 1, 2)$. Also, assume that the sample $(\mathbf{X}_{11}, \ldots, \mathbf{X}_{1n_1})$ are independent of the sample $(\mathbf{X}_{21}, \ldots, \mathbf{X}_{2n_2})$. Then it is known by the Central Limit theorem that $\bar{\mathbf{X}}_1 - \bar{\mathbf{X}}_2 \sim N_p(\mu_1 - \mu_2, \frac{\mathbf{\Sigma}_1}{n_1} + \frac{\mathbf{\Sigma}_2}{n_2})$ approximately. When n_1 and n_2 are large, $\mathbf{\Sigma}_i$ will be close to $\mathbf{S}_i$ with high probability. Hence,

$$(\bar{\mathbf{X}}_1 - \bar{\mathbf{X}}_2 - (\mu_1 - \mu_2))'(\frac{\mathbf{S}_1}{n_1} + \frac{\mathbf{S}_2}{n_2})^{-1}(\bar{\mathbf{X}}_1 - \bar{\mathbf{X}}_2 - (\mu_1 - \mu_2)) \sim \chi^2_{(p)} \qquad (5.7.16)$$

approximately. A large sample test for $H_0(\mu_1 - \mu_2 = \delta_0)$ against $H(\mu_1 - \mu_2 \neq \delta_o)$ is to reject H_0 if

$$(\bar{\mathbf{x}}_1 - \bar{\mathbf{x}}_2 - \delta_0)'(\frac{\mathbf{S}_1}{n_1} + \frac{\mathbf{S}_2}{n_2})^{-1}(\bar{\mathbf{x}}_1 - \bar{\mathbf{x}}_2 - \delta_0) > \chi^2_{p;\alpha}. \qquad (5.7.17)$$

The simultaneous confidence intervals for $\mathbf{a}'(\mu_1 - \mu_2)$ is given by

$$[\mathbf{a}'(\bar{\mathbf{x}}_1 - \bar{\mathbf{x}}_2) - \sqrt{\mathbf{a}'(\frac{\mathbf{S}_1}{n_1} + \frac{\mathbf{S}_2}{n_2})\mathbf{a}\chi^2_{p;\alpha}},$$

$$\mathbf{a}'(\bar{\mathbf{x}}_1 - \bar{\mathbf{x}}_2) + \sqrt{\mathbf{a}'(\frac{\mathbf{S}_1}{n_1} + \frac{\mathbf{S}_2}{n_2})\mathbf{a}\chi^2_{p;\alpha}}]. \qquad (5.7.18)$$

EXAMPLE 5.7.2 : For the data in Example 5.7.1, large-sample simultaneous 95% percent confidence interval for $\mu_{11} - \mu_{21}$ is

$$(\bar{x}_1 - \bar{y}_1) \mp \sqrt{(\frac{s_{11}^{(1)}}{n_1} + \frac{s_{11}^{(2)}}{n_2})\chi^2_{3;.05}}$$

$$= (-3.93,\ 1.83). \qquad (xiii)$$

where $((s_{ij}^{(k)})) = \mathbf{S}^{(k)}, k = 1, 2$. Similarly, large-sample confidence interval for $(\mu_{11} - \mu_{21}) - (\mu_{12} - \mu_{22})$ is

$$(\bar{x}_1 - \bar{y}_1) - (\bar{x}_2 - \bar{y}_2) \mp \sqrt{\left(\frac{s_{11}^{(1)} - 2s_{12}^{(1)} + s_{22}^{(1)}}{n_1} + \frac{s_{11}^{(2)} - 2s_{12}^{(2)} + s_{22}^{(2)}}{n_2}\right)\chi^2_{3;.05}}$$

$$= (-3.23,\ 2.27). \qquad (xiv)$$

5.7.4 *Paired comparison*

Experiments are often conducted on the same set of units under two different sets of conditions. One of the problems one may be interested in such situations is to see if there is a significant change in average response due to change in experimental conditions. Thus, measurements may be taken on n subjects, before ($\mathbf{X}_1$) and after ($\mathbf{X}_2$) subjecting them to a certain treatment (diet). It is required to test if $\mu_1 - \mu_2 = \delta_0$.

It is obvious that $\mathbf{X}_1$ and $\mathbf{X}_2$ are not independent in this case and hence the test in (5.7.2) does not apply. Instead we make a paired T^2 test as follows.

Assume that $\mathbf{D}_i = \mathbf{X}_{1i} - \mathbf{X}_{2i}$, $(i = 1, \ldots, n)$ is a random sample from a p-variate normal population with mean δ and variance $\mathbf{\Sigma}_\mathbf{d}$. Under $H_0, \delta = \delta_0$.

The sample covariance matrix calculated from $\mathbf{D}_1, \ldots, \mathbf{D}_n$ is

$$\mathbf{S}_\mathbf{d} = \frac{1}{n-1}\sum_{i=1}^{n}(\mathbf{D}_i - \bar{\mathbf{D}})(\mathbf{D}_i - \bar{\mathbf{D}})'$$

where $\bar{\mathbf{D}} = \sum_{i=1}^{n} \mathbf{D}_i/n$. Under $H_0, \sqrt{n}\bar{\mathbf{D}} \cap N_p(\delta_0, \mathbf{\Sigma}_\mathbf{d})$. Also, $(n-1)\mathbf{S}_\mathbf{d} \cap W_p(n-1, \mathbf{\Sigma}_\mathbf{d})$ and $\bar{\mathbf{D}}$ and $\mathbf{S}_\mathbf{d}$ are independently distributed. Therefore, under H_0,

$$T^2 = n(\bar{\mathbf{D}} - \delta_0)'\mathbf{S}_\mathbf{d}^{-1}(\bar{\mathbf{D}} - \delta_0) \qquad (5.7.19)$$

is distributed as Hotelling's $T^2(p, n-1)$ distribution. A test for $H_0(\mu_1 - \mu_2 = \delta_0)$ against $H(\mu_1 - \mu_2 \neq \delta_0)$ is to reject H_0 if

$$T_0^2 = n(\bar{\mathbf{d}} - \delta_0)'\mathbf{S}_\mathbf{d}^{-1}(\bar{\mathbf{d}} - \delta_0) \geq \frac{(n-1)p}{n-p}F_{p,n-p;\alpha}. \qquad (5.7.20)$$

A $100(1-\alpha)\%$ confidence region for $\delta = \mu_1 - \mu_2$ is given by those value of δ for which

$$n(\bar{\mathbf{d}} - \delta)'\mathbf{S}_\mathbf{d}^{-1}(\bar{\mathbf{d}} - \delta) \leq \frac{(n-1)p}{n-p}F_{p,n-p;\alpha}. \qquad (5.7.21)$$

Also, $100(1-\alpha)\%$ simultaneous confidence interval for the linear function of components of δ is given by

$$\left[\mathbf{a}'\bar{\mathbf{d}} - \sqrt{\frac{\mathbf{a}'\mathbf{S_d}\mathbf{a}}{n}}\sqrt{\frac{(n-1)p}{n-p}F_{p,n-p;\alpha}},\right.$$

$$\left.\mathbf{a}'\bar{\mathbf{d}} + \sqrt{\frac{\mathbf{a}'\mathbf{S_d}\mathbf{a}}{n}}\sqrt{\frac{(n-1)p}{n-p}F_{p,n-p;\alpha}}\right]. \qquad (5.7.22)$$

The Bonferroni $100(1-\alpha)\%$ simultaneous confidence interval for the p individual mean differences are

$$\delta_i \in \mu_{1i} - \mu_{2i} \in [\bar{d}_i - t_{n-1;\alpha/2p}\sqrt{\frac{s^2_{d_i}}{n}}, \bar{d}_i + t_{n-1;\alpha/2p}\sqrt{\frac{s^2_{d_i}}{n}}]. \qquad (5.7.23)$$

If $(n-p)$ is large T^2 follows approximately a $\chi^2_{(p)}$ distribution and normality need not be assumed. In this case, large sample test for $H_0(\delta = \delta_0)$, confidence region for δ and simultaneous confidence intervals for $\mathbf{a}'\delta$ can be developed similarly as in Subsection 5.7.3.

Example 5.7.3: Consider the blood-glucose data in Table 11.E.3 where x_1, x_2, x_3 give measurements of blood-glucose level of 52 workers at three occasions on fasting and y_1, y_2, y_3 the same one hour after sugar-intake. Test the null hypothesis $H_0(\mu_1 - \mu_2 = \delta_0 = (-30,\ -30,\ -30)'$ against the alternative hypothesis $H(\mu_1 - \mu_2 \neq \delta_0)$, where μ_1 is the mean of the population from which $\mathbf{x}_i (i = 1, \ldots, 52)$ is a sample. Obtain 95% confidence ellipsoid for $\delta = (\mu_1 - \mu_2)$. Derive simultaneous T^2 confidence intervals and Bonferroni's simultaneous confidence interval with $m = 3$ for $\delta_1 = (\mu_{11} - \mu_{21})$.

We make the usual assumptions of a paired T^2 test. Let $d_i = x_i - y_i, i = 1, \ldots, 52$. Here

$$\bar{d}_1 = -38.96,\ \bar{d}_2 = -30.54,\ \bar{d}_3 = -34.69,$$

$$\mathbf{S}_d = \begin{bmatrix} 763.449 & 280.825 & 121.243 \\ 280.825 & 432.449 & 103.483 \\ 121.243 & 103.483 & 466.139 \end{bmatrix},$$

$$\mathbf{S}_d^{-1} = \begin{bmatrix} .0017407 & -.0010794 & -.002131 \\ -.0010794 & .0031115 & -.0004100 \\ -.0002131 & -.0004100 & .0022917 \end{bmatrix}.$$

The eigenvalues and eigenvectors of $\mathbf{S}_d$ are

$$\lambda_1 = 972.474,\ \lambda_2 = 423.014,\ \lambda_3 = 266.50,$$

$$\begin{aligned}\mathbf{e}_1 &= (.823006,\ .484730,\ .296138),\\ \mathbf{e}_2 &= (-.346164,\ .014676,\ .938060),\\ \mathbf{e}_3 &= (.450368,\ -.874451,\ .179850)'.\end{aligned}$$

Now $T^2 = 52(\bar{\mathbf{d}} - \delta_0)'\mathbf{S}_d^1(\bar{\mathbf{d}} - \delta_0) = 8.35281$. Also,

$$\frac{(n-1)p}{(n-p)}F_{p,n-p;.05} = \frac{(51)(3)}{49}F_{3,49;.05} = 8.74286.$$

Hence, H_0 is accepted. Confidence ellipsoid for δ is

$$(\delta - \bar{\mathbf{d}})'\mathbf{S}_d^{-1}(\delta - \bar{\mathbf{d}}) \leq \frac{(n-1)p}{n(n-p)}F_{p,n-p;.05} = .168132 = c_0^2 \ \text{(say)}\ .$$

The length of its semi-axes are

$$\sqrt{\lambda_i}c_0 = 12.787,\ 8.433,\ 6.6938$$

and its direction cosines are given by $\mathbf{e}_1, \mathbf{e}_2, \mathbf{e}_3$, respectively when plotted with $\bar{\mathbf{d}}$ as the origin.

The T^2-simultaneous confidence interval for $\delta_1 = (\mu_{11} - \mu_{12})$ is

$$\bar{d}_1 \underset{+}{-} c_0 s_{d_1} = \sqrt{.168132 \times 763.449} = -38.96 \underset{+}{-} 11.33 = (-50.29,\ -27.63).$$

For $m = 3$, Bonferroni's simultaneous confidence interval for δ_1 is

$$\bar{d}_1 \underset{+}{-} t_{51;.025/6}\sqrt{\frac{s_{d_1}^2}{n}} = -38.76 \underset{+}{-} 10.55 = (-49.51,\ -28.41),$$

since $t_{51,.00417} = 2.7539$.

5.7.5 *Testing that all components of μ are equal*

Suppose in a clinical trial we wish to test $(p-1)$ different drugs and along with the placebo we apply these p drugs at different points of time at random order to each of n patients. Denoting the measurements on the patients as $\mathbf{x}_1, \ldots, \mathbf{x}_n$ (where each $\mathbf{x}_i$ is a $p \times 1$ vector), we want to test the null hypothesis that all the drugs have the same effect. Assume that each $\mathbf{x}_i$ is a sample from $N_p(\mu, \Sigma)$, this is equivalent to testing $H_0(\mu_1 = \mu_2 = \ldots = \mu_p)$. The hypothesis may be written as

$$H_{01} : \mu_1 - \mu_2 = \mu_2 - \mu_3 = \ldots = \mu_{p-1} - \mu_p = 0,$$

or

$$H_{02} = \mu_1 - \mu_p = \mu_2 - \mu_p = \ldots = \mu_{p-1} - \mu_p = 0. \tag{5.7.24}$$

The former version may be written as

$$\mathbf{C}_1\mu = \begin{bmatrix} 1 & -1 & 0 & \ldots & 0 & 0 \\ 0 & 1 & -1 & \ldots & 0 & 0 \\ . & . & . & \ldots & . & . \\ 0 & 0 & 0 & \ldots & 1 & -1 \end{bmatrix} \mu = \mathbf{0} \tag{5.7.25}$$

while the latter version as

$$\mathbf{C}_2\mu = \begin{bmatrix} 1 & 0 & \ldots & 0 & -1 \\ 0 & 1 & \ldots & 0 & -1 \\ 0 & 0 & \ldots & 1 & -1 \end{bmatrix} \mu = \mathbf{0}. \tag{5.7.26}$$

Note that each row of $\mathbf{C}_1\mu$ or $\mathbf{C}_2\mu$ is a contrast among μ_i's, because this is of the form $\sum_i c_i\mu_i$ with $\sum_i c_i = 0$. Also, all the rows of $\mathbf{C}_1$ (and $\mathbf{C}_2$) are linearly independent, rank $(\mathbf{C}_i) = p - 1 (i = 1, 2)$. Such a matrix is called a *contrast matrix* and has the property $\mathbf{C1} = \mathbf{0}$. As is obvious, there are many such choices of $\mathbf{C}$.

The above example is a special case of the *repeated measurement design* in which p similar measurements are made on the same sampling unit.

Under $H_0, \mathbf{C}\mu = \mathbf{0}$. We will, therefore, test the hypothesis on the basis of the observations $\mathbf{Cx}_1, \ldots, \mathbf{Cx}_n$ of the transformed variable $\mathbf{Cx}$. Clearly, $E(\mathbf{C\bar{X}}) = \mathbf{C}\mu = \mathbf{0}, V(\mathbf{C\bar{X}}) = \mathbf{C\Sigma C}'/n$. Also, $\mathbf{C\bar{X}} \cap N_{p-1}(\mathbf{0}, \mathbf{C\Sigma C}'/n)$ under H_0. Again $(n-1)\mathbf{CSC}' \cap W_{p-1}(n-1, \mathbf{C\Sigma C}')$ and is independent of $\mathbf{C\bar{X}}$. Hence a test for $H_0(\mu_1 = \ldots = \mu_p)$ against $H(\mu_i$'s are not all equal) is to reject H_0 if

$$T^2 = n(\mathbf{C\bar{x}})'(\mathbf{CSC}')^{-1}(\mathbf{C\bar{x}}) > \frac{(n-1)(p-1)}{n-p+1} F_{p-1,n-p+1;\alpha}. \tag{5.7.27}$$

It can be shown that T^2 does not depend on the particular form of $\mathbf{C}$. A confidence region for $\mathbf{C}\mu$ is the set of all μ's such that

$$n(\mathbf{C\bar{x}} - \mathbf{C}\mu)'(\mathbf{CSC}')^{-1}(\mathbf{C\bar{x}} - \mathbf{C}\mu) \leq \frac{(n-1)(p-1)}{n-p+1} F_{p-1,n-p+1;\alpha}. \tag{5.7.28}$$

Simultaneous confidence intervals for linear functions of components of $\mathbf{C}\mu, \mathbf{a}'\mathbf{C}\mu$ are

$$\mathbf{a}'\mathbf{C\bar{x}} \mp \sqrt{\frac{\mathbf{a}'(\mathbf{CSC}')\mathbf{a}}{n}} \sqrt{\frac{(n-1)(p-1)}{n-p+1} F_{p-1,n-p+1;\alpha}}. \tag{5.7.29}$$

Bonferroni simultaneous confidence intervals for the individual contrast $\mathbf{c}'\mu$, where $\mathbf{c}'$ is a row vector of $\mathbf{C}$ is

$$\mathbf{c}'\bar{\mathbf{x}} \mp \sqrt{\frac{\mathbf{c}'\mathbf{S}\mathbf{c}}{n}} t_{n-1;\alpha/2(p-1)}. \tag{5.7.30}$$

EXAMPLE 5.7.4: Consider the blood-glucose data of 52 workers at three occasions on fasting, x_1, x_2, x_3 as provided in Table 11.E.3. Assuming that the data come from a 3-variate normal population we want to test the null hypothesis $H_0(\mu_1 = \mu_2 = \mu_3)$.

Here

$$\bar{x}_1 = 70.12,\ \bar{x}_2 = 73.62,\ \bar{x}_3 = 75.15,$$

$$\mathbf{S} = \begin{bmatrix} 93.7511 & 19.5943 & 13.0995 \\ 19.5943 & 69.3002 & 14.3152 \\ 13.0995 & 14.3152 & 73.3484 \end{bmatrix},\ \mathbf{CSC}' = \begin{bmatrix} 123.863 & 75.372 \\ 75.372 & 140.900 \end{bmatrix},$$

$$(\mathbf{CSC}')^{-1} = \begin{bmatrix} .01197 & -.00640 \\ -.00640 & .01052 \end{bmatrix},\quad \text{where } \mathbf{C} = \begin{bmatrix} 1 & -1 & 0 \\ 1 & 0 & -1 \end{bmatrix}.$$

The value of T^2 as given by formula (5.7.27) is 9.74506. Also, $\{(51)(2)/(50)\}F_{2,50;.05} = 6.5076$. Hence, H_0 is rejected at .05-point level of significance.

The 95% simultaneous confidence interval for $\mu_1 - \mu_2$ is, by formula (5.7.29),

$$(\bar{x}_1 - \bar{x}_2) \mp \sqrt{\frac{s_{11} - 2s_{12} + s_{22}}{n} \cdot \frac{(n-1)(p-1)}{n-p+1} F_{p-1,n-p+1;.05}}$$

$$= (-6.54,\ -0.46).$$

EXAMPLE 5.7.5: *Randomized Block Design (RBD) with Correlated Observations*

Consider a RBD with N blocks and p treatments. Let y_{ij} be the yield of the ith treatment in the jth block ($i = 1, \ldots, p; j = 1, \ldots, N$). Assume that

$$E(Y_{ij}) = \mu + \tau_i + \beta_j,$$

$$\operatorname{cov}(Y_{ij}, Y_{i'j'}) = \begin{cases} \sigma_{ii'} & \text{if } j = j' \\ 0, & \text{otherwise} \end{cases} \tag{5.7.31}$$

where τ_i is the effect of the ith treatment and β_j is the effect of the jth block. Let

$$\mathbf{Y}_j' = (Y_{1j}, Y_{2j}, \ldots, Y_{pj}), i = 1, \ldots, p; j = 1, \ldots, N.$$

Then $\text{Cov}(\mathbf{Y}_j) = \mathbf{\Sigma}$. Assume that $\mathbf{Y}_j$'s are independently distributed as p-variate normal variables. Now

$$(\mathbf{C}_1\mathbf{Y}_j)' = (Y_{1j} - Y_{2j}, Y_{2j} - Y_{3j}, \ldots, Y_{p-1j} - Y_{pj}) = \mathbf{z}_j' \text{ (say)}.$$

Let $\mathbf{Z}(N \times (p-1)) = (\mathbf{z}_1, \mathbf{z}_2, \ldots, \mathbf{z}_N)'$. Then

$$E(\mathbf{z}_j) = E(\mathbf{C}_1\mathbf{Y}_j) = \begin{bmatrix} \mu_1 - \mu_2 \\ \mu_2 - \mu_3 \\ \cdot \\ \cdot \\ \mu_{p-1} - \mu_p \end{bmatrix}$$

where $\mu_i = \mu + \tau_i (i = 1, \ldots, p)$. Under H_0, the vectors $\mathbf{z}_1, \mathbf{z}_2, \ldots, \mathbf{z}_N$ are independently and identically distributed as $N_{p-1}(\mathbf{0}, \mathbf{C}_1\mathbf{\Sigma}\mathbf{C}_1')$. Therefore, by Theorem 4.2.1(c), under $H_0, (N-1)\mathbf{S} \cap \mathbf{W}_{p-1}(N-1, \mathbf{C}_1\mathbf{\Sigma}\mathbf{C}_1')$ where

$$\begin{aligned} \mathbf{S} &= \textstyle\sum_{j=1}^N (\mathbf{z}_j - \bar{\mathbf{z}})(\mathbf{z}_j - \bar{\mathbf{z}})'/(N-1) \\ &= \textstyle\sum_{j=1}^N (\mathbf{C}_1\mathbf{y}_j - \mathbf{C}_1\bar{\mathbf{y}})(\mathbf{C}_1\mathbf{y}_j - \mathbf{C}_1\bar{\mathbf{y}})'/(N-1) \\ &= \mathbf{C}_1\mathbf{S}_y\mathbf{C}_1' \end{aligned} \tag{5.7.32}$$

and $\bar{\mathbf{y}} = \sum_{j=1}^N \mathbf{y}_j/N$. Also, under $H_0, \sqrt{N}\mathbf{C}_1\bar{\mathbf{y}} \cap N_{p-1}(\mathbf{0}, \mathbf{C}_1\mathbf{\Sigma}\mathbf{C}_1')$ and is distributed independently of $\mathbf{S}$. Therefore, under H_0, the statistic

$$T_0^2 = N(\mathbf{C}_1\bar{\mathbf{y}})'\mathbf{S}^{-1}(\mathbf{C}_1\bar{\mathbf{y}}) = N(\mathbf{C}_1\bar{\mathbf{y}})'(\mathbf{C}_1\mathbf{S}_y\mathbf{C}_1)^{-1}(\mathbf{C}_1\bar{\mathbf{y}}) \tag{5.7.33}$$

follows Hotelling's T^2 distribution with $(p-1, N-1)$ d.f. Under the alternative hypothesis T_0^2 follows Hotelling's T^2 distribution with non-centrality parameter

$$\delta = N(\mu_1 - \mu_2, \mu_2 - \mu_3, \ldots, \mu_{p-1} - \mu_p)'(\mathbf{C}_1\mathbf{\Sigma}\mathbf{C}_1')^{-1}$$

$$(\mu_1 - \mu_2, \mu_2 - \mu_3, \ldots, \mu_{p-1} - \mu_p). \tag{5.7.34}$$

5.7.6 *Testing that two subvectors have equal means*

Suppose that $\mathbf{X}_i$ is a $2p \times 1$ vector in which the first p elements represent measurements on the left side of the ith person and the second p elements represent the same measurements on the right side. In another example, the elements may be measurements of the twins. A natural problem is to test $H_0 : \mu_i = \mu_{i+p}$ $(i = 1, \ldots, p)$, which represents the equality of the measurements on the left and right sides of an individual.

The null hypothesis can be expressed in the form $H_0 : \mathbf{A}\mu = \mathbf{0}$, where $\mathbf{A} = [\mathbf{I}_p, \ -\mathbf{I}_p]$. If we partition in a similar manner,

$$\bar{\mathbf{x}} = \begin{bmatrix} \bar{\mathbf{x}}_1 \\ \bar{\mathbf{x}}_2 \end{bmatrix}, \mathbf{S} = \begin{bmatrix} \mathbf{S}_{11} & \mathbf{S}_{12} \\ \mathbf{S}_{21} & \mathbf{S}_{22} \end{bmatrix}, \tag{5.7.35}$$

where $\mathbf{S} = (n-1)^{-1} \sum (\mathbf{x}_i - \bar{\mathbf{x}})(\mathbf{x}_i - \bar{\mathbf{x}})'$. Thus, $\mathbf{A}\bar{\mathbf{X}} \cap N_p(\mathbf{0}, \mathbf{A}\Sigma\mathbf{A}'/n)$. Also, $\mathbf{ASA}' \cap W_p(n-1, \mathbf{A}\Sigma\mathbf{A}')$. Hence, under H_0,

$$n(\mathbf{A}\bar{\mathbf{x}})'(\mathbf{ASA}')^{-1}(\mathbf{A}\bar{x})$$

$$= n(\bar{\mathbf{x}}_1 - \bar{\mathbf{x}}_2)'(\mathbf{S}_{11} - \mathbf{S}_{12} - \mathbf{S}_{21} + \mathbf{S}_{22})^{-1}(\bar{\mathbf{x}}_1 - \bar{\mathbf{x}}_2) \ \cap T^2(p, n-1). \tag{5.7.36}$$

5.8 Testing for the Variance of a Single Population

Let $\mathbf{x}_1, \ldots, \mathbf{x}_n$ constitute a random sample from $N_p(\mu, \Sigma)$. We want to test $H_0(\Sigma = \Sigma_0)$, μ being unknown. Here $\Theta_0 = \{\mu, \Sigma_0\}, \mu \in \mathcal{R}^p$. The m.l.e. of μ under H_0 is $\hat{\mu} = \bar{\mathbf{x}}$. When the parameters lie in Θ, the m.l.e.'s are $\hat{\mu} = \bar{\mathbf{x}}, \hat{\Sigma} = \mathbf{S}_n$. From (3.6.6),

$$\ln L = l(\mu, \Sigma : \mathbf{x}) = -\frac{n}{2} \ln |2\pi\Sigma| - \frac{n}{2} tr\Sigma^{-1}\mathbf{S}_n - \frac{n}{2}(\bar{\mathbf{x}} - \mu)\Sigma^{-1}(\bar{\mathbf{x}} - \mu).$$

Hence

$$l_0^* = \ln(\sup{}_{\theta \in \Theta_0} L(\theta)) = -\frac{n}{2} \ln |2\pi\Sigma_0| - \frac{n}{2} tr(\Sigma_0^{-1}\mathbf{S}_n), \tag{5.8.1}$$

$$l^* = \ln(\sup{}_{\theta \in \Theta} L(\theta)) = -\frac{n}{2} \ln |2\pi\mathbf{S}_n| - \frac{np}{2}. \tag{5.8.2}$$

Hence,

$$\begin{aligned} -2\ln\lambda &= -2\ln \frac{\sup{}_{\theta\in\Theta_0} L(\theta)}{\sup{}_{\theta\in\Theta} L(\theta)} \\ &= 2l^* - 2l_0^* \\ &= n \ln |2\pi\Sigma_0| + ntr(\Sigma_0^{-1}\mathbf{S}_n) - n \ln |2\pi\mathbf{S}_n| - np \\ &= ntr(\Sigma_0^{-1}\mathbf{S}_n) - n \ln |\Sigma_0^{-1}\mathbf{S}_n| - np. \end{aligned} \tag{5.8.3}$$

Let a and g denote, respectively, the arithmetic mean (AM) and geometric mean (GM) of the eigenvalues of $\Sigma_0^{-1}\mathbf{S}_n$. then,

$$tr(\Sigma_0^{-1}\mathbf{S}_n) = ap, \quad |\Sigma_0^{-1}\mathbf{S}_n| = g^p.$$

Hence, from (5.8.3),

$$-2\ln\lambda = np(a - \ln g - 1), \tag{5.8.4}$$

which in large samples follows approximately a $\chi^2_{(\nu)}$ distribution with $\nu = p(p+1)/2$ degrees of freedom.

EXAMPLE 5.8.1: For the data in Example 5.7.4, test the hypothesis

$$H_0 : \mathbf{\Sigma} = \mathbf{\Sigma}_0 = \begin{bmatrix} 5.00 & 1.00 & 0.75 \\ 1.00 & 3.75 & 0.75 \\ 0.75 & 0.75 & 4.00 \end{bmatrix}.$$

Here

$$\mathbf{S}_n = \begin{bmatrix} 91.9482 & 19.2175 & 12.8476 \\ 19.2175 & 67.9675 & 14.0399 \\ 12.8476 & 14.0399 & 71.9378 \end{bmatrix}.$$

Note that in Example 7.8.4 we calculated $\mathbf{S} = (n\mathbf{S}_n)/(n-1)$. To calculate the eigenvalues of $\mathbf{\Sigma}_0^{-1}\mathbf{S}_n$ we calculate the eigenvalues of $\mathbf{\Sigma}_0^{-1/2}\mathbf{S}_n\mathbf{\Sigma}_0^{-1/2}$, since both have the same eigenvalues (vide Section 4.5). Now,

$$\mathbf{\Sigma}_0^{-1} = \begin{bmatrix} .21483 & -.05115 & -.03069 \\ -.05115 & .29924 & -.04464 \\ -0.03069 & -.04464 & .26413 \end{bmatrix},$$

$$\mathbf{\Sigma}_0^{-1/2} = \begin{bmatrix} .45919 & -.05306 & -.03402 \\ -.05306 & .53333 & -.04448 \\ -.03402 & -.04448 & .51087 \end{bmatrix},$$

$$\mathbf{\Sigma}_0^{-1/2}\mathbf{S}_n\mathbf{\Sigma}_0^{-1/2} = \begin{bmatrix} 18.3751 & .2444 & -.2186 \\ .2444 & 18.0407 & .1395 \\ -.2186 & .1395 & 17.9888 \end{bmatrix}.$$

Its eigenvalues are

$$\lambda_1 = 18.5391, \lambda_2 = 18.1527, \lambda_3 = 17.7128.$$

Hence,

$$a = 18.1349, \ g = 18.1265.$$

Therefore, from (5.8.3),

$$-2\ln\lambda = (52)(3)(18.1349 - 2.8937 - 1) = 2230.41.$$

The null hypothesis is rejected.

5.8.1 $H_0(\Sigma = k\Sigma_0), \Sigma_0$ *is known, but* k *is unknown*

From (3.6.28), the *m.l.e.* of k is

$$\hat{k} = tr(\mathbf{\Sigma}_0^{-1}\mathbf{S}_n)/p. \tag{5.8.5}$$

Hence, from (5.8.4),

$$-2\log\lambda = np(a - \log g - 1)$$

where a and g are, respectively, the AM and GM of the eigenvalues of $(\hat{k}\mathbf{\Sigma}_0)^{-1}\mathbf{S}_n$. Taking a_0 and g_0 as the AM and the GM of the eigenvalues of $\mathbf{\Sigma}_0^{-1}\mathbf{S}_n$,

$$a = \frac{a_0}{\hat{k}}, \quad g = \frac{g_0}{\hat{k}}.$$

Hence, from (5.8.5),

$$\hat{k} = a_0. \tag{5.8.6}$$

Therefore, $a = 1, g = g_0/a_0$ and

$$-2\log\lambda = np\log(a_0/g_0). \tag{5.8.7}$$

In large samples, $-2\log\lambda$ follows approximately, a $\chi^2_{(\nu)}$ distribution with $\nu = p(p+1)/2 - 1 = (p-1)(p+2)/2$ d.f.

5.8.2 $H_0(\Sigma_{12} = 0)$

Suppose $\mathbf{X} = (\mathbf{X}'_{(1)}, \mathbf{X}'_{(2)})$, where $\mathbf{X}_{(i)}$ is of order $p_i \times 1 (p_1 + p_2 = p)$, follows $N_p(\mu, \mathbf{\Sigma})$ where

$$\mu = (\mu'_{(1)}, \mu'_{(2)})', \ \mathbf{\Sigma} = \begin{bmatrix} \mathbf{\Sigma}_{11} & \mathbf{\Sigma}_{12} \\ \mathbf{\Sigma}_{21} & \mathbf{\Sigma}_{22} \end{bmatrix},$$

μ_1 being of order $p_1 \times 1$ and $\mathbf{\Sigma}_{11}$ of order $p_1 \times p_1$. We want to test $\mathbf{\Sigma}_{12} = \mathbf{0}$, i.e., variables in $\mathbf{X}_{(1)}$ and $\mathbf{X}_{(2)}$ are mutually independent. Samples are denoted as $\mathbf{x}'_i = (\mathbf{x}'_{(1)i}, \mathbf{x}'_{(2)i})$ $(i = 1, \ldots, n)$.

Under H_0 the m.l.e.'s are

$$\hat{\mu}_1 = \bar{\mathbf{x}}_{(1)}, \quad \hat{\mu}_2 = \bar{\mathbf{x}}_{(2)},$$

$$\hat{\mathbf{\Sigma}}_{11} = \mathbf{Q}_{11}/n, \quad \hat{\mathbf{\Sigma}}_{22} = \mathbf{Q}_{22}/n,$$

where

$$\mathbf{Q}_{11} = \sum_{i=1}^{n}(\mathbf{x}_{(1)i} - \bar{\mathbf{x}}_{(1)})(\mathbf{x}_{(1)i} - \bar{\mathbf{x}}_{(1)})',$$

$$\mathbf{Q}_{22} = \sum_{i=1}^{n} (\mathbf{x}_{(2)i} - \bar{\mathbf{x}}_{(2)})(\mathbf{x}_{(2)i} - \bar{\mathbf{x}}_{(2)})'.$$

In the unrestricted case,

$$\hat{\mu} = \bar{\mathbf{x}} = (\bar{\mathbf{x}}'_{(1)},\ \bar{\mathbf{x}}_{(2)})',$$

$$\hat{\boldsymbol{\Sigma}} = \mathbf{Q}/n, \quad \text{where } \mathbf{Q} = \sum_{i=1}^{n} (\mathbf{x}_i - \bar{\mathbf{x}})(\mathbf{x}_i - \bar{\mathbf{x}})',\ \bar{\mathbf{x}} = (\bar{\mathbf{x}}'_1,\ \bar{\mathbf{x}}'_2).$$

If H_0 is true, variables $\mathbf{X}_{(1)}, \mathbf{X}_{(2)}$ are mutually independent and hence the maximum value of the likelihood is

$$\sup{}_{(\theta \in \boldsymbol{\Theta}_0)} L(\mu, \boldsymbol{\Sigma}) = \sup{}_{(\mu_{(1)}, \boldsymbol{\Sigma}_{11})} L_1(\mu_{(1)}, \boldsymbol{\Sigma}_{11}) \sup{}_{(\mu_{(2)}, \boldsymbol{\Sigma}_{22})} L_2(\mu_{(2)} \boldsymbol{\Sigma}_{22}).$$

The likelihood ratio is

$$\begin{aligned} \lambda &= \frac{L_1(\hat{\mu}_{(1)}, \hat{\boldsymbol{\Sigma}}_{11}) L_2(\hat{\mu}_2, \hat{\boldsymbol{\Sigma}}_{22})}{L(\hat{\mu}, \hat{\boldsymbol{\Sigma}})} \\ &= \frac{|\hat{\boldsymbol{\Sigma}}_{11}|^{-n/2} |\hat{\boldsymbol{\Sigma}}_{22}|^{-n/2}}{|\hat{\boldsymbol{\Sigma}}|^{-n/2}}. \end{aligned}$$

Hence,

$$U = \lambda^{2/n} = \frac{|\hat{\boldsymbol{\Sigma}}|}{|\hat{\boldsymbol{\Sigma}}_{11}||\hat{\boldsymbol{\Sigma}}_{22}|} = \frac{|\mathbf{Q}|}{|\mathbf{Q}_{11}||\mathbf{Q}_{22}|}. \tag{5.8.8}$$

Since $\mathbf{Q} > 0$ with probability one, $\mathbf{Q}_{11}$ is non-singular and from (A.5.7),

$$|\mathbf{Q}| = |\mathbf{Q}_{11}| . |\mathbf{Q}_{22} - \mathbf{Q}_{21} \mathbf{Q}_{11}^{-1} \mathbf{Q}_{12}|.$$

Setting $\mathbf{Q}_{22} - \mathbf{Q}_{21}\mathbf{Q}_{11}^{-1}\mathbf{Q}_{12} = \mathbf{Q}_{22.1} = \mathbf{E}$ and

$$\mathbf{Q}_{21}\mathbf{Q}_{11}^{-1}\mathbf{Q}_{12} = \mathbf{H},$$

we have

$$U = \frac{|\mathbf{Q}_{22.1}|}{|\mathbf{Q}_{22}|} = \frac{|\mathbf{E}|}{|\mathbf{H} + \mathbf{E}|}. \tag{5.8.9}$$

By Theorem 4.2.7, $\mathbf{E} \cap W_{p_2}(n - 1 - p_1, \boldsymbol{\Sigma}_{22.1})$ where $\boldsymbol{\Sigma}_{22.1} = \boldsymbol{\Sigma}_{22} - \boldsymbol{\Sigma}_{21}\boldsymbol{\Sigma}_{11}^{-1}\Sigma_{12}$. If $\boldsymbol{\Sigma}_{12} = \mathbf{0}, \mathbf{H}$ has $W_{p_2}(p_1, \boldsymbol{\Sigma}_{22})$ distribution and $\mathbf{H}, \mathbf{E}$ are mutually independent. It then follows that under H_0, U follows Wilk's $\Lambda(p_2; n - 1 - p_1, p_1)$ distribution.

Since $\Lambda(p_2; n-1-p_1, p_1)$ has the same distribution as $\Lambda(p_1 : n-1-p_2, p_2)$ (vide Theorem 4.4.3) we arrive at the same distribution for U in (5.8.9), if we write

$$U = \frac{|\mathbf{Q}_{11.2}|}{|\mathbf{Q}_{11}|}.$$

We note that H_0 is rejected, if λ is too small or equivalently, U is too small. For large n, $-2\ln\lambda$ follows a $\chi^2_{(\nu)}$ distribution, where $\nu = p(p+1)/2 - \sum_{j=1}^{2} p_j(p_j+1)/2 = (p^2 - p_1^2 - p_2^2)/2 = p_1p_2$.

Again, from (5.8.9), by A.8(3),

$$U = \Pi_{j=1}^{p_2}(1-\theta_j)$$

where $\theta_j$are ordered roots of $|\mathbf{H} - \theta(\mathbf{E}+\mathbf{H})| = 0$, i.e., θ_j's are the ordered eigenvalues of $\mathbf{H}(\mathbf{E}+\mathbf{H})^{-1} = \mathbf{Q}_{21}\mathbf{Q}_{11}^{-1}\mathbf{Q}_{12}\mathbf{Q}_{22}^{-1}$, i.e., the roots of

$$|\mathbf{S}_{21}\mathbf{S}_{11}^{-1}\mathbf{S}_{12}\mathbf{S}_{22}^{-1} - \theta\mathbf{I}_{p_2}| = 0$$

where $\mathbf{S}_{ij} = \mathbf{Q}_{ij}/(n-1), \mathbf{Q}_{12} = \sum_{i=1}^{n}(\mathbf{x}_{(1)i} - \bar{\mathbf{x}}_{(1)})(\mathbf{x}_{(2)i} - \bar{\mathbf{x}}_{(2)})'$. This test has also been considered in Section 10.4.

5.8.3 *Blockwise independence*

Let $\mathbf{X} = (\mathbf{X}'_{(1)}, \ldots, \mathbf{X}'_{(b)})'$ where $\mathbf{X}_{(i)}$ is of order $p_i \times 1, \sum_i p_i = p$, follow $N_p(\mu, \mathbf{\Sigma})$ distribution with $\mu = (\mu'_{(1)}, \ldots, \mu'_{(b)})'$,

$$\mathbf{\Sigma} = \begin{bmatrix} \mathbf{\Sigma}_{11} & \mathbf{\Sigma}_{12} & \ldots & \mathbf{\Sigma}_{1b} \\ . & . & \ldots & . \\ \mathbf{\Sigma}_{b1} & \mathbf{\Sigma}_{b2} & \ldots & \mathbf{\Sigma}_{bb} \end{bmatrix},$$

μ_j being of order $p_j \times 1$ and $\mathbf{\Sigma}_{jk}$ of order $p_j \times p_k (j,k = 1, \ldots b)$. We want to test

$$H_{01} : \mathbf{\Sigma} = \mathbf{\Sigma}_{01} = \begin{bmatrix} \mathbf{\Sigma}_{11} & \mathbf{0} & \ldots & \mathbf{0} \\ 0 & \mathbf{\Sigma}_{22} & \ldots & .\mathbf{0} \\ . & . & \ldots & . \\ \mathbf{0} & \mathbf{0} & \ldots & \mathbf{\Sigma}_{bb} \end{bmatrix},$$

i.e., the variables in $\mathbf{X}_{(1)}, \ldots, \mathbf{X}_{(b)}$ are mutually independent.

Sample observations are $(\mathbf{x}'_{(1)i}, \ldots, \mathbf{x}'_{(b)i})', i = 1, \ldots, n$. Under H_{01}, the *m.l.e.*'s are

$$\begin{aligned} \hat{\mu}_{(j)} &= \bar{\mathbf{x}}_{(j)} = \textstyle\sum_{i=1}^{n} \mathbf{x}_{(j)i}/n, \\ n\hat{\mathbf{\Sigma}}_{jj} &= \textstyle\sum_{i=1}^{n}(\mathbf{x}_{(j)i} - \bar{\mathbf{x}}_{(j)})(\mathbf{x}_{(j)i} - \bar{\mathbf{x}}_{(j)})', j = 1, \ldots, b. \end{aligned}$$

Hence the LR test statistic is

$$\begin{aligned} U = \lambda^{2/n} &= \frac{|\hat{\mathbf{\Sigma}}|}{|\hat{\mathbf{\Sigma}}_{11}|....|\hat{\mathbf{\Sigma}}_{bb}|} \\ &= \frac{|\mathbf{Q}|}{|\mathbf{Q}_{11}...|\mathbf{Q}_{bb}|} \end{aligned} \tag{5.8.10}$$

The special method used for the case $b = 2$ is not available here. The number of parameters in Θ is $p(p+1)/2 + p$. The number of parameters in Θ_0 is $\sum_{j=1}^{b}\{p_j(p_j+1)/2+p_j\} = \sum_{j=1}^{b} p_j^2/2 + (3/2)p$. Hence, for large n, $-2\ln\lambda$ follows approximately a $\chi^2_{(\nu)}$ distribution with $\nu = (p^2 - \sum_j p_j^2)/2$ degrees of freedom.

Diagonal Dispersion Matrix

Here the null hypothesis is $H_{02} : \mathbf{\Sigma} =$ Diag. $(\sigma_{11}, \ldots, \sigma_{pp})$. This is a special case of the null hypothesis H_{02} with $b = p, \mathbf{x}_{(j)k} = x_{jk}, \mu_{(j)} = \mu_j, j = 1, \ldots, p$. Here

$$U = \lambda^{2/n} = \frac{|\mathbf{Q}|}{q_{11}\cdots q_{pp}} \tag{5.8.11}$$

where $q_{jj} = \sum_{i=1}^{n}(x_{ji} - \bar{x}_j)^2 = n\hat{\sigma}_{jj}(j = 1, \ldots, p)$. Here $\nu = p(p-1)/2$.

In practice, we may be interested in testing

$$H_{03} : \mathbf{\Sigma} = \sigma^2 \mathbf{I}_p,$$

where σ^2 is known. This is the *hypothesis of sphericity*. Under $H_{03} : |\mathbf{\Sigma}| = \sigma^{2p}$ and *m. l. e* of σ^2 is $\hat{\sigma}^2 =$ tr $(\mathbf{S}_n)/p$. Therefore,

$$\lambda = \frac{|\mathbf{S}_n|^{n/2}}{[\text{ tr }(\mathbf{Q})/p]^p}. \tag{5.8.12}$$

To test

$$H'_{03} : \mathbf{\Sigma} = \mathbf{I}_p$$

we have the LR statistic

$$\lambda = (\frac{e}{n})^{np/2}|\mathbf{Q}|^{n/2}\exp(-\frac{1}{2}\text{ tr }\mathbf{Q}). \tag{5.8.13}$$

When H'_{03} is true, $-2\ln\lambda$ is approximately a $\chi^2_{(\nu)}$ variate with $\nu = p(p+1)/2$ d.f.

5.8.4 *Hypothesis of equicorrelation matrix*

Here the null hypothesis is

$$H_4 : \mathbf{\Sigma} = \sigma^2 \begin{bmatrix} 1 & \rho & \ldots & \rho & \rho \\ \rho & 1 & \ldots & \rho & \rho \\ . & . & \ldots & . & . \\ \rho & \rho & \ldots & \rho & 1 \end{bmatrix}. \tag{5.8.14}$$

Under H_0, the *m.l.e.* of σ^2 and ρ are $\hat{\sigma}^2 = \sum_{j=1}^{p} s_{jjn}/p$ and

$$\hat{\sigma}^2\hat{\rho} = \sum_{j\neq k=1}^{p} s_{jkn}/p(p-1) \tag{5.8.15}$$

where $s_{jkn} = n^{-1}\sum_{i=1}^{n}(x_{ij}-\bar{x}_j)(x_{ik}-\bar{x}_k), j,k=1,\ldots,p$. The L.R. test is given by

$$U = \lambda^{2/n} = \frac{|\mathbf{S}_n|}{(\hat{\sigma}^2)^p(1-\hat{\rho})^{p-1}[1+(p-1)\hat{\rho}]}. \tag{5.8.16}$$

Box (1949) showed that under H_{04},

$$-[n-1-\frac{p(p+1)^2(2p-3)}{6(p-1)(p^2+p-4)}]\ln\lambda \tag{5.8.17}$$

is asymptotically distributed as $\chi^2_{(\nu)}$ where $\nu = p(p+1)/2-2$.

5.9 Test for Equality of Two Dispersion Matrices

Suppose we have a random sample $(\mathbf{x}_1,\ldots,\mathbf{x}_{n_1})$ from a $N_p(\mu_1,\mathbf{\Sigma}_1)$ population and another independent sample $(\mathbf{y}_1,\ldots,\mathbf{y}_{n_2})$ from another $N_p(\mu_2,\mathbf{\Sigma}_2)$ population. We want to test

$$H_0 : \mathbf{\Sigma}_1 = \mathbf{\Sigma}_2.$$

The likelihood function is

$$L(\mu_1,\mu_2,\mathbf{\Sigma}_1,\mathbf{\Sigma}_2) = L_1(\mu_1,\mathbf{\Sigma}_1)L(\mu_2,\mathbf{\Sigma}_2).$$

Maximizing L is equivalent to maximizing each $L_i(i=1,2)$. Thus L is maximized at the *m.l.e.*'s $\hat{\mu}_I = \bar{\mathbf{x}}, \hat{\mu}_2 = \bar{\mathbf{y}}, \hat{\mathbf{\Sigma}} = \mathbf{Q}_1/n_1, \hat{\mathbf{\Sigma}}_2 = \mathbf{Q}_2/n_2$. The maximum value of L is

$$\begin{aligned} L(\hat{\mu}_1,\hat{\mu}_2,\hat{\mathbf{\Sigma}}_1,\hat{\mathbf{\Sigma}}_2) &= L_1(\hat{\mu}_1,\hat{\mathbf{\Sigma}}_1)L_2(\hat{\mu}_2,\hat{\mathbf{\Sigma}}_2) \\ &= (2\pi)^{-np/2}|\mathbf{\Sigma}_1|^{-n_1/2}|\mathbf{\Sigma}_2|^{-n_2/2}e^{-np/2}, \end{aligned}$$

where $n = n_1+n_2$. Under H_0, the log-likelihood is

$$\begin{aligned} \ln L_0 &= \ln L_1(\mu_1,\mathbf{\Sigma}_1) + \ln L_2(\mu_2,\mathbf{\Sigma}_2) \\ &= K - \tfrac{n}{2}\ln|\mathbf{\Sigma}| - \tfrac{1}{2}\operatorname{tr}\{\mathbf{\Sigma}^{-1}\mathbf{Q}\} \\ &\quad -\tfrac{1}{2}\operatorname{tr}\{\mathbf{\Sigma}^{-1}[n_1(\bar{\mathbf{x}}-\mu_1)(\bar{\mathbf{x}}-\mu_1)' + n_2(\bar{\mathbf{y}}-\mu_2)(\bar{\mathbf{y}}-\mu_2)']\} \end{aligned} \tag{5.9.1}$$

where $K = -\frac{np}{2}\log 2\pi$ and $\mathbf{Q} = \mathbf{Q}_1+\mathbf{Q}_2$. Since, $\operatorname{tr}[\mathbf{\Sigma}^{-1}\{n_1(\bar{\mathbf{x}}-\mu_1)(\bar{\mathbf{x}}-\mu_1)'\}] = n_1(\bar{\mathbf{x}}-\mu_1)'\mathbf{\Sigma}^{-1}(\bar{\mathbf{x}}-\mu_1) \geq 0$, (5.9.1) is maximized for any $\mathbf{\Sigma} > 0$ for $\mu_1 = \bar{\mathbf{x}}, \mu_2 = \bar{\mathbf{y}}$. Now

$$\ln L_0(\hat{\mu}_1,\hat{\mu}_2,\mathbf{\Sigma},\mathbf{\Sigma}) = K - \frac{n}{2}\ln|\mathbf{\Sigma}| - \frac{n}{2}\operatorname{tr}[\mathbf{\Sigma}^{-1}\mathbf{Q}/n] \tag{5.9.2}$$

which is to be maximized with respect to $\mathbf{Q}$ subject to the condition $\mathbf{\Sigma} > 0$. Since, each $\mathbf{Q}_i > 0, \mathbf{Q} > 0$ and (5.9.2) is maximized when $\hat{\mathbf{\Sigma}} = \mathbf{Q}/n$ (vide Theorem 3.6.1). Thus

$$\log L_0(\hat{\mu}_1,\hat{\mu}_2,\hat{\mathbf{\Sigma}},\hat{\mathbf{\Sigma}}) = K - \frac{n}{2}\log|\hat{\mathbf{\Sigma}}| - \frac{n}{2}p$$

so that

$$L_0(\hat{\mu}_1, \hat{\mu}_2, \hat{\mathbf{\Sigma}}, \hat{\mathbf{\Sigma}}) = \frac{1}{(2\pi)^{np/2}|\hat{\mathbf{\Sigma}}|^{n/2}} e^{-np/2}.$$

The likelihood ratio statistic is therefore,

$$\lambda = \frac{L_0(\hat{\mu}_1, \hat{\mu}_2, \hat{\mathbf{\Sigma}}, \hat{\mathbf{\Sigma}})}{L(\hat{\mu}_1, \hat{\mu}_2, \hat{\mathbf{\Sigma}}_1, \hat{\mathbf{\Sigma}}_2)} = \frac{|\hat{\mathbf{\Sigma}}|^{-n/2}}{|\hat{\mathbf{\Sigma}}_1|^{-n_1/2}|\hat{\mathbf{\Sigma}}_2|^{-n_2/2}}$$

$$= \frac{n^{np/2}}{n_1^{n_1 p/2} n_2^{n_2 p/2}} \cdot \frac{|\mathbf{Q}_1|^{n_1/2}|\mathbf{Q}_2|^{n_2/2}}{|\mathbf{Q}_1 + \mathbf{Q}_2|^{n/2}}. \tag{5.9.3}$$

In large samples, under H_0, $-2\ln\lambda$ follows $\chi^2_{(\nu)}$ with $\nu = p(p+1)/2$.

Note 5.9.1: If we are to test for the equality of two normal populations $N_p(\mu_1, \mathbf{\Sigma}_1)$ and $N_p(\mu_2, \mathbf{\Sigma}_2)$, we first test for the hypotheses $\mathbf{\Sigma}_1 = \mathbf{\Sigma}_2$. If this hypothesis is accepted, then we test for the hypothesis $\mu_1 = \mu_2$, given that $\mathbf{\Sigma}_1 = \mathbf{\Sigma}_2$ by the method in Section 5.7.

EXAMPLE 5.9.1: For the data in Example 5.7.1 test the null hypothesis $H_0 : \mathbf{\Sigma}_1 = \mathbf{\Sigma}_2$.

Here

$$\mathbf{Q}_1 = \begin{bmatrix} 220.952 & 182.000 & 31.133 \\ 182.000 & 350.000 & 38.200 \\ 31.133 & 38.200 & 10.627 \end{bmatrix},$$

$$\mathbf{Q} = \begin{bmatrix} 406.857 & 464.143 & 60.557 \\ 464.143 & 878.857 & 91.743 \\ 60.557 & 91.743 & 19.667 \end{bmatrix},$$

$$|\mathbf{Q}_1| = 241037, \ |\mathbf{Q}_2| = 1304942, \ |\mathbf{Q}| = 5187793,$$

$$-2\ln\lambda = 2.70808 < \chi_{6;.05} = 12.59.$$

Hence the null hypothesis is accepted.

5.10 Profile Analysis

Suppose that a battery of p treatments is administered to a group of n_1 subjects selected randomly from a given population and let $\mathbf{x}_1, \ldots, \mathbf{x}_{n_1}$ be the vector of responses to these treatments by these n_1 subjects. Assume that $\mathbf{x}_1, \ldots, \mathbf{x}_{n_1}$ are *iid* $N_p(\mu, \mathbf{\Sigma})$. The graph obtained by joining the points

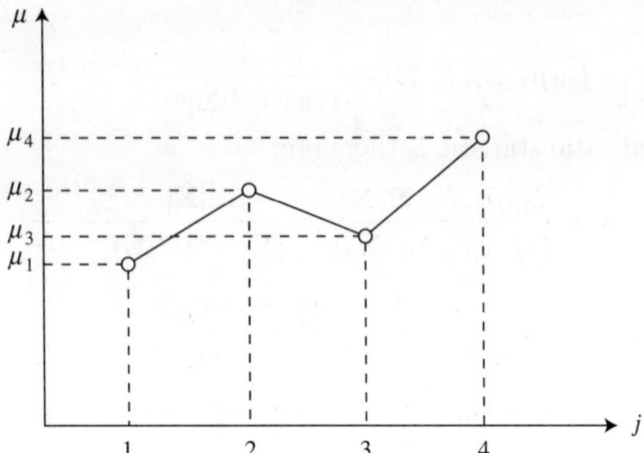

Fig. 5.3a: Profile of a 4-variate population

$(j, \mu_j)(j = 1, \ldots, p)$ successively is called the *profile* of the population (Fig. 5.3a).

We draw a sample of n_2 subjects from another population and the above battery of treatments is administered to them. Let $\mathbf{y}_1, \ldots, \mathbf{y}_{n_2}$ be the responses. Assume that $\mathbf{y}_1, \ldots, \mathbf{y}_{n_2}$ are *iid* $N_p(\nu, \mathbf{\Sigma})$. We wish to compare the two profiles $(j, \mu_j), (j, \nu_j)(j = 1, \ldots, p)$ using the sample data. Let $\bar{\mathbf{x}} = (\bar{x}_1, \ldots, \bar{x}_p)', \bar{\mathbf{y}} = (\bar{y}_1, \ldots, \bar{y}_p)'$ be the sample mean vectors.

The first hypothesis of interest is: Are the two profiles parallel? or, expressed mathematically, we wish to test

$$H_{01} : \mu_k - \mu_{k-1} = \nu_k - \nu_{k-1} (k = 2, \ldots, p)$$

or

$$\mu_2 - \mu_1 = \nu_2 - \nu_1, \mu_3 - \mu_2 = \nu_3 - \nu_2, \ldots,$$

$$\mu_p - \mu_{p-1} = \nu_p - \nu_{p-1};$$

or

$$\mathbf{C}_1 \mu = \mathbf{C}_1 \nu$$

where

$$\mathbf{C}_1 = \begin{bmatrix} -1 & 1 & 0 & \ldots & 0 & 0 \\ 0 & -1 & 1 & \ldots & 0 & 0 \\ . & . & . & \ldots & . & . \\ 0 & 0 & 0 & \ldots & -1 & 1 \end{bmatrix} \tag{5.10.1}$$

is the $(p-1)\times p$ contrast matrix. In the terminology of experimental designs, H_{01} is the hypothesis of no interaction between groups (populations) and treatments (Fig. 5.3b).

The hypothesis is tested by considering the transformed observations $\mathbf{C}_1\mathbf{x}_i(i=1,\ldots,n_1), \mathbf{C}_1\mathbf{y}_i(i=1,\ldots,n_2)$. Under H_{01}

$$\mathbf{C}_1(\bar{\mathbf{x}}-\bar{\mathbf{y}}) \cap N_{p-1}(\mathbf{0}, \mathbf{C}_1\boldsymbol{\Sigma}\mathbf{C}_1'(\frac{1}{n_1}+\frac{1}{n_2}))$$

or

$$\sqrt{\frac{n_1 n_2}{n}}\mathbf{C}_1(\bar{\mathbf{x}}-\bar{\mathbf{y}}) \cap N_{p-1}(\mathbf{0}, \mathbf{C}_1\boldsymbol{\Sigma}\mathbf{C}_1') \tag{5.10.2}$$

where $n = n_1 + n_2$. Also, defining

$$\mathbf{S}_p = \frac{(n_1-1)\mathbf{S}_1 + (n_2-1)\mathbf{S}_2}{n_1+n_2-2}$$

where

$$\mathbf{S}_1 = \frac{1}{n_1-1}\sum_{i=1}^{n_1}(\mathbf{x}_i-\bar{\mathbf{x}})(\mathbf{x}_i-\bar{\mathbf{x}})', \ \mathbf{S}_2 = \frac{1}{n_2-2}\sum_{i=1}^{n_2}(\mathbf{y}_i-\bar{\mathbf{y}})(\mathbf{y}_i-\bar{\mathbf{y}})',$$

we note that

$$(n_1+n_2-2)\mathbf{S}_p \cap W_p(n_1+n_2-2, \boldsymbol{\Sigma}).$$

Here

$$(n_1+n_2-2)\mathbf{C}_1\mathbf{S}_p\mathbf{C}_1' \cap W_{p-1}(n_1+n_2-2), \mathbf{C}_1\boldsymbol{\Sigma}\mathbf{C}_1'). \tag{5.10.3}$$

Also, the statistics in (5.10.2) and (5.10.3) are independently distributed. Hence, under H_{01},

$$T_{01}^2 = \frac{n_1 n_2}{n}(\bar{\mathbf{x}}-\bar{\mathbf{y}})'\mathbf{C}_1'(\mathbf{C}_1\mathbf{S}_p\mathbf{C}_1')^{-1}\mathbf{C}_1(\bar{\mathbf{x}}-\bar{\mathbf{y}}) \cap T^2(p-1, n_1+n_2-2). \tag{5.10.4}$$

Therefore, we reject H_{01} if

$$T_{01}^2 > \frac{(n_1+n_2-2)(p-1)}{n_1+n_2-p}F_{p-1,n_1+n_2-p;\alpha}. \tag{5.10.5}$$

If H_{01} is accepted, we may wish to test if the two population profiles are coincident. If the two profiles are parallel, they will be coincident *iff* $\mu_1 + \ldots + \mu_p = \nu_1 + \ldots + \nu_p$. The null hypothesis at stage 2 can, therefore, be written as

$$H_{02} : \mathbf{1}'\mu = \mathbf{1}'\nu.$$

This hypothesis is of the same type as H_{01}. Hence, under H_{02},

$$\begin{aligned} T_{02}^2 &= \frac{n_1 n_2}{n}(\bar{\mathbf{x}}-\bar{\mathbf{y}})'\mathbf{1}[\mathbf{1}'\mathbf{S}_p\mathbf{1}]^{-1}\mathbf{1}'(\bar{\mathbf{x}}-\bar{\mathbf{y}}) \\ &= \frac{n_1 n_2}{n}\frac{[\mathbf{1}'(\bar{\mathbf{x}}-\bar{\mathbf{y}})]^2}{\mathbf{1}'\mathbf{S}_p\mathbf{1}} \\ &= \left\{\frac{\mathbf{1}'(\bar{\mathbf{x}}-\bar{\mathbf{y}})}{[\mathbf{1}'\mathbf{S}_p\mathbf{1}(\frac{1}{n_1}+\frac{1}{n_2})]^{1/2}}\right\}^2 \\ &= t_0^2 \ \text{(say)} . \end{aligned} \tag{5.10.6}$$

The statistic t_0^2 is the ordinary $t-$ statistic for testing equality of two mean of two variables $\mathbf{1}'\mathbf{x}$ and $\mathbf{1}'\mathbf{y}$ whose sample means are $\mathbf{1}'\bar{\mathbf{x}}$ and $\mathbf{1}\mathbf{y}$, respectively. Under $H_{02}, t_0 \cap t_{(n_1+n_2-2)}$ and $T_{02}^2 \cap F_{(1,n_1+n_2-2)}$. Therefore, H_{02} is rejected if

$$|t_0| > t_{n_1+n_2-2;\alpha/2}$$

or

$$T_{02}^2 > F_{1,n_1+n_2-2;\alpha}.$$

The hypothesis H_{02} may be regarded as the hypothesis of no population 'main effects'. Note that H_{02} may be true even if H_{01} is false.

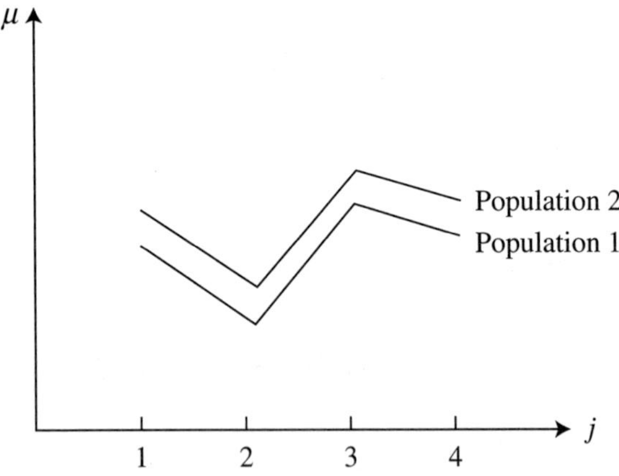

Fig. 5.3b: Parallel profiles of two populations

Similarly, we can test the hypothesis that all the treatments have the same average effects, i.e.,

$$H_{03} : \frac{1}{2}(\mu_1 + \nu_1) = \frac{1}{2}(\mu_2 + \nu_2) = \ldots = \frac{1}{2}(\mu_p + \nu_p)$$

or

$$\frac{1}{2}(\mu_i + \nu_i) - \frac{1}{2}(\mu_{i-1} + \nu_{i-1})) = 0, i = 2, \ldots, p$$

or

$$\mathbf{C}_1(\mu + \nu) = \mathbf{0}. \tag{5.10.7}$$

If H_{01} is true (i.e., $\mathbf{C}_1(\mu - \nu) = \mathbf{0}$), H_{03} implies $\mathbf{C}_1\mu = \mathbf{0}, \mathbf{C}_1\nu = \mathbf{0}$, i.e.,

$$\mu_1 = \mu_2 = \ldots = \mu_p$$

and

$$\nu_1 = \nu_2 = \ldots = \nu_p,$$

i.e. the two profiles are horizontal parallel lines (Fig. 5.3c). Here, there are population main effects but no differential effects due to treatments.

If all of H_{01}, H_{02}, H_{03} are true, it means $\mu_1 = \ldots = \mu_p = \nu_1 = \ldots = \nu_p$, i.e., the two horizontal parallel lines coincide. This means, there are no population main effects, no test mean effects.

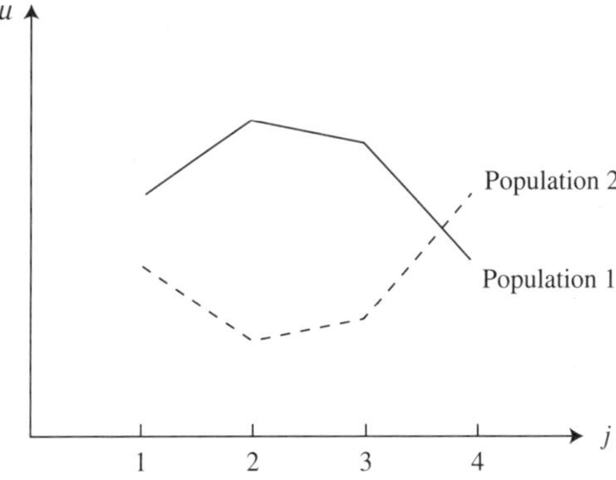

Fig. 5.3c: Profiles with same the average effects

Let

$$\bar{\mathbf{z}} = \frac{n_1\bar{\mathbf{x}} + n_2\bar{\mathbf{y}}}{n_1 + n_2} \text{ and } \mathbf{Q} = (n_1 + n_2 - 2)\mathbf{S}_p.$$

Then

$$\bar{\mathbf{z}} \cap N_p(E(\bar{\mathbf{z}}), \mathbf{\Sigma}/n);$$

$$\mathbf{C}_1\bar{\mathbf{z}} \cap N_{p-1}(\mathbf{C}_1 E(\bar{\mathbf{z}}), \mathbf{C}_1\mathbf{\Sigma}\mathbf{C}_1'/n).$$

Also,

$$\mathbf{Q} \cap W_p(n_1+n_2-2, \mathbf{\Sigma}),\ \mathbf{C}_1\mathbf{Q}\mathbf{C}_1' \cap W_{p-1}(n_1+n_2-2, \mathbf{C}_1\mathbf{\Sigma}\mathbf{C}_1').$$

Also, $\mathbf{C}_1\bar{\mathbf{z}}$ and $\mathbf{C}_1\mathbf{Q}\mathbf{C}_1'$ are statistically independent. If H_{01} is true, $E(\mathbf{C}_1\bar{\mathbf{z}}) = \mathbf{C}_1\mu$. If, further, H_{03} is true, $E(\mathbf{C}_1\bar{\mathbf{z}}) = \mathbf{0}$. Therefore, given that H_{01} is true,

$$n(\mathbf{C}_1\bar{\mathbf{z}})'(\mathbf{C}_1\mathbf{S}_p\mathbf{C}_1')^{-1}(\mathbf{C}_1\bar{\mathbf{z}}) \cap T^2(p-1, n_1+n_2-2)$$

under H_{02}.

Profile analysis in the case of $K(\geq 2)$ populations will be considered in Section 6.7.

EXAMPLE 5.10.1: For the blood-glucose data of Table 11.E.3, test the hypothesis that the profiles of fasting-sugar and after-glucose intake sugar on each of three occasions are parallel. If this hypothesis is accepted, test if these profiles are coincident.

Here $\bar{\mathbf{x}} = (70.12,\ 73.52,\ 75.15)'$, $\bar{\mathbf{y}} = (109.08,\ 104.15,\ 109.85)'$,

$$\mathbf{S}_p = \begin{bmatrix} 445.343 & 162.527 & 111.085 \\ 162.527 & 269.089 & 68.683 \\ 111.085 & 68.660 & 295.054 \end{bmatrix},$$

$$\mathbf{C}_1 = \begin{bmatrix} -1 & 1 & 0 \\ 0 & -1 & 1 \end{bmatrix},\ (\mathbf{C}_1'\mathbf{S}_p\mathbf{C}_1)^{-1} = \begin{bmatrix} 0.0029640 & 0.0010345 \\ 0.0010347 & 0.00270412 \end{bmatrix}.$$

By formula (5.10.4), $T_{01}^2 = 4.688$. Also, the right side of (5.10.5) is $\{(102)(2)/101\}F_{2,101;.05} = 6.201$. Hence the null hypothesis H_{01} of the parallelism of two profiles is accepted.

We therefore test if the two population profiles are coincident. By formula (5.10.6) $T_{02}^2 = .09854 < 3.92 = F_{1,102;.05}$. Therefore the data support this hypothesis.

5.11 Multi-Sample Hypotheses

Suppose we have independent data matrices $\mathbf{X}_1, \ldots, \mathbf{X}_k$ where each row of $\mathbf{X}_i(n_i \times p)$ is an independent sample from $N_p(\mu_i, \mathbf{\Sigma}_i)(i = 1, \ldots, k)$-population. Here $\mu_i = (\mu_{1i}, \mu_{2i}, \ldots, \mu_{pi})'$. A typical observation is denoted as $\mathbf{x}_{ri} = (x_{ri1}, x_{ri2}, \ldots, x_{rip})', r = 1, \ldots, n_i; i = 1, \ldots, k$. Thus $\mathbf{X}_i = (\mathbf{x}_{1i}, \mathbf{x}_{2i}, \ldots, x_{n_i i})'$. We denote $\bar{\mathbf{x}}_i = \sum_{r=1}^{n_i} \mathbf{x}_{ri}/n_i, (i = 1, \ldots, k), \bar{\mathbf{x}} = \sum_{i=1}^{k} n_i\bar{\mathbf{x}}_i/n, n = \sum_{i=1}^{k} n_i$.

5.11.1 *Hypothesis* $H_a(\mu_1 = \ldots = \mu_k = \mu)$ *(unknown), given* $\Sigma_1 = \ldots = \Sigma_k = \Sigma$ *(unknown)*

This is the model for the multivariate one-way analysis of variance where we want to test the hypothesis that the k treatments have the same effects on the p-variate responses. We shall consider the likelihood ratio test.

The parameter space Θ and its subspace Θ_a under H_0 are, therefore,

$$\Theta = \begin{cases} \mathbf{\Sigma} = ((\sigma_{ij})) \text{ such that } & \mathbf{\Sigma} \text{ is positive definite,} \\ -\infty < \mu_{ri} < \infty, & r = 1, \ldots, p; i = 1, \ldots, k. \end{cases}$$

$$\Theta_a = \{ \text{ subspace of } \Theta \text{ for which } \mu_1 = \ldots = \mu_k = \mu \}.$$

From (3.6.6), the likelihood function under Θ is

$$L(\mu_1, \ldots, \mu_k; \mathbf{\Sigma}) = (2\pi)^{-np/2} |\mathbf{\Sigma}|^{-n/2} \exp[-\frac{1}{2} \text{ tr } \mathbf{\Sigma}^{-1}$$

$$\{\sum_{i=1}^{k} \sum_{r=1}^{n_i} (\mathbf{x}_{ri} - \bar{\mathbf{x}}_i)(\mathbf{x}_{ri} - \bar{\mathbf{x}}_i)' + \sum_{i=1}^{k} n_i (\bar{\mathbf{x}}_i - \mu_i)(\bar{\mathbf{x}}_i - \mu_i)'\}]$$

$$= (2\pi)^{-np/2} |\mathbf{\Sigma}|^{-n/2} \exp[-\frac{1}{2} \text{ tr } \mathbf{\Sigma}^{-1} \{\mathbf{W} + \sum_{i=1}^{k} n_i (\bar{\mathbf{x}}_i - \mu_i)(\bar{\mathbf{x}}_i - \mu_i)'\}] \tag{5.11.1}$$

where

$$\mathbf{W} = \sum_{i=1}^{k} \sum_{r=1}^{n_i} (\mathbf{x}_{ri} - \bar{\mathbf{x}}_i)(\mathbf{x}_{ri} - \bar{\mathbf{x}}_i)', \tag{5.11.2}$$

the within sum of squares and product (SSP) matrix. Under Θ, the m.l.e.'s are

$$\hat{\mu}_i = \bar{\mathbf{x}}_i, \ \hat{\mathbf{\Sigma}} = \frac{\mathbf{W}}{n}. \tag{5.11.3}$$

Hence,

$$\sup{}_{\mu, \mathbf{\Sigma}} L(\mu, \mathbf{\Sigma} | H_a) = l_0^* \text{ (say) } = (2\pi)^{-np/2} |\frac{\mathbf{W}}{n}|^{-n/2} e^{-np/2}. \tag{5.11.4}$$

Under H_a, the likelihood-ratio function is

$$L(\mu, \mathbf{\Sigma} | H_a) = (2\pi)^{-np/2} |\mathbf{\Sigma}|^{n/2} \exp[-\frac{1}{2} \text{ tr } \mathbf{\Sigma}^{-1} \sum_{i=1}^{k} \{\sum_{r=1}^{n_i} (\mathbf{x}_{ri} - \bar{\mathbf{x}})(\mathbf{x}_{ri} - \bar{\mathbf{x}})'$$

$$+ n_i (\bar{\mathbf{x}} - \mu)(\bar{\mathbf{x}} - \mu)'\}]. \tag{5.11.5}$$

Under H_a, the m.l.e.'s are

$$\hat{\mu} = \bar{\mathbf{x}},$$

$$\hat{\boldsymbol{\Sigma}} = \mathbf{S}_0 = \textstyle\sum_{i=1}^{k}\sum_{r=1}^{n_i}(\mathbf{x}_{ri} - \bar{\mathbf{x}})(\mathbf{x}_{ri} - \bar{\mathbf{x}})'/n = \frac{\mathbf{T}}{n} \tag{5.11.6}$$

where $\mathbf{T}$ is the total SSP matrix. Hence,

$$\sup{}_{\mu,\boldsymbol{\Sigma}} L(\mu, \boldsymbol{\Sigma}|H_a) = l_0^* \text{ (say) } = (2\pi)^{-np/2}|\mathbf{S}_0|^{-n/2}e^{-np/2}. \tag{5.11.7}$$

Therefore, the LR statistic is

$$\lambda = \frac{l_0^*}{l^*} = \{\frac{|\mathbf{W}|}{n|\mathbf{S}_0|}\}^{n/2} = |\mathbf{T}^{-1}\mathbf{W}|^{n/2}. \tag{5.11.8}$$

We note that

$$\mathbf{T} - \mathbf{W} = \sum_{i=1}^{k} n_i(\bar{\mathbf{x}}_i - \bar{\mathbf{x}})(\bar{\mathbf{x}}_i - \bar{\mathbf{x}})' = \mathbf{B} \text{ (say)} \tag{5.11.9}$$

can be regarded as the 'between-groups' SSP matrix. Hence, from (5.11.8)

$$\lambda^{2/n} = \frac{|\mathbf{W}|}{|\mathbf{B} + \mathbf{W}|} = |\mathbf{I} + \mathbf{W}^{-1}\mathbf{B}|^{-1}. \tag{5.11.10}$$

To find the distribution of (5.11.10) we write the k sample as constituting the single data matrix

$$\mathbf{X} = \begin{bmatrix} \mathbf{X}_1 \\ \mathbf{X}_2 \\ . \\ . \\ \mathbf{X}_k \end{bmatrix}_{n\times p}.$$

Let $\mathbf{1}_i(n\times 1) = (0, \dots, 0, 1, \dots, 1, 0, \dots, 0)'$, the n-vector with 1 in the places corresponding to the ith sample and 0 otherwise,

$$\mathbf{I}_i = \text{ Diag. } (0, \dots, 0, 1, \dots, 1, 0, \dots, 0)_{n\times n},$$

$$\mathbf{H}_i = \mathbf{I}_i - n_i^{-1}\mathbf{1}_i\mathbf{1}_i'. \tag{5.11.11}$$

Then

$$\mathbf{1}_n = \sum_{i=1}^{k} \mathbf{1}_{n_i}, \quad \mathbf{I}_n = \sum_{i=1}^{k} \mathbf{I}_i.$$

Then

$$\begin{aligned} n_i\mathbf{S}_{n_i} &= \textstyle\sum_{r=1}^{n_i}(\mathbf{x}_{ri}-\bar{\mathbf{x}}_i)(\mathbf{x}_{ri}-\bar{\mathbf{x}}_i)' \\ &= \textstyle\sum_{r=1}^{n_i}\mathbf{x}_{ri}\mathbf{x}_{ri}' - n_i\bar{\mathbf{x}}_i\bar{\mathbf{x}}_i' \\ &= \mathbf{X}_i'(\mathbf{I}_{n_i} - n_i^{-1}\mathbf{1}_{n_i}\mathbf{1}_{n_i}')\mathbf{X}_i \\ &= \mathbf{X}'(\mathbf{I}_i - n_i^{-1}\mathbf{1}_i\mathbf{1}_i')\mathbf{X} \\ &= \mathbf{X}'\mathbf{H}_i\mathbf{X}. \end{aligned} \tag{5.11.12}$$

Let

$$\mathbf{C}_1 = \sum_{i=1}^{k}\mathbf{H}_i,\ \mathbf{C}_2 = \sum_{i=1}^{k} n_i^{-1}\mathbf{1}_i\mathbf{1}_i' - n^{-1}\mathbf{1}_n\mathbf{1}_n'. \tag{5.11.13}$$

Then,

$$\begin{aligned} \mathbf{W} &= \mathbf{X}'\mathbf{C}_1\mathbf{X}, \\ \mathbf{B} &= \textstyle\sum_{i=1}^{k} n_i\bar{\mathbf{x}}_i\bar{\mathbf{x}}_i' - n\bar{\mathbf{x}}\bar{\mathbf{x}}' \\ &= \textstyle\sum_{i=1}^{k} n_i^{-1}\mathbf{X}'\mathbf{1}_i\mathbf{1}_i'\mathbf{X} - n^{-1}\mathbf{X}'\mathbf{1}_n\mathbf{1}_n'\mathbf{X} \\ &= \mathbf{X}'\{\textstyle\sum_{i=1}^{k} n_i^{-1}\mathbf{1}_i\mathbf{1}_i' - n^{-1}\mathbf{1}_n\mathbf{1}_n'\}\mathbf{X} \\ &= \mathbf{X}'\mathbf{C}_2\mathbf{X}. \end{aligned} \tag{5.11.14}$$

Also, $\mathbf{C}_1, \mathbf{C}_2$ are symmetric idempotent matrices with ranks $n-k$ and $k-1$ respectively. Therefore, under $H_a, \mathbf{W} \cap W_p(n-k, \mathbf{\Sigma})$ and $\mathbf{B} \cap W_p(k-1, \mathbf{\Sigma})$ (vide Theorem 4.2.10(b)). Hence, provided $n \geq p+k, |\mathbf{I}+\mathbf{W}^{-1}\mathbf{B}|^{-1} \cap \Lambda(p; k-1, n-k)$ under H_a (Section 4.4).

One can use Bartlett's approximation, Rao's approximation to the distribution of Λ as given in (4.4.9) and (4.4.10) respectively.

Following Section 4.5 we can also use Roy's largest root statistic θ_1 which is the largest eigenvalue of $|\mathbf{B}|/(|\mathbf{B}+\mathbf{W}|)$, Lawley-Hotelling Statistic $T_g^2 = (n-k)\sum_{j=1}^{s}\lambda_j$ where $\lambda_1 > \lambda_2 > \ldots > \lambda_s$ are the non-zero eigenvalues of $\mathbf{W}^{-1}\mathbf{B}, s =$ min $(p, k-1)$ and Pillai's trace statistic $V^{(s)} = \sum_{j=1}^{s}\theta_j$. When $k=2$, the L.R. statistic (5.11.8) can be simplified in terms of two-sample Hotelling T^2 statistic. Here

$$\mathbf{B} = n_1(\bar{\mathbf{x}}_1 - \bar{\mathbf{x}})(\bar{\mathbf{x}}_1 - \bar{\mathbf{x}})' + n_2(\bar{\mathbf{x}}_2 - \bar{\mathbf{x}})(\bar{\mathbf{x}}_2 - \bar{\mathbf{x}})'.$$

However,

$$\bar{\mathbf{x}}_1 - \bar{\mathbf{x}} = \frac{n_2}{n}\mathbf{d},\ \bar{\mathbf{x}}_2 - \bar{\mathbf{x}} = -\frac{n_1}{n}\mathbf{d}$$

where $\mathbf{d} = \bar{\mathbf{x}}_1 - \bar{\mathbf{x}}_2$. Therefore,

$$\mathbf{B} = \frac{n_1 n_2}{n}\mathbf{d}\mathbf{d}'. \tag{5.11.15}$$

Hence,

$$\begin{aligned}|\mathbf{I}+\mathbf{W}^{-1}\mathbf{B}|^{-1} &= |\mathbf{I}+(\tfrac{n_1n_2}{n})\mathbf{W}^{-1}\mathbf{dd}'| \\ &= 1+\tfrac{n_1n_2}{n}\mathbf{d}'\mathbf{W}^{-1}\mathbf{d} \quad \text{(by A.5.9)} \\ &= 1+\tfrac{T^2}{n-2}.\end{aligned} \tag{5.11.16}$$

EXAMPLE 5.11.1: The following data were obtained from the analysis of an experiment with $n_1 = 22, n_2 = 24, n_3 = 26$,

$$\bar{\mathbf{x}}_1 = (36.231,\ 56.754,\ 12.802),\ \bar{\mathbf{x}}_2 = (30.791,\ 64.725,\ 9.784),$$

$$\bar{\mathbf{x}}_3 = (43.582,\ 75.893,\ 24.986),$$

$$\mathbf{S}_1 = \begin{bmatrix} 12.50 & 9.35 & 2.38 \\ 9.35 & 18.50 & 1.92 \\ 2.38 & 1.92 & 0.54 \end{bmatrix},$$

$$\mathbf{S}_2 = \begin{bmatrix} 15.07 & 18.25 & 2.24 \\ 18.25 & 32.56 & 4.40 \\ 2.24 & 4.40 & 0.76 \end{bmatrix},$$

$$\mathbf{S}_3 = \begin{bmatrix} 18.09 & 20.52 & 4.76 \\ 20.52 & 28.75 & 3.82 \\ 4.76 & 3.82 & 1.24 \end{bmatrix}.$$

Test the hypothesis $H_a(\mu_1 = \mu_2 = \mu_3)$.

Here,

$$\begin{aligned}\bar{\mathbf{x}} &= \tfrac{22\bar{\mathbf{x}}_1+24\bar{\mathbf{x}}_2+26\bar{\mathbf{x}}_3}{72} \\ &= (37.0722,\ 66.3223,\ 16.1958)',\end{aligned}$$

$$\begin{aligned}\mathbf{W} &= (n_1-1)\mathbf{S}_1 + (n_2-1)\mathbf{S}_2 + (n_3-1)\mathbf{S}_3 \\ &= \begin{bmatrix} 1061.36 & 1129.10 & 220.50 \\ 1129.10 & 1856.13 & 237.02 \\ 220.50 & 237.02 & 59.82 \end{bmatrix},\end{aligned}$$

$$\mathbf{B} = \begin{bmatrix} 2064.27 & 2037.75 & 2517.16 \\ 2037.75 & 4456.94 & 3147.54 \\ 2517.16 & 3147.54 & 3249.02 \end{bmatrix},$$

$$\mathbf{T} = \begin{bmatrix} 3125.63 & 3166.86 & 2737.66 \\ 3166.86 & 6313.07 & 3384.56 \\ 2737.66 & 3384.56 & 3308.84 \end{bmatrix}.$$

Thus,

$$\lambda^{2/n} = \frac{|\mathbf{W}|}{|\mathbf{T}|} = .0012681,$$

which follows under $H_0, \Lambda(3; 2, 69)$ distribution. By (4.4.6)

$$\frac{(1-\sqrt{\Lambda})(m_E - p + 1)}{p(\sqrt{\Lambda})} = \frac{(67)(1-\sqrt{.0012681})}{3(\sqrt{.0012681})} = 604.822.$$

Since this exceeds $F_{6,134;.05} = 2.10$ the hypothesis H_a is strongly rejected. Also, $-2\ln\lambda = 480.257$. It far exceeds $\chi^2_{36;.o5} = 12.59$. Bartlett's approximation (4.4.9) gives $W = -f\ln\Lambda$ where $f = m_W - (p - m_B + 1)/2 = 68$ follows a $\chi^2_{(6)}$ distribution. Here $W = -68\ln\Lambda = 453.576$.

5.11.2 *Hypothesis* $H_b(\Sigma_1 \ldots = \Sigma_k)$ *(test of homogeneity of covariances)*

Here the unrestricted space Θ is the parameter space in which $\mathbf{\Sigma}_i$ is positive definite and μ_i is any vector of real numbers $(i = 1, \ldots, k)$. The restricted space Θ_b is the subspace of Θ in which $\mathbf{\Sigma}_1 = \ldots = \mathbf{\Sigma}_k$.

Under both Θ_b and Θ, the *m.l.e.* of μ_i is $\hat{\mu}_i = \bar{\mathbf{x}}_i$. Under H_b, the m.l.e. of $\mathbf{\Sigma}_i$ is $\hat{\mathbf{\Sigma}}_i = \mathbf{W}/n$ and under the alternative hypothesis the m.l.e. of $\mathbf{\Sigma}_i$ is $\tilde{\mathbf{\Sigma}}_i = \mathbf{S}_{ni} = n_i^{-1}(\sum_{r=1}^{n_i}(\mathbf{x}_{ri} - \bar{\mathbf{x}}_i)(\mathbf{x}_{ri} - \bar{\mathbf{x}}_i)'(i = 1, \ldots, k)$. Therefore, the likelihood ratio is (vide (5.9.3))

$$\lambda_b = \frac{\Pi_{i=1}^k |\mathbf{S}_{ni}|^{n_i/2}}{|\frac{\mathbf{W}}{n}|^{n/2}}. \tag{5.11.17}$$

Hence,

$$\begin{aligned} -2\ln\lambda_b &= n\ln|\bar{\mathbf{S}}| - \textstyle\sum_{i=1}^k n_i \ln|\mathbf{S}_{n_i}| \\ &= \textstyle\sum_{i=1}^k n_i[\ln|\bar{\mathbf{S}}| - \ln|\mathbf{S}_{ni}|] \\ &= \textstyle\sum_{i=1}^k n_i \ln|\mathbf{S}_{ni}^{-1}\bar{\mathbf{S}}|, \end{aligned} \tag{5.11.18}$$

where $\bar{\mathbf{S}} = \mathbf{W}/n$. The statistic (5.11.18) has an asymptotic χ^2 distribution with $p(p+1)(k-1)/2$ degrees of freedom.

Box (1949) proposed the test statistic M in place of (5.11.18)

$$M = \gamma\sum_{i=1}^{k}(n_i - 1)\ln|\mathbf{S}_{ui}^{-1}\mathbf{S}_u|, \tag{5.11.19}$$

where

$$\gamma = 1 - \frac{2p^2 + 3p - 1}{6(p+1)(k-1)}(\sum_i \frac{1}{n_i - 1} - \frac{1}{n-k})$$

and $\mathbf{S}_u, \mathbf{S}_{ui}$ are unbiased estimates

$$\mathbf{S}_u = \frac{n}{n-k}\mathbf{S}, \ \mathbf{S}_{ui} = \frac{n_i}{n_i - 1}\mathbf{S}_i.$$

Box's M also has an asymptotic $\chi^2_{(Q)}$ distribution with $Q = p(p+1)(k-1)/2$. Box's approximation seems to be good if each n_i exceeds 20 and p and k does not exceed 5.

For $p = 1$, (5.11.18) reduces to Bartlett's statistic for test of homogeneity,

$$n \ln s^2 - \sum_{i=1}^{k} n_i \ln s_i^2$$

where s_i^2 is the variance of the ith sample and s^2 is the pooled estimate of the variance.

EXAMPLE 5.11.2: For the data of Example 5.11.1 test the hypothesis $H_b(\mathbf{\Sigma}_1 = \mathbf{\Sigma}_2 = \mathbf{\Sigma}_3)$.

Here,

$$\mathbf{S}_{n_1} = (21/22)\mathbf{S}_1 = \begin{bmatrix} 11.9318 & 8.9250 & 2.2718 \\ 8.9250 & 17.6591 & 1.8327 \\ 2.2718 & 1.8327 & 0.5155 \end{bmatrix},$$

$$\mathbf{S}_{n_2} = (23/24)\mathbf{S}_2 = \begin{bmatrix} 14.4421 & 17.4896 & 2.1467 \\ 17.4896 & 31.2033 & 4.2167 \\ 2.1467 & 4.2167 & 0.7283 \end{bmatrix},$$

$$\mathbf{S}_{n_3} = (25/26)\mathbf{S}_3 = \begin{bmatrix} 17.3942 & 19.7308 & 4.5769 \\ 19.7308 & 27.6442 & 3.6731 \\ 4.5769 & 3.6731 & 1.1923 \end{bmatrix},$$

$$\bar{\mathbf{S}} = \frac{\mathbf{W}}{n} = \begin{bmatrix} 14.7411 & 15.6820 & 3.0625 \\ 15.6820 & 25.7796 & 3.2919 \\ 3.0625 & 3.2919 & 0.8308 \end{bmatrix}.$$

The eigenvalues of $\mathbf{S}_{n_1}$ are $\lambda_1 = 24.5004, \lambda_2 = 5.5273, \lambda_3 = .0787$. Hence,

$$|\mathbf{S}_{n_1}| = \lambda_1 \times \lambda_2 \times \lambda_3 = 10.6576.$$

Similarly

$$|\mathbf{S}_{n_2}| = 21.4779, \ \mathbf{S}_{n_3}| = -41.2127, \ |\bar{\mathbf{S}}| = 26.0722.$$

Since, $|\mathbf{S}_{n_3}|^{26} = \{-|\mathbf{S}_{n_3}|\}^{26}$, we have considered in λ_b in (5.11.17), $\{-|\mathbf{S}_{n_3}|\}$ in place of $|\mathbf{S}_{n_3}|$. Thus, from (5.11.8),

$$\begin{aligned} -2\ln\lambda_b &= (72)\ln(26.0722) - 22\ln(10.6576) - \\ &\quad -(24)\ln(21.4779) - (26)\ln(41.2127) \\ &= 12.4286. \end{aligned}$$

This is less than $\chi^2_{12;.05} = 21.05$. Hence the hypothesis of homogeneity of covariance matrices is accepted.

5.11.3 *Hypothesis that the k multinormal populations are equal*

In Subsection 5.11.1 we considered the null hypothesis

$$H_a : \mu_1 = \ldots = \mu_k$$

given that $\mathbf{\Sigma}_1 = \ldots = \mathbf{\Sigma}_k$. In Subsection 5.11.2 we considered the null hypothesis

$$H_b : \mathbf{\Sigma}_1 = \ldots = \mathbf{\Sigma}_k.$$

Now, we consider the null hypothesis

$$H_c : \mu_1 = \ldots = \mu_k,\ \mathbf{\Sigma}_1 = \ldots = \mathbf{\Sigma}_k \tag{5.11.20}$$

which is a combination of H_a and H_b. As before, let $\mathbf{X}_i$ be the $n_i \times p$ data matrix whose rows are independent samples from $N_p(\mu_i, \mathbf{\Sigma}_i)(i = 1, \ldots, k)$. Thus Θ is the unrestricted parameter space $\{\mu_i, \mathbf{\Sigma}_i\}(i = 1, \ldots, k)$ where $\mathbf{\Sigma}_i$ is positive definite and Θ_c is the subspace of Θ restricted by (5.11.20).

The hypothesis H_b is then $\theta \in \Theta_b$ given that $\theta \in \Theta$. The hypothesis H_a is then $\theta \in \Theta_a \cap \Theta_b$ given that $\theta \in \Theta_b$. The hypothesis H_c is then $\theta \in \Theta_a \cap \Theta_b$ given that $\theta \in \Theta$.

We now consider the following Lemma.

Lemma 5.11.1: Let $L(\theta|\mathbf{y})$ be the likelihood of θ based on the observations $\mathbf{y}, \theta \in \Omega$. Let H_{01} be the hypothesis that $\theta \in \Omega_{01} \subset \Omega$. Let H_{02} be the hypothesis that $\theta \in \Omega_{01} \cap \Omega_{02}$ given that $\theta \in \Omega_{01}$. Let H_{012} be the hypothesis that $\theta \in \Omega_{01} \cap \Omega_{02}$ given that $\theta \in \Omega$. If $\lambda_1, \lambda_2, \lambda_{12}$ and the likelihood criteria for testing H_{01}, H_{02}, H_{012} are uniquely determined, than

$$\lambda_{12} = \lambda_1 \lambda_2. \tag{5.11.21}$$

Proof. We have

$$\lambda_1 = \frac{\text{Sup}_{\ \theta \in \Omega_{01}} L(\theta|\mathbf{y})}{\text{Sup}_{\ \theta \in \Omega} L(\theta|\mathbf{y})},$$

$$\lambda_2 = \frac{\text{Sup}_{\ \theta \in \Omega_{01} \cap \Omega_{02}} L(\theta|\mathbf{y})}{\text{Sup}_{\ \theta \in \Omega_{01}} L(\theta|\mathbf{y})},$$

$$\lambda_3 = \frac{\text{Sup}_{\ \theta \in \Omega_{01} \cap \Omega_{02}} L(\theta|\mathbf{y})}{\text{Sup}_{\ \theta \in \Omega} L(\theta|\mathbf{y})}.$$

Hence, equality (5.11.21) follows. □

For H_c, the likelihood function is, therefore,

$$\lambda_c = \lambda_a \lambda_b = |\mathbf{I} + \mathbf{W}^{-1}\mathbf{B}|^{-n/2} \Pi_{i=1}^k |\mathbf{S}_{n_i}^{-1} \bar{\mathbf{S}}|^{-n_i/2}.$$

Hence,

$$\begin{aligned} -2\ln\lambda_c &= n\ln|\tfrac{\mathbf{T}}{\mathbf{W}}| - \textstyle\sum_{i=1}^k n_i \ln|\mathbf{S}_{n_i}| + n\ln|\bar{\mathbf{S}}| \\ &= n\ln|\tfrac{\mathbf{S}_0}{\bar{\mathbf{S}}}| - \textstyle\sum_{i=1}^k n_i \ln|\mathbf{S}_{n_i}| + n\ln|\bar{\mathbf{S}}| \\ &= n\ln|\mathbf{S}_0| - \textstyle\sum_{i=1}^k n_i \ln|\mathbf{S}_{n_i}| \end{aligned} \tag{5.11.22}$$

which follows $\chi^2_{(Q)}$ where $Q = p(p+3)(k-1)/2$.

5.11.4 *Union-intersection method*

We want to test the hypothesis $H_a(\mu_1 = \ldots = \mu_k)$ by union-intersection method of Roy (1957). For this we define linear transformation $z_{ri} = \mathbf{a}'\mathbf{x}_{ri}, i = 1, \ldots, k; r = 1, \ldots, n_i$, where $\mathbf{a}$ is a $p \times 1$ vector of constants, thus reducing observation vectors to scalars. To test the univariate k-sample hypothesis $H_{aa}(\mathbf{a}'\mu_1 = \ldots = \mathbf{a}'\mu_k)$ we use the usual univariate F ratio

$$F(\mathbf{a}) = \frac{\sum_{i=1}^k (\bar{z}_i - \bar{z})^2/(k-1)}{\sum_{i=1}^k \sum_{r=1}^{n_i} (z_{ri} - \bar{z}_i)^2/(n-k)} \tag{5.11.23}$$

where $n = \sum_{i=1}^k n_i$, $\bar{z}_i = \sum_{r=1}^{n_i} z_{ri}/n_i$, $\bar{z} = \sum_{i=1}^k \sum_{r=1}^{n_i} z_{ri}/n$. We reject H_{aa} if $F(\mathbf{a}) \geq c$ where c is such that the size of the test α is attained. Note that value of c does not depend on $\mathbf{a}$. We therefore, reject $H_a(\mu_1 = \ldots = \mu_k)$ if $\max_{\mathbf{a}} F(\mathbf{a}) \geq c$. Now,

$$F(\mathbf{a}) = \frac{\mathbf{a}'\mathbf{B}\mathbf{a}/(k-1)}{\mathbf{a}'\mathbf{W}\mathbf{a}/(n-k)}. \tag{5.11.24}$$

Maximizing $F(\mathbf{a})$ is equivalent to maximizing the ratio $\mathbf{a}'\mathbf{B}\mathbf{a}/\mathbf{a}'\mathbf{W}\mathbf{a} = \lambda$. By Theorem A.14.2 the maximum value of λ and the vector $\mathbf{a}$ that produces this maximum value are given by the largest eigenvalue λ_1 of $\mathbf{W}^{-1}\mathbf{B}$ and the associated eigenvector. This may also be seen directly as follows.

To find $\mathbf{a}$ we differentiate λ with respect to $\mathbf{a}$ and set the resulting equation to $\mathbf{0}$.

$$\frac{\partial\lambda}{\partial\mathbf{a}} = \frac{\mathbf{a}'\mathbf{W}\mathbf{a}(2\mathbf{B}\mathbf{a}) - \mathbf{a}'\mathbf{B}\mathbf{a}(2\mathbf{W}\mathbf{a})}{(\mathbf{a}'\mathbf{W}\mathbf{a})^2} = \mathbf{0}.$$

Multiplying by $(\mathbf{a}'\mathbf{W}\mathbf{a})$ we obtain,

$$\mathbf{B}\mathbf{a} - (\frac{\mathbf{a}'\mathbf{B}\mathbf{a}}{\mathbf{a}'\mathbf{W}\mathbf{a}})\mathbf{W}\mathbf{a} = \mathbf{0}$$

i.e.,

$$\mathbf{Ba} - \lambda \mathbf{Wa} = \mathbf{0}$$

or

$$(\mathbf{B} - \lambda \mathbf{W})\mathbf{a} = \mathbf{0}. \tag{5.11.25}$$

Pre-multiplying by $\mathbf{W}^{-1}$, this reduces to

$$(\mathbf{W}^{-1}\mathbf{B} - \lambda \mathbf{I})\mathbf{a} = \mathbf{0}. \tag{5.11.26}$$

The union-intersection test uses only λ_1, the largest eigenvalue of $\mathbf{W}^{-1}\mathbf{B}$. Since the rank of $\mathbf{W}^{-1}\mathbf{B}$ is the same as the rank of $\mathbf{B}$, there are $s = \min(p, \nu_B)$ non-zero eigenvalues $\lambda_1 > \lambda_2 > \ldots > \lambda_s > 0$, where ν_B = rank of $\mathbf{B}$.

The function $z = \mathbf{a}'\mathbf{x}$ where $\mathbf{a}$ is the eigenvector corresponding to λ_1 is called the discriminant function for k populations $N_p(\mu_1, \mathbf{\Sigma}), \ldots, N_p(\mu_k, \mathbf{\Sigma})$ (Chapter 11).

To test $H_0(\mu_1 = \ldots = \mu_k)$ using λ_1, we consider two cases.

(a): $\nu_B = 1, p \geq 2$. Here we use

$$F = \frac{\nu_W - p + 1}{p} \lambda_1$$

which has the exact F distribution with $|\nu_B - p| + 1$ and $\nu_W - p + 1$ d.f.

(b): $\nu_B \geq 2, p \geq 2$. Here, we use Roy's test statistic

$$\theta = \frac{\lambda_1}{1 + \lambda_1}.$$

Critical values of θ are given in Table B.12. We reject $H_a : (\mu_1 = \ldots = \mu_k)$ if $\theta \geq \theta_{\alpha, s, m, N}$, where

$$s = \min\,(\nu_B, p), m = \frac{1}{2}(|\nu_B - p| - 1), N = \frac{1}{2}(\nu_W - p - 1).$$

5.12 Some Further Tests for $H_0(\mu_1 = \ldots \mu_k)$ Assuming Unequal Dispersion Matrices

As in Section 5.11, assume that $\mathbf{x}_{ri}$ is a random sample from $N_p(\mu_i, \mathbf{\Sigma}_i), r = 1, \ldots, n_i; i = 1, \ldots, k$. We want to test the null hypothesis $H_0(\mu_1 = \ldots = \mu_k)$ but do not assume equality of $\mathbf{\Sigma}_i$'s. Clearly, $\mathbf{x}_{ri}$ and $\mathbf{x}_{sj}$ are independent, $i \neq j = 1, \ldots, k$.

Scheffe' (1943) solved the problem for the univariate case for $k = 2$. This was generalized to multivariate case by Bennette (1951) and Anderson (1963a). Eaton (1969) gave a more general solution which covers all previous solutions as special cases.

5.12.1 *Eaton's test*

Let $n_0 = \min_{i=1,\ldots,k} n_i$ and suppose that for each i we randomly select n_0 of the n_i observations. Without loss of generality, assume that they are the first n_0 observations, $\mathbf{x}_{ri}, r = 1, \ldots, n_0; i = 1, \ldots, k$. Let

$$\mathbf{y}_r = (\mathbf{x}'_{r1}, \mathbf{x}'_{r2}, \ldots, \mathbf{x}'_{rk})'.$$

If $\tilde{\mathbf{x}}_i$ be the mean of the observations $\mathbf{x}_{ri} (r = 1, \ldots, n_0)$ let

$$\bar{\mathbf{y}} = (\tilde{\mathbf{x}}'_1, \tilde{\mathbf{x}}'_2, \ldots, \tilde{\mathbf{x}}'_k)'$$

be the sample mean of $\mathbf{y}$. Let also

$$\theta = (\mu'_1, \mu'_2, \ldots, \mu'_k)'.$$

Then

$$\bar{\mathbf{y}} \cap N_{kp}(\theta, \mathbf{\Sigma}) \text{ where } \mathbf{\Sigma} = \text{ Diag. } (\mathbf{\Sigma}/n_1, \mathbf{\Sigma}/2, \ldots, \mathbf{\Sigma}_k/n_k).$$

Define

$$\mathbf{v}_r = (n_1^{-1/2}\mathbf{x}'_{r1}, n_2^{-1/2}\mathbf{x}'_{r2}, \ldots, n_k^{-1/2}\mathbf{x}'_{rk})', r = 1, \ldots, n_0;$$

$$\bar{\mathbf{v}} = \sum_{r=1}^{n_0} \mathbf{v}_r/n_0, \quad \mathbf{Q}_0 = \sum_{r=1}^{n_0} (\mathbf{v}_r - \bar{\mathbf{v}})(\mathbf{v}_r - \bar{\mathbf{v}})'.$$

Then $\mathbf{Q}_0 \cap W_{kp}(n_0 - 1, \mathbf{\Sigma})$, because, $\mathbf{v}_r (r = 1, \ldots, n_0)$ are *iid np*-variate normal with covariance matrix $\mathbf{\Sigma}$ (Corollary 4.2.10.1). Also $\mathbf{Q}_0$ is statistically independent of $\bar{\mathbf{v}}$ and hence of $\bar{\mathbf{y}}$.

Now, H_0 is true *iff*

$$\mu_1 - \mu_2 = \mu_2 - \mu_3 = \ldots = \mu_{k-1} - \mu_k = \mathbf{0},$$

i.e.,

$$\begin{bmatrix} \mathbf{I}_p & -\mathbf{I}_p & \mathbf{0} & \ldots & \mathbf{0} \\ \mathbf{0} & -\mathbf{I}_p & -\mathbf{I}_p & \ldots & \mathbf{0} \\ . & . & . & \ldots & . \\ \mathbf{0} & \mathbf{0} & \mathbf{0} & \ldots & -\mathbf{I}_p \end{bmatrix} \begin{bmatrix} \mu_1 \\ \mu_2 \\ . \\ . \\ \mu_k \end{bmatrix} = \mathbf{0} \tag{5.12.1}$$

or

$$\mathbf{C}\theta = \mathbf{0}. \tag{5.12.2}$$

We use the transformed observations $\mathbf{z}_r = \mathbf{C}\mathbf{y}_r, r = 1, \ldots, n_0$. Then the mean $\bar{\mathbf{z}} = \mathbf{C}\bar{\mathbf{y}}$ follows, under H_0, $N_{(k-1)p}(\mathbf{0}, \mathbf{C\Sigma C}')$. Also, $\mathbf{C}\mathbf{Q}_0\mathbf{C}' \cap W_{p(k-1)}(n_0 - 1, \mathbf{C\Sigma C}')$. Hence, under H_0,

$$T_0^2 = (n_0 - 1)(\mathbf{C}\bar{\mathbf{z}})'[\mathbf{C}\mathbf{Q}_0\mathbf{C}']^{-1}\mathbf{C}\bar{\mathbf{z}} \tag{5.12.3}$$

follows T^2 distribution with $(p(k-1), n_0-1)$ degrees of freedom, which can be used for testing H_0.

The above procedure holds for any matrix $\mathbf{C}$ of rank r, provided $r \le \min(pk, n_0-1)$.

The test suffers from the following disadvantages: (i) when sample sizes are unequal, a part of the data is ignored in calculating T_0^2; (ii) the test involves estimates of known elements of $\mathbf{\Sigma}$, namely, the off-diagonal blocks of zeros, so that the test is inefficient; (iii) randomization is required for selecting n_0 observations from each group. For these reasons the test is seldom recommended in practice.

5.12.2 *James's test*

James (1954) gave an approximate test for $H_0(\mu_1 = \ldots = \mu_k)$ when the $\mathbf{\Sigma}_i$'s may be unequal. Let

$$\mathbf{z}_i = \bar{\mathbf{x}}_i - \bar{\mathbf{x}}_k, i = 1, \ldots, k-1.$$

Then,

$$\text{Cov }(\mathbf{z}_i, \mathbf{z}_h) = V(\bar{\mathbf{x}}_k) = \mathbf{\Sigma}_k / n_k = \mathbf{A}_k \quad (\text{say}) \, , i \neq h;$$

$$V(\mathbf{z}_i) = \mathbf{A}_i + \mathbf{A}_k, i = 1, \ldots, k-1.$$

Writing $\mathbf{z}' = (\mathbf{z}_1', \mathbf{z}_2', \ldots, \mathbf{z}_{k-1})'$,

$$V[\mathbf{z}] = \begin{bmatrix} \mathbf{A}_1 + \mathbf{A}_k & \mathbf{A}_k & \ldots & \mathbf{A}_k \\ \mathbf{A}_k & \mathbf{A}_2 + \mathbf{A}_k & \ldots & \mathbf{A}_k \\ . & . & \ldots & . \\ \mathbf{A}_k & \mathbf{A}_k & \ldots & \mathbf{A}_{k-1} + \mathbf{A}_k \end{bmatrix} = \mathbf{V} \quad (\text{say}) \, .$$

Hence, under $H_0, \mathbf{z} \cap N_{(k-1)p}(\mathbf{0}, \mathbf{V})$ and $\mathbf{z}'\mathbf{V}^{-1}\mathbf{z} \cap \chi^2_{(k-1)p}$. If $\mathbf{V}$ is estimated by $\hat{\mathbf{V}}$, which is obtained by replacing $\mathbf{A}_i$ by $\mathbf{S}_i/n_i$ where $\mathbf{S}_i = (n_i-1)^{-1}\sum_{r=1}^{n_i}(\mathbf{x}_{ri} - \bar{\mathbf{x}}_i)(\mathbf{x}_{ri} - \bar{\mathbf{x}}_i)'$, then $\mathbf{z}'\hat{\mathbf{V}}^{-1}\mathbf{z}$ follows, under H_0, approximately a $\chi^2_{(k-1)p}$ distribution and this can be used for testing the null hypothesis.

5.13 Exercises and Complements

5.1 Let $\mathbf{x}_1, \ldots, \mathbf{x}_n$ be a random sample from $N_p(\mu, \mathbf{\Sigma})$. Show that the likelihood ratio statistic for $H_0(\mu = \mu_0, \mathbf{\Sigma} = \mathbf{\Sigma}_0)$ is given by

$$\lambda = (\frac{e}{n})^{np/2} |\mathbf{Q}\mathbf{\Sigma}_0^{-1}|^{n/2} \exp(-\frac{1}{2}[\text{ tr } (\mathbf{Q}\mathbf{\Sigma}_0^{-1})$$

$$+n(\bar{\mathbf{x}}-\mu_0)'\boldsymbol{\Sigma}_0^{-1}(\bar{\mathbf{x}}-\mu_0)]\}.$$

5.2 Let $\mathbf{x}_1, \ldots, \mathbf{x}_n$ be *iid* $N_p(\mu, \boldsymbol{\Sigma})$ where $\boldsymbol{\Sigma}$ is of the intraclass correlation form (5.8.14). Show that the maximum likelihood estimates of σ^2 and ρ are obtained from (5.8.15).

5.3 The unbiased estimates of the mean vector μ and variance-covariance matrix $\boldsymbol{\Sigma}$ of a tri-variate normal distribution are given by

$$\bar{\mathbf{x}} = (3.41,\ 2.23,\ 1.40)',$$

$$\mathbf{S} = \begin{bmatrix} 16.90 & 22.01 & 12.35 \\ 22.01 & 51.72 & 19.46 \\ 12.35 & 19.46 & 28.73 \end{bmatrix}.$$

Test the following hypotheses at 5% level of significance: (i) $\mu_1 = \mu_2$; (ii) $\mu_1 - \mu_2 = \mu_2 - \mu_3$; (iii) $\mu_1 = \mu_2 = \mu_3 = 3$.

5.4 Let $\mathbf{x}_1, \ldots, \mathbf{x}_n (n > p)$ be a random sample from $N_p(\mu, \boldsymbol{\Sigma})$ where μ and $\boldsymbol{\Sigma}$ are unknown. Show that $T^2 = n(\bar{\mathbf{x}}'\mathbf{S}^{-1}\bar{\mathbf{x}} - \frac{(\bar{\mathbf{x}}'\mathbf{S}^{-1}\mathbf{1})^2}{\mathbf{1}'\mathbf{S}^{-1}\mathbf{1}}]$ can be used for testing $H_0(\mu_1 = \ldots = \mu_p)$ against $H($ not all μ's are equal), where $\bar{\mathbf{x}}$ and $\mathbf{S}$ are sample mean vector and the unbiased estimates of $\boldsymbol{\Sigma}$ respectively and $\mathbf{1} = (1, \ldots, 1)'$. Derive the non-null distribution of the test statistic.

5.5 Let $\bar{\mathbf{x}}$ and $\mathbf{A}$ represent the mean vector and the matrix of corrected sum of squares (SS) and the sum of products (SP) of observations in a sample of size n from a $N_p(\mu, \boldsymbol{\Sigma})$ respectively. Show that Hotelling's T^2 for testing $H_0(\mu = \mu_0)$, given by

$$\frac{T^2}{n-1} = n(\bar{\mathbf{x}} - \mu_0)'\mathbf{A}^{-1}(\bar{\mathbf{x}} - \mu_0)$$

is obtained as

$$\max{}_{\mathbf{a}} n(n-1)\frac{[\mathbf{a}'(\bar{\mathbf{x}} - \mu_0)]^2}{\mathbf{a}'\mathbf{A}\mathbf{a}}$$

where $\mathbf{a}$ is a non-null $p \times 1$ vector.

A statistician, by mistake, uses the statistic

$$\frac{T_0^2}{n-1} = n(\bar{\mathbf{x}} - \mu_0)\mathbf{A}_0^{-1}(\bar{\mathbf{x}} - \mu_0)$$

where $\mathbf{A}_0$ is the matrix of SS and SP of the observations measured about the true mean μ_0.

Find the relation between T^2 and T_0^2 and derive the distribution of T_0^2, assuming the distribution of T^2.

5.6 Let $\bar{\mathbf{x}} = (\bar{x}_1, \ldots, \bar{x}_p)'$ and $\mathbf{S} = ((s_{ij}))(i, j = 1, \ldots, p)$ be based on n observations from $N_p(\mu, \mathbf{\Sigma})$. Let $\mathbf{x} = (x_1, \ldots, x_p)'$ be an additional observation from the same distribution. Show that $(\mathbf{x} - \bar{\mathbf{x}})$ is distributed as $N_p(\mathbf{0}, \frac{n+1}{n}\mathbf{\Sigma})$. Verify that $\frac{n}{n+1}(\mathbf{x} - \bar{\mathbf{x}})'\mathbf{S}^{-1}(\mathbf{x} - \bar{\mathbf{x}})$ has T^2 distribution with $(n-1)$ degrees of freedom.

5.7 The following table gives the estimated dispersion matrices of two anthropometric characters, maximum head length x_1 and maximum head breadth x_2 for Muslims in two districts of Bangladesh:

$$\text{Dacca:} \quad \begin{bmatrix} 41.00 & 7.05 \\ 7.05 & 27.72 \end{bmatrix}$$

$$\text{Mymensingh:} \quad \begin{bmatrix} 40.37 & 4.16 \\ 4.16 & 25.37 \end{bmatrix}$$

Sample sizes for Dacca and Mymensingh are 357 and 299 respectively. Are the two dispersion matrices significantly different?

5.8 In the bird data of Table 2.E.1 consider two groups of birds: Group S consisting of birds 1-21 who survived a certain treatment and Group P comprising of the remaining birds. Test the hypothesis (i) the two groups have the same body measurements (ii) the dispersion matrices of the measurements for the groups are the same (iii) the two groups come from the same population.

5.9 For the data on agricultural land and household income as given in Table 2.E.2 test the hypotheses (i) $\mu_1 = \mu_2$, (ii) $\mathbf{\Sigma}_1 = \mathbf{\Sigma}_2$, (iii) $\mu_1 = \mu_2, \mathbf{\Sigma}_1 = \mathbf{\Sigma}_2$.

5.10 For the data on irises in Table 11.E.2 test the hypotheses that all the three groups setosa, versicolor and verginica have the same measurements.

Chapter 6

Multivariate Regression Analysis

6.1 Introduction

Consider the model defined by

$$\mathbf{Y} = \mathbf{X}\mathbf{B} + \mathbf{U} \tag{6.1.1}$$

where $\mathbf{Y}(n \times p)$ is an observed matrix of p response variables on each of n individuals, $\mathbf{X}(n \times q)$ is a known matrix of rank $r(\leq q), \mathbf{B}(q \times p)$ is a matrix of unknown regression parameters and $\mathbf{U}(n \times p)$ is a matrix of unobserved random disturbances. We shall assume that for given $\mathbf{X}$, rows of $\mathbf{U}$ are uncorrelated, each with mean $\mathbf{0}$ and common variance matrix $\mathbf{\Sigma}$. Later, for studying the properties of the estimators of the parameters and testing of hypotheses we shall assume that the rows of $\mathbf{U}$ are normally distributed.

If the elements of $\mathbf{X}$ are quantitative, for example, measurements on controlled regressor or predictor variables, then (6.1.1) is the *multiple regression model* and generally, $r = q$. However, if the elements of $\mathbf{X}'$ are 1 or 0 so that $\mathbf{X}$ represents underlying qualitative factors, then we have an *analysis of variance* model. In this case, $\mathbf{X}$ is sometimes called the *design matrix*, as it expresses the structure imposed by the underlying experimental design, and here, usually $r < q$. When $\mathbf{X}$ is a mixture of qualitative and quantitative elements, we refer to the model as an *analysis of covariance model.*

If $\mathbf{X}$ is a random matrix, then the distribution of $\mathbf{X}$ is assumed to be uncorrelated with $\mathbf{U}$. If $\mathbf{X}$ is random then the likelihood of all parameters and expectation of all functions of $\mathbf{Y}$ are to be interpreted as conditional on $\mathbf{X}$.

We shall suppose that $\mathbf{X} = (\mathbf{1}, \mathbf{X}_1)$ meaning that the first column of $\mathbf{X}$ is $\mathbf{1}$ to allow for an overall mean effect. For simplicity, we shall treat $\mathbf{X}$ as a fixed matrix throughout.

The columns of $\mathbf{Y}$ represent observations on dependent variables, which are to be explained in terms of the independent variables, whose values on different units are given by the columns of $\mathbf{X}$. We have,

$$\mathbf{Y} = \begin{bmatrix} \mathbf{y}_1' \\ \mathbf{y}_2' \\ . \\ . \\ \mathbf{y}_n' \end{bmatrix}, \quad \mathbf{X} = \begin{bmatrix} \mathbf{x}_1' \\ \mathbf{x}_2' \\ . \\ . \\ \mathbf{x}_n' \end{bmatrix}, \quad \mathbf{U} = \begin{bmatrix} \mathbf{u}_1' \\ \mathbf{u}_2' \\ . \\ . \\ \mathbf{u}_n' \end{bmatrix}$$

where $\mathbf{y}_i = (y_{i1}, y_{i2}, \dots y_{ip})', y_{ij}$ denoting the value of the jth response variable y_j on the ith unit, $i = 1, \dots, n$ and similarly for $\mathbf{x}_i$ and $\mathbf{u}_i$. Also

$$\mathbf{B} = (\beta_{(1)}, \beta_{(2)}, \dots, \beta_{(p)}) = \begin{bmatrix} \beta_1' \\ . \\ . \\ \beta_q' \end{bmatrix}$$

where $\beta_{(j)}(q \times 1) = (\beta_{1j}, \beta_{2j}, \dots, \beta_{qj})'(j = 1, \dots, p), \beta_i = (\beta_{i1}\ \beta_{i2}, \dots \beta_{ip})'$ $(i = 1, \dots, q)$. Note that

$$E(y_{ij}) = \mathbf{x}_i'\beta_{(j)}$$

so that the expected value of y_{ij} depends on the ith row of $\mathbf{X}$ and the jth column of the matrix of regression coefficients. Note also that the model (6.3.1) means

$$[\mathbf{y}_{(1)}\ \mathbf{y}_{(2)} \cdots \mathbf{y}_{(p)}] = \mathbf{X}[\beta_{(1)}\ \beta_{(2)}\ \dots \beta_{(p)}] + [\mathbf{u}_{(1)}\ \mathbf{u}_{(2)}\ \dots \mathbf{u}_{(p)}]$$

where $\mathbf{y}_{(j)} = (y_{1j}, \dots, y_{nj})'$ denotes the vector of values of y_j and similarly for $\mathbf{u}_{(j)}$. Thus,

$$\mathbf{y}_{(j)} = \mathbf{X}\beta_{(j)} + \mathbf{u}_{(j)}, \quad j = 1 \dots p.$$

We shall assume that $\mathbf{u}_i$ are normally distributed so that $\mathbf{U}$ is a data matrix from

$$N_p(\mathbf{0}, \mathbf{\Sigma}). \tag{6.1.2}$$

Sometimes a row representation is useful.

$$\mathbf{y}_i' = \mathbf{x}_i'\mathbf{B} + \mathbf{u}_i'$$

or

$$\mathbf{y}_i = \mathbf{B}'\mathbf{x}_i + \mathbf{u}_i, \quad i = 1, \dots, n.$$

The log-likelihood for $\mathbf{B}$ and $\mathbf{\Sigma}$ for given $\mathbf{Y}$ under the assumption (6.1.2) is given by

$$l(\mathbf{B}, \mathbf{\Sigma}) = -\frac{n}{2} \log |2\pi\Sigma| - \frac{1}{2} \text{ tr } [(\mathbf{Y} - \mathbf{XB})\mathbf{\Sigma}^{-1}(\mathbf{Y} - \mathbf{XB})']. \qquad (6.1.3)$$

This can be seen as follows. The probability density function of $\mathbf{Y}$ is

$$f_{\mathbf{Y}}(\mathbf{y}; \mathbf{B}, \mathbf{\Sigma}) = \frac{1}{(2\pi)^{np/2}|\mathbf{\Sigma}|^{n/2}} \exp[-\frac{1}{2}\sum_{i=1}^{n}(\mathbf{y}_i - \mathbf{B}'\mathbf{x}_i)'\mathbf{\Sigma}^{-1}(\mathbf{y}_i - \mathbf{B}'\mathbf{x}_i)]. \qquad (6.1.4)$$

The exponent is

$$\begin{aligned} &= -\tfrac{1}{2}\textstyle\sum_{i=1}^{n} \text{ tr } [\mathbf{\Sigma}^{-1}(\mathbf{y}_i - \mathbf{B}'\mathbf{x}_i)(\mathbf{y}_i - \mathbf{B}'\mathbf{x}_i)'] \quad \text{(by (A.9.7))} \\ &= -\tfrac{1}{2} \text{ tr } [\mathbf{\Sigma}^{-1}\textstyle\sum_{i=1}^{n}(\mathbf{y}_i - \mathbf{B}'\mathbf{x}_i)(\mathbf{y}_i - \mathbf{B}'\mathbf{x}_i)'] \\ &= -\tfrac{1}{2} \text{ tr } [\mathbf{\Sigma}^{-1}(\mathbf{Y} - \mathbf{XB})'(\mathbf{Y} - \mathbf{XB})] \\ &= -\tfrac{1}{2} \text{ tr } [(\mathbf{Y} - \mathbf{XB})\mathbf{\Sigma}^{-1}(\mathbf{Y} - \mathbf{XB})'] \quad \text{(by (A.9.1))}\ . \end{aligned} \qquad (6.1.5)$$

In Section 6.2 we shall find the maximum likelihood estimates (*m.l.e.*'s) of parameters $\mathbf{B}$ and $\mathbf{\Sigma}$ and study their properties. The next section addresses their least square estimation, offers geometrical interpretation of $\hat{\mathbf{B}}$ and reconciles between the ordinary and generalized least square estimates under the multivariate regression model. Forecast of an observation along with producing simultaneous confidence intervals are the subject matter of the subsequent section. Afterwards we consider the likelihood ratio test of the general linear hypothesis and restricted least square estimation of $\hat{\mathbf{B}}$. The results are elucidated by examples drawn from the field of comparison of two mean vectors, missing observations, Profile analysis for $K(\geq 2)$ populations, etc. The test of general linear hypothesis by Union-Intersection principle and simultaneous confidence intervals for linear functions of regression parameters are considered. Subsequently, the mean-centered multivariate regression model is looked into, the classical linear regression model ($p = 1$) is visited, the notion of multiple correlation coefficient is examined and the problem of selection of variables is investigated. Section 6.11 assesses the proportion of variation in $\mathbf{Y}$ explained by the multivariate regression model. Subsequently the problem of selection of variables in the multivariate model is investigated and different growth curve models are studied.

6.2 Maximum Likelihood Estimators

Assume that $n \geq p+q$. Also assume that $\mathbf{X}$ is of full rank q so that $(\mathbf{X}'\mathbf{X})^{-1}$ exists. This is because, $\mathbf{X}'\mathbf{X}$ has rank $q(< n)$. If $\mathbf{X}'\mathbf{X}$ is not of full rank,

$\mathbf{X'Xa} = \mathbf{0}$ for some $\mathbf{a} \neq \mathbf{0}$. But then $\mathbf{a'X'Xa} = 0$ or $\mathbf{Xa} = \mathbf{0}$. This contradicts the assumption that $\mathbf{X}$ is of full rank. Hencs, $\mathbf{X'X}$ has full rank q.

Let

$$\mathbf{P} = \mathbf{I}_n - \mathbf{X}(\mathbf{X'X})^{-1}\mathbf{X'}. \tag{6.2.1}$$

Thus $\mathbf{P}(n \times n)$ is a symmetric idempotent matrix of rank $n - q$ which projects any vector (matrix) onto the subspace of $\mathcal{R}^n$ which is orthogonal to the column space of $\mathbf{X}$. Note that $\mathbf{PX} = \mathbf{0}$. Also $\mathbf{Pz} = \mathbf{z}$ *iff* $\mathbf{z}$ is orthogonal to the range-space of $\mathbf{X}$.

Theorem 6.2.1 For the log likelihood function (6.1.3), the *m.l.e.* of $\mathbf{B}$ and $\mathbf{\Sigma}$ are given by

$$\hat{\mathbf{B}} = (\mathbf{X'X})^{-1}\mathbf{X'Y}, \tag{6.2.2}$$

$$\hat{\mathbf{\Sigma}} = n^{-1}\mathbf{Y'PY}. \tag{6.2.3}$$

Proof. Assuming that the results (6.2.2) and (6.2.3) are true, the 'fitted values' of $\mathbf{Y}$ and the 'estimated error' matrix or residual matrix are, respectively,

$$\hat{\mathbf{Y}} = \mathbf{X}\hat{\mathbf{B}} = \mathbf{X}(\mathbf{X'X})^{-1}\mathbf{X'Y}, \tag{6.2.4}$$

$$\hat{\mathbf{U}} = \mathbf{Y} - \hat{\mathbf{Y}} = \mathbf{PY}. \tag{6.2.5}$$

Then,

$$\mathbf{Y} - \mathbf{XB} = \hat{\mathbf{U}} + \mathbf{X}(\hat{\mathbf{B}} - \mathbf{B}).$$

Therefore, from (6.1.5), the second term on the right hand side of (6.1.3) can be written as

$$-\frac{1}{2}\ \text{tr}\ [\mathbf{\Sigma}^{-1}\{\hat{\mathbf{U}} + \mathbf{X}(\hat{\mathbf{B}} - \mathbf{B})\}'\{\hat{\mathbf{U}} + \mathbf{X}(\hat{\mathbf{B}} - \mathbf{B})\}]$$

$$= -\frac{1}{2}\ \text{tr}\ [\mathbf{\Sigma}^{-1}\hat{\mathbf{U}}'\hat{\mathbf{U}} + \mathbf{\Sigma}^{-1}(\hat{\mathbf{B}} - \mathbf{B})'\mathbf{X'X}(\hat{\mathbf{B}} - \mathbf{B})],$$

because $\hat{\mathbf{U}}'\mathbf{X} = \mathbf{Y'PX} = \mathbf{0}$. Now,

$$\hat{\mathbf{\Sigma}} = n^{-1}\mathbf{Y'PY} = n^{-1}\hat{\mathbf{U}}'\hat{\mathbf{U}}.$$

Hence, from (6.1.3),

$$\begin{aligned} l(\mathbf{B}, \mathbf{\Sigma}) &= -\tfrac{n}{2}\log|2\pi\mathbf{\Sigma}| - \tfrac{1}{2}\ \text{tr}\ \{\mathbf{\Sigma}^{-1}[\hat{\mathbf{U}}'\hat{\mathbf{U}} + (\hat{\mathbf{B}} - \mathbf{B})'\mathbf{X'X}(\hat{\mathbf{B}} - \mathbf{B})]\} \\ &= -\tfrac{n}{2}\log|2\pi\mathbf{\Sigma}| - \tfrac{n}{2}\ \text{tr}\ [\mathbf{\Sigma}^{-1}\hat{\mathbf{\Sigma}}] - \tfrac{1}{2}\ \text{tr}\ \mathbf{\Sigma}^{-1}(\hat{\mathbf{B}} - \mathbf{B})'\mathbf{X'X}(\hat{\mathbf{B}} - \mathbf{B}). \end{aligned} \tag{6.2.6}$$

Only the last term of (6.2.6) involves $\mathbf{B}$ and this is maximized when $\mathbf{B} = \hat{\mathbf{B}}$. Therefore, the reduced log-likelihood function is given by

$$\begin{aligned} l(\hat{\mathbf{B}}, \mathbf{\Sigma}) &= -\tfrac{n}{2} \log |2\pi\mathbf{\Sigma}| - \tfrac{n}{2} \text{ tr } (\mathbf{\Sigma}^{-1}\hat{\mathbf{\Sigma}}) \\ &= -\tfrac{np}{2} \log 2\pi - \tfrac{n}{2} \log |\mathbf{\Sigma}| - \tfrac{n}{2} \text{ tr } (\mathbf{\Sigma}^{-1}\hat{\mathbf{\Sigma}}). \end{aligned} \tag{6.2.7}$$

Now we recall

Theorem 3.6.1: For any fixed matrix $\mathbf{A} > 0$,

$$f(\mathbf{\Sigma}) = \frac{1}{|\mathbf{\Sigma}|^{n/2}} \exp[-\frac{1}{2} \text{ tr } (\mathbf{\Sigma}^{-1}\mathbf{A})]$$

is maximized over $\mathbf{\Sigma} > 0$ by $\mathbf{\Sigma} = n^{-1}\mathbf{A}$ and

$$f(n^{-1}\mathbf{A}) = |n^{-1}\mathbf{A}|^{-n/2} \exp[-\frac{np}{2}].$$

Applying this theorem, from (6.2.7), the log-likelihood function $l(\hat{\mathbf{B}}, \mathbf{\Sigma})$ is maximized over $\mathbf{\Sigma} > 0$ for $\mathbf{\Sigma} = \hat{\mathbf{\Sigma}}$. Hence the theorem. □

The maximum value of the log-likelihood function is given by

$$l(\hat{\mathbf{B}}, \hat{\mathbf{\Sigma}}) = -\frac{n}{2} \log |2\pi\hat{\mathbf{\Sigma}}| - \frac{np}{2}. \tag{6.2.8}$$

From (6.2.6) we note that the statistic $(\hat{\mathbf{B}}, \hat{\mathbf{\Sigma}})$ are sufficient for $(\mathbf{B}, \mathbf{\Sigma})$. Also, the estimated value of the error matrix is

$$\hat{\mathbf{U}} = \mathbf{PY}. \tag{6.2.9}$$

The estimated value of the $\mathbf{Y}$-matrix is

$$\hat{\mathbf{Y}} = \mathbf{X}\hat{\mathbf{B}} = (\mathbf{I} - \mathbf{P})\mathbf{Y}. \tag{6.2.10}$$

Again,

$$\hat{\mathbf{U}}'\hat{\mathbf{U}} = (\mathbf{Y} - \mathbf{X}\hat{\mathbf{B}})'(\mathbf{Y} - \mathbf{X}\hat{\mathbf{B}}) = \mathbf{Y}'\mathbf{Y} - \hat{\mathbf{Y}}'\hat{\mathbf{Y}}. \tag{6.2.11}$$

Again,

$$n\hat{\mathbf{\Sigma}} = \mathbf{Y}'\mathbf{PY} = \hat{\mathbf{U}}'\hat{\mathbf{U}}. \tag{6.2.12}$$

Thus, the estimated error or the residual sum of squares and the product (SSP) matrix $\hat{\mathbf{U}}'\hat{\mathbf{U}}$ equals the observed SSP matrix $\mathbf{Y}'\mathbf{Y}$ minus the fitted SSP matrix $\hat{\mathbf{Y}}\hat{\mathbf{Y}}$.

Also,

$$\mathbf{X}\hat{\mathbf{U}} = \mathbf{X}'\mathbf{PY} = \mathbf{0}, \; \hat{\mathbf{Y}}'\hat{\mathbf{U}} = \mathbf{Y}'\mathbf{QPY} = \mathbf{0}.$$

EXAMPLE 6.2.1: Suppose

$$\mathbf{Y} = \begin{bmatrix} 8 & 12 \\ 7 & 9 \\ 10 & 5 \\ 6 & 8 \\ 8 & 8 \\ 7 & 12 \\ 9 & 12 \\ 5 & 21 \\ 7 & 11 \\ 8 & 13 \end{bmatrix}, \quad \mathbf{X} = \begin{bmatrix} 1 & 98 & 8 \\ 1 & 107 & 5 \\ 1 & 103 & 6 \\ 1 & 88 & 15 \\ 1 & 91 & 10 \\ 1 & 90 & 12 \\ 1 & 84 & 15 \\ 1 & 72 & 14 \\ 1 & 82 & 11 \\ 1 & 64 & 9 \end{bmatrix}.$$

Here, the m.l.e. of $\mathbf{B}$ is

$$\begin{aligned} \hat{\mathbf{B}} &= (\mathbf{X}'\mathbf{X})^{-1}\mathbf{X}'\mathbf{Y} \\ &= \begin{bmatrix} 12.0364 & -0.1001 & -0.2985 \\ -0.1001 & 0.0009 & 0.0019 \\ -0.2985 & 0.0019 & 0.0126 \end{bmatrix} \begin{bmatrix} 75 & 111 \\ 6651 & 9420 \\ 767 & 1227 \end{bmatrix} \\ &= \begin{bmatrix} 7.7620 & 26.4756 \\ 0.0148 & -0.1918 \\ -0.1485 & 0.1410 \end{bmatrix}. \end{aligned}$$

Also, the m.l.e. of $\mathbf{\Sigma}$ is

$$\begin{aligned} \hat{\mathbf{\Sigma}} &= n^{-1}\mathbf{Y}'\{\mathbf{I}_n - \mathbf{X}(\mathbf{X}'\mathbf{X})^{-1}\mathbf{X}'\}\mathbf{Y} \\ &= \begin{bmatrix} 1.45912 & -1.93907 \\ -1.93907 & 9.16208 \end{bmatrix}. \end{aligned}$$

The matrix of estimated y-values is

$$\hat{\mathbf{Y}} = \mathbf{X}\hat{\mathbf{B}} = \begin{bmatrix} 8.0205 & 8.8106 \\ 8.5990 & 6.6616 \\ 8.3914 & 7.5697 \\ 6.8330 & 11.7154 \\ 7.6200 & 10.4350 \\ 7.3082 & 10.9088 \\ 6.7740 & 12.4825 \\ 6.7454 & 14.6427 \\ 7.3386 & 12.3019 \\ 7.3700 & 15.4717 \end{bmatrix},$$

and the estimated error matrix is

$$\hat{\mathbf{U}} = \begin{bmatrix} -0.02047 & 3.18940 \\ -1.59898 & 2.33838 \\ 1.60862 & -2.56972 \\ -0.83302 & -3.71542 \\ 0.37996 & -2.43501 \\ -0.30818 & 1.09118 \\ 2.22603 & -0.48249 \\ -1.74535 & 6.35732 \\ -0.33862 & -1.30194 \\ 0.63003 & -2.47170 \end{bmatrix}.$$

6.2.1 *Properties of maximum likelihood estimators*

The following theorem gives the properties of the *m.l.e.*'s $\hat{\mathbf{B}}$ and $\hat{\mathbf{\Sigma}}$.

Theorem 6.2.2: Under the model defined by (6.1.1) along with the assumption of normally distributed errors,

(a) $\hat{\mathbf{B}}$ is unbiased for $\mathbf{B}$;
(b) $E(\hat{\mathbf{U}}) = \mathbf{0}$;
(c) $\hat{\mathbf{B}}$ is distributed independently of $\hat{\mathbf{\Sigma}}$.

Proof. (a) We have

$$\begin{aligned}\hat{\mathbf{B}} &= (\mathbf{X}'\mathbf{X})^{-1}\mathbf{X}'\mathbf{Y} = (\mathbf{X}'\mathbf{X})^{-1}\mathbf{X}'(\mathbf{X}\mathbf{B} + \mathbf{U}) \\ &= \mathbf{B} + (\mathbf{X}'\mathbf{X})^{-1}\mathbf{X}'\mathbf{U}.\end{aligned} \tag{6.2.13}$$

Hence, $E(\hat{\mathbf{B}}) = \mathbf{B}$, since $E(\mathbf{U}) = \mathbf{0}$.

(b) $\hat{\mathbf{U}} = \mathbf{Y} - \hat{\mathbf{Y}} = \mathbf{Y} - \mathbf{X}\hat{\mathbf{B}}$. Hence, $E(\hat{\mathbf{U}}) = \mathbf{XB} - \mathbf{XB} = \mathbf{0}$.

(c) Consider Theorem 3.3.3. If $\mathbf{Z}$ is a data matrix from $N_p(\nu, \mathbf{S})$ and if $\mathbf{U} = \mathbf{AZC}$ and $\mathbf{W} = \mathbf{DZE}$, then the elements of $\mathbf{U}$ are independent of the elements $\mathbf{W}$ *iff* either (a) $\mathbf{C}'\mathbf{SE} = \mathbf{0}$ or (b) $\mathbf{AD}' = \mathbf{0}$.

Consider $\hat{\mathbf{B}} = (\mathbf{X}'\mathbf{X})^{-1}\mathbf{X}'\mathbf{Y}$and $\hat{\mathbf{U}} = \mathbf{PY}$. Here we take $\mathbf{A} = \mathbf{P}, \mathbf{D} = (\mathbf{X}'\mathbf{X})^{-1}\mathbf{X}', \mathbf{C} = \mathbf{E} = \mathbf{I}_p, \mathbf{Z} = \mathbf{Y}$. Hence, $\mathbf{AD}' = \mathbf{0}$, since $\mathbf{PX} = \mathbf{0}$. Therefore, $\hat{\mathbf{B}}$ is independent of $\hat{\mathbf{U}}$ and hence of $\hat{\mathbf{\Sigma}} = \hat{\mathbf{U}}'\hat{\mathbf{U}}/n$.

Theorem 6.2.3: (a) The covariance matrix between $\hat{\beta}_{(j)}$ and $\hat{\beta}_{(k)}$ is $\text{Cov}\,(\hat{\beta}_{(j)}, \hat{\beta}_{(k)}) = \sigma_{jk}(\mathbf{X}'\mathbf{X})^{-1},\ j = 1, \ldots, p.$

(b) $n\hat{\mathbf{\Sigma}} = \hat{\mathbf{U}}'\hat{\mathbf{U}} \sim W_p(n - q, \mathbf{\Sigma})$ and hence $E(\hat{\mathbf{U}}'\hat{\mathbf{U}}/(n - q)) = \mathbf{\Sigma}$. Thus, $\hat{\mathbf{\Sigma}} = \hat{\mathbf{U}}'\hat{\mathbf{U}}/n$, the *m.l.e.* of $\mathbf{\Sigma}$ is a biased estimator of $\mathbf{\Sigma}, E(\hat{\mathbf{\Sigma}}) = \frac{n-q}{n}\mathbf{\Sigma}$.

Proof. We have Cov $(\mathbf{y}_j, \mathbf{y}_k) = \delta_{jk}\boldsymbol{\Sigma}$ where $\delta_{jk} = 1(0)$ for $j = (\neq)k = 1, \ldots, n$. Also, Cov $(\mathbf{y}_{(j)}, \mathbf{y}_{(k)}) = \sigma_{jk}\mathbf{I}_n$. Hence,

$$\begin{aligned}\text{Cov.}(\hat{\beta}_{(j)}, \hat{\beta}_{(k)})) &= \text{Cov. } ((\mathbf{X}'\mathbf{X})^{-1}\mathbf{X}'\mathbf{y}_{(j)}, (\mathbf{X}'\mathbf{X})^{-1}\mathbf{X}'\mathbf{y}_{(k)}) \\ &= \sigma_{jk}(\mathbf{X}'\mathbf{X})^{-1}.\end{aligned}$$

Hence, $\hat{\mathbf{B}}$, thought of as a column vector $\hat{\mathbf{B}}^V = (\hat{\beta}'_{(1)}, \hat{\beta}'_{(2)}, \ldots, \hat{\beta}'_{(p)})'$ (vide Definition A.10.1) has Cov $(\hat{\mathbf{B}}^v) = \boldsymbol{\Sigma} \otimes (\mathbf{X}'\mathbf{X})^{-1}$ (Exercise 6.3).

(b) We have $n\hat{\boldsymbol{\Sigma}} = \hat{\mathbf{U}}'\hat{\mathbf{U}} = \mathbf{U}'\mathbf{P}\mathbf{U}$ where $\mathbf{U}$ is a data matrix from $N_p(\mathbf{0}, \boldsymbol{\Sigma})$ and $\mathbf{P}$ given by (6.2.1) is an idempotent matrix of rank $n - q$. Hence the result follows by Theorem 4.2.10(b). Also,

$$E(n\hat{\boldsymbol{\Sigma}}) = (n - q)\boldsymbol{\Sigma}.$$

A proof without the assumption of normality is given in Theorem 6.3.1 (d).)

Corollary 6.2.3.1: Writing $(\mathbf{X}'\mathbf{X})^{-1} = ((g_{ij})) = \mathbf{G}$ we have

(i) Cov $(\hat{\beta}_{ij}, \hat{\beta}_{kl}) = g_{ik}\sigma_{jl}$, Corr $(\hat{\beta}_{ij}, \hat{\beta}_{kl}) = \rho_{jl}g_{ik}/\sqrt{g_{ii}g_{kk}}$ where $\rho_{jl} = \sigma_{jl}/\sqrt{\sigma_{jj}\sigma_{ll}}$.

(ii) Cov $(\hat{\beta}_i, \hat{\beta}_k) = g_{ik}\boldsymbol{\Sigma}$.

6.3 Least Square Estimators

Let $\mathbf{XB} = (\theta_{(1)}, \theta_{(2)}, \ldots, \theta_{(p)}) = \theta$ where $\theta_{(j)} = \mathbf{X}\beta_{(j)}$. Clearly, $\theta_{(j)}$ belongs to $\boldsymbol{\Omega} = \mathcal{R}(\mathbf{X})$, the column-space of $\mathbf{X}, j = 1, \ldots, p$. Assume that $\mathbf{X}$ is of full rank.

To estimate $\mathbf{B}$ we shall estimate $\mathbf{XB} = \theta$ by minimizing

$$(\mathbf{Y} - \theta)'(\mathbf{Y} - \theta) \tag{6.3.1}$$

with respect to θ subject to the conditions that the columns of θ belong to the range-space of $\mathbf{X}$. This will give least square estimate (l.s.e.) of $\mathbf{B}$.

Let $\mathbf{Q}_{\boldsymbol{\Omega}} = \mathbf{Q}$ be the orthogonal projection matrix of $\mathcal{R}_n$ onto $\boldsymbol{\Omega}$, i.e., $\mathbf{Q}$ projects any vector in $\mathcal{R}_n$ onto $\boldsymbol{\Omega}$. Then

$$\mathbf{Q}\theta_{(j)} = \theta_{(j)}, \ \mathbf{Q}\theta = \theta. \tag{6.3.2}$$

Also,

$$\mathbf{Q}_{\boldsymbol{\Omega}} = \mathbf{I}_n - \mathbf{P}_{\boldsymbol{\Omega}} = \mathbf{I}_n - \mathbf{P} = \mathbf{X}(\mathbf{X}'\mathbf{X})^{-1}\mathbf{X}' \tag{6.3.3}$$

where $\mathbf{P}$ has been defined in Section 6.2. $\mathbf{Q}$ is a symmetric idempotent matrix of rank q.

We set $\hat{\theta} = \mathbf{X}\tilde{\mathbf{B}} = \mathbf{QY}$. Now for all θ with columns in $\mathbf{\Omega}$,

$$\begin{aligned}\mathbf{S}(\theta) &= (\mathbf{Y} - \theta)'(\mathbf{Y} - \theta) \\ &= (\mathbf{Y} - \hat{\theta})'(\mathbf{Y} - \hat{\theta}) + (\hat{\theta} - \theta)'(\hat{\theta} - \theta) \\ &\geq (\mathbf{Y} - \hat{\theta})'(\mathbf{Y} - \hat{\theta}) = \mathbf{S}(\hat{\theta}) \end{aligned} \tag{6.3.4}$$

since,

$$\begin{aligned}(\mathbf{Y} - \hat{\theta})'(\hat{\theta} - \theta) &= (\mathbf{Y} - \mathbf{QY})'(\mathbf{QY} - \mathbf{Q}\theta) \\ &= \mathbf{Y}'(\mathbf{I} - \mathbf{Q})'\mathbf{Q}(\mathbf{Y} - \theta) = \mathbf{0}.\end{aligned}$$

Therefore, $\mathbf{S}(\theta)$ is minimized for $\theta = \hat{\theta}$. Also, the minimum value of $\mathbf{S}(\theta)$ is

$$\begin{aligned}\mathbf{S}(\hat{\theta}) &= (\mathbf{Y} - \hat{\theta})'(\mathbf{Y} - \hat{\theta}) = (\mathbf{Y} - \mathbf{QY})'(\mathbf{Y} - \mathbf{QY}) \\ &= \mathbf{Y}'(\mathbf{I} - \mathbf{Q})(\mathbf{I} - \mathbf{Q})\mathbf{Y} = \mathbf{Y}'(\mathbf{I} - \mathbf{Q})\mathbf{Y} = \mathbf{Y}'\mathbf{PY}.\end{aligned} \tag{6.3.5}$$

The above result shows that $\hat{\theta} = \mathbf{QY}$ where $\mathbf{Q}$ is the orthogonal projection of $\mathcal{R}^n$ onto the column space $\mathbf{\Omega} = \mathcal{R}(\mathbf{X})$ of $\mathbf{X}$ is the least square estimator of θ. Thus, the l.s.e. of θ is

$$\hat{\theta} = \mathbf{X}\tilde{\mathbf{B}} = \mathbf{X}(\mathbf{XX}')^{-1}\mathbf{X}'\mathbf{Y}$$

or the l.s.e. of $\mathbf{B}$ is

$$\tilde{\mathbf{B}} = (\mathbf{X}'\mathbf{X})^{-1}\mathbf{X}'\mathbf{Y}. \tag{6.3.6}$$

Alternatively, consider the normal equations for $\mathbf{B}$,

$$\frac{\partial \mathbf{S}}{\partial \mathbf{B}} = \mathbf{0}$$

or

$$\mathbf{X}'(\mathbf{Y} - \hat{\theta}) = \mathbf{0}. \tag{6.3.7}$$

Taking $\hat{\theta} = \mathbf{X}(\mathbf{X}'\mathbf{X})^{-1}\mathbf{X}'\mathbf{Y}$ we see that $\hat{\theta}$ satisfies (6.3.7). Conversely, suppose that $\tilde{\mathbf{B}}$ satisfies (6.3.7). Then $\mathbf{X}'(\mathbf{Y} - \mathbf{X}\tilde{\mathbf{B}}) = \mathbf{0}$. But $\mathbf{X}'(\mathbf{Y} - \hat{\theta}) = \mathbf{0}$, because, $\hat{\theta} \in \mathbf{\Omega}$ (since each $\hat{\theta}_{(j)} \in \mathbf{\Omega}$) and $\mathbf{Y} - \hat{\theta}$ belongs to the space which is orthogonal to $\mathcal{R}(\mathbf{X})$. Hence,

$$\mathbf{X}'(\mathbf{Y} - \mathbf{X}\tilde{\mathbf{B}}) - \mathbf{X}'(\mathbf{Y} - \hat{\theta}) = \mathbf{X}'(\hat{\theta} - \mathbf{X}\tilde{\mathbf{B}}) = \mathbf{0},$$

that is, $\hat{\theta} - \mathbf{X}\tilde{\mathbf{B}} \in \mathbf{\Omega}^{\perp}$. Again, $\mathbf{X}\tilde{\mathbf{B}} \in \mathbf{\Omega}$ and $\hat{\theta} \in \mathbf{\Omega}$. Therefore, $\hat{\theta} - \mathbf{X}\tilde{\mathbf{B}} \in \mathbf{\Omega}$. Hence, $\mathbf{X}\tilde{\mathbf{B}} - \hat{\theta} = \mathbf{0}$ or

$$\mathbf{X}\tilde{\mathbf{B}} = \hat{\theta}$$

or

$$\tilde{\mathbf{B}} = (\mathbf{X}'\mathbf{X})^{-1}\mathbf{X}'\mathbf{Y}, \tag{6.3.8}$$

using $\hat{\theta} = \mathbf{X}(\mathbf{X}'\mathbf{X})^{-1}(\mathbf{X}'\mathbf{X})$.

Thus, we note that the *m.l.e.* and *l.s.e.* of $\mathbf{B}$ are identical and

$$\tilde{\mathbf{B}} = \hat{\mathbf{B}} = (\mathbf{X}'\mathbf{X})^{-1}\mathbf{X}'\mathbf{Y}. \tag{6.3.9}$$

However, normality assumptions are not needed for the derivation of the *l.s.e.*.

Note 6.3.1: Extracting the jth column from each side of (6.3.2),

$$\hat{\beta}_{(j)} = (\mathbf{X}'\mathbf{X})^{-1}\mathbf{X}'\mathbf{y}_{(j)},$$

which is the same as the estimator obtained from the univariate model

$$\mathbf{y}_{(j)} = \mathbf{X}\beta_{(j)} + \mathbf{u}_{(j)}.$$

Thus for least square estimation we can treat each of the p response variables separately, even though $\mathbf{y}_{(j)}$'s are correlated. Therefore, any (e.g., *l.s.e.*, *m.l.e.*) method for finding $\hat{\beta}$ in the univariate case can be applied to find $\hat{\beta}_{(j)}$. The univariate computational techniques extend naturally to the multivariate cases. When $\mathbf{X}$ is ill-conditioned so that $\mathbf{X}'\mathbf{X}$ is nearly singular, the ridge-regression biased estimator in the univariate case can be generalized to the multivariate case. For details in this area some references are Seber (1977, Chapters 11, 12), Berk (1978), Thompson (1978 a, b), Brown and Zidek (1980).

Note 6.3.2: We have

$$\begin{aligned}\mathbf{Y} &= \mathbf{QY} + (\mathbf{I}_n - \mathbf{Q})\mathbf{Y} = \mathbf{QY} + \mathbf{PY}\\ &= \hat{\theta} + (\mathbf{Y} - \hat{\theta}) = \mathbf{X}\hat{\mathbf{B}} + (\mathbf{Y} - \mathbf{X}\hat{\mathbf{B}})\end{aligned}$$

where $\mathbf{X}\hat{\mathbf{B}} \subset \mathbf{\Omega}$ and $(\mathbf{Y} - \mathbf{X}\hat{\mathbf{B}}) \subset \mathbf{\Omega}^{\perp}$. The decomposition is unique.

Note 6.3.3: The least square estimator $\hat{\theta} = \mathbf{X}\hat{\mathbf{B}}$ is the orthogonal projection of $\mathbf{Y}$ onto $\mathbf{\Omega}$, the column space of $\mathbf{X}$. By (A.17.1.2), $\mathbf{X}\hat{\mathbf{B}}$ has the property that

$$\text{Min}_{\ \theta \subset \mathbf{\Omega}} ||\mathbf{Y} - \theta|| = ||\mathbf{Y} - \mathbf{X}\hat{\mathbf{B}}||$$

or

$$\text{Min}_{\ \theta \subset \mathbf{\Omega}} (\mathbf{Y} - \theta)'(\mathbf{Y} - \theta) = (\mathbf{Y} - \mathbf{X}\hat{\mathbf{B}})'(\mathbf{Y} - \mathbf{X}\hat{\mathbf{B}})$$

where $\mathbf{X}\hat{\mathbf{B}} = \mathbf{QY} = \hat{\theta}$.

6.3.1 *Geometrical interpretation of* $\hat{\mathrm{B}}$

Each column of $\mathbf{Y}$ can be split up into two orthogonal parts. The first part $\hat{\mathbf{y}}_{(j)}$ is the projection of $\mathbf{y}_{(j)}$ onto the space spanned by the columns of $\mathbf{X}, \hat{\mathbf{y}}_{(j)} = \mathbf{Q}\mathbf{y}_{(j)}$. The second part $\hat{\mathbf{u}}_{(j)} = \mathbf{P}\mathbf{y}_{(j)}$ is the projection of $\mathbf{y}_{(j)}$ onto the space orthogonal to the columns of $\mathbf{X}$. This property is illustrated in Fig. 6.1.

6.3.2 *Properties of the least square estimators*

Theorem 6.3.1 Under the model (6.1.1), the l.s.e $\hat{\mathbf{B}}$ has the following properties:
(a) $E(\hat{\mathbf{B}}) = \mathbf{B}$; (b) $E(\hat{\mathbf{U}}) = \mathbf{0}$;
(c) Cov $(\hat{\beta}_{(j)}, \hat{\beta}_{(k)}) = \sigma_{jk}(\mathbf{X}'\mathbf{X})^{-1}$
(d) $E(\hat{\mathbf{U}}'\hat{\mathbf{U}}) = (n-q)\mathbf{\Sigma}$. Thus, the l.s.e. of $\mathbf{\Sigma}$ is

$$\tilde{\mathbf{\Sigma}} = \frac{\hat{\mathbf{U}}'\hat{\mathbf{U}}}{n-q}$$

and is unbiased for $\mathbf{\Sigma}$.

Proofs of (a), (b), (c) follow from the proof of Theorems 6.2.2 and 6.2.3. Now

$$\begin{aligned}\hat{\mathbf{U}}'\hat{\mathbf{U}} &= \mathbf{Y}'\mathbf{P}\mathbf{Y} = \mathbf{Y}'\mathbf{P}\mathbf{P}\mathbf{Y} = \mathbf{Y}'(\mathbf{I}-\mathbf{Q})(\mathbf{I}-\mathbf{Q})\mathbf{Y}\\ &= (\mathbf{Y}-\mathbf{Q}\mathbf{Y})'(\mathbf{Y}-\mathbf{Q}\mathbf{Y})\\ &= (\mathbf{Y}-\hat{\theta})'(\mathbf{I}-\mathbf{Q})(\mathbf{Y}-\hat{\theta})\\ &= \hat{\mathbf{U}}'(\mathbf{I}-\mathbf{Q})\hat{\mathbf{U}} = \mathbf{U}'(\mathbf{I}-\mathbf{Q})\mathbf{U}(\text{ because } \mathbf{P}\mathbf{X} = \mathbf{0})\\ &= \textstyle\sum_h \sum_i (\mathbf{I}-\mathbf{Q})_{hi}\mathbf{u}_h\mathbf{u}_i'\end{aligned}$$

where $(\mathbf{I}-\mathbf{Q})_{hi}$ denotes the (h,i)th element of $\mathbf{I}-\mathbf{Q}$. Hence,

$$\begin{aligned}E(\hat{\mathbf{U}}'\hat{\mathbf{U}}) &= \textstyle\sum_h \sum_i (\mathbf{I}-\mathbf{Q})_{hi}\delta_{hi}\mathbf{\Sigma}\\ &= [\textstyle\sum_h (\mathbf{I}-\mathbf{Q}_h)_{hh}]\mathbf{\Sigma}\\ &= \{\text{ tr } (\mathbf{I}_n - \mathbf{Q})\}\mathbf{\Sigma}\\ &= (n-q)\mathbf{\Sigma}.\square\end{aligned}$$

Note 6.3.4: The m.l.e. of $\mathbf{\Sigma}, \hat{\mathbf{\Sigma}} = \hat{\mathbf{U}}\hat{\mathbf{U}}'/n$ is biased, while the l.s.e. $\tilde{\mathbf{\Sigma}}$ is unbiased for $\mathbf{\Sigma}$.

Note 6.3.5: We have $\hat{\mathbf{U}}'\hat{\mathbf{U}} = \hat{\mathbf{U}}'(\mathbf{I}-\mathbf{Q})\hat{\mathbf{U}}$. Hence, if $\mathbf{\Sigma} > 0$ and $n - q = \text{rank}(\mathbf{I}-\mathbf{Q}) \geq p$, then $\hat{\mathbf{U}}'\mathbf{U}$ is positive definite with probability 1 (by result A.11.13).

Note 6.3.6: Note that

$$E(\hat{\theta}) = E(\mathbf{Q}_{\Omega}\mathbf{Y}) = \mathbf{Q}_{\Omega}\theta = \theta.$$

Thus, $\hat{\theta}$ is an unbiased estimator of θ.

Note 6.3.7: Note that $\hat{\mathbf{U}}'\hat{\mathbf{U}} = \mathbf{Y}'\mathbf{P}\mathbf{Y} = \mathbf{Y}'(\mathbf{I}-\mathbf{Q})\mathbf{Y} = \mathbf{Y}'(\mathbf{Y}-\mathbf{Q}) = (\mathbf{Y}-\theta)'(\mathbf{Y}-\theta) = \mathbf{U}'\mathbf{U}$, because, $\theta \in \boldsymbol{\Omega}, \mathbf{Q}\theta = \theta$ and $\theta'\mathbf{P} = \theta'\mathbf{Q}\mathbf{P} = \theta'\mathbf{Q}(\mathbf{I}-\mathbf{Q}) = \mathbf{0}$.

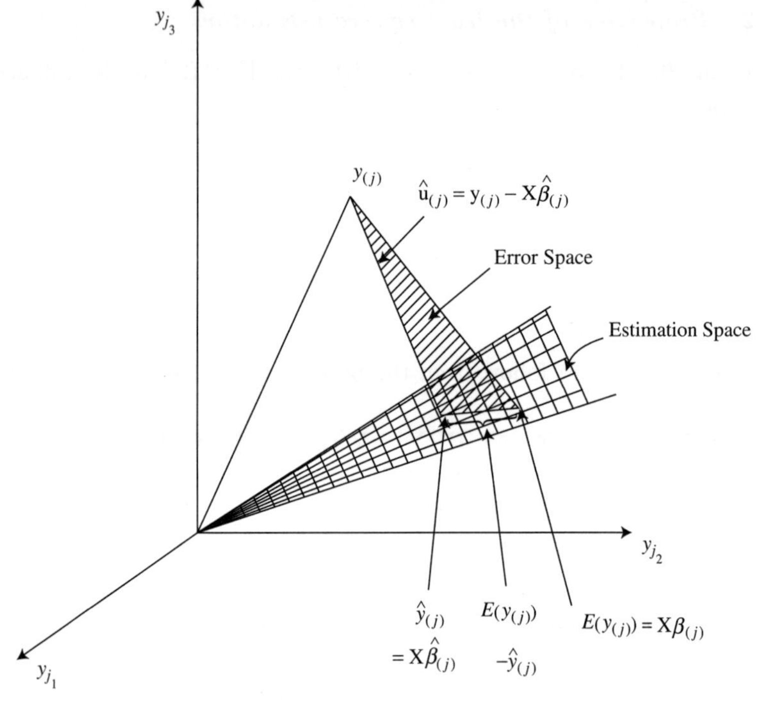

Fig. 6.1: Geometry of the multiple regression model $\mathbf{y}_{(j)} = \mathbf{X}\beta_{(j)} + \mathbf{u}_{(j)}$

We have obtained above the l.s.e. $\hat{\theta}_{(j)}$ of each component $\theta_{(j)}$ and of θ. We shall now obtain an important property of the estimator of a linear function of $\theta_{(j)}$ obtained from such estimators. This is given in the following Theorem.

Theorem 6.3.2: *A multivariate generalization of Gauss-Markov Theorem*

(Seber, 1984) Consider the model $\mathbf{Y} = \theta + \mathbf{U}$ where the rows $\mathbf{u}_i$ of $\mathbf{U}$ are uncorrelated with mean $\mathbf{0}$ and a common variance $\mathbf{\Sigma}$ and $\theta = (\theta_{(1)}, \ldots, \theta_{(p)})$. Let $\theta \subset \mathbf{\Omega}$ and $\hat{\theta} = (\hat{\theta}_{(1)}, \ldots, \hat{\theta}_{(p)})$ be the *l.s.e.* of θ subject to the condition $\hat{\theta}$ belongs to $\mathbf{\Omega}$. Let $\psi = \sum_{j=1}^{p} \mathbf{a}_j' \theta_{(j)}$ where $\mathbf{a}_j$ is a $q \times 1$ vector of known constants. Then $\hat{\psi} = \sum_{j=1}^{p} \mathbf{a}_j' \hat{\theta}_{(j)}$ is the best linear unbiased estimator (BLUE) of ψ in the sense of minimum variance.

Proof. We have $\hat{\theta} = \mathbf{QY}$ where $\mathbf{Q}$ is the orthogonal projection of $\mathcal{R}^n$ onto $\mathbf{\Omega}$. Clearly, $\mathbf{Q}\theta = \theta, \mathbf{Q}\theta_{(j)} = \theta_{(j)}$. Hence, $E(\hat{\theta}) = \mathbf{Q}E(\mathbf{Y}) = \mathbf{Q}\theta = \theta$.

Similarly, $E(\hat{\theta}_{(j)}) = \mathbf{Q}E(\mathbf{y}_{(j)}) = \mathbf{Q}\theta_{(j)} = \theta_{(j)}$. Thus, $\hat{\psi}$ is unbiased for ψ.

Let $\hat{\psi}^* = \sum_{j=1}^{p} \mathbf{c}_{(j)}' \mathbf{y}_{(j)}$ be any other unbiased estimator of ψ. Hence,

$$E(\hat{\psi}^*) = \sum_{j=1}^{p} \mathbf{a}_{(j)}' \theta_{(j)},$$

for all $\theta_{(j)} \subset \mathbf{\Omega}$. Hence,

$$\sum_j (\mathbf{c}_{(j)} - \mathbf{a}_j)' \theta_{(j)} = \mathbf{0} \ \forall \ \theta_{(j)} \subset \mathbf{\Omega}.$$

Hence, $\mathbf{c}_{(j)} - \mathbf{a}_j$ is perpendicular to $\mathbf{\Omega}$ and its projection onto $\mathbf{\Omega}$ is zero. Thus,

$$\mathbf{Q}(\mathbf{c}_{(j)} - \mathbf{a}_j) = \mathbf{0}$$

or

$$\mathbf{Qc}_{(j)} = \mathbf{Qa}_j, \quad (j = 1, \ldots, p). \tag{6.3.10}$$

Now,

$$\begin{aligned} V(\hat{\psi}) &= V[\textstyle\sum_j \mathbf{a}_j' \mathbf{Qy}_{(j)}] = V[\sum_j \mathbf{c}_{(j)}' \mathbf{Qy}_{(j)}] \quad \text{(by (6.3.10))} \\ &= \textstyle\sum_j \sum_k \mathbf{c}_{(j)}' \mathbf{Q} \text{ Cov } (\mathbf{y}_{(j)}, \mathbf{y}_{(k)}) \mathbf{Qc}_{(k)} \\ &= \textstyle\sum_j \sum_k \mathbf{c}_{(j)}' \mathbf{Q} \sigma_{jk} \mathbf{I}_n \mathbf{Qc}_{(k)} \\ &= \sigma_{jk} \textstyle\sum_j \sum_k \mathbf{c}_{(j)}' \mathbf{Qc}_{(k)}. \end{aligned}$$

Similarly,

$$V(\hat{\psi}^*) = V(\sum_j \mathbf{c}_{(j)}' \mathbf{y}_{(j)}) = \sum_j \sum_k \mathbf{c}_{(j)}' \mathbf{c}_{(k)} \sigma_{jk}.$$

Writing $\mathbf{C}_{n \times p} = [\mathbf{c}_{(1)} \ \mathbf{c}_{(2)} \ \ldots \mathbf{c}_{(p)}]$ and $\mathbf{\Sigma} = \mathbf{RR}'$ where $\mathbf{R}$ is non-singular,

$$\begin{aligned} V(\hat{\psi}^*) - V(\hat{\psi}) &= \textstyle\sum_j \sum_k \mathbf{c}_{(j)}' \mathbf{c}_{(k)} \sigma_{jk} - \sum_j \sum_k \mathbf{c}_{(j)}' \mathbf{Qc}_{(k)} \sigma_{jk} \\ &= \textstyle\sum_j \sum_k \mathbf{c}_{(j)}' (\mathbf{I}_n - \mathbf{Q}) \mathbf{c}_{(k)} \sigma_{jk} \\ &= \text{tr } [\mathbf{C}'(\mathbf{I}_n - \mathbf{Q})\mathbf{C\Sigma}] = \text{tr } [\mathbf{C}'(\mathbf{I}_n - \mathbf{Q})\mathbf{CRR}'] \\ &= \text{tr } [\mathbf{R}'\mathbf{C}'(\mathbf{I}_n - \mathbf{Q})\mathbf{CR}] \\ &= \text{tr } [\mathbf{R}'\mathbf{C}'(\mathbf{I}_n - \mathbf{Q})(\mathbf{I}_n - \mathbf{Q})\mathbf{CR}] \\ &= \text{tr } [\mathbf{D}'\mathbf{D}] \geq 0, \quad \text{where } \mathbf{D} = (\mathbf{I}_n - \mathbf{Q})\mathbf{CR}, \end{aligned} \tag{6.3.11}$$

since $\mathbf{D}'\mathbf{D}$ is positive semi-definite and its trace is the sum of its eigenvalues. Equality holds *iff* $\mathbf{D}'\mathbf{D} = \mathbf{0}$ or $\mathbf{D} = \mathbf{0}$, i.e. if $(\mathbf{I}_n - \mathbf{Q})\mathbf{C} = \mathbf{0}$ or $\mathbf{c}_{(j)} = \mathbf{Q}\mathbf{c}_{(j)} = \mathbf{Q}\mathbf{a}_j$ (vide (6.3.10)). In this case, $\hat{\psi}^* = \sum_j \mathbf{c}'_{(j)}\mathbf{y}_{(j)} = \sum_j \mathbf{a}'_j\mathbf{Q}\mathbf{y}_{(j)} = \sum_j \mathbf{a}'_j\hat{\theta}_j = \psi$. Therefore, $V(\hat{\psi}^*) \geq V(\hat{\psi})$, with equality *iff* $\hat{\psi}^* = \hat{\psi}$ and $\hat{\psi}$ is the unique unbiased estimator with the minimum variance. □

Corollary 6.3.2.1 If $\hat{\theta} = ((\hat{\theta}_{ij}))$, then $\hat{\theta}_{ij}$ is the BLUE of θ_{ij}.

We note an important definition.

DEFINITION 6.3.1 *Estimability of a Parameter Function* : A function $\psi = \sum_{j=1}^{p} \mathbf{a}'_j\beta_{(j)}$ is estimable if it has a linear unbiased estimator of the form $\sum_{j=1}^{p} \mathbf{c}'_j\mathbf{y}_{(j)}$. Then

$$\sum_{j=1}^{p} \mathbf{a}'_j\beta_{(j)} = E(\sum_{j=1}^{p} \mathbf{c}'_j\mathbf{y}_{(j)}) = \sum_{j=1}^{p} \mathbf{c}'_j\mathbf{X}\beta_{(j)}$$

identically in $\beta_{(j)}$. This implies

$$\mathbf{a}_j = \mathbf{X}'\mathbf{c}_j (j = 1, \ldots, p), \tag{6.3.12}$$

i.e. $\mathbf{a}_j$ is linearly dependent on the rows of $\mathbf{X}$, i.e. $\mathbf{a}_j$ belongs to the row-space of $\mathbf{X}$. Thus ψ is estimable *iff* all the univariate functions $\mathbf{a}'_{(j)}\beta_{(j)}$ are estimable.

Again, (6.3.12) implies

$$\mathbf{a} = \mathbf{X}'\mathbf{C}$$

where $\mathbf{a} = (\mathbf{a}_1, \mathbf{a}_2, \ldots \mathbf{a}_p)$. Pre-multiplying both sides by a matrix $\mathbf{H}'$ we have

$$\mathbf{H}'\mathbf{a} = \mathbf{H}'\mathbf{X}\mathbf{C}$$

which means that

$$\mathbf{X}\mathbf{H} = \mathbf{0} \text{ implies } \mathbf{a}'\mathbf{H} = \mathbf{0}. \tag{6.3.13}$$

6.3.3 *Ordinary and generalized least square estimator of* B

In this subsction we shall show that for multivariate regression model (6.1.1), the ordinary least square estimator (*OLS*) and the generalized least square (*GLS*) estimator are the same. To clarify the distinction between the

OLS and GLS estimators we first consider the simple multiple regression model ($p = 1$),

$$\mathbf{y} = \mathbf{X}\beta + \mathbf{u} \tag{6.3.14}$$

where $\mathbf{u}(n \times 1)$ is a vector of disturbances, $\mathbf{X}$ is a known $n \times q$ matrix of rank q. We assume that

$$E(\mathbf{u}) = \mathbf{0}, \quad V(\mathbf{u}) = \sigma^2\mathbf{V} \tag{6.3.15}$$

but do not assume any distribution of $\mathbf{u}$. The model (6.3.14) is also called the *Classical Linear Regression Model* and has been considered in detail in Section 6.10. The OLS estimator of β is

$$\hat{\beta} = (\mathbf{X}'\mathbf{X})^{-1}\mathbf{X}'\mathbf{y} \tag{6.3.16}$$

with $E(\hat{\beta}) = \beta,\ V(\hat{\beta}) = \sigma^2(\mathbf{X}'\mathbf{X})^{-1}(\mathbf{X}'\mathbf{V}\mathbf{X})(\mathbf{X}'\mathbf{X})^{-1}$. In particular, when $\mathbf{V} = \mathbf{I}$, it is known by Gauss-Markov theorem that the OLS estimator $\hat{\beta}$ is the BLUE of β.

When $V(\mathbf{u})$ does not have the structure $\sigma^2\mathbf{I}$, then in general OLS estimator is not BLUE. In this case, by a simple transformation we can find BLUE of β. Consider the transformed model obtained from (6.3.14):

$$\mathbf{z} = \mathbf{V}^{-1/2}\mathbf{X}\beta + \mathbf{v} \tag{6.3.17}$$

where $\mathbf{z} = \mathbf{V}^{-1/2}\mathbf{y}, \mathbf{v} = \mathbf{V}^{-1/2}\mathbf{u}$. Since $V(\mathbf{v}) = \mathbf{I}$, the condition of Gauss-Markov theorem is satisfied and the BLUE of β is

$$\hat{\beta}_G = (\mathbf{X}'\mathbf{V}^{-1}\mathbf{X})^{-1}\mathbf{X}'\mathbf{V}^{-1}\mathbf{y} \tag{6.3.18}$$

with covariance matrix, $V(\hat{\beta}_G) = (\mathbf{X}'\mathbf{V}^{-1}\mathbf{X})^{-1}$. The estimator defined by (6.3.18) is called the generalized least square (*GLS*) estimator of β.

We now come back to the multivariate multiple regression model (6.1.1). Writing the vec-matrices

$$\mathbf{Y}^V = [\mathbf{y}'_{(1)}, \mathbf{y}'_{(2)}, \ldots, \mathbf{y}'_{(p)}]'_{np\times 1}$$

$$\mathbf{B}^V = [\beta'_{(1)}, \beta'_{(2)}, \ldots, \beta'_{(p)}]'_{pq\times 1}$$

$$\mathbf{U}^V = [\mathbf{u}'_{(1)}, \mathbf{u}'_{(2)}, \ldots, \mathbf{u}'_{(p)}]'_{np\times 1}$$

and

$$\mathbf{X}^* = \mathbf{I}_p \otimes \mathbf{X}_{n\times q},$$

the model (6.1.1) can be written as

$$\mathbf{Y}^V = \mathbf{X}^*\mathbf{B}^V + \mathbf{U}^V \tag{6.3.19}$$

where

$$E(\mathbf{U}^V) = \mathbf{0},\ V(\mathbf{U}^V) = \mathbf{\Sigma} \otimes \mathbf{I}_n = \mathbf{\Omega}\ \text{(say)}\ .$$

Then the *GLS* estimator of $\mathbf{B}^V$ is

$$\begin{aligned}\hat{\mathbf{B}}^V &= (\mathbf{X}^{*'}\mathbf{\Omega}^{-1}\mathbf{X}^*)^{-1}(\mathbf{X}^{*'}\mathbf{\Omega}^{-1}\mathbf{Y}^V)\\ &= [\mathbf{I}_p \otimes (\mathbf{X}'\mathbf{X})^{-1}\mathbf{X}']\mathbf{Y}^V\end{aligned} \tag{6.3.20}$$

by simplification, using the properties of direct product matrices (items (5), (6), (8) of Section A.10).

Note that $\hat{\mathbf{B}}^V$ does not depend on $\mathbf{\Sigma}$ and hence defines an estimator whether $\mathbf{\Sigma}$ is known or unknown. The *GLS* estimator is obtained when $\mathbf{\Sigma} = \mathbf{I}_p$, so that the *OLS* and the *GLS* are the same in this case. Also, (6.3.20) coincides with the *m.l.e.* of $\mathbf{B}$ obtained under Theorem 6.2.1.

6.4 Forecasting an Observation

Consider the problem of estimating the mean response $E(\mathbf{Y})$ when the predictor variables have values $\mathbf{x}_0 = (x_{01}, x_{02}, \ldots, x_{0q})'$. The vector of the mean of response variables corresponding to $\mathbf{x}_0$ is $\mathbf{B}'\mathbf{x}_0$ and is estimated unbiasedly by

$$\hat{\mathbf{B}}'\mathbf{x}_0 = \begin{bmatrix} \hat{\beta}'_{(1)}\mathbf{x}_0 \\ \hat{\beta}'_{(2)}\mathbf{x}_0 \\ \cdot \\ \cdot\ \hat{\beta}'_{(p)}\mathbf{x}_0 \end{bmatrix}. \tag{6.4.1}$$

Covariance between estimation errors $\mathbf{x}'_0\hat{\beta}_{(i)} - \mathbf{x}'_0\beta_{(i)}$ and $\mathbf{x}'_0\hat{\beta}_{(k)} - \mathbf{x}'_0\beta_{(k)}$ is

$$\begin{aligned}&\mathbf{x}'_0\ \text{Cov}\ (\hat{\beta}_{(i)}, \hat{\beta}_{(k)})\mathbf{x}_0\\ &= \sigma_{ik}\mathbf{x}'_0(\mathbf{X}'\mathbf{X})^{-1}\mathbf{x}_0.\end{aligned} \tag{6.4.2}$$

Under normality assumptions,

$$\hat{\mathbf{B}}'\mathbf{x}_0 \sim N_p(\mathbf{B}'\mathbf{x}_0, \mathbf{x}'_0(\mathbf{X}'\mathbf{X})^{-1}\mathbf{x}_0\mathbf{\Sigma})$$

and is independent of $n\hat{\mathbf{\Sigma}}$ which follows $W_p(n-q, \mathbf{\Sigma})$ (Theorem 6.2.3(b)). Therefore,

$$T^2 = \frac{(n-q)}{n}\left(\frac{\hat{\mathbf{B}}'\mathbf{x}_0 - \mathbf{B}'\mathbf{x}_0}{\sqrt{\mathbf{x}'_0(\mathbf{X}'\mathbf{X})^{-1}\mathbf{x}_0}}\right)'(\hat{\mathbf{\Sigma}})^{-1}\left(\frac{\hat{\mathbf{B}}'\mathbf{x}_0 - \mathbf{B}'\mathbf{x}_0}{\sqrt{\mathbf{x}'_0(\mathbf{X}'\mathbf{X})^{-1}\mathbf{x}_0}}\right) \tag{6.4.3}$$

follows Hotelling's T^2 distribution with $(p, n-q)$ d.f. The $100(1-\alpha)\%$ confidence ellipsoid for $\mathbf{B}'\mathbf{x}_0$ is given by the inequality

$$\frac{n-q}{n}(\mathbf{B}'\mathbf{x}_0 - \hat{\mathbf{B}}'\mathbf{x}_0)'\hat{\mathbf{\Sigma}}^{-1}(\mathbf{B}'\mathbf{x}_0 - \hat{\mathbf{B}}'\mathbf{x}_0) \leq \mathbf{x}_0'(\mathbf{X}'\mathbf{X})^{-1}\mathbf{x}_0 c^2 \tag{6.4.4}$$

where

$$c^2 = \frac{p(n-q)}{n-q-p+1} F_{p,n-q-p+1;\alpha}. \tag{6.4.5}$$

The center of the ellipsoid is $\hat{\mathbf{B}}'\mathbf{x}_0$ and beginning at the center, the axes of the ellipsoid are

$$\pm\{c\sqrt{\lambda_i}\sqrt{\mathbf{x}_0'(\mathbf{X}'\mathbf{X})^{-1}\mathbf{x}_0\frac{n}{n-q}}\}\mathbf{e}_i \tag{6.4.6}$$

where $(\lambda_i, \mathbf{e}_i)(i = 1, \ldots, p)$ are the (eigenvalue-eigenvector) pairs of $\hat{\mathbf{\Sigma}}$. If a point ν_0 lies inside the ellipse then the hypothesis $H_0(\mathbf{B}'\mathbf{x}_0 = \nu_0)$ is accepted.

To find simultaneous confidence interval for $\mathbf{a}'\mathbf{B}'\mathbf{x}_0$, following Theorem 5.5.1, we maximize

$$\Psi(\mathbf{a}; \hat{\mathbf{B}}, \mathbf{x}_0) = \frac{\{\mathbf{a}'(\hat{\mathbf{B}}'\mathbf{x}_0 - \mathbf{B}'\mathbf{x}_0)\}^2}{\mathbf{a}'\hat{V}(\hat{\mathbf{B}}'\mathbf{x}_0)\mathbf{a}}$$

with respect to $\mathbf{a}$ where $\hat{V}$ is an estimate of the variance. Considering Theorem A.14.1, if we put $\mathbf{x}$ (in the Theorem's notation) $= \mathbf{a}, \mathbf{b} = (\hat{\mathbf{B}}'\mathbf{x}_0 - \mathbf{B}'\mathbf{x}_0), \mathbf{C} = \hat{V}(\hat{\mathbf{B}}\mathbf{x}_0)$, then

$$\max_{\mathbf{a}} \Psi(\mathbf{a}; \hat{\mathbf{B}}, \mathbf{x}_0) = (\hat{\mathbf{B}}'\mathbf{x}_0 - \mathbf{B}'\mathbf{x}_0)'\{\hat{V}(\hat{\mathbf{B}}'\mathbf{x}_0)\}^{-1}(\hat{\mathbf{B}}'\mathbf{x}_0 - \mathbf{B}'\mathbf{x}_0) = \frac{n}{n-q}T^2 \tag{6.4.7}$$

where $T^2 = T^2(p, n-q)$ is given in (6.4.3). Hence,

$$P[\Psi(\mathbf{a}; \hat{\mathbf{B}}, \mathbf{x}_0) \leq c^2\frac{n}{n-q} \ \forall \mathbf{a}] \geq 1-\alpha. \tag{6.4.8}$$

Therefore, simultaneous confidence interval for $\mathbf{a}'\mathbf{B}'\mathbf{x}_0$ is

$$\mathbf{a}'\hat{\mathbf{B}}\mathbf{x}_0 \mp c\sqrt{\mathbf{x}_0'(\mathbf{X}'\mathbf{X}')^{-1}\mathbf{x}_0\mathbf{a}'\hat{\mathbf{\Sigma}}\mathbf{a}}\sqrt{\frac{n}{n-q}}. \tag{6.4.9}$$

In particular, $100(1-\alpha)\%$ simultaneous confidence interval for $E(Y_i) = \mathbf{x}_0'\beta_{(i)}$ is

$$\mathbf{x}_0'\hat{\beta}_{(i)} \pm c\sqrt{\mathbf{x}_0'(\mathbf{X}'\mathbf{X})^{-1}\mathbf{x}_0(\frac{n}{n-q}\hat{\sigma}_{ii})} \tag{6.4.10}$$

where $\hat{\sigma}_{ii}$ is the ith diagonal element of $\hat{\mathbf{\Sigma}}$.

We may want to forecast a new observation $\mathbf{y}_0 = (y_{01}, y_{02}, \ldots, y_{0p})'$ at $\mathbf{x}_0$. By the model (6.1.1), $y_{0i} = \mathbf{x}_0'\beta_{(i)} + u_{0i}$ where the error vector corresponding to $\mathbf{y}_0$ is $\mathbf{u}_0 = (u_{01}, u_{02}, \ldots, u_{0p})'$ and is independent of $\mathbf{u}_1, \ldots, \mathbf{u}_n$. It is assumed, as in model (6.1.1), that $E\mathbf{u}_0 = \mathbf{0}$ and Cov $(u_{0i}, u_{0k}) = \sigma_{ik}$. The forecast error for the ith component of $\mathbf{y}_0$ is

$$\begin{aligned}\mathbf{y}_{0i} - \mathbf{x}_0'\hat{\beta}_{(i)} &= \mathbf{y}_{0i} - \mathbf{x}_0'\beta_{(i)} + \mathbf{x}_0'\beta_{(i)} - \mathbf{x}_0'\hat{\beta}_{(i)} \\ &= u_{0i} - \mathbf{x}_0'(\hat{\beta}_{(i)} - \beta_{(i)})\end{aligned}$$

so that

$$E(y_{0i} - \mathbf{x}_0'\hat{\beta}_{(i)}) = 0.$$

Also, covariance among forecast errors is

$$\text{Cov }\{(y_{0i} - \mathbf{x}_0'\hat{\beta}_{(i)})(y_{0k} - \mathbf{x}_0'\hat{\beta}_{(k)})\}$$

$$\begin{aligned}&= E[u_{0i} - \mathbf{x}_0'(\hat{\beta}_{(i)} - \beta_{(i)})][u_{0k} - \mathbf{x}_0'(\hat{\beta}_{(k)} - \beta_{(k)})] \\ &= E(u_{0i}u_{0k}) + \mathbf{x}_0'E(\hat{\beta}_{(i)} - \beta_{(i)})(\hat{\beta}_{(k)} - \beta_{(k)})\mathbf{x}_0 \\ &= \sigma_{ik}[1 + \mathbf{x}_0'(\mathbf{X}'\mathbf{X})^{-1}\mathbf{x}_0],\end{aligned} \tag{6.4.11}$$

because $E(\hat{\beta}_{(i)} - \beta_{(i)})u_{0k} = E(\mathbf{X}'\mathbf{X})^{-1}\mathbf{X}'\mathbf{u}_{(i)}u_{0k} = 0$. Under normality assumption, $\mathbf{y}_0 - \hat{\mathbf{B}}'\mathbf{x}_0 = (\mathbf{B} - \hat{\mathbf{B}})'\mathbf{x}_0 + \mathbf{u}_0$ is distributed as $N_p(\mathbf{0}, \{1 + \mathbf{x}_0'(\mathbf{X}'\mathbf{X})^{-1}\mathbf{x}_0\}\boldsymbol{\Sigma})$ independently of $\hat{\boldsymbol{\Sigma}}$. Therefore, following derivations in (6.4.3) and (6.4.4), $100(1-\alpha)\%$ prediction ellipsoid for $\mathbf{y}_0$ is given by

$$\frac{n-q}{n}(\mathbf{y}_0 - \hat{\mathbf{B}}'\mathbf{x}_0)'\hat{\boldsymbol{\Sigma}}^{-1}(\mathbf{y}_0 - \hat{\mathbf{B}}'\mathbf{x}_0) \le \{1 + (\mathbf{x}_0'\mathbf{X}'\mathbf{X})^{-1}\mathbf{x}_0\}c^2. \tag{6.4.12}$$

The $100(1-\alpha)\%$ simultaneous prediction intervals for the individual response y_{oi} are

$$\mathbf{x}_0'\hat{\beta}_{(i)} \pm \sqrt{\frac{n}{n-q}c^2\{1 + \mathbf{x}_0'(\mathbf{X}'\mathbf{X})^{-1}\mathbf{x}_0\}\hat{\sigma}_{ii}},\ i = 1, \ldots, p. \tag{6.4.13}$$

Example 6.4.1: For the data of Example 6.2.1, predict the expected value $\mathbf{y}_0$ when $\mathbf{x} = \mathbf{x}_0 = (1,\ 100,\ 10)'$. Obtain 95% confidence ellipsoid for this expected value and for $\mathbf{y}_0$. Also obtain simultaneous confidence interval for $E(y_{10})$ and y_{10}.

We have

$$\begin{aligned}E(\mathbf{y}_0) = \hat{\mathbf{B}}'\mathbf{x}_0 &= \begin{bmatrix} 7.7620 & 0.0148 & -0.1485 \\ 26.4756 & -0.1918 & 0.1410 \end{bmatrix}\begin{bmatrix} 1 \\ 100 \\ 10 \end{bmatrix} \\ &= \begin{bmatrix} 7.75291 \\ 8.70911 \end{bmatrix}.\end{aligned}$$

Also

$$\mathbf{x}_0'(\mathbf{X}'\mathbf{X})^{-1}\mathbf{x}_0 = 0.214101,$$

$$c^2 = \frac{(2)(7)}{6}F_{2,6;.05} = 11.9933.$$

The 95% confidence ellipsoid for $E(\mathbf{y}_0)$ is

$$(E(\mathbf{y}_0) - \hat{\mathbf{B}}'\mathbf{x}_0)'\hat{\mathbf{\Sigma}}^{-1}(E(\mathbf{y}_0) - \hat{\mathbf{B}}'\mathbf{x}_0) \leq (0.214101)(11.9933)(\frac{10}{7}) = 3.64378. \quad (i)$$

The eigenvalues and eigenvectors of $\hat{\mathbf{\Sigma}}$ are

$$\lambda_1 = 9.62266,\ \lambda_2 = 0.99853,$$

$$\mathbf{e}_1 = (-.231098,\ .972931)',\ \mathbf{e}_2 = (.972931,\ .231098)'.$$

The center of the ellipsoid at (i) is $\hat{\mathbf{B}}'\mathbf{x}_0$ and beginning at the center, the axes of the ellipsoid are

$$\mp 5.94124\mathbf{e}_1,\ \mp 1.91386\mathbf{e}_2.$$

Similarly, the confidence ellipsoid for $\mathbf{y}_0$ is the ellipse with center at $\hat{\mathbf{B}}'\mathbf{x}_0$ and axes

$$\mp 14.1480\mathbf{e}_1,\ \mp 4.5575\mathbf{e}_2.$$

By formula (6.4.10) simultaneous confidence interval for $E(y_{10})$ is

$$7.757 \mp 2.31353 = (5.443,\ 10.071).$$

Similarly, by formula (6.4.13), simultaneous confidence interval for y_{10} is

$$7.757 \mp 5.510 = (2.247,\ 13.267).$$

6.5 Likelihood Ratio Tests for Regression Parameters

Part of the regression analysis is concerned with assessing the effects of different dependent variables on the response variables. We therefore consider different hypotheses concerning the regression parameters $\mathbf{B}$.

6.5.1 *The matrix* $\mathbf{X}$ *is of full rank*

(a) H_0 : A submatrix of $\mathbf{B}$ is null.

Let

$$\mathbf{B} = \begin{bmatrix} \mathbf{B}_1(q_1 \times p) \\ \mathbf{B}_2(q_2 \times p) \end{bmatrix}, \; q_1 + q_2 = q.$$

We want to tst $H_0 : \mathbf{B}_2 = \mathbf{0}$, i.e. the responses do not depend on the regressors $x_{q_1+1}, \ldots, x_{q_1+q_2}$. Setting

$$\mathbf{X}(n \times q) = [\mathbf{X}_1(n \times q_1) \;\; \mathbf{X}_2(n \times q_2)],$$

we can write the model (6.1.1) as

$$E(\mathbf{Y}) = \mathbf{X}_1\mathbf{B}_1 + \mathbf{X}_2\mathbf{B}_2.$$

Under $H_0, \mathbf{B}_2 = \mathbf{0}$ and hence the model (6.1.1) reduces to

$$\mathbf{Y} = \mathbf{X}_1\mathbf{B}_1 + \mathbf{U}. \tag{6.5.1}$$

The LR statistic for H_0 can be written as

$$\begin{aligned}\lambda &= \frac{\max_{(\mathbf{B}_1,\mathbf{\Sigma})} L(\mathbf{B}_1,\mathbf{\Sigma})}{\max_{(\mathbf{B},\mathbf{\Sigma})} L(\mathbf{B},\mathbf{\Sigma})} \\ &= \frac{L(\hat{\mathbf{B}}_1,\hat{\mathbf{\Sigma}}_1)}{L(\hat{\mathbf{B}},\hat{\mathbf{\Sigma}})}\end{aligned} \tag{6.5.2}$$

where $\hat{\mathbf{B}}_1, \hat{\mathbf{\Sigma}}_1$, are, respectively, m.l.e.'s of $\mathbf{B}_1$ and $\mathbf{\Sigma}$ under model (6.5.1), in association with the multivariate normality assumptions on the row vectors of $\mathbf{\Sigma}$. Following (6.2.8), we have λ in (6.5.2) as

$$\lambda = (\frac{|\hat{\mathbf{\Sigma}}|}{|\hat{\mathbf{\Sigma}}_1|})^{n/2}, \tag{6.5.3}$$

where, as in (6.2.11),

$$\hat{\mathbf{\Sigma}}_1 = n^{-1}(\mathbf{Y} - \mathbf{X}_1\hat{\mathbf{B}}_1)'(\mathbf{Y} - \mathbf{X}_1\hat{\mathbf{B}}_1) \tag{6.5.4}$$

and

$$\hat{\mathbf{\Sigma}} = n^{-1}(\mathbf{Y} - \mathbf{X}\hat{\mathbf{B}})'(\mathbf{Y} - \mathbf{X}\hat{\mathbf{B}}). \tag{6.5.5}$$

Here

$$\hat{\mathbf{B}} = (\mathbf{X}'\mathbf{X})^{-1}\mathbf{X}'\mathbf{Y},$$

$$\hat{\mathbf{B}}_1 = (\mathbf{X}_1'\mathbf{X}_1)^{-1}\mathbf{X}_1'\mathbf{Y}. \tag{6.5.6}$$

The LR test of H_0 is equivalent to rejecting H_0 for large values of

$$-2\ln\lambda = -n\ln(\frac{|\hat{\mathbf{\Sigma}}|}{|\hat{\mathbf{\Sigma}}_1|}) \tag{6.5.7}$$

which follows χ^2 with pq_2 d.f.

For large n, the modified statistic

$$-[n-q-\frac{1}{2}(p-q_2+1)]ln(\frac{|\hat{\mathbf{\Sigma}}|}{|\hat{\mathbf{\Sigma}}_1|}) \tag{6.5.8}$$

follows closely a χ^2 distribution with pq_2 d.f.

EXAMPLE 6.5.1: For the data of Example 6.4.1 let us test the hypothesis $H_0 : \mathbf{B}_2 = \mathbf{0}$ where $\mathbf{B}_2(1 \times 2)$ is the last row of $\mathbf{B}$.

Here $\mathbf{X}_1$ consists of the first two columns of $\mathbf{X}$.

$$\begin{aligned}\hat{\mathbf{B}}_1 = (\mathbf{X}_1'\mathbf{X}_1)^{-1}\mathbf{X}_1\mathbf{Y} &= \begin{bmatrix} 4.98117 & -0.05553 \\ -0.05553 & 0.00063 \end{bmatrix}\begin{bmatrix} 75 & 111 \\ 6651 & 9420 \end{bmatrix} \\ &= \begin{bmatrix} 4.2514 & 29.8084 \\ 0.0370 & -0.2128 \end{bmatrix},\end{aligned}$$

$$\hat{\mathbf{\Sigma}}_1 = (10)^{-1}(\mathbf{Y}-\mathbf{X}_1\hat{\mathbf{B}}_1)'(\mathbf{Y}-\mathbf{X}_1\hat{\mathbf{B}}) = \begin{bmatrix} 1.63380 & -2.10490 \\ -2.10490 & 9.31951 \end{bmatrix}.$$

Also,

$$|\hat{\mathbf{\Sigma}}| = 9.60851,\ |\hat{\mathbf{\Sigma}}_1| = 10.7959.$$

Hence,

$$-2\ln\lambda = -10\ln(9.60851/10.7959) = 1.1656.$$

Also, the modified statistic (6.5.8) has the value $-6\ln(|\hat{\mathbf{\Sigma}}|/|\hat{\mathbf{\Sigma}}_1|) = .6994$. Since $\chi^2_{2;.05} = 5.9915$, the hypothesis $H_0(\mathbf{B}_2 = \mathbf{0})$ is accepted. The responses do not significantly depend on x_3.

(b) The General Linear Hypothesis:

Suppose we want to test the null hypothesis, $H_G : \mathbf{A}_1\mathbf{B} = \mathbf{D}$ where $\mathbf{A}_1$ is a $g \times q$ matrix of rank g.

To do this, we define some matrices. Let $\mathbf{A}_2$ be a $(q-g) \times q$ matrix such that $\mathbf{A}' = (\mathbf{A}_1',\ \mathbf{A}_2')$ is a non-singular $q \times q$ matrix. Also, let

$$\mathbf{A}^{-1} = (\mathbf{A}^{(1)}(q \times g),\ \mathbf{A}^{(2)}(q \times (q-g))).$$

Let $\mathbf{B}_0$ be a $q \times p$ matrix such that $\mathbf{A}_1\mathbf{B}_0 = \mathbf{D}$.

The model (6.1.1) can be written as

$$\mathbf{Y} = \mathbf{X}\mathbf{B}_0 + \mathbf{X}(\mathbf{B} - \mathbf{B}_0) + \mathbf{U}$$

or

$$\mathbf{Y} - \mathbf{X}\mathbf{B}_0 = \mathbf{X}(\mathbf{B} - \mathbf{B}_0) + \mathbf{U}$$

or

$$\mathbf{Y}_+ = (\mathbf{X}\mathbf{A}^{-1})\mathbf{A}(\mathbf{B} - \mathbf{B}_0) + \mathbf{U}$$

or

$$\mathbf{Y}_+ = \mathbf{Z}\boldsymbol{\Delta} + \mathbf{U} \tag{6.5.9}$$

where

$$\mathbf{Y}_+ = \mathbf{Y} - \mathbf{X}\mathbf{B}_0,\ \mathbf{Z} = \mathbf{X}\mathbf{A}^{-1},\ \boldsymbol{\Delta} = \mathbf{A}(\mathbf{B} - \mathbf{B}_0).$$

Now,

$$\mathbf{Z} = \mathbf{X}(\mathbf{A}^{(1)},\ \mathbf{A}^{(2)}) = ((\mathbf{Z}_1(n \times g),\ \mathbf{Z}_2(n \times (q-g))$$

where $\mathbf{Z}_1 = \mathbf{X}\mathbf{A}^{(1)}$. Also,

$$\boldsymbol{\Delta} = \mathbf{A}(\mathbf{B} - \mathbf{B}_0) = \begin{bmatrix} \mathbf{A}_1(\mathbf{B} - \mathbf{B}_0)(g \times q) \\ \mathbf{A}_2(\mathbf{B} - \mathbf{B}_0)((q-g) \times q) \end{bmatrix} = \begin{bmatrix} \boldsymbol{\Delta}_1 \\ \boldsymbol{\Delta}_2 \end{bmatrix}.$$

Under $H_G, \boldsymbol{\Delta}_1 = \mathbf{0}$. Therefore under H_G, the model (6.5.9) reduces to

$$\mathbf{Y}_+ = \mathbf{Z}_2\boldsymbol{\Delta}_2 + \mathbf{U}. \tag{6.5.10}$$

The technique used in testing H_0 in (6.5.3) can, therefore, be used. Note that we can write $\mathbf{P}$ defined in (6.2.1) as

$$\mathbf{P} = \mathbf{I}_n - \mathbf{Z}(\mathbf{Z}'\mathbf{Z})^{-1}\mathbf{Z}' \tag{6.5.11}$$

which represents the projection matrix onto the subspace which is orthogonal to the column-space of $\mathbf{Z} = \mathbf{X}\mathbf{A}^{-1}$. We now define

$$\mathbf{P}_1 = \mathbf{I}_n - \mathbf{Z}_2(\mathbf{Z}_2'\mathbf{Z}_2)^{-1}\mathbf{Z}_2' \tag{6.5.12}$$

which represents the projection onto the subspace orthogonal to the columns of $\mathbf{Z}_2 = \mathbf{X}\mathbf{A}^{(2)}$.

Following (6.2.8), the maximum value of the likelihood under under the model (6.5.10) is

$$|2\pi n^{-1}\mathbf{Y}_+'\mathbf{P}_1\mathbf{Y}_+|^{-n/2}\exp(-np/2). \tag{6.5.13}$$

Therefore, from (6.2.12),

$$\begin{aligned} n\hat{\mathbf{\Sigma}} &= \mathbf{Y}'\mathbf{PY} = \text{ SSP matrix due to error under full model (6.5.4)} \\ &= \mathbf{E} \text{ (say)}, \end{aligned}$$

and following (6.2.12),

$$\begin{aligned} n\hat{\mathbf{\Sigma}}_G &= \mathbf{Y}'_+\mathbf{P}_1\mathbf{Y}_+ = \text{ SSP matrix due to error under model (6.5.10)} \\ &= \text{ SSP matrix due to error under } H_G \\ &= \mathbf{Y}'\mathbf{PY} + \mathbf{Y}'_+\mathbf{P}_2\mathbf{Y}_+ \text{ where } P_2 = P_1 - P(\text{as } \mathbf{Y}'\mathbf{PY} = \mathbf{Y}'_+\mathbf{PY}_+, \\ &\quad \text{because,} (\mathbf{XB}_0)'\mathbf{P} = \mathbf{B}'_0\mathbf{X}'\mathbf{P} = \mathbf{B}'_0\mathbf{A}'\mathbf{Z}'\mathbf{P} = \mathbf{0}) \\ &= \mathbf{E} + \text{ Extra SSP matrix due to } H_G \\ &= \mathbf{E} + \mathbf{H} \text{ (say)} \\ &= \mathbf{E}_H \text{ (say)}. \end{aligned} \tag{6.5.14}$$

Here, $\hat{\mathbf{\Sigma}}$ and $\hat{\mathbf{\Sigma}}_G$ are estimates of $\mathbf{\Sigma}$ under full model (6.5.4) and restricted model (6.5.10) respectively. The likelihood ratio statistic for testing H_G is, therefore, given by

$$\begin{aligned} \lambda^{2/n} &= \frac{|\hat{\mathbf{\Sigma}}|}{|\hat{\mathbf{\Sigma}}_G|} \\ &= \frac{|\mathbf{Y}'\mathbf{PY}|}{|\mathbf{Y}'_+\mathbf{P}_1\mathbf{Y}_+|} \\ &= \frac{|\mathbf{E}|}{|\mathbf{E}_H|} \\ &= \frac{|\mathbf{E}|}{|\mathbf{E}+\mathbf{H}|} \\ &= |\mathbf{I}_p - \mathbf{V}| \\ &= \Pi_{j=1}^p(1-\theta_j) \\ &= \Pi_{j=1}^p(1+\phi_j)^{-1}, \end{aligned} \tag{6.5.15}$$

where $\mathbf{V} = \mathbf{E}_H^{-1}\mathbf{H}$ and θ_j's are the ordered eigenvalues of $\mathbf{V}$, i.e., the ordered eigenvalues of $\mathbf{E}_H^{-1/2}\mathbf{H}\mathbf{E}_H^{-1/2}$. Also, ϕ_js are the ordered eigenvalues of $\mathbf{HE}^{-1}$.

It can be shown that

$$\mathbf{P}_2 = \mathbf{P}_1 - \mathbf{P} = \mathbf{X}(\mathbf{X}'\mathbf{X})^{-1}\mathbf{A}'_1[\mathbf{A}_1(\mathbf{X}'\mathbf{X})^{-1}\mathbf{A}'_1]^{-1}\mathbf{A}_1(\mathbf{X}'\mathbf{X})^{-1}\mathbf{X}'. \tag{6.5.16}$$

(Exercise 6.6). Here, the SSP matrix due to H_G is

$$\begin{aligned} \mathbf{H} &= \mathbf{E}_H - \mathbf{E} \\ &= \mathbf{Y}'_+\mathbf{P}_2\mathbf{Y}_+ \\ &= \mathbf{Y}'\mathbf{P}_2\mathbf{Y} \\ &= \mathbf{Y}'[\mathbf{X}(\mathbf{X}'\mathbf{X})^{-1}\mathbf{A}'_1\{\mathbf{A}_1(\mathbf{X}'\mathbf{X})^{-1}\mathbf{A}'_1\}^{-1} \\ &\quad \mathbf{A}_1(\mathbf{X}'\mathbf{X})^{-1}\mathbf{X}']\mathbf{Y} \end{aligned} \tag{6.5.17}$$

Theorem 6.5.1: The likelihood ratio test statistic for the hypothesis H_0 : $\mathbf{A}_1\mathbf{B} = \mathbf{D}$,

$$\lambda^{2/n} = \frac{|\mathbf{Y}'\mathbf{P}\mathbf{Y}|}{|\mathbf{Y}'\mathbf{P}\mathbf{Y} + \mathbf{Y}'_+\mathbf{P}_2\mathbf{Y}_+|} \tag{6.5.18}$$

follows $\Lambda(p; n-q, g)$ distribution under the null hypothesis.

Proof. It follows that both under the full model (6.5.4) and the hypothesized model (6.5.5) $\mathbf{Y}'\mathbf{P}\mathbf{Y} = \mathbf{U}'\mathbf{P}\mathbf{U}$. Since $\mathbf{P}$ is idempotent matrix of rank $n-q$, $\mathbf{Y}'\mathbf{P}\mathbf{Y}$ follows $W_p(n-q, \mathbf{\Sigma})$ distribution (Theorem 4.2.10). Again, under H_0,

$$\begin{aligned} \mathbf{Y}'_+\mathbf{P}_2\mathbf{Y}_+ &= (\mathbf{Z}_2\mathbf{\Delta}_2 + \mathbf{U})'(\mathbf{P}_1 - \mathbf{P})(\mathbf{Z}_2\mathbf{\Delta}_2 + \mathbf{U}) \\ &= \mathbf{U}'\mathbf{P}_2\mathbf{U}. \end{aligned} \tag{6.5.19}$$

follows a $W_p(g, \mathbf{\Sigma})$ distribution since $\mathbf{P}_2$ is a symmetric idempotent matrix of rank g. Also, $\mathbf{P}\mathbf{P}_2 = \mathbf{0}$. Hence, by Theorem 4.2.11, these two Wishart distributions are independent. Therefore, by (4.4.2), the statistic $\lambda^{2/n}$ follows Wilk's $\Lambda(p; n-q, g)$ distribution. By Theorem 4.4.2,

$$\Lambda(p; n-q, g) = \Pi_{i=1}^{g} u_i$$

where $u_1, \ldots, u_g$ are g independent variables and $u_i \cap B((n-q+i-p)/2, p/2), i = 1, \ldots, g$.

Consider now the hypothesis H_{G1} : $\mathbf{A}_1\mathbf{B}\mathbf{L}_1 = \mathbf{D}_1$ for the model (6.1.1), where $\mathbf{L}_1(p \times r)$ is of rank r. This is equivalent to studying the hypothesis under the transformed model

$$\mathbf{Y}\mathbf{L}_1 = \mathbf{X}\mathbf{L}_1 + \mathbf{U}\mathbf{L}_1 \tag{6.5.20}$$

where $\mathbf{U}\mathbf{L}_1$ is a data matrix from $N_r(\mathbf{0}, \mathbf{L}_1'\mathbf{\Sigma}\mathbf{L}_1)$ and $\mathbf{B}\mathbf{L}_1$ contains qr parameters. We have the following result.

Theorem 6.5.2 : For testing the hypothesis H_{G1} : $\mathbf{A}_1\mathbf{B}\mathbf{L}_1 = \mathbf{D}_1$ under the model (6.1.1), where $\mathbf{A}_1(g \times q)$ has rank g and $\mathbf{L}_1(p \times r)$ has rank r, the likelihood ratio test statistic is given by

$$\lambda^{2/n} = \frac{|\mathbf{E}|}{|\mathbf{E} + \mathbf{H}|} \tag{6.5.21}$$

where

$$\mathbf{E} = \mathbf{L}_1'\mathbf{Y}'\mathbf{P}\mathbf{Y}\mathbf{L}_1 \tag{6.5.22}$$

$$\mathbf{H} = \mathbf{L}_1'\mathbf{Y}'_+\mathbf{P}_2\mathbf{Y}_+\mathbf{L}_1. \tag{6.5.23}$$

It follows by (A.8.3) that if $\phi_1, \ldots, \phi_r$ are the ordered eigenvalues of $\mathbf{HE}^{-1}$, then the LR statistic in (6.5.18) can be written as

$$\frac{|\mathbf{E}|}{|\mathbf{H}+\mathbf{E}|} = \Pi_{i=1}^{r}(1+\phi_i)^{-1}. \tag{6.5.24}$$

As discussed in Section 4.4, other statistics that can be used for testing H_G are Roy's maximum root statistic θ_1, Lawley-Hotelling statistic and Pillai's statistic.

Note 6.5.1: The least square estimate of $\mathbf{\Delta}_2$ under H_G is given by

$$\mathbf{Z}_2\hat{\mathbf{\Delta}}_2 = \mathbf{Q}_1\mathbf{Y}_+, \quad \text{where } \mathbf{Q}_1 = \mathbf{I}_n - \mathbf{P}_1, \tag{6.5.25}$$

from which the restricted least square estimate of $\mathbf{B}$ under $H_G, \hat{\mathbf{B}}_H$ can be found out. For a complete expression of restriced least square estimate of $\mathbf{B}$ under different hypotheses, see Section 6.6.

6.5.2 *The matrix* X *not of full rank*

When the design matrix $\mathbf{X}$ is not of full rank, interpretation of the matrix $\mathbf{B}$ becomes difficult. The simplest way of taking care of this situation is to confine to a subset of regressor variables.

Suppose $\mathbf{X}(n \times q)$ is of rank $k(< q)$ and it is possible to write $\mathbf{X}$ as $\mathbf{X} = (\mathbf{X}_1(n \times k),\ \mathbf{X}_2(n \times (q-k))$ such that $\mathbf{X}_1$ is of full rank and $\mathbf{X}_2$ being dependent on $\mathbf{X}_1$ can be written as $\mathbf{X}_2 = \mathbf{X}_1\mathbf{C}$ where $\mathbf{C}$ is of order $k\times(q-k)$. Thus

$$\mathbf{X} = (\mathbf{X}_1,\ \mathbf{X}_2) = (\mathbf{X}_1,\ \mathbf{X}_1\mathbf{C}).$$

Write

$$\mathbf{B} = \begin{bmatrix} \mathbf{B}_1(k \times p) \\ \mathbf{B}_2((q-k) \times p) \end{bmatrix},$$

where $\mathbf{B}_1$ is the matrix of the regression coefficients corresponding to $\mathbf{X}_1$. The model (6.1.1) then reduces to

$$\begin{aligned} \mathbf{Y} &= \mathbf{X}_1\mathbf{B}_1 + \mathbf{X}_2\mathbf{B}_2 + \mathbf{U} \\ &= \mathbf{X}_1\mathbf{B}_1 + \mathbf{X}_1\mathbf{C}\mathbf{B}_2 + \mathbf{U} \\ &= \mathbf{X}_1\mathbf{B}_1^* + \mathbf{U}, \end{aligned} \tag{6.5.26}$$

where $\mathbf{B}_1^* = \mathbf{B}_1 + \mathbf{C}\mathbf{B}_2$. Suppose we want to test $H_G : \mathbf{A}_1\mathbf{B} = \mathbf{D}$ where $\mathbf{A}_1(g \times q)$ is of rank g. Write $\mathbf{A}_1 = (\mathbf{A}_{11}(g \times k),\ \mathbf{A}_{12}(g \times (q-k))$. Then

$$\mathbf{A}_1\mathbf{B} = \mathbf{A}_{11}\mathbf{B}_1 + \mathbf{A}_{12}\mathbf{B}_2.$$

One has to investigate the conditions, if H_0 is testable, i.e. if $\mathbf{A}_1\mathbf{B}$ is estimable. It is known that if $\mathbf{A}_1\mathbf{B}$ is estimable, there exists a statistic $\mathbf{FY}$ such that $E(\mathbf{FY}) = \mathbf{A}_1\mathbf{B}$, i.e. $\mathbf{FX} = \mathbf{A}_1$. Now,

$$\mathbf{FX} = (\mathbf{FX}_1,\ \mathbf{FX}_2) = (\mathbf{A}_{11},\ \mathbf{A}_{12})$$

so that $\mathbf{FX}_1 = \mathbf{A}_{11}$, $\mathbf{FX}_2 = \mathbf{A}_{12}$. Therefore, H_{01} reduces to

$$\mathbf{A}_1\mathbf{B} = \mathbf{FX}_1\mathbf{B}_1 + \mathbf{FX}_1\mathbf{CB}_2 = \mathbf{D}$$

or

$$\mathbf{FX}_1\mathbf{B}_1^* = \mathbf{A}_{11}\mathbf{B}_1^* = \mathbf{D}. \tag{6.5.27}$$

Therefore, one has to test H_{01} given by (6.5.27) under the model (6.5.26). If $\hat{\mathbf{B}}_1^*$ is the m.l.e. of $\mathbf{B}_1^*$ and if $\mathbf{A}_1$ is testable, then the unique m.l.e. of $\mathbf{A}_1\mathbf{B}$ is given by

$$\begin{aligned}\mathbf{FX}_1\hat{\mathbf{B}}_1^* &= \mathbf{A}_{11}\hat{\mathbf{B}}_1^* \\ &= \mathbf{A}_{11}(\mathbf{X}_1'\mathbf{X}_1)^{-1}\mathbf{X}_1'\mathbf{Y}.\end{aligned} \tag{6.5.28}$$

Roy (1957) has shown that H_G is testable *iff*

$$\mathbf{A}_{12} = \mathbf{A}_{11}(\mathbf{X}_1'\mathbf{X}_1)^{-1}\mathbf{X}_1'\mathbf{X}_2. \tag{6.5.29}$$

EXAMPLE 6.5.2: *One-way Multivariate Analysis of Variance* Consider k, $p-$variate normal samples $\mathbf{y}_1, \mathbf{y}_2, \ldots, \mathbf{y}_{n_1}, \mathbf{y}_{n_1+1}, \ldots, \mathbf{y}_n$, where $n = n_1 + \ldots + n_k$. The ith sample consists of n_i observations from $N_p(\mu + \tau_i, \mathbf{\Sigma}), i = 1, \ldots, k$. We can write the model as

$$E\begin{bmatrix}\mathbf{y}_1' \\ \mathbf{y}_2' \\ \cdot \\ \cdot \\ \mathbf{y}_n'\end{bmatrix} = \begin{bmatrix}\mathbf{1}_{n_1} & \mathbf{0} & \ldots & \mathbf{0} & \mathbf{1}_{n_1} \\ \mathbf{0} & \mathbf{1}_{n_2} & \ldots & \mathbf{0} & \mathbf{1}_{n_2} \\ \cdot & \cdot & \ldots & \cdot & \cdot \\ \cdot & \cdot & \ldots & \cdot & \cdot \\ \mathbf{0} & \mathbf{0} & \ldots & \mathbf{1}_{n_k} & \mathbf{1}_{n_k}\end{bmatrix}\begin{bmatrix}\tau_1' \\ \tau_2' \\ \cdot \\ \cdot \\ \tau_k' \\ \mu'\end{bmatrix}. \tag{6.5.30}$$

Here $\mathbf{X}(n\times(k+1))$ is of rank k. Let us write $\mathbf{X} = (\mathbf{X}_1, \mathbf{1}_n)$ so that $\mathbf{X}_1(n\times k)$ is of rank k and $\mathbf{X}_2 = \mathbf{X}_1\mathbf{1}_k$ so that $\mathbf{A} = \mathbf{1}_k$. Therefore,

$$\mathbf{B}_1^* = \begin{bmatrix}\tau_1' \\ \cdot \\ \cdot \\ \tau_k'\end{bmatrix} + \mathbf{1}_k\mu' = \begin{bmatrix}\tau_1^{*'} \\ \cdot \\ \cdot \\ \tau_k^{*'}\end{bmatrix},$$

where $\tau^* = \tau + \mu$. We want to test $\tau_1 = \dots \tau_k$. This hypothesis can be written as

$$\begin{bmatrix} 1 & 0 & \dots & 0 & -1 & 0 \\ 0 & 1 & \dots & 0 & -1 & 0 \\ \cdot & \cdot & \dots & \cdot & \cdot & \cdot \\ 0 & 0 & \dots & 1 & -1 & 0 \end{bmatrix}_{(k-1)\times(k+1)} \mathbf{B} = \mathbf{0},$$

i.e.

$$\mathbf{C}_1\mathbf{B} = \mathbf{0}$$

where $\mathbf{B} = [\tau_1 \ \tau_2 \dots \tau_k \ \mu]'$ and $\mathbf{C}_1 = (\mathbf{A}_{11} \ \mathbf{0}_{(k-1)\times 1})$. The hypothesis can be written as

$$\mathbf{A}_{11}\mathbf{B}_1^* = \mathbf{0}.$$

It can be easily checked that Roy's condition (6.5.18) is satisfied so that the null hypothesis $\mathbf{A}_1\mathbf{B} = \mathbf{0}$ is testable. The *m.l.e.* of $\mathbf{B}_1^*$ is given by

$$(\hat{\tau}_1^* \dots \hat{\tau}_k^*)' = (\mathbf{X}_1'\mathbf{X}_1)^{-1}\mathbf{X}_1'\mathbf{Y} = (\bar{\mathbf{y}}_1, \dots, \bar{\mathbf{y}}_k)'$$

where $\bar{\mathbf{y}}_i$ is the sample mean for the ith multinormal sample. A more detailed solution of the problem has been given in Section 7.2.

6.6 Restricted Least Square Estimator of B

We shall now consider an alternative approach to the problems discussed in Subsections 6.5.1 and 6.5.2 by using projection-matrices on the relevant subspaces. This will simplify some calculations. We assume that $\mathbf{X}$ is of full rank.

Consider the null hypothesis

$$H_{01} : \mathbf{AB} = \mathbf{0} \tag{6.6.1}$$

where $\mathbf{A}$ is a $g \times q$ matrix of rank g.

To find the restricted least square estimator $\hat{\mathbf{B}}_H$ of $\mathbf{B}$ under H_{01} we are to minimize $(\mathbf{Y} - \mathbf{XB})'(\mathbf{Y} - \mathbf{XB})$ subject to H_{01}. Instead of $\mathbf{B}$ we shall work with $\theta = \mathbf{XB}$.

Now

$$\mathbf{B} = (\mathbf{X}'\mathbf{X})^{-1}\mathbf{X}'\mathbf{XB} = (\mathbf{X}'\mathbf{X})^{-1}\mathbf{X}'\theta. \tag{6.6.2}$$

Hence H_{01} becomes

$$H_{01} : \mathbf{A}_1\theta = \mathbf{0} \tag{6.6.3}$$

where $\mathbf{A}_1 = \mathbf{A}(\mathbf{X}'\mathbf{X})^{-1}\mathbf{X}'$. We therefore want to minimize $(\mathbf{Y}-\theta)'(\mathbf{Y}-\theta)$ subject to conditions (i) $\theta \in \Omega = \mathcal{R}(\mathbf{X})$, the column-space of $\mathbf{X}$, (ii) $\theta \in \mathcal{N}[\mathbf{A}_1]$, the null-space of $\mathbf{A}_1$; that is, to $\theta \in \omega = \Omega \cap \mathcal{N}[\mathbf{A}_1]$.

Hence, least square estimate of θ is

$$\hat{\theta}_H = \mathbf{Q}_\omega \mathbf{Y} \tag{6.6.4}$$

where $\mathbf{Q}_\omega$ is the orthogonal projection operator onto ω. (Clearly, $\mathbf{Q}_\omega = \mathbf{Q}_1 = \mathbf{I}_n - \mathbf{P}_1$ in the notation of (6.5.25).)

Now,

$$\begin{aligned} \mathbf{Q}_\Omega \mathbf{A}_1' &= \mathbf{X}(\mathbf{X}'\mathbf{X})^{-1}\mathbf{X}'\mathbf{A}_1' \quad \text{(by A.17.1.8)} \\ &= \mathbf{X}(\mathbf{X}'\mathbf{X})^{-1}\mathbf{A}' \end{aligned} \tag{6.6.5}$$

is of rank g, because $\mathcal{R}(\mathbf{A}') \subset \Omega = \mathcal{R}(\mathbf{X})$ (by condition of estimability there exists an estimator $\mathbf{FY}$ such that $E(\mathbf{FY}) = \mathbf{AB}$ or $\mathbf{FX} = \mathbf{A}$ or $\mathbf{A}' = \mathbf{X}'\mathbf{F}'$) and $\Omega^\perp \cap \mathcal{R}(\mathbf{A}') = \mathbf{0}$ (vide A.17.3.4).

Also,

$$\begin{aligned} \mathbf{X}\hat{\mathbf{B}}_H &= \mathbf{Q}_\omega \mathbf{Y} = \mathbf{Q}_\Omega \mathbf{Y} + (\mathbf{Q}_\omega - \mathbf{Q}_\Omega)\mathbf{Y} \\ &= \hat{\theta} - \mathbf{Q}_{\omega^\perp \cap \Omega}\mathbf{Y} \quad \text{(by A.17.3.2)} \end{aligned} \tag{6.6.6}$$

where $\hat{\theta} = \mathbf{X}\hat{\mathbf{B}}, \hat{\mathbf{B}}$ being the least square estimate of $\mathbf{B}$ under full model (6.1.1). Again,

$$\begin{aligned} \mathbf{Q}_{\omega^\perp \cap \Omega} &= \mathbf{Q}_{\mathcal{R}[\mathbf{Q}_\Omega \mathbf{A}_1']} \quad \text{(by A.17.3.3)} \\ &= (\mathbf{Q}_\Omega \mathbf{A}_1')\{\mathbf{A}_1 \mathbf{Q}_\Omega^2 \mathbf{A}_1'\}^{-1}(\mathbf{Q}_\Omega \mathbf{A}_1')' \\ &= \mathbf{X}(\mathbf{X}'\mathbf{X})^{-1}\mathbf{A}'[\mathbf{A}(\mathbf{X}'\mathbf{X})^{-1}\mathbf{A}']^{-1}\mathbf{A}(\mathbf{X}'\mathbf{X})^{-1}\mathbf{X}'. \end{aligned} \tag{6.6.7}$$

By (6.6.5), $\mathbf{Q}_{\omega^\perp \cap \Omega}$ is a symmetric idempotent matrix of rank g.

Premultiplying both sides of (6.6.6) by $\mathbf{X}'$ and using (6.6.7),

$$\begin{aligned} \mathbf{X}'\mathbf{X}\hat{\mathbf{B}}_H &= \mathbf{X}'\hat{\theta} - \mathbf{A}'[\mathbf{A}(\mathbf{X}'\mathbf{X})^{-1}\mathbf{A}']^{-1}\mathbf{A}(\mathbf{X}'\mathbf{X})^{-1}\mathbf{X}'\mathbf{Y} \\ &= \mathbf{X}'\mathbf{Y} - \mathbf{A}'[\mathbf{A}(\mathbf{X}'\mathbf{X})^{-1}\mathbf{A}']^{-1}\mathbf{A}\hat{\mathbf{B}} \end{aligned} \tag{6.6.8}$$

from which $\hat{\mathbf{B}}_H$ can be estimated. (Clearly, $\mathbf{Q}_{\omega \perp \cap \Omega} = \mathbf{Q}_\Omega - \mathbf{Q}_\omega = \mathbf{P}_\omega - \mathbf{P}_\Omega = \mathbf{P}_1 - \mathbf{P} = \mathbf{P}_2$ in the notation of (6.5.16). Also, compare (6.5.16) with (6.6.7).)

We now consider hypotheses

$$H_{02} : \mathbf{AB} = \mathbf{D}. \tag{6.6.9}$$

Let $\mathbf{B}_0$ be any solution satisfying $\mathbf{AB}_0 = \mathbf{D}$. Let $\tilde{\mathbf{Y}} = \mathbf{Y} - \mathbf{XB}_0$. Now

$$\begin{aligned} \tilde{\mathbf{Y}} &= \mathbf{X}(\mathbf{B} - \mathbf{B}_0) + \mathbf{U} \\ &= \mathbf{X}\mathbf{\Lambda} + \mathbf{U} \\ &= \mathbf{\Phi} + \mathbf{U} \end{aligned} \tag{6.6.10}$$

where $\Lambda = \mathbf{B} - \mathbf{B}_0, \mathbf{\Phi} = \mathbf{X\Lambda}$. The hypothesis H_{02} becomes

$$H_{02} : \mathbf{A\Lambda} = \mathbf{0}.$$

As before, we shall work with $\mathbf{\Phi}$ instead of $\mathbf{\Lambda}$.

Now,

$$\begin{aligned} \mathbf{A\Lambda} &= \mathbf{A}(\mathbf{X}'\mathbf{X})^{-1}\mathbf{X}'\mathbf{\Phi} \\ &= \mathbf{A}_1\mathbf{\Phi}. \end{aligned}$$

Therefore, from (6.6.6),

$$\begin{aligned} \hat{\mathbf{\Phi}}_H &= \mathbf{Q}_\Omega\tilde{\mathbf{Y}} + (\mathbf{Q}_\omega - \mathbf{Q}_\Omega)\tilde{\mathbf{Y}} \\ &= \mathbf{Q}_\Omega(\mathbf{Y} - \mathbf{XB}_0) - \mathbf{Q}_{\omega^\perp\cap\Omega}\tilde{\mathbf{Y}} \\ &= \hat{\theta} - \mathbf{X}(\mathbf{X}'\mathbf{X})^{-1}\mathbf{X}'\mathbf{XB}_0 - \mathbf{Q}_{\omega^\perp\cap\Omega}\tilde{\mathbf{Y}} \\ &= \hat{\theta} - \mathbf{XB}_0 - \mathbf{Q}_{\omega^\perp\cap\Omega}\tilde{\mathbf{Y}}. \end{aligned} \tag{6.6.11}$$

Again,

$$\theta = (\mathbf{\Phi} + \mathbf{XB}_0) = \mathbf{X}(\mathbf{B} - \mathbf{B}_0) + \mathbf{XB}_0.$$

Therefore,

$$\begin{aligned} \hat{\theta}_H &= \mathbf{X}(\hat{\mathbf{B}}_H - \mathbf{B}_0) + \mathbf{XB}_0 \\ &= \hat{\mathbf{\Phi}}_H + \mathbf{XB}_0 \end{aligned} \tag{6.6.12}$$

where $\hat{\theta}_H$ is the least square estimator of θ under H_{02} and $\hat{\mathbf{\Phi}}_H$ is the least square estimator of $\mathbf{\Phi}$ under H_{02}.

Hence, from (6.6.11) and (6.6.12), since we are interested in estimating $\theta = \mathbf{XB}$,

$$\hat{\theta}_H = \hat{\theta} - \mathbf{Q}_{\omega^\perp\cap\Omega}\tilde{\mathbf{Y}}, \tag{6.6.13}$$

the same expression as (6.6.6), only change is that $\mathbf{Y}$ is replaced by $\tilde{\mathbf{Y}}$. Here $\hat{\theta}$ is the l.s.e. of θ under model (6.1.1).

Proceeding exactly as in (6.6.7) and (6.6.8) we shall have

$$\mathbf{X}'\mathbf{X}\hat{\mathbf{B}}_H = \mathbf{X}'\mathbf{Y} - \mathbf{A}'[\mathbf{A}(\mathbf{X}'\mathbf{X})^{-1}\mathbf{A}']^{-1}(\mathbf{A}\hat{\mathbf{B}} - \mathbf{D}) \tag{6.6.14}$$

from which $\hat{\mathbf{B}}_H$, the restricted least square estimator of $\mathbf{B}$ under H_{02} can be obtained.

If the $g \times q$ matrix $\mathbf{A}$ has rank less than g, then we shall use a generalized inverse $\mathbf{A}'[\mathbf{A}(\mathbf{X}'\mathbf{X})^{-1}\mathbf{A}']^-$ in (6.6.14).

When $\mathbf{X}$ is not of full rank $r(< q)$ (and $\mathbf{A}$ is of rank g), since the constraints $\mathbf{a}_i'\beta_{(j)} (i = 1, \ldots, g; j = 1, \ldots, p)$ must be estimable, there should be a $g \times n$ matrix $\mathbf{M}$ of rank g such that $\mathbf{A} = \mathbf{MX}$.

The hypothesis H_{01} then reduces to H_{01} : $\mathbf{MXB} = \mathbf{0}$ or $\mathbf{M}\theta = \mathbf{0}$. Hence, $\theta \in \omega = \Omega \cap \mathcal{N}(\mathbf{M})$. Also,

$$\begin{aligned} \mathbf{Q}_\Omega \mathbf{M}' &= \mathbf{X}(\mathbf{X}'\mathbf{X})^- \mathbf{X}'\mathbf{M}' \\ &= \mathbf{X}(\mathbf{X}'\mathbf{X})^- \mathbf{A}' \end{aligned} \tag{6.6.15}$$

has rank g. This can be seen as follows. Suppose the $n \times g$ matrix $\mathbf{Q}_\Omega\mathbf{M}'$ has linearly dependent columns so that for some $\mathbf{c}(\neq \mathbf{0})$, $\mathbf{Q}_\Omega\mathbf{M}'\mathbf{c} = \mathbf{0}$. Thus $\mathbf{c}'\mathbf{M} \in \Omega^\perp$. Also, $\mathbf{A}'\mathbf{c} = \mathbf{X}'\mathbf{M}'\mathbf{c} = \mathbf{0}$ which is a contradiction since the rows of $\mathbf{A}$ are linearly independent. Hence, $\mathbf{Q}_\Omega\mathbf{M}'$ is a $n \times g$ matrix of rank g.

Also,

$$\begin{aligned} \mathbf{Q}_{\omega^\perp \cap \Omega} &= \mathbf{Q}_{\mathcal{R}[\mathbf{Q}_\Omega \mathbf{M}']} = \mathbf{Q}_\Omega \mathbf{M}'[\mathbf{M}\mathbf{Q}_\Omega^2 \mathbf{M}']^{-1}(\mathbf{Q}_\Omega\mathbf{M}')' \\ &= \mathbf{X}(\mathbf{X}'\mathbf{X})^- \mathbf{A}'[\mathbf{A}(\mathbf{X}'\mathbf{X})^- \mathbf{A}']^{-1}\mathbf{A}(\mathbf{X}'\mathbf{X})^- \mathbf{A}'. \end{aligned} \tag{6.6.16}$$

It follows that equation (6.6.8) till holds with $(\mathbf{X}'\mathbf{X})^{-1}$ replaced by $(\mathbf{X}'\mathbf{X})^-$ and $\hat{\mathbf{B}}$ any solution of the normal equation (6.3.7).

6.6.1 *SSP matrices and their distributions*

If H_{02} in (6.6.9) holds,

$$\begin{aligned} (\mathbf{Y} - \hat{\theta}_H)'(\hat{\theta}_H - \theta) &= (\tilde{\mathbf{Y}} - \hat{\mathbf{\Phi}}_H)'(\hat{\mathbf{\Phi}}_H - \mathbf{\Phi}) \quad \text{(in (6.6.12))} \\ &= (\tilde{\mathbf{Y}} - \mathbf{Q}_\omega \tilde{\mathbf{Y}})'(\mathbf{Q}_\omega \tilde{\mathbf{Y}} - \mathbf{Q}_\omega \mathbf{\Phi}) \\ &\quad [\text{since } \mathbf{Q}_\omega \mathbf{\Phi} = \mathbf{\Phi}] \\ &= \tilde{\mathbf{Y}}'(\mathbf{I} - \mathbf{Q}_\omega)'\mathbf{Q}_\omega(\tilde{\mathbf{Y}} - \mathbf{\Phi}) \\ &= \mathbf{0}. \end{aligned} \tag{6.6.17}$$

Therefore,

$$(\mathbf{Y} - \theta)'(\mathbf{Y} - \theta) = (\mathbf{Y} - \hat{\theta}_H)'(\mathbf{Y} - \hat{\theta}_H) + (\hat{\theta}_H - \theta)'(\hat{\theta}_H - \theta). \tag{6.6.18}$$

The SSP-matrix due to error under the full model (6.6.10) is

$$\begin{aligned} \mathbf{E} &= (\tilde{\mathbf{Y}} - \hat{\mathbf{\Phi}})'(\tilde{\mathbf{Y}} - \hat{\mathbf{\Phi}}) \\ &= (\tilde{\mathbf{Y}} - \mathbf{Q}_\Omega \tilde{\mathbf{Y}})'(\tilde{\mathbf{Y}} - \mathbf{Q}_\Omega \tilde{\mathbf{Y}}) \\ &= \tilde{\mathbf{Y}}'\mathbf{P}_\Omega \tilde{\mathbf{Y}} \\ &= (\mathbf{Y} - \mathbf{X}\mathbf{B}_0)'\mathbf{P}_\Omega(\mathbf{Y} - \mathbf{X}\mathbf{B}_0) \\ &= \mathbf{U}'\mathbf{P}_\Omega \mathbf{U} \end{aligned} \tag{6.6.19}$$

where $\hat{\mathbf{\Phi}}$ is the *l.s.e.* of $\mathbf{\Phi}$ under model (6.6.10).

The SSP matrix due to error when H_{02} is true, is

$$\begin{aligned} \mathbf{E}_H &= (\tilde{\mathbf{Y}} - \hat{\mathbf{\Phi}}_H)'(\tilde{\mathbf{Y}} - \hat{\mathbf{\Phi}}_H) \\ &= (\tilde{\mathbf{Y}} - \mathbf{Q}_\omega \tilde{\mathbf{Y}})'(\tilde{\mathbf{Y}} - \mathbf{Q}_\omega \tilde{\mathbf{Y}}) \\ &= \tilde{\mathbf{Y}}'\mathbf{P}_\omega \tilde{\mathbf{Y}} \\ &= (\mathbf{\Phi} + \mathbf{U})'\mathbf{P}_\omega(\mathbf{\Phi} + \mathbf{U}) \quad \text{(by (6.6.19))} \\ &= \mathbf{U}'\mathbf{P}_\omega \mathbf{U}. \end{aligned} \tag{6.6.20}$$

Therefore, SSP matrix due to H_{02} is

$$\begin{aligned}\mathbf{H} &= \mathbf{E}_H - \mathbf{E}\\ &= \mathbf{U}'(\mathbf{P}_\omega - \mathbf{P}_\Omega)\mathbf{U}\\ &= \mathbf{U}'(\mathbf{Q}_\Omega - \mathbf{Q}_\omega)\mathbf{U}\\ &= \mathbf{U}'\mathbf{Q}_{\omega^\perp\cap\Omega}\mathbf{U}\end{aligned} \tag{6.6.21}$$

where $\mathbf{Q}_{\omega^\perp\cap\Omega}$ is a symmetric idempotent matrix of rank g (vide (6.6.7)). Again, $\mathbf{H}$ can be written as

$$\begin{aligned}\mathbf{H} &= \mathbf{E}_H - \mathbf{E} = \tilde{\mathbf{Y}}'(\mathbf{P}_\omega - \mathbf{P}_\Omega)\\ &= \tilde{\mathbf{Y}}'(\mathbf{Q}_\Omega - \mathbf{Q}_\omega)\tilde{\mathbf{Y}}\\ &= (\mathbf{Y} - \mathbf{X}\mathbf{B}_0)'(\mathbf{Q}_\Omega\mathbf{A}_1')\{\mathbf{A}_1\mathbf{Q}_\Omega\mathbf{A}_1'\}^{-1}(\mathbf{Q}_\Omega\mathbf{A}_1')'(\mathbf{Y} - \mathbf{X}\mathbf{B}_0)\\ &\quad \text{(vide (6.6.7))}\\ &= (\mathbf{A}_1\mathbf{Q}_\Omega\mathbf{Y} - \mathbf{A}_1\mathbf{Q}_\Omega\mathbf{X}\mathbf{B}_0)'\{\mathbf{A}_1\mathbf{Q}_\Omega\mathbf{A}_1\}^{-1}(\mathbf{A}_1\mathbf{Q}_\Omega\mathbf{Y} - \mathbf{A}_1\mathbf{Q}_\Omega\mathbf{X}\mathbf{B}_0)\\ &= (\mathbf{A}\hat{\mathbf{B}} - \mathbf{D})'[\mathbf{A}(\mathbf{X}'\mathbf{X})^{-1}\mathbf{A}']^{-1}(\mathbf{A}\hat{\mathbf{B}} - \mathbf{D}).\end{aligned} \tag{6.6.22}$$

We have, therefore, the following theorem.

Theorem 6.6.1: When $H_{02} : \mathbf{AB} = \mathbf{D}$ is true, $\mathbf{E}$ and $\mathbf{H} = \mathbf{E}_H - \mathbf{E}$ are independently distributed as $\mathbf{W}_p(n - q, \mathbf{\Sigma})$ and $\mathbf{W}_p(g, \mathbf{\Sigma})$ respectively.

Proof. Since $\mathbf{E} = \mathbf{U}'\mathbf{P}_\Omega\mathbf{U}$ and $\mathbf{H} = \mathbf{U}'\mathbf{Q}_{\omega^\perp\cap\Omega}\mathbf{U}$ where $\mathbf{P}_\Omega$ is a symmetric idempotent matrix of rank $n - q$ and $\mathbf{Q}_{\omega^\perp\cap\Omega}$ is a symmetric idempotent matrix of rank g and since

$$\mathbf{P}_\Omega'\mathbf{Q}_{\omega^\perp\cap\Omega} = (\mathbf{I}_n - \mathbf{Q}_\Omega)'(\mathbf{Q}_\Omega - \mathbf{Q}_\omega) = \mathbf{0},$$

the result follows by Theorem 4.2.10(b). □

6.6.2 *A generalized linear hypothesis*

Consider the model (6.1.1) with rank $(\mathbf{X}) = q$. Suppose we want to test the hypothesis

$$H_{03} : \mathbf{ABL} = \mathbf{0} \tag{6.6.23}$$

where $\mathbf{A}(g \times q)$ is of rank $g, \mathbf{B}(q \times p)$ is of rank p and $\mathbf{L}(p \times v)$ is of rank $v(\leq p)$.

We shall write $\mathbf{Y}_L = \mathbf{YL}$. The model (6.1.1) reduces to

$$\begin{aligned}\mathbf{Y}_L &= \mathbf{XBL} + \mathbf{UL}\\ &= \mathbf{X\Lambda} + \mathbf{V}\end{aligned} \tag{6.6.24}$$

where rows of $\mathbf{V}(n \times v)$

$$\mathbf{V} = \begin{bmatrix} \mathbf{v}_1' = (\mathbf{L}'\mathbf{u}_1)' \\ \mathbf{v}_2' = (\mathbf{L}'\mathbf{u}_2)' \\ \cdot \\ \cdot \\ \mathbf{v}_n' = (\mathbf{L}'\mathbf{u}_n)' \end{bmatrix}$$

are *iid* $N_v(\mathbf{0}, \mathbf{L}'\mathbf{\Sigma}\mathbf{L})$. The hypothesis H_{03} in (6.6.23) reduces to $H_{03} : \mathbf{A}\mathbf{\Lambda} = \mathbf{0}$ when the theory considered in connection with (6.6.3) can be applied.

The SSP matrix due to hypothesis in (6.6.23) is, by (6.6.22),

$$\begin{aligned} \mathbf{H} &= (\mathbf{A}\hat{\mathbf{\Lambda}})'[\mathbf{A}(\mathbf{X}'\mathbf{X})^{-1}\mathbf{A}']^{-1}(\mathbf{A}\hat{\mathbf{\Lambda}}) \\ &= \mathbf{Y}_L'\mathbf{X}(\mathbf{X}'\mathbf{X})^{-1}\mathbf{A}'[\mathbf{A}(\mathbf{X}'\mathbf{X})^{-1}\mathbf{A}']^{-1}\mathbf{A}'(\mathbf{X}'\mathbf{X})^{-1}\mathbf{X}'\mathbf{Y}_L \\ &= (\mathbf{A}\hat{\mathbf{B}}\mathbf{L})'[\mathbf{A}(\mathbf{X}'\mathbf{X})^{-1}\mathbf{A}']^{-1}(\mathbf{A}\hat{\mathbf{B}}\mathbf{L}) \\ &= \mathbf{V}'\mathbf{Q}_{\omega^\perp \cap \Omega}\mathbf{V} \quad \text{(following (6.6.21))} \end{aligned} \tag{6.6.25}$$

and

$$\begin{aligned} \mathbf{E} &= \mathbf{Y}_L'(\mathbf{I}_n - \mathbf{Q}_\Omega)\mathbf{Y}_L \\ &= \mathbf{L}'\mathbf{Y}'(\mathbf{I}_n - \mathbf{Q}_\Omega)\mathbf{Y}\mathbf{L}. \end{aligned} \tag{6.6.26}$$

Thus $\mathbf{E} \cap \mathbf{W}_v(n-q, \mathbf{L}'\mathbf{\Sigma}\mathbf{L})$. By Theorem 6.6.1, when H_{03} is true, $\mathbf{H}$ in (6.6.25) follows $\mathbf{W}_v(g, \mathbf{L}'\mathbf{\Sigma}\mathbf{L})$. One can therefore test H_{03} by using $\Lambda = |\mathbf{E}|/|\mathbf{E}+\mathbf{H}|$.

If $\mathbf{X}$ has rank $r(< q)$, then the above theory still holds with $(\mathbf{X}'\mathbf{X})^{-1}$ replaced by $(\mathbf{X}'\mathbf{X})^{-}$.

6.7 Some Examples

We now consider some examples of testing of hypotheses, applying the theory developed in Sections 6.5 and 6.6.

EXAMPLE 6.7.1: *Test for the Constrained Mean of a Normal Population*: Let $\mathbf{z}_1, \ldots, \mathbf{z}_n$ be a random sample from a $N_p(\mu, \mathbf{\Sigma})$ population. We want to test $H_0 : \mathbf{L}'\mu = \mathbf{0}$ where $\mathbf{L}'$ is a $v \times p$ matrix of rank v. Setting

$$\mathbf{Y} = [\mathbf{z}_1, \ldots, \mathbf{z}_n]', \; \mathbf{X}\mathbf{B} = [\mu, \ldots, \mu]' = \mathbf{1}_n\mu'$$

we have the linear model

$$\mathbf{Y} = \mathbf{X}\mathbf{B} + \mathbf{U}$$

where the rows $\mathbf{u}_1', \ldots, \mathbf{u}_n'$ of $\mathbf{U}$ are *iid* $N_p(\mathbf{0}, \mathbf{\Sigma})$. Then H_0 becomes

$$H_0 : \mu'\mathbf{L} = 0' \;\text{ or }\; \mathbf{B}\mathbf{L} = \mathbf{0}$$

which is a special case of the hypothesis H_0 : $\mathbf{ABL} = \mathbf{0}$ in (6.6.23) with $\mathbf{A} = 1$. Hence the test statistic is $\Lambda = |\mathbf{E}|/|\mathbf{E}+\mathbf{H}|$ where

$$\begin{aligned} \mathbf{H} &= (\hat{\mathbf{B}}L)'[(\mathbf{X}'\mathbf{X})^{-1}]^{-1}(\hat{\mathbf{B}}\mathbf{L}) \\ &= \mathbf{L}'\mathbf{Y}'\mathbf{X}(\mathbf{X}'\mathbf{X})^{-1}\mathbf{X}'\mathbf{Y}\mathbf{L} \\ &= n^{-1}\mathbf{L}'\mathbf{Y}'\mathbf{1}_n\mathbf{1}_n'\mathbf{Y}\mathbf{L} \\ &= n^{-1}\mathbf{L}'\bar{\mathbf{y}}\bar{\mathbf{y}}'\mathbf{L} \end{aligned} \tag{6.7.1}$$

where $\bar{\mathbf{y}} = (\bar{y}_1, \ldots, \bar{y}_p)'$ and

$$\begin{aligned} \mathbf{E} &= \mathbf{L}'\mathbf{Y}'\{\mathbf{I}_n - \mathbf{1}_n(\mathbf{1}_n'\mathbf{1}_n)^{-1}\mathbf{1}_n'\}\mathbf{Y}\mathbf{L} \\ &= \mathbf{L}'\textstyle\sum_i(\mathbf{z}_i - \bar{\mathbf{z}})(\mathbf{z}_i - \bar{\mathbf{z}})'\mathbf{L} \\ &= \mathbf{L}'\mathbf{Q}\mathbf{L} \quad \text{(say)} . \end{aligned} \tag{6.7.2}$$

Under $H_0, \mathbf{H}$ follows a Wishart distribution $\mathbf{W}_v(1, \mathbf{L}'\mathbf{\Sigma}\mathbf{L})$ and $\mathbf{E}$ follows $\mathbf{W}_v(n-1, \mathbf{L}'\mathbf{\Sigma}\mathbf{L})$. Since rank $(\mathbf{H}) = 1$, we can use Lawley-Hotelling trace statistics (vide Section 4.5)

$$\begin{aligned} T_g^2 &= m_E \text{ tr } (\mathbf{H}\mathbf{E}^{-1}) \\ &= (n-1) \text{ tr } [(\mathbf{L}'\mathbf{Q}\mathbf{L})^{-1}n\mathbf{L}'\bar{\mathbf{y}}\bar{\mathbf{y}}'\mathbf{L}] \\ &= n(\mathbf{L}'\bar{\mathbf{y}})'(\mathbf{L}'\mathbf{S}\mathbf{L})^{-1}(\mathbf{L}'\bar{\mathbf{y}}) \end{aligned} \tag{6.7.3}$$

(by (A.9 (7))), where $\mathbf{S} = (n-1)^{-1}\mathbf{Q}$.

EXAMPLE 6.7.2: *Comparing Means of Two Multivariate Normal Populations*: Let $\mathbf{w}_1, \ldots, \mathbf{w}_{n_1}$ and $\mathbf{z}_1, \ldots, \mathbf{z}_{n_2}$ be two independent random samples from $N_p(\mu_1, \mathbf{\Sigma})$ and $N_p(\mu_2, \mathbf{\Sigma})$ respectively. Setting $\mathbf{y}_i = \mathbf{w}_i (i = 1, \ldots, n_1)$ and $\mathbf{y}_{n_1+j} = \mathbf{z}_j (j = 1, \ldots, n_2)$ we have the linear model

$$\begin{bmatrix} \mathbf{y}_1' \\ \cdot \\ \mathbf{y}_{n_1+1}' \\ \cdot \\ \mathbf{y}_n' \end{bmatrix} = \begin{bmatrix} \mathbf{1}_{n_1} & \mathbf{0} \\ \mathbf{0} & \mathbf{1}_{n_2} \end{bmatrix} \begin{bmatrix} \mu_1' \\ \mu_2' \end{bmatrix} + \begin{bmatrix} \mathbf{u}_1' \\ \cdot \\ \mathbf{u}_{n_1+1}' \\ \cdot \\ \mathbf{u}_n' \end{bmatrix}$$

or

$$\mathbf{Y} = \mathbf{XB} + \mathbf{U}$$

where $n = n_1 + n_2$. The hypothesis $H_0 : \mu_1 = \mu_2$ can be written in the form

$$[1, -1] \begin{bmatrix} \mu_1' \\ \mu_2' \end{bmatrix} = \mathbf{0}$$

or

$$\mathbf{AB} = \mathbf{0}.$$

This is a special case of (6.6.1). The least square estimate of $\mathbf{B}$ is

$$\hat{\mathbf{B}} = (\mathbf{X}'\mathbf{X})^{-1}\mathbf{X}'\mathbf{Y} = \begin{bmatrix} n_1^{-1} & 0 \\ 0 & n_2^{-1} \end{bmatrix} \begin{bmatrix} \sum_i \mathbf{w}_i \\ \sum_j \mathbf{z}_j \end{bmatrix} = \begin{bmatrix} \bar{\mathbf{w}}' \\ \bar{\mathbf{z}}' \end{bmatrix}.$$

Therefore,

$$\begin{aligned} \mathbf{E} &= (\mathbf{Y} - \mathbf{X}\hat{\mathbf{B}})'(\mathbf{Y} - \mathbf{X}\hat{\mathbf{B}}) \\ &= \textstyle\sum_{i=1}^{n_1} (\mathbf{w}_i - \bar{\mathbf{w}})(\mathbf{w}_i - \bar{\mathbf{w}})' + \sum_j (\mathbf{z}_j - \bar{\mathbf{z}})(\mathbf{z}_j - \bar{\mathbf{z}})' \\ &= (n_1 + n_2 - 2)\mathbf{S}_p \quad \text{(say)} . \end{aligned}$$

Also,

$$\mathbf{A}\hat{\mathbf{B}} = \bar{\mathbf{w}}' - \bar{\mathbf{z}}'.$$

Hence,

$$\begin{aligned} \mathbf{H} &= (\mathbf{A}\hat{\mathbf{B}})'\{\mathbf{A}(\mathbf{X}'\mathbf{X})^{-1}\mathbf{A}'\}^{-1}(\mathbf{A}\hat{\mathbf{B}}) \\ &= \tfrac{n_1 n_2}{n_1 + n_2}(\bar{\mathbf{w}} - \bar{\mathbf{z}})(\bar{\mathbf{w}} - \bar{\mathbf{z}})'. \end{aligned}$$

For testing H_0 we use $|\mathbf{E}|/|\mathbf{E}+\mathbf{H}|$. Since $\mathbf{H}$ has rank 1, we can use Lawley-Hotelling trace statistic (vide Section 4.5) with $m_E = n_1 + n_2 - 2$,

$$\begin{aligned} T_g^2 &= m_E \text{ tr } [\mathbf{E}^{-1}\mathbf{H}] \\ &= m_E \text{ tr } [\tfrac{1}{m_E}\mathbf{S}_p^{-1}(\bar{\mathbf{w}} - \bar{\mathbf{z}})(\bar{\mathbf{w}} - \bar{\mathbf{z}})'\tfrac{n_1 n_2}{n_1 + n_2}] \\ &= \tfrac{n_1 n_2}{n}(\bar{\mathbf{w}} - \bar{\mathbf{z}})'\mathbf{S}_p^{-1}(\bar{\mathbf{w}} - \bar{\mathbf{z}}) \quad [\text{ by A.9(7)}]. \end{aligned}$$

EXAMPLE 6.7.3: *Missing Observations*: Suppose that in an experiment we have complete observations on $\mathbf{y}$ only on $n-m$ units, while the observations on the remaining m units are missing. Assume without any loss of generality

$$\mathbf{Y} = \begin{bmatrix} \mathbf{Y}_{(1)}(n-m) \times p \\ \mathbf{Y}_{(2)} m \times p \end{bmatrix}, \quad \mathbf{X} = \begin{bmatrix} \mathbf{X}_{(1)}(n-m) \times q \\ \mathbf{X}_{(2)} m \times q \end{bmatrix}$$

where observations on $\mathbf{Y}_{(2)}$ are missing. However, we assume that both $\mathbf{X}_{(1)}$ and $\mathbf{X}_{(2)}$ are available. For the full observations we have the model (6.1.1).

The least square estimate $\hat{\mathbf{B}}$ from the available data is a solution of the equation

$$\mathbf{X}_{(1)}'\mathbf{X}_{(1)}\mathbf{B} = \mathbf{X}_{(1)}'\mathbf{Y}_{(1)}. \tag{6.7.4}$$

The corresponding error SSP matrix is

$$\mathbf{E} = (\mathbf{Y} - \mathbf{X}\hat{\mathbf{B}})'(\mathbf{Y} - \mathbf{X}\hat{\mathbf{B}}). \tag{6.7.5}$$

However, a part of $\mathbf{Y}$ is unknown.

Now,

$$(\mathbf{Y} - \mathbf{XB})'(\mathbf{Y} - \mathbf{XB}) = (\mathbf{Y}_{(1)} - \mathbf{X}_{(1)}\mathbf{B})'(\mathbf{Y}_{(1)} - \mathbf{X}_{(1)}\mathbf{B}) + (\mathbf{Y}_{(2)} - \mathbf{X}_{(2)}\mathbf{B})'(\mathbf{Y}_{(2)} - \mathbf{X}_{(2)}\mathbf{B}). \quad (6.7.6)$$

Let $\hat{\mathbf{Y}}_{(2)} = \mathbf{X}_{(2)}\hat{\mathbf{B}}$ and $\hat{\mathbf{Y}} = (\mathbf{Y}'_{(1)}, \hat{\mathbf{Y}}'_{(2)})'$. Then from (6.7.6),

$$\begin{aligned}(\hat{\mathbf{Y}} - \mathbf{X}\hat{\mathbf{B}})'(\hat{\mathbf{Y}} - \mathbf{X}\hat{\mathbf{B}}) &= (\mathbf{Y}_{(1)} - \mathbf{X}_{(1)}\hat{\mathbf{B}})'(\mathbf{Y}_{(1)} - \mathbf{X}_{(1)}\hat{\mathbf{B}}) + (\hat{\mathbf{Y}}_{(2)} - \mathbf{X}_{(2)}\hat{\mathbf{B}})'(\hat{\mathbf{Y}}_{(2)} - \mathbf{X}_{(2)}\hat{\mathbf{B}}) \\ &= (\mathbf{Y}_{(1)} - \mathbf{X}_{(1)}\hat{\mathbf{B}})'(\mathbf{Y}_{(1)} - \mathbf{X}_{(1)}\hat{\mathbf{B}}).\end{aligned} \quad (6.7.7)$$

The matrix in (6.7.5) can, therefore, be taken as error matrix if $\mathbf{Y}$ is replaced by $\hat{\mathbf{Y}}$. Again, adding $\mathbf{X}'_{(2)}\mathbf{X}_{(2)}\hat{\mathbf{B}} = \mathbf{X}'_{(2)}\hat{\mathbf{Y}}_{(2)}$ to both sides of (6.7.4), we have

$$(\mathbf{X}'_{(1)}\mathbf{X}_{(1)} + \mathbf{X}'_{(2)}\mathbf{X}_{(2)})\hat{\mathbf{B}} = \mathbf{X}'_{(1)}\mathbf{Y}_{(1)} + \mathbf{X}'_{(2)}\hat{\mathbf{Y}}_{(2)}$$

i.e.

$$\mathbf{X}'\mathbf{X}\hat{\mathbf{B}} = \mathbf{X}'\hat{\mathbf{Y}} \quad (6.7.8)$$

so that $\hat{\mathbf{B}}$ is obtained by minimizing $(\hat{\mathbf{Y}} - \mathbf{X}\hat{\mathbf{B}})'(\hat{\mathbf{Y}} - \mathbf{X}\hat{\mathbf{B}})$. The procedure, therefore, reduces to completing the data set by replacing $\mathbf{Y}$ by $\hat{\mathbf{Y}}$ and then finding $\hat{\mathbf{B}}$ from (6.7.8) and $\mathbf{E}$ for the augmented model. One tests for the null hypothesis $H_0 : \mathbf{AB} = \mathbf{L}$ (where $\mathbf{A}$ has rank g), using $\mathbf{H}$ and $\mathbf{E}$ with g and $n - m - r$ degrees of freedom respectively (where $r =$ rank $(\mathbf{X})$).

EXAMPLE 6.7.4: **Profile Analysis** In section 5.2 we considered a profile analysis for $K = 2$ populations with mean μ and ν. Writing μ_1 and μ_2 in place of μ and ν respectively, the hypothesis of parallelism of profiles is

$$H_0 : \mathbf{C}_1\mu_1 = \mathbf{C}_1\mu_2$$

or

$$\begin{aligned}\mathbf{0}' &= (\mu_1' - \mu_2')\mathbf{C}_1' \\ &= (1, -1)\begin{bmatrix}\mu_1' \\ \mu_2'\end{bmatrix}\mathbf{C}_1'\end{aligned} \quad (6.7.9)$$

where $\mathbf{C}_1$ is given in (5.10.1),

$$\mathbf{C}_1 = \begin{bmatrix} -1 & 1 & 0 & \dots & 0 \\ 0 & -1 & 1 & \dots & 0 \\ . & . & . & \dots & . \\ 0 & 0 & 0 & \dots & 1 \end{bmatrix}.$$

We now consider profile analysis for $K \geq 2$ populations. Let $\mathbf{y}_{i1}, \ldots, \mathbf{y}_{in_i}$ be K independent random samples from $N_p(\mu_i, \mathbf{\Sigma})$-population $(i = 1, \ldots, K)$. We have the linear model

$$\begin{bmatrix} \mathbf{y}'_{11} \\ \cdot \\ \cdot \\ \mathbf{y}'_{1n_1} \\ \mathbf{y}'_{21} \\ \cdot \\ \cdot \\ \mathbf{y}'_{Kn_K} \end{bmatrix} = \begin{bmatrix} \mathbf{1}_{n_1} & \mathbf{0} & \mathbf{0} & \ldots & \mathbf{0} \\ \mathbf{0} & \mathbf{1}_{n_2} & \mathbf{0} & \ldots & \mathbf{0} \\ \cdot & \cdot & \cdot & \ldots & \cdot \\ \mathbf{0} & \mathbf{0} & \mathbf{0} & \ldots & \mathbf{1}_{n_K} \end{bmatrix} \begin{bmatrix} \mu'_1 \\ \mu'_2 \\ \cdot \\ \cdot \\ \mu'_K \end{bmatrix} + \begin{bmatrix} \mathbf{u}'_{11} \\ \cdot \\ \cdot \\ \mathbf{u}'_{1n_1} \\ \mathbf{u}'_{21} \\ \cdot \\ \cdot \\ \mathbf{u}'_{Kn_K} \end{bmatrix}$$

or

$$\mathbf{Y} = \mathbf{XB} + \mathbf{U}. \tag{6.7.10}$$

In case of K populations, hypothesis of parallelism of profiles is

$$H_{01K} : \mathbf{C}_1\mu_1 = \mathbf{C}_1\mu_2 = \ldots = \mathbf{C}_1\mu_K$$

or

$$\mathbf{0}'(1 \times (p-1)) = (\mu'_1 - \mu'_2)(1 \times p)\mathbf{C}'_1(p \times (p-1)),$$

$$\mathbf{0}' = (\mu'_2 - \mu'_3)\mathbf{C}'_1,$$

$$\ldots$$

$$\mathbf{0}' = (\mu'_{K-1} - \mu'_K)\mathbf{C}'_1,$$

or

$$\begin{bmatrix} \mathbf{0}' \\ \mathbf{0}' \\ \cdot \\ \cdot \\ \mathbf{0}' \end{bmatrix} = \begin{bmatrix} 1 & -1 & 0 & \ldots 0 \\ 0 & 1 & -1 & \ldots 0 \\ \cdot & \cdot & \cdot & \ldots . \\ \cdot & \cdot & \cdot & \ldots . \\ 0 & 0 & 0 & \ldots -1 \end{bmatrix} \begin{bmatrix} \mu'_1 \\ \mu'_2 \\ \cdot \\ \cdot \\ \mu'_K \end{bmatrix} \mathbf{C}'_1$$

or

$$\mathbf{0}((K-1) \times (p-1)) = \mathbf{A}((K-1) \times K)\mathbf{B}(K \times p)\mathbf{C}'_1(p \times (p-1)). \tag{6.7.11}$$

We find that (6.7.9) and (6.7.11) are special cases of (6.6.23).

In Section 5.10, the hypothesis H_{02} that the profiles are coincident given that they are parallel was given as

$$H_{02} : \mathbf{1}'_p\mu = \mathbf{1}'_p\nu$$

or

$$(\mu' - \nu')\mathbf{1}_p = 0. \tag{6.7.12}$$

Generalizing to the case of K populations, the null hypothesis of coincidence of profiles, given that they are parallel is,

$$H_{02K} : \mathbf{1}'_p\mu_1 = \mathbf{1}'_p\mu_2 = \ldots = \mathbf{1}'_p\mu_K$$

or expanding similarly as in (6.7.11),

$$H_{02K} : \mathbf{0}_{(K-1)\times 1} = \mathbf{AB1}_p. \tag{6.7.13}$$

For the case of two populations the hypothesis that all the treatments have the same mean effect was

$$H_{03} : \mathbf{C}_1(\mu_1 + \nu) = \mathbf{0}$$

or

$$\mathbf{0}' = (1,1)\begin{bmatrix}\mu' \\ \nu'\end{bmatrix}\mathbf{C}'_1. \tag{6.7.14}$$

Generalizing this to K populations we have the hypothesis

$$H_{03K} : \frac{1}{2}(\mu_{11} + \mu_{21}) = \ldots = \frac{1}{2}(\mu_{1p} + \mu_{2p}) \text{ or } \mathbf{C}_1(\mu_1 + \mu_2) = \mathbf{0},$$

$$\frac{1}{2}(\mu_{21} + \mu_{31}) = \ldots = \frac{1}{2}(\mu_{2p} + \mu_{3p}) \text{ or } \mathbf{C}_1(\mu_2 + \mu_3) = \mathbf{0},$$

$$\ldots$$

$$\frac{1}{2}(\mu_{K-11} + \mu_{K1}) = \ldots = \frac{1}{2}(\mu_{K-1p} + \mu_{Kp}) \text{ or } \mathbf{C}_1(\mu_{K-1} + \mu_K) = \mathbf{0}$$

or

$$\begin{bmatrix}\mathbf{0}' \\ \mathbf{0}' \\ \cdot \\ \cdot \\ \mathbf{0}'\end{bmatrix} = \begin{bmatrix}1 & 1 & 0 & \ldots & 0 & 0 \\ 0 & 1 & 1 & \ldots & 0 & 0 \\ \cdot & \cdot & \cdot & \ldots & \cdot & \cdot \\ \cdot & \cdot & \cdot & \ldots & \cdot & \cdot \\ 0 & 0 & 0 & \ldots & 1 & 1\end{bmatrix}\begin{bmatrix}\mu'_1 \\ \mu'_2 \\ \\ \mu'_K\end{bmatrix}\mathbf{C}'_1$$

or

$$\mathbf{0}((K-1)\times(p-1)) = \mathbf{A}_1((K-1)\times K)\mathbf{BC}'_1 \text{ (say)}. \tag{6.7.15}$$

If H_{01K} is true, H_{03K} implies $\mathbf{C}_1\mu_1 = \mathbf{C}_1\mu_2 = \ldots = \mathbf{C}_1\mu_K = \mathbf{0}$, i.e.,

$$\mu_{11} = \mu_{12} = \ldots = \mu_{1p},$$

$$\mu_{21} = \mu_{22} = \ldots = \mu_{2p},$$

$$\cdots$$

$$\mu_{K1} = \mu_{K2} = \ldots = \mu_{Kp},$$

i.e., all the profiles are horizontal parallel lines. Here, there are population main effects, but no differential effects due to treatments.

If all the hypotheses $H_{01K}, H_{02K}, H_{03K}$ are true, it means $\mu_{11} = \ldots = \mu_{1p} = \ldots = \mu_{K1} = \ldots = \mu_{Kp}$, i.e., all the K profiles are horizontal lines and they coincide. This means there are no differential population main effects, no differential test mean effects.

All the hypotheses $H_{01K}, H_{02K}, H_{03K}$ are special cases of the general linear hypothesis (6.6.23) and can be tested following the theory developed therein.

We indicate here another method of testing these hypotheses. Let

$$\bar{\mathbf{y}} = \frac{1}{n}(n_1\bar{\mathbf{y}}_1 + n_2\bar{\mathbf{y}}_2 + \ldots + n_K\bar{\mathbf{y}}_K), \ \mathbf{Q} = (n-K)\mathbf{S}_p$$

where $n = \sum_{i=1}^{K} n_i, \mathbf{S}_p = \sum_{i=1}^{n_1}(\mathbf{y}_{i1} - \bar{\mathbf{y}}_1)(\mathbf{y}_{i1} - \bar{\mathbf{y}}_1)' + \ldots + \sum_{i=1}^{n_K}(\mathbf{y}_{iK} - \bar{\mathbf{y}}_K)(\mathbf{y}_{iK} - \bar{\mathbf{y}}_K)'$, all the symbols having usual meanings. Then

$$\bar{\mathbf{y}} \cap N_p(E(\bar{\mathbf{y}}), \mathbf{\Sigma}/n);$$

$$\mathbf{C}_1\bar{\mathbf{y}} \cap N_{p-1}(\mathbf{C}_1E(\bar{\mathbf{y}}), \mathbf{C}_1\mathbf{\Sigma}\mathbf{C}_1'/n_1).$$

Also,

$$\mathbf{Q} \cap \mathbf{W}_p(n-K, \mathbf{\Sigma}),$$

$$\mathbf{C}_1\mathbf{Q}\mathbf{C}_1' \cap \mathbf{W}_{p-1}(n-K, \mathbf{C}_1\mathbf{\Sigma}\mathbf{C}_1').$$

Also, $\mathbf{C}_1\bar{\mathbf{y}}$ and $\mathbf{C}_1\mathbf{Q}\mathbf{C}_1'$ are independently distributed. Therefore,

$$n(\mathbf{C}_1\bar{\mathbf{y}} - \mathbf{C}_1E(\bar{\mathbf{y}}))'(\mathbf{C}_1\mathbf{Q}\mathbf{C}_1)^{-1}(\mathbf{C}_1\bar{\mathbf{y}} - \mathbf{C}_1E(\bar{\mathbf{y}})) \cap T^2(p-1, n-K). \quad (6.7.16)$$

If H_{01K} is true, $E(\mathbf{C}_1\bar{\mathbf{y}}) = \mathbf{C}_1\mu$. If, further, H_{03K} is true, $E(\mathbf{C}_1\bar{\mathbf{y}}) = \mathbf{0}$, because $\mathbf{C}_1\mu = \ldots = \mathbf{C}_1\mu_K = \mathbf{0}$. Hence, under $H_{01K} \cap H_{03K}$,

$$n(\mathbf{C}_1\bar{\mathbf{y}})'(\mathbf{C}_1\mathbf{Q}\mathbf{C}_1')^{-1}(\mathbf{C}_1\bar{\mathbf{y}}) \cap T^2(p-1, n-K). \quad (6.7.17)$$

6.8 Testing H_0 by Union-Intersection Principle

We have introduced Roy's (1959) union-intersection principle in Section 5.9. Here we use this principle to test $H_0 : \mathbf{AB} = \mathbf{D}$ where, as before, $\mathbf{A}$ is a known $g \times q$ matrix of rank g and $\mathbf{D}$ is a known $g \times p$ matrix.

We may thus wish to test simultaneously p hypotheses, $H_{0j} : \mathbf{A}\beta_{(j)} = \mathbf{d}_{(j)}, j = 1, \ldots, p$ where $\mathbf{D} = (\mathbf{d}_{(1)}, \ldots, \mathbf{d}_{(p)})$. The hypothesis H_0 is

$$H_0 = \cap_{j=1}^{p} H_{0j}. \tag{6.8.1}$$

We reduce the multivariate model to a linear model by the transformation

$$\mathbf{y} = \mathbf{Yl},\ \beta = \mathbf{Bl},\ \mathbf{u} = \mathbf{Ul} \tag{6.8.2}$$

where $\mathbf{l}(\neq \mathbf{0})$ is a $p \times 1$ vector of arbitrary constants. Then the model (6.1.1) along with the normality assumption reduces to the model

$$\mathbf{y} = \mathbf{X}\beta + \mathbf{u},$$

$$\mathbf{u} \sim N(\mathbf{0}, \sigma_l^2 \mathbf{I}_n),\ \sigma_l^2 = \mathbf{l}'\boldsymbol{\Sigma}\mathbf{l}. \tag{6.8.3}$$

The hypothesis H_0 reduces to

$$H_{0l} : \mathbf{A}\beta = \mathbf{d} \text{ where } \mathbf{d} = \mathbf{Dl}. \tag{6.8.4}$$

For testing H_{0l} we calculate E_l and E_{Hl} where

$$\begin{aligned} E_l &= (\mathbf{y} - \mathbf{X}\hat{\beta})'(\mathbf{y} - \mathbf{X}\hat{\beta}) \\ &= \mathbf{l}'\mathbf{E}\mathbf{l}, \end{aligned} \tag{6.8.5}$$

the error SSP matrix under full (transformed) model (6.8.3) where $\mathbf{E} = (\mathbf{Y} - \mathbf{X}\hat{\mathbf{B}})'(\mathbf{X} - \hat{\mathbf{B}}) = \mathbf{Y}'\mathbf{P}\mathbf{Y}$, the error SSP matrix under full model (6.1.1);

$$\begin{aligned} E_{Hl} &= (\mathbf{y} - \mathbf{X}\hat{\beta}_{Hl})'(\mathbf{y} - \mathbf{X}\hat{\beta}_{Hl}) \\ &= \mathbf{l}'\mathbf{E}_H\mathbf{l}, \end{aligned} \tag{6.8.6}$$

(because, $\hat{\beta}_{Hl} = \hat{\mathbf{B}}_H\mathbf{l}$) the SSP matrix under H_{0l} for the transformed model (6.8.3), when $\mathbf{E}_H = (\mathbf{Y} - \mathbf{X}\hat{\mathbf{B}}_H)'(\mathbf{Y} - \mathbf{X}\hat{\mathbf{B}}_H) = \tilde{\mathbf{Y}}'\mathbf{P}_\omega\tilde{\mathbf{Y}}$, the error SSP matrix under $H_0 : \mathbf{AB} = \mathbf{D}$ for the model (6.1.1) (vide (6.6.20)). Here, $\hat{\beta}_{Hl}$ is the least square estimate of β under model (6.8.3) subject to H_{0l}.

Since $E_{Hl} - E_l = H_l = \mathbf{l}'\mathbf{H}\mathbf{l}$ (where $\mathbf{H} = \mathbf{E}_H - \mathbf{E}$ is the SSP matrix due to $H_0 : \mathbf{AB} = \mathbf{D}$) is distributed as $\sigma_l^2\chi^2$ with g degrees of freedom when H_0 is true and E_l is distributed as $\sigma_l^2\chi^2$ with $(n - r)$ degrees of freedom (where rank $(\mathbf{X}) = r$) and since these χ^2's are independent (vide Theorems 6.5.2 and 4.2.4) we can test H_{0l} using the F ratio

$$F_l = \frac{(E_{Hl} - E_l)/g}{E_l/(n - r)} \tag{6.8.7}$$

which follows a $F_{(g,n-r)}$ distribution under H_{ol}.

Now, $H_0 = \cap_l H_{ol}$, so that using the union-intersection principle, a test for H_0 has the acceptance region

$$\begin{aligned}\cap_l\{\mathbf{Y} : F_l \le k\} &= \{\mathbf{Y} : \text{Sup }_l F_l \le k\} \\ &= \{\mathbf{Y} : \text{Sup }_l \tfrac{\mathbf{l'Hl}}{\mathbf{l'El}} \le \tfrac{gk}{n-r} = k_1\} \\ &= \{\mathbf{Y} : \Phi_{max} \le k_1\}\end{aligned} \tag{6.8.8}$$

where Φ_{max} is the largest eigenvalue of $\mathbf{HE}^{-1}$ (by A.8.14), i.e., the largest root of $|\mathbf{H} - \Phi\mathbf{E}| = 0$. We reject H_0 if Φ_{max} is too large.

Critical values of $\theta_{max} = \Phi_{max}/(1 + \Phi_{max})$, the maximum eigenvalue of $\mathbf{H}(\mathbf{H}+\mathbf{E})^{-1}$ [i.e. the largest root of $|\mathbf{H} - \theta(\mathbf{H}+\mathbf{E})| = 0$] have been given in Table B.11. We reject H_0 if $\theta_{max} \ge \theta_\alpha$, where θ_α is obtained from Table B.11 with $s = \min(g,p), \nu_1 = (|g-p|-1)/2, \nu_2 = (n-r-p-1)/2$. When $s = 1$, we use tables of F distribution.

6.8.1 *Simultaneous confidence intervals*

Suppose we want to find a set of simultaneous confidence intervals for all linear combinations of the form $\mathbf{a'ABb}$, where $\mathbf{a}, \mathbf{b}$ are vectors of arbitrary constants. This is obtained by considering the acceptance region given by the largest root test of hypothesis $H_0 : \mathbf{AB} = \mathbf{0}$.

The hypothesis H_0 is true *iff* $H_{0ab} : \mathbf{a'ABb} = \mathbf{0}$ is true for all $\mathbf{a}, \mathbf{b}$. Thus

$$H_0 = \cap_a \cap_b H_{0ab}.$$

Set

$$\mathbf{y} = \mathbf{Yb},\ \beta = \mathbf{Bb},\ \mathbf{u} = \mathbf{Ub}.$$

Then the model (6.1.1) reduces to

$$\mathbf{y} = \mathbf{Xb} + \mathbf{u} \tag{6.8.9}$$

where $\mathbf{u} \cap N_n(\mathbf{0}, \sigma_b^2\mathbf{I}_n), \sigma_b^2 = \mathbf{b'\Sigma b}$. The hypothesis H_{0ab} reduces to H_{0ab} : $\mathbf{a}_0'\beta = 0$ where $\mathbf{a}_0' = \mathbf{a'A}$. We assume $\mathbf{X}$ is of full rank q. As in (6.8.7), H_{0ab} can be tested by using the F ratio

$$F_{(a,b)} = \frac{Q_H - Q}{Q/(n-q)}$$

where $Q_H - Q$ = SS due to H_{0ab} and is given by

$$(\mathbf{a}_0'\hat{\beta})'\{\mathbf{a}_0'(\mathbf{X'X})^{-1}\mathbf{a}_0]^{-1}\mathbf{a}_0'\hat{\beta} \quad \text{(compare with (6.6.22))}$$

$$\begin{aligned} &= \mathbf{y}'\mathbf{X}(\mathbf{X}'\mathbf{X})^{-1}\mathbf{A}'\mathbf{a}[\mathbf{a}'\mathbf{A}(\mathbf{X}'\mathbf{X})^{-1}\mathbf{A}'\mathbf{a}]^{-1}\mathbf{a}'\mathbf{A}(\mathbf{X}'\mathbf{X})^{-1}\mathbf{X}'\mathbf{y} \\ &= \{\mathbf{a}'\mathbf{A}(\mathbf{X}'\mathbf{X})^{-1}\mathbf{X}'\mathbf{Y}\mathbf{b}\}^2/\{\mathbf{a}'\mathbf{A}(\mathbf{X}'\mathbf{X})^{-1}\mathbf{A}'\mathbf{a}\} \\ &= (\mathbf{a}'\mathbf{L}\mathbf{b})^2/\mathbf{a}'\mathbf{M}\mathbf{a} \ \ \text{(say)} \end{aligned} \tag{6.8.10}$$

and

$$\begin{aligned} Q &= SS \text{ due to error under the full model (6.8.9)} \\ &= \mathbf{y}'\mathbf{P}_\Omega\mathbf{y} \\ &= \mathbf{b}'\mathbf{Y}'\mathbf{P}_\Omega\mathbf{Y}\mathbf{b} \\ &= \mathbf{b}'\mathbf{E}\mathbf{b}. \end{aligned} \tag{6.8.11}$$

Using the union-intersection principle, a test for H_0 has the acceptance region

$$\begin{aligned} \cap_a \cap_b \{\mathbf{Y} : F_{(a,b)} \leq k\} &= \{\mathbf{Y} : \ \sup{}_a \sup{}_b \frac{(\mathbf{a}'\mathbf{L}\mathbf{b})^2}{(\mathbf{a}'\mathbf{M}\mathbf{a})(\mathbf{b}'\mathbf{E}\mathbf{b})} \leq \frac{k}{n-q} = k_1\} \\ &= \{\mathbf{Y} : \Phi_{max} \leq k_1\} \end{aligned} \tag{6.8.12}$$

where Φ_{max} is the maximum eigenvalue of $\mathbf{M}^{-1}\mathbf{L}\mathbf{E}^{-1}\mathbf{L}'$ (by Theorem A.14.3), i.e., of

$$\begin{aligned} \mathbf{L}'\mathbf{M}^{-1}\mathbf{L}\mathbf{E}^{-1} &= \mathbf{Y}'\mathbf{X}(\mathbf{X}'\mathbf{X})^{-1}\mathbf{A}'[\mathbf{A}(\mathbf{X}'\mathbf{X})^{-1}\mathbf{A}']^{-1}\mathbf{A}(\mathbf{X}'\mathbf{X})^{-1}\mathbf{X}'\mathbf{Y}\mathbf{E}^{-1} \\ &= (\mathbf{A}\hat{\mathbf{B}})'[\mathbf{A}(\mathbf{X}'\mathbf{X})^{-1}\mathbf{A}']^{-1}\mathbf{A}\hat{\mathbf{B}}\mathbf{E}^{-1} \\ &= \mathbf{H}\mathbf{E}^{-1} \end{aligned} \tag{6.8.13}$$

where $\mathbf{H}$ is SSP matrix due to $H_0 : \mathbf{AB} = \mathbf{0}$ (by (6.6.22)). We have therefore arrived at Roy's union-intersection test.

Using (6.8.12) with $\hat{\mathbf{B}}$ replaced by $\hat{\mathbf{B}} - \mathbf{B}$ we have the following confidence statement:

$$\begin{aligned} 1-\alpha &= Pr[\Phi_{max} \leq \Phi_\alpha] \ \forall \mathbf{a}, \mathbf{b} \\ &= Pr\left[\frac{|\mathbf{a}'\mathbf{A}(\hat{\mathbf{B}}-\mathbf{B})\mathbf{b}|}{\sqrt{\mathbf{a}'\mathbf{A}(\mathbf{X}'\mathbf{X})^{-1}\mathbf{A}'\mathbf{a}}\sqrt{\mathbf{b}'\mathbf{E}\mathbf{b}}} \leq \sqrt{\Phi_\alpha}\right] \ \ \text{(by (6.8.12), (6.8.13))} \\ &= Pr[|\mathbf{a}'\mathbf{A}(\hat{\mathbf{B}}-\mathbf{B})\mathbf{b}| \leq [\Phi_\alpha\{\mathbf{a}'\mathbf{A}(\mathbf{X}'\mathbf{X})^{-1}\mathbf{A}'\mathbf{a}\}(\mathbf{b}'\mathbf{E}\mathbf{b})]^{1/2} \end{aligned} \tag{6.8.14}$$

$\forall \mathbf{a}, \mathbf{b}$ where Φ_α is the quantile value of the maximum eigenvalue of $\mathbf{H}\mathbf{E}^{-1}$. Thus a set of simultaneous confidence intervals for all linear combinations $\mathbf{a}'\mathbf{A}\mathbf{B}\mathbf{b}$ is given by

$$\mathbf{a}'\mathbf{A}\hat{\mathbf{B}}\mathbf{b} \pm [\Phi_\alpha(\mathbf{a}'\mathbf{A}(\mathbf{X}'\mathbf{X})^{-1}\mathbf{A}'\mathbf{a})(\mathbf{b}'\mathbf{E}\mathbf{b})]^{1/2}. \tag{6.8.15}$$

The set has overall confidence coefficient $100(1-\alpha)\%$. The largest root test of $H_0 : \mathbf{AB} = \mathbf{0}$ will be significant at $100\alpha\%$ level if at least one of the intervals in (6.8.15) does not contain zero.

Now

$$\begin{aligned} V(\mathbf{a}'\mathbf{A}\hat{\mathbf{B}}\mathbf{b}) &= \mathbf{a}'\mathbf{A}V(\hat{\mathbf{B}}\mathbf{b})\mathbf{A}'\mathbf{a} \\ &= \mathbf{a}'\mathbf{A}\mathbf{b}'V(\hat{\mathbf{B}})\mathbf{b}\mathbf{A}'\mathbf{a} \\ &= \mathbf{a}'\mathbf{A}\mathbf{b}'\{\boldsymbol{\Sigma}\otimes(\mathbf{X}'\mathbf{X})^{-1}\}\mathbf{b}\mathbf{A}'\mathbf{a} \quad \text{(vide Theorem 6.2.3)} \\ &= \mathbf{a}'\mathbf{A}(\mathbf{b}'\boldsymbol{\Sigma}\mathbf{b})(\mathbf{X}'\mathbf{X})^{-1}\mathbf{A}'\mathbf{a} \\ &= \mathbf{a}'\mathbf{A}(\mathbf{X}'\mathbf{X})^{-1}\mathbf{A}'\mathbf{a}\sigma_b^2 \quad \text{where } \sigma_b^2 = \mathbf{b}'\boldsymbol{\Sigma}\mathbf{b}. \end{aligned} \tag{6.8.16}$$

Hence the intervals (6.8.15) may be written as

$$\mathbf{a}'\mathbf{A}\hat{\mathbf{B}}\mathbf{b} \pm [\Phi_\alpha \hat{V}\{\mathbf{a}'\mathbf{A}\hat{\mathbf{B}}\mathbf{b}\}]^{1/2} \tag{6.8.17}$$

where $\hat{V}(\mathbf{a}'\mathbf{A}\hat{\mathbf{B}}\mathbf{b})$ is $V(\mathbf{a}'\mathbf{A}\hat{\mathbf{B}}\mathbf{b})$ with $\boldsymbol{\Sigma}$ replaced by $\mathbf{E}$.

If we set $\mathbf{A} = \mathbf{I}_q (g = q)$ then we get the set of intervals for $\mathbf{a}'\mathbf{B}\mathbf{b}$

$$\mathbf{a}'\hat{\mathbf{B}}\mathbf{b} \pm [\Phi_\alpha\{\mathbf{a}'(\mathbf{X}'\mathbf{X})^{-1}\mathbf{a}\}\{\mathbf{b}'\mathbf{E}\mathbf{b}\}]^{1/2}. \tag{6.8.18}$$

If we want to find confidence interval for β_{ij}, we set vectors $\mathbf{a}$ and $\mathbf{b}$ with 1 in the ith position and 1 in the jth position respectively, and zeros elsewhere.

However, it is known that for univariate models Scheffe's simultaneous intervals are far too wide. Hence, for multivariate models, intervals (6.8.15) may be far too wider and may not be of much use.

6.9 Mean Centered Model

Suppose the model (6.1.1) is modified as

$$\mathbf{Y} = \mathbf{1}_n\mu' + \mathbf{X}\mathbf{B} + \mathbf{U} \tag{6.9.1}$$

where $(\mathbf{1}_n, \mathbf{X})$ is a $n\times(q+1)$ matrix of rank $(q+1)$, $\mu = (\mu_1, \ldots, \mu_p)'$ and $\mathbf{B}$ ia $q\times p$ matrix of regression coefficients. In model (6.9.1) each $\mathbf{y}_{(j)}$ contains an intercept term $\mu_j\mathbf{1}_n$. (We no longer assume that $\mathbf{X}$ contains $\mathbf{1}_n$ as its first column.) Let

$$\tilde{\mathbf{X}} = \mathbf{X} - \mathbf{1}_n\bar{\mathbf{x}}'.$$

In $\tilde{\mathbf{X}}$ the column $\tilde{\mathbf{x}}_{(j)} = \mathbf{x}_{(j)} - \bar{x}_j\mathbf{1}_n, (j = 1, \ldots, p)$, so that the columns have mean zeros. Thus, $\mathbf{1}_n'\tilde{\mathbf{X}} = \mathbf{1}_n'\mathbf{X} - \mathbf{1}_n'\mathbf{1}_n\bar{\mathbf{x}}' = \mathbf{0}$. We can write the model (6.9.1) as

$$\begin{aligned} \mathbf{Y} &= \mathbf{1}_n\mu' + \tilde{\mathbf{X}}\mathbf{B} + \mathbf{1}_n\bar{\mathbf{x}}'\mathbf{B} + \mathbf{U} \\ &= \mathbf{1}_n(\mu' + \textstyle\sum_{j=1}^q \beta_j'\bar{x}_j) + \tilde{\mathbf{X}}\mathbf{B} + \mathbf{U} \\ &= \mathbf{1}_n\nu' + \tilde{\mathbf{X}}\mathbf{B} + \mathbf{U}. \end{aligned} \tag{6.9.2}$$

Thus, in (6.9.1) if μ is replaced by $\nu = \mu + \sum_{j=1}^{q} \beta_j \bar{x}_j$ then (6.9.1) remains valid with $\mathbf{X}$ replaced by $\tilde{\mathbf{X}}$.

We shall find the least square estimates of ν and $\mathbf{B}$ under model (6.9.2). Now,

$$\begin{bmatrix} \mathbf{1}'_n \\ \tilde{\mathbf{X}}' \end{bmatrix} [\mathbf{1}_n, \tilde{\mathbf{X}}] = \begin{bmatrix} n & \mathbf{0} \\ \mathbf{0} & \tilde{\mathbf{X}}'\tilde{\mathbf{X}} \end{bmatrix}.$$

Therefore, the *l.s.e.*'s and also the *m.l.e.*'s under normality assumptions are given by

$$\begin{bmatrix} \nu' \\ \mathbf{B} \end{bmatrix} = \begin{bmatrix} n^{-1} & \mathbf{0} \\ \mathbf{0} & (\tilde{\mathbf{X}}\tilde{\mathbf{X}})^{-1} \end{bmatrix} \begin{bmatrix} \mathbf{1}'_n \mathbf{Y} \\ \tilde{\mathbf{X}}'\mathbf{Y} \end{bmatrix}$$

$$= \begin{bmatrix} \bar{\mathbf{Y}}' \\ (\tilde{\mathbf{X}}'\tilde{\mathbf{X}})^{-1}\tilde{\mathbf{X}}'\mathbf{Y} \end{bmatrix}.$$

Again,

$$\text{Cov} \begin{bmatrix} \hat{\nu}' \\ \hat{\mathbf{B}} \end{bmatrix}^V = \mathbf{\Sigma} \otimes \begin{bmatrix} n^{-1} & \mathbf{0} \\ \mathbf{0} & (\tilde{\mathbf{X}}'\tilde{\mathbf{X}})^{-1} \end{bmatrix} \tag{6.9.3}$$

(Theorem 6.2.3). Hence, $\hat{\nu}$ and $\hat{\mathbf{B}}$ are independent. From (6.9.2),

$$\bar{\mathbf{Y}}' = \nu' + \bar{\mathbf{u}}',$$

where $\bar{\mathbf{u}}' = n^{-1}\mathbf{1}'_n\mathbf{U}$. Hence,

$$\mathbf{Y} - \mathbf{1}_n\bar{\mathbf{Y}}' = \tilde{\mathbf{X}}\mathbf{B} + \mathbf{W} \tag{6.9.4}$$

where

$$\mathbf{W} = \mathbf{U} - \mathbf{1}_n\bar{\mathbf{u}}' = \mathbf{H}\mathbf{U}, \quad \mathbf{H} = \mathbf{I}_n - n^{-1}\mathbf{1}_n\mathbf{1}'_n.$$

Now,

$$\mathbf{W} = \mathbf{H}\mathbf{U} = (\mathbf{H}\mathbf{u}_{(1)},\ \mathbf{H}\mathbf{u}_{(2)},\ \ldots, \mathbf{H}\mathbf{u}_{(p)}).$$

Writing

$$Vec(\mathbf{W}) = \mathbf{W}^V = \begin{bmatrix} \mathbf{H}\mathbf{u}_{(1)} \\ \mathbf{H}\mathbf{u}_{(2)} \\ \cdot \\ \cdot \\ \mathbf{H}\mathbf{u}_{(p)} \end{bmatrix},$$

$\text{Var}(\mathbf{W}^V) = \mathbf{\Sigma} \otimes \mathbf{\Sigma}$. Hence, the distribution of $\mathbf{W}^V$ is $N_{np}(\mathbf{0}, \mathbf{\Sigma} \otimes \mathbf{H})$.

6.10 Classical Linear Regression Model

In this section we shall review the results for the Classical Linear Model, i.e., the model (6.1.1) with $p = 1$,

$$\mathbf{Y} = \mathbf{X}\beta + \mathbf{U} \tag{6.10.1}$$

where $\mathbf{Y}$ is a $n \times 1$ vector of observed vector of response variables on each of n units, $\mathbf{X}(n \times q)$ is a known matrix of rank $q, \beta(q \times 1)$ is a vector of unknown regression parameters and $\mathbf{U}(n \times 1)$ is a vector of unobserved random disturbances. We assume that

$$E(\mathbf{u}) = \mathbf{0},\ V(\mathbf{u}) = \sigma^2\mathbf{I}_n \tag{6.10.2}$$

where $\sigma^2(> 0)$ is an unknown parameter of the model. As before, we assume that $\mathbf{X} = (\mathbf{1}_n, \mathbf{X}_1)$ and $\mathbf{X}$ is a fixed matrix. We shall later add the assumption of joint normality of $\mathbf{U}$ for making confidence statements and testing hypotheses.

We shall state the results without proof which follow as special cases of the corresponding results for multivariate multiple regression model.

Theorem 6.10.1: Assume $n \geq q + 1$. Assume that $\mathbf{U} \cap N_n(\mathbf{0}, \sigma^2\mathbf{I}_n)$. The m.l.e. of β and σ^2 are given by

$$\hat{\beta} = (\mathbf{X}'\mathbf{X})^{-1}\mathbf{X}'\mathbf{Y}, \tag{6.10.3}$$

$$\begin{aligned}\hat{\sigma^2} &= n^{-1}\mathbf{Y}'\mathbf{P}\mathbf{Y}\\ &= n^{-1}\hat{\mathbf{U}}'\hat{\mathbf{U}}\\ &= n^{-1}(\mathbf{Y} - \mathbf{X}\hat{\beta})'(\mathbf{Y} - \mathbf{X}\hat{\beta})\\ &= \tfrac{1}{n}\textstyle\sum_{i=1}^{n}(y_i - \mathbf{x}_i'\hat{\beta})^2\end{aligned} \tag{6.10.4}$$

where $\mathbf{P}$ is given in (6.2.1). Also,

(i) $E(\hat{\beta}) = \beta$;
(ii) $E(\hat{\mathbf{U}}) = \mathbf{0}$; Cov $(\hat{\mathbf{U}}) = \sigma^2\mathbf{P}$;
(iii) $\hat{\beta}$ is distributed independently of $\hat{\sigma^2}$;
(iv) Cov $(\hat{\beta}) = \sigma^2(\mathbf{X}'\mathbf{X})^{-1}$;
(v) $n\hat{\sigma^2} = \sum_{i=1}^{n}(y_i - \mathbf{x}_i'\hat{\beta})^2 \cap \sigma^2\chi^2_{(n-q)}$; however, $\hat{\sigma^2}$ is a biased estimator of σ^2, $E(\hat{\sigma^2}) = \frac{n-q}{n}\sigma^2$.

Proofs follow from Theorems 6.2.1, 6.2.2 and 6.2.3.

Theorem 6.10.2: The least square estimator of β obtained by minimizing $\mathbf{U}'\mathbf{U} = (\mathbf{Y}-\mathbf{X}\beta)'(\mathbf{Y}-\mathbf{X}\beta)$ is the same as the maximum likelihood estimator

$\hat{\beta}$ given in (6.10.3). All the properties (i), (ii) of Theorem 6.10.1 hold. Also, $E(\hat{\mathbf{U}}'\hat{\mathbf{U}}) = E[(\sum_{i=1}^{n}(y_i - \mathbf{x}_i'\hat{\beta})^2] = (n-q)\sigma^2$. Therefore,

$$E(s^2) = E(\frac{\sum_{i=1}^{n}(y_i - \mathbf{x}_i'\hat{\beta})^2}{n-q}) = \sigma^2. \tag{6.10.5}$$

Hence, the l.s.e of σ^2, s^2 is unbiased for σ^2. Also, $\hat{\beta}$ and $\hat{\mathbf{U}}$ are uncorrelated.

Theorem 6.10.3: *Gauss-Makov Theorem* Let $\mathbf{Y} = \mathbf{X}\beta + \mathbf{U}$, where $E(\mathbf{U}) = \mathbf{0}$, Cov $(\mathbf{U}) = \sigma^2\mathbf{I}_n$ and $\mathbf{X}$ has full rank q. For any vector $\mathbf{c}$ of constants, the estimator $\mathbf{c}'\hat{\beta}$ has the smallest possible variance among all linear estimator of the form $\mathbf{a}'\mathbf{Y}$ that are unbiased for $\mathbf{c}'\beta$, that is, $\mathbf{c}'\hat{\beta}$ is the best linear unbiased estimator (BLUE) of its expectation in the sense of minimum variance.

Theorem 6.10.4: Let $\mathbf{Y} = \mathbf{X}\beta + \mathbf{U}$, where $\mathbf{X}$ has full rank q and $\mathbf{U}$ is $N_n(\mathbf{0}, \sigma^2\mathbf{I}_n)$. Then a $100(1-\alpha)\%$ confidence-ellipsoid for β is given by

$$(\beta - \hat{\beta})'(\mathbf{X}'\mathbf{X})(\beta - \hat{\beta}) \leq qs^2 F_{q,n-q;\alpha}. \tag{6.10.6}$$

Also, simultaneous $100(1-\alpha)\%$ confidence intervals for the β_i are given by

$$\hat{\beta}_i \mathop{-}_{+} \sqrt{s^2(\mathbf{X}'\mathbf{X})^{-1}}\sqrt{qF_{q,n-q;\alpha}},\ i = 0, 1, \ldots, r. \tag{6.10.7}$$

Proof. We have $\hat{\beta} \cap N_p(\beta, \sigma^2(\mathbf{X}'\mathbf{X})^{-1})$. Hence

$$\frac{1}{\sigma^2}(\hat{\beta} - \beta)'(\mathbf{X}'\mathbf{X})(\hat{\beta} - \beta) \cap \chi^2_{(q)}.$$

Also, $(n-q)s^2/\sigma^2 \cap \chi^2_{(n-q)}$ and these two distributions are independent. Therefore,

$$(\hat{\beta} - \beta)'(\mathbf{X}'\mathbf{X})(\hat{\beta} - \beta) \cap qs^2 F_{(q,n-q)}.$$

Hence the result (6.10.6). To find the simultaneous confidence interval for $\mathbf{a}'\beta$, following Theorem 5.5.1, we maximize

$$\frac{[\mathbf{a}'(\hat{\beta} - \beta)]^2}{\mathbf{a}'(\mathbf{X}'\mathbf{X})^{-1}\mathbf{a}s^2}$$

with respect to $\mathbf{a}$. By Theorem A.14.1, the maximum value of this statistic is

$$(\hat{\beta} - \beta)'(\mathbf{X}'\mathbf{X})(\hat{\beta} - \beta)/s^2$$

which follows $qF_{(q,n-q)}$. Hence for any vector $\mathbf{a}$,

$$P[\frac{[\mathbf{a}'(\hat{\beta} - \beta)]^2}{\mathbf{a}'(\mathbf{X}'\mathbf{X})^{-1}\mathbf{a}s^2} \leq qF_{q,n-q;\alpha}] \geq 1 - \alpha. \tag{6.10.8}$$

Therefore, $100(1-\alpha)\%$ simultaneous confidence interval for $\mathbf{a}'\beta$ ia

$$\mathbf{a}'\hat{\beta} \underset{+}{-} \sqrt{\mathbf{a}'(\mathbf{X}'\mathbf{X})^{-1}\mathbf{a}s^2}\sqrt{qF_{q,n-q;\alpha}}.$$

The result (6.10.7) follows when $\mathbf{a} = (0, 0, \ldots, 0, 1, 0, \ldots, 0)'$. □

The one-at-a-time t-interval for β_i is

$$\hat{\beta}_i \underset{+}{-} t_{n-q;\alpha/2}\sqrt{\hat{\text{Var}}\ (\hat{\beta}_i)} \tag{6.10.9}$$

where $\hat{\text{Var}}$ denotes estimated variance obtained by using s^2 in place of σ^2.

6.10.1 *Forecasting a new observation*

Suppose we want to estimate the mean response, $E(Y)$ when $\mathbf{x} = \mathbf{x}_0$. This mean response is estimated unbiasedly by $\mathbf{x}_0'\hat{\beta}$. Under normality assumption

$$\mathbf{x}_0'\hat{\beta} \cap N(\mathbf{x}_0'\beta, \sigma^2\mathbf{x}_0'(\mathbf{X}'\mathbf{X})^{-1}\mathbf{x}_0)$$

and is independent of $(n-q)s^2 = \sum_{i=1}^{n}(Y_i - \mathbf{x}_i'\hat{\beta})^2$ which follows $\sigma^2\chi^2_{(n-q)}$ distribution. Therefore,

$$\frac{\mathbf{x}_0'\hat{\beta} - \mathbf{x}_0'\beta}{\sqrt{s^2\mathbf{x}_0'(\mathbf{X}'\mathbf{X})^{-1}\mathbf{x}_0}}$$

follows the t_{n-q} distribution. Hence, one-at-a-time t confidence interval for $\mathbf{x}_0'\beta$ is

$$\mathbf{x}_0'\hat{\beta} \underset{+}{-} t_{n-q;\alpha/2}\sqrt{s^2\mathbf{x}_0'(\mathbf{X}'\mathbf{X})^{-1}\mathbf{x}_0}. \tag{6.10.10}$$

Similarly, if we want to forecast a new observation $Y_0 = \mathbf{x}_0'\beta + u_0$ corresponding $\mathbf{x}_0$, the unbiased predictor $\mathbf{x}_0'\hat{\beta}$ has the forecast error

$$Y_0 - \mathbf{x}_0'\hat{\beta} = \mathbf{x}_0'(\beta - \hat{\beta}) + u_0$$

with variance of the forecast error

$$\sigma^2[\mathbf{x}_0'(\mathbf{X}'\mathbf{X})^{-1}\mathbf{x}_0 + 1].$$

Therefore $100(1-\alpha)\%$ prediction interval for Y_0 is

$$\mathbf{x}_0'\hat{\beta} \underset{+}{-} t_{n-q;\alpha/2}\sqrt{s^2[1 + \mathbf{x}_0'(\mathbf{X}'\mathbf{X})^{-1}\mathbf{x}_0]}. \tag{6.10.11}$$

The simultaneous confidence (prediction) intervals for $E(Y)$ at $\mathbf{x} = \mathbf{x}_0$ and for Y_0 can similarly be found out.

6.10.2 *Tests of hypotheses regarding regression parameters*

Consider the null hypotheses $H_0 : \beta_{(2)} = (\beta_{q_1+1}, \beta_{q_1+2}, \ldots, \beta_q)' = \mathbf{0}$, i.e., the variables $x_{q_1+1}, \ldots, x_q$, where $q = q_1 + q_2$, do not influence the response.

The overall regression hypothesis that none of the x's predict y can be expressed as $H_0' : \beta^0 = (\beta_2, \ldots, \beta_q)' = \mathbf{0}$. We do not include β_1, because β_1 represents an general effect and the hypothesis $(\beta_1, \ldots, \beta_q)' = \mathbf{0}$ would restrict Y to have a mean of zero.

Setting $\mathbf{X} = (\mathbf{X}_1(n \times q_1),\ \mathbf{X}_2(n \times q_2)), \beta = (\beta_{(1)}', \beta_{(2)}')'$, the general linear model (6.10.1) can be written as

$$\mathbf{Y} = \mathbf{X}_1\beta_{(1)} + \mathbf{X}_2\beta_{(2)} + \mathbf{U}.$$

Under H_0, the model reduces to

$$\mathbf{Y} = \mathbf{X}_1\beta_{(1)} + \mathbf{U}.$$

The likelihood ratio statistic for H_0 can be written as

$$\begin{aligned} \lambda &= \frac{\max_{(\beta_{(1)},\sigma^2)} L(\beta_{(1)},\sigma^2)}{\max_{(\beta,\sigma^2)}} \\ &= \frac{L(\hat{\beta}_{(1)},\hat{\sigma_1^2})}{L(\hat{\beta},\hat{\sigma^2})} \\ &= (\frac{\hat{\sigma^2}}{\hat{\sigma_1^2}})^{n/2} \end{aligned} \tag{6.10.12}$$

where

$$\begin{aligned} n\hat{\sigma}_1^2 &= (\mathbf{Y} - \mathbf{X}_1\hat{\beta}_{(1)})'(\mathbf{Y} - \mathbf{X}_1\hat{\beta}_{(1)}) = SS_{res}(\mathbf{X}_1) \quad \text{(say)} \\ n\hat{\sigma}^2 &= (\mathbf{Y} - \mathbf{X}\hat{\beta})'(\mathbf{X} - \mathbf{X}\hat{\beta}) = SS_{res}(\mathbf{X}) \quad \text{(say)} \\ \hat{\beta}_{(1)} &= (\mathbf{X}_1'\mathbf{X}_1)^{-1}(\mathbf{X}_1\mathbf{Y}),\ \hat{\beta} = (\mathbf{X}'\mathbf{X})^{-1}(\mathbf{X}'\mathbf{Y}). \end{aligned}$$

The LR test for testing H_0 is equivalent to rejecting H_0 for large value of

$$-2\ln\lambda = -n\ln(\frac{\hat{\sigma^2}}{\hat{\sigma_1^2}}) \tag{6.10.13}$$

which follows $\chi^2_{(q_2)}$ under H_0.

For large n, the modified statistic

$$-(n - q_1 - q_2/2 - 1)\ln(\frac{\hat{\sigma^2}}{\hat{\sigma_1^2}}) \cap \chi^2_{(q_2)}.$$

Alternatively, we can use the test based on

$$\begin{aligned} H &= \text{ Sum of Squares (SS) due to } H_0 = \text{ Extra SS} \\ &= SS_{res}(\mathbf{X}_1) - SS_{res}(\mathbf{X}) \\ &= (\mathbf{Y} - \mathbf{X}_1\hat{\beta}_{(1)})'(\mathbf{Y} - \mathbf{X}_1\hat{\beta}_{(1)}) - (\mathbf{Y} - \mathbf{X}\hat{\beta})'(\mathbf{Y} - \mathbf{X}\hat{\beta}) \\ &= \hat{\beta}'\mathbf{X}'\mathbf{Y} - \hat{\beta}_{(1)}'\mathbf{X}_{(1)}'\mathbf{Y} \\ &= SS_{reg}(\mathbf{X}) - SS_{reg}(\mathbf{X}_1) \end{aligned} \tag{6.10.14}$$

where SS_{res} and SS_{reg} denote respectively the SS due to residual (error) and SS due to regression. By Theorem 6.6.1, H follows $\sigma^2\chi^2_{(q_2)}$ distribution under H_0. Therefore, H_0 is rejected if

$$\frac{H/q_2}{SS_{res}(\mathbf{X})/(n-q)} = \frac{SS_{res}(\mathbf{X}_1) - SS_{res}(\mathbf{X})}{q_2 s^2} > F_{q_2,n-q;\alpha}. \tag{6.10.15}$$

A test for the individual β_i above and beyond the other β's is readily obtained using (6.10.14). To test $H_{0i}(\beta_i = 0)$ we arrange β_i last in β,

$$\beta = \begin{bmatrix} \beta_r \\ \beta_i \end{bmatrix},$$

where $\beta_r' = (\beta_1, \ldots, \beta_q)$ contains all β's except β_i. By (6.10.15), the test statistic is

$$F = \frac{\hat{\beta}'\mathbf{X}'\mathbf{Y} - \hat{\beta}_r'\mathbf{X}_r'\mathbf{Y}}{SS_{res}(\mathbf{X})/(n-q)} \tag{6.10.16}$$

which follows $F_{(1,n-q)}$. The test for $H_{0i} : \beta_i = 0$ made by the F statistic above is called the *partial F test.*

A test for the significance of the overall regression using multiple correlation coefficient has been considered in (6.10.29).

Consider now the general linear hypotheses $H_G : \mathbf{A}\beta = \mathbf{0}$ where $\mathbf{A}(g \times q)$ is of rank g. Under the full model (6.10.1)

$$\mathbf{A}\hat{\beta} \cap N_g(\mathbf{A}\beta,\ \sigma^2\mathbf{A}(\mathbf{X}'\mathbf{X})^{-1}\mathbf{A}').$$

Therefore, under H_G,

$$(\mathbf{A}\hat{\beta})'\{\mathbf{A}(\mathbf{X}'\mathbf{X})^{-1}\mathbf{A}'\}^{-1}(\mathbf{A}\hat{\beta}\} \cap \sigma^2\chi^2_{(g)}.$$

Also, $(n-q)s^2/\sigma^2 \cap \chi^2_{(n-q)}$ and the two distributions are independent. Therefore, H_G is rejected if

$$\frac{(\mathbf{A}\hat{\beta})'\{\mathbf{A}(\mathbf{X}'\mathbf{X})^{-1}\mathbf{A}'\}^{-1}(\mathbf{A}\hat{\beta})}{s^2} > gF_{g,n-q;\alpha}. \tag{6.10.17}$$

6.10.3 *Multiple correlation coefficient*

With a little change of notation, let us write $q = r+1$, the regressor variables as $x_0, x_1, \ldots, x_r$ and $\beta = (\beta_0, \beta_1, \ldots, \beta_r)' = (\beta_0,\ \tilde{\beta}')', \mathbf{X}_1 = ((x_{ij}, i = 1, \ldots, n; j = 1, \ldots, r))$. Then the model (6.10.1) can be written in the form of a mean-centered model,

$$\mathbf{Y} - \mathbf{1}_n\bar{Y} = \tilde{\mathbf{X}}\tilde{\beta} + \mathbf{W} \tag{6.10.18}$$

where

$$\tilde{\mathbf{X}}_{n\times r} = \mathbf{X}_1 - \mathbf{1}_n\bar{\mathbf{x}}', \mathbf{W} = \mathbf{U} - \mathbf{1}_n\bar{U},$$

and $\bar{\mathbf{x}} = (\bar{x}_1, \bar{x}_2, \ldots, \bar{x}_r)'$. Note that the errors $w_1, \ldots, w_n$ in (6.10.18) are no longer independent.

Define

$$\begin{aligned}\mathbf{S} &= (n-1)^{-1}\begin{bmatrix}(\mathbf{Y}-\bar{Y}\mathbf{1}_n)' \\ \tilde{\mathbf{X}}'\end{bmatrix}[\mathbf{Y}-\bar{Y}\mathbf{1}_n \;\; \tilde{\mathbf{X}}] \\ &= \begin{bmatrix} s_{yy} & \mathbf{S}_{y\mathbf{x}} \\ \mathbf{S}_{\mathbf{x}y} & \mathbf{S}_{\mathbf{xx}}\end{bmatrix}.\end{aligned} \tag{6.10.19}$$

Let $\mathbf{R}$ denote the corresponding matrix of correlation coefficients,

$$\mathbf{R} = \begin{bmatrix} 1 & \mathbf{R}_{y\mathbf{x}} \\ \mathbf{R}_{y\mathbf{x}} & \mathbf{R}_{\mathbf{xx}}\end{bmatrix}. \tag{6.10.20}$$

From (6.10.18), the ordinary least square estimator of $\tilde{\beta}$ is (vide equation (6.3.7)),

$$\begin{aligned}\hat{\tilde{\beta}} &= (\tilde{\mathbf{X}}'\tilde{\mathbf{X}})^{-1}\tilde{\mathbf{X}}'(\mathbf{Y} - \mathbf{1}_n\bar{Y}) \\ &= \mathbf{S}_{\mathbf{xx}}^{-1}\mathbf{S}_{\mathbf{x}y}.\end{aligned} \tag{6.10.21}$$

The simple correlation coefficient between $\mathbf{Y}$ and $\hat{\mathbf{Y}} = \tilde{\mathbf{X}}\hat{\tilde{\beta}}$ is called the multiple correlation coefficient of y on $x_1, \ldots .x_r$.

$$\begin{aligned}R_{y.\mathbf{x}} &= \text{Corr. } (\mathbf{Y}, \tilde{\mathbf{X}}\mathbf{S}_{\mathbf{xx}}^{-1}\mathbf{S}_{\mathbf{x}y}) \\ &= \sqrt{\frac{\mathbf{S}_{y\mathbf{x}}\mathbf{S}_{\mathbf{xx}}^{-1}\mathbf{S}_{\mathbf{x}y}}{s_{yy}}},\end{aligned} \tag{6.10.22}$$

because,

$$\begin{aligned}V(\tilde{\mathbf{X}}\mathbf{S}_{\mathbf{xx}}^{-1}\mathbf{S}_{\mathbf{x}y}) &= \mathbf{S}_{y\mathbf{x}}\mathbf{S}_{\mathbf{xx}}^{-1}\mathbf{S}_{\mathbf{x}y} \\ \text{Cov } (\mathbf{Y}, \tilde{\mathbf{X}}\mathbf{S}_{\mathbf{xx}}^{-1}\mathbf{S}_{\mathbf{x}y}) &= \mathbf{S}_{y\mathbf{x}}\mathbf{S}_{\mathbf{xx}}^{-1}\mathbf{S}_{\mathbf{x}y}.\end{aligned}$$

The coefficient $R_{y.\mathbf{x}}$ can also be defined in terms of correlation as

$$R_{y.\mathbf{x}}^2 = R^2 = R_{y\mathbf{x}}R_{\mathbf{xx}}^{-1}R_{\mathbf{x}y}. \tag{6.10.23}$$

R^2 is called the squared multiple correlation, also, the *coefficient of multiple determination.* Multiple correlation coefficient measures the efficacy of the variables $\mathbf{x}_1, \ldots, \mathbf{x}_r$ for predicting $\mathbf{y}$ through a linear combination of $\mathbf{x}_1, \ldots, \mathbf{x}_r$. Note that all the above results hold even if the elements of $\mathbf{W}$ are uncorrelated.

Considering the model (6.10.18) again, we have

$$\hat{\mathbf{W}} = \mathbf{Y} - \bar{Y}\mathbf{1}_n - \tilde{\mathbf{X}}\hat{\tilde{\beta}},$$

the estimated residual. Then

$$(\mathbf{Y} - \bar{Y}\mathbf{1}_n)'(\mathbf{Y} - \bar{Y}\mathbf{1}_n) = (\hat{\mathbf{W}} + \tilde{\mathbf{X}}\hat{\tilde{\beta}})'(\hat{\mathbf{W}} + \tilde{\mathbf{X}}\hat{\tilde{\beta}})$$
$$= \hat{\mathbf{W}}'\hat{\mathbf{W}} + \hat{\tilde{\mathbf{Y}}}'\hat{\tilde{\mathbf{Y}}} + \hat{\mathbf{W}}'\tilde{\mathbf{X}}\hat{\tilde{\beta}} + \hat{\tilde{\beta}}'\tilde{\mathbf{X}}'\hat{\mathbf{W}}$$
$$= \hat{\mathbf{W}}'\hat{\mathbf{W}} + \hat{\tilde{\mathbf{Y}}}'\hat{\tilde{\mathbf{Y}}} \tag{6.10.24}$$
$$= \text{SS due to residual (SSE)} + \text{SS due to regression (SSR)}$$

where $\tilde{\mathbf{Y}} = \mathbf{Y} - \bar{Y}\mathbf{1}_n$, because $\hat{\mathbf{W}}'\tilde{\mathbf{X}}\hat{\tilde{\beta}} = 0$. Now,

$$\begin{aligned} \hat{\tilde{\mathbf{Y}}}'\hat{\tilde{\mathbf{Y}}} &= \tilde{\mathbf{Y}}'\tilde{\mathbf{X}}(\tilde{\mathbf{X}}'\tilde{\mathbf{X}})^{-1}\tilde{\mathbf{X}}'\tilde{\mathbf{Y}} \\ &= (\mathbf{Y} - \mathbf{1}_n\bar{Y})'\tilde{\mathbf{X}}(\tilde{\mathbf{X}}'\tilde{\mathbf{X}})^{-1}\tilde{\mathbf{X}}'(\mathbf{Y} - \mathbf{1}_n\bar{Y}) \\ &= n\mathbf{S}_{y\mathbf{x}}\mathbf{S}_{\mathbf{xx}}^{-1}\mathbf{S}_{\mathbf{x}y} \\ &= ns_{yy}R^2_{y.\mathbf{x}}. \end{aligned} \tag{6.10.25}$$

Therefore,

$$R^2_{y.\mathbf{x}} = \frac{\hat{\tilde{\mathbf{Y}}}'\hat{\tilde{\mathbf{Y}}}}{\mathbf{Y}'\mathbf{Y} - n\bar{Y}^2} \tag{6.10.26}$$

is the proportion of the total (corrected) sum of squares due to y which is explained by its regression on $x_1, \ldots, x_r$.

From (6.10.24) and (6.10.25),

$$\hat{\mathbf{W}}\hat{\mathbf{W}} = ns_{yy}(1 - R^2_{y.\mathbf{x}})$$

or

$$1 - R^2_{y.\mathbf{x}} = \frac{\hat{\mathbf{W}}'\hat{\mathbf{W}}}{\tilde{\mathbf{Y}}'\tilde{\mathbf{Y}}}. \tag{6.10.27}$$

We can test for the significance of the overall regression, $H_0(\tilde{\beta} = \mathbf{0})$ by means of the statistic

$$\begin{aligned} F &= \frac{SSR/r}{SSE/(n-r-1)} \\ &= \frac{n-r-1}{r}\frac{R^2}{1-R^2} \end{aligned} \tag{6.10.28}$$

which is distributed as $F(r, n-r-1)$ under H_0. This is, because, SSR $= \hat{\mathbf{Y}}\hat{\mathbf{Y}} = \mathbf{Y}'\mathbf{Q}\mathbf{Y}$ where $\mathbf{Q} = \tilde{\mathbf{X}}(\tilde{\mathbf{X}}'\tilde{\mathbf{X}})^{-1}\tilde{\mathbf{X}}'$ is a symmetric idempotent matrix of rank r and SSE $= \hat{\mathbf{W}}'\hat{\mathbf{W}} = (\mathbf{HU})'\mathbf{P}(\mathbf{HU}) = \mathbf{U}'\mathbf{HPU}$ where $\mathbf{H} = \mathbf{I}_n - n^{-1}\mathbf{1}_n\mathbf{1}_n'$, $\mathbf{P} = \mathbf{I}_n - \mathbf{Q}$ and $\mathbf{HP} = \mathbf{H} - \tilde{\mathbf{X}}(\tilde{\mathbf{X}}'\tilde{\mathbf{X}})^{-1}\tilde{\mathbf{X}}'$ is a symmetric idempotent matrix of rank $n-1-r$.

We now consider an important property of the multiple correlation coefficient.

Theorem 6.10.5: Consider the univariate multiple regression model (6.10.18) (where the x values and y values are centered about the means). The largest sample correlation between $\mathbf{y}$ and a linear combination of the columns of $\tilde{\mathbf{X}}$ is given by $R_{y.\mathbf{x}}$ and is attained by the linear combination $\mathbf{X}\hat{\tilde{\beta}}$ where $\hat{\tilde{\beta}} = (\tilde{\mathbf{X}}'\tilde{\mathbf{X}})^{-1}\tilde{\mathbf{X}}'\mathbf{Y}$ is the regression coefficient.

Proof. Let $\mathbf{X}\beta$ be any linear combination of the columns of $\tilde{\mathbf{X}}$. Then

$$\text{Corr. }^2(\mathbf{Y}, \tilde{\mathbf{X}}\beta) = \frac{(\beta'\mathbf{S}_{\mathbf{x}y})^2}{s_{yy}\beta'\mathbf{S}_{\mathbf{xx}}\beta}. \qquad (6.10.29).$$

By Theorem A.14.1, this is maximized for $\beta = \mathbf{S}_{\mathbf{xx}}^{-1}\mathbf{S}_{\mathbf{x}y}$ and the maximum value is $\mathbf{S}_{y\mathbf{x}}\mathbf{S}_{\mathbf{xx}}^{-1}\mathbf{S}_{\mathbf{x}y}/s_{yy} = R^2_{y.\mathbf{x}}$. □

6.10.4 *Selection of variables*

In practice, one has at one's disposal more predictor variables x's than can be conveniently or economically accommodated in predicting y. Many of these variables may be redundant and could be discarded. Consider the classical linear regression model (6.10.1) and suppose that some of the columns of $\mathbf{X}(n \times q)$ are nearly collinear (i.e. some of the columns of $\mathbf{X}$ are nearly linearly related). Now, rank of $\mathbf{X}$ equals the number of independent columns of $\mathbf{X}$ and is also the number of non-zero eigenvalues of $\mathbf{X}'\mathbf{X}$ (vide A.7 (6), A.7(11)). Hence, in this situation some of the eigenvalues of $\mathbf{X}'\mathbf{X}$ will be nearly zeros. Therefore, some of the eigenvalues of $(\mathbf{X}'\mathbf{X})^{-1}$ will be very large. Since variance of the ordinary least square (OLS) estimator of β is $\sigma^2(\mathbf{X}'\mathbf{X})^{-1}$, at least some of the regression estimators will have large variances.

Because of the multi-collinearity some of the explanatory variables add very little to explain the response variable to what has already been explained by the other predictor variables. Our aim is to retain those variables which nontrivially explain the response variables and discard those variables that contribute only marginally to explain the total variation in response variable. This will increase the precision of the regression estimates based on the retained variables and would also reduce the number of measurements needed on similar data.

We shall examine in this section the problem of selection of a suitable subset of variables. There are two problems, - determining the size of the subset and determining the variables to form the optimum subset of the chosen size.

The two most popular approaches to subset selection are: (i) to examine all possible subsets (ii) to use a stepwise technique.

6.10.4.1 *Examining all possible subsets*

In many cases we may take advantage of algorithms that find the optimum subset of each size. The 'leaps-and-bounds' technique of Furnival and Wilson (1974) is an example of this technique.

We discuss three techniques for comparing subsets. Our objective is to find a subset of $k-1$ regressor variables out of an available number of $q-1$ variables, the intercept term being always included in the model. (The established notation in the literature considers this as a problem of choice of $p-1$ regressors out of an available set of $k-1$ predictor variables, the overall general effect being always included in the model.)

(a) R_k^2: By definition, R^2, the squared multiple correlation coefficient is a measure of model fit. Since R^2 can only increase if a new regressor variable is added to the model, the best choice for k is obviously q. The usual procedure is, therefore, to find the subset with the largest R_k^2 for each chosen $k = 2, 3, \ldots$ and then choose a value of k beyond which the increase in the value of R_k^2 is only marginal.

(b) s_k^2: Another criterion is to examine the smallest value of the variance estimator for each subset of size k,

$$s_k^2 = \frac{SSE_k}{n-k} \tag{6.10.30}$$

for $k = 2, 3, \ldots$ where SSE_k denotes SS due to residual (error) in fitting a model with k regressor variables. If q is fairly large, the minimal value of s_k^2 will gradually decrease to a value less than s_q^2 as k approaches q and then increase. The minimum value of s_k^2 can be less than s_q^2 if the decrease in SSE_k with an additional variable does not affect the loss of a d.f. It is often suggested that the subset with absolute minimum value of s_k^2 be chosen. An alternative suggestion is to choose the subset with the smallest k such that $s_k^2 < s_q^2$.

(c) C_k: The C_k (C_p in usual notation) criterion is due to Mallows (1964, 1973). Let a model be fitted to the data and let $\hat{Y}_i$ be the fitted value of Y_i. If a correct model is fitted, $E(Y_i) = E(\hat{Y}_i)$. If the fitted model is incorrect, $E_1(\hat{Y}_i) \neq E(Y_i)$ where expectation E_1 is taken with respect to working model and E is with respect to the true model. In this case,

$E_1(\hat{Y}_i) - E(Y_i) =$ Bias $(\hat{Y}_i)$. Now,

$$\begin{aligned} E_1[\hat{Y}_i - E(Y_i)]^2 &= E_1[\hat{Y}_i - E_1(\hat{Y}_i)]^2 + [E_1(\hat{Y}_i) - E(Y_i)]^2 \\ &= \text{Var}_1(\hat{Y}_i) + \{\text{ Bias } (\hat{Y}_i)\}^2. \end{aligned} \quad (6.10.31)$$

The total mean squared errors for the n observations standardized by division by σ^2 is

$$\frac{1}{\sigma^2}\sum_{i=1}^{n} E_1[\hat{Y}_i - E_1(\hat{Y}_i)]^2 = \frac{1}{\sigma^2}\sum_{i=1}^{n} \text{ Var}_1(\hat{Y}_i) + \frac{1}{\sigma^2}\sum_{i=1}^{n}\{\text{ Bias } (\hat{Y}_i)\}^2 \quad (6.10.32)$$

where σ^2 is $\text{Var}(Y_i)$ under the true model.

Let the working model which is fitted to the data be denoted as

$$\mathbf{Y} = \mathbf{X}^{(k)}\beta^{(k)} + \epsilon \quad (6.10.33)$$

where $\mathbf{X}^{(k)}$ is a $n \times k$ matrix and $\beta^{(k)}$ a $k \times 1$ vector of regression coefficients. Then $\hat{Y}_i = \mathbf{x}_i^{(k)'}\hat{\beta}^{(k)}$. We have

$$\begin{aligned} \tfrac{1}{\sigma^2}\textstyle\sum_{i=1}^{n} \text{ Var}_1(\hat{Y}_i) &= \tfrac{1}{\sigma^2}\textstyle\sum_{i=1}^{n} \mathbf{x}_i^{(k)'} \text{ Var}_1(\hat{\beta}^{(k)})\mathbf{x}_i^{(k)} \\ &= \textstyle\sum_{i=1}^{n} \mathbf{x}_i^{(k)'}(\mathbf{X}^{(k)'}\mathbf{X}^{(k)})^{-1}\mathbf{x}_i^{(k)} \\ &= \textstyle\sum_{i=1}^{n} \text{ tr } [(\mathbf{X}^{(k)'}\mathbf{X}^{(k)})^{-1}\mathbf{x}_i^{(k)}\mathbf{x}_i^{(k)'}] \text{ (by (A.9(7))} \\ &= \text{ tr } [(\mathbf{X}^{(k)'}\mathbf{X}^{(k)^{-1}})(\mathbf{X}^{(k)'}\mathbf{X}^{(k)})] \\ &= \text{ tr } (\mathbf{I}_k) = k. \end{aligned} \quad (6.10.34)$$

It can be shown (Myers 1990, pp 178-179) that the second term on the right side of (6.10.34) can be expressed as $(n-k)E_1(s_k^2 - \sigma^2)/\sigma^2$. Therefore,

$$\frac{1}{\sigma^2}\sum_{i=1}^{n} E_1[\hat{Y}_i - E_1(Y_i)]^2 = k + \frac{n-k}{\sigma^2}E_1(s_k^2 - \sigma^2). \quad (6.10.35)$$

In particular, σ^2 is estimated by s_q^2, the mean square error (MSE) for the full model. We therefore estimate (6.10.35) by

$$C_k = k + \frac{n-k}{s_q^2}(s_k^2 - s_q^2) \quad (6.10.36)$$

or, alternatively,

$$C_k = \frac{SSE_k}{s_q^2} - (n - 2k). \quad (6.10.37)$$

From (6.10.36) we see that if for a particular model, the bias is small, C_k will be close to k. For this reason, the line $C_k = k$ is commonly plotted along with the C_k values of the several competing models. We look for values of k for which C_k are very close to this line. This gives the optimum subset.

6.10.4.2 *Stepwise selection*

We shall now consider procedures that add or delete variables one at a time.

In the *forward selection* procedure, at the first step y is regressed on each x_i alone and the x with the largest F-value, x_1 (say), is entered into the model. At the second step we calculate the partial-F value

$$F_{i|1} = \frac{SS_{reg}(x_1, x_i) - SS_{reg}(x_1)}{MSE(x_1, x_i)} \tag{6.10.38}$$

for each of the variables $x_i (\neq x_1)$, using the formula (6.10.16) and the variable with the largest $F_{i|1}$ value is entered in the model. If x_2 is the variable entered, we calculate at the second step, the partial F,

$$F_{j|1,2} = \frac{SS_{reg}(1,2,j) - SS_{reg}(1,2)}{MSE(x_1, x_2, x_j)} \tag{6.10.39}$$

and enter in the model the variable with the largest partial-F value. This procedure continues at each step until the largest partial F for an entering variable falls below a particular pre-specified threshold value, F_{IN} or until the corresponding p-value exceeds some predetermined value. The variables are entered one at a time and are not tested for removal once they are brought in.

The *backward* selection procedure begins with all the available x's in the model and deletes one variable at a time. The partial F-statistic for each variable in the presence of the other variables is calculated and the variable with the smallest F is eliminated. The process continues until the smallest partial F at some step exceeds a pre-specified threshold value, F_{OUT}.

The stepwise selection procedure is a combination of the forward and backward approaches. At any given step the variable with the largest partial F is added to the current subset if its partial F value is larger than F_{IN}. After a variable has been entered, all the variables in the subset are reexamined and the one with the smallest F is deleted, if its F value is less than F_{OUT}. At each step we do not try to eliminate the variable that was just entered or add the variable that was just eliminated. This can be ensured by choosing $F_{IN} > F_{OUT}$.

Since these step-by-step procedures do not examine all subsets, there is no guarantee that the optimum subset will be selected, specially if q is large. Moreover, in deciding a k-regressor variable model, the denominator s_k^2 in (6.10.39) (etc.) is based on a under-specified model, and is, therefore, generally biased upwards, thus artificially reducing the partial F value. In addition, since we are dealing with maximal F values, it is clear that the partial F statistic for an entering variable does not have an F distribution. Because of these reasons, the result obtained from stepwise selection should be taken with some caution. In practice, however, the criteria, R_{k}^2; s_k^2 or C_k may not show substantial differences between the optimum subset and the one found by stepwise selection.

We shall now consider two criteria based on the squared multiple correlation for choice of a suitable subset of a pres-specified number of k regressor variables out of a set of available q variables.

6.10.4.3 *Dependence analysis*

Suppose from other considerations we have decided to retain $k(< q)$ variables. Then we should retain a set of variables $x_{i_1}, x_{i_2}, \ldots, x_{i_k}$ which maximize the squared multiple correlation coefficient

$$R^2_{y.x_1,x_2,\ldots,x_k} \tag{6.10.40}$$

among all choices of k variables from the set of q. The subset $(x_{i_1}, x_{i_2}, \ldots, x_{i_k})$ best explains y among all subsets of k variables. An efficient algorithm for the search for the best subset of k variables has been developed by Beale *et al.* (1967), Beale (1970).

Again, though there is usually just one subset of k variables maximizing (6.10.40), in practice there may remain more than one subset of variables which are nearly optimal.

The choice of k is somewhat arbitrary. From the common sense point of view, one should retain enough variables so that the squared multiple correlation with y using k variables is at least $90 - 95\%$ of the squared multiple correlation coefficient using all the q variables.

One can use principal component analysis (Chapter 8) to obtain a set of derived variables (linear combinations of variables $x_1, x_2, \ldots, x_q$) which explain the variations in $x_1, \ldots, x_q$ almost completely and take these derived variables as predictors in the regression analysis.

6.10.4.4 *Interdependence analysis*

We nw consider the data matrix $\mathbf{X}(n \times q)$ and investigate the problem of finding which $(q - k)$ of the q variables $x_1, \ldots, x_q$ can be discarded without sacrificing much of the information contained in $\mathbf{X}$. This will reduce the number of measurements needed to describe the data effectively. We, therefore, need a measure of how well the retained set of k variables $x_{i_1}, \ldots, x_{i_k}$ explain the data set.

A measure of how well the set of retained variables $x_{i_1}, \ldots, x_{i_k}$ explain a rejected variable x_j is given by the squared multiple correlation $R^2_{x_j \cdot x_{i_1}, x_{i_2}, \ldots, x_{i_k}}$. An overall measure of how well the set of retained variables $x_{i_1}, \ldots, x_{i_k}$ explain the rejected variables is, therefore, given by

$$\min R^2_{x_h \cdot x_{i_1}, \ldots, x_{i_k}} \tag{6.10.41}$$

where the minimization is with respect to all rejected variables x_h. One should, therefore, choose a set of variables $x_{j_1}, \ldots, x_{j_k}$ such that

$$\min{}_h R^2_{x_h \cdot x_{j_1}, \ldots, x_{j_k}} = \max \{ \min{}_h R^2_{x_h \cdot x_{i_1}, \ldots, x_{i_k}} \}$$

where the maximum is with respect to choice of all subsets of k variables among $x_1, \ldots, x_q$.

As before, there may be several nearly optimal choices of k variables making the selection non-unique for all practical purposes.

The choice of k is somewhat arbitrary. A rule of thumb is to retain enough variables so that min R^2 with any rejected variable is at least 0.50.

6.11 Proportion of Variation in Y Explained by the Multivariate Model

In this section we considers measures of variations among y's accounted for by the multivariate regression model. These are also the measures of association between y's and x's. Analogous with (6.10.27) in the univariate multiple regression case ($p = 1$), we consider the statistic

$$\mathbf{D} = (\tilde{\mathbf{Y}}'\tilde{\mathbf{Y}})^{-1}(\hat{\mathbf{U}}'\hat{\mathbf{U}}) \tag{6.11.1}$$

where the observations are centered about their means, $\tilde{\mathbf{Y}}'\tilde{\mathbf{Y}} = ((\sum_i (y_{ij} - \bar{y}_j)(y_{ik} - \bar{y}_k)))$. The matrix $\mathbf{D}$ is a generalization of $1 - R^2$ in the univariate case. Now $\hat{\mathbf{U}}'\hat{\mathbf{U}}$ varies between the zero matrix (when all the variations in $\mathbf{Y}$ are explained by the model) and $\tilde{\mathbf{Y}}'\tilde{\mathbf{Y}}$ (when no part of the variation is explained by the model). Therefore, $\mathbf{I} - \mathbf{D}$ varies between the identity matrix and the null matrix. Any measure of multivariate multiple correlation should range between zero and one and this property is satisfied by the following two measures.

Trace correlation, r_T where $r_T^2 = p^{-1} \text{ tr } (\mathbf{I} - \mathbf{D})$;

Determinant correlation, r_D where $r_D^2 = |\mathbf{I} - \mathbf{D}|$.

If $d_1, \ldots, d_p$ are eigenvalues of $(\mathbf{I} - \mathbf{D})$,

$$r_T^2 = p^{-1} \sum_{i=1}^{p} d_i, \; r_D^2 = \Pi_{i=1}^p d_i.$$

Hotelling (1936) suggested the *vector alienation coefficient*

$$r_D = |\mathbf{D}| = \Pi_{i=1}^p (1 - d_i).$$

These measures of vector correlations are invariant under commutation, i.e., they would remain the same if we define $\mathbf{D} = (\hat{\mathbf{U}}'\hat{\mathbf{U}})(\tilde{\mathbf{Y}}'\tilde{\mathbf{Y}})^{-1}$.

6.12 Subset Selection for the Multivariate Regression Model

In the case of multivariate regression there are two distinct problems on subset selection, - selection of x's and selection of y's. As in the univariate response situation, there may be more potential predictor variables (x's) than can be economically accommodated. Similarly, some of the y's may be deleted if they do not significantly depend on x's.

We present two approaches, - (i) stepwise selection, (ii) all possible subsets.

6.12.1 *Stepwise procedures*

6.12.1.1 *Subset of x's*

We begin with a forward selection procedure based on Wilk's Λ. At the first step we regress $\mathbf{y} = (y_1, y_2, \ldots, y_p)'$ on each $x_j (j = 2, 3, \ldots, q)$, the variable x_1 which takes the value 1 on all the units being always included in the model.

If only, x_1 is included in the model, SSP matrix due to error is $(\mathbf{Y} - \mathbf{1}_n\bar{\mathbf{Y}}')'(\mathbf{Y} - \mathbf{1}_n\bar{\mathbf{Y}}') = \mathbf{Y}'\mathbf{Y} - n\bar{\mathbf{Y}}\bar{\mathbf{Y}}'$, where $\bar{\mathbf{Y}} = (\bar{Y}_1, \ldots, \bar{Y}_p)'$.

If only x_1, x_j are included in the model, SSP matrix due to error is

$$(\mathbf{Y} - \mathbf{X}_{(j)}\hat{\mathbf{B}}_{(j)})'(\mathbf{Y} - \mathbf{X}_{(j)}\hat{\mathbf{B}}_{(j)}) = \mathbf{Y}'\mathbf{Y} - \hat{\mathbf{B}}_{(j)}'\mathbf{X}_{(j)}'\mathbf{Y} \tag{6.12.1}$$

where

$$\mathbf{X}_{(j)} = (\mathbf{1}_n, \; \mathbf{x}_{(j)}), \; \mathbf{B}_{(j)} = \begin{bmatrix} \beta_{11} & \beta_{12} & \ldots & \beta_{1p} \\ \beta_{j1} & \beta_{j2} & \ldots & \beta_{jp} \end{bmatrix} = \begin{bmatrix} \beta_1' \\ \beta_j' \end{bmatrix}$$

and $\mathbf{x}'_{(j)} = (x_{1j}, x_{2j}, \ldots, x_{nj})'$. Therefore, the hypothesis $H_{0j}(\beta_j = \mathbf{0})$ is tested by using

$$\begin{aligned}\Lambda(x_j|x_1) &= \frac{|(\mathbf{Y}-\mathbf{X}_{(j)}\hat{\mathbf{B}}_{(j)})'(\mathbf{Y}-\mathbf{X}_{(j)}\hat{\mathbf{B}}_{(j)})|}{|\mathbf{Y}'\mathbf{Y}-n\bar{\mathbf{Y}}\bar{\mathbf{Y}}'|} \\ &= \frac{\Lambda(x_1,x_j)}{\Lambda(x_1)} \text{ (say)}\end{aligned} \tag{6.12.2}$$

(vide (6.5.2)). The statistic $\Lambda(x_j|x_1)$ is distributed as Wilk's $\Lambda(p; n-2, 1)$, because $\mathbf{Y}'\mathbf{Y} - \hat{\mathbf{B}}'_{(j)}\mathbf{X}'_{(j)}\mathbf{Y} \cap W_p(n-2, \mathbf{\Sigma})$ and $(\mathbf{Y}'\mathbf{Y} - n\bar{\mathbf{Y}}\bar{\mathbf{Y}}') - (\mathbf{Y}'\mathbf{Y} - \hat{\mathbf{B}}'_{(j)}\mathbf{X}'_{(j)}\mathbf{Y}) = \hat{\mathbf{B}}'_{(j)}\mathbf{X}'_{(j)}\mathbf{Y} - n\bar{\mathbf{Y}}\bar{\mathbf{Y}}' \cap W_p(1, \mathbf{\Sigma})$ and these distributions are independent. We choose the variable with the minimum value of $\Lambda(x_j|x_1)(j \neq 1)$.

Suppose the variable x_2 has been selected. We calculate the partial Λ,

$$\begin{aligned}\Lambda(x_j|x_1, x_2) &= \frac{|(\mathbf{Y}-\mathbf{X}_{(2,j)}\hat{\mathbf{B}}_{(2,j)})'(\mathbf{Y}-\mathbf{X}_{(2,j)}\hat{\mathbf{B}}_{(2,j)})|}{|(\mathbf{Y}-\mathbf{X}_{(j)}\hat{\mathbf{B}}_{(j)})'(\mathbf{Y}-\mathbf{X}_{(j)}\hat{\mathbf{B}}_{(j)})|} \\ &= \frac{\Lambda(x_1,x_2,x_j)}{\Lambda(x_1,x_2)} \text{ (say)}\end{aligned} \tag{6.12.3}$$

where

$$\mathbf{X}_{(2,j)} = (\mathbf{1}_n,\ \mathbf{x}_{(2)},\ \mathbf{x}_{(j)}),\ \mathbf{B}'_{(2,j)} = (\beta_1,\ \beta_2,\ \beta_j).$$

The statistic $\Lambda(x_j|x_1, x_2)$ is distributed as $\Lambda(p; n-3, 1)$ and we enter in the model the variable with the smallest value of $\Lambda(x_j|x_1, x_2), j \neq (1, 2)$.

After m variables $x_1, x_2, \ldots, x_m$ have been selected, the partial Λ at the next step would be

$$\Lambda(x_j|x_1, \ldots, x_m) = \frac{\Lambda(x_1, \ldots, x_m, x_j)}{\Lambda(x_1, \ldots, x_m)}$$

and would be distributed as $\Lambda(p; n-m+1, 1)$.

The procedure is continued till a step is reached in which the minimum partial Λ exceeds a predetermined threshold value Λ_{IN} or equivalently, the associated partial F falls below a pre-specified value, F_{IN}.

The backward elimination procedure starts with all the x's in the model and calculate for each $x_j (j \neq 1)$, the partial Λ,

$$\begin{aligned}\Lambda(x_j|x_1, \ldots, x_{j-1}, x_{j+1}, \ldots, x_q) &= \frac{\Lambda(x_1,x_2,\ldots,x_q)}{\Lambda(x_1,\ldots,x_{j-1},x_{j+1},\ldots,x_q)} \\ &= \frac{|(\mathbf{Y}-\mathbf{X}\hat{\mathbf{B}})'(\mathbf{Y}-\mathbf{X}\hat{\mathbf{B}})|}{|(\mathbf{Y}-\mathbf{X}_{-j}\hat{\mathbf{B}}_{-j})'(\mathbf{Y}-\mathbf{X}_{-j}\hat{\mathbf{B}}_{-j})|}\end{aligned} \tag{6.12.4}$$

where $\mathbf{X}_{-j}$ is the $n \times (q-1)$ matrix $\mathbf{X}$ with $\mathbf{x}_{(j)}$ deleted and $\mathbf{B}_{-j}$ is the matrix $\mathbf{B}$ with the jth row β'_j deleted. The statistic $\Lambda(x_j|x_1, \ldots, x_{j-1}, x_{j+1}, \ldots, x_q)$ follows $\Lambda(p; n-q, 1)$ distribution. The variable x_j with the largest Λ in (6.12.4) or smallest F is removed from the model. At the next step, a partial Λ or partial F is calculated for each of the remaining $q-2$ variables (x_1 being never deleted) and the variable with the largest partial Λ or smallest partial F is eliminated. The process is continued until at some step the partial Λ falls below a predetermined threshold value Λ_{OUT} or the partial F exceeds a pre-specified value F_{OUT}.

The variables can also be selected by stepwise procedures. Here each time a variable is entered by a forward selection process, all the variables that have already been entered are re-examined by a partial Λ or F test to see if the least significant one has become redundant and and be removed.

Since these step-by-step procedures do not examine all the possible subsets, there is no guarantee that the optimum subset will be achieved. Again in forward selection, the error SSP matrix occurring in the denominator of partial Λ (as in (6.12.3)) may be inflated being based on an under-specified model and as such the true predictors of $\mathbf{y}$ may be overlooked. Also, some spurious x's may appear non-redundant because of their chance correlations in the particular sample.

6.12.1.2 *Subset of y's*

After a subset of x's has been found, we may want to know if these variables predict all the y's. A subset of y's may be selected using forward selection, backward elimination or stepwise procedures using x's as the dependent variables and y's as independent variables. Thus, in forward selection at the third step we calculate the partial Λ as in (6.12.3),

$$\Lambda(y_i|y_1, y_2) = \frac{\Lambda(y_1, y_2, y_i)}{\Lambda(y_1, y_2)},$$

assuming that the variables y_1, y_2 have already been entered in the model. The statistic $\Lambda(y_i|y_1, y_2)$ is distributed as $\Lambda(q; n-3, 1) = \Lambda(1; n-2-q, q)$ (Theorem 4.4.3). We choose the variable y_i with the minimum value of the partial Λ. Backward selection and stepwise selection can similarly be carried out.

6.12.2 *All possible subsets*

We now consider matrix analogues of R_k^2, s_k^2 and C_k discussed in Subsection 6.10.4 for the univariate response case. The extended statistics are denoted as $\mathbf{R}_k^2, \mathbf{S}_k$ and $\mathbf{C}_k$ respectively and are due to Mallows (1973) and Sparkes *et al.* (1983).

(i) $\mathbf{R}_k^2$: In the univariate response case, R_k^2 for a k-variate model $\mathbf{X}$ (including x_1) was defined as, vide (6.10.26),

$$R_k^2 = \frac{\hat{\beta}_k'\mathbf{X}_k'\mathbf{Y} - n\bar{Y}^2}{\mathbf{Y}'\mathbf{Y} - n\bar{Y}^2}.$$

A direct extension of R_k^2 in the multivariate case is given by

$$\mathbf{R}_k^2 = (\mathbf{Y}'\mathbf{Y} - n\bar{\mathbf{Y}}\bar{\mathbf{Y}}')^{-1}(\hat{\mathbf{B}}_k\mathbf{X}_k'\mathbf{Y} - n\bar{\mathbf{Y}}\bar{\mathbf{Y}}') \tag{6.12.5}$$

where $\mathbf{B}_k$ is of order $k \times p$. Converting $\mathbf{R}_k^2$ to its scalar form we use

$$\psi(p;k) = \frac{1}{p}[\text{ tr } (\mathbf{R}_k^2)]. \tag{6.12.6}$$

Clearly, $0 \le \psi(p;k) \le 1$. As in the univariate case, we identify the subset which maximizes $\psi(p;k)$ for each value of $k = 2, 3, \ldots, q$. The statistic $\psi(p;k)$ reaches its maximum value at $k = q$, but we look at the value of k beyond which increase is only marginal. This gives the best subset. We could also use $|\mathbf{R}_k^2|$ in place of $\psi(p;k)$.

(ii) $\mathbf{S}_k$: A direct extension of the univariate $s_k^2 = MSE_k = SSE/(n-k)$ (vide (6.10.30)) is provided by

$$\mathbf{S}_k = \frac{(\mathbf{Y} - \mathbf{X}_k\hat{\mathbf{B}}_k)'(\mathbf{Y} - \mathbf{X}_k\hat{\mathbf{B}}_k)}{n-k}. \tag{6.12.7}$$

We can use the measure tr $(\mathbf{S}_k)$ or $|\mathbf{S}_k|$, each of which behaves in a fashion similar to s_k^2. We select the subset with the minimum value of tr $(\mathbf{S}_k)$ or with the smallest k such that tr $(\mathbf{S}_k) <$ tr $(\mathbf{S}_q)$. A similar treatment may be given to $|\mathbf{S}_k|$.

(iii) $\mathbf{C}_k$: In using model (6.1.1) with $\mathbf{X}_k$, the matrix of the estimated y-values is $\hat{\mathbf{Y}} = \mathbf{X}_k\hat{\mathbf{B}}_k$ which gives

$$\hat{\mathbf{Y}}_i' = \mathbf{x}_{i(k)}'\hat{\mathbf{B}}_k$$

where $\mathbf{x}_{i(k)}' = (x_{i1}, x_{i2}, \ldots, x_{ik})$, assuming that the variables included in the model are $x_1, \ldots, x_k$. If a correct model is fitted to the data $E(\hat{\mathbf{Y}}_i) = E(\mathbf{Y}_i)$. If the fitted model is incorrect, $E_1(\hat{\mathbf{Y}}_i) \ne E(\mathbf{Y}_i)$, where E_1 is with respect to the working model and E is with respect to the true model. In this case $E_1(\hat{\mathbf{Y}}_i) - E(\mathbf{Y}_i) = \text{Bias } (\hat{\mathbf{Y}}_i)$. Now

$$\begin{aligned} E_1[\hat{\mathbf{Y}}_i - E(\mathbf{Y}_i)][\hat{\mathbf{Y}}_i - E(\mathbf{Y}_i)]' &= E_1[\hat{\mathbf{Y}}_i - E_1(\hat{\mathbf{Y}}_i)][\hat{\mathbf{Y}}_i - E_1(\hat{\mathbf{Y}}_i)]' + \\ &\quad [E_1(\hat{\mathbf{Y}}_i) - E(\mathbf{Y}_i)][E_1(\hat{\mathbf{Y}}_i) - E(\mathbf{Y}_i)]' \\ &= \text{Cov}_1(\hat{\mathbf{Y}}_i) + [\text{ Bias } (\hat{\mathbf{Y}}_i)][\text{ Bias } (\hat{\mathbf{Y}}_i)]'. \end{aligned} \tag{6.12.8}$$

Again,

$$\begin{aligned}
\text{Cov}_1(\hat{\mathbf{Y}}_i') &= \text{Cov}_1(\mathbf{x}_{i(k)}'\hat{\mathbf{B}}_k)\\
&= \text{Cov}_1(\mathbf{x}_{i(k)}'\hat{\beta}_{k(1)},\ \mathbf{x}_{i(k)}'\hat{\beta}_{k(2)},\ldots,\mathbf{x}_{i(k)}'\hat{\beta}_{k(p)})\\
&= \begin{bmatrix} \sigma_{11}\mathbf{x}_{i(k)}'(\mathbf{X}_k'\mathbf{X}_k)^{-1}\mathbf{x}_{i(k)} & \cdots & \sigma_{1p}\mathbf{x}_{i(k)}'(\mathbf{X}_k'\mathbf{X}_k)^{-1}\mathbf{x}_{i(k)}\\ \cdot & \cdots & \cdot \\ \cdot & \cdots & \cdot \\ \sigma_{p1}\mathbf{x}_{i(k)}'(\mathbf{X}_k'\mathbf{X}_k)^{-1}\mathbf{x}_{i(k)} & \cdots & \sigma_{pp}\mathbf{x}_{i(k)}'(\mathbf{X}_k'\mathbf{X}_k)^{-1}\mathbf{x}_{i(k)} \end{bmatrix}\\
&= \mathbf{x}_{i(k)}'(\mathbf{X}_k'\mathbf{X}_k)^{-1}\mathbf{x}_{i(k)}\mathbf{\Sigma}.
\end{aligned} \tag{6.12.9}$$

Therefore,

$$\begin{aligned}
\textstyle\sum_{i=1}^n \text{Cov}_1(\hat{\mathbf{Y}}_i') &= \textstyle\sum_{i=1}^n \mathbf{x}_{i(k)}'(\mathbf{X}_k'\mathbf{X}_k)^{-1}\mathbf{x}_{i(k)}\mathbf{\Sigma}\\
&= \mathbf{\Sigma}\ \text{tr}\ \textstyle\sum_{i=1}^n \mathbf{x}_{i(k)}\mathbf{x}_{i(k)}'(\mathbf{X}_k'\mathbf{X}_k)^{-1}\\
&= k\mathbf{\Sigma}.
\end{aligned} \tag{6.12.10}$$

It can be shown that

$$\sum_{i=1}^n [\text{ Bias }(\hat{\mathbf{Y}}_i)][\text{ Bias }(\hat{\mathbf{Y}}_i)]' = (n-k)E(\mathbf{S}_k - \mathbf{\Sigma}). \tag{6.12.11}$$

Therefore, from (6.12.8) - (6.12.11),

$$\begin{aligned}
\mathbf{\Sigma}^{-1}\textstyle\sum_{i=1}^n E_1[\hat{\mathbf{Y}}_i - E(\mathbf{Y}_i)][\hat{\mathbf{Y}}_i - E(\mathbf{Y}_i)]' &= \mathbf{\Sigma}^{-1}[k\mathbf{\Sigma} + (n-k)E(\mathbf{S}_k - \mathbf{\Sigma})]\\
&= k\mathbf{I}_p + (n-k)\mathbf{\Sigma}^{-1}E(\mathbf{S}_k - \mathbf{\Sigma}).
\end{aligned} \tag{6.12.12}$$

Using $\mathbf{S}_q = \mathbf{E}_q/(n-q)$ where $\mathbf{E}_q$ is the SSP matrix due to residuals based on all the q auxiliary variables, as an estimate of $\mathbf{\Sigma}$, we can estimate (6.12.12) by

$$\mathbf{C}_k = k\mathbf{I}_p + (n-k)\mathbf{S}_q^{-1}(\mathbf{S}_k - \mathbf{S}_q) \tag{6.12.13}$$

$$= \mathbf{S}_q^{-1}\mathbf{E}_k + (2k-n)\mathbf{I}_p. \tag{6.12.14}$$

To reduce $\mathbf{C}_k$ to a scalar we can use tr $(\mathbf{C}_k)$ or $|\mathbf{C}_k|$. If $(2k-n)$ is negative, $|\mathbf{C}_k|$ can be negative. Sparkes *et al.*, therefore, suggested an alternative measure

$$|\mathbf{E}_q^{-1}\mathbf{E}_k| \tag{6.12.15}$$

which is always positive.

To find the optimum subset of x's, we look for each k, the subset with the smallest $\mathbf{C}_k$ matrix. We note from (6.12.12) that when the bias is $\mathbf{0}$), the population $\mathbf{C}_k = k\mathbf{I}_p$. Thus we seek a subset for which $\mathbf{C}_k$ is small and also close to $k\mathbf{I}_p$. In terms of trace we seek tr$(\mathbf{C}_k)$ close to kp.

Now, from (6.12.15),

$$\mathbf{E}_q^{-1}\mathbf{E}_k = \frac{\mathbf{C}_k + (n-2k)\mathbf{I}_p}{n-q}. \tag{6.12.16}$$

When the bias is $\mathbf{0}, \mathbf{C}_k = k\mathbf{I}_p$ and (6.12.16) reduces to

$$\mathbf{E}_q^{-1}\mathbf{E}_k = \frac{n-k}{n-q}\mathbf{I}_p$$

or

$$|\mathbf{E}_q^{-1}\mathbf{E}_k| = (\frac{n-k}{n-q})^p. \tag{6.12.17}$$

Choosing a subset with tr $(\mathbf{C}_k) \leq kp$ is equivalent to choosing a subset with $|\mathbf{E}_q^{-1}\mathbf{E}_l| \leq (\frac{n-k}{n-q})^p$.

All these methods hold for choosing an optimum subset of y's with the role of x and y interchanged.

6.13 Growth Curve Models

We now consider a variant of the model (6.1.1), the so-called 'growth curve models'. First we consider some examples.

6.13.1 *Examples*

6.13.1.1 *Single growth curve*

Suppose we have a sample of n units (animals, plants, etc.) all subject to the same condition and the size, weight, etc of these units are observed at p points of time $t_1, t_2, \ldots, t_p$, the observation on the ith unit at time t_r being denoted as y_{ir}. Clearly, the p observations on the ith unit, $\mathbf{y}_i = (y_{i1}, \ldots, y_{ip})'$ will be correlated having a covariance matrix $\mathbf{\Sigma}$. We assume that each observation vector $\mathbf{y}_i$ is *iid* $N_p(\mu, \mathbf{\Sigma})$ where $\mu = (\mu_1, \ldots, \mu_p)'$. Here μ_r represents the average size of the animal at time t_r. The elements of $\mathbf{y}_i$ plotted against time gives the observed growth for the ith individual. A curve joining the points $(r, \mu_r)(r = 1, \ldots, p)$ gives the growth curve of the population of the animals from which the sample has been drawn. We assume that the growth curve is a polynomial of degree $k-1$, i.e.

$$\mu_r = \gamma_0 + \gamma_1 t_r + \gamma_2 t_r^2 + \ldots + \gamma_{k-1} t_r^{k-1}, r = 1, \ldots, p. \tag{6.13.1}$$

Written in the matrix form the relation (6.13.1) is

$$\begin{bmatrix} \mu_1 \\ \mu_2 \\ \cdot \\ \cdot \\ \mu_p \end{bmatrix} = \begin{bmatrix} 1 & t_1 & t_1^2 & \dots & t_1^{k-1} \\ 1 & t_2 & t_2^2 & \dots & t_2^{k-1} \\ \cdot & \cdot & \cdot & \dots & \cdot \\ \cdot & \cdot & \cdot & \dots & \cdot \\ 1 & t_p & t_p^2 & \dots & t_p^{k-1} \end{bmatrix} \begin{bmatrix} \gamma_0 \\ \gamma_1 \\ \gamma_2 \\ \cdot \\ \cdot \\ \gamma_{k-1} \end{bmatrix}$$

or

$$\mu = \mathbf{K}\boldsymbol{\Gamma} \tag{6.13.2}$$

where $\mathbf{K}$ is $p \times k$ of rank $k(< p)$. Writing

$$\mathbf{Y} = \begin{bmatrix} \mathbf{y}_1' \\ \mathbf{y}_2' \\ \cdot \\ \cdot \\ \mathbf{y}_n' \end{bmatrix}, \; E(\mathbf{Y}) = \begin{bmatrix} \mu' \\ \mu' \\ \cdot \\ \cdot \\ \mu' \end{bmatrix},$$

the model is, therefore, by (6.13.2),

$$E(\mathbf{Y}) = \mathbf{1}_n \mu' = \mathbf{1}_n \boldsymbol{\Gamma}' \mathbf{K}'. \tag{6.13.3}$$

Given that the model (6.13.2) is acceptable we may wish to test the hypothesis that the growth curve is actually a polynomial of lower degree, i.e. $H_0 : \gamma_{k-1} = \gamma_{k-2} = \dots - \gamma_{k-v} = 0$. The hypothesis H_0 can be written as

$$H_0 : \boldsymbol{\Gamma}'\mathbf{L} = \mathbf{0} \tag{6.13.4}$$

where

$$\mathbf{L} = (0, \dots, 0, 1, \dots, 1)'.$$

This is a special case of the hypothesis

$$\mathbf{A}\boldsymbol{\Delta}\mathbf{L} = \mathbf{0} \tag{6.13.5}$$

for the model

$$\mathbf{Y} = \mathbf{X}\boldsymbol{\Delta}\mathbf{K}' + \mathbf{U} \tag{6.13.6}$$

where $\boldsymbol{\Delta} = \boldsymbol{\Gamma}', \mathbf{X} = \mathbf{1}_n, \mathbf{A} = 1$.

Note that in model (6.13.2) there are two design matrices, $\mathbf{1}_n$ which expresses $E(\mathbf{Y})$ in terms of μ and an internal design matrix $\mathbf{K}$ which expresses μ in terms of fewer parameters in $\boldsymbol{\Gamma}$. All the hypotheses should relate to the elements of $\boldsymbol{\Gamma}$ only. For the general model (6.13.6), $\mathbf{X}$ is the external design matrix and $\mathbf{K}$ is the internal design matrix.

A general theory for testing for the model (6.13.6) is given in Subsection 6.13.2.

6.13.1.2 *Two growth curves*

Suppose we have two populations. A sample of size m is drawn from the first population resulting in the observation vectors $\mathbf{v}_1, \ldots, \mathbf{v}_m, \mathbf{v}_i$ being the observations on the size of the ith unit at points of time $t_1, \ldots, t_p$. It is assumed that $\mathbf{v}_1, \ldots, \mathbf{v}_m$ are *iid* $N_p(\nu, \mathbf{\Sigma})$. Likewise, let $\mathbf{w}_1, \ldots, \mathbf{w}_n$ be the observation vectors for samples from the second population, with the assumption that $\mathbf{w}_1, \ldots, \mathbf{w}_n$ are *iid* $N_p(\eta, \mathbf{\Sigma})$. Assuming the growth curves are cubic,

$$\begin{aligned} \nu_r &= \gamma_0 + \gamma_1 t_r + \gamma_2 t_r^2 + \gamma_3 t_r^3, \\ \eta_r &= \lambda_0 + \lambda_1 t_r + \lambda_2 t_r^2 + \lambda_3 t_r^3, \ (r = 1, \ldots, p) \end{aligned} \tag{6.13.7}$$

we want to test the hypothesis that the growth curves are parallel, that is, $\gamma_i = \lambda_i, i = 1, 2, 3$.

Writing

$$\mathbf{Y} = \begin{bmatrix} \mathbf{v}_1' \\ \mathbf{v}_2' \\ . \\ \mathbf{v}_m' \\ \mathbf{w}_1' \\ \mathbf{w}_2' \\ \\ \mathbf{w}_n' \end{bmatrix}, \ \mathbf{X} = \begin{bmatrix} \mathbf{1}_m & \mathbf{0} \\ \mathbf{0} & \mathbf{1}_n \end{bmatrix}, \ \mathbf{B} = \begin{bmatrix} \nu' \\ \eta' \end{bmatrix}, \tag{6.13.8}$$

the model is

$$\mathbf{Y} = \mathbf{XB} + \mathbf{U} \tag{6.13.9}$$

where rows of $\mathbf{U}$ are *iid* $N_p(\mathbf{0}, \mathbf{\Sigma})$. As in (6.13.2),

$$\begin{aligned} \nu &= \mathbf{K\Gamma} \\ \eta &= \mathbf{K\Lambda}, \end{aligned} \tag{6.13.10}$$

where

$$\mathbf{K} = \begin{bmatrix} 1 & t_1 & t_1^2 & t_1^3 \\ 1 & t_2 & t_2^2 & t_2^3 \\ . & . & . & . \\ . & . & . & . \\ 1 & t_p & t_p^2 & t_p^3 \end{bmatrix},$$

$\mathbf{\Gamma} = (\gamma_0, \gamma_1, \gamma_2, \gamma_3)', \mathbf{\Lambda} = (\lambda_0, \lambda_1, \lambda_2, \lambda_3)'$. Hence we can write

$$\mathbf{B} = \begin{bmatrix} \mathbf{\Gamma}'\mathbf{K}' \\ \mathbf{\Lambda}'\mathbf{K}' \end{bmatrix} = \begin{bmatrix} \mathbf{\Gamma}' \\ \mathbf{\Lambda}' \end{bmatrix} \mathbf{K}' = \mathbf{\Delta K}'. \tag{6.13.11}$$

By (6.13.9) the model reduces to the form

$$\mathbf{Y} = \mathbf{X}\boldsymbol{\Delta}\mathbf{K}' + \mathbf{U}.$$

As before, here $\mathbf{X}$ is the external design matrix and $\mathbf{K}$ the internal design matrix. The hypothesis can be written as

$$\begin{aligned}\mathbf{0}' &= (\gamma_1 - \lambda_1,\ \gamma_2 - \lambda_2\ \gamma_3 - \lambda_3)\\ &= (1,\ -1)\begin{bmatrix}\gamma_0 & \gamma_1 & \gamma_2 & \gamma_3\\ \lambda_0 & \lambda_1 & \lambda_2 & \lambda_3\end{bmatrix}\begin{bmatrix}0 & 0 & 0\\ 1 & 0 & 0\\ 0 & 1 & 0\\ 0 & 0 & 1\end{bmatrix}\\ &= \mathbf{A}\boldsymbol{\Delta}\mathbf{L}\end{aligned}$$

where $\mathbf{A} = (1,\ -1)$ and $\mathbf{L}$ is the last matrix in the right side of (6.13.11).

6.13.1.3 *Growth curve in a randomized block design*

Consider a randomized block design with a treatments and b blocks. The suitable general linear model is

$$\mathbf{y}_{ij} = \theta_{ij} + \mathbf{u}_{ij},\ i = 1, \ldots, a;\ j = 1, \ldots, b \tag{6.13.12}$$

where

$$\theta_{ij} = \mu + \alpha_i + \beta_j \tag{6.13.13}$$

with usual notations. The side restrictions are

$$\sum_i \alpha_i = \mathbf{0},\ \sum_j \beta_j = \mathbf{0}. \tag{6.13.14}$$

As before, θ_{ijr} is the mean value of the response for the (i, j)th unit at time $t_r, r = 1, \ldots, p$ where $\theta_{ij} = (\theta_{ij1}, \ldots, \theta_{ijp})'$. Assuming a growth curve of degree $k-1$, we have

$$\theta_{ijr} = \lambda_{0ij} + \lambda_{1ij}t_r + \lambda_{2ij}t_r^2 + \ldots + \lambda_{(k-1)ij}t_r^{k-1},\ r = 1, \ldots, p$$

or written in the matrix form

$$\theta_{ij} = \mathbf{K}\lambda_{ij},\ i = 1, \ldots, a;\ j = 1, \ldots, b \tag{6.13.15}$$

where $\mathbf{K}$ is obtained from (6.13.2) and $\lambda_{ij} = (\lambda_{0ij}, \lambda_{1ij}, \ldots, \lambda_{(k-1)ij})'$.

Now, by relation (6.13.15),

$$\begin{aligned}\mu &= \bar{\theta}_{00} = \mathbf{K}\bar{\lambda}_{00},\\ \alpha_i &= \bar{\theta}_{i0} - \bar{\theta}_{00} = \mathbf{K}(\bar{\lambda}_{i0} - \bar{\lambda}_{00}),\\ \beta_j &= \bar{\theta}_{0j} - \bar{\theta}_{00} = \mathbf{K}(\bar{\lambda}_{0j} - \bar{\lambda}_{00}),\end{aligned} \tag{6.13.16}$$

where the symbols have the usual meanings. Therefore

$$\begin{aligned}\theta_{ij} &= \mu + \alpha_i + \beta_j \\ &= \mathbf{K}[\bar{\lambda}_{00} + (\bar{\lambda}_{i0} - \bar{\lambda}_{00}) + (\bar{\lambda}_{0j} - \bar{\lambda}_{00}) \\ &= \mathbf{K}[\mathbf{\Phi} + \mathbf{\Psi}_i + \xi_j] \quad \text{(say)}\end{aligned} \tag{6.13.17}$$

where $\sum_i \mathbf{\Psi}_i = \xi_j = \mathbf{0}$ (by (6.13.14)). Hence,

$$\theta = \begin{bmatrix} \theta'_{11} \\ \cdot \\ \theta'_{1b} \\ \cdot \\ \cdot \\ \theta'_{a1} \\ \cdot \\ \theta'_{ab} \end{bmatrix} = \begin{bmatrix} \mathbf{\Phi}' + \mathbf{\Psi}'_{11} + \xi'_{11} \\ \cdot \\ \mathbf{\Phi}' + \mathbf{\Psi}'_1 + \xi'_b \\ \cdot \\ \cdot \\ \mathbf{\Phi}' + \mathbf{\Psi}'_a + \xi'_1 \\ \cdot \\ \mathbf{\Phi}' + \mathbf{\Psi}'_a + \xi'_b \end{bmatrix} \mathbf{K}' \tag{6.13.18}$$
$$= \mathbf{X}\mathbf{\Delta}\mathbf{K}'$$

where $\mathbf{\Delta} = (\mathbf{\Phi}, \mathbf{\Psi}_1, \ldots, \mathbf{\Psi}_a, \ldots, \xi_1 \ldots, \xi_b)'$ and $\mathbf{X}$ is a suitably chosen external design matrix. The identification restrictions (6.13.14) reduce to

$$\mathbf{F}\mathbf{\Delta} = \mathbf{0}$$

where

$$\mathbf{F} = \begin{bmatrix} 0 & 1 & \ldots 1 & 0 & \ldots & 0 \\ 0 & 0 & \ldots 0 & 1 & \ldots & 1 \end{bmatrix}.$$

Therefore, the same model (6.13.6) is obtained. All the hypotheses must relate to $\mathbf{\Delta}$.

6.13.1.4 *Two-dimensional growth curve*

Suppose we observe two variables, say, height and weight at points of time $t_1, \ldots, t_p$ for each of m individuals and assume that the growth curve for the first variable is cubic, while that for the second is cubic. We have therefore $2p$ observations for each individual, the first p being on height and the second set of p being on weight. Denoting the observations as $\mathbf{v}_1, \ldots, \mathbf{v}_m$ we assume that $\mathbf{v}_i$ are *iid* $N_{2p}(\mu, \mathbf{\Sigma})$ where

$$\mu' = (\mu_1^{(1)}, \ldots, \mu_p^{(1)}, \mu_1^{(2)}, \ldots, \mu_p^{(2)}) = (\mu^{(1)'}, \mu^{(2)'})$$

with suffixes (1) and (2) referring to variables height and weight respectively. Now,

$$\begin{aligned}\mu_r^{(1)} &= \gamma_0^{(0)} + \gamma_1^{(1)} t_r + \gamma_2^{(1)} t_r^2 + \gamma_3^{(1)} t_r^3, \\ \mu_r^{(2)} &= \gamma_0^{(2)} + \gamma_1^{(2)} t_r + \gamma_2^{(2)} t_r^2.\end{aligned} \tag{6.13.19}$$

Thus

$$
\begin{aligned}
\mu^{(1)} &= \begin{bmatrix} 1 & t_1 & t_1^2 & t_1^3 \\ \cdot & \cdot & \cdot & \cdot \\ \cdot & \cdot & \cdot & \cdot \\ 1 & t_p & t_p^2 & t_p^3 \end{bmatrix} \begin{bmatrix} \gamma_0^{(1)} \\ \gamma_1^{(1)} \\ \gamma_2^{(1)} \\ \gamma_3^{(1)} \end{bmatrix} = \mathbf{K}_1 \mathbf{\Gamma}^{(1)}, \\
\mu^{(2)} &= \mathbf{K}_2 \mathbf{\Gamma}^{(2)}, \\
\mu &= \begin{bmatrix} \mu^{(1)} \\ \mu^{(2)} \end{bmatrix} = \begin{bmatrix} \mathbf{K}_1 & \mathbf{0} \\ \mathbf{0} & \mathbf{K}_2 \end{bmatrix} \begin{bmatrix} \gamma^{(1)} \\ \gamma^{(2)} \end{bmatrix} \\
&= \mathbf{K}\mathbf{\Gamma} \ \text{(say)} \ .
\end{aligned}
\tag{6.13.20}
$$

Similarly if we have another population from which a sample of size n is drawn and the above two variables are measured at points of time $t_1, \ldots, t_p$, then assuming the observations $\mathbf{w}_i (i = 1, \ldots, n)$ are *iid* $N_{2p}(\nu, \mathbf{\Sigma})$, we shall have the model $\nu = \mathbf{K}\mathbf{\Lambda}$. Putting the two samples together we again arrive at the model (6.13.6) where

$$
\mathbf{X} = \begin{bmatrix} \mathbf{1}_m & \mathbf{0} \\ \mathbf{0} & \mathbf{1}_n \end{bmatrix}, \ \mathbf{\Delta} = \begin{bmatrix} \mathbf{\Gamma}' \\ \mathbf{\Lambda}' \end{bmatrix} = \begin{bmatrix} \gamma^{(1)'} & \gamma^{(2)'} \\ \lambda^{(1)} & \lambda^{(2)} \end{bmatrix}. \tag{6.13.21}
$$

Various generalizations to this model are available (see for example, Potthoff and Roy, 1964).

6.13.2 *A general solution*

Consider (as in (6.13.6)) the linear model $\mathbf{Y} = \mathbf{X\Delta K}' + \mathbf{U}$ where $\mathbf{X}(n \times q)$ is of rank q, $\mathbf{\Delta}$ is $q \times k$ and $\mathbf{K}'$ is $k \times p$ of rank $k(< p)$ and the rows of $\mathbf{U}$ are *iid* $N_p(\mathbf{0}, \mathbf{\Sigma})$. We want to develop a test for the hypothesis $H_0(\mathbf{A\Delta L} = \mathbf{0})$ where $\mathbf{A}(g \times q)$ is of rank g, and $\mathbf{L}(p \times v)$ is of rank v. The following solution is due to Potthoff and Roy (1964).

Choose a nonsingular $p \times p$ matrix $\mathbf{G}$ such that the $k \times k$ matrix $\mathbf{K}'\mathbf{G}^{-1}\mathbf{K}$ is of rank $k(< p)$. Let

$$
\mathbf{C} = \mathbf{G}^{-1}\mathbf{K}(\mathbf{K}'\mathbf{G}^{-1}\mathbf{K})^{-1}. \tag{6.13.22}.
$$

Then,

$$
\mathbf{K}'\mathbf{C} = \mathbf{I}_k. \tag{6.13.23}
$$

Consider the transformation

$$
\mathbf{z}_i = \mathbf{C}'\mathbf{y}_i. \tag{6.13.24}
$$

Then $\mathbf{z}_i$ are *iid* with dispersion matrix $\mathbf{C}'\mathbf{\Sigma}\mathbf{C} = \mathbf{\Sigma}_1$. Now

$$\mathbf{Z} = \begin{bmatrix} \mathbf{z}_1' \\ \mathbf{z}_2' \\ \cdot \\ \cdot \\ \mathbf{z}_n' \end{bmatrix} = \begin{bmatrix} \mathbf{y}_1'\mathbf{C} \\ \mathbf{y}_2'\mathbf{C} \\ \cdot \\ \cdot \\ \mathbf{y}_n'\mathbf{C} \end{bmatrix} = \mathbf{YC},$$

$$E(\mathbf{Z}) = E(\mathbf{YC}) = \mathbf{X\Delta K}'\mathbf{C} = \mathbf{X\Delta}. \tag{6.13.25}$$

We have, therefore, reduced the original model (6.13.6) to the model

$$\mathbf{Z} = \mathbf{X\Delta} + \mathbf{U}_1 \tag{6.13.26}$$

where $\mathbf{U}_1 = \mathbf{CU}$ and the rows of $\mathbf{U}_1$ are *iid* $N_k(\mathbf{0}, \mathbf{\Sigma}_1)$. The hypothesis $H_0(\mathbf{A\Delta L} = \mathbf{0})$ can therefore be tested by using the general theory in Subsection 6.6.2, setting k for $p, \mathbf{Z}$ for $\mathbf{Y}, \mathbf{\Delta}$ for $\mathbf{B}$ and $\mathbf{ZL}$ for $\mathbf{Y}_L$. By (6.6.25), the SSP matrix due to H_0 is

$$\mathbf{H}_1 = (\mathbf{A}\hat{\mathbf{\Delta}}\mathbf{L})'[\mathbf{A}(\mathbf{X}'\mathbf{X})^{-1}\mathbf{A}']^{-1}(\mathbf{A}\hat{\mathbf{\Delta}}\mathbf{L}) \tag{6.13.27}$$

and the SSP matrix due to error is

$$\mathbf{E}_1 = \mathbf{L}'\mathbf{Z}'(\mathbf{I}_n - \mathbf{X}(\mathbf{X}'\mathbf{X})^{-1}\mathbf{X}')\mathbf{ZL} \tag{6.13.28}$$

where

$$\begin{aligned} \hat{\mathbf{\Delta}} &= (\mathbf{X}'\mathbf{X})^{-1}\mathbf{X}'\mathbf{Z} \\ &= (\mathbf{X}'\mathbf{X})^{-1}\mathbf{X}'\mathbf{Y}\mathbf{G}^{-1}\mathbf{K}(\mathbf{K}'\mathbf{G}^{-1}\mathbf{K})^{-1}. \end{aligned} \tag{6.13.29}$$

The matrices $\mathbf{H}_1$ and $\mathbf{E}_1$ are independently distributed as $W_v(g, \mathbf{L}'\mathbf{\Sigma}_1\mathbf{L})$ and $W_v(n-q, \mathbf{L}'\mathbf{\Sigma}_1\mathbf{L})$ respectively, when H_0 is true. One can, therefore, test H_0 by using $\Lambda = |\mathbf{E}_1|/|\mathbf{E}_1 + \mathbf{H}_1|$.

If $k = p$, then $\mathbf{K}$ is nonsingular and we use the transformation $\mathbf{Z} = \mathbf{YK}^{-1}$. If $k < p$, there are many choices of $\mathbf{G}$ which may be a matrix of constants or a random matrix whose distribution is independent of $\mathbf{Y}$. The simplest choice is $\mathbf{G} = \mathbf{I}_p$ when $\mathbf{Z} = \mathbf{YK}(\mathbf{K}'\mathbf{K})^{-1}$ so that

$$\mathbf{z}_i = (\mathbf{K}'\mathbf{K})^{-1}\mathbf{K}'\mathbf{y}_i. \tag{6.13.30}$$

In this case $\mathbf{z}_i$ is the vector of regression coefficients γ_i in the polynomial regression model $\mathbf{y}_i = \mathbf{K}\gamma_i$. If orthogonal polynomials are used in $\mathbf{K}$, then $\mathbf{K}'\mathbf{K} = \mathbf{I}_p$ and $\mathbf{Z} = \mathbf{YK}$.

Using a minimum-variance criterion it has been shown that the optimal choice of $\mathbf{G}$ is $\mathbf{\Sigma}$. In practice, $\mathbf{\Sigma}$ is unknown and is to be replaced by it maximum likelihood estimate

$$\begin{aligned} \mathbf{S}_e &= \frac{\mathbf{Y}'\mathbf{PY}}{n-p} \\ &= \frac{\mathbf{Y}'(\mathbf{I}_n - \mathbf{X}(\mathbf{X}'\mathbf{X})^{-1}\mathbf{X}')\mathbf{Y}}{n-p}. \end{aligned} \tag{6.13.31}$$

However, $\mathbf{S}$ depends on $\mathbf{Y}$ and the above-mentioned distributional results of $\mathbf{H}_1$ and $\mathbf{E}_1$ no longer hold.

Rao (1966, 1967) and Khatri (1966) proposed another method based on the analysis of covariance, the details of which are omitted. The multivariate analysis of covariance is considered in Chapter 7.

6.14 Exercises and Complements

6.1 Show that the $n \times n$ matrix $\mathbf{P}$ defined in (6.2.1) is symmetric and idempotent. Show that $\mathbf{Pz} = \mathbf{z}$ *iff* $\mathbf{z}$ is orthogonal to the column-space of $\mathbf{X}$ and that $\mathbf{Pz} = \mathbf{0}$ *iff* $\mathbf{z}$ is a linear combination of columns of $\mathbf{X}$. Hence, show that $\mathbf{P}$ is a projection onto the subspace of $\mathcal{R}^n$ orthogonal to the columns of $\mathbf{X}$.

6.2 Using (6.2.6) show that $l(\mathbf{B}, \hat{\mathbf{\Sigma}})$ depends on $\mathbf{Y}$ only through $\hat{\mathbf{B}}$ and $\hat{\mathbf{\Sigma}}$ and hence show that $(\hat{\mathbf{B}}, \hat{\mathbf{\Sigma}})$ is jointly sufficient for $(\mathbf{B}, \mathbf{\Sigma})$. Is $\hat{\mathbf{B}}$ sufficient for $\mathbf{B}$ when $\mathbf{\Sigma}$ is known? Is $\hat{\mathbf{\Sigma}}$ sufficient for $\mathbf{\Sigma}$ when $\mathbf{B}$ is known? Justify your answer.

6.3 Show that $\hat{\mathbf{B}} - \mathbf{B} = \mathbf{AU}$ where $\mathbf{A} = (\mathbf{X}'\mathbf{X})^{-1}\mathbf{X}'$ and $\mathbf{AA}' = (\mathbf{X}'\mathbf{X})^{-1}$. Hence show that $\hat{\mathbf{B}} - \mathbf{B}$ thought of as a column vector has the distribution

$$(\hat{\mathbf{B}} - \mathbf{B})^V \sim N_{np}(\mathbf{0}, \mathbf{\Sigma} \otimes (\mathbf{X}'\mathbf{X})^{-1}).$$

6.4 Prove the results of Corollary 6.2.3.1.

6.5 Consider the matrix $\mathbf{Q}$ in (6.3.3). Prove that $\mathrm{tr}(\mathbf{Q}) = q$.

6.6 Check the relation (6.5.16).

6.7 Check the equation (6.9.3).

6.8 Considering the model (6.9.4), show that $E(\mathbf{w}_i) = \mathbf{0}, i = 1, \ldots, n,$

$$\mathrm{Cov}\ (\mathbf{w}_i, \mathbf{w}_j) = \begin{cases} (1 - \frac{1}{n})\mathbf{\Sigma} & i = j \\ -\frac{1}{n}\mathbf{\Sigma} & i \neq j. \end{cases}$$

Hence, writing $\mathbf{W}$ as a vector, deduce that

$$\mathbf{W}^V \sim N_{np}(\mathbf{0}, \mathbf{\Sigma} \otimes \mathbf{H}).$$

6.9 *Population multiple correlation coefficient*: If $\mathbf{x} = (x_1, \ldots, x_p)'$ is a random p-vector with covariance matrix $\mathbf{\Sigma}$, show that the largest correlation between x_1 and a linear combination of $x_2, \ldots, x_p$ equals $\mathbf{\Sigma}_{12}\mathbf{\Sigma}_{22}^{-1}\mathbf{\Sigma}_{21}/\sigma_{11}$.

6.10 If $\mathbf{R}$ is a 3×3 correlation matrix, show that

$$R_{1.23}^2 = (r_{12}^2 + r_{13}^2 - 2r_{12}r_{13}r_{23})/(1 - r_{23}^2).$$

6.11 If the random variable $\mathbf{X} = (X_1, \ldots, X_4)'$ has the covariance matrix

$$\mathbf{\Sigma} = \begin{bmatrix} \sigma^2 & \sigma_{12} & \sigma_{13} & \sigma_{14} \\ \sigma_{12} & \sigma^2 & \sigma_{14} & \sigma_{13} \\ \sigma_{13} & \sigma_{14} & \sigma^2 & \sigma_{12} \\ \sigma_{14} & \sigma_{13} & \sigma_{12} & \sigma^2 \end{bmatrix}$$

show that the four multiple correlation coefficients between one variable and the other three variables are equal.

6.12 Let $\mathbf{x} = (\mathbf{x}_1', \mathbf{x}_2')'$ be a random $(p+q)$-vector with known mean μ and covariance matrix $\mathbf{\Sigma}$. Suppose $\mathbf{x}_2$ is regressed on $\mathbf{x}_1$ using

$$E[(\mathbf{x}_2 - \mu_2|\mathbf{x}_1] = \mathbf{\Sigma}_{21}\mathbf{\Sigma}_{11}^{-1}(\mathbf{x}_1 - \mu_1).$$

Then show that $\mathbf{\Delta}$ (the population quantity corresponding to $\mathbf{D}$ in (6.11.1)) is

$$\mathbf{\Delta} = \mathbf{\Sigma}_{22}^{-1}\mathbf{\Sigma}_{22.1} = \mathbf{I} - \mathbf{\Sigma}_{22}^{-1}\mathbf{\Sigma}_{21}\mathbf{\Sigma}_{11}^{-1}\mathbf{\Sigma}_{12}.$$

Show also that the corresponding population value of ρ_T, the trace correlation and ρ_D, the determinant correlation are given by

$$\rho_T^2 = \frac{1}{q}\text{ tr }(\mathbf{I} - \mathbf{\Delta}) = \frac{1}{q}\text{ tr }(\mathbf{\Sigma}_{22}^{-1}\mathbf{\Sigma}_{21}\mathbf{\Sigma}_{11}^{-1}\mathbf{\Sigma}_{12}$$

and

$$\rho_D^2 = |\mathbf{I} - \mathbf{D}| = |\mathbf{\Sigma}_{21}\mathbf{\Sigma}_{11}^{-1}\mathbf{\Sigma}_{12}|/|\mathbf{\Sigma}_{22}|.$$

6.13 Show that when the hypothesis H_{02} in Section 6.6 is false,

$$\begin{aligned} E(\mathbf{H}) &= g\mathbf{\Sigma} + (\mathbf{AB} - \mathbf{D})'[\mathbf{A}(\mathbf{X}'\mathbf{X})^{-}\mathbf{A}']^{-1}(\mathbf{AB} - \mathbf{D}) \\ &= g\mathbf{\Sigma} + \mathbf{G} \quad \text{(say)} \end{aligned}$$

and $\mathbf{H}$ has a noncentral Wishart distribution $W_p(g, \mathbf{\Sigma}; \mathbf{\Delta})$ with noncentrality parameter $\mathbf{\Delta} = \mathbf{\Sigma}^{-1/2}\mathbf{G}\mathbf{\Sigma}^{-1/2}$.

Chapter 7

Multivariate Analysis of Variance and Covariance

7.1 Introduction

In this chapter we apply the theory of multivariate regression models to the problems of multivariate analysis of variance and covariance. After reviewing the univariate model, Section 7.2 considers multivariate one-way fixed effects analysis of variance. This section also discusses tests for multivariate treatment contrasts and makes a comparison among them. Two-way fixed effects models are addressed in the next section. The last two sections display applications of the regression theory to the multivariate analysis of covariance and a conditional hypothesis model.

7.2 Multivariate One-Way Analysis of Variance

We first recall univariate one-way analysis of variance model for better understanding of the results under the multivariate model.

7.2.1 *Univariate model*

Let y_{ij} $(j = 1, \dots n_i)$ be a sample of size n_i from a Normal Population $N(\mu_i, \sigma^2), i = 1, \dots, k$. We want to test the null hypothesis $H_0(\mu_1 = \dots = \mu_k)$. The sample mean for the ith population can be written as $\bar{y}_{i0} = \sum_{j=1}^{n_i} y_{ij}/n_i$, the grand mean is $\bar{y}_{00} = \sum_{i=1}^{k} \sum_{j=1}^{n_i} y_{ij}/n = \sum_{i=1}^{k} n_i \bar{y}_{io}/n$, where $n = \sum_{i=1}^{k} n_i$.

The observation y_{ij} can be written as

$$y_{ij} = \mu_i + \epsilon_{ij} \tag{7.2.1}$$

where ϵ_{ij} are *iid* $N(0, \sigma^2)$. Hence we can write the vector of observations as

$$\begin{bmatrix} y_{11} \\ \cdot \\ y_{1n_1} \\ \cdot \\ \cdot \\ y_{k1} \\ \cdot \\ y_{kn_k} \end{bmatrix} = \begin{bmatrix} \mathbf{1}_{n_1} & \mathbf{0} & \dots & \mathbf{0} \\ \mathbf{0} & \mathbf{1}_{n_2} & \dots & \mathbf{0} \\ \mathbf{0} & \mathbf{0} & \dots & \mathbf{0} \\ \cdot & \cdot & \dots & \cdot \\ \cdot & \cdot & \dots & \cdot \\ \mathbf{0} & \mathbf{0} & \dots & \mathbf{1}_{r_k} \end{bmatrix} \begin{bmatrix} \mu_1 \\ \mu_2 \\ \cdot \\ \cdot \\ \mu_k \end{bmatrix} + \begin{bmatrix} \epsilon_{11} \\ \cdot \\ \epsilon_{1n_1} \\ \cdot \\ \cdot \\ \epsilon_{k1} \\ \cdot \\ \epsilon_{kn_k} \end{bmatrix} \tag{7.2.2}$$

or

$$\mathbf{Y} = \mathbf{XB} + \epsilon \tag{7.2.3}$$

where each row of ϵ is $N(0, \sigma^2)$.

We have to calculate E, the sum of squares (SS) due to error under the full model (7.2.2) and E_H, the SS due to error under the restricted model , i.e., the full model (7.2.1) subject to the condition that $\mu_1 = \ldots = \mu_k$.

Under H_0, the model (7.2.1) reduces to

$$\mathbf{Y} = \tilde{\mathbf{X}} B_H + \epsilon \tag{7.2.4}$$

where $\tilde{\mathbf{X}} = \mathbf{1}_n$ and $B_H = \mu$, the common value of $\mu_1, \ldots, \mu_k$.

Now, the least square estimate (*l.s.e*) of μ under (7.2.2) is

$$\hat{\mathbf{B}} = \hat{\mu} = (\mathbf{X}'\mathbf{X})^{-1}\mathbf{X}'\mathbf{Y},$$

i.e.

$$\hat{\mu} = (\bar{y}_{10}, \bar{y}_{20}, \ldots, \bar{y}_{k0})'. \tag{7.2.5}$$

Also, the *l.s.e.* of μ under (7.2.4) is

$$\hat{B}_H = \hat{\mu} = \bar{y}_{C0} \tag{7.2.6}$$

Therefore,

$$\begin{aligned} E &= (\mathbf{Y} - \mathbf{X}\hat{\mathbf{B}})'(\mathbf{Y} - \mathbf{X}\hat{\mathbf{B}}) \\ &= \textstyle\sum_{i=1}^{k} \sum_{j=1}^{n_i} (y_{ij} - \bar{y}_{i0})^2. \end{aligned} \tag{7.2.7}$$

Similarly,

$$\begin{aligned} E_H &= (\mathbf{Y} - \tilde{\mathbf{X}}\hat{B}_H)'(\mathbf{X} - \tilde{\mathbf{X}}\hat{B}_H) \\ &= \textstyle\sum_{i=1}^{k} \sum_{j=1}^{n_i} (y_{ij} - \bar{y}_{00})^2 \end{aligned} . \tag{7.2.8}$$

Therefore, SS due to H_0 is

$$\begin{aligned} H &= E_H - E \\ &= \sum_{i=1}^{k}\sum_{j=1}^{n_i}(y_{ij} - \bar{y}_{00})^2 - \sum_{i=1}^{k}\sum_{j=1}^{n_i}(\bar{y}_{ij} - \bar{y}_{i0})^2 \\ &= \sum_{i=1}^{k} n_i(\bar{y}_{i0} - \bar{y}_{00})^2. \end{aligned} \tag{7.2.9}$$

Now, for fixed i, y_{ij}'s are independent normal with common mean and common variance σ^2. Hence $\sum_{j=1}^{n_i}(y_{ij}-\bar{y}_{i0})^2$ is distributed as $\sigma^2\chi^2_{(n_i-1)}$. Hence, E is distributed as $\sigma^2\chi^2_{(n-k)}$, as the samples $y_{ij}, j = 1, \ldots, n_i$ are independent, $i = 1, \ldots, k$. This is true both under the full model and the restricted model.

Again, under the restricted model, each of $y_{ij}(i = 1, \ldots, k; j = 1, \ldots, n_i)$ are normal with mean μ and variance σ^2. Hence, under H_0, E_H is distributed as $\sigma^2\chi^2_{(n-1)}$.

Also, since $\sum_{j=1}^{n_i}(y_{ij} - \bar{y}_{i0})^2$ is distributed independently of $\bar{y}_{i0}$ and hence of $(\bar{y}_{i0} - \bar{y}_{00})^2, H = \sum_{i=1}^{k} n_i(\bar{y}_{i0} - \bar{y}_{00})^2$ is distributed independently of $E = \sum_{i=1}^{k}\sum_{j=1}^{n_i}(y_{ij} - \bar{y}_{i0})^2$. Hence, by Cochran's Theorem, under H_0, H follows $\chi^2_{(k-1)}$ distribution. Also, under H_0, $\frac{H/(k-1)}{E/(n-k)}$ follows $F_{(k-1,n-k)}$ distribution, which can be used in testing H_0.

The quantity H is popularly known as SS due to treatment or between SS or SSA and E as SS due to error or within SS or SSE. The quantity E_H is the total SS, TSS. Writing $\mu_i = \mu + \alpha_i$, under H_0,

$$E(MSA) = \sigma^2 + \tfrac{1}{k-1}\sum_{i=1}^{k} n_i\alpha_i^2, \text{ where } MSA = SSA/(k-1),$$
$$E(MSE) = \sigma^2 \text{ where } MSE = SSE/(n-k).$$

The maximum likelihood estimates (*mle*) of σ^2 under full model (7.2.2) and restricted model [(7.2.2) subject to H_0] are $\hat{\sigma}^2 = SSE/n$ and $\tilde{\sigma}^2 = E_H/n$ respectively. These are biased estimates.

If the null hypothesis H_0 is rejected, we may want to test for the equality of two class means $H_{ab}(\mu_a = \mu_b)$. This is done by using the test statistic

$$t = \frac{\hat{\mu}_a - \hat{\mu}_b}{s.\hat{e}.(\hat{\mu}_a - \hat{\mu}_b)} = \frac{\bar{y}_{a0} - \bar{y}_{b0}}{\sqrt{MSE(\frac{1}{n_a} + \frac{1}{n_b})}} \tag{7.2.10}$$

which follows, under H_{ab}, Student's t distribution with $n - k$ d.f. If the observed $|t| > t_{n-k;\alpha/2}$, then H_{ab} is rejected at $100\alpha\%$ level of significance. If $n_1 = \ldots = n_k = m$, the value of the *critical difference* or *the least significant difference* between two class means is

$$\sqrt{\frac{2}{m}MSE}t_{n-k;\alpha/2} = d_0 \;\; \text{(say)} \; ,$$

so that if $|\bar{y}_{a0} - \bar{y}_{b0}| > d_\alpha$, then the class means μ_a, μ_b differ significantly at $100\alpha\%$ level of significance.

Sch'effe's method of multiple comparison gives the value of critical difference as

$$d'_\alpha = \sqrt{\frac{2MSE}{m}(k-1)F_{k-1,n-k;\alpha}}. \tag{7.2.11}$$

Another procedure is to use Bonferroni's t-statistic. To test H_{ab}, one rejects the null hypothesis if

$$|\bar{y}_{ao} - \bar{y}_{b0}| > \sqrt{MSE(\frac{1}{n_a} + \frac{1}{n_b})}t'_{n-k;\alpha/2M} = t'_0 \text{ (say)} \tag{7.2.12}$$

where $M = \binom{k}{2}$ and $t'_{n-k;\alpha/2M}$ is the upper $100\alpha/2M$-point of Bonferroni's t-statistics, available from tables by Miller (1981). The above critical region ensures that the probability that at least one of the $\binom{k}{2}$ values $|\bar{y}_{a0} - \bar{y}_{b0}| (a \neq b = 1, \ldots, k)$ will exceed t'_0 is less than α.

Table 7.2.1: Univariate ANOVA Table for One-Way Classified Data

Source of variation	d.f.	SS	MS	F
Treatments (or between) classes)	k-1	$SSA = \sum_{i=1}^k n_i(\bar{y}_{i0} - \bar{y}_{00})^2$	MSA = SSA/(k-1)	F = MSA/MSE
Error (or within classes)	n-k	SSE = $\sum_{i=1}^k \sum_{j=1}^{n_i}(y_{ij} - \bar{y}_{i0})^2$	MSE = SSE/(n-k)	
Total	n-1	TSS = $\sum_{i=1}^k \sum_{j=1}^{n_i}(y_{ij} - \bar{y}_{00})^2$		

With this background we now consider multivariate one-way analysis of variance.

7.2.2 *Multivariate one-way fixed effects model*

Let $\mathbf{y}_{ij}$ be the jth sample observation from the ith multivariate normal population $N_p(\mu_i, \mathbf{\Sigma})(i = 1, \ldots, k; j = 1, \ldots, n_i)$. The sample mean for the ith population is $\bar{\mathbf{y}}_{i0} = \sum_{j=1}^{n_i} \mathbf{y}_{ij}/n_i$, the grand mean is $\bar{\mathbf{y}}_{00} = \sum_{i=1}^k \sum_{j=1}^{n_i} \mathbf{y}_{ij}/n = \sum_{i=1}^k n_i \bar{\mathbf{y}}_{i0}/n$ where $n = \sum_{i=1}^k n_i$. We want to test the null hypothesis $H_0(\mu_1 = \ldots = \mu_k)$. The problem has already been considered in Section 5.11 and the likelihood ratio test and the union-intersection test for the same have been developed.

We shall now apply the general theory of multivariate regression considered in Chapter 6 to this problem. For this, we see that the vector-observation $\mathbf{y}_{ij}$ can be written as

$$\mathbf{y}_{ij} = \mu_i + \mathbf{u}_{ij} \tag{7.2.13}$$

where $\mathbf{u}_{ij}(i = 1, \ldots, k; j = 1, \ldots, n_i)$ are *iid* $N_p(\mathbf{0}, \mathbf{\Sigma})$. Hence, we can write the matrix of observations as

$$\begin{bmatrix} \mathbf{y}'_{11} \\ \cdot \\ \cdot \\ \mathbf{y}'_{1n_1} \\ \cdot \\ \cdot \\ \mathbf{y}'_{k1} \\ \cdot \\ \cdot \\ \mathbf{y}'_{kn_k} \end{bmatrix} = \begin{bmatrix} \mathbf{1}_{n_1} & \mathbf{0} & \ldots & \mathbf{0} \\ \mathbf{0} & \mathbf{1}_{n_2} & \ldots & \mathbf{0} \\ \cdot & \cdot & \ldots & \cdot \\ \cdot & \cdot & \ldots & \cdot \\ \mathbf{0} & \mathbf{0} & \ldots & \mathbf{1}_{n_k} \end{bmatrix} \begin{bmatrix} \mu'_1 \\ \mu'_2 \\ \cdot \\ \cdot \\ \mu'_k \end{bmatrix} + \begin{bmatrix} \mathbf{u}'_{11} \\ \cdot \\ \cdot \\ \mathbf{u}'_{1n_1} \\ \cdot \\ \cdot \\ \mathbf{u}'_{k1} \\ \cdot \\ \cdot \\ \mathbf{u}'_{kn_k} \end{bmatrix} \tag{7.2.14}$$

or

$$\mathbf{Y} = \mathbf{XB} + \mathbf{U} \tag{7.2.15}$$

where each row of $\mathbf{U}$ is $N_p(\mathbf{0}, \mathbf{\Sigma})$, i.e., $\mathbf{U}$ is a data-matrix from $N_p(\mathbf{0}, \mathbf{\Sigma})$. Here, rank $(\mathbf{X}) = k$.

The hypothesis H_0 of equality of means can be written as

$$H_0 : \begin{bmatrix} 1 & 0 & \ldots & 0 & -1 \\ 0 & 1 & \ldots & 0 & -1 \\ \cdot & \cdot & \ldots & \cdot & \cdot \\ \cdot & \cdot & \ldots & \cdot & \cdot \\ 0 & 0 & \ldots & 1 & -1 \end{bmatrix} \begin{bmatrix} \mu'_1 \\ \mu'_2 \\ \cdot \\ \cdot \\ \mu'_k \end{bmatrix} = \mathbf{0}$$

or

$$H_0 : \mathbf{AB} = \mathbf{0}. \tag{7.2.16}$$

The hypothesis H_0 can then be tested using the general theory of Section 6.5.1 or 6.6.

We have to calculate $\mathbf{E}$, the sum of squares and product (SSP) matrix due to error under the full model (7.2.15) and $\mathbf{E}_H$, the SSP matrix due to error under the restricted model, i.e., the full model (7.2.15) subject to (7.2.16).

By Theorem 6.2, the *l.s.e* of $\mathbf{B}$ under (7.2.15) is

$$\begin{aligned}\hat{\mathbf{B}} &= (\mathbf{X}'\mathbf{X})^{-1}\mathbf{X}'\mathbf{Y} \\ &= (\bar{\mathbf{y}}_{10}, \ldots, \bar{\mathbf{y}}_{k0})'.\end{aligned} \tag{7.2.17}$$

Also, the restricted model is

$$\mathbf{Y} = \tilde{\mathbf{X}}\mathbf{B}_H + \mathbf{U} \tag{7.2.18}$$

where $\tilde{\mathbf{X}} = \mathbf{1}_n, \mathbf{B}_H = \mu'$, the common value of $\mu_i'(i = 1, \ldots, k)$. Hence, the *l.s.e.* of $\mathbf{B}_H$ is

$$\hat{\mathbf{B}}_H = \hat{\mu}' = (\tilde{\mathbf{X}}'\tilde{\mathbf{X}})^{-1}\tilde{\mathbf{X}}'\mathbf{Y} = \bar{\mathbf{y}}_{00}'. \tag{7.2.19}$$

Therefore,

$$\begin{aligned}\mathbf{E} &= (\mathbf{Y} - \mathbf{X}\hat{\mathbf{B}})'(\mathbf{Y} - \mathbf{X}\hat{\mathbf{B}}) \\ &= \textstyle\sum_{i=1}^{k}\sum_{j=1}^{n_i}(\mathbf{y}_{ij} - \bar{\mathbf{y}}_{i0})(\mathbf{y}_{ij} - \bar{\mathbf{y}}_{i0})',\end{aligned} \tag{7.2.20}$$

and

$$\begin{aligned}\mathbf{E}_H &= (\mathbf{Y} - \tilde{\mathbf{X}}\hat{\mathbf{B}}_H)'(\mathbf{Y} - \tilde{\mathbf{X}}\hat{\mathbf{B}}_H) \\ &= \textstyle\sum_{i=1}^{k}\sum_{j=1}^{n_i}(\mathbf{y}_{ij} - \bar{\mathbf{y}}_{00})(\mathbf{y}_{ij} - \bar{\mathbf{y}}_{00})'.\end{aligned} \tag{7.2.21}$$

Hence, SSP matrix due to H_0 is

$$\begin{aligned}\mathbf{H} &= \mathbf{E}_H - \mathbf{E} \\ &= \textstyle\sum_{i=1}^{k} n_i(\bar{\mathbf{y}}_{i0} - \bar{\mathbf{y}}_{00})(\bar{\mathbf{y}}_{i0} - \bar{\mathbf{y}}_{00})'.\end{aligned} \tag{7.2.22}$$

Also, $\mathbf{E} = \mathbf{U}'\mathbf{P}\mathbf{U}$ where $\mathbf{P} = \mathbf{I}_n - \mathbf{X}(\mathbf{X}'\mathbf{X})^{-1}\mathbf{X}'$ has rank $n - k$. Similarly, $\mathbf{E}_H = \mathbf{U}'\mathbf{P}_1\mathbf{U}$ where $\mathbf{P}_1 = \mathbf{I}_n - \tilde{\mathbf{X}}(\tilde{\mathbf{X}}'\tilde{\mathbf{X}})^{-1}\tilde{\mathbf{X}}'$ has rank $n - 1$. Hence, $\mathbf{E}$ follows $W_p(n - k, \mathbf{\Sigma})$ and when H_0 is true, $\mathbf{E}_H$ follows $W_p(n - 1, \mathbf{\Sigma})$ (Theorem 4.2.10(b)). Also, under $H_0, \mathbf{H} = \mathbf{U}'(\mathbf{P}_1 - \mathbf{P})\mathbf{U}$ follows $W_p(k - 1, \mathbf{\Sigma})$ since $\mathbf{P}_2 = \mathbf{P}_1 - \mathbf{P}$ is a symmetric idempotent matrix of rank $k - 1$. Again, $\mathbf{P}\mathbf{P}_2 = \mathbf{0}$. Hence, when H_0 is true,

$$\Lambda = \frac{|\mathbf{E}|}{|\mathbf{E} + \mathbf{H}|} \tag{7.2.23}$$

follows Wilk's $\Lambda(p, k - 1, n - k)$ distribution.

We can also use Roy's maximum root statistic θ_1 where θ_1 is the largest eigenvalue of $(\mathbf{E} + \mathbf{H})^{-1/2}\mathbf{E}(\mathbf{E} + \mathbf{H})^{-1/2}$, Lawley-Hotelling statistic T_g^2 in (4.5.6) and Pillai's statistic $V^{(s)}$ in (4.5.8) and use the corresponding percentage points for the acceptance or rejection of H_0 (vide Section 4.5).

To estimate $\mathbf{\Sigma}$ we use

$$\mathbf{S}_{pl} = \mathbf{E}/(n - k)$$

which is an unbiased estimator both under H_0 and full model.

The above problem has also been considered in Example 6.5.1 in a slightly different setting.

Table 7.2.2: Multivariate One-Way Analysis of Variance

Source	Sum of Squares and Product Matrix	d.f.
Treatments	$\mathbf{H} = \sum_{i=1}^{k} n_i(\bar{\mathbf{y}}_{i0} - \bar{\mathbf{y}}_{00})(\bar{\mathbf{y}}_{i0} - \bar{\mathbf{y}}_{00})'$	$k-1$
Error	$\mathbf{E} = \sum_{i=1}^{k}\sum_{j=1}^{n_i}(\mathbf{y}_{ij} - \bar{\mathbf{y}}_{i0})(\mathbf{y}_{ij} - \bar{\mathbf{y}}_{i0})'$	$n-k$
Total	$\mathbf{E}_H = \sum_{i=1}^{k}\sum_{j=1}^{n_i}(\mathbf{y}_{ij} - \bar{\mathbf{y}}_{00})(\bar{\mathbf{y}}_{ij} - \bar{\mathbf{y}}_{00})'$	$n-1$

EXAMPLE 7.2.1: The following data give the values of 4 characteristics on samples taken from 3 groups. Test the hypothesis that the group means are equal.

Group I

$$\begin{bmatrix}19.2\\26.7\\12.5\\0.9\end{bmatrix}, \begin{bmatrix}18.4\\27.2\\13.5\\1.8\end{bmatrix}, \begin{bmatrix}16.5\\24.6\\12.6\\1.2\end{bmatrix}, \begin{bmatrix}15.8\\24.7\\13.5\\0.9\end{bmatrix}, \begin{bmatrix}14.7\\21.8\\12.9\\2.1\end{bmatrix};$$

Group II

$$\begin{bmatrix}16.5\\28.2\\11.4\\1.6\end{bmatrix}, \begin{bmatrix}15.4\\22.3\\9.8\\0.8\end{bmatrix}, \begin{bmatrix}19.2\\27.5\\8.4\\2.1\end{bmatrix};$$

Group III

$$\begin{bmatrix}12.5\\18.2\\9.7\\0.5\end{bmatrix}, \begin{bmatrix}13.8\\20.4\\6.9\\1.5\end{bmatrix}, \begin{bmatrix}11.7\\20.9\\8.3\\0.9\end{bmatrix}, \begin{bmatrix}15.2\\21.7\\7.5\\1.2\end{bmatrix}.$$

Here

$$\mathbf{H} = \begin{bmatrix}35.794 & 49.778 & 34.250 & 3.932\\ 49.778 & 70.557 & 42.507 & 5.638\\ 34.250 & 42.507 & 52.490 & 3.108\\ 3.932 & 5.638 & 3.108 & 0.454\end{bmatrix},$$

$$\mathbf{E} = \begin{bmatrix}28.455 & 26.280 & -8.757 & 2.122\\ 26.290 & 45.750 & -1.840 & 3.050\\ -8.757 & -1.840 & 10.337 & -2.085\\ 2.122 & 3.050 & -2.085 & 2.596\end{bmatrix};$$

$m_H = k - 1 = 2, m_E = n - k = 9$. The eigenvalues of $\mathbf{E}$ are 65.8211, 15.5367, 4.2540, and 1.5162. Also, the eigenvalues of $\mathbf{E}+\mathbf{H}$ are 189.441, 45.012, 9.670 and 2.291. Hence

$$|\mathbf{E}| = 6595.96, \; |\mathbf{E}+\mathbf{H}| = 188910.$$

From (4.4.2), Wilk's statistic

$$\Lambda(p; m_H, m_E) = \Lambda(4; 2, 9) = \frac{|\mathbf{E}|}{|\mathbf{E} + \mathbf{H}|} = 0.391.$$

Since the observed value of $\Lambda <$ the tabulated value $\Lambda_{.05}(4; 2, 9) = .119$, the hypothesis $H_0(\mu_1 = \mu_2 = \mu_3)$ is rejected at 5% level of significance.

From (4.4.10) and (4.4.11) using Rao's F-approximation to Λ-distribution, the above value of $\Lambda(4; 2, 9)$ is approximated by $F_{(8,9)} = 6.07805$ (here $t = 2, f = 7.5, 4 = 3$). Since the observed value of F exceeds the tabulated value $F_{8,9;.os} = 3.23$, the null hypothesis H_0 is rejected.

From Table B.12 (Table of percentage points of θ_{max}), $s = \min(p, m_H) = 2, m = (|p - m_|) - 1)/2 = 0.5, N = (m_E - p - 1)/2 = 2$. Also $\theta_1 =$ the largest eigenvalue of $\mathbf{H}(\mathbf{E} + \mathbf{H})^{-1} = .94157$.

Also, from (4.5.13), Pillai's statistic $V^{(2)} = \sum_{j=1}^{2} \theta_j = 1.34418$ where θ_j are the roots of $\mathbf{H}(\mathbf{E}+\mathbf{H})^{-1}$. Since the observed value exceeds the (estimated) tabulated value $V_{.05}(2; .5, 2) = 1.145$, the null hypothesis is rejected.

Following (4.5.12)

$$U^{(s)} = \sum_{j=1}^{2} \phi_j = 16.78714$$

where ϕ_j are the eigenvectors of $\mathbf{HE}^{-1}$. In Table B.13, the Lawley-Hotelling test statistic is

$$\tilde{T}_g^2 = \frac{m_E U^{(s)}}{m_H}.$$

We reject H_0 if the observed value of $\tilde{T}_g^2$ exceeds the tabulated value. Here the observed value of $\tilde{T}_g^2$ is 75.54213. This far exceeds the tabulated value $T_{.05}(p; m_H, m_E) = T_{.05}(4; 2, 9) = T_{.05}(m_H; p, m_E+m_H-p) = T_{.05}(2; 4, 7) = 9.1694$. Hence the null hypothesis is rejected.

7.2.2.1 *Multiple comparison*

For the general linear model (6.1.1) we have seen that the confidence interval for the linear function $\mathbf{a}'\mathbf{ABb}$ is given by (vide (6.8.15))

$$1 - \alpha = \text{ Pr. } [\mathbf{a}'\mathbf{ABb} \in \mathbf{a}'\mathbf{A}\hat{\mathbf{B}}\mathbf{b} \pm \sqrt{\Phi_\alpha\{\mathbf{a}'\mathbf{A}(\mathbf{X}'\mathbf{X})^{-1}\mathbf{A}'\mathbf{a}\}\{\mathbf{b}'\mathbf{Eb}\}}\ \forall \mathbf{a}, \mathbf{b}] \tag{7.2.24}$$

where $\mathbf{E} = (\mathbf{Y} - \mathbf{X}\hat{\mathbf{B}})'(\mathbf{Y} - \mathbf{X}\hat{\mathbf{B}})$ is the SSP matrix due to error under the model (6.1.1). In our case,

$$\mathbf{AB} = \begin{bmatrix} 1 & 0 & 0 & \dots & -1 \\ 0 & 1 & 0 & \dots & -1 \\ . & . & . & \dots & . \\ . & . & . & \dots & . \\ 0 & 0 & 0 & \dots & -1 \end{bmatrix} = \begin{bmatrix} (\mu_1 - \mu_k)' \\ (\mu_2 - \mu_k)' \\ . \\ . \\ (\mu_{k-1} - \mu_k)' \end{bmatrix}.$$

Hence,

$$\begin{aligned} \mathbf{a}'\mathbf{AB} &= a_1(\mu_1 - \mu_k)' + a_2(\mu_2 - \mu_k)' + \ldots + a_{k-1}(\mu_{k-1} - \mu_k)' \\ &= \textstyle\sum_{i=1}^{k-1} a_i\mu_i' - \mu_k' \sum_{i=1}^{k-1} a_i \\ &= \textstyle\sum_{i=1}^{k} c_i\mu_i' \end{aligned} \tag{7.2.25}$$

where

$$\begin{aligned} c_i &= a_i,\ i = 1, \ldots, k-1 \\ c_k &= -\textstyle\sum_{i=1}^{k-1} a_i = -\sum_{i=1}^{k-1} c_i. \end{aligned} \tag{7.2.26}$$

Therefore, $\sum_{i=1}^{k} c_i = 0$. Hence the class of all functions $\mathbf{a}'\mathbf{AB}$ is the same as the class of all linear contrasts among $\mu_1', \mu_2', \ldots, \mu_k'$.

For this model, $\hat{\mathbf{B}} = (\bar{\mathbf{y}}_{10}', \bar{\mathbf{y}}_{20}', \ldots, \bar{\mathbf{y}}_{k0}')'$. Hence,

$$\mathbf{a}'\mathbf{A}\hat{\mathbf{B}}\mathbf{b} = \sum_{i=1}^{k} c_i\bar{\mathbf{y}}_{i0}'\mathbf{b},$$

which is a contrast among linear combinations $\bar{\mathbf{y}}_{i0}'\mathbf{b}$. Also,

$$\mathbf{a}'\mathbf{A}(\mathbf{X}'\mathbf{X})^{-1}\mathbf{A}'\mathbf{a} = \sum_{i=1}^{k} \frac{c_i^2}{n_i}. \tag{7.2.27}$$

Therefore, from (7.2.24)

$$1 - \alpha = \Pr\left[\sum_{i=1}^{k} c_i\mu_i'\mathbf{b} \in \sum_{i=1}^{k} c_i\bar{\mathbf{y}}_{i0}'\mathbf{b} \pm \sqrt{\Phi_\alpha\left(\sum_{i=1}^{k} \frac{c_i^2}{n_i}\right)\{\mathbf{b}'\mathbf{E}\mathbf{b}\}}\right] \tag{7.2.28}$$

simultaneously for all contrasts and all $\mathbf{b}$.

If we put $c_i = 1, c_{i'} = 0 (i' \neq i = 1, \ldots, k)$ and $b_j = 1, b_{j'} = 0\ (j' \neq j = 1, \ldots, p)$, then $\sum_{i=1}^{k} c_i\mu_i'\mathbf{b} = \mu_{ij}$. Therefore,

$$1 - \alpha = \Pr.\ \left[\mu_{ij} \in \bar{y}_{ij0} \pm \sqrt{\frac{\Phi_\alpha}{n_i} e_{jj}}\right] \tag{7.2.29}$$

where $\bar{\mathbf{y}}_{i0} = (\bar{y}_{i10}, \ldots, \bar{y}_{ip0})'$ and $\mathbf{E} = \sum_{i=1}^{k}\sum_{j=1}^{n_i}(\mathbf{y}_{ij} - \bar{\mathbf{y}}_{i0})(\mathbf{y}_{ij} - \bar{\mathbf{y}}_{i0})' = ((e_{jj'}))$.

If we are interested in the contrsat $\mu_r - \mu_s$ so that $c_r = 1, c_s = -1, c_t = 0(t \neq r, s = 1, \ldots, k)$, then

$$(\mu_r - \mu_s)'\mathbf{b} \in (\bar{\mathbf{y}}_{r0} - \bar{\mathbf{y}}_{s0})'\mathbf{b} \pm \sqrt{\Phi_\alpha(\frac{1}{n_r} + \frac{1}{n_s})\mathbf{b}'\mathbf{E}\mathbf{b}}. \tag{7.2.30}$$

However, it is well known that for univariate models, Scheffe's simultaneous intervals are very wide. Hence, for multivariate models, Scheffe's simultaneous intervals may be far too wide.

Note that when $k = 2, m_h = 1, s = 1$ and all the five statistics, Hotelling's T^2, Wilk's Λ, Roy's θ_1, T_g^2 and $V^{(1)}$ are related (vide Subsection 4.5.1). In particular, in this case,

$$U^{(1)} = \frac{T^2}{n_1 + n_2 - 2} \tag{7.2.31}$$

where T^2 follows $T^2(p, n_1 + n_2 - 2)$ (Exercise 7.1). Because of this relationship, the Lawley-Hotelling statistic $U^{(s)}$ is often called the generalized T^2 statistic.

Bonferroni's confidence intervals are also available for r specified comparisons. We simply replace $\mathbf{E}$ by $\mathbf{E}/m_E$ and $\Phi_\alpha^{1/2}$ by $t_{m_E;\alpha/(2r)}$ with $m_E = n - k$.

7.2.3 *Comparison among MANOVA tests*

When H_0 is true, all the four tests, - $\Lambda, \theta_1, U^{(s)}$ and $V^{(s)}$ have the same type-I error rate α. However, if H_0 is false, the tests have different probabilities of rejection.

Wilk's Λ test is not always the most powerful among the four tests. When $p = 1, \mu_1, \ldots, \mu_k$ all lie in one dimension and the usual F-test is the most powerful test. In the multivariate case ($p > 1$), the mean vectors $\mu_1, \ldots, \mu_k$ lie in $s =$ min (p, m_H) dimensions and none of the four tests is uniformly most powerful. For some configuration of $\mu_1, \ldots, \mu_k$, a given test will be the most powerful, while for other configuration another test will be the most powerful.

An indication of the pattern of the configuration of $\mu_1, \ldots, \mu_k$ is given by the eigenvalues of $\mathbf{E}^{-1}\mathbf{H}$. If there is one large eigenvalue and the rest are small, the mean vectors lie close in a line in the s-dimensional space. If there are two large eigenvalues, the mean vectors lie almost in a two-dimensional space.

Roy's θ_1 test which uses the largest eigenvalue of $\mathbf{E}^{-1}(\mathbf{E}+\mathbf{H})$ is more powerful than the other three tests if the mean vectors lie close to a line (almost collinear). In the collinear case, $\theta_1 \succeq U^{(s)} \succeq \Lambda \succeq V^{(s)}$ where $t_1 \succ t_2$ means t_1 is more powerful than t_2. In the diffuse case and for intermediate structure between diffuse and collinear, the ordering of the power is reversed: $V^{(s)} \succeq \Lambda \succeq U^{(s)} \succeq \theta_1$.

If the sample sizes are equal, all the tests generally perform well even under the heterogeneity of covariance matrices, i.e., the tests are robust with respect to heterogeneity of covariance matrices. If the group sizes are unequal and the covariances matrices are found unequal by Box's M test (say) (p. 173), then the α level of the tests may be affected. If the larger value of $|\mathbf{\Sigma}_i|$ is associated with the larger sample size n_i, then the actual value of α may be lower than the nominal value, i.e., the tests become conservative. On the other hand if larger $|\mathbf{\Sigma}_i|$ is associated with smaller n_i, the true value of α may be higher so that the test becomes liberal.

In short, θ_1 test is not to be recommended except in collinear case under the standard assumptions. If the data come from non-normal populations exhibiting skewness or positive kurtosis, any of the other three tests perform acceptably well. When there is heterogeneity of covariance matrices, $V^{(s)}$ is superior to the other two followed by Λ. It appears Λ is the most popular test because it has well-known χ^2 and F approximation.

7.2.4 *Testing a contrast*

If the hypothesis $H_0(\mu_1 = \ldots = \mu_p)$ is rejected we may wish to test hypothesis about a contrast. We consider only the balanced case $n_1 = n_2 = \ldots = n_k$.

First we consider the univariate case for some definitions.

A contrast among the parameters $\mu_1, \ldots, \mu_k$ is $\sum_{i=1}^{k} c_i\mu_i$ with known constant coefficients subject to the condition $\sum_{i=1}^{k} c_i = 0$. For example, the functions $\mu_i - \mu_{i'}, \mu_i - 2\mu_{i'} + \mu_{i''}$ are contrasts.

Similarly, we can define the contrasts among the observations $y_i (i = 1, \ldots, n)$. The linear function $C = \sum_{i=1}^{n} l_i y_i$ where l_i are given numbers such that $\sum_{i=1}^{n} l_i = 0$ is called a contrast of y_i's. The sum of squares due to the contrast C is $C^2/\sum_{i=1}^{n} l_i^2$. If y_i's follow independent $N(\mu, \sigma^2)$ distribution, then SS due to C follows $\sigma^2\chi^2_{(1)}$ distribution.

Two contrasts $C_1 = \sum l_i y_i$ and $C_2 = \sum m_i y_i$ are said to be orthogonal if $\sum_i l_i m_i = 0$. Thus, if the observations y_i's are independent, Cov $(C_1, C_2) = \sigma^2 \sum_i l_i m_i = 0$. A set of contrasts are mutually orthogonal, if the contrasts are pairwise orthogonal.

The SS due to a set of contrasts is the SS due to the set of mutually orthogonal contrasts which span this set of contrasts and carries as many degrees of freedom as the number of orthogonal contrasts. The SS due to r mutually orthogonal contrasts $C_i = \sum_{j=1}^{n} l_{ij} y_j, (i = 1, \ldots, r)$ is

$$\sum_{i=1}^{r} \frac{C_i^2}{\sum_{j=1}^{n} l_{ij}^2}.$$

We now come back to the multivariate case. A contrast among the population means $\mu_1, \ldots, \mu_k$ is

$$\delta = c_1\mu_1 + \ldots + c_k\mu_k \tag{7.2.32}$$

where $c_1 + \ldots + c_k = 0$. An unbiased estimate of δ is given by the corresponding contrasts among the sample means,

$$\hat{\delta} = c_1\bar{\mathbf{y}}_{10} + \ldots + c_k\bar{\mathbf{y}}_{k0}. \tag{7.2.33}$$

The sample observations $\mathbf{y}_{ij}$ are assumed to be independent with common covariance matrix $\mathbf{\Sigma}$. Hence,

$$\text{cov }(\hat{\delta}) = \frac{\mathbf{\Sigma}}{n}\sum_{i=1}^{k} c_i^2 \tag{7.2.34}$$

which is estimated by

$$\frac{\mathbf{S}_p}{n}\sum_{i=1}^{n} c_i^2 \tag{7.2.35}$$

where the pooled estimate $\mathbf{S}_p = \mathbf{E}/m_E$. Here n is the sample size for each population.

Two contrasts $\delta = \sum_{i=1}^{k} c_i\mu_i$ and $\gamma = \sum_{i=1}^{k} d_i\mu_i$ are said to be orthogonal, if $\sum_{i=1}^{k} c_i d_i = 0$.

The SSP matrix due to a contrast

$$\mathbf{L} = c_1\sum_{j=1}^{n} l_{1j}\mathbf{y}_{1j} + c_2\sum_{j=1}^{n} l_{2j}\mathbf{y}_{2j} + \ldots + c_k\sum_{j=1}^{n} l_{kj}\mathbf{y}_{kj} \tag{7.2.36}$$

is

$$\frac{\mathbf{LL}'}{c_1^2\sum_{j=1}^{n} l_{1j}^2 + \ldots + c_k^2\sum_{j=1}^{n} l_{kj}^2} = \frac{\mathbf{LL}'}{\text{Coefficient of } \mathbf{\Sigma} \text{ in V } (\mathbf{L})}. \tag{7.2.37}$$

When two contrasts are orthogonal, the two corresponding SSP matrices are independent. In fact, for k treatments, we can find $(k-1)$ orthogonal contrasts among the treatment means $\mu_1, \dots, \mu_k$ and their unbiased estimates are also orthogonal contrasts among observations $\mathbf{y}_{ij}(i = 1, \dots, k; j = 1, \dots, n)$. The SSP matrices due to these observational contrasts are independent each carrying one degree of freedom. Thus the SSP matrix due to treatments (i.e. due to hypothesis $H_0(\mu_1 = \dots = \mu_k)$) can be partitioned into $(k-1)$ independent SSP matrices, each carrying one degree of freedom. In the unbalanced case, orthogonal contrasts no longer partition SSP matrix due to H_0 into $(k-1)$ independent SSP matrices.

Suppose we want to test $H_\delta(\sum_{i=1}^k c_i\mu_i = \mathbf{0})$. For example, $\mu_1 - 2\mu_2 + \mu_3 = \mathbf{0}$ is equivalent to $\mu_2 = (\mu_1 + \mu_3)/2$. Since $\mathbf{E} \cap W_p(m_E, \mathbf{\Sigma})$ and $\hat{\delta} \cap N_p(\mathbf{0}, (\mathbf{\Sigma}/n)\sum_{i=1}^k c_i^2)$ and since these variables are independent

$$\begin{aligned} T^2 &= \frac{n\hat{\delta}'}{\sum_{i=1}^k c_i^2}\left(\frac{\mathbf{S}_p}{n}\right)^{-1}\frac{n\hat{\delta}}{\sum_{i=1}^k c_i^2} \\ &= \frac{n}{\sum_{i=1}^n c_i^2}(\textstyle\sum_{i=1}^k c_i\bar{\mathbf{y}}_{i0})'\left(\frac{\mathbf{E}}{m_E}^{-1}\right)(\textstyle\sum_{i=1}^k c_i\bar{\mathbf{y}}_{i0}) \end{aligned} \tag{7.2.38}$$

is distributed as $T^2(p, m_E)$. This can be used for testing H_δ.

The hypothesis H_δ can be tested equivalently using Wilk's $\mathbf{\Lambda}$. The SSP matrix due to hypothesis is

$$\mathbf{H}_\delta = \frac{n}{\sum_{i=1}^k c_i^2}(\sum_{i=1}^k c_i\bar{\mathbf{y}}_{i0})(\sum_{i=1}^k c_i\bar{\mathbf{y}}_{i0})' \tag{7.2.39}$$

(by (6.6.22)) with rank 1. Hence, the test statistic is

$$\Lambda = \frac{|\mathbf{E}|}{|\mathbf{E} + \mathbf{H}_\delta|} \tag{7.2.40}$$

which is distributed as $\Lambda(p; 1, m_E)$. Since $m_H = 1$, all four MANOVA statistics and T^2 give the same result.

If the multivariate hypothesis $H_0(\mu_1 = \dots = \mu_k)$ is rejected, one should make univariate test for each of the hypothesis $H_{0r}(\mu_{1r} = \dots = \mu_{kr}), r = 1, \dots, p$. The hypothesis H_{or} can be tested by an ordinary univariate F test.

If to test the multivariate null hypothesis $H_0(\mu_1 = \mu_2 = \dots = \mu_k)$, one carries p tests on each of the univariate hypothesis H_{0r}, each at level α, then the overall experimental error rate will be $1-(1-\alpha)^p$ which is greater than the nominal value α. The overall α or the experimental error rate is defined as the probability of rejecting one or more of the univariate tests when $H_0(\mu_1 = \dots = \mu_p)$ is true.

The correct procedure is, therefore, to first test H_0 and if this hypothesis is rejected, then to test $H_{0r}(r = 1, \ldots, p)$ to find out which of the variables contributed to the rejection of H_0.

EXAMPLE 7.2.2: For the data of Example 7.2.1 test the hypothesis H_δ : $\mu_1 - 2\mu_2 + \mu_3 = \mathbf{0}$. Assume that the sample size is $n = 3$ for each group.

Here

$$\mathbf{H}_\delta = \frac{3}{1+4+1}(\bar{\mathbf{y}}_{10} - 2\bar{\mathbf{y}}_{20} + \bar{\mathbf{y}}_{30})(\bar{\mathbf{y}}_{10} - 2\bar{\mathbf{y}}_{20} + \bar{\mathbf{y}}_{30})'$$

$$= \begin{bmatrix} 7.3728 & 12.8640 & -2.8992 & 1.1424 \\ 12.8640 & 22.4450 & -5.0585 & 1.9292 \\ -2.8992 & -5.0585 & 1.1400 & -.4492 \\ 1.1424 & 1.9932 & -.4492 & .1770 \end{bmatrix}.$$

The test statistic is

$$\Lambda(4; 1, 9) = \frac{|\mathbf{E}|}{|\mathbf{E} + \mathbf{H}_\delta|} = \frac{6595.96}{10804.4} = 0.6105.$$

Since this value is greater than $\Lambda_{.05}(4; 1, 9) = 0.249$, the hypothesis H_δ is accepted.

7.3 Multivariate Two-Way Fixed Effects Model

7.3.1 *Univariate case: One observation per cell*

We first consider the univariate case. This situation corresponds to that of randomized block design (RBD).

Let y_{ij} be the observation corresponding to the ith treatment (ith level of factor A) and jth block (j level of factor B) of a RBD. Then the linear model for this design is

$$y_{ij} = \mu + \alpha_i + \beta_j + \epsilon_{ij}, \ i = 1, \ldots, a; j = 1, \ldots, b \qquad (7.3.1)$$

where μ is an overall effect, α_i is the effect of the ith treatment, β_j the effect of the jth block and ϵ_{ij} are independently distributed $N(0, \sigma^2)$ variables. We impose the side conditions

$$\sum_{i=1}^{a} \alpha_i = 0, \ \sum_{j=1}^{b} \beta_j = 0. \qquad (7.3.2)$$

We call these assumptions as Ω. We define the means as $\bar{y}_{i0} = \sum_{j=1}^{b} y_{ij}/b$, $\bar{y}_{0j} = \sum_{i=1}^{a} y_{ij}/a$, $\bar{y}_{00} = \sum_{i=1}^{a}\sum_{j=1}^{b} y_{ij}/ab$. The least square estimates of the parameters obtained by minimizing the residual sum of squares

$$S = \sum_{i=1}^{a}\sum_{j=1}^{b}(y_{ij} - \mu - \alpha_i - \beta_j)^2$$

subject to Ω are

$$\hat{\mu} = \bar{y}_{00},\ \hat{\alpha}_i = \bar{y}_{i0} - \bar{y}_{00},\ \hat{\beta}_j = \bar{y}_{0j} - \bar{y}_{00}. \tag{7.3.3}$$

Substituting these values in (7.3.1), we have

$$y_{ij} = \bar{y}_{00} + (\bar{y}_{i0} - \bar{y}_{00}) + (\bar{y}_{0j} - \bar{y}_{00}) + (y_{ij} - \bar{y}_{i0} - \bar{y}_{0j} + \bar{y}_{00}), \tag{7.3.4}$$

where the last bracketed term in the right side is written so as to keep the two sides equal. Hence,

$$y_{ij} - \bar{y}_{00} = (\bar{y}_{i0} - \bar{y}_{00}) + (\bar{y}_{0j} - \bar{y}_{00}) + (y_{ij} - \bar{y}_{i0} - \bar{y}_{0j} + \bar{y}_{00}).$$

Squaring both sides and summing over all i and j, and noting that the cross-product terms vanish,

$$\begin{aligned}\textstyle\sum_i\sum_j(y_{ij} - \bar{y}_{00})^2 = b\sum_{i=1}^{a}(\bar{y}_{i0} - \bar{y}_{00})^2 + a\sum_j(\bar{y}_{0j} - \bar{y}_{00})^2 +\\ \textstyle\sum_i\sum_j(y_{ij} - \bar{y}_{i0} - \bar{y}_{0j} + \bar{y}_{00})^2,\end{aligned} \tag{7.3.5}$$

i.e.,

Total SS (TSS) = SS due to factor A (SSA) + SS due to factor B (SSB) + SS due to error (SSE). (7.3.6)

The partition of the total degrees of freedom into the corresponding groups are

$$ab - 1 = (a-1) + (b-1) + (a-1)(b-1). \tag{7.3.7}$$

The SSA carries $(a-1)$ d.f., because it consists of the set of linear functions $\{\bar{y}_{i0} - \bar{y}_{00}, i = 1, \ldots, a\}$ of which $(a-1)$ functions are linearly independent (as $\sum_{i=1}^{a}\bar{y}_{i0} = a\bar{y}_{00}$ and the coefficient vectors of y_{ij}'s in any two functions $\bar{y}_{i0} - \bar{y}_{00}$ and $\bar{y}_{i'0} - \bar{y}_{00}$ are mutually orthogonal). Similar results follow for SSB and SSE. Note that the linear functions in the sets $\{\bar{y}_{i0} - \bar{y}_{00}, i = 1, \ldots, a\}$, $\{\bar{y}_{0j} - \bar{y}_{00}, j = 1, \ldots, b\}$ and $\{y_{ij} - \bar{y}_{i0} - \bar{y}_{0j} + \bar{y}_{00}, i = 1, \ldots, a; j = 1, \ldots, b\}$ are mutually orthogonal. Hence, the degrees of freedom occurring in the sums of squares SSA, SSB, SSE are mutually orthogonal.

By dividing each SS by its d.f. we get the corresponding MS. It can be shown that

$$\begin{aligned}E(MSA) &= \sigma^2 + \tfrac{b}{a-1}\textstyle\sum_{i=1}^{a}\alpha_i^2\\ E(MSB) &= \sigma^2 + \tfrac{a}{b-1}\textstyle\sum_{j=1}^{b}\beta_j^2\\ E(MSE) &= \sigma^2.\end{aligned}$$

We are now interested in testing two null hypotheses:

$$H_{0A} : \alpha_1 = \alpha_2 = \ldots = \alpha_a = 0, \tag{7.3.8}$$

$$H_{0b} : \beta_1 = \beta_2 = \ldots = \beta_b = 0. \tag{7.3.9}$$

We call the assumptions in (7.3.8) as Ω_A. To test H_{0A} we minimize the residual SS S subject to conditions $\Omega \cap \Omega_A$. The value of SS so obtained is S_A^2 (say). The SS due to H_{0A} is then $S_A^2 - SSE$. This gives SS due to H_{0A} as

$$b \sum_{i=1}^{a} (\bar{y}_{i0} - \bar{y}_{00})^2 = SSA \tag{7.3.10}$$

which follows under H_{0A}, a $\sigma^2\chi^2_{(a-1)}$ distribution and is distributed independently of SSE, which again, follows a $\sigma^2\chi^2_{(a-1)(b-1)}$ distribution. Hence, under $H_{0A}, MSA/MSE \sim F_{(a-1,(a-1)(b-1))}$ which can be used for testing H_{0A}. The case of H_{0B} can be similarly dealt with.

Table 7.3.1: Univariate ANOVA Table for Two-Way Classified Data

(one observation per cell)

Sources	d.f.	SS	MS	F
Between the levels of A	$a-1$	$b\sum_{i=1}^{a}(\bar{y}_{i0} - \bar{y}_{00})^2 = SSA$	MSA	$\frac{MSA}{MSE}$
Between the levels of B	$b-1$	$a\sum_{j=1}^{b}(\bar{y}_{0j} - \bar{y}_{00})^2 = SSB$	MSB	$\frac{MSB}{MSE}$
Error	(a-1)(b-1)	$\sum_i \sum_j (y_{ij} - \bar{y}_{i0} - \bar{y}_{0j} + \bar{y}_{00})^2 = SSE$	MSE	-
Total	$ab-1$			

The diagonal elements of **H** are the between sum of squares for each of the p variables. The off-diagonal elements are the corresponding sum of products for each of the $\binom{p}{2}$ pairs of variables. The sth diagonal element of **H** is

$$\sum_{i=1}^{k} n_i(\bar{y}_{ijs} - \bar{y}_{00s})^2 = \sum_i \frac{y_{i0s}^2}{n_i} - \frac{y_{00s}^2}{n}, s = 1, \ldots, p.$$

The (s,t)

$$\sum_{i=1}^{k} (\bar{y}_{i0s} - \bar{y}_{00s})(\bar{y}_{i0t} - \bar{y}_{00t}) = \sum_i \frac{y_{i0s}y_{i0t}}{n_i} - \frac{y_{00s}y_{00t}}{n}, s \neq t = 1, \ldots, p.$$

Here $\bar{\mathbf{y}}_{i0} = (\bar{y}_{i01}, \ldots, \bar{y}_{i0p})', \bar{\mathbf{y}}_{00} = (\bar{y}_{001}, \ldots, \bar{y}_{00p})', y_{i0s} = \sum_j y_{ijs}, y_{00s} = \sum_i \sum_j y_{ijs}, \mathbf{y}_{ij} = (y_{ij1}, \ldots, y_{ijp})'$.

7.3.2 *Multivariate two-way fixed effects model (one observation per cell)*

A two-way fixed effects model for p dependent variables can be expressed by substituting vectors in (7.3.1):

$$\mathbf{y}_{ij} = \mu + \alpha_i + \beta_j + \mathbf{u}_{ij},\ i = 1, \ldots, a; j = 1, \ldots, b \tag{7.3.11}$$

where $\mathbf{y}_{ij}$ is the vector of observations on the unit receiving the ith level of factor A and the jth level of the factor B, μ is an overall effect, α_i is the effect of the ith level of A on each of the p variables, β_j is the effect of the jth level of B and $\mathbf{u}_{ij}$ are *iid* $N_p(\mathbf{0}, \mathbf{\Sigma})$ variables. As before, we assume that there is only one observation for each (i, j) level combination. We impose the side conditions

$$\sum_{i=1}^{a} \alpha_i = \mathbf{0},\ \sum_{j=1}^{b} \beta_j = \mathbf{0}. \tag{7.3.12}$$

Here, $\bar{\mathbf{y}}_{i0} = \sum_{j=1}^{b} \mathbf{y}_{ij}/b$, $\bar{y}_{0j} = \sum_{i=1}^{a} \mathbf{y}_{ij}/a$, $\bar{\mathbf{y}}_{00} = \sum_{i=1}^{a} \sum_{j=1}^{b} \mathbf{y}_{ij}/ab$. We have the multivariate analysis of variance (MANOVA) table as follows.

Table 7.3.2: MANOVA Table for Two-Way Classified Data
(one observation per cell)

Source	SSP Matrix	d.f.
Factor A	$\mathbf{H}_A = b\sum_{i=1}^{a}(\bar{\mathbf{y}}_{i0} - \bar{\mathbf{y}}_{00})(\bar{\mathbf{y}}_{i0} - \bar{\mathbf{y}}_{00})'$	$m_1 = a - 1$
Factor B	$\mathbf{H}_B = a\sum_{j=1}^{b}(\bar{\mathbf{y}}_{0j} - \bar{\mathbf{y}}_{00})(\bar{\mathbf{y}}_{0j} - \bar{\mathbf{y}}_{00})'$	$m_2 = b - 1$
Error	$\mathbf{E} = \sum_{i=1}^{a}\sum_{j=1}^{b}(\mathbf{y}_{ij} - \bar{\mathbf{y}}_{i0} - \bar{\mathbf{y}}_{0j} + \bar{\mathbf{y}}_{00})$ $(\mathbf{y}_{ij} - \bar{\mathbf{y}}_{i0} - \bar{\mathbf{y}}_{0j} + \bar{\mathbf{y}}_{00})'$	$m_E = (a-1)(b-1)$

For testing $H_{0A}(\alpha_1 = \ldots = \alpha_a)$ we use one of the four statistics, $\Lambda, \theta_1, T_g^2, V^{(s)}$. For example, the likelihood ratio test for this hypothesis is

$$\Lambda_A = \frac{|\mathbf{E}|}{|\mathbf{E} + \mathbf{H}_A|} \tag{7.3.13}$$

where $\Lambda_A \sim U_{p,m_A,m_E}$ when H_{0A} is true. Here, θ_A is the maximum eigenvalue of $(\mathbf{E} + \mathbf{H}_A)^{-1}\mathbf{H}_A$. Similar is the case for $H_{0B}(\beta_1 = \ldots = \beta_b)$.

The multiple comparison among contrasts can be made following the procedure in the case of one-way effects model in Section 7.2. Suppose we want to find simultaneous confidence intervals for $\sum_i c_i \alpha_i' \mathbf{b}_i$ where $\sum_i c_i = 0$. For this consider the derived model

$$\bar{\mathbf{y}}_{i0} = \mu + \alpha_i + \bar{\mathbf{u}}_{i0} \tag{7.3.14}$$

where $\bar{\mathbf{u}}_{i0} = \sum_{j=1}^{b} \mathbf{u}_{ij}/b$. Since, $\sum_i c_i \hat{\alpha}_i = \sum_i c_i(\bar{\mathbf{y}}_{i0} - \bar{\mathbf{y}}_{00}) = \sum_i c_i \bar{\mathbf{y}}_{i0}$, by (7.2.28),

$$1 - \alpha = \text{ Pr. } [\sum_i c_i \alpha_i' \mathbf{b} \in \sum_i c_i \bar{\mathbf{y}}_{i0}' \mathbf{b} \pm \sqrt{\Phi_\alpha \frac{\sum_i c_i^2}{b} \mathbf{b}'\mathbf{E}\mathbf{b}}] \tag{7.3.15}$$

where Φ_α is the quantile value of the maximum eigenvalue of $\mathbf{H}_1\mathbf{E}^{-1}$. Similarly, the probability is $1 - \alpha$ that

$$(\alpha_r - \alpha_s)'\mathbf{b} \in (\bar{\mathbf{y}}_{r0} - \bar{\mathbf{y}}_{s0})'\mathbf{b} \pm \sqrt{\Phi_\alpha \frac{2}{b} \mathbf{b}'\mathbf{E}\mathbf{b}}$$

simultaneously for all contrasts and all $\mathbf{b}$. Similar comparisons can be made among the contrasts of β_j's.

Bonferroni confidence intervals are also available for m prespecified comparisons. We replace $\mathbf{E}$ by $\mathbf{E}/m_E$ and $\sqrt{\Phi_\alpha}$ by $t_{m_E;\alpha/(2m)}$, with $m_E = (a-1)(b-1)$.

7.3.3 *Univariate case: r observations per cell*

We first review the univariate case.

We first consider some population parameters. Assume that the kth observation in the (i, j) cell, y_{ijk} is a random sample from the population of all such possible observations in that population with (true) mean η_{ij} and variance σ^2. Thus,

$$y_{ijk} = \eta_{ij} + \epsilon_{ijk}, \ \ i = 1, \ldots, a; j = 1, \ldots, b; k = 1, \ldots, r \tag{7.3.16}$$

where ϵ_{ijk} is a random error variable with

$$E(\epsilon_{ijk}) = 0, V(\epsilon_{ijk}) = \sigma^2.$$

In order to test different hypotheses we assume that ϵ_{ijk} are independent $N(0, \sigma^2)$ variables.

Writing $\bar{\eta}_{i0} = \sum_{j=1}^{b} \eta_{ij}/b$ and $\bar{\eta}_{0j}, \bar{\eta}_{00}$ similarly, we define μ, the overall population mean; α_i, the effect (on y_{ijk}) of the ith level of A; β_j, the effect of the jth level of B; γ_{ij}, the corresponding interaction effect, as follows:

$$\begin{aligned} \mu &= \bar{\eta}_{00} \\ \alpha_i &= \bar{\eta}_{i0} - \bar{\eta}_{00}, \\ \beta_j &= \bar{\eta}_{0j} - \bar{\eta}_{00}, \\ \gamma_{ij} &= \eta_{ij} - \bar{\eta}_{i0} - \bar{\eta}_{0j} + \bar{\eta}_{00}. \end{aligned} \tag{7.3.17}$$

It follows therefore, that

$$\sum_i \alpha_i = 0,\ \sum_j \beta_j = 0,\ \sum_i \gamma_{ij}(\forall\ j) = \sum_j \gamma_{ij}(\forall\ i) = 0.$$

Thus

$$\eta_{ij} = \mu + \alpha_i + \beta_j + \gamma_{ij}. \tag{7.3.18}$$

These conditions together with the model (7.3.16) uniquely determine $\mu, \alpha_i, \beta_j, \gamma_{ij}$.

Before considering the case of an equal number of observations per cell we consider the case where some cell frequencies are empty. Let n_{ij} be the number of observations in the cell (i, j) and D be the set of pairs (i, j) which label the non-empty cells. The assumptions are

$$y_{ijk} = \eta_{ij} + \epsilon_{ijk},\quad k = 1, \ldots, n_{ij}; (i, j) \in D,$$

$$E(\epsilon_{ijk}) = 0,\ V(\epsilon_{ijk}) = \sigma^2. \tag{7.3.19}$$

The number of unknown parameters η_{ij} in the model (7.3.19), apart from σ^2, is the number of non-empty cell in D, equal to ν (say) and their least square estimates are obtained by minimizing

$$\sum_{(i,j)\in D} \sum_{k=1}^{n_{ij}} (y_{ijk} - \eta_{ij})^2 \tag{7.3.20}$$

with respect to η_{ij}'s. It can be easily checked from the theory of linear estimation that writing the model in the form

$$E(\mathbf{Y}) = \mathbf{X}\beta$$

where β is the vector of unknown parameters η_{ij}, rank $(\mathbf{X}) = \nu$ so that this is a case of full-rank model. The least square estimates are

$$\hat{\eta}_{ij} = \bar{y}_{ij0},\ (i, j) \in D, \tag{7.3.21}$$

where

$$\bar{y}_{ij0} = \frac{1}{n_{ij}} \sum_{k=1}^{n_{ij}} y_{ijk}.$$

Since, this is a case of full rank model, all linear functions of η_{ij} corresponding to the non-empty cells are estimable. However, if there is a single empty cell, then the general mean, main effects and interaction effects are not estimable, since all these quantities involve in their definitions the η_{ij}'s of the empty cell which do not enter the observations.

From now on, we shall consider the balanced case, $\eta_{ij} = r(> 1)\ \forall\ (i,j)$. The model is, therefore,

$$y_{ijk} = \mu + \alpha_i + \beta_j + \gamma_{ij} + \epsilon_{ijk} \tag{7.3.22}$$

where the parameters satisfy the side-conditions (7.3.17).The least-square estimates obtained by minimizing

$$\sum_i \sum_j \sum_k (y_{ijk} - \mu - \alpha_i - \beta_j - \gamma_{ij})^2$$

are

$$\hat{\mu} = \bar{y}_{000},\ \hat{\alpha}_i = \bar{y}_{i00} - \bar{y}_{000},\ \hat{\beta}_j = \bar{y}_{0j0} - \bar{y}_{000},$$

$$\hat{\gamma}_{ij} = \bar{y}_{ij0} - \bar{y}_{i00} - \bar{y}_{0j0} + \bar{y}_{000}, \tag{7.3.23}$$

where

$$\bar{y}_{i00} = \sum_j \sum_k y_{ijk}/(br),\ \bar{y}_{ojo} = \sum_i \sum_k y_{ijk}/(ar),$$
$$\bar{y}_{000} = \sum_i \sum_j \sum_k y_{ijk}/(abr).$$

Substituting these values in (7.3.22) we have

$$y_{ijk} - \bar{y}_{000} = (\bar{y}_{i00} - \bar{y}_{000}) + (\bar{y}_{0j0} - \bar{y}_{000}) +$$

$$(\bar{y}_{ij0} - \bar{y}_{i00} - \bar{y}_{0j0} + \bar{y}_{000}) + (y_{ijk} - \bar{y}_{ij0}).$$

Squaring both sides and summing over i, j, k,

$$\begin{aligned}\sum_i \sum_j \sum_k (y_{ijk} - \bar{y}_{000})^2 &= br \sum_{i=1}^a (\bar{y}_{i00} - \bar{y}_{000})^2 + ar \sum_{j=1}^b (\bar{y}_{0j0} - \bar{y}_{000})^2 \\ &+ r \sum_i \sum_j (y_{ijk} - \bar{y}_{i00} - \bar{y}_{0j0} + \bar{y}_{000})^2 \\ &+ \sum_i \sum_j \sum_k (y_{ijk} - \bar{y}_{ij0})^2,\end{aligned} \tag{7.3.24}$$

i.e.,

$$\text{Total SS} = \text{SSA} + \text{SSB} + \text{SSAB} + \text{SSE}.$$

The corresponding partitioning of d.f. is

$$abr - 1 = a - 1) + (b - 1) + (a - 1)(b - 1) + ab(r - 1).$$

The hypotheses of interest are

$$H_A(\alpha_1 = \ldots = \alpha_a = 0)$$

$$H_B(\beta_1 = \ldots = \beta_b = 0)$$

$$H_{AB}(\gamma_{ij} = 0\ \forall\ i, j). \tag{7.3.25}$$

The hypothesis H_{AB} relates to the independence of factors A and B. The appropriate tests are suggested by considering the following expected values of mean squares.

$$\begin{aligned} E(MSA) &= \sigma^2 + \tfrac{br}{a-1}\textstyle\sum_i \alpha_i^2 \\ E(MSB) &= \sigma^2 + \tfrac{ar}{b-1}\textstyle\sum_j \beta_j^2 \\ E(MS(AB)) &= \sigma^2 + \tfrac{r}{(a-1)(b-1)}\textstyle\sum_i\sum_j \gamma_{ij}^2 \\ E(MSE) &= \sigma^2. \end{aligned}$$

If H_{AB} is true, $E(MS(AB)) = \sigma^2$ and hence, $MS(AB)/MSE$ gives the test for H_{03}. Under H_{03}, the ratio follows an F distribution with $(a-1)(b-1)$ and $ab(r-1)$ degrees of freedom. If the observed value $F > F_{(a-1)(b-1),ab(r-1);\alpha}$, the hypothesis is rejected at $100\alpha\%$ level of significance.

If H_{03} is rejected, then the tests for H_A and H_B need not be done, because in that case there is significant interaction effects between A and B and hence the α_i's must differ among themselves significantly and so the β_j's. If H_{AB} is rejected, but H_A, H_B are accepted, the correct conclusion is that there are significant differences among the cell means, but when the effects of levels of one factor are averaged over the levels of the other, no appropriate differences of these average effects are demonstrated.

Thus, when significant interaction exists and H_A is accepted, the levels of A will differ significantly in the presence of a particular level of factor B. This can be investigated by considering one-way classified data with a fixed level of B , but all levels of A and testing whether levels of A differ significantly. Such test should be done for each level of B. Similarly, if significant interaction exists and H_B is accepted, one may test for equality of effects of various levels of factor B for each fixed level of factor A by considering a number of one-way classified data.

If H_{AB} is accepted, i.e., if additivity holds, then rejection (acceptance) of H_A implies that the effects α_i differ (do not differ) significantly, irrespective of the levels of B. In this case, if $\alpha_1 - \alpha_2$ is significantly positive, it means that the effects of A_1 is significantly higher than that of A_2, irrespective of levels of B.

If the overall F test for the hypothesis $H_0(\alpha_1 = \ldots = \alpha_a = 0)$ is rejected one may choose to test hypotheses regarding contrasts. Such contrasts should be chosen prior to seeing the data. A contrast among the main effects of A is given by $\sum_i c_i\alpha_i$ with $\sum_i c_i = 0$. This is estimated by $\sum_i c_i\bar{y}_{i0}$ with variance $\sigma^2 \sum_{i=1}^{a} c_i^2/br$. To test the hypothesis $H_0 : \sum_i c_i\alpha_i = 0$ we can use the statistic

$$F = \frac{br(\sum_i c_i \bar{y}_{i0})^2 / \sum_i c_i^2}{MSE} \tag{7.3.26}$$

which is distributed as F with 1 and $ab(r-1)$ d.f. Contrasts $\sum_j d_j \beta_j$ with $\sum_j d_j = 0$ may be tested in an analogous manner.

Table 7.3.3: Univariate ANOVA Table for Two-Way Classified Data
(r observations per cell)

Sources of Variation	d.f.	SS	MS	F
Between the levels of A	$a-1$	$br\sum_i(\bar{y}_{i00} - \bar{y}_{000})^2$ $= \text{SSA}$	MSA	$\frac{MSA}{MSE}$
Between the levels of B	$b-1$	$ar\sum_j(\bar{y}_{0j0} - \bar{y}_{000})^2$ $= \text{SSB}$	MSB	$\frac{MSB}{MSE}$
Interaction $(A \times B)$	$(a-1)(b-1)$	$r\sum_i\sum_j(\bar{y}_{ij0} - \bar{y}_{000}$ $-\bar{y}_{0j0} + \bar{y}_{000})^2$ $= SS(AB)$	MS(AB)	$\frac{MS(AB)}{MSE}$
Error	$ab(r-1)$	$\sum_i\sum_j\sum_k(y_{ijk} - \bar{y}_{ij0})^2$ $= SSE$	MSE	-
Total	$abr-1$	$\sum_i\sum_j\sum_k(y_{ijk} - \bar{y}_{000})^2$	-	-

7.3.4 *Multivariate two-way fixed effects model (r observations per cell)*

We now come to the multivariate case. A balanced two-way fixed effects model for p dependent variables can be expressed by substituting vectors in (7.3.14) and (7.3.18):

$$\begin{aligned} \mathbf{y}_{ijk} &= \mu + \alpha_i + \beta_j + \gamma_{ij} + \mathbf{u}_{ijk} \\ &= \eta_{ijk} + \mathbf{u}_{ijk} \\ &\quad i = 1, \ldots, a; j = 1, \ldots, b; k = 1, \ldots, r \end{aligned} \tag{7.3.27}$$

where $\alpha_i, \beta_j, \gamma_{ij}$ have the usual meaning. The random variables ϵ_{ijk} are independently distributed $N_p(\mathbf{0}, \mathbf{\Sigma})$ variables. We also assume side conditions

$$\sum_i \alpha_i = \sum_j \beta_j = \sum_i \gamma_{ij}\ (\forall\ i) = \sum_j \gamma_{ij}\ (\forall\ i) = \mathbf{0}. \tag{7.3.28}$$

Under these conditions,

$$\alpha_i = \bar{\eta}_{i0} - \bar{\eta}_{00}, \beta_j = \bar{\eta}_{0j} - \bar{\eta}_{00},$$

$$\gamma_{ij} = \eta_{ij} - \bar{\eta}_{i0} - \bar{\eta}_{0j} + \bar{\eta}_{00} \tag{7.3.29}$$

where $\bar{\mu}_{i0} = \sum_j \mu_{ij}/b$ and all other symbols have usual meanings. The least square estimate of α_i is $\bar{\mathbf{y}}_{i00} - \bar{\mathbf{y}}_{000}$where $\bar{\mathbf{y}}_{i00} = \sum_j \sum_k \mathbf{y}_{ijk}/(br), \bar{\mathbf{y}}_{000} = \sum_i \sum_j \sum_k \mathbf{y}_{ijk}/(abr)$ and similar results hold for β_j, γ_{ij}.

Table 7.3.4: Multivariate Table for Two-Way Classified Data

Source	SSP Matrix	d.f.
A	$\mathbf{H}_A = br \sum_i (\bar{\mathbf{y}}_{i00} - \bar{\mathbf{y}}_{000})(\bar{\mathbf{y}}_{i00} - \bar{\mathbf{y}}_{000})'$	$a-1$
B	$\mathbf{H}_B = ar \sum_j (\bar{\mathbf{y}}_{0j0} - \bar{\mathbf{y}}_{000})(\bar{\mathbf{y}}_{0j0} - \bar{\mathbf{y}}_{000})'$	$b-1$
AB	$\mathbf{H}_{AB} = r \sum_i \sum_j (\bar{\mathbf{y}}_{ij0} - \bar{\mathbf{y}}_{i00} - \bar{\mathbf{y}}_{0j0} + \bar{\mathbf{y}}_{000})$ $(\bar{\mathbf{y}}_{ij0} - \bar{\mathbf{y}}_{i00} - \bar{\mathbf{y}}_{0j0} + \bar{\mathbf{y}}_{000})'$	$(a-1)(b-1)$
Error	$\mathbf{E} = \sum_i \sum_j \sum_k (\mathbf{y}_{ijk} - \bar{\mathbf{y}}_{ij0})(\mathbf{y}_{ijk} - \bar{\mathbf{y}}_{ij0})'$	$ab(r-1)$
Total	$\mathbf{T} = \sum_i \sum_j \sum_k (\mathbf{y}_{ijk} - \bar{\mathbf{y}}_{000})(\mathbf{y}_{ijk} - \bar{\mathbf{y}}_{000})'$	$abr-1$

The sth diagonal element of $\mathbf{H}_A$ is

$$br \sum_{i=1}^{a} (\bar{y}_{i00s} - \bar{y}_{000s})^2 = \sum_{i=1}^{a} \frac{y_{i00s}^2}{br} - \frac{y_{000s}^2}{abr}.$$

The (s,t)th element of $\mathbf{H}_A$ is

$$br \sum_{i=1}^{a} (\bar{y}_{i00s} - \bar{y}_{000s})(\bar{y}_{i00t} - \bar{y}_{000t})$$

$$= \sum_{i=1}^{a} \frac{y_{i00s} y_{i00t}}{br} - \frac{y_{000s} y_{000t}}{abr}.$$

Here $\bar{\mathbf{y}}_{i00} = (\bar{y}_{i001}, \ldots, \bar{y}_{i00p})', \bar{\mathbf{y}}_{000} = (\bar{y}_{0001}, \ldots, \bar{y}_{000p})', y_{i00s} = \sum_j \sum_k y_{ijks}/(br), y_{000s} = \sum_i \sum_j \sum_k y_{ijks}/(abr), \mathbf{y}_{ijk} = (y_{ijk1}, \ldots, y_{ijkp})'$. Similar results hold for all the SSP matrices.

The interaction and the main effects can be tested by using any of the four statistics - Wilk's Λ, Roy's largest root statistics, Lawley-Hotelling statistics, and Pillai's statistics. Thus for Wilk's Λ, we use

$$\Lambda_A = \frac{|\mathbf{E}|}{|\mathbf{E} + \mathbf{H}_A|} \cap \Lambda(p; a-1, ab(r-1)) \text{ under } H_A, \text{ for testing } H_A;$$

$$\Lambda_B = \frac{|\mathbf{E}|}{|\mathbf{E} + \mathbf{H}_a|} \cap \Lambda(p; b-1, ab(r-1)) \text{ under } H_B, \text{ for testing } H_B;$$

$$\Lambda_{AB} = \frac{|\mathbf{e}|}{|\mathbf{E} + \mathbf{H}_B|} \cap \Lambda(p; (a-1)(b-1), ab(r-1)) \text{ under } H_{AB}, \text{ for testing } H_{AB}.$$

We test the interaction hypothesis H_{AB} first One must be careful in interpretation while testing the main effects and the interaction and the comments made in case of univariate models in such situations in Subsection 7.3.3 also apply in this case.

A contrast among the levels of factor A can be defined as $\sum_{i=1}^{a} c_i \bar{\eta}_{i00}$ where $\sum_{i=1}^{a} c_i = 0$. Similarly, a contrast among the levels of factor B can be defined as $\sum_{j=1}^{b} d_j \bar{\eta}_{0j}$ with $\sum_{j=1}^{b} d_j = 0$. The hypothesis $H_\delta : \sum_{i=1}^{a} c_i \bar{\mathbf{eta}}_{i0} = \mathbf{0}$ can be tested by using T^2 or any of the four MANOVA test statistics, as in (7.2.36) - (7.2.38). Here

$$T^2 = \frac{br}{\sum_{i=1}^{a} c_i^2} (\sum_{i=1}^{a} c_i \bar{\mathbf{y}}_{i00})' (\frac{\mathbf{E}}{m_E})^{-1} (\sum_{i=1}^{a} c_i \bar{\mathbf{y}}_{i00}) \tag{7.3.30}$$

is distributed as $T^2(p, m_E)$. Alternatively, H_δ can be tested using Wilk's Λ. The SSP matrix due to H_δ is

$$\mathbf{H}_\delta = \frac{br}{\sum_{i=1}^{a} c_i^2} (\sum_{i=1}^{a} c_i \bar{\mathbf{y}}_{i00}) (\sum_{i=1}^{a} c_i \bar{\mathbf{y}}_{i00})' \tag{7.3.31}$$

with d.f. 1. Hence the test statistic is

$$\Lambda_\delta = \frac{|\mathbf{E}|}{|\mathbf{E} + \mathbf{H}_\delta|} \tag{7.3.32}$$

which is distributed as $\Lambda(p; 1, ab(r-1))$.

If the multivariate hypothesis H_A is rejected, one should make univariate test for each of the hypotheses $H_{As}(\alpha_{1s} = \alpha_{2s} = \ldots = \alpha_{ps}), s = 1, \ldots, p$. The hypothesis H_{As} can be tested by using an ordinary F test. Similar action can be taken with respect to H_B.

The above procedures are expected to be reasonably robust to small departures from the assumption of equal dispersion matrices.

EXAMPLE 7.3.1: Table 7.3.5 gives fictitious data for three variables y_1, y_2, y_3 on each of three units for two levels of factor A and three levels of factor B. Perform analysis of variance.

Table 7.3.5: Data for a Two-Way Table

	A_1B_1			A_1B_2			A_1B_3	
y_1	y_2	y_3	y_1	y_2	y_3	y_1	y_2	y_3
23.2	37.5	2.5	28.5	34.7	2.1	28.7	36.5	1.9
25.4	33.4	1.8	32.8	41.7	2.2	29.7	35.8	2.0
28.6	35.9	2.7	30.5	39.4	2.3	30.2	34.3	2.2

	A_2B_1			A_2B_2			A_2B_3	
y_1	y_2	y_3	y_1	y_2	y_3	y_1	y_2	y_3
27.5	29.3	1.5	30.2	34.6	2.8	26.7	30.4	3.1
22.4	28.4	1.9	28.9	33.5	2.6	29.7	35.6	3.2
25.6	29.8	2.1	31.2	36.5	2.9	28.4	38.2	3.4

Here $a = 2, b = 3, r = 3, m_E = ab(r-1) = 12$.

$$\mathbf{H}_A = \begin{bmatrix} 2.722 & 12.794 & -1.478 \\ 12.794 & 60.134 & -6.946 \\ -1.478 & -6.946 & .802 \end{bmatrix},$$

$$\mathbf{H}_B = \begin{bmatrix} 76.030 & 66.245 & 7.280 \\ 66.245 & 58.050 & 6.025 \\ 7.280 & 6.025 & 0.970 \end{bmatrix},$$

$$\mathbf{H}_{AB} = \begin{bmatrix} 0.541 & -3.006 & -0.802 \\ -3.006 & 23.814 & 7.157 \\ -0.802 & 7.157 & 2.214 \end{bmatrix},$$

$$\mathbf{E} = \begin{bmatrix} 45.647 & 24.737 & 0.910 \\ 24.737 & 73.687 & 3.363 \\ 0.910 & 3.363 & 0.793 \end{bmatrix}.$$

The eigenvalues of $\mathbf{E}$ are 88.230, 31.260 and 0.637. Hence $|\mathbf{E}| = 1756.82$.

First we consider H_{AB}. Here $s = 2, \nu_{1AB} = 0, \nu_{2AB} = 4$ (notation of Section 4.5).

$$\mathbf{E}^{-1/2}\mathbf{H}_{AB}\mathbf{E}^{-1/2} = \begin{bmatrix} 0.06490 & -0.13998 & -0.36474 \\ -0.13998 & 0.31626 & 0.87201 \\ -0.36474 & 0.87201 & 2.55595 \end{bmatrix}.$$

Its eigenvalues are $\phi_{1AB} = 2.90901, \phi_{2AB} = .02811, \phi_{3AB} = 0$. Hence,

$$\text{Wilk's Statistic } \Lambda_{AB} = \frac{|\mathbf{E}|}{|\mathbf{E} + \mathbf{H}_{AB}|} = \frac{1756.82}{7057.32}$$
$$= .24894 < \Lambda_{.05}(3; 2, 12) = .316;$$

$$\text{Lawley-Hotelling Statistic } = U^{(2)} = \phi_1 + \phi_2 = 2.93712.$$

$$\tilde{T}_g^2 = \frac{m_E U^{(2)}}{m_H} = 17.62272 >$$

$$\tilde{T}^2_{g;.05}(p; m_H, m_E) = \tilde{T}^2_{g;.05}(m_h; p, m_E + m_H - p) = \tilde{T}^2_{g;.05}(2; 3, 11) = 6.8763;$$

$$\text{Pillai's Statistic } V^{(2)} = \theta_1 + \theta_2 = .77152 < .782 = V^{(2)}_{.05}(2; 0, 4);$$

$$\text{Roy's Statistic} = \theta_1 = .74418 > .565.$$

All the tests (except Pillai's) reject the null hypothesis H_{AB} at 5% level of significance. Pillai's statistic is on the border line and almost rejects the hypothesis. The hypothesis of additivity of effects is not tenable.

We however test the hypotheses H_A, H_B. Here

$$\mathbf{E}^{-1/2}\mathbf{H}_A\mathbf{E}^{-1/2} = \begin{bmatrix} 0.00020 & 0.01384 & -0.02157 \\ 0.01384 & 0.96720 & -1.50487 \\ -0.02157 & -1.50487 & 2.34100 \end{bmatrix}.$$

Its eigenvalues are 3.3085, 0.000, 0.000. Also, $s_A = \min(p, m_A) = 1, \nu_{1A} = 0.5, \nu_{2A} = 4$. Hence,

$$\text{Wilk's } \Lambda_A = \frac{|\mathbf{E}|}{|\mathbf{E} + \mathbf{H}_A|} = \frac{1756.82}{7563.74} = 0.2323 < \Lambda_{.05}(3; 1, 12) = 0.473;$$

$$\text{Lawley-Hotelling Statistic }, U^{(1)} = \phi_1 = 3.3085; \frac{m_E}{m_A} U^{(1)} = 39.702 >$$

$$\tilde{T}^2_{g;.05}(p; m_A, m_E) = \tilde{T}^2_{g;.05}(m_A; p, m_A + m_E - p) = \tilde{T}^2_{g;.05}(1; 3, 10) = 7.2243;$$

$$\text{Pillai's Statistic, } V^{(1)} = \theta_1 = \frac{\phi_1}{1 + \phi_1} = 0.7679 > V^{(1)}_{.05}(1, .5, 4) = .4905;$$

$$\text{Roy 's Statistic } = \theta_1 = 0.7679.$$

The hypothesis H_A is, therefore, rejected.

Similarly, for testing H_B, the eigenvalues of $\mathbf{E}^{-1/2}\mathbf{H}_B\mathbf{E}^{-1/2}$ are 2.3305, .38524, 0. Also eigenvalues of $\mathbf{E} + \mathbf{H}_A$ are 218.564, 35.612 and 1.041. Also $s_B = 2, \nu_{1B} = 0, \nu_{2B} = 4$. Hence,

$$\text{Wilk's Statistics } \Lambda_B = \frac{\mathbf{E}|}{|\mathbf{E} + \mathbf{H}_B|} = .216821 < \Lambda_{.05}(3; 2, 12) = .316;$$

$$\text{Lawley-Hotelling Statistic } U^{(2)} = 2.71574; \frac{m_E}{m_H} U^{(2)} = 16.29444 >$$

$$\tilde{T}^2_{g;.05}(2; 3, 11) = 6.87635;$$

$$\text{Pillai's Statistic } V^{(2)} = .97784 > .782 = V^{(2)}_{.05}(2; 0, 4);$$

$$\text{Roy's Statistic } \theta_1 = .69974 > .565 = \theta_{.05}(2; 0, 4).$$

All the tests reject H_B.

7.4 Analysis of Covariance

We first consider a special form of the general linear model.

7.4.1 *A univariate general linear model*

Consider the model

$$\begin{aligned} \mathbf{y} &= \mathbf{X}\beta + \mathbf{Z}\gamma + \epsilon \\ &= (\mathbf{X}\ \mathbf{Z}) \begin{bmatrix} \beta \\ \gamma \end{bmatrix} + \epsilon \end{aligned} \tag{7.4.1}$$

where $\mathbf{y}$ is $n \times 1, \mathbf{X} = ((x_{ij}))$ is $n \times q, \mathbf{Z} = ((z_{ij}))$ is $n \times t, \beta$ ia $q \times 1$, and γ is $t \times 1$. It is assumed that the columns of $\mathbf{Z}$ are linearly independent of the columns of $\mathbf{X}$ and $\epsilon \sim N_n(\mathbf{0}, \sigma^2\mathbf{I}_n)$. Generally, $\mathbf{X}$ is a design matrix, its elements are 0 or 1, $\mathbf{Z}$ is a matrix of values of auxiliary or concomitant variables, $z_1, \ldots, z_t$. In this case the model (7.4.1) is a univariate analysis of covariance model. As is well-known, the analysis of covariance is a technique for increasing the precision of an experiment (Mukhopadhyay, 2005, p.331). (When the concomitant variables are absent, this becomes a analysis of variance model.) We wish to analyze this model and test different hypotheses relating to β and γ. When the parameters are estimated using the covariance model (like (7.4.1)) we shall use the suffix $'C'$ and when the parameters are estimated using the analysis of variance model (like $\mathbf{y} = \mathbf{X}\beta + \epsilon$ we shall omit the suffix.

The least square estimate of β and γ are

$$\begin{bmatrix} \hat{\beta}_C \\ \hat{\gamma}_C \end{bmatrix} = \begin{bmatrix} \mathbf{X'X} & \mathbf{X'Z} \\ \mathbf{Z'Z} & \mathbf{Z'Z} \end{bmatrix}^{-1} \begin{bmatrix} \mathbf{X'Y} \\ \mathbf{Z'Y} \end{bmatrix}. \tag{7.4.2}$$

Now

$$\begin{bmatrix} \mathbf{X'X} & \mathbf{X'Z} \\ \mathbf{Z'X} & \mathbf{Z'Z} \end{bmatrix}^{-1} = \begin{bmatrix} (\mathbf{X'X})^{-1} + \mathbf{LML'} & -\mathbf{LM} \\ -\mathbf{ML'} & \mathbf{M} \end{bmatrix} \tag{7.4.3}$$

where

$$\mathbf{L} = (\mathbf{X'X})^{-1}\mathbf{X'Z},\ \mathbf{M} = (\mathbf{Z'PZ})^{-1},\ \mathbf{P} = \mathbf{I}_n - \mathbf{X}(\mathbf{X'X})^{-1}\mathbf{X'}$$

(Exercise 7.2). Hence,

$$\begin{aligned} \hat{\beta}_C &= \{(\mathbf{X'X})^{-1} + \mathbf{LML'}\}\mathbf{X'Y} = \mathbf{LMZ'Y} \\ \hat{\gamma}_C &= -\mathbf{ML'X'Y} + \mathbf{MZ'Y}. \end{aligned} \tag{7.4.4}$$

Now,

$$\begin{aligned}\hat{\gamma}_C &= -(\mathbf{Z}'\mathbf{P}\mathbf{Z})^{-1}\mathbf{Z}'\mathbf{Z}(\mathbf{Z}'\mathbf{X})^{-1}\mathbf{X}'\mathbf{Y} + (\mathbf{Z}'\mathbf{P}\mathbf{Z})^{-1}\mathbf{Z}'\mathbf{Y} \\ &= (\mathbf{Z}'\mathbf{P}\mathbf{Z})^{-1}\mathbf{Z}'[\mathbf{I}_n - \mathbf{X}(\mathbf{X}'\mathbf{X})^{-1}\mathbf{X}']\mathbf{Y} \\ &= (\mathbf{Z}'\mathbf{P}\mathbf{Z})^{-1}\mathbf{Z}'\mathbf{P}\mathbf{Y}.\end{aligned} \tag{7.4.5}$$

Also,

$$\begin{aligned}\hat{\beta}_C &= (\mathbf{X}'\mathbf{X})^{-1}\mathbf{X}'\mathbf{Y} + \mathbf{L}(\mathbf{M}\mathbf{L}'\mathbf{X}' - \mathbf{M}\mathbf{Z}')\mathbf{Y} \\ &= (\mathbf{X}'\mathbf{X})^{-1}\mathbf{X}'\mathbf{Y} - \mathbf{L}\hat{\gamma}_C \\ &= (\mathbf{X}'\mathbf{X})^{-1}\mathbf{X}'[\mathbf{Y} - \mathbf{Z}\hat{\gamma}_C].\end{aligned} \tag{7.4.6}$$

The residual sum of squares under the model (7.4.1) is

$$\begin{aligned}RSS_C &= \{(\mathbf{y} - \mathbf{Z}\hat{\gamma}_C) - \mathbf{X}\hat{\beta}_C\}'\{(\mathbf{y} - \mathbf{Z}\hat{\gamma}_C) - \mathbf{X}\hat{\beta}_C\} \\ &= (\mathbf{y} - \mathbf{Z}\hat{\gamma}_C)'(\mathbf{y} - \mathbf{Z}\hat{\gamma}_C) - 2(\mathbf{y} - \mathbf{Z}\hat{\gamma}_C)'\mathbf{X}\hat{\beta}_C + \hat{\beta}_C'\mathbf{X}'\mathbf{X}\hat{\beta}_C \\ &= (\mathbf{y} - \mathbf{Z}\hat{\gamma}_C)'\mathbf{P}(\mathbf{y} - \mathbf{Z}\hat{\gamma}_C) \\ &= \mathbf{y}'\mathbf{P}\mathbf{y} - \hat{\gamma}_C'\mathbf{Z}'\mathbf{P}\mathbf{y}.\end{aligned} \tag{7.4.7}$$

If we want to test $H_{0\gamma} : \gamma = \mathbf{0}$, i.e. the concomitant variables have no effect, we calculate the residual sum of squares under $H_{0\gamma}$,

$$\begin{aligned}E_{H_\gamma} &= (\mathbf{y} - \mathbf{X}\hat{\beta}_G)'(\mathbf{y} - \mathbf{X}\hat{\beta}_G) \\ &= \mathbf{y}'\mathbf{P}\mathbf{y},\end{aligned} \tag{7.4.8}$$

since $\hat{\beta}_C$ now reduces to $\hat{\beta}$, the least square estimate of β under the model $\mathbf{y} = \mathbf{X}\beta + \epsilon$ (vide Section 6.2). Hence the sum of squares due to the hypothesis $H_{0\gamma}$ is

$$H_\gamma = E_{H_\gamma} - RSS_C = \hat{\gamma}_C'\mathbf{Z}'\mathbf{P}\mathbf{y} \tag{7.4.9}$$

and carries t degrees of freedom. The appropriate test statistic is, therefore,

$$F_\gamma = \frac{(E_{H_\gamma} - RSS_C)/t}{RSS_C/(n-q-t)} = \frac{\hat{\gamma}_C'\mathbf{Z}'\mathbf{P}\mathbf{Y}}{ts^2} \quad \text{(say)} \tag{7.4.10}$$

which follows $F_{(t,n-q-t)}$ under $H0\gamma$.

Suppose we want to test $H_{0\beta} : \mathbf{A}\beta = \mathbf{0}$ where $\mathbf{A}$ is a $g \times q$ matrix of rank g. We can write the model (7.4.1) as

$$\mathbf{y} = \mathbf{W}\mathbf{\Phi} + \epsilon \tag{7.4.11}$$

where

$$\mathbf{W} = [\mathbf{X},\ \mathbf{Z}],\ \mathbf{\Phi} = \begin{bmatrix}\beta \\ \gamma\end{bmatrix},$$

$\mathbf{W}$ is a $n \times (q+t)$ matrix of rank $(q+t)$. Now,

$$\mathbf{A}\beta = [\mathbf{A}_{g\times q},\ \mathbf{0}_{g\times t}]\begin{bmatrix}\beta \\ \gamma\end{bmatrix} = \mathbf{A}_1\mathbf{\Phi} \quad \text{(say)}\ . \tag{7.4.12}$$

Hence, by (6.6.22), the sum of squares due to hypothesis $\mathbf{A}_1\mathbf{\Phi} = \mathbf{0}$ is

$$\begin{aligned}&(\mathbf{A}_1\hat{\mathbf{\Phi}})'[\mathbf{A}_1(\mathbf{W}'\mathbf{W})^{-1}\mathbf{A}_1']^{-1}\mathbf{A}_1\hat{\mathbf{\Phi}} \\ &= (\mathbf{A}\hat{\beta}_C)'[\mathbf{A}(\mathbf{X}'\mathbf{X})^{-1}\mathbf{A}' + \mathbf{ALML}'\mathbf{A}']^{-1}\mathbf{A}\hat{\beta}_C \\ &= H_\beta \quad \text{(say)}.\end{aligned} \tag{7.4.13}$$

(Exercise 7.3). The F-statistic for testing $H_{0\beta}$ is, therefore,

$$\frac{H_\beta}{gs^2} \tag{7.4.14}$$

which follows $F_{g,n-q-t}$ under $H_{0\beta}$.

If $\mathbf{X}$ has less than full rank, we replace q by rank $(\mathbf{X})$ and $(\mathbf{X}'\mathbf{X})^{-1}$ by a generalized inverse $(\mathbf{X}'\mathbf{X})^{-}$.

Note that

$$\text{Cov }(\hat{\gamma}_C, \hat{\beta}) = \mathbf{0} \tag{7.4.15}$$

(Exercise 7.4). Also,

$$\text{cov} \begin{bmatrix}\hat{\beta}_C \\ \hat{\gamma}_C\end{bmatrix} = \sigma^2(\mathbf{W}'\mathbf{W})^{-1} = \sigma^2\begin{bmatrix}(\mathbf{X}'\mathbf{X})^{-1} + \mathbf{LML}' & -\mathbf{LM} \\ -\mathbf{ML}' & \mathbf{M}\end{bmatrix}. \tag{7.4.16}$$

7.4.2 *Univariate analysis of covariance: Two-way model with one covariate*

Consider a randomized block design (RBD) with b blocks and v treatments and suppose that observations on an auxiliary variate $'z'$, closely related to the main variable $'y'$ are available along with those of y on each unit. Consider the model

$$y_{ij} = \mu + \beta_i + \tau_j + \gamma z_{ij} + \epsilon_{ij} \; i = 1, \ldots, b; \; j = 1, \ldots, v, \tag{7.4.17}$$

where $y_{ij}(z_{ij})$ is the yield of treatment j in block i (the corresponding value of z), β_i the block-effect, τ_j the treatment effect, γ is the unknown regression coefficient of y on x and ϵ_{ij}'s are random error components. It is assumed that ϵ_{ij} are *iid* $N(0, \sigma^2)$.

The least square estimates of the parameters $\mu, \beta_i, \tau_j, \gamma$ are obtained by minimizing the error sum of squares

$$\sum_{i=1}^{b}\sum_{j=1}^{v}(y_{ij} - \mu - \beta_i - \tau_j - \gamma z_{ij})^2$$

with respect to $\mu, \beta_i, \tau_j, \gamma$. The estimates are

$$\begin{aligned}
\hat{\mu}_C &= \tfrac{[G-\hat{\gamma}_C G(z)]}{bv} \\
\hat{\beta}_{iC} &= \tfrac{[B_i-\hat{\gamma}_C B_i(z)]}{v} - \hat{\mu}_C, \\
\hat{\tau}_{jC} &= \tfrac{[T_j-\hat{\gamma}_C T_j(z)]}{b} - \hat{\mu}_C \\
\hat{\gamma}_C &= \tfrac{E_{yz}}{E_{zz}},
\end{aligned} \tag{7.4.18}$$

where

$$B_i = \sum_{j=1}^{v} y_{ij}, \quad B_i(z) = \sum_{j=1}^{v} z_{ij},$$

$$T_j = \sum_{i=1}^{b} z_{ij}, \quad T_j(z) = \sum_{i=1}^{b} z_{ij},$$

$$G = \sum_{i=1}^{b} B_i = \sum_{j=1}^{v} T_j, \; G(z) = \sum_{i=1}^{b} B_i(z) = \sum_{j=1}^{v} T_j(z).$$

Now, the total sum of squares due to variable $z = SS_Z = \sum_{i=1}^{a}\sum_{j=1}^{b}(z_{ij} - \bar{z}_{00})^2 = \sum_i\sum_j z_{ij}^2 - \frac{G(z)^2}{bv}$.

The sum of squares due to blocks with respect to variable $z = B_{zz} = v\sum_{i=1}^{b}(\bar{z}_{i0} - \bar{z}_{00})^2 = \sum_{i=1}^{a} \frac{B_i^2(z)}{v} - \frac{G(z)^2}{bv}$.

The sum of squares due to treatments with respect to variable $z = T_{zz} = b\sum_{j=1}^{v}(\bar{z}_{0j} - \bar{z}_{00})^2 = \sum_{j=1}^{b} \frac{T_j^2}{b} - \frac{G(z)^2}{bv}$.

The sum of squares due to error with respect to variable $z = E_{zz} = \sum_{i=1}^{a}\sum_{j=1}^{b}(z_{ij} - \bar{z}_{i0} - \bar{z}_{0j} + \bar{z}_{00})^2 = SS_z - B_{zz} - T_{zz}$. In the matrix notation $E_{zz} = \mathbf{z}'\mathbf{P}\mathbf{z}$.

The sums of squares due to variable $y, SS_y, B_{yy}, T_{yy}, E_{yy}$ can be defined similarly.

The total sum of products due to y and $z = SP(y,z) = \sum_{i=1}^{b}\sum_{j=1}^{v}(y_{ij} - \bar{y}_{00})(z_{ij} - \bar{z}_{00}) = \sum_{i=1}^{b}\sum_{j=1}^{v} y_{ij}z_{ij} - \frac{GG(z)}{bv}$.

The
sum of products due to blocks with respect to variables y and $z = B_{yz} = v\sum_{i=1}^{b}(\bar{y}_{i0} - \bar{y}_{00})(\bar{z}_{i0} - \bar{z}_{00}) = \sum_{i=1}^{b} \frac{B_iB_i(z)}{v} - \frac{GG(z)}{bv}$.

The sum of products due to treatments with respect to variables y and $z = T_{yz} = b\sum_{j=1}^{v}(\bar{y}_{0j} - \bar{y}_{00})(\bar{z}_{0j} - \bar{z}_{00}) = \sum_{j=1}^{v} \frac{T_jT_j(z)}{b} - \frac{GG(z)}{bv}$.

The sum of products due to error with respect to variables y and $z = E_{yz} = SP(y,z) - B_{yz} - T_{yz}$. In the matrix notation $E_{yz} = \mathbf{y}'\mathbf{P}\mathbf{z}$.

The error sum of squares under model (7.5.17) is

$$\begin{aligned} SSE &= \sum_i \sum_j (y_{ij} - \hat{\mu}_C - \hat{\beta}_{iC} - \sum_j \tau_{jC} - \hat{\gamma}_C z_{ij})^2 \\ &= \sum_i \sum_j y_{ij}^2 - \sum_i \tfrac{B_i^2}{v} - \sum_j \tfrac{T_j^2}{b} + \tfrac{G^2}{bv} - \hat{\gamma}_C E_{yz} \\ &= SS_y - B_{yy} - T_{yy} - \hat{\gamma}_C E_{yz} \\ &= E_{yy} - \hat{\gamma}_C E_{yz} = E \ \ \text{(say)} \ . \end{aligned} \tag{7.4.19}$$

The quantity E carries $(b-1)(v-1)-1$ d.f., the reduction of one d.f. over the one in a RBD without using the concomitant variable is due to estimation of the parameter γ.

To test the null hypothesis $H_{0\tau}(\tau_1 = \tau_2 = \ldots = \tau_v)$, we calculate the error sum of squares due to $H_{0\tau}$ after adjusting for the variation in z. This is given by

$$\sum_i \sum_j (y_{ij} - \tilde{\mu} - \tilde{\beta}_i - \tilde{\gamma} z_{ij})^2 = E_1 \ \ \text{(say)} \ , \tag{7.4.20}$$

where $\tilde{\mu}, \tilde{\beta}_i, \tilde{\gamma}$ are the least square estimates of the corresponding parameters under $H_{0\tau}$. It is seen that

$$\begin{aligned} E_1 &= \sum_i \sum_j y_{ij}^2 - \sum_i \tfrac{B_i^2}{v} - \tilde{\gamma} E'_{yz} \\ &= SS_y - B'_{yy} - \tilde{\gamma} E'_{yz} \end{aligned} \tag{7.4.21}$$

where

$$\begin{aligned} E'_{yy} &= E_{yy} + T_{yy} \\ E'_{yz} &= E_{yz} + T_{yz} \\ E'_{zz} &= E_{zz} + T_{zz} \\ \tilde{\gamma} &= \tfrac{E'_{yz}}{E'_{zz}} . \end{aligned} \tag{7.4.22}$$

The sum of squares due to $H_{0\tau}$ after adjusting for the variation in z is given by $E_1 - E$ and carries $v - 1$ d.f.

Table 7.4.1: Analysis of a Covariance Table for a RBD

Source	d.f	SS due to z	SP due to y, z	SS due to y
Blocks	$b-1$	B_{zz}	B_{yz}	B_{yy}
Treatments	$v-1$	T_{zz}	T_{yz}	T_{yy}
Error	$(b-1)(v-1)-1$	E_{zz}	E_{yz}	E_{yy}
Total	$bv-2$	SS_z	$SP(y,z)$	SS_z

Table 7.4.2: Table Showing Calculation of SS and SP under $H_{0\tau}$

Source	SS due to z	SP of y, z	SS due to y	Regression Coefficient	Adjusted SS	Adjusted d.f.
Treatment + Error	$T_{zz}+E_{zz} = E'_{zz}$	$T_{yz}+E_{yz} = E'_{yz}$	$T_{yy}+E_{yy}$ E'_{yy}	$\tilde{\gamma}=\frac{E'_{yz}}{E'_{zz}}$	$E'_{yy}-\tilde{\gamma}E'_{yz} = E_1$	$b(v-1)$ -1
Treatment	T_{zz}	T_{yz}	T_{yy}		E_1-E	$v-1$
Error	E_{zz}	E_{yz}	E_{yy}	$\hat{\gamma}=\frac{E_{yz}}{E_{zz}}$	$E_{yy}-\hat{\gamma}E_{yz} = E$	$(b-1)$ $(v-1)-1$

The mean square errors due to treatment is

$$M_T = \frac{(E_1 - E)}{(v-1)}$$

and that due to error is

$$M_E = \frac{E}{\{(b-1)(v-1)-1\}}.$$

The hypothesis $H_{0\tau}$ is tested by the ratio M_T/M_E which follows $F_{((v-1),(b-1)(v-1)-1)}$ under the null hypothesis.

From (7.4.18),

$$\begin{aligned}\hat{\tau}_{jC} &= \tfrac{T_j}{b} - \tfrac{G}{bv} - \hat{\gamma}_C(\tfrac{T_j(z)}{b} - \tfrac{G(z)}{bv}) \\ &= \bar{y}_{0j} - \bar{y}_{00} - \hat{\gamma}_C(\bar{z}_{0j} - \bar{z}_{00}).\end{aligned} \tag{7.4.23}$$

Hence an estimate of treatment contrast $\sum_j l_j\tau_j$ is

$$\sum_j l_j\hat{\tau}_{jC} = \sum_j l_j(\bar{y}_{0j} - \hat{\gamma}_C\bar{z}_{0j}). \tag{7.4.24}$$

Also,

$$V(\sum_j l_j\hat{\tau}_{jC}) = \sigma^2[\sum_j \frac{l_j^2}{b} + \frac{(\sum_j l_j\bar{z}_{0j})^2}{E_{zz}}], \tag{7.4.25}$$

because

$$V(\hat{\gamma}_C) = \frac{\sigma^2}{E_{zz}}.$$

Thus,

$$V(\hat{\tau}_{jC} - \hat{\tau}_{j'C}) = \sigma^2(\frac{2}{b} + \frac{(\bar{z}_{0j} - \bar{z}_{0j'})^2}{E_{zz}}). \tag{7.4.26}$$

An unbiased estimate of σ^2 is given by the error mean square M_E.

A disturbing feature is that the variance of the estimate of the difference between any pair of treatments is not a constant, unlike the design without using the concomitant variable. To make the critical difference a constant for each pair of treatments, sometimes $(\bar{z}_{0j}-\bar{z}_{0j'})^2$ is replaced by $T_{zz}/(v-1)$.

In the above example we have taken one observation per cell and hence the interaction term was absent in the model. In case there are $m(>1)$ observations per cell we can include an interaction effect in the model (7.4.17) and the analysis will follow similarly.

7.4.3 *Multivariate analysis of covariance (MANCOV)*

Consider the multivariate analysis of covariance model

$$\mathbf{Y} = \mathbf{XB} + \mathbf{Z\Gamma} + \mathbf{U} \tag{7.4.27}$$

where $\mathbf{Y}(n\times p)$ is a matrix of n observations on each of p response variables, $\mathbf{X}(n\times q)$ is a known matrix of rank q, $\mathbf{Z}(n\times t)$ is a matrix of n observations on each of q concomitant variables with rank t and the rows of $\mathbf{U}(n\times p)$ are *iid* $N_p(\mathbf{0},\mathbf{\Sigma})$.

The least square estimates are, following (7.4.5) and (7.4.6),

$$\begin{aligned}\hat{\mathbf{\Gamma}}_C &= (\mathbf{Z}'\mathbf{PZ})^{-1}\mathbf{Z}'\mathbf{PY}\\ \hat{\mathbf{B}}_C &= (\mathbf{X}'\mathbf{X})^{-1}\mathbf{X}'[\mathbf{Y}-\mathbf{Z}\hat{\mathbf{\Gamma}}_C] = \hat{\mathbf{B}} - (\mathbf{X}'\mathbf{X})^{-1}\mathbf{X}'\mathbf{Z}\hat{\mathbf{\Gamma}}_C\end{aligned} \tag{7.4.28}$$

Following (7.4.7), the residual SSP matrix for the model (7.4.27) is

$$\begin{aligned}\mathbf{RSSP}_C &= (\mathbf{Y}-\mathbf{X}\hat{\mathbf{B}}_C-\mathbf{Z}\hat{\mathbf{\Gamma}}_C)'(\mathbf{Y}-\mathbf{X}\hat{\mathbf{B}}_C-\mathbf{Z}\hat{\mathbf{\Gamma}}_C)\\ &= \mathbf{Y}'\mathbf{PY} - \hat{\mathbf{\Gamma}}_C'\mathbf{Z}'\mathbf{PY},\end{aligned} \tag{7.4.29}$$

carrying $(n-q-t)$ d.f.

If we are to test $H_{0\mathbf{\Gamma}}(\mathbf{\Gamma}=\mathbf{0})$, then the error SSP matrix under the null hypothesis is

$$\begin{aligned}\mathbf{RSSP}_{\mathbf{\Gamma}} &= (\mathbf{Y}-\mathbf{X}\hat{\mathbf{B}})'(\mathbf{Y}-\mathbf{X}\hat{\mathbf{B}})\\ &= \mathbf{Y}'\mathbf{PY},\end{aligned} \tag{7.4.30}$$

(because the least square estimate of $\mathbf{B}$ under model $Y=\mathbf{XB}+\mathbf{U}$ is $\hat{\mathbf{B}}$) carrying $n-q$ d.f. Therefore, SSP matrix due to $H_{0\mathbf{\Gamma}}$ is

$$\mathbf{H}_{\mathbf{\Gamma}} = \mathbf{RSSP}_{\mathbf{\Gamma}} - \mathbf{RSSP}_C = \hat{\mathbf{\Gamma}}_C'\mathbf{Z}'\mathbf{PY} \tag{7.4.31}$$

and it carries t d.f.

When $H_{0\mathbf{\Gamma}}$ is true, $\mathbf{RSSP}_C$ and $\mathbf{H}_{\mathbf{\Gamma}}$ are independently distributed as $W(p; n-q-t, \mathbf{\Sigma})$ and $W(p; t, \mathbf{\Sigma})$ respectively. One can test $H_{0\mathbf{\Gamma}}$ by using

$$\Lambda = \frac{|\mathbf{RSSP}_C|}{|\mathbf{RSSP}_C + \mathbf{H}_{\mathbf{\Gamma}}|} \tag{7.4.32}$$

which follows $\Lambda(p; t, n-q-t)$ distribution under the null hypothesis. Likewise, other three multivariate tests,- Roy's, Lawley-Hotelling and Pillai's - can be used.

If we want to test $H_{0\mathbf{B}}(\mathbf{AB} = \mathbf{0})$ where $\mathbf{A}$ is a $g \times q$ matrix of rank g, the SSP matrix due to $H_{0\mathbf{B}}$, following (7.5.11), is

$$\mathbf{H}_{0\mathbf{B}} = (\mathbf{A}\hat{\mathbf{B}}_C)'[\mathbf{A}(\mathbf{X}'\mathbf{X})^{-1}\mathbf{A}' + \mathbf{ALML}'\mathbf{A}']^{-1}\mathbf{A}\hat{\mathbf{B}}_C \tag{7.4.33}$$

and it carries g d.f. under null hypothesis, $\mathbf{H}_{0\mathbf{B}} \cap W(p; g, \mathbf{\Sigma})$. The hypothesis $H_{0\mathbf{B}}$ is tested by using

$$\begin{aligned} \Lambda &= \frac{|\mathbf{RSSP}_C|}{|\mathbf{RSSP}_C + \mathbf{H}_{0\mathbf{B}}|} \\ &= \frac{|\mathbf{Y}'\mathbf{PY} - \hat{\mathbf{\Gamma}}_C'\mathbf{Z}'\mathbf{PY}|}{|\{(\mathbf{Y} - \mathbf{Z}\tilde{\Gamma}) - \mathbf{X}\tilde{\mathbf{B}}_H\}'\{(\mathbf{Y} - \mathbf{Z}\tilde{\Gamma}) - \mathbf{X}\tilde{\mathbf{B}}_H\}|} \end{aligned} \tag{7.4.34}$$

where $\tilde{\mathbf{\Gamma}}, \tilde{\mathbf{B}}_H$, respectively, are the least square estimates of $\mathbf{\Gamma}$ and $\mathbf{B}$ under the model (7.5.27) subject to the hypothesis $H_{0\mathbf{B}}$ (vide (6.6.20)). Under null hypothesis Λ in (7.4.34) follows $\Lambda(p; g. n-q-t)$ distribution.

For testing $H'_{0\mathbf{B}}(\mathbf{ABD} = \mathbf{0})$ where $\mathbf{D}$ is a $p \times v$ matrix of rank v, the SSP matrix due to the null hypothesis, following equation (6.6.25), is

$$\tilde{\mathbf{H}}_{0\mathbf{B}} = (\mathbf{A}\hat{\mathbf{B}}_C\mathbf{D})'[\mathbf{A}(\mathbf{X}'\mathbf{X})^{-1}\mathbf{A}' + \mathbf{ALML}'\mathbf{A}']^{-1}(\mathbf{A}\hat{\mathbf{B}}_C\mathbf{D}) \tag{7.4.35}$$

which under null hypothesis follows $W(v; g, \mathbf{D}'\mathbf{\Sigma}\mathbf{D})$. Also, in this case, the SSP matrix due to error is (vide equation (6.6.26))

$$\tilde{\mathbf{E}} = \mathbf{D}'\mathbf{Y}'\mathbf{PYL} \tag{7.4.36}$$

which follows $W(v; n-q-t, \mathbf{D}'\mathbf{\Sigma}\mathbf{D})$ distribution. The hypothesis $H'_{0\mathbf{B}}$ can be tested by using the statistic

$$\Lambda = \frac{|\tilde{\mathbf{E}}|}{|\tilde{\mathbf{E}} + \tilde{\mathbf{H}}_{0\mathbf{B}}|} \tag{7.4.37}$$

which follows $\Lambda(v; g, n-q-t)$ distribution if the hypothesis is true.

Simultaneous confidence intervals for linear combinations $\alpha'\mathbf{\Delta\Phi f}$ where $\mathbf{\Delta}$ is a $g \times (p+t)$ matrix of rank g, α and $\mathbf{f}$ are constant vectors of order $g \times 1$ and $p \times 1$, respectively, can be constructed using the general theory of Subsection 6.8.1 applied to the model (7.4.11), suitably modified.

The analysis for testing the hypothesis $H_{0\mathbf{B}}(\mathbf{AB} = \mathbf{0})$ can be simplified through the following steps of calculations. We have seen that

$$\begin{aligned} \mathbf{E}_{YY} &= \mathbf{Y}'\mathbf{PY}, \\ \mathbf{E}_{ZZ} &= \mathbf{Z}'\mathbf{PZ}, \\ \mathbf{E}_{YZ} &= \mathbf{Y}'\mathbf{PZ}. \end{aligned} \tag{7.4.38}$$

The different steps are as follows:

(1) Find $\hat{\mathbf{B}} = (\mathbf{X}'\mathbf{X})^{-1}\mathbf{X}'\mathbf{Y}$ and the error SSP matrix $\mathbf{E}_{YY} = \mathbf{Y}'\mathbf{PY}$ from the MANOVA model $E(\mathbf{Y}) = \mathbf{XB}$.

(2) Find

$$\begin{aligned} \hat{\Gamma}_C &= (\mathbf{Z}'\mathbf{PZ})^{-1}\mathbf{Z}'\mathbf{PZ} \\ &= \mathbf{E}_{ZZ}^{-1}\mathbf{E}_{ZY}. \end{aligned} \tag{7.4.39}$$

(3) Calculate the SSP error matrix due to the analysis of covariance model (7.4.27),

$$\begin{aligned} \mathbf{E} &= \{(\mathbf{Y} - \mathbf{Z}\hat{\mathbf{\Gamma}}_C) - \mathbf{X}\hat{\mathbf{B}}_C\}'\{(\mathbf{Y} - \mathbf{Z}\hat{\mathbf{\Gamma}}_C) - \mathbf{X}\hat{\mathbf{B}}_C\} \\ &= (\mathbf{Y} - \mathbf{Z}\hat{\Gamma}_C)'\mathbf{P}(\mathbf{Y} - \mathbf{Z}\hat{\mathbf{\Gamma}}_C) \\ &= \mathbf{Y}'\mathbf{PY} - \hat{\mathbf{\Gamma}}_C'\mathbf{Z}'\mathbf{PY} \quad \text{(vide (7.4.29))} \\ &= \mathbf{E}_{YY} - \mathbf{E}_{YZ}\mathbf{E}_{ZZ}^{-1}\mathbf{E}_{ZY}. \end{aligned} \tag{7.4.40}$$

(4) Replacing $\mathbf{Y}$ by $\mathbf{Y} - \mathbf{Z}\hat{\mathbf{\Gamma}}_C$ in $\hat{\mathbf{B}}$ obtain

$$\hat{\mathbf{B}}_C = (\mathbf{X}'\mathbf{X})^{-1}\mathbf{X}'(\mathbf{Y} - \mathbf{Z}\hat{\Gamma}_C). \tag{7.4.41}$$

(5) We have seen that the idempotent matrix $\mathbf{P}$ occurring in $\mathbf{E}_{YY}$ is crucial for calculations. To obtain the error SSP matrix under the covariance model (7.4.27) subject to the hypothesis $H_{0\mathbf{B}}$ we first obtain the error SSP matrix under the model $E(\mathbf{Y}) = \mathbf{XB}$ subject to the hypothesis $H_{0\mathbf{B}}(\mathbf{AB} = \mathbf{0})$. This is

$$\begin{aligned} \mathbf{E}_{HYY} &= (\mathbf{Y} - \mathbf{X}\hat{\mathbf{B}}_H)'(\mathbf{Y} - \mathbf{X}\hat{\mathbf{B}}_H) \\ &= \mathbf{Y}'\mathbf{P}_H\mathbf{Y}, \end{aligned} \tag{7.4.42}$$

where $\hat{\mathbf{B}}_H$ is the least square of $\mathbf{B}$ under the MANOVA model subject to the condition $\mathbf{AB} = \mathbf{0}$. $\mathbf{E}_{HYY}$ is given by (6.6.20). Thus P_H is suitably defined.

(6) Find least square estimate of $\mathbf{\Gamma}$ under the model (7.4.27) subject to the null hypothesis $H_{0\mathbf{B}}$,

$$\tilde{\mathbf{\Gamma}} = (\mathbf{Z}'\mathbf{P}_H\mathbf{Z})^{-1}\mathbf{Z}'\mathbf{P}_H\mathbf{Z}.$$

(7) Obtain the SSP matrix due to error under the model (7.4.27) subject to the hypothesis H_{0B},

$$\begin{aligned}\mathbf{E}_H &= \mathbf{E}_{HYY} - \tilde{\mathbf{\Gamma}}'\mathbf{Z}'\mathbf{R}_H\mathbf{Y} \\ &= \mathbf{Y}'\mathbf{R}_H\mathbf{Y} - \tilde{\mathbf{\Gamma}}'_G\mathbf{Z}'\mathbf{R}_H\mathbf{Y} \\ &= \mathbf{Y}'\mathbf{R}_H\mathbf{Y} - \mathbf{Y}'\mathbf{R}_H\mathbf{Z}(\mathbf{Z}'\mathbf{R}_H\mathbf{Z})^{-1}\mathbf{Z}'\mathbf{R}_H\mathbf{Y} \\ &= \mathbf{E}_{HYY} - \mathbf{E}'_{HZY}(\mathbf{E}_{HZZ})^{-1}\mathbf{E}_{HZY} \\ &= \mathbf{E}_{HYY} - \mathbf{E}_{HYZ}(\mathbf{E}_{HZZ})^{-1}\mathbf{E}_{HZY}.\end{aligned} \tag{7.4.43}$$

Following $\mathbf{E}_{HYY}$ we shall write expressions for $\mathbf{E}_{HYZ}, \mathbf{E}_{HYZ}$.

We shall also write

$$\mathbf{E}_{HYY} = \mathbf{E}_{YY} + \mathbf{H}_{YY},$$

where $\mathbf{H}_{YY}$ is the SSP matrix due to the hypothesis $\mathbf{AB} = \mathbf{0}$ under the MANOVA model $E(\mathbf{Y}) = \mathbf{XB}$. Following $\mathbf{H}_{YY}$ we shall also write expressions for $\mathbf{H}_{ZZ}, \mathbf{H}_{YZ}$.

The hypothesis $H_{0\mathbf{B}}$ can, therefore, be tested by the Wilk's likelihood ratio,

$$\Lambda = \frac{|\mathbf{E}|}{|\mathbf{E}_H|} = \frac{|\mathbf{E}_{YY} - \mathbf{E}_{YZ}\mathbf{E}_{ZZ}^{-1}\mathbf{E}_{ZY}|}{|\mathbf{E}_{HYY} - \mathbf{E}_{HYZ}(\mathbf{E}_{HZZ})^{-1}\mathbf{E}_{HZY}|} \tag{7.4.44}$$

which follows $\Lambda(p; g, n-q-t)$ if the hypothesis is true. This test may be compared with the corresponding test for the null hypothesis $\mathbf{AB} = \mathbf{0}$ for the MANOVA model $E(\mathbf{Y}) = \mathbf{XB}$, which is

$$\Lambda = \frac{|\mathbf{E}_{YY}|}{|\mathbf{E}_{HYY}|} = \frac{|\mathbf{E}_{YY}|}{|\mathbf{E}_{YY} + \mathbf{H}_{YY}|}, \tag{7.4.45}$$

where $\mathbf{H}_{YY}$ is given by (6.6.22).

We note that the SSP matrix due to hypothesis $H_{0\mathbf{B}}$ under the covariance model (7.5.27) is

$$\begin{aligned}\mathbf{H} &= \mathbf{E}_H - \mathbf{E} \\ &= \mathbf{E}_{HYY} - \mathbf{E}_{HYZ}\mathbf{E}_{HZZ}^{-1}\mathbf{E}_{HZZ} - \mathbf{E}_{YY} + \mathbf{E}_{YZ}\mathbf{E}_{ZZ}^{-1}\mathbf{E}_{ZY} \\ &= (\mathbf{E}_{HYY} - \mathbf{E}_{YY}) - \mathbf{E}_{HYZ}\mathbf{E}_{HZZ}^{-1}\mathbf{E}_{HZY} + \mathbf{E}_{YZ}\mathbf{E}_{ZZ}^{-1}\mathbf{E}_{ZY} \\ &= \mathbf{H}_{YY} - (\mathbf{E}_{YZ} + \mathbf{H}_{YZ})(\mathbf{E}_{ZZ} + \mathbf{H}_{ZZ})^{-1}(\mathbf{E}_{YZ} + \mathbf{H}_{YZ}) \\ &\quad + \mathbf{E}_{YZ}\mathbf{E}_{ZZ}^{-1}\mathbf{E}_{ZY}.\end{aligned} \tag{7.4.46}$$

EXAMPLE 7.4.1: Consider the one-way multivariate analysis of covariance model with one concomitant variable,

$$\mathbf{y}_{ij} = \mu_i + \gamma z_{ij} + \mathbf{u}_{ij},\ i = 1, \ldots, a; j = i, \ldots, b. \tag{7.4.47}$$

The model can be written as

$$\begin{bmatrix} \mathbf{y}'_{11} \\ \mathbf{y}'_{12} \\ \\ \cdot \\ \cdot \\ \mathbf{y}'_{1b} \\ \cdot \\ \cdot \\ \mathbf{y}'_{ab} \end{bmatrix}_{ab\times p} = \begin{bmatrix} \mathbf{1}_b & \mathbf{0} & \dots & \mathbf{0} \\ \mathbf{0} & \mathbf{1}_b & \dots & \mathbf{0} \\ \mathbf{0} & \mathbf{0} & \dots & \mathbf{0} \\ \cdot & \cdot & \dots & \cdot \\ \cdot & \cdot & \dots. & \\ \mathbf{0} & \mathbf{0} & \dots & \mathbf{0} \end{bmatrix} \begin{bmatrix} \mu'_1 \\ \mu'_2 \\ \cdot \\ \cdot \\ \mu'_a \end{bmatrix} + \mathbf{Z\Gamma}_{1\times p} + \mathbf{U} \tag{7.4.48}$$

where $\gamma = (\gamma_1, \dots, \gamma_p), \mathbf{\Gamma} = \gamma'$ and $\mathbf{U}$ is defined accordingly. Writing $\mathbf{y}_{ij} = (y_{ij1}, \dots, y_{ijp})'$ and $\mathbf{u}_{ij} = (u_{ij1}, \dots, u_{ijp})'$, we have from model (7.4.47),

$$y_{ijt} = \mu_{it} + z_{ij}\gamma_t + u_{ijt}, \; t = 1, \dots, p. \tag{7.4.49}$$

Thus each element of the observation vector $\mathbf{y}_{ij}$ has the same value of the concomitant variable z_{ij} but is affected differently by the different values of the regression coefficient γ_t $(t = 1, \dots, p)$.

For example, if i is the type of factory, $i = 1, 2$ according as the factory was commissioned before or after the year 2000, $j = 1, 2, 3$ and y_1 = monthly output (value), y_2 = average value of raw materials consumed in a month, z = number of employees, the model is

$$\begin{aligned} y_{ij1} &= \mu_{i1} + \gamma_1 z_{ij} + u_{ij1}, \\ y_{ij2} &= \mu_{i2} + \gamma_2 z_{ij} + u_{ij2}, \; i = 1, 2; j = 1, ., 3. \end{aligned}$$

Suppose we want to test the null hypothesis $H_{0\mu} : \mu_1 = \dots \mu_a$ for the model (7.4.47). Here

$$\begin{aligned} \mathbf{E} = &\text{ Residual SSP matrix for the MANOVA model } E(\mathbf{Y}) = \mathbf{XB} \\ = &\textstyle\sum_{i=1}^a \sum_{j=1}^b (\mathbf{y}_{ij} - \bar{\mathbf{y}}_{i0})(\mathbf{y}_{ij} - \bar{\mathbf{y}}_{i0})' \end{aligned}$$

$$\begin{aligned} \mathbf{E}_{HYY} &= \text{Residual SSP matrix due to hypthesis } H_{0\mu} \\ &\quad\;\text{for the MANOVA model} \\ &= \text{ Total SSP matrix for the MANOVA model} \\ &= \textstyle\sum_{i=1}^a \sum_{j=1}^b (\mathbf{y}_{ij} - \bar{\mathbf{y}}_{00})(\mathbf{y}_{ij} - \bar{\mathbf{y}}_{00})'. \end{aligned}$$

$$\begin{aligned} \mathbf{H}_{YY} &= \text{ SSP matrix due to } H_{0\mu} \text{ for the MANOVA model} \\ &= \mathbf{E}_{HYY} - \mathbf{E}_{YY} \\ &= b\textstyle\sum_{i=1}^a (\bar{\mathbf{y}}_{i0} - \bar{\mathbf{y}}_{00})(\bar{\mathbf{y}}_{i0} - \bar{\mathbf{y}}_{00})'. \end{aligned}$$

Also,

$$\begin{aligned}\mathbf{E}_{ZZ} &= \sum_{i=1}^{a}\sum_{j=1}^{b}(z_{ij}-\bar{z}_{i0})(z_{ij}-\bar{z}_{i0})',\\ \mathbf{E}_{YZ} &= \sum_{i=1}^{a}\sum_{j=1}^{b}(z_{ij}-\bar{z}_{i0})(\mathbf{y}_{ij}-\bar{\mathbf{y}}_{i0}).\end{aligned}$$

Hence, the least square estimate of $\mathbf{\Gamma}$ under the full model (7.4.47)is

$$\hat{\mathbf{\Gamma}}_G = \mathbf{E}_{ZZ}^{-1}\mathbf{E}_{YZ} = \frac{\sum_{i=1}^{a}\sum_{j=1}^{b}(z_{ij}-\bar{z}_{i0})(\mathbf{y}_{ij}-\bar{\mathbf{y}}_{i0})}{\sum_{i=1}^{a}\sum_{j=1}^{b}(z_{ij}-\bar{z}_{i0})^2}.$$

Therefore,

$$\hat{\gamma}_t = \frac{\sum_{i=1}^{a}\sum_{j=1}^{b}(y_{ijt}-\bar{y}_{i0t})(z_{ij}-\bar{z}_{i0})}{\sum_{i=1}^{a}\sum_{j=1}^{b}(z_{ij}-\bar{z}_{i0})^2}.$$

We can, therefore, calculate $\mathbf{E} = \mathbf{E}_{YY} - \mathbf{E}_{YZ}\mathbf{E}_{ZZ}^{-1}\mathbf{E}_{ZY}$, the error SSP matrix under the full model (7.4.47) (vide (7.4.29)) and the SSP matrix due to hypothesis $H_{0\mu}$ by formula (7.4.33). Here

$$\mathbf{A} = \begin{bmatrix} 1 & 0 & \dots & -1\\ 0 & 1 & \dots & -1\\ . & . & \dots & .\\ . & . & \dots & .\\ 0 & 0 & \dots & -1\end{bmatrix}$$

whose rank is $(a-1)$. The hypothesis $H_{0\mu}$ can be tested by the Wilk's formula

$$\Lambda = \frac{|\mathbf{E}|}{|\mathbf{E}+\mathbf{H}|}$$

which follows $\Lambda(p; a-1, n-a-1)$ if the hypothesis is true.

The least square estimate of $\mathbf{B}$ under the covariance model (7.4.47) is

$$\hat{\mathbf{B}}_G = \begin{bmatrix}\hat{\mu}_{1G}\\ \hat{\mu}_{2G}\\ .\\ .\\ \hat{\mu}_{aG}\end{bmatrix} = (\mathbf{X}'\mathbf{X})^{-1}\mathbf{X}'(\mathbf{Y}-\mathbf{Z}\hat{\mathbf{\Gamma}}_G) = \begin{bmatrix}\bar{\mathbf{y}}'_{10}-\bar{z}_{10}\hat{\gamma}'\\ \bar{\mathbf{y}}'_{20}-\bar{z}_{20}\hat{\gamma}'\\ .\\ .\\ \bar{\mathbf{y}}'_{a0}-\bar{z}_{a)}\hat{\gamma}'\end{bmatrix}.$$

As in equation (7.4.11) we can express the model as

$$\mathbf{Y} = \mathbf{W\Phi} + \mathbf{U}$$

and the hypothesis $\mathbf{AB} = \mathbf{0}$ as the hypothesis $H_0 : \mathbf{A}_1\mathbf{\Phi} = \mathbf{0}$. From the theory of general linear model in Chapter 6,

$$\begin{aligned}\mathbf{H} &= (\mathbf{A}_1\hat{\mathbf{\Phi}})'[\mathbf{A}_1(\mathbf{W}'\mathbf{W})^{-1}\mathbf{A}_1']^{-1}\mathbf{A}_1\hat{\mathbf{\Phi}}\\ \mathbf{E} &= \mathbf{Y}'[\mathbf{I}_n - \mathbf{W}(\mathbf{W}'\mathbf{W})^{-1}\mathbf{W}']\mathbf{Y}.\end{aligned}$$

EXAMPLE 7.4.2: The following table gives the monthly output (in \$ thousand), raw materials consumed (in \$ thousand) and the number of workers in six factories classified according to their year of installation - A, installed before 1990, and B, installed after 1990. Test the hypothesis that the average output and amount of raw materials consumed in these two sectors are the same, taking the number of workers as the concomitant variable.

Table 7.4.3: Output of Factories

Type of Factory	Unit	Output y_1	Raw Material y_2	Number of workers, z
	1	251	102	72
A	2	378	157	93
	3	108	52	64
	1	504	207	78
B	2	380	125	65
	3	482	198	97

The model is

$$\mathbf{Y} = \mathbf{XB} + \mathbf{Z}\Gamma + \mathbf{U}$$

where

$$\mathbf{X} = \begin{bmatrix} \mathbf{1}_3 & \mathbf{0} \\ \mathbf{0} & \mathbf{1}_3 \end{bmatrix}, \ \mathbf{B} = \begin{bmatrix} \mu_1' \\ \mu_2' \end{bmatrix}.$$

The hypothesis $H_0(\mu_1 = \mu_2)$ can be written as $H_{0B}(\mathbf{AB} = \mathbf{0})$ where $\mathbf{A} = [1, \ -1]$. Here

$$\begin{aligned}
\mathbf{E}_{YY} &= \mathbf{Y}'\mathbf{PY} \\
&= \textstyle\sum_{i=1}^2 \sum_{j=1}^3 (\mathbf{y}_{ij} - \bar{\mathbf{y}}_{i0})(\mathbf{y}_{ij} - \bar{\mathbf{y}}_{i0})' \\
&= \textstyle\sum_{i=1}^2 \sum_{j=1}^3 \begin{bmatrix} y_{ij1} - \bar{y}_{i01} \\ y_{ij2} - \bar{y}_{i02} \end{bmatrix} [y_{ij1} - \bar{y}_{i01}, \ y_{ij2} - \bar{y}_{i02}] \\
&= \begin{bmatrix} 45247.3 & 18961.3 \\ 18961.3 & 9561.3 \end{bmatrix},
\end{aligned}$$

$$\begin{aligned}
\mathbf{E}_{YZ} &= \mathbf{Y}'\mathbf{PZ} \\
&= \textstyle\sum_{i=1}^2 \sum_{j=1}^3 (\mathbf{y}_{ij} - \bar{\mathbf{y}}_{i0})(z_{ij} - \bar{z}_{i0}) \\
&= \begin{bmatrix} y_{ij1} - \bar{y}_{i01} \\ y_{ij2} - \bar{y}_{i02} \end{bmatrix} (z_{ij} - \bar{z}_{i0}) \\
&= \begin{bmatrix} 4654.23 \\ 2731.65 \end{bmatrix},
\end{aligned}$$

$$\begin{aligned}\mathbf{E}_{ZZ} &= \mathbf{Z}'\mathbf{P}\mathbf{Z}\\ &= \textstyle\sum_{i=1}^{2}\sum_{j=1}^{3}(z_{ij}-\bar{z}_{i0})^2\\ &= 966.667,\end{aligned}$$

$$\begin{aligned}\hat{\mathbf{\Gamma}}_G &= \mathbf{E}_{ZZ}^{-1}\mathbf{E}_{ZY}\\ &= [4.81472,\ 2.82535],\end{aligned}$$

$$\begin{aligned}\mathbf{E} &= \mathbf{E}_{YY} - \mathbf{E}_{YZ}\mathbf{E}_{ZZ}^{-1}\mathbf{E}_{ZY}\\ &= \begin{bmatrix} 22838.1 & 5808.9\\ 5808.9 & 1842.0\end{bmatrix},\end{aligned}$$

$$\begin{aligned}\hat{\mathbf{B}} &= (\mathbf{X}'\mathbf{X})^{-1}\mathbf{X}'\mathbf{Y}\\ &= \begin{bmatrix} 245.67 & 103.67\\ 455.33 & 276.67\end{bmatrix},\end{aligned}$$

$$\begin{aligned}\hat{\mathbf{B}}_G &= \hat{\mathbf{B}} - (\mathbf{X}'\mathbf{X})^{-1}\mathbf{X}'\mathbf{Z}\hat{\Gamma}_G\\ &= \begin{bmatrix} -121.838 & -112.027\\ 70.132 & 50.602\end{bmatrix}.\end{aligned}$$

The SSP matrix due to $H_0(\mu_1 = \mu_2)$ is

$$\mathbf{H} = (\mathbf{A}\hat{\mathbf{B}}_G)'[\mathbf{A}(\mathbf{X}'\mathbf{X})^{-1}\mathbf{A}' + \mathbf{ALML}'\mathbf{A}']^{-1}(\mathbf{A}\hat{\mathbf{B}}_G)$$

where $\mathbf{L} = (\mathbf{X}'\mathbf{X})^{-1}\mathbf{X}' = [76.33,\ 80]', \mathbf{M} = \mathbf{E}_{ZZ}^{-1}$. Hence

$$\mathbf{H} = \begin{bmatrix} 53883.1 & 17573.4\\ 17573.4 & 5731.4\end{bmatrix}.$$

The test statistic is

$$\Lambda = \frac{|\mathbf{E}|}{|\mathbf{E}+\mathbf{H}|} = .002426.$$

Under null hypothesis Λ follows $\Lambda(2;1,3)$. Since the observed value of Λ is less than $\Lambda_{.05}(2;1,3) = .05$, the null hypothesis is rejected.

If we are to test the hypothesis $H_{0\mathbf{\Gamma}}(\mathbf{\Gamma} = \mathbf{0})$, then the error SSP matrix under $H_{0\mathbf{\Gamma}}$ is $\mathbf{E}_{YY}$. Hence the SSP matrix due to $H_{0\mathbf{\Gamma}}$ is

$$\mathbf{H}_{\mathbf{\Gamma}} = \mathbf{E}_{YY} - \mathbf{E} = \mathbf{E}_{YZ}\mathbf{E}_{ZZ}^{-1}\mathbf{E}_{ZY} = \begin{bmatrix} 22409.2 & 13152.4\\ 13152.4 & 7719.3\end{bmatrix}.$$

Hence the test statistic for testing $H_{0\mathbf{\Gamma}}$ is

$$\Lambda_{\mathbf{\Gamma}} = \frac{|\mathbf{E}|}{|\mathbf{E}_{YY}}| = .0011388.$$

Under the null hypothesis Λ_{Γ} follows $\Lambda(2;1,3)$. Since the observed value is less than $\Lambda_{.05}(2;1,3) = .05$ the null hypothesis $H_{0\mathbf{\Gamma}}$ is rejected.

7.5 A Conditional Hypothesis

Consider the general linear model in (6.1.1), $\mathbf{Y} = \mathbf{XB} + \mathbf{U}$. Suppose $\mathbf{Y} = (\mathbf{Y}_1,\ \mathbf{Y}_2)$ where $\mathbf{Y}_1(n \times p_1)$ represents n observations on each of p_1 response variables and $\mathbf{Y}_2(n \times p_2), n$ observations on each of p_2 response variables, $p_1 + p_2 = p$. We partition $\mathbf{B}$ as $\mathbf{B} = (\mathbf{B}_1,\ \mathbf{B}_2)$ where $\mathbf{B}_1$ is $q \times p_1, \mathbf{B}_2, q \times p_2$. We have, therefore,

$$\mathbf{Y} = \mathbf{XB} + \mathbf{U}$$

or

$$(\mathbf{Y}_1,\ \mathbf{Y}_2) = (\mathbf{XB}_1,\ \mathbf{XB}_2) + (\mathbf{U}_1,\ \mathbf{U}_2) \tag{7.5.1}$$

where $\mathbf{U}_i$ is $n \times p_i$ and rows of $\mathbf{U}_i$ are *iid* $N_{p_i}(\mathbf{0}, \mathbf{\Sigma}_{ii}), i = 1, 2$. The matrices $\mathbf{\Sigma}_{ii}, i = 1, 2$ are given by

$$\mathbf{\Sigma} = \begin{bmatrix} \mathbf{\Sigma}_{11} & \mathbf{\Sigma}_{12} \\ \mathbf{\Sigma}_{21} & \mathbf{\Sigma}_{22} \end{bmatrix}. \tag{7.5.2}$$

We examine the following marginal model,

$$\mathbf{Y}_1 = \mathbf{XB}_1 + \mathbf{U}_1 \tag{7.5.3}$$

and a conditional model as follows. Consider a row of $\mathbf{Y}_2$, say, $\mathbf{y}_i^{(2)'} = (y_{ip_1+1}, \ldots, y_{ip_2})'$. Its conditional distribution given $\mathbf{Y}_1$ is affected only by $\mathbf{y}_i^{(1)'} = (y_{i1}, \ldots, y_{ip_1})'$ and not by $y_j^{(1)} (j \neq i)$. Hence

$$\begin{aligned} E(\mathbf{y}_i^{(2)}|\mathbf{Y}_1) &= E(\mathbf{y}_i^{(2)}|\mathbf{y}_i^{(1)}) \\ &= E(\mathbf{y}_i^{(2)} + \text{ Cov } (\mathbf{y}_i^{(2)}, \mathbf{y}_i^{(1)})[\text{ Cov } (\mathbf{y}_i^{(1)}]^{-1}(\mathbf{y}_i^{(1)} - E(\mathbf{y}_i^{(1)})) \end{aligned}$$

or

$$E(\mathbf{y}_i^{(2)'}|\mathbf{Y}_1) = E(\mathbf{y}_i^{(1)'}) + (\mathbf{y}_i^{(1)} - E(\mathbf{y}_i^{(1)})'[\text{ Cov } (\mathbf{y}_i^{(1)})]^{-1} \text{ Cov } (\mathbf{y}_i^{(2)}, \mathbf{y}_i^{(1)})'.$$

Hence,

$$\begin{aligned} E(\mathbf{Y}_2|\mathbf{Y}_1) &= E(\mathbf{Y}_2) + (\mathbf{Y}_1 - E(\mathbf{Y}_1))\mathbf{\Sigma}_{11}^{-1}\mathbf{\Sigma}_{12} \\ &= \mathbf{XB}_2 + (\mathbf{Y}_1 - \mathbf{XB}_1)\mathbf{\Sigma}_{11}^{-1}\mathbf{\Sigma}_{12} \\ &= \mathbf{X}(\mathbf{B}_2 - \mathbf{B}_1\mathbf{\Sigma}_{11}^{-1}\mathbf{\Sigma}_{12}) + \mathbf{Y}_1\mathbf{\Sigma}_{11}^{-1}\mathbf{\Sigma}_{12} \\ &= \mathbf{X\Delta} + \mathbf{Y}_1\mathbf{\Gamma} \ \text{(say)}\ . \end{aligned} \tag{7.5.4}$$

This is a special case of the covariance model (7.4.27) with $p = p_2, t = p_1, \mathbf{B} = \mathbf{\Delta} = \mathbf{B}_2 - \mathbf{B}_1\mathbf{\Sigma}_{11}^{-1}\mathbf{\Sigma}_{12}$ and $\mathbf{Z} = \mathbf{Y}_1$.

We now consider the likelihood ratio tests for the following three hypotheses:

(i) $H_0 : \mathbf{AB} = \mathbf{0}$

(ii) $H_{01} : \mathbf{AB}_1 = \mathbf{0}$
(iii) $H_{02|1} : \mathbf{A\Delta} = \mathbf{0}$ given that H_{01} is true.

where $\mathbf{A}$ is a $g \times q$ matrix of rank g. We note that if H_0 is true then both H_{01} and $H_{02|1}$ are true. Conversely, if both H_{01} and $H_{02|1}$ are true, then H_0 is true. Hence, $H_0 \Leftrightarrow H_{02} \cap H_{02|1}$. Therefore, $H_0 = H_{01} \cap H_{02|1}$.

Noting that we are considering here the general linear model $\mathbf{Y} = \mathbf{XB} + \mathbf{U}$, the statistics used for testing H_0 are:

$$\begin{aligned} \mathbf{E} &= (\mathbf{Y} - \mathbf{X}\hat{\mathbf{B}})'(\mathbf{Y} - \mathbf{X}\hat{\mathbf{B}}) = \mathbf{Y}'\mathbf{PY}, \text{ d.f. } = n - q, \\ \mathbf{E}_H &= (\mathbf{Y} - \mathbf{X}\hat{\mathbf{B}}_H)'(\mathbf{Y} - \mathbf{X}\hat{\mathbf{B}}_H), \text{ d.f } = n - q + g, \\ \mathbf{H} &= (\mathbf{A}\hat{\mathbf{B}})'[\mathbf{A}(\mathbf{X}'\mathbf{X})^{-1}\mathbf{A}']^{-1}\mathbf{A}\hat{\mathbf{B}}, \text{ d.f. } = g, \end{aligned} \tag{7.5.5}$$

where $\hat{\mathbf{B}}_H$ is the least square estimate of $\mathbf{B}$ under the hypothesis H_0 and is given by (6.6.14). The expression for $\mathbf{H}$ is given by (6.6.22).

For testing H_{01} we start with the marginal model (7.5.3). Hence the error SSP matrix is

$$\mathbf{E}_{11} = (\mathbf{Y}_1 - \mathbf{X}\hat{\mathbf{B}}_1)'(\mathbf{Y}_1 - \mathbf{X}\hat{\mathbf{B}}_1) = \mathbf{Y}_1'\mathbf{PY}_1. \tag{7.5.6}$$

Now, from the property of the least square estimates, $\hat{\mathbf{B}} = (\hat{\mathbf{B}}_1,\ \hat{\mathbf{B}}_2)$. Hence, $\mathbf{E}_{11}$ is a submatrix of $\mathbf{E}$ and is obtained from

$$\mathbf{E} = \begin{bmatrix} \mathbf{E}_{11} & \mathbf{E}_{12} \\ \mathbf{E}_{21} & \mathbf{E}_{22} \end{bmatrix}. \tag{7.5.7}$$

Similarly $\mathbf{H}_{11}$, the SSP matrix due to H_{01} will be a submatrix of $\mathbf{H}$,

$$\mathbf{H} = \begin{bmatrix} \mathbf{H}_{11} & \mathbf{H}_{12} \\ \mathbf{H}_{21} & \mathbf{H}_{22} \end{bmatrix}.$$

Now, from (6.6.14) $\hat{\mathbf{B}}_H$ is given by

$$\hat{\mathbf{B}}_H = \hat{\mathbf{B}} - (\mathbf{X}'\mathbf{X})^{-1}\mathbf{A}'[\mathbf{A}(\mathbf{X}'\mathbf{X})^{-1}\mathbf{A}']^{-1}\mathbf{A}\hat{\mathbf{B}}'.$$

Hence,

$$\hat{\mathbf{B}}_{H1} = \hat{\mathbf{B}}_1 - (\mathbf{X}'\mathbf{X}){-}1(\mathbf{A}'[\mathbf{A}(\mathbf{X}'\mathbf{X})^{-1}\mathbf{A}']^{-1}\mathbf{A}\hat{\mathbf{B}}_1 \tag{7.5.8}$$

where $\hat{\mathbf{B}}_{H1}$ is the least square estimate of $\mathbf{B}_1$ under the hypothesis $\mathbf{H}_{01}$. Hence, $\hat{\mathbf{B}}_{H1}$ is a submatrix of $\hat{\mathbf{B}}_H$ corresponding to $\mathbf{B}_1$,

$$\hat{\mathbf{B}}_H = (\hat{\mathbf{B}}_{H1},\ \hat{\mathbf{B}}_{H2}). \tag{7.5.9}$$

Hence, the error SSP matrix under the hypothesis H_{01} is

$$\mathbf{E}_{H11} = (\mathbf{Y}_1 - \mathbf{X}\hat{\mathbf{B}}_{H1})'(\mathbf{Y} - \mathbf{X}\hat{\mathbf{B}}_{H1}) \tag{7.5.10}$$

will be a submatrix of $\mathbf{E}_H$,

$$\mathbf{E}_H = \begin{bmatrix} \mathbf{E}_{H11} & \mathbf{E}_{H12} \\ \mathbf{E}_{H21} & \mathbf{E}_{H22} \end{bmatrix}. \tag{7.5.11}$$

The likelihood tests for H_0 and H_{01} are, therefore, respectively,

$$\Lambda_0 = \frac{|\mathbf{E}|}{|\mathbf{E}_H|}, \ \Lambda_{01} = \frac{|\mathbf{E}_{11}|}{|\mathbf{E}_{H11}|} \tag{7.5.12}$$

which follow under the respective null hypotheses $\Lambda(p; g, n-q)$ and $\Lambda(p_1 : g, n-q)$ distributions.

We now consider the hypothesis $H_{02|1}$ which is to be tested under the covariance model (7.5.4). Since $\mathbf{Y}_1$ here plays the role of $\mathbf{Z}$ in (7.4.27), the likelihood ratio test for testing $H_{02|1}$ is, by (7.4.43),

$$\Lambda_{02|1} = \frac{|\mathbf{E}_{22} - \mathbf{E}_{21}\mathbf{E}_{11}^{-1}\mathbf{E}_{12}|}{|\mathbf{E}_{H22} - \mathbf{E}_{H21}\mathbf{E}_{H11}^{-1}\mathbf{E}_{H21}|}. \tag{7.5.13}$$

The statistic $\Lambda_{02|1}$ follows $\Lambda(p_2; g, n-q-p_1)$, because, rank$(\mathbf{Y}_1) = p_1$. Note that $\mathbf{E}_{H22}$ is the error SSP matrix under the MANOVA model $E(\mathbf{Y}) = \mathbf{XB}$ subject to the hypothesis $H_{02}(\mathbf{AB}_2) = \mathbf{0}$.This is the same as the error SSP matrix under the model $E(\mathbf{3}_2|\mathbf{Y}_1) = \mathbf{X\Delta}$ subject to the hypothesis $\mathbf{A\Delta} = \mathbf{0}$ given that the hypothesis H_{01} is true. By (A.5.6) and (A.5.7), $\Lambda_0 = \Lambda_{01}\Lambda_{02|1}$. Therefore, when H_0 is true, we can write this relation as

$$\Lambda(p; g, n-q) = \Lambda(p_1; g, n-q)\Lambda(p_2; g, n-q). \tag{7.5.14}$$

The test $H_{02|1}$ is the test for the hypothesis $\mathbf{AB}_2 = \mathbf{0}$ given that $\mathbf{AB}_1 = \mathbf{0}$. As such $H_{02|1}$ may be regarded as the test for additional information as supplied by the variables in $\mathbf{Y}_2$.

7.6 Exercises and Complements

7.1 Prove the relation (7.2.31). Hence, or otherwise find the relation between T^2, Wilk's Λ, Roy's θ_1, T_g^2 and Pillai's statistic when $k = 2$.

7.2 Prove the relation (7.4.3).

[*Hints*: Use the relations A.6.9 - A.6.12 and the Binomial Inversion Theorem.]

7.3 Verify the equation (7.4.13).

7.4 Verify equation (7.4.15).

Table 7.E.1

Plot No.	HT Pt.I	Pt.II	Pt.III	LT Pt.I	Pt.II	Pt.III	Control Pt.I	Pt.II	Pt.III
1	15	16	13	20	22	21	25	24	24
	32	41	34	52	47	51	38	39	40
2	20	15	18	30	29	28	40	35	38
	35	37	42	54	49	51	40	38	40
3	16	14	15	18	16	17	20	18	19
	37	43	39	51	49	49	40	45	41
4	18	15	16	20	22	23	22	26	25
	37	45	38	53	49	48	39	39	41
5	21	22	21	20	19	18	30	28	27
	42	39	39	48	47	52	39	38	43
6	10	08	09	16	14	15	16	15	15
	45	39	34	57	48	52	39	40	42
7	14	12	13	17	10	16	16	20	18
	34	42	38	52	49	52	39	40	43
8	15	11	12	20	18	19	19	17	16
	35	45	38	54	49	52	39	42	45
9	16	15	14	14	15	15	25	22	24
	37	43	40	57	51	54	39	32	37
10	17	14	15	15	14	13	14	18	16
	37	43	38	54	49	54	39	40	43

7.5 The following experiment was conducted to see whether subjecting seeds to high temperature and low temperature before planting has any effect on germination time and yield. Ten plots were taken and at each 9 seeds were sown at random. Three of the seeds were subjected to high temperature (HT), three to low temperature (LT) and three to control. Time to germination (in hours) and the yield of individual plants (in grammes) are recorded in Table 7.E.1. (first and second entry in a cell respectively). Perform an analysis of variance.

Chapter 8

Principal Component Analysis

8.1 Introduction

The principal component analysis is concerned with explaining the variations among a number of variables, as reflected in the dispersion matrix of the vector of variables, more precisely, the total variance of the variables, through a few linear combination of the original variables. Data on multivariate populations often involve repeated observations, n in number, on p, a large number of correlated variables. It will, therefore, be convenient for analysis and interpretation of the data, if a smaller number of new independent variables or principal components (which are linear combinations of original variables) can be found which account for most of the variations in the original variables. Although p components are required to represent the total variability of the original variables, often most of this variability can be accounted for by a smaller number $m(<< p)$ of principal components. There is then almost as much information in the m components as there is in the original p variables. The m principal components can then replace the p original variables and the original data set, consisting of n observations on p variables can be replaced by a set consisting of n observations on the principal components.

An analysis of principal components often reveals relationship among original variables that were not previously suspected and thereby allows a new insight into the data. Again, in replacing the original variables by a fewer number of variables, some information have to be necessarily sacrificed. Principal components are optimum in the sense that the amount of lost information is kept at a minimum among all the similar processes.

It may be noted that the analysis of principal components is more a means to an end rather than an end in itself. The principal component analysis often serves as an intermediate step in much larger investigations with data analysis. For example, in regression analysis if the number of independent variables is large relative to the number of observations, a test may be ineffective. Also, if the independent variables are highly correlated, the estimates of regression coefficients may be unstable. In such situations the number of independent variables may be reduced to a smaller number of principal components that will yield a better test and/or more stable regression coefficients.

The principal component analysis was initiated by Hotelling (1933) who developed the technique for his work in educational psychlogy. Its mathematical theory has been developed by Girshick (1939), Anderson (1963a), Geisser (1965), among others.

8.2 Population Principal Components

Consider a random variable $\mathbf{X} = (X_1, \ldots, X_p)'$ with mean $\mu = (\mu_1, \ldots, \mu_p)', \mu_i < \infty$ $(i = 1, \ldots, p)$ and variance $\mathbf{\Sigma} = ((\sigma_{ij})), \sigma_{ij} < \infty (i, j = 1, \ldots, p)$. Assume that the rank of $\mathbf{\Sigma}$ is p and

$$\lambda_1 \geq \lambda_2 \geq \ldots \geq \lambda_p \geq 0 \tag{8.2.1}$$

are the p eigenvalues of $\mathbf{\Sigma}$.

In the principal component analysis we want to find uncorrelated linear functions of $X_1, \ldots, X_p$, say, $Z_1, \ldots, Z_m, (m \leq p)$, such that variances $V(Z_1), \ldots, V(Z_m)$ account for most of the total variances among $X_1, \ldots, X_p$. Also, we require $V(Z_1) > V(Z_2) > \ldots > V(Z_m)$. Algebraically, principal components are particular linear combinations of $X_1, \ldots, X_p$. Geometrically, the principal components represent a new coordinate system obtained by rotating the original axes $X_1, \ldots, X_p$. The new axes represent the directions with maximum variability.

Let $\alpha = (\alpha_1, \ldots, \alpha_p)'$ be a $p \times 1$ vector of weights for the respective components of $\mathbf{X}$ and consider the linear function

$$Z_1 = \alpha'\mathbf{X} = \sum_{i=1}^{p} \alpha_i X_i \tag{8.2.2}$$

Our aim is to find α such that $V(Z_1)$ is maximum subject to the condition $\alpha'\alpha = 1$. It is clear that $V(Z_1)$ can be increased by multiplying α by some constant. To eliminate this arbitrariness we restrict our attention to coefficient vectors of unit lengths.

Now,

$$V(Z_1) = \alpha'\mathbf{\Sigma}\alpha.$$

Hence, we are required to find α such that

$$\alpha'\mathbf{\Sigma}\alpha \tag{8.2.3}$$

is maximum subject to the condition $\alpha'\alpha = 1$.

It is known from matrix algebra that

$$\max{}_{\beta} \frac{\beta'\mathbf{\Sigma}\beta}{\beta'\beta} = \lambda_1 \tag{8.2.4}$$

and the vector β which gives this maximum is the eigenvector corresponding to λ_1. Hence, the maximum value in (8.2.3) is λ_1 and this is attained by $\alpha = \alpha_1$, the normalized eigenvector corresponding to λ_1.

Alternatively, for finding (8.2.3), consider the Lagrangian,

$$\psi = \alpha'\mathbf{\Sigma}\alpha - \theta(\alpha'\alpha - 1) \tag{8.2.5}$$

where θ is a Lagrangian constant. Differentiating both sides with respect to α and equating these to zero,

$$\frac{\partial\psi}{\partial\alpha} = 2\mathbf{\Sigma}\alpha - 2\theta\alpha = \mathbf{0}$$

i.e.

$$(\mathbf{\Sigma} - \theta\mathbf{I})\alpha = \mathbf{0}. \tag{8.2.6}$$

Since, $\alpha \neq \mathbf{0}$, there can be a solution only if $\mathbf{\Sigma} - \theta\mathbf{I}$ is singular, i.e. if

$$\mid \mathbf{\Sigma} - \theta\mathbf{I} \mid = 0$$

i.e. if θ is a latent root of $\mathbf{\Sigma}$ and α is its corresponding normalized latent vector. Now,

$$\begin{aligned} V(Z) &= \alpha'\mathbf{\Sigma}\alpha = \alpha'\theta\alpha \quad \text{(by (8.2.6))} \\ &= \theta \end{aligned}$$

and it is maximum if $\theta = \lambda_1$. Therefore, $\alpha = \alpha_1$, the normalized latent vector corresponding to λ_1. The first principal component is, therefore,

$$Z_1 = \alpha_1'\mathbf{X}$$

with

$$V(Z_1) = \lambda_1. \tag{8.2.7}$$

The second principal component is given by the linear function

$$Z_2 = \alpha' \mathbf{X}$$

where α is such that $V(Z_2)$ is maximum subject to the condition $\alpha'\alpha = 1$ and Z_2 is orthogonal to Z_1 (i.e. $\alpha'\alpha_1 = 0$, since we require Z_1, Z_2 to be stochastically independent). It is known from matrix algebra that

$$\max_{\alpha:\alpha'\alpha=1,\alpha'\alpha_1=0} \ \alpha'\mathbf{\Sigma}\alpha = \lambda_2 \tag{8.2.8}$$

and the maximum is attained for $\alpha = \alpha_2$, the normalized eigenvector corresponding to λ_2.

Alternatively, consider the Lagrangian

$$\phi = \alpha'\mathbf{\Sigma}\alpha - \theta_1(\alpha'\alpha - 1) - \theta_2(\alpha'\alpha_1 - 0)$$

where θ_1, θ_2 are Lagrangian multipliers. Differentiating both sides with respect to α and equating the result to zero,

$$\frac{\partial \phi}{\partial \alpha} = 2(\mathbf{\Sigma} - \theta_1 \mathbf{I})\alpha - \theta_2 \alpha_1 = \mathbf{0}. \tag{8.2.9}$$

Premultiplying by α_1'

$$2\alpha_1'\mathbf{\Sigma}\alpha - \theta_2 = \mathbf{0}. \tag{8.2.10}$$

Again, premultiplying the relation

$$(\mathbf{\Sigma} - \lambda_1 \mathbf{I})\alpha_1 = \mathbf{0}$$

by α',

$$\alpha'\mathbf{\Sigma}\alpha_1 = 0.$$

Hence, from (8.2.10), $\theta_2 = 0$. Therefore, by (8.2.9),

$$(\mathbf{\Sigma} - \theta_1 \mathbf{I})\alpha = \mathbf{0} \tag{8.2.11}$$

and it follows similarly (as in (8.2.6) and subsequent statements) $\theta_1 = \lambda_2$ and $\alpha = \alpha_2$, the corresponding eigenvector. The second principal component is, therefore,

$$Z_2 = \alpha_2'\mathbf{X}$$

with

$$V(Z_2) = \lambda_2. \tag{8.2.12}$$

Clearly,

$$\begin{aligned} \text{Cov}\,(Z_1, Z_2) &= \text{Cov}\,(\alpha_1'X, \alpha_2'X) = \alpha_1'\mathbf{\Sigma}\alpha_2 \\ &= \alpha'\lambda_2\alpha_2 \ \ (\text{by } (8.2.11)) \\ &= 0. \end{aligned}$$

To find the kth principal component, $Z_k = \alpha'\mathbf{X}$, we are to find α such that $V(Z_k)$ is maximum subject to the conditions

$$\begin{aligned} \alpha'\alpha &= 1 \\ \alpha'\alpha_i &= 0 \quad (i = 1, \ldots, k-1). \end{aligned}$$

It follows that

$$Z_k = \alpha_k'\mathbf{X}$$

with

$$V(Z_k) = \lambda_k, \quad k = 1, \ldots, p, \tag{8.2.13}$$

where α_k is the normalized eigenvector corresponding to λ_k. Clearly,

$$\begin{aligned} \text{Cov}\ (Z_k, Z_{k'}) &= \text{Cov}\ (\alpha_k'\mathbf{X}, \alpha_{k'}'\mathbf{X}) \\ &= \alpha_k'\mathbf{\Sigma}\alpha_{k'} \\ &= \alpha_k'\lambda_{k'}\alpha_{k'} \\ &= 0, \ (k \neq k'). \end{aligned} \tag{8.2.14}$$

Note that if $\lambda_i = \lambda_{i+1}$ for some i, the orthogonal vectors α_i, α_{i+1} can be chosen in an infinite number of ways.

Hence, we have the following theorem.

Theorem 8.2.1: Let $\mathbf{\Sigma}$ have the eigenvalue-eigenvector pair $(\lambda_1, \alpha_1), \ldots, (\lambda_p, \alpha_p)$ where $\lambda_1 \geq \lambda_2 \geq \ldots \lambda_p \geq 0$. The kth principal component of $\mathbf{X}$ is given by

$$Z_k = \alpha_k'\mathbf{X}$$

with

$$V(Z_k) = \lambda_k.$$

Also,

$$\text{Cov}\ (Z_k, Z_{k'}) = 0 \quad (k \neq k').$$

If some λ_k's are equal, the corresponding principal components may not be uniquely chosen. □

By Spectral Decomposition Theorem (Theorem A.12.1), we can write

$$\mathbf{\Sigma} = \mathbf{A\Lambda A}'$$

where

$$\mathbf{A} = (\alpha_1, \ldots, \alpha_p), \quad \mathbf{\Lambda} = \text{ Diag. } (\lambda_1, \ldots, \lambda_p).$$

Note that some of the λ_i's may be zeroes. Therefore, the total population variance among $X_1, \ldots, X_p$ is

$$\begin{aligned} \textstyle\sum_{i=1}^p V(X_i) = \text{ tr } \mathbf{\Sigma} = &\text{ tr } (\mathbf{A\Lambda A'}) = \text{ tr } (\mathbf{\Lambda A A'}) \text{ (by A.9(1))} \\ = &\text{ tr } (\mathbf{\Lambda}) \text{ (since } \mathbf{AA'} = \mathbf{I}) \\ = &\textstyle\sum_{i=1}^p \lambda_i = \sum_{i=1}^p V(Z_i). \end{aligned} \tag{8.2.15}$$

The total population variance among $Z_1, \ldots, Z_p$ is the same as the total population variance among $X_1, \ldots, X_p$. The proportion of the total variance accounted for by the kth P.C. is $\lambda_k / \sum_{i=1}^p \lambda_i$. The first m P.C.'s with the m largest variances account for

$$\sum_{i=1}^m \lambda_i / \sum_{i=1}^p \lambda_i$$

proportion of the total variance of $\mathbf{X}$. If, therefore, most ($80\% - 90\%$) of the total variance in $\mathbf{X}$ ia accounted for by the first m components $Z_1, \ldots, Z_m$, then, for large p, these components can replace the p original variables $X_1, \ldots, X_p$ for explaining the variability among the variables and the subsequent components $Z_{m+1}, \ldots Z_p$ can be discarded with.

Theorem 8.2.2: If $Z_1 = \alpha_1' \mathbf{X}, \ldots, Z_p = \alpha_p' \mathbf{X}$ are the principal components, then

$$\rho_{Z_i, X_k} = \frac{\alpha_{ik}\sqrt{\lambda_i}}{\sqrt{\sigma_{kk}}}, \quad i, k = 1, \ldots, p \tag{8.2.16}$$

where $\alpha_i = (\alpha_{i1}, \alpha_{i2}, \ldots, \alpha_{ip})'$.

Proof. Let $X_k = \mathbf{e}_k' \mathbf{X}$, where $\mathbf{e}_k = (0, \ldots, 0, 1, 0, \ldots, 0)'$. Hence,

$$\begin{aligned} \text{Cov } (X_k, Z_i) = &\text{ Cov } (\mathbf{e}_k' \mathbf{X}, \alpha_i' \mathbf{X}) \\ = &\ \mathbf{e}_k' \mathbf{\Sigma} \alpha_i = \mathbf{e}_k' \lambda_i \alpha_i \\ = &\ \lambda_i \alpha_{ik}. \end{aligned} \tag{8.2.17}$$

Therefore,

$$\rho_{Z_i, X_k} = \frac{\lambda_i \alpha_{ik}}{\sqrt{\lambda_i}\sqrt{\sigma_{kk}}} = \frac{\alpha_{ik}\sqrt{\lambda_i}}{\sqrt{\sigma_{kk}}}.$$

□

Note 8.2.1: The correlation ρ_{Z_i, X_k} measures the univariate correlation between the variable X_k and the principal component Z_i, i.e. it does not measure the contribution of the variable X_k to Z_i in the presence of other variables as is done by the coefficient α_{ik}. However, it is often found that variable X_k with relatively large coefficient (in absolute value) α_{ik} in Z_i will have also relatively large (in absolute value) correlation ρ_{Z_i,X_k}.

Note 8.2.2: We have, from (8.2.14),

$$\begin{aligned}\Sigma &= \mathbf{A}\sqrt{\mathbf{\Lambda}}\sqrt{\mathbf{\Lambda}}\mathbf{A}' \\ &= \mathbf{L}\mathbf{L}'\end{aligned} \tag{8.2.18}$$

where

$$\mathbf{L} = \mathbf{A}\sqrt{\mathbf{\Lambda}},\ \sqrt{\mathbf{\Lambda}} = \text{ Diag } (\sqrt{\lambda_1} \ldots, \sqrt{\lambda_p}).$$

Also, by Spectral Decomposition Theorem,

$$\mathbf{\Sigma} = \lambda_1\alpha_1\alpha_1' + \ldots + \lambda_r\alpha_r\alpha_r'$$

where $r(\leq p)$ is the rank of $\mathbf{\Sigma}$. As the principal components are formed, the matrix $\lambda_i\alpha_i\alpha_i'$ can be calculated and their running sums show how well the matrix $\mathbf{\Sigma}$ is approximated by a smaller number of variables.

The principal components are not scale-invariant in that they depend on the scales on which the variables are measured (vide Section 8.4). For example, changes in measurement of length from foot to meter, of time from hour to second will produce different eigenvalues. Again, if the variables are measured in widely different units (e.g age in years, weight in kilograms, price in Rupees), their linear combinations would have little meanings. To get rid of these difficulties, one should work with the standardized variables

$$Y_i = \frac{X_i - \mu_i}{\sqrt{\sigma_{ii}}},\ i = 1, \ldots, p.$$

Thus

$$\mathbf{Y} = (\sqrt{\mathbf{V}})^{-1}(\mathbf{X} - \mu)$$

where $\mathbf{Y} = (Y_1, \ldots, Y_p)', \sqrt{\mathbf{V}} = \text{ Diag } (\sqrt{\sigma_{11}}, \ldots, \sqrt{\sigma_{pp}})$. The covariance matrix of $\mathbf{Y}$ is the correlation matrix $\rho = ((\rho_{ij}))$ of $\mathbf{X}$. The principal components of $\mathbf{Y}$ may be calculated from the eigenvectors of ρ and its variances from its eigenvalues.

The eigenvalues and eigenvectors of ρ will not generally be equal to the eigenvalues and eigenvectors of $\mathbf{\Sigma}$, from which the covariance matrix is derived. However, we will use the same notation (λ_i, α_i) for the eigenvalue-eigenvector pair of both $\mathbf{\Sigma}$ and ρ, their actual meanings being clear from the context.

Theorem 8.2.3: Let $(\lambda_i, \alpha_i)(i = 1, \ldots, p)$ be the eigenvalue-eigenvector pair of ρ. The ith principal component of the standardized variable $\mathbf{Y}$ with covariance matrix $V(\mathbf{Y}) = \rho$ is given by

$$W_i = \alpha_i'\mathbf{Y} = \alpha_i'(\sqrt{\mathbf{V}})^{-1}(\mathbf{X} - \mu) \tag{8.2.19}$$

with

$$V(W_i) = \lambda_i.$$

Again,

$$\sum_{i=1}^{p} V(Y_i) = \text{ tr } (\rho) = \text{ tr } (\mathbf{\Lambda}) = \sum_{i=1}^{p} \lambda_i = \sum_{i=1}^{p} V(W_i) = p,$$

$$\rho_{W_i, Y_k} = \alpha_{ik}\sqrt{\lambda_i}, \ i, k = 1, \ldots, p.$$

Proof. Follows easily from Theorems 8.2.1 and 8.2.2.

In this case, the proportion of total variance of Y explained by the kth principal component is λ_k/p.

One should decide whether the components are to be extracted from the covariance matrix $\mathbf{\Sigma}$ or the correlation matrix ρ. The proportion of variation explained by the corresponding components using each method will be different. Generally, the covariance matrices produce components with large variances because of the scale of measurement of $\mathbf{X}$.

EXAMPLE 8.2.1: Suppose the random variables X_1, X_2, X_3 have covariance matrix

$$\mathbf{\Sigma} = \begin{bmatrix} 8 & 0 & 1 \\ 0 & 8 & 3 \\ 1 & 3 & 5 \end{bmatrix}.$$

It can be seen that

$$\begin{aligned} \lambda_1 &= 10, \quad \alpha_1 = (-.267, -.802, -.535)', \\ \lambda_2 &= 8, \quad \alpha_2 = (.949, -.316, 0)', \\ \lambda_3 &= 3, \quad \alpha_3 = (-.169, -.507, .845)'. \end{aligned}$$

Therefore, the principal components are

$$\begin{aligned} Z_1 &= -.267X_1 - .802X_2 - .535X_3 \\ Z_2 &= .949X_1 - .316X_2 \\ Z_3 &= -.169X_1 - .507X_2 + .845X_3. \end{aligned}$$

It is readily seen that $\sum \sigma_{ii} = \sum \lambda_i$. The proportion of total variance accounted for by the first PC is $\lambda_1 / \sum \lambda_i = .48$. The first two components account for .86 of the total variance. Again,

$$\rho_{Z_1,X_1} = \frac{\alpha_{11}\sqrt{\lambda_1}}{\sqrt{\sigma_{11}}} = -.299,$$

$$\rho_{Z_1,X_2} = \frac{\alpha_{12}\sqrt{\lambda_1}}{\sqrt{\sigma_2}} = -.897,$$

$$\rho_{Z_1,X_3} = \frac{\alpha_{13}\sqrt{\lambda_1}}{\sqrt{\sigma_{33}}} = -.757.$$

The principal component Z_1 depends most on X_2, then on X_3 and X_1 in that order. This is also reflected on the absolute values of $\alpha_{11}, \alpha_{12}, \alpha_{13}$. The principal component Z_2 does not depend on X_3. It is a contrast between X_1 and X_2. The principal component Z_3 depends mostly on X_3, then on X_2 and X_1. This is a contrast between X_3 on one hand and X_1, X_2 on the other. Similarly, the other correlations can be found out and interpreted.

If $\mathbf{\Sigma}$ is converted into the correlation matrix,

$$\rho = \begin{bmatrix} 1.000 & 0 & .158 \\ 0 & 1.000 & .474 \\ .158 & .474 & 1.000 \end{bmatrix},$$

its eigenvalues and eigenvectors are,

$$\begin{aligned} \lambda_1 &= 1.50, \quad \alpha_1 = (-.224, -.671, -.707)', \\ \lambda_2 &= 1.0, \quad \alpha_2 = (.949, -.316, 0)' \\ \lambda_3 &= .50, \quad \alpha_3 = (-.224, -.671, .707)'. \end{aligned}$$

Therefore, the PC's are

$$\begin{aligned} Z_1 &= -.224X_1 - .671X_2 - .707X_3 \\ Z_2 &= .949X_1 - .316X_2 \\ Z_3 &= -.224X_1 - .671X_2 + .707X_3 \end{aligned} .$$

Also,

$$\rho(Z_1, X_1) = -.274, \ \rho(Z_1, X_2) = -.671, \ \rho(Z_1, X_3) = -.5.$$

The eigenvalues are obviously different. The eigenvectors are also different, though, surprisingly, α_2 remains the same. Here Z_1 accounts for .50 and Z_1, Z_2 together for .83 of the total variance. The rank of absolute values of $\rho(Z_1, X_i)(i = 1, 2, 3)$ remains the same as for $\mathbf{\Sigma}$, though the rank of absolute values of coefficients $\alpha_{1i}(i = 1, 2, 3)$ has changed, now X_3 has the highest contribution to the first principal component. All other correlations can be worked out and interpreted similarly.

8.3 Principal Components of a Multivariate Normal Distribution

Suppose $\mathbf{X}$ is distributed as $N_p(\mu, \mathbf{\Sigma})$. Then it is known that the probability distribution $f(\mathbf{x}; \mu, \mathbf{\Sigma})$ is a constant on the ellipsoid defined by

$$(\mathbf{x}-\mu)'\mathbf{\Sigma}^{-1}(\mathbf{x}-\mu) = c^2(\text{ a constant }). \quad (8.3.1)$$

This ellipsoid has center μ and axes $\pm c\sqrt{\lambda_i}\alpha_i$ where (λ_i, α_i) are eigenvalue-eigenvector pairs of $\mathbf{\Sigma}, (i = 1, \ldots, p), \lambda_1 \geq \lambda_2 \geq \ldots \geq \lambda_p > 0$. Any point lying on the ith axis of the ellipsoid (8.3.1) has coordinates proportional to $\alpha_i = (\alpha_{i1}, \ldots, \alpha_{ip})'$ in a coordinate system having origin μ and axes that are parallel to the axes $X_1, \ldots, X_p$. We assume without loss of generality that $\mu = \mathbf{0}$. Now

$$\begin{aligned} c^2 &= \mathbf{x}'\mathbf{\Sigma}^{-1}\mathbf{x} \\ &= \mathbf{x}'(\textstyle\sum_{i=1}^p \lambda_i^{-1}\alpha_i\alpha_i')\mathbf{x}(\text{by spectral decomposition of } \mathbf{\Sigma}^{-1}) \\ &= \textstyle\sum_{i=1}^p \frac{\mathbf{x}'\alpha_i\alpha_i'\mathbf{x}}{\lambda_i} \\ &= \textstyle\sum_{i=1}^p (\alpha_i'\mathbf{x})^2/\lambda_i, \ \lambda_i > 0. \end{aligned}$$

Note that $\alpha_1'\mathbf{X}, \ldots, \alpha_i'\mathbf{X}_p$ are the principal components of $\mathbf{X}$. Letting $z_i = \alpha_i'\mathbf{x}$, we have

$$c^2 = \sum_{i=1}^p \frac{z_i^2}{\lambda_i}, \ \lambda_i > 0. \quad (8.3.2)$$

The equation (8.3.2) represents an ellipsoid whose axes are in the direction of $Z_1, \ldots, Z_p$, i.e. in the direction $\alpha_1, \ldots, \alpha_p$, respectively, with respect to the original coordinate system in $X_1, \ldots, X_p$. Any point in the ith principal axis has $\mathbf{x}$-coordinates proportional to $\alpha_i = (\alpha_{i1}, \ldots, \alpha_{ip})'$ and $\mathbf{z}$-coordinate $(0, \ldots, 0, z_i, 0, \ldots, 0)$ with respect to the changed coordinate system in $Z_1, \ldots, Z_p$. If λ_1 is the largest eigenvalue, then the major axis lies in the direction of α_1. The remaining axes lie in the direction defined by $\alpha_2, \ldots, \alpha_p$, respectively.

Fig. 8.1 shows the principal components Z_1, Z_2 for a bivariate normal random vector $\mathbf{X} = (X_1, X_2)'$ having mean vector $\mathbf{0}$. The principal components are obtained by rotating the original coordinate axes through an angle θ until they coincide with the axes of the ellipse.

8.4 Sample Principal Components

Suppose a sample of size n is drawn from a population with mean μ and dispersion matrix $\mathbf{\Sigma}$, the sample observations being represented by the data matrix $\mathbf{X} = ((x_{ij})), x_{ij}$, being the ith observation on the variable $X_j(i = 1, \ldots, n; j = 1, \ldots, p)$. Assume for simplicity that $\mathbf{X}$ is of full rank p. If some of the variables are linearly dependent on the other variables, rank of $\mathbf{X}$ will be reduced and we do not envisage such situations.

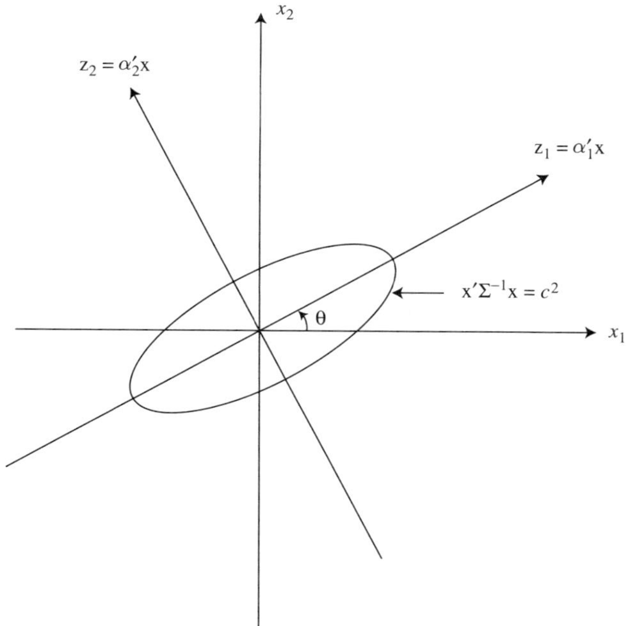

Fig. 8.1: The constant density ellipse $\mathbf{x}'\mathbf{\Sigma}^{-1}\mathbf{x} = c^2$ and the principal components Z_1, Z_2 for a bivariate normal random vector $\mathbf{X}$ having mean $\mathbf{0}$.

The results obtained in Section 8.2 will be directly applicable only if $\mathbf{\Sigma}$ is known. However, it will be unknown and has to be estimated from the sample data.

The estimate of $\mathbf{\Sigma}$ will be the usual sample covariance matrix

$$\mathbf{S} = ((s_{jj'})) = \frac{1}{n-1} \sum_{k=1}^{n} (\mathbf{x}_k - \bar{\mathbf{x}})(\mathbf{x}_k - \bar{\mathbf{x}})' \tag{8.4.1}$$

where

$$\mathbf{x}_k = (x_{k1}, \ldots, x_{kp})', \quad \bar{\mathbf{x}} = (\bar{x}_1, \ldots, \bar{x}_p)', \quad \bar{x}_j = \frac{1}{n}\sum_{k=1}^{n} x_{kj},$$

$$s_{jj'} = \frac{1}{n-1}\sum_{i=1}^{n}(x_{ij} - \bar{x}_j)(x_{ij'} - \bar{x}_{j'}).$$

If $\mathbf{X}$ follows a $N_p(\mu, \mathbf{\Sigma})$ distribution, $\bar{\mathbf{x}}$ and $\mathbf{S}_n = \frac{n-1}{n}\mathbf{S}$ are the maximum likelihood estimators of μ and $\mathbf{\Sigma}$ respectively.

As noted before, a decision has to be made as to whether one should work with the covariance matrix of the original observations x_{ij} or should work with the covariance matrix of the transformed observations

$$y_{ij} = \frac{x_{ij} - \bar{x}_j}{\sqrt{s_{jj}}}.$$

If the responses are in widely different units, linear components of original observations would have little meaning and the standardized observations y_{ij} and the correlation matrix $\mathbf{R}$ should be used. Conversely, if the responses are in reasonably commensurable units, the covariance form $\mathbf{S}$ has a more realistic statistical justification. Furthermore, the sampling theory of components extracted from the sample correlation matrix is exceedingly more complex than that of the covariance-matrix components (Anderson, 1963a).

The following theorem is obtained following Theorems 8.2.1 and 8.2.2.

Theorem 8.4.1: Let $(l_1, \mathbf{a}_1), \ldots, (l_p, \mathbf{a}_p)$ be eigenvalue-eigenvector pairs of $\mathbf{S}$. The kth Sample principal component is given by

$$\xi_k = \mathbf{a}_k'\mathbf{x} = a_{k1}x_1 + a_{k2}x_2 + \ldots + a_{kp}x_p \tag{8.4.2}$$

with sample variance

$$\begin{aligned} V(\xi_k) &= \tfrac{1}{n-1}\textstyle\sum_{i=1}^{n}(a_{k1}x_{i1} + a_{k2}x_{i2} + \ldots + a_{kp}x_{ip} - \\ &\quad a_{k1}\bar{x}_1 - a_{k2}\bar{x}_2 - \ldots - a_{kp}\bar{x}_p)^2 \\ &= \textstyle\sum_j \sum_{j'=1}^{p} a_{kj}a_{kj'}s_{jj'} \\ &= \mathbf{a}_k'\mathbf{S}\mathbf{a}_k = \mathbf{a}_k' l_k \mathbf{a}_k \quad (\text{ since } \mathbf{S}\mathbf{a}_k = l_k\mathbf{a}_k) \\ &= l_k \end{aligned} \tag{8.4.3}$$

and sample covariance

$$\text{Cov }(\xi_k, \xi_{k'}) = 0 \quad (k \neq k').$$

The total sample variance of X is

$$\sum_{i=1}^{p} s_{ii} = \sum_{i=1}^{p} l_i.$$

The contribution of the kth principal component to the total sample variance is, therefore, given by

$$l_k / \sum_{i=1}^{p} l_i.$$

Also, the sample covariance $s_{\xi_k, x_j} = l_k a_{kj}$ and sample correlation

$$r_{\xi_k, X_j} = \frac{a_{kj}\sqrt{l_k}}{\sqrt{s_{jj}}}. \tag{8.4.4}$$

Note that if $l_i = l_{i+1}$, the elements of $\mathbf{a}_i$ and $\mathbf{a}_{i+1}$ can be chosen to be orthogonal in an infinite number of ways.

Sometimes, the observations are centered at the sample mean $\bar{\mathbf{x}}$. In this case, the sample covariance $\mathbf{S}_n$ remains the same and the kth PC is

$$\xi_k = \mathbf{a}'_k(\mathbf{x} - \bar{\mathbf{x}}), \ k = 1, \ldots, p.$$

For the ith observation $\mathbf{x}_i = (x_{i1}, \ldots, x_{ip})'$, the value of the kth P.C. is

$$\xi_{ik} = a'_k(\mathbf{x}_i - \bar{\mathbf{x}}), \ i = 1, \ldots, n.$$

Hence,

$$\bar{\xi}_k = \frac{1}{n}\sum_{i=1}^{n} \xi_{ik} = \frac{1}{n} a'_k \sum_{i=1}^{n} (\mathbf{x}_i - \bar{\mathbf{x}}) = 0.$$

We note below some more properties of sample principal components.

(i) The following theorem shows that $r^2_{\xi_k, X_j}$ is a component of the squared multiple correlation of X_j on ξ's.

Theorem 8.4.2: Let r_{ξ_k, X_j} be the correlation between the variable X_j and the kth Principal Component ξ_k. Then

$$r^2_{\xi_1, X_j} + r^2_{\xi_2, X_j} + \ldots + r^2_{\xi_t, X_j} = R^2_{X_j \cdot \xi_1, \ \ldots, \xi_t} \tag{8.4.5}$$

where t is the number of components retained and $R^2_{X_j \cdot \xi_1, \ldots, \xi_t}$ is the squared multiple correlation of X_j with the ξ_k's.

Proof. We have

$$R^2_{X_j \cdot \xi_1, \ldots, \xi_k} = \frac{\mathbf{S}'_{X_j \xi} \mathbf{S}^{-1}_{\xi\xi} \mathbf{S}_{X_j \xi}}{s_{jj}}$$

where $\mathbf{S}_{X_j\xi}$ is the vector covariances of X_j with $\xi = (\xi_1, \ldots, \xi_t)', \mathbf{S}_{\xi\xi}$ is the covariance matrix of ξ and s_{jj} is the variance of X_j (vide (6.10.22)). Now, by (8.4.4),

$$\mathbf{S}'_{X_j\xi} = (l_1 a_{1j}, l_2 a_{2j}, \ldots, l_t a_{tj}).$$

Again,

$$\mathbf{S}_{\xi\xi} = \text{ Diag } (l_1, \ldots, l_t).$$

Hence,

$$\begin{aligned} R^2_{X_j.\xi_1,\ldots,\xi_t} &= \textstyle\sum_{k=1}^t \frac{l_k a_{kj}^2}{s_{jj}} \\ &= \textstyle\sum_{k=1}^t r^2_{X_j,\xi_t} \quad \text{(by (8.4.4))} . \end{aligned}$$

Corollary 8.4.2.1: If $t = p$, i.e. all the p components are used, then $R^2_{X_j.\xi_1,\ldots,\xi_p} = 1$.

Proof.

$$\begin{aligned} R^2_{X_j.\xi_1,\ldots,\xi_p} &= \textstyle\sum_{k=1}^p r^2_{X_j\xi_t} \\ &= \frac{1}{s_{jj}} \textstyle\sum_{k=1}^p l_k a_{kj}^2. \end{aligned} \tag{8.4.6}$$

From the spectral decomposition

$$\mathbf{S}_n = \sum_{k=1}^p l_k \mathbf{a}_k \mathbf{a}'_k,$$

the sum $\sum_{k=1}^t l_k a_{kj}$ in (8.4.5) is s_{jj}. Hence the result.

(ii) The principal components are not scale-invariant, i.e., the eigenvalues and eigenvectors of the original variables do not remain the same as the eigenvalues and eigenvectors of the transformed variables.

Suppose $\mathbf{X}$ is transformed to $\mathbf{Y}$ by $\mathbf{Y} = \mathbf{CX}$ where $\mathbf{C}$ is nonsingular. The sample covariance matrix of $\mathbf{Y}$ is $\mathbf{S}(y) = \mathbf{CSC}'$ where $\mathbf{S} = \mathbf{S}(x) =$ Cov $(\mathbf{X})$.

A principal component of $\mathbf{Y}$ is of the form $\mathbf{b}'_j\mathbf{Y} = \mathbf{b}'_j\mathbf{CX}$ where $\mathbf{b}_j$ is a eigenvector of $\mathbf{S}(y) = \mathbf{CSC}'$. A principal component of $\mathbf{X}$ is of the form

$$\mathbf{a}'_j\mathbf{X}$$

where $\mathbf{a}_j$ is a eigenvector of $\mathbf{S}$. Moreover,

$$\mathbf{b}'_j\mathbf{CX} \neq \mathbf{a}'_j\mathbf{X}$$

because $(\mathbf{C}')^{-1}\mathbf{a}_j$ is not a eigenvector of $C\mathbf{S}C'$.

To see that scaling has an effect on eigenvalues, note that

$$\sum_{j=1}^{p} s_{jj} = \text{ tr } (\mathbf{S}) = \sum_{j=1}^{p} l_j.$$

If an s_{jj} is increased by a scale change, then one or more of the l_j's will also increase.

(iii) Note that if one of the variables, say, X_p is uncorrelated with other variables, then its variance s_{pp} is one of the eigenvalues of $\mathbf{S}$ and the corresponding eigenvector is $(0, \ldots, 0, 1)'$. Hence, X_p itself is a principal component. In general, if all the p-variables are uncorrelated, the principal component analysis will reproduce these variables. Hence, if correlations are small, there is little to gain from a principal component analysis.

(iv) Again, note that in addition to explaining a portion of the total variance, trace (S), the first few principal components provide the best approximation (in some sense) to the covariance matrix $\mathbf{S}$ as follows. By spectral decomposition

$$\mathbf{S} = l_1 \mathbf{a}_1 \mathbf{a}_1' + l_2 \mathbf{a}_2 \mathbf{a}_2' + \ldots + l_p \mathbf{a}_p \mathbf{a}_p'. \tag{8.4.7}$$

Hence the first k terms in (8.4.6) provide an approximation to $\mathbf{S}$:

$$\tilde{\mathbf{S}} = l_1 \mathbf{a}_1 \mathbf{a}_1' + \ldots + l_k \mathbf{a}_k \mathbf{a}_k'. \tag{8.4.8}$$

This is the best approximation to $\mathbf{S}$ of rank r in the sense that the sum of squares of elements of $\mathbf{S} - \tilde{\mathbf{S}}$ is minimum (Rao, 1964).

(v) The principal components of $\mathbf{R}$ and $\mathbf{R}_k$ are same, where

$$\mathbf{R} = \begin{bmatrix} 1 & r_{12} & \ldots & r_{1p} \\ r_{21} & 1 & \ldots & r_{2p} \\ . & . & \ldots & . \\ r_{p1} & r_{p2} & \ldots & 1 \end{bmatrix} \quad \text{and} \quad \mathbf{R}_k = \begin{bmatrix} 1 & kr_{12} & \ldots & kr_{1p} \\ kr_{21} & 1 & \ldots & kr_{2p} \\ . & . & \ldots & . \\ kr_{p1} & kr_{p2} & \ldots & 1 \end{bmatrix},$$

$$-1 \leq kr_{ij} \leq 1.$$

To see this note that

$$\mathbf{R}_k = k\mathbf{R} - (k-1)\mathbf{I}.$$

The eigenvalues θ_k and eigenvectors $\mathbf{b}_k$ of $\mathbf{R}_k$ are defined as:

$$(\mathbf{R}_k - \theta_k \mathbf{I})\mathbf{b}_k = \mathbf{0},$$

i.e.

$$[k\mathbf{R} - (k-1)\mathbf{I} - \theta_k \mathbf{I}]\mathbf{b}_k = \mathbf{0},$$

i.e.

$$[\mathbf{R} - (\frac{k-1+\theta_k}{k})\mathbf{I}]\mathbf{b}_k = \mathbf{0}. \tag{8.4.9}$$

Again, the eigenvector for $\mathbf{R}$ are:

$$[\mathbf{R} - l\mathbf{I}]\mathbf{a} = \mathbf{0}. \tag{8.4.10}$$

Comparing (8.4.9) and (8.4.10) we see that eigenvectors $\mathbf{b}_k$ and $\mathbf{a}$ are the same and the eigenvalues are related by

$$l = \frac{k-1+\theta_k}{k}$$

or

$$\theta_k = k(l-1)+1.$$

Hence, the principal components of $\mathbf{R}$ and $\mathbf{R}_k$ are the same.

The following theorems by Girshick (1936) and Anderson (1958) give the estimates of population principal components.

Theorem 8.4.3: Let $\hat{\mu}, \hat{\mathbf{\Sigma}}$ denote moment estimators of μ and $\mathbf{\Sigma}$ (maximum likelihood estimates for μ and $\mathbf{\Sigma}$, if $\mathbf{X}$ follows $N_p(\mu, \mathbf{\Sigma})), \mathbf{\Sigma} > 0)$. The moment estimators (maximum likelihood estimators) of the principal components of $\mathbf{X}$ and of the variances of principal components of $\mathbf{X}$ are obtained by substituting $\hat{\mathbf{\Sigma}}$ for $\mathbf{\Sigma}$ in Theorem 8.2.1.

Note 8.4.1: Unusually small values of the last few eigenvalues of the sample covariance or correlation matrix indicate almost linear dependency of some variables on the remaining variables in the data set. The multi-collinearity in the data set is indicated by small eigenvalues. In such situations some of the variables are redundant and should be deleted. Thus, although large eigenvalues and the corresponding eigenvectors are important in principal component analysis, the eigenvalues close to zero should not be overlooked. The eigenvectors associated with these eigenvalues may indicate linear dependency in the data set which should be considered for interpretation.

8.5 Principal Components of Covariance Matrices with Special Structures

In this section we consider principal components of $\mathbf{X}$ when $\mathbf{\Sigma}$ has some special forms.

(a) Suppose,

$$\mathbf{\Sigma} = \text{Diag}\ (\sigma_{11}, \ldots, \sigma_{pp}).$$

Here,

$$\lambda_k = \sigma_{kk},\ \alpha_k = (0, \ldots, 0, 1, 0, \ldots, 0)',$$

$$Z_k = \alpha_k' \mathbf{X} = X_k. \tag{8.5.1}$$

Again,

$$\rho = \mathbf{I},\ \rho\alpha_k = 1\alpha_k \quad (k = 1, \ldots p)$$

so that for the correlation matrix ρ, eigenvalue 1 has multiplicity p and $\alpha_k (k = 1, \ldots p)$ are the convenient choices for its eigenvectors. The standardized variables are

$$Y_i = \frac{X_i - \mu_i}{\sqrt{\sigma_{ii}}},\ i = 1, \ldots, p$$

and the corresponding principal components are

$$W_k = \alpha_k' \mathbf{Y} = Y_k,\ k = 1, \ldots, p.$$

(b) Suppose $\mathbf{\Sigma}$ has intraclass correlation structure

$$\mathbf{\Sigma} = \mathbf{\Sigma}_0 \text{ (say) } = \sigma^2 \begin{bmatrix} 1 & \rho & \ldots & \rho \\ \rho & 1 & \ldots & \rho \\ \cdot & \cdot & \ldots & \cdot \\ \rho & \rho & \ldots & 1 \end{bmatrix} = \sigma^2 \rho_0 \text{ (say)}. \tag{8.5.2}$$

We shall require $0 < \rho < 1$. The p eigenvalues of ρ_0 are, when $\rho > 0$,

$$\lambda_1 = 1 + (p-1)\rho,$$

$$\lambda_2 = \ldots = \lambda_p = 1 - \rho.$$

We have

$$\alpha_1 = (\frac{1}{\sqrt{p}}, \ldots, \frac{1}{\sqrt{p}})'.$$

The orthogonal vectors $\alpha_2, \ldots, \alpha_p$ $(\alpha_i'\alpha_1 = 0, i = 2, \ldots, p)$ can be chosen in an infinite number of ways. One set of choices for the eigenvectors are the elements of Helmert's matrix:

$$\alpha_2 = (\tfrac{1}{\sqrt{1.2}},\ -\tfrac{1}{\sqrt{1.2}},\ 0, \ldots, 0)',$$
$$\alpha_3 = (\tfrac{1}{\sqrt{2.3}},\ \tfrac{1}{\sqrt{2.3}},\ -\tfrac{2}{\sqrt{2.3}}, \ldots, 0),'$$
$$\ldots$$
$$\alpha_i = (\tfrac{1}{\sqrt{(i-1)i}},\ \tfrac{1}{\sqrt{(i-1)i}},\ \ldots,\ \tfrac{1}{\sqrt{(i-1)i}}, -\tfrac{(i-1)}{\sqrt{(i-1)i}}, 0, \ldots, 0)',$$
$$\ldots$$
$$\alpha_p = (\tfrac{1}{\sqrt{(p-1)p}}, \tfrac{1}{\sqrt{(p-1)p}}, \ldots, \tfrac{1}{\sqrt{p-1}}, -\tfrac{p}{\sqrt{(p-1)p}})'.$$

The first component is

$$Z_1 = \alpha_1'\mathbf{X} = \frac{1}{\sqrt{p}}\sum_{i=1}^{p} X_i$$

which explains a proportion

$$\frac{\lambda_1}{p} = \frac{1+(p-1)\rho}{p} = \rho + \frac{1-\rho}{p}$$

of the total population variance. Note that

$$\frac{\lambda_1}{p} \approx \rho \quad \text{for} \quad \rho \approx 1 \quad \text{or } p \quad \text{large.}$$

For example, if $\rho = .80$ and $p = 4$, the first component explains 85% of the total variance. When ρ is very close to one, the last $(p-1)$ components contribute very little to the total variance.

The equal variance-covariance structure matrix has, therefore, a single component which explains a major proportion of the total variation. The remaining $(p-1)$ principal components have isotopic variation (i.e. equal variances) and are not uniquely defined.

(c) *The equiprobability covariance matrix*:

Bock (1960) considered the following patterned covariance matrix due to Bergmann (1957):

$$\mathbf{\Sigma} = \begin{bmatrix} \sigma^2 & \sigma_{12} & \sigma_{13} & \sigma_{14} \\ \sigma_{12} & \sigma^2 & \sigma_{14} & \sigma_{13} \\ \sigma_{13} & \sigma_{14} & \sigma^2 & \sigma_{12} \\ \sigma_{14} & \sigma_{13} & \sigma_{12} & \sigma^2 \end{bmatrix}. \tag{8.5.3}$$

Such matrices have the property that the multiple correlation coefficient of one variable on the remaining variables is always the same, whatever be the dependent variable.

It can be verified that the matrix in (8.5.3) will reduce to its canonical form under pre and post multiplication by the orthogonal matrix

$$\begin{bmatrix} 1 & 1 & 1 & 1 \\ 1 & 1 & -1 & -1 \\ 1 & -1 & 1 & -1 \\ 1 & -1 & -1 & 1 \end{bmatrix}.$$

For this matrix, the principal component structure is:

Characteristic root	*Characteristic vector*
$\sigma^2 + \sigma_{12} + \sigma_{13} + \sigma_{14}$	(1/2 1/2 1/2 1/2)'
$\sigma^2 + \sigma_{12} - \sigma_{13} - \sigma_{14}$	(1/2 1/2 -1/2 -1/2)'
$\sigma^2 - \sigma_{12} + \sigma_{13} - \sigma_{14}$	(1/2 -1/2 1/2 -1/2)'
$\sigma^2 - \sigma_{12} - \sigma_{13} + \sigma_{14}$	(1/2 -1/2 -1/2 1/2)'

Note 8.5.1: It has been shown (Brauer, 1953) that for a general covariance matrix,

$$\lambda_1 \leq \max_{i(=1,\ldots,p)} (\sum_{j=1}^{p} \sigma_{ij}). \tag{8.5.4}$$

If $\mathbf{\Sigma}$ has the intraclass correlation pattern $\mathbf{\Sigma}_0$, the upper bound in (8.5.4) is attained. As $\mathbf{\Sigma}$ departs from the structure Σ_0, the upper bound becomes less sensitive. However, the upper bound in (8.5.4) gives an idea of the maximum amount of total variation ascribable to the first principal component.

Note 8.5.2: Another pattern for $\mathbf{R}$ or $\mathbf{S}$ is that all correlations or covariances are positive, but not equal. In this situation, a lower bound for the first eigenvalue of $\mathbf{R}$ is

$$\lambda_1 \geq 1 + (p-1)\tilde{r}$$

where $\tilde{r}$ is the average of the off-diagonal elements of $\mathbf{R}$ (Meyer, 1975).

EXAMPLE 8.5.1: Consider the correlation matrix

$$\mathbf{R} = \begin{bmatrix} 1.000 & .7234 & .7089 & .6958 \\ .7234 & 1.000 & .7184 & .6574 \\ .7089 & .7184 & 1.000 & .7832 \\ .6958 & .6574 & .7832 & 1.000 \end{bmatrix}.$$

The eigenvalues, eigenvectors of this matrix are:

$$\begin{aligned} l_1 &= 3.144,\ a_1 = (-.497, -.492, -.512, -.499)', \\ l_2 &= .370,\ a_2 = (.380, .599. - .344, -.616)', \\ l_3 &= .281,\ a_3 = (.768, -.564, -.290, .088)', \\ l_4 &= .205,\ a_4 = (.135, -.286, .732, -.603)'. \end{aligned}$$

We note that $l_1 = 3.144 \approx 1 + (p-1)\bar{r} = 1 + 3 \times .7145 = 3.1435$, where $\bar{r}$ is the average of the off-diagonal elements of $\mathbf{R}$. The remaining eigenvalues are small and nearly equal. It appears that the corresponding population correlation matrix of the intraclass correlation structure ρ_0. The first principal component is

$$W_1 = -.497Y_1 - .497Y_2 - .512Y_3 - .499Y_4$$

and accounts for $l_1/4 = .79$ of the total variance. This example will be further investigated in Section 8.7.

8.6 Geometrical Interpretation of Sample Principal Components

We shall illustrate with observations on a 3-dimensional vector of variables $\mathbf{X} = (X_1, X_2, X_3)', x_{ij}$ being the ith observation on variable $X_j (j = 1, 2, 3)$. The original observations are measured in co-ordinate system with X_1, X_2, X_3 as the axes. The cluster of observations $\mathbf{X} = ((x_{ij}))$ will often look like one as constituting an ellipsoid with major axis in the direction of ξ_1 and the remaining two less well-defined orthogonal axes ξ_2, ξ_3 as shown in Figure 8.2. Let us first interpret the first principal component.

Let ξ_1 make angles $\theta_1, \theta_2, \theta_3$ with the axes X_1, X_2, X_3 respectively. The ξ_1-value (projection) of observation $\mathbf{x}_i = (x_{i1}, x_{i2}, x_{i3})'$ on the new co-ordinate axis ξ_1 will be

$$\xi_{i1} = a_{11}(x_{i1} - \bar{x}_1) + a_{12}(x_{i2} - \bar{x}_2) + a_{13}(x_{i3} - \bar{x}_3) \qquad (8.6.1)$$

where

$$a_{11} = \cos\theta_1,\ a_{12} = \cos\theta_2,\ a_{13} = \cos\theta_3, \qquad (8.6.2)$$

and

$$a_{11}^2 + a_{12}^2 + a_{13}^2 = 1.$$

Here we have assumed that the origin of the new coordinate system (ξ_1, ξ_2, ξ_3) is at the point $(\bar{x}_1, \bar{x}_2, \bar{x}_3)$ of the old co-ordinate system. Then, the mean of ξ_1 is

$$\bar{\xi}_1 = \frac{1}{n}\sum_{i=1}^{n} \xi_{i1} = \sum_{j=1}^{3} a_{1j} \sum_{i=1}^{n} (x_{ij} - \bar{x}_j) = 0.$$

We now define the concept of major axis of a cluster of points as that axis which is in the direction of maximum variation among observations. Thus, this major axis is defined as the axis with respect to which the transformed observations $\xi_{i1} (i = 1, \ldots, n)$ has the maximum variation. Since the choice of $a_{1j}(j = 1, \ldots, 3)$ uniquely determines the orientation of ξ_1 this is obtained by minimizing $V(\xi_1)$ with respect to a_{11}, a_{21}, a_{31}. Now, the solution would be the characteristic vector of the largest latest root of the sample covariance matrix of x_{ij} and the first principal axis would be

$$\xi_1 = a_{11}(X_1 - \bar{x}_1) + a_{12}(X_2 - \bar{x}_2) + a_{13}(X_3 - \bar{x}_3).$$

We now prove the result for the case of p response variables $X_1, \ldots, X_p$. Let the direction cosines of first principal axis be $\mathbf{a}_1 = (a_{11}, a_{12}, \ldots, a_{1p})'$ where we must have

$$\mathbf{a}_1'\mathbf{a}_1 = 1. \qquad (8.6.3)$$

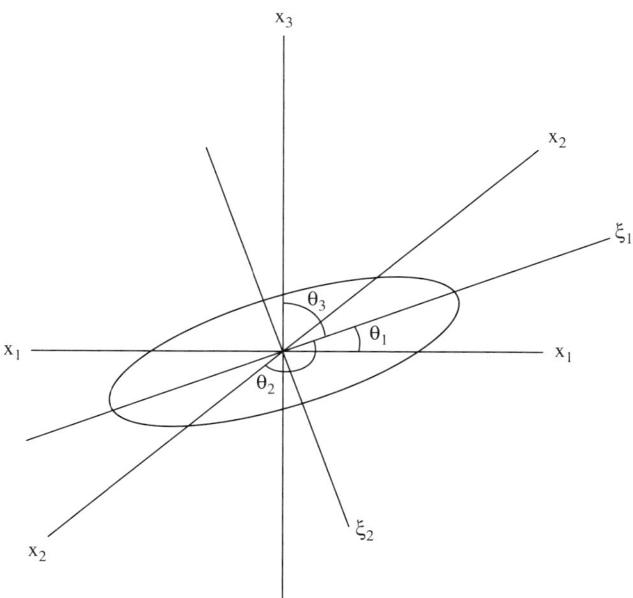

Fig. 8.2: An ellipsoid with three principal components

The variance of the transformed observations (8.6.1) on the ξ_1 axis is

$$\begin{aligned} s_{\xi_1}^2 &= \tfrac{1}{n-1} \textstyle\sum_{i=1}^n (\xi_{i1} - \bar{\xi}_1)^2 = \tfrac{1}{n-1} \sum_{i=1}^n \xi_{i1}^2 \\ &= \tfrac{1}{n-1} \textstyle\sum_{i=1}^n [\sum_{j=1}^p a_{1j}(x_{ij} - \bar{x}_j)]^2 \\ &= \tfrac{1}{n-1} \textstyle\sum_{i=1}^n [(x_i - \bar{x})'\mathbf{a}_1]^2 \\ &= \tfrac{1}{n-1} \textstyle\sum_{i=1}^n a_1'(x_i - \bar{x})(x_i - \bar{x})'\mathbf{a}_1 \\ &= \mathbf{a}_1'\mathbf{S}\mathbf{a}_1. \end{aligned}$$

To maximize $s^2_{\xi_1}$ with respect to $\mathbf{a}_1$ subject to the condition (8.6.3) we have to maximize the Lagrangian

$$\mathbf{a}_1'\mathbf{S}\mathbf{a}_1 + t_1(1 - \mathbf{a}_1'\mathbf{a}_1)$$

where t_1 is Lagrangian multiplier. This is precisely similar to the equation (8.2.5). The direction cosines of the first principal axis are the elements of the first characteristic vector of $\mathbf{S}$. The maximized variance is the largest characteristic root which determines the length of the first principal axis. It is found that $t_1 = l_1$.

Similarly, the remaining characteristic roots and vectors of $\mathbf{S}$ determine the length and orientation of second and higher component axes. If two successive roots l_i and l_{i+1} are equal, the corresponding major axes are not uniquely determined and the cluster of points in those directions look more like circular than elliptical. Such dispersion is called *isotopic* in those dimensions with equal l_i.

8.7 Large Sample Properties of Sample Principal Components

Assume that $\mathbf{X}_1, \mathbf{X}_2, \ldots, \mathbf{X}_n$ is a random sample from a $N_p(\mu, \mathbf{\Sigma})$ distribution. It is also assumed that

$$\lambda_1 > \lambda_2 > \ldots > \lambda_p > 0.$$

Let

$$\mathbf{\Lambda} = \text{ Diag } (\lambda_1, \ldots, \lambda_p).$$

The following results hold in large sample. The results are due to Girshick (1936, 1939), Anderson (1951, 1963 a), Bartlett (1954), Lawley (1956, 1963), among others. The tests based on these large sample are generally conservative in the sense that the true p-values will be frequently higher than those obtained from these asymptotic tests.

Let $(l_i, \mathbf{a}_i)(i = 1, \ldots, p)$ denote the eigenvalue-eigenvector pairs of $\mathbf{S}$.

(i)

$$\sqrt{n}(\mathbf{l} - \lambda) \sim N_p(\mathbf{0}, 2\mathbf{\Lambda}^2),$$

where $\mathbf{l} = (l_1, \ldots, l_p)', \lambda = (\lambda_1, \ldots, \lambda_p)'$.

(ii) Let

$$E_i = \lambda_i \sum_{k(\neq i)=1}^{p} \frac{\lambda_k}{(\lambda_k - \lambda_i)^2} \alpha_k \alpha_{\mathbf{k}}'.$$

Then

$$\sqrt{n}(\mathbf{a}_i - \alpha_i) \sim N_p(\mathbf{0}, \mathbf{E}_i).$$

(iii) Each l_i is distributed independently of the elements of $\mathbf{a}_i$.

(iv) The covariance of the rth element of $\mathbf{a}_i$ and sth element of $\mathbf{a}_j$ is

$$-\frac{\lambda_i \lambda_j \alpha_{ir} \alpha_{js}}{n(\lambda_i - \lambda_j)^2}, \quad i \neq j.$$

Anderson (1963 a) pointed out that the result (ii) requires that only λ_i be distinct from the other $(p-1)$ roots which may have any number of multiplicities. The result (i), (iii), (iv) strictly require that all the roots must be distinct.

The result in (i) shows that for large n,

$$l_i \sim N(\lambda_i, 2\lambda_i^2/n) \tag{8.7.1}$$

and l_i is distributed independently of $l_j (i \neq j)$.

Hence,

$$P\{[l_i - \lambda_i| < \lambda_i \sqrt{\frac{2}{n}} \tau_{\alpha/2}] = 1 - \alpha, \tag{8.7.2}$$

where $\tau_{\alpha/2}$ is the $100(1-\alpha/2)\%$ point of the standard normal distribution. A large sample $100(1-\alpha)\%$ confidence interval for λ_i is, therefore,

$$P[-\sqrt{\frac{2}{n}} \lambda_i \tau_{\alpha/2} \leq l_i - \lambda_i \leq \sqrt{\frac{2}{n}} \lambda_i \tau_{\alpha/2}] = 1 - \alpha$$

i.e.

$$P[\lambda_i \{1 - \sqrt{\frac{2}{n}} \tau_{\alpha/2}\} \leq l_i \leq \lambda_i \{1 + \sqrt{\frac{2}{n}} \tau_{\alpha/2}\}] = 1 - \alpha$$

i.e.

$$P[\frac{l_i}{1 + \sqrt{\frac{2}{n}} \tau_{\alpha/2}} \leq \lambda_i \leq \frac{l_i}{1 - \sqrt{\frac{2}{n}} \tau_{\alpha/2}}] = 1 - \alpha. \tag{8.7.3}$$

Bonferroni-type simultaneous $100(1-\alpha)\%$ confidence interval for λ_i is obtained by replacing $\tau_{\alpha/2}$ by $\tau_{\alpha/2m}$. We have,

$$P[\frac{l_i}{1 + \sqrt{\frac{2}{n}} \tau_{\alpha/2m}} \leq \lambda_i \leq \frac{l_i}{1 - \sqrt{\frac{2}{n}} \tau_{\alpha/2m}}] \geq 1 - \alpha \tag{8.7.4}$$

where m is the number of λ_i's in which we are interested.

It is also possible to generalize the asymptotic confidence interval (8.7.3) in the case of multiple roots. If $\lambda_{q+1} = \ldots = \lambda_{q+r} = \lambda_j$ (say) the $100(1-\alpha)\%$ asymptotic confidence interval for λ_j is

$$\frac{\tilde{l}_j}{1+\tau_{\alpha/2}\sqrt{\frac{2}{nr}}} \leq \lambda_j \leq \frac{\tilde{l}_j}{1-\tau_{\alpha/2}\sqrt{\frac{2}{nr}}} \tag{8.7.5}$$

where

$$\tilde{l}_j = \frac{1}{r}(l_{q+1} + \ldots + l_{q+r}).$$

Result (ii) implies that the $\mathbf{a}_i$'s are normally distributed about mean vector α_i in large samples. The elements of each $\mathbf{a}_i$ are correlated and the correlation depends to a large extent on the separation of eigenvalues $\lambda_1, \ldots, \lambda_p$ and the sample size n. Approximate standard error for the elements a_{ij} of $\mathbf{a}_i$ is given by the square root of the jth diagonal element of $\hat{\mathbf{E}}_i$ divided by n, where $\hat{\mathbf{E}}_i$ is obtained from $\mathbf{E}_i$ by substituting l_i for λ_i and $\mathbf{a}_i$ for α_i.

For the equi-correlation matrix ρ_0 of (8.5.2) we have

$$\begin{aligned} \text{Var }(a_{1r}) &= \tfrac{[1+(p-1)\rho](p-1)(1-\rho)}{np^3\rho^2}, \\ \text{Cov }(a_{1r}, a_{1s}) &= -\tfrac{[1+(p-1)\rho](1-\rho)}{np^3\rho^2}. \end{aligned}$$

Hence,

$$\text{Corr. }(a_{1r}, a_{1s}) = -\frac{1}{p-1}.$$

Another problem of interest is one concerning a population correlation matrix with two different characteristic roots λ_1 and λ_2 of respective multiplicities q_1 and q_2, $q_1 + q_2 = p$. Let the characteristic roots of the sample correlation matrix $\mathbf{R}$ be $l_1 > l_2 > \ldots > l_{q_1} > \ldots > l_p$. We take

$$\hat{\lambda}_1 = \frac{l_1 + \ldots + l_{q_1}}{q_1}, \; \hat{\lambda}_2 = \frac{l_{q_1+1} + \ldots + l_p}{q_2}. \tag{8.7.6}$$

It has been shown (Anderson, 1963a) that in large sample $\hat{\lambda}_2$ is approximately normally distributed about mean λ_2 and variance

$$\frac{2\lambda_2^2(p - q_2\lambda_2)^2}{npq_1q_2}. \tag{8.7.7}$$

Hence, $100(1-\alpha)\%$ confidence interval for λ_2 is given by the fact:

$$P\{\frac{(\hat{\lambda}_2 - \lambda_2)^2}{\lambda_2^2(p - q_2\lambda_2)^2}\frac{npq_1q_2}{2} \leq \tau_{\alpha/2}^2\} = 1 - \alpha. \tag{8.7.8}$$

EXAMPLE 8.7.1: For Jackson's data of Exercise 8.8 compute confidence interval for λ_2. Take $n = 100$.

Here $l_1 = 335.335, l_2 = 48.034, l_3 = 29.330, l_4 = 16.410$. The 95% confidence interval for λ_2 is (37.609, 66.454).

8.7.1 *Tests of hypotheses*

We now consider tests of different hypotheses relating to λ_i's and α_i's.

(a) To test the hypothesis $H_0(\lambda_i = \lambda_{i0})$ against $H_1(\lambda_i \neq \lambda_{i0})$, we compute

$$\tau = \frac{l_i - \lambda_{i0}}{\lambda_{i0}\sqrt{\frac{2}{n}}} \tag{8.7.9}$$

and reject H_0 if $|\tau| > \tau_{\alpha/2}$.

(b) Suppose we want to test the hypothesis

$$H_{01} : \lambda_{j+1} = \ldots = \lambda_p.$$

Bartlett (1951) gave the following test statistic

$$u = (n - \frac{2p+11}{6})\{(p-j)\ln \bar{l}_j - \sum_{k=j+1}^{p} \ln l_k\} \tag{8.7.10}$$

where

$$\bar{l}_j = \frac{1}{p-j}\sum_{k=j+1}^{p} l_k.$$

The statistic u follows under the null hypothesis χ^2 with $(s-1)(s+2)/2$ degrees of freedom where $s = p - j$.

For the hypothesis

$$H_0'(\lambda_{j+1} = \ldots = \lambda_{j+r}), j + r \leq p, \tag{8.7.11}$$

Anderson (1963a) gave the following likelihood ratio statistic

$$Q = r(n-1)\{\ln \bar{l}_j - \frac{1}{r}\sum_{t=1}^{r} \ln l_{j+t}\} \tag{8.7.12}$$

which under the null hypothesis follows asymptotically a χ^2 with $[\frac{r(r+1)}{2} - 1]$ degrees of freedom.

If $j + r = p$, so that $r = p - j$, then from (8.7.12),

$$Q = (n-1)\{(p-j)\ln \bar{l}_j - \sum_{k=j+1}^{p} \ln l_k\}$$

which is almost equal to the test statistic obtained from (8.7.10).

(c) For testing the hypothesis

$$H_{02}: \ \alpha_i = \alpha_{i0}, \tag{8.7.13}$$

that the characteristic vector associated with the distinct root λ_i of $\mathbf{\Sigma}$ is equal to some specified value, the test statistic is (Anderson, 1963a),

$$U = n(l_i\alpha'_{i0}\mathbf{S}^{-1}\alpha_{i0} + \frac{1}{l_i}\alpha'_{i0}\mathbf{S}\alpha_{i0} - 2) \tag{8.7.14}$$

which is asymptotically distributed as χ^2 with $(p-1)$ degrees of freedom.

EXAMPLE 8.7.2: Suppose we want to test

$$H_{03}: \ \mathbf{\Sigma} = \mathbf{\Sigma}_0 = \sigma^2 \begin{bmatrix} 1 & \rho & \dots & \rho \\ \rho & 1 & \dots & \rho \\ . & . & \dots. & . \\ \rho & \rho & \dots & 1 \end{bmatrix}, \ -\frac{1}{p-1} < \rho < 1, \tag{8.7.15}$$

against the alternative hypothesis H_3 : $\mathbf{\Sigma}$ is a positive definite matrix. This is equivalent to testing $H_{03} : \lambda_2 = \ldots = \lambda_p$ against the alternative the last $(p-1)$ roots are not equal. The test statistic is, using Bartlett's test,

$$u = (p-1)(n - \frac{2p+11}{6})\{\ln \bar{l}_1 - \frac{1}{p-1}\sum_{j=2}^{p} \ln l_j\}. \tag{8.7.16}$$

Here, u follows central χ^2 with $[\frac{p(p-1)}{2} - 1]$ degrees of freedom under the null hypothesis.

An *alternative test* for the hypothesis H_{03} based on maximum likelihood ratio is given below.

(c) Consider the null hypothesis H_{03} : $\mathbf{\Sigma} = \mathbf{\Sigma}_0$ given in (8.7.15). The maximum likelihood estimators (adjusted for bias) for σ^2 and $\sigma^2\rho$ under H_{03} are:

$$\tilde{s}^2 = \frac{1}{p}\sum_{j=1}^{p} s_{jj}, \quad \tilde{s}^2 r = \frac{1}{p(p-1)}\sum_{j \neq k} s_{jk}. \tag{8.7.17}$$

Hence, under H_{03}, an estimate of $\mathbf{\Sigma}$ is

$$\mathbf{S}_0 = \tilde{s}^2[(1-r)\mathbf{I} + r\mathbf{J}] \tag{8.7.18}$$

where r can be found from (8.7.18). A test statistic based on the likelihood ratio (LR) is

$$\begin{aligned} u = (LR)^{2/n} &= \frac{\mathbf{S}|}{|\mathbf{S}_C|} \\ &= \frac{|\mathbf{S}|}{(\tilde{s}^2)^p(1-r)^{p-1}[1+(p-1)r]}. \end{aligned} \tag{8.7.19}$$

Box (1949) showed that under H_{03},

$$u' = -[\nu - \frac{p(p+1)^2(2p-3)}{6(p-1)(p^2+p-4)}] \ln u \tag{8.7.20}$$

is approximately distributed as $\chi^2_{(\frac{p(p+1)}{2}-2)}$, where ν is the degrees of freedom of $\mathbf{S}$. Nagarsenkar (1975) gave the exact distribution of u and provided a table for $p = 4, 5, \ldots, 10$. A nonparametric test was obtained by Choi (1977).

Box (1950) gave the following approximation which is more precise for larger p or smaller ν.

$$F = -\frac{(\eta_2 - \eta_2 c_1 - \eta_1)}{\eta_1 \eta_2} \ln u \tag{8.7.21}$$

which follows a F-distribution with (η_1, η_2) degrees of freedom, where

$$c_1 = \frac{p(p+1)^2(2p-3)}{6\nu(p-1)(p^2+p-4)}, \quad c_2 = \frac{p(p^2-1)(p+2)}{6\nu^2(p^2+p-4)},$$

$$\eta_1 = \frac{p(p+1)}{2} - 2, \quad \eta_2 = \frac{\eta_1 + 2}{c_2 - c_1^2}.$$

(d) Consider

$$H_{04} : \rho = \rho_0 = (1-\rho)\mathbf{I} + \rho\mathbf{J} \tag{8.7.22}$$

where $\mathbf{J}$ is a $p \times p$ matrix with 1 in all the places, against the alternative $H_4 : \rho$ is of a general form. The likelihood ratio is intractable and Lawley (1963) proposed the following procedure on heuristic arguments. Compute

$$\bar{r}_k = \frac{1}{p-1} \sum_{i(\neq k)=1}^{p} r_{ik}, \; k = 1, \ldots, p,$$

$$\bar{r} = \frac{2}{p(p-1)} \sum\sum_{i<k} r_{ik},$$

$$\hat{\gamma} = \frac{(p-1)^2[1-(1-\bar{r})^2]}{p-(p-2)(1-\bar{r})^2},$$

where r_{ik} is the sample correlation coefficient between X_i and X_k. The large sample approximate α-level test has the form : reject H_{04} in favor of H_4 if

$$T^2 = \frac{n-1}{(1-\bar{r})^2}[\sum\sum_{i<k}(r_{ik}-\bar{r})^2 - \hat{\gamma}\sum_{k=1}^{p}(\bar{r}_k - \bar{r})^2]$$

$$> \chi^2_{\nu;\alpha}$$

where $\nu = (p+1)(p-2)/2$ and $\chi^2_{\nu;\alpha}$ is the upper $(100)(1-\alpha)$th percentile of a chi-square distribution with ν d.f.

Note 8.7.1: Under $H_{03} : \mathbf{\Sigma} = \mathbf{\Sigma}_0$, maximum likelihood estimate of λ_2 is

$$\hat{\lambda}_2 = \frac{1}{p-1} \sum_{k=1}^{p-1} l_{1+k}.$$

Again, asymptotic distribution of $\hat{\lambda}_2$ is given by

$$\sqrt{n-1}(\hat{\lambda}_2 - \lambda_2) \sim N(0, \frac{2\lambda_2^2}{p-1})$$

From this result large sample confidence interval for λ_2 can be found out.

EXAMPLE 8.7.3: For the data of Example 8.5.1 test the null hypothesis H_{04}. Take $n = 100$.

Here $\bar{r}_1 = .7094, \bar{r}_2 = .6997, \bar{r}_3 = .7368, \bar{r}_4 = .7121, \bar{r} = .7145$.

$$\sum\sum_{i<k}(r_{ik}-\bar{r})^2 = .00845,\ \sum_{k=1}^{4}(r_k-\bar{r})^2 = .00075,\ \bar{\gamma} = 2.15440,\ T^2 = 8.2955$$

Now, $\chi^2_{5;.05} = 11.07$. Hence, hypothesis H_{04} is accepted.

EXAMPLE 8.7.4: For the data of Example 8.5.1 test the hypothesis $H_0(\lambda_2 = \lambda_3 = \lambda_4)$. Take $n = 100$.

Using Bartlett's test $u = 8.33799 < \chi^2_{5;.05} = 11.07$. Hence the null hypothesis is accepted. By Anderson's test $Q = 8.52108$. The null hypothesis is accepted.

EXAMPLE 8.7.5: For Jackson's data of Exercise 8.8 test the null hypothesis $H_0(\alpha_2 = \alpha'_{20} = (.7,\ .1,\ -.1,\ .7)'$. Take $n = 100$.

Here

$$\begin{aligned}
\alpha_1 &= (-0.468143,\ 0.621474,\ 0.571554,\ -0.260649)' \\
\alpha_2 &= (-0.607927,\ 0.178760,\ -0.759463,\ -0.147258)' \\
\alpha_3 &= (-0.458988,\ -0.138745,\ 0.167733,\ 0.861363)' \\
\alpha_4 &= (-0.447880,\ -0.750043,\ 0.261548,\ -0.410404)'.
\end{aligned}$$

Also $l_2 = 48.034$. The test statistic is

$$\begin{aligned}
U &= 100(l_2\alpha'_{20}\mathbf{S}^{-1}\alpha_{20} + \tfrac{1}{l_2}\alpha'_{20}\mathbf{S}\alpha_{20} - 2) \\
&= 38.831.
\end{aligned}$$

This is greater than $7.815 = \chi^2_{3;.05}$. Hence the hypothesis is rejected.

8.8 Last Few Principal Components

The kth eigenvalue of $\mathbf{S}$ is the variance of the kth principal component ($k = 1, \ldots, p$). Hence, the last few principal components have smallest variances. A principal component with a very small variance defines a linear relationship among the variables such that the observed values of this linear relationship remains more or less constant over the observations. Thus, an extremely small eigenvalue signals a collinearity among the variables. The last few principal components, therefore, provide approximately linear relationships which exist among the variables.

This result may be used in research for detection of outliers that deviate significantly from the above (approximately) linear relationships. Suppose that X_7 is highly linearly correlated with $X_1, \ldots, X_6$ so that $X_7 = c_1X_1 + \ldots + c_6X_6 + \epsilon$ where the variables are measured against their means. Thus the linear combination $X_7 - c_1X_1 - \ldots - c_6X_6$ will have a very small variance and this fact will be reflected in one of the principal components. A plot of the values of this principal component will show outlier values of X, if any, that deviate markedly from this observed relationship.

Now,

$$\mathbf{a}_1\mathbf{a}_1' + \ldots + \mathbf{a}_p\mathbf{a}_p' = \mathbf{I}_p. \tag{8.8.1}$$

Multiplying both sides by the observation vector $\mathbf{x}_i$,

$$\mathbf{a}_1\mathbf{a}_1'\mathbf{x}_i + \ldots + \mathbf{a}_p\mathbf{a}_p'\mathbf{x}_i = \mathbf{I}_p\mathbf{x}_i.$$

Now,

$$\mathbf{a}_1'\mathbf{x}_i = \xi_{i1},$$

the value of the first principal component ξ_1 for the observed vector $\mathbf{x}_i$. Similarly,

$$\mathbf{a}_2'\mathbf{x}_i = \xi_{i2},$$

etc. Thus,

$$\mathbf{a}_1\xi_{i1} + \ldots + \mathbf{a}_p\xi_{ip} = \mathbf{x}_i,\ i = 1, \ldots, n. \tag{8.8.2}$$

If the first k principal components are only retained, we can use the residual

$$\mathbf{d}_i = \mathbf{a}_{k+1}\xi_{ik+1} + \ldots + \mathbf{a}_p\xi_{ip} \tag{8.8.3}$$

to see how well the first k components fit the observations. Now,

$$\mathbf{d}_i'\mathbf{d}_i = \xi_{ik+1}^2 + \ldots + \xi_{ip}^2. \tag{8.8.4}$$

We compute the residuals $\mathbf{d}_i'\mathbf{d}_i (i = 1, \ldots, n)$. Thus, an observation with an unusually large value of $\mathbf{d}_i'\mathbf{d}_i$ will indicate a poor performance by the first k principal components which may be due to the fact that the observation $\mathbf{x}_i$ is an outlier with respect to the underlying correlation structure.

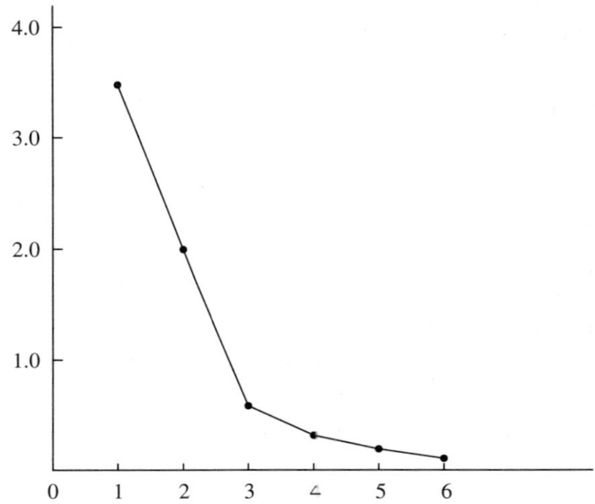

Fig. 8.3: A Scree Plot. An elbow occurs at $i = 2$ or perhaps 3

The quantities $\xi_{i1}, \xi_{i2}, \ldots, \xi_{ik},\ k < m$ are called the *component scores* for the ith observation and the Matrix of component scores can be written as

$$\begin{bmatrix} \xi_{11} & \xi_{12} & \cdots & \xi_{1k} \\ \xi_{21} & \xi_{22} & \cdots & \xi_{2k} \\ \cdot & \cdot & \cdots & \cdot \\ \xi_{n1} & \xi_{n2} & \cdots & \xi_{nk} \end{bmatrix} = \begin{bmatrix} \mathbf{x}_1' \\ \mathbf{x}_2' \\ \cdot \\ \mathbf{x}_n' \end{bmatrix} (\mathbf{a}_1\ \mathbf{a}_2\ \ldots \mathbf{a}_k)$$

or

$$\xi = \mathbf{XA}, \tag{8.8.5}$$

where ξ is of order $n \times k, \mathbf{X}$ of order $n \times p$ and $\mathbf{A}$ of order $p \times k$.

The plot of the first two principal components may reveal some important features of the data set. One of the objectives of plotting is to check for departures from normality, such as outliers or nonlinearity. It has been pointed out (Gnanadesikan, 1997; p.308) that, in general, the first few principal components are sensitive to outliers that inflate variances or distort covariances, and the last few are sensitive to outliers that introduce the artificial dimensions. One should examine the bivariate plots of at least the first two and the last two principal components in search for outliers.

8.8.1 *Number of principal components to retain*

There is no hard and fast rule for determining the number of principal components to be retained. If a variance-covariance matrix $\mathbf{S}$ is analyzed, only those components whose associated eigenvalues are significantly different from zero, should be retained. However, even among these components, some components may make only marginal contribution to the total variance. Such components may be ignored in the final analysis. Another criterion may be to look into the the values of $\theta_m = \sum_{j=1}^{m} l_j / \sum_{i=1}^{p} l_i$, $m = 1, 2, \ldots$ and retain only the first m PC's for which $\theta_m \approx 1$.

The above rules do not apply if a correlation matrix is analyzed. Here, one suggestion is to retain those components whose eigenvalues are greater than one. Another visual approach proposed by Cattell (1966) is to draw a *scree plot.* The eigenvalues l_i are plotted against i, $(l_1 > l_2 > \ldots)$ and are joined by straight lines. To determine the appropriate number of components to be retained, one looks for an elbow (bend) in the scree plot. The number of components to be retained is taken to be the point at which the elbow is found (Figure 8.3). Other suggestions are due to Cortella and Jespers (1967), Horn (1965), among others.

8.9 Exercises and Complements

8.1 Determine the population principal component Y_1, Y_2 for the covariance matrix $\begin{bmatrix} 5 & 7 \\ 7 & 2 \end{bmatrix}$. Also, calculate the proportion of the total variance explained by the first principal component.

8.2. Reduce the covariance matrix in Exercise 8.1 to a correlation matrix ρ. Determine the Principal Component Y_1 and Y_2 from ρ and compute the proportion of total variance explained by Y_1. Compare the components with those obtained before. Compute the correlations $\rho_{Y_1,Z_1}, \rho_{Y_1,Z_2}$ and ρ_{Y_2,Z_1} where Z_1, Z_2 are the standardized variables.

8.3 The following table produces observations on x_1 and x_2.

Observation	x_1	x_2
1	2	3
2	3	5
3	1	4
4	4	4
5	6	2
6	8	5
6	2	7
8	3	10

(a) Compute the variance of each variable and percentage of total variance accounted for by each variable.

(b) Let Z_1 be any axis in a two-dimensional space making an angle θ with X_1. Projection of the observations on Z_1 gives the coordinates z_1 of the observations with respect to Z_1. Express z_1 as a function of θ, x_1, x_2.

(c) For what values of θ does Z_1^* have maximum variance? What percentage of the total variance is accounted for by Z_1^*?

8.4 Let

$$\mathbf{\Sigma} = \begin{bmatrix} 8 & 0 & 1 \\ 0 & 8 & 3 \\ 1 & 3 & 5 \end{bmatrix}.$$

(a) Derive the Principal Components.

(b) Show that $\lambda_1 + \lambda_2 + \lambda_3 =$ trace (Σ).

(c) Show that $\lambda_1 \times \lambda_2 \times \lambda_3 = \mid \Sigma \mid$.

8.5 Suppose the dispersion matrix of $\mathbf{Y} = (Y_1, \ldots, Y_p)'$ is

$$\mathbf{\Sigma} = c\mathbf{I} + \mathbf{J}.$$

Show that p is the non-zero eigenvalue of $\mathbf{J}$. Show also that $c+p, c, \ldots, c$ are the eigenvalues of $\mathbf{\Sigma}$ and that $Z_1 = \sqrt{p}\bar{Y}$ is the first principal component of $\mathbf{\Sigma}$, where $\bar{Y} = \sum_{j=1}^{p} Y_j/p$.

8.6 Show that if the variable X_p is uncorrelated with all other variables, its variance σ_{pp} is an eigenvalue of $\mathbf{\Sigma}$ with the corresponding eigenvector $(0, \ldots, 0, 1)'$.

8.7 Consider the data in Table 11.E.3 where y_1, y_2, y_3 and x_1, x_2, x_3 represent measurements of blood glucose levels on three occasions, on fasting and after one hour of sugar-intake, respectively, for 52 women. Carry out a principal component analysis using all the variables $y_1, y_2, y_3, x_1, x_2, x_3$ and using $\mathbf{S}$ and $\mathbf{R}$. On the basis of the average eigenvalue or a scree-plot, decide on the number of components to retain. Interpret the components of $\mathbf{S}$ and $\mathbf{R}$. Which set of components, you think, is more appropriate for these data ? Also, find the scores for the first two principal components for each observation and display them in a scatter plot.

8.8 In an experiment for testing the ballistic missiles for total impulse, two identical thrust gages were attached to the head of a rocket during static testing. Each of these gages was connected to a separate recording system consisting of (1) an electronic integration (2) an oscilloscope and camera. The variables are: x_1 = Gage (1), integration reading; x_2 = Gage (2), planimeter measurement; x_3 = Gage (2), integration reading; x_4 = Gage (2), planimeter measurement. The sample covariances for these variables were as follows (Jackson, 1959).

$$\mathbf{S} = \begin{bmatrix} 102.74 & 88.67 & 67.04 & 54.06 \\ 88.67 & 142.74 & 86.56 & 80.03 \\ 67.04 & 86.56 & 84.57 & 69.42 \\ 54.06 & 80.03 & 69.42 & 99.06 \end{bmatrix}.$$

Perform a principal component analysis.

8.9 Consider the correlation matrix:

$$\mathbf{R} = \begin{bmatrix} 1.000 & .920 & .875 & .625 \\ .920 & 1.000 & .589 & .750 \\ .875 & .589 & 1.000 & .425 \\ .625 & .750 & .425 & 1.000 \end{bmatrix}.$$

Test the null hypothesis that ρ is a equi-correlation matrix $\rho_0 = (1-\rho)\mathbf{I}+\rho\mathbf{J}$.

8.10 The lengths of the humerus, ulma, tibia and femur bones of 276 leghorn fowl were found to have the following correlation matrix (Wright, 1954):

$$\mathbf{R} = \begin{bmatrix} 1 & .940 & .875 & .878 \\ .940 & 1 & .877 & .886 \\ .875 & .877 & 1 & .924 \\ .878 & .886 & .924 & 1 \end{bmatrix}.$$

Perform a principal component analysis.

8.7 Consider the d[illegible]
resent measurements [illegible]
and after the door [illegible]
on principal [illegible]
analysing S and R [illegible]
decide on the [illegible]
of S and R [illegible]
these data [illegible]
[illegible]

8.8 In an experiment [illegible]
identical [illegible]
testing. Each [illegible]
consisting of [illegible]
The variables [illegible]
planimeter [illegible]
12), planimeter [illegible]
were as follows [illegible]

[illegible]

Perform a principal [illegible]

8.9 Consider the [illegible]

[illegible]

Test the null [illegible]

8.10 The length [illegible]
[illegible]

[illegible]

Perform a principal [illegible]

Chapter 9

Factor Analysis

9.1 Introduction

Like Principal Component analysis, factor analysis also attempts to explain the covariance (or correlation) among a large number of variables in terms of a smaller number of factors. However, unlike principal components, these factors cannot be observed and are unobservable random variables. Such a situation is particularly suitable for studies in subjects like psychology where it is not possible to measure exactly the concepts one is interested in, eg., intelligence, kindness, devotion.

Basically, the factor analysis is motivated by the following consideration. Suppose, variables can be grouped by their correlations. That is, all variables within a particular group are highly correlated among themselves, but have relatively small correlations with variables in a different group. It can then be conceived that each group of variables represents a single underlying factor that is responsible for correlations. For example, correlations from the test scores on literature, English language, Mathematics, Statistics, Physics, Music, Painting, a 200-meter race, boxing, table-tennis may support three underlying factors, - intelligence, aptitude for art and body-fitness, each of which will produce high correlations among scores on tests in its own group.

The latent factor analysis was first suggested by Galton (1888). The mathematical model for factor analysis was initiated by Spearman (1904) in the context of analyzing a set of intelligence-test scores and subsequently developed by Thurstone and Thurstone (1941), Thurstone (1947), among others.

9.2 The Orthogonal Factor Model

Let $\mathbf{X} = (X_1, X_2, \ldots, X_p)'$ be a observable $p \times 1$ random vector with mean $\mu = (\mu_1, \mu_2, \ldots, \mu_p)'$ and covariance matrix $\mathbf{\Sigma}$. The factor model postulates that $\mathbf{X}$ is linearly dependent upon a few unobservable random factor $F_1, \ldots, F_m$ and p additional sources of variations, called *errors* or *specific factors.* The factor $F_1, \ldots, F_m$ are common to all the variable $X_1, \ldots, X_m$ and are called *common factors.* The error ϵ_i is specific to variable X_i and therefore, $\epsilon_1, \ldots, \epsilon_p$ are called specific factors.

Thus the factor analysis model is:

$$\begin{aligned} X_1 - \mu_1 &= l_{11}F_1 + l_{12}F_2 + \ldots + l_{1m}F_m + \epsilon_1 \\ X_2 - \mu_2 &= l_{21}F_1 + l_{22}F_2 + \ldots + l_{2m}F_m + \epsilon_2 \\ .. &= ...... \\ X_p - \mu_p &= l_{p1}F_1 + l_{p2}F_2 + \ldots + l_{pm}F_m + \epsilon_p \end{aligned} \tag{9.2.1}$$

or in matrix notation

$$\mathbf{X} - \mu = \mathbf{LF} + \epsilon \tag{9.2.2}$$

where $\mathbf{L} = ((l_{ij})), \mathbf{F} = (F_1, F_2, \ldots, F_m)'$ and $\epsilon = (\epsilon_1, \ldots, \epsilon_p)'$. The coefficient l_{ij} is called loading of the variable X_i on the factor F_j. We assume that

$$E(\mathbf{F}) = \mathbf{0}, \quad \text{Cov } (\mathbf{F}) = \mathbf{I},$$

$$E(\epsilon) = \mathbf{0}, \quad \text{Cov } (\epsilon) = \text{ Diag. } (\psi_1, \ldots, \psi_p) = \mathbf{\Psi},\ \psi_i > 0, \tag{9.2.3}$$

$$\text{Cov } (\epsilon, \mathbf{F}) = \mathbf{0}.$$

The relation (9.2.2) along with the assumption (9.2.3) constitute the *orthogonal factor model.*

From (9.2.2),

$$\text{Cov } (\mathbf{X}) = \mathbf{\Sigma} = \text{ Cov } (\mathbf{LF} + \mathbf{\Psi}) = \mathbf{LL}' + \mathbf{\Psi}, \tag{9.2.4}$$

$$\text{Cov } (\mathbf{X}, \mathbf{F}) = \text{ Cov } (\mathbf{LF} + \epsilon, \mathbf{F}) = \mathbf{L}. \tag{9.2.5}$$

The set of equations (9.2.4) is called the *factor pattern* and the $p \times m$ matrix $\mathbf{L}$ is called the *pattern matrix* or *loading matrix.* From (9.2.4) it follows ,therefore,

$$\sigma_{ii} = \sum_{j=1}^{m} l_{ij}^2 + \psi_i = l_{i1}^2 + \ldots l_{im}^2 + \psi_i, i = 1, \ldots, p$$

$$\sigma_{ik} = \sum_{j=1}^{m} l_{ij} l_{kj} = l_{i1} l_{k1} + \ldots + l_{im} l_{km}, i \neq k = 1, \ldots, p. \tag{9.2.6}$$

From (9.2.5),

$$\text{Cov }(X_i, F_j) = l_{ij}.$$

The portion of $\sigma_{ii} = V(X_i)$ accounted for by the m common factors $F_1, \ldots, F_m$ is called the ith *communality*, often denoted as h_i^2. The remaining part of σ_{ii} is a measure of the extent to which the common factors fail to account for the variance of X_i and is called the *specific variance.* Thus

$$\begin{aligned} \sigma_{ii} &= (l_{i1}^2 + \ldots l_{im}^2) + \psi_i \\ &= h_i^2 + \psi_i = \text{ Communality + Specific Variance.} \end{aligned} \tag{9.2.7}$$

Clearly, in any factor analysis problem one should choose $m < p$ as, otherwise, there is no gain in terms of summarization of data using smaller number of variables as factors. Also, one should make h_i^2 as large as possible, so that variable X_i is well-explained by the common factors.

The total contribution of the factor F_j to the total variance of $\mathbf{X} = \sum_{i=1}^{p} \sigma_{ii}$ is

$$\begin{aligned} V_j &= \textstyle\sum_{i=1}^{p} l_{ij}^2 \\ &= \mathbf{l}_{(j)}' \mathbf{l}_{(j)} \end{aligned} \tag{9.2.8}$$

where $\mathbf{l}_{(j)} = (l_{1j}, \ldots, l_{pj})'$ is the jth column of $\mathbf{L}$.

The total contribution of all the common factors $F_1, \ldots, F_m$ to the total variance is

$$\begin{aligned} V &= \textstyle\sum_{j=1}^{m} V_j = \sum_{j=1}^{m} \sum_{i=1}^{p} l_{ij}^2 \\ &= \textstyle\sum_{i=1}^{p} h_i^2. \end{aligned} \tag{9.2.9}$$

Therefore, the proportion of the contribution of the factor F_j to that part of the total variance of $\mathbf{X}$ that is accounted for by the common factors $F_1, \ldots, F_m$ is

$$W_j = \frac{V_j}{V}. \tag{9.2.10}$$

The factor model assumes that the $\frac{p(p+1)}{2}$ parameters of $\mathbf{\Sigma}$ can be expressed in terms of $(pm + p)$ factor loadings and specific variances. Specifically, it assumes that $\mathbf{\Sigma}$ can be factored as $\mathbf{\Sigma} = \mathbf{FF}' + \mathbf{\Psi}$. When $p = m$, every covariance matrix $\mathbf{\Sigma}$ can be written in the form $\mathbf{\Sigma} = \mathbf{LL}'$ with $\mathbf{\Psi} = \mathbf{0}$ (vide A.11(6)). However, cases of interest are $m < p$.

We are actually required to determine $pm + p$ unknown parameters l_{ij} and ψ by solving $p(p+1)/2$ equations. The requirement of identifiability is that the number of parameters should be less than the number of equations. Hence, we must have

$$pm + p < \frac{p(p+1)}{2}, \text{ or } m < \frac{p-1}{2}.$$

Thus, m must be fairly small. However, this does not guarantee that a solution will exist (vide Example 9.2.2).

Note 9.2.1: Note that unlike the multivariate regression model where the independent variables (regressors) are observable, the factors $F_1, \ldots, F_m$ in the factor model are unobservable random variables.

Note 9.2.2: The model (9.2.2) is linear in common factors $F_1, \ldots, F_m$. If the p responses are, in fact, related to the underlying factors in a nonlinear way, like $X_1 - \mu_1 = l_{11}F_1 + l_{12}F_2F_3 + \epsilon_1, X_2 - \mu_2 = l_{21}F_1F_2 + \epsilon_2$, etc., then the covariance structure given in (9.2.4) would not hold.

EXAMPLE 9.2.1: Consider the following covariance matrix,

$$\mathbf{\Sigma} = \begin{bmatrix} 41 & 33 & 47 & 85 \\ 33 & 50 & 45 & 93 \\ 47 & 45 & 67 & 117 \\ 85 & 93 & 117 & 225 \end{bmatrix}.$$

Now,

$$\begin{aligned} \mathbf{\Sigma} &= \begin{bmatrix} 5 & 3 \\ 3 & 6 \\ 7 & 4 \\ 11 & 10 \end{bmatrix} \begin{bmatrix} 5 & 3 & 7 & 11 \\ 3 & 6 & 4 & 10 \end{bmatrix} + \text{ Diag. } (7, 5, 2, 4) \\ &= \mathbf{LL}' + \mathbf{\Psi}. \end{aligned}$$

Therefore, $\mathbf{\Sigma}$ can be factorized as (9.2.4) with $m = 2$. We can, therefore, write

$$\begin{aligned} X_1 - \mu_1 &= 5F_1 + 3F_2 + 7 \\ X_2 - \mu_2 &= 3F_1 + 6F_2 + 5 \\ X_3 - \mu_3 &= 7F_1 + 4F_2 + 2 \\ X_4 - \mu_4 &= 11F_1 + 10F_2 + 4\,. \end{aligned}$$

For X_1, communality is $h_1^2 = l_{11}^2 + l_{12}^2 = 5^2 + 3^2 = 34$ and specific variance $\psi_1 = \sigma_{11} - h_1^2 = 7$. Similarly the other communalities and specific variances can be calculated.

EXAMPLE 9.2.2: The following example shows that a factor model does not always exist. Let

$$\rho = \begin{bmatrix} 1 & .4 & .9 \\ .4 & 1 & .7 \\ .9 & .7 & 1 \end{bmatrix}.$$

Consider a one-factor model

$$Z_1 = l_{11}F_1 + \epsilon_1,$$
$$Z_2 = l_{21}F_1 + \epsilon_2,$$
$$Z_3 = l_{31}F_1 + \epsilon_3.$$

Hence,

$$l_{11}^2 + \psi_1 = 1, \; l_{11}l_{21} = .4, \; l_{11}l_{31} = .9,$$
$$l_{22}^2 + \psi_2 = 1, \; l_{21}l_{31} = .7, \; l_{31}^2 + \psi_3 = 1.$$

Hence, $l_{11}/l_{21} = .9/.7$, or $l_{21} = (.7l_{11})/.9$. Again, $l_{11}l_{21} = (.7l_{11}^2)/.9 = .4$ or $l_{11} = \underline{+}(.6)/\sqrt{.7}$. Also, $l_{31} = .9/l_{11} = \underline{+}(.9\sqrt{.7})/.6$ or $l_{31}^2 = 1.575$. This gives $\psi_3 = -.575$. But ψ_3 cannot be negative. Hence, there is no one-factor solution in this case.

The case when h_i^2 exceeds 1 (when factoring ρ) and hence $\psi_i < 0$ is known as a *Heywood* case (Heywood, 1931).

9.2.1 *Scale-invariance of factor model*

Suppose a m-factor model (9.2.2) holds for $\mathbf{X}$. If the variables are standardized so that $\mathbf{Z} = \mathbf{V}^{-1/2}(\mathbf{X} - \mu)$ is the vector of standardized variables, where $\mathbf{V} =$ Diag. $(\sigma_{11}, \ldots, \sigma_{pp})$, then the covariance matrix ρ of $\mathbf{Z}$ is

$$\begin{aligned} \rho &= \text{Cov}\ (\mathbf{Z}) \\ &= \text{Cov}\ [\mathbf{V}^{-1/2}(\mathbf{LF} + \epsilon)] \\ &= (\mathbf{V}^{-1/2}\mathbf{L})(\mathbf{L}'\mathbf{V}^{-1/2}) + \mathbf{V}^{-1/2}\mathbf{\Psi}\mathbf{V}^{-1/2} \\ &= \mathbf{L}_z\mathbf{L}_{z'} + \mathbf{\Psi}_z \text{ (say)} \end{aligned} \tag{9.2.11}$$

where

$$\mathbf{L}_z = \mathbf{V}^{-1/2}\mathbf{L}, \mathbf{\Psi}_z = \mathbf{V}^{-1/2}\mathbf{\Psi}\mathbf{V}^{-1/2}.$$

Also

$$\text{Cov}\ (\mathbf{Z}, \mathbf{F}) = \mathbf{V}^{-1/2}\mathbf{L} = \mathbf{L}_z.$$

Therefore, the orthogonal factor model (9.2.2), (9.2.3) also holds for $\mathbf{Z}$ with factor loading matrix $\mathbf{L}_z = \mathbf{V}^{-1/2}\mathbf{L}$ and specific factor $\mathbf{V}^{-1/2}\epsilon$ and hence specific variance $\mathbf{\Psi}_z = \mathbf{V}^{-1}\mathbf{\Psi} =$ Diag. $(\Psi/\sigma_{ii}, i = 1, \ldots, p)$. In other words, factor analysis, unlike principal component analysis, remains unaffected by rescaling of the variables.

9.2.2 *Non-uniqueness of factor loadings*

Let $\mathbf{T}$ be any $m \times m$ orthogonal matrix so that $\mathbf{TT}' = \mathbf{T}'\mathbf{T} = \mathbf{I}$. Now, we can write

$$\mathbf{X} - \mu = \mathbf{LF} + \epsilon = \mathbf{LTT}'\mathbf{F} + \epsilon = \mathbf{L}^*\mathbf{F}^* + \epsilon \tag{9.2.12}$$

where

$$\mathbf{LT} = \mathbf{L}^*, \quad \mathbf{T}'\mathbf{F} = \mathbf{F}^*.$$

We have,

$$E(\mathbf{F}^*) = \mathbf{0}, \quad \text{Cov }(\mathbf{F}^*) = \mathbf{T}' \text{ Cov }(\mathbf{F})\mathbf{T} = \mathbf{T}'\mathbf{T} = \mathbf{I},$$

$$\text{Cov }(\mathbf{F}^*, \epsilon) = \mathbf{0}.$$

Hence, all the properties (9.2.4), (9.2.5) are satisfied by the loading matrix $\mathbf{L}^*$ and factor $\mathbf{F}^*$. Thus, if $(\mathbf{L}, \mathbf{F})$ is a solution so is $(\mathbf{L}^*, \mathbf{F}^*)$, both generate the same covariance matrix

$$\mathbf{\Sigma} = \mathbf{LL}' + \mathbf{\Psi} = \mathbf{L}^*\mathbf{L}^{*'} + \mathbf{\Psi}.$$

In fact, for fixed $\mathbf{\Psi}$ and $m > 1$, this rotation is the only indeterminate step in the decomposition of $\mathbf{\Sigma}$ in terms of $\mathbf{L}$ and $\mathbf{\Psi}$ i.e. if $\mathbf{\Sigma} = \mathbf{LL}' + \mathbf{\Psi} = \mathbf{L}^*\mathbf{L}^{*'} + \mathbf{\Psi}$, then $\mathbf{L}^* = \mathbf{LT}$ for some orthogonal matrix $\mathbf{T}$.

Note that the communalities $h_i^2 = l_{i1}^2 + l_{i2}^2 + \ldots + l_{im}^2, i = 1, \ldots, p$ are also unaffected by the transformation $\mathbf{L}^* = \mathbf{LT}$. This can be seen as follows. The communality h_i^2 is the sum of squares of the ith row of $\mathbf{L}$. If we denote the ith row of $\mathbf{L}$ by $\mathbf{l}_i'$, then $h_i^2 = \mathbf{l}_i'\mathbf{l}_i$. The ith row of $\mathbf{L}^*$ is $\mathbf{l}_i^{*'} = \mathbf{l}_i'\mathbf{T}$ and the corresponding communality is

$$h_i^{*2} = \mathbf{l}_i^{*'}\mathbf{l}_i^* = \mathbf{l}_i'\mathbf{TT}'\mathbf{l}_i = \mathbf{l}_i'\mathbf{l}_i = h_i^2.$$

In general, the loading matrices $\mathbf{L}, \mathbf{L}^*$ will be different. The factors $\mathbf{F}, \mathbf{F}^* = \mathbf{T}'\mathbf{F}$ will have different interpretations, though, they will have the same statistical properties. Thus, there are an infinite number of alternative solutions for the factor analysis model. Since orthogonal matrices correspond to the rotation and reflection of the coordinate system $\mathbf{X}$, this is called *factor rotation.*

The analysis of factor model is done by imposing conditions that allows one to uniquely determine $\mathbf{L}$ and $\mathbf{\Psi}$. The loading matrix is then rotated (multiplied by some orthogonal matrix) until easily interpretable factors are determined. Once the loadings and specific variances are obtained and factors are identified, estimated values of factors for different observations, also called *factor scores* are constructed.

To uniquely determine $\mathbf{L}$, $\mathbf{\Psi}$, the following arbitrary constraints are, generally, imposed:

$$\mathbf{L}'\mathbf{\Psi}^{-1}\mathbf{L} = \text{ A diagonal matrix;} \tag{9.2.13a}$$

$$\mathbf{L}'\mathbf{V}^{-1}\mathbf{L} = \text{ A diagonal matrix.} \tag{9.2.13b}$$

Both the constraints are scale invariant. Note that when $\psi_i = 0$ for some i, the constraint (9.2.13a) can not be used.

Now, $mp+p$ parameters in $\mathbf{L}$ and $\mathbf{\Psi}$ become subject to $\frac{m(m-1)}{2}$ constraints if conditions (9.2.13a) or (9.2.13b) are imposed. Therefore, the number of unknown parameters in $\mathbf{L}$ and $\mathbf{\Psi}$ reduces to $p(m+1) - \frac{m(m-1)}{2}$. In the unconstrained $\mathbf{\Sigma}$, there are $\frac{p(p+1)}{2}$ parameters. Therefore, for obtaining advantage in factor model we must have

$$t = \frac{p(p+1)}{2} - \{p(m+1) - \frac{m(m-1)}{2}\} = \frac{1}{2}(p-m)^2 - \frac{1}{2}(p+m) > 0. \tag{9.2.14}$$

9.2.3 *Interpretation of factors*

As noted in (9.2.5), l_{ij} is the covariance between the variable X_i and the factor F_j. In some sense, l_{ij}'s tell us which variables involve which factors and to what extent. Thus by comparing the l_{ij}'s the factors which are most important for interpreting a variable can be identified. However, in large-scale problems, the pattern matrices may be complex and the interpretation of factors may be very difficult. The following procedure simplifies the problem considerably in many cases.

Starting with the first variable and the first factor in the loading matrix $\mathbf{L}$ and proceeding horizontally from left to right we circle the loading with the highest absolute value, then circle the loading with the next highest absolute value and so on for a practically acceptable number of factors. Repeat the procedure for each variable.

After all the variables have been considered, we examine each circled loading for significance. This can be done by tests of statistical significance of l_{ij}'s or on the basis of practical significance, eg., by setting a minimum value of l_{ij}^2/σ_{ii}, the proportion of the variable's variance that must be accounted for by the factor. With regard to statistical significance, it is found that for most applications with sample size less than 100, correlation between X_i

and F_j, i.e. $l_{ij}/\sqrt{\sigma_{ii}}$ should be greater then .30 in absolute value in order to be considered significant. Significant loading should be underlined and the corresponding factors should be accommodated in the factor model.

The variables which do not have any significant loading on any factor should be reviewed as to whether they should be dropped from further consideration or should be retained for their sensitivity to the research problem.

We then try to give some meaning to the retained factors on the basis of their (significant) loadings on the variables and the meaning of the variables.

If the interpretation of the factors are difficult we frequently rotate the factors by orthogonal (or oblique) transformations until some easily interpretable set of factors are attained. In practical applications, interpretation of the loading matrix is often difficult as many variables will have a number of factor loadings all of which are significant. If each variable is mainly dependent on one or two factors, interpretation becomes simple. Ideally one would like to minimize the number of significant loadings for each variable and maximize the number of insignificant loadings. This notion is related to the concepts of *simple structure* and *factor rotation* and is discussed in Section 9.5.

9.3 Estimation of Model-Parameters

Before undergoing factor analysis, the validity of factor analysis model (9.2.2), (9.2.3) should be assessed with respect to the data. Many populations have covariance matrices that do not approach the pattern $\mathbf{\Sigma} = \mathbf{LL}' + \mathbf{\Psi}$ unless m is large. On the other hand, for a population in which $\mathbf{\Sigma}$ is close to this pattern for small m, the sample covariance matrix $\mathbf{S}$ (from which $\mathbf{L}$ and $\mathbf{\Psi}$) will be estimated) may not follow this pattern due to sampling fluctuations.

A basic requirement for a factor analysis model is that the variables are not independent. This can be done by testing the hypothesis $H_0 : \rho = \mathbf{I}$ (by using the test in (5.8.11)).

It has been suggested that $\mathbf{R}^{-1}$ should be close to a diagonal matrix in order to successfully fit a factor analysis model. To this aim Kaiser (1970) proposed a measure of sampling adequacy

$$S_A = \frac{\sum\sum_{i\neq j} r_{ij}^2}{\sum\sum_{i\neq j} r_{ij}^2 + \sum\sum_{i\neq j} q_{ij}^2}$$

where $\mathbf{Q} = ((q_{ij})) = \mathbf{DR}^{-1}\mathbf{D}, \mathbf{D} = [(\text{ Diag } \mathbf{R}^{-1})^{1/2}]^{-1}$. As $\mathbf{R}^{-1}$ approaches a diagonal matrix, S_A approaches 1. Kaiser and Rice (1974) suggested that S_A should be greater than 0.8 for good results to be expected.

We shall now consider methods of estimating the factor-loadings l_{ij} and the specific variances $\psi_i (i = 1, \ldots, p; j = 1, \ldots, m)$. Let $\mathbf{X}_1, \ldots, \mathbf{X}_n$ be a random sample of size n from a population with mean vector μ and covariance matrix $\mathbf{\Sigma}$. The sample mean vector $\bar{\mathbf{x}}$ and sample covariance matrix $\mathbf{S}$ are estimates of μ and $\mathbf{\Sigma}$, respectively.

If the off-diagonal elements of $\mathbf{S}$ or those of the sample correlation matrix $\mathbf{R}$ are nearly zeros, the variables $X_1, \ldots, X_p$ are nearly unrelated and in this case, there is little point in carrying out a factor analysis, as then, $m \approx p$. If $\mathbf{\Sigma}$ appears to be significantly different from a diagonal matrix, the factor model can be entertained.

There are two popular methods of estimation: (1) Principal Component Method and its modification, Principal Factor method (2) Maximum Likelihood Method. The solution from either method can be rotated, as noted in Subsection 9.2.2, until a suitable interpretation of the factors are arrived at. Also, it is always wise to try more than one method of solution. If the data support a factor model, all the solutions will be consistent. We shall discuss these below.

9.3.1 *Principal component method*

Let $(\lambda_i, \mathbf{e}_i), i = 1, \ldots, p, \lambda_1 \geq \lambda_2 \geq \ldots \geq \lambda_p \geq 0$ be the (eigenvalue, eigenvector) pairs of $\mathbf{\Sigma}$. By spectral decomposition (Theorem A.12.1) we can write

$$\begin{aligned}\mathbf{\Sigma} &= \textstyle\sum_{i=1}^{p} \lambda_i \mathbf{e}_i \mathbf{e}_i' \\ &= (\sqrt{\lambda_1}\mathbf{e}_1, \sqrt{\lambda_2}\mathbf{e}_2, \ldots, \sqrt{\lambda_p}\mathbf{e}_p) \begin{bmatrix} \sqrt{\lambda_1}\mathbf{e}_1' \\ \sqrt{\lambda_2}\mathbf{e}_2' \\ \cdot \\ \cdot \\ \cdot \\ \sqrt{\lambda_p}\mathbf{e}_p' \end{bmatrix} = \mathbf{L}^{(0)}\mathbf{L}^{(0)'} \text{ (say)} . \end{aligned} \quad (9.3.1)$$

If we decompose $\mathbf{\Sigma}$ in this way, in a factor model, $\mathbf{L} = \mathbf{L}^0, m = p$ and $\mathbf{\Psi} = \mathbf{0}$. However, we should employ $m < p$ factors. Retaining, therefore, only contributions from the first m eigenvalues, we can write

$$\mathbf{\Sigma} \approx (\sqrt{\lambda_1}\mathbf{e}_1, \sqrt{\lambda_2}\mathbf{e}_2, \ldots \sqrt{\lambda_m}\mathbf{e}_m) \begin{bmatrix} \sqrt{\lambda_1}\mathbf{e}_1 \\ \sqrt{\lambda_2}\mathbf{e}_2' \\ \cdot \\ \cdot \\ \cdot \\ \sqrt{\lambda_m}\mathbf{e}_m' \end{bmatrix} + \tilde{\mathbf{\Psi}} \tag{9.3.2}$$

$$= \tilde{\mathbf{L}}\tilde{\mathbf{L}}' + \tilde{\mathbf{\Psi}} \quad \text{(say)}, \tag{9.3.3}$$

where $\tilde{\mathbf{L}} = ((\tilde{l}_{ij})) = (\sqrt{\lambda_1}\mathbf{e}_1, \ldots, \sqrt{\lambda_m}\mathbf{e}_m), \tilde{\mathbf{\Psi}} = \text{ Diag } (\tilde{\psi}_1, \ldots, \tilde{\psi}_p), \tilde{\psi}_i = \sigma_{ii} - \sum_{j=1}^m \tilde{l}_{ij}^2, i = 1, \ldots, p$. Clearly, in the factorization (9.3.2), the loading-vector of the jth factor $\tilde{\mathbf{l}}_{(j)} = (\tilde{l}_{1j}, \ldots, \tilde{l}_{pj})'$ is given by $\sqrt{\lambda j}\mathbf{e}_j$, the jth column of $\tilde{\mathbf{L}}(j = 1, \ldots, m)$.

In the above analysis, if we want to extend the number of factors to (m+1), the loadings of the first m factors $\tilde{\mathbf{l}}_{(j)}$ will remain unchanged.

We now examine the number of factor to be retained. If the off-diagonal elements of $\mathbf{\Sigma} - \tilde{\mathbf{L}}\tilde{\mathbf{L}}' - \tilde{\mathbf{\Psi}}$ are nearly zeros, then m should be the appropriate number of factors.

Now $\mathbf{\Sigma} - \tilde{\mathbf{L}}\tilde{\mathbf{L}}' - \tilde{\mathbf{\Psi}}$ have zeros as diagonal elements. Therefore, the sum of squared elements of $\mathbf{\Sigma} - \tilde{\mathbf{L}}\tilde{\mathbf{L}}' - \tilde{\mathbf{\Psi}} \leq$ sum of squared elements of $\mathbf{\Sigma} - \tilde{\mathbf{L}}\tilde{\mathbf{L}}'$. Again,

$$\begin{aligned} \mathbf{\Sigma} - \tilde{\mathbf{L}}\tilde{\mathbf{L}}' &= \lambda_{m+1}\mathbf{e}_{m+1}\mathbf{e}_{m+1}' + \ldots + \lambda_p\mathbf{e}_p\mathbf{e}_p' \\ &= \mathbf{PAP}' \end{aligned} \tag{9.3.4}$$

where

$$\mathbf{P} = (\mathbf{e}_{m+1}, \ldots, \mathbf{e}_p), \mathbf{A} = \text{ Diag. } (\lambda_{m+1}, \ldots, \lambda_p).$$

Again, for any squared matrix $\mathbf{B}$, the sum of squared entries of $\mathbf{B}$ is trace($\mathbf{BB}'$). Therefore, sum of squared entries of $\mathbf{\Sigma} - \tilde{\mathbf{L}}\tilde{\mathbf{L}}' =$ tr $(\mathbf{PAP}'\mathbf{PAP}')$ = tr $(\mathbf{PAAP}')$ = tr $(\mathbf{AA})$ (vide A.9(7)) $= \lambda_{m+1}^2 + \ldots + \lambda_p^2$. Hence the sum of squared elements of the residual matrix

$$\mathbf{\Sigma} - \tilde{\mathbf{L}}\tilde{\mathbf{L}}' - \mathbf{\Psi} \leq \lambda_{m+1}^2 + \ldots + \lambda_p^2. \tag{9.3.5}$$

Therefore, if $\sum_{j=m+1}^p \lambda_j^2$ is small, m can be taken as the appropriate number of factors.

As noted in (9.2.8), the contribution of the factor F_j to the total variance is $\mathbf{l}'_{(j)}\mathbf{l}_{(j)}$. For principal component method, this is given by,

$$(\sqrt{\lambda_j}\mathbf{e}_j)'(\sqrt{\lambda_j}\mathbf{e}_j) = \lambda_j \mathbf{e}'_j \mathbf{e}_j = \lambda_j, j = 1, \ldots, m. \tag{9.3.6}$$

Also, total contribution of all factors to the total variance $= V = \sum_{j=1}^{m} \lambda_j$. Thus, in the principal component method, proportion of the total communality due to the jth factor is

$$\frac{V_j}{V} = \frac{\lambda_j}{\sum_{i=1}^{p} \lambda_i}, \quad (j = 1, \ldots, p).$$

Therefore, if the contribution of the first m factors, $\sum_{j=1}^{m} \lambda_j / \sum_{i=1}^{p} \lambda_i$ is substantial (.8 or more), we use a m-factor model.

To implement the method in practice we consider the data set $\mathbf{x}_j, j = 1, \ldots, n$. It is customary to consider the centered data set $\mathbf{x}_j - \bar{\mathbf{x}}, \bar{\mathbf{x}} = (\bar{x}_1, \ldots, \bar{x}_p)', \bar{x}_i = \sum_{j=1}^{n} x_{ji}/n, i = 1, \ldots, p$. If the variables are in different units of measurement, one should consider the standardized observations

$$\mathbf{z}_j = (\frac{x_{j1} - \bar{x}_1}{\sqrt{s_{11}}}, \ldots, \frac{x_{jp} - \bar{x}_p}{\sqrt{s_{pp}}})', j = 1, \ldots, n.$$

Let $(\hat{\lambda}_i, \hat{\mathbf{e}}_i)$ be the eigenvalue-eigenvector pair of $\mathbf{S}$ or $\mathbf{R}$, as the case may be (we use the same symbol for both $\mathbf{S}$ and $\mathbf{R}$). Then a m-factor solution is obtained by writing

$$\mathbf{S} \text{ (or } \mathbf{R}) \approx \hat{\hat{\mathbf{L}}}\hat{\hat{\mathbf{L}}}' + \hat{\hat{\mathbf{\Psi}}} \tag{9.3.7}$$

where

$$\hat{\hat{\mathbf{L}}} = ((\hat{\hat{l}}_{ij})) = (\sqrt{\hat{\lambda}_1}\hat{\mathbf{e}}_1, \ldots, \sqrt{\hat{\lambda}_m}\hat{\mathbf{e}}_m), \tag{9.3.8}$$

and

$$\hat{\hat{\psi}}_i = s_{ii} - \sum_{j=1}^{m} \hat{\hat{l}}_{ij}^2.$$

The diagonal elements of the residual matrix $\mathbf{S} - (\hat{\hat{\mathbf{L}}}\hat{\hat{\mathbf{L}}}' + \hat{\hat{\mathbf{\Psi}}})$ will be zero. If the off-diagonal elements are also nearly zero, we may take the m-factor model to be appropriate.

Analytically, as seen above, if $\hat{\lambda}_1, \ldots, \hat{\lambda}_m$ account for a substantial portion $\sum_{j=1}^{m} \hat{\lambda}_j / \sum_{i=1}^{p} \hat{\lambda}_i$ of the total communality, and hence of the total sampling variance $\sum_{i=1}^{p} s_{ii}$, then an m-factor model is appropriate.

EXAMPLE 9.3.1: Consider the sample correlation matrix

$$\mathbf{R} = \begin{bmatrix} 1 & .75 & .45 & .40 \\ .75 & 1 & .60 & .35 \\ .45 & .60 & 1 & .80 \\ .40 & .35 & .80 & 1 \end{bmatrix}.$$

Its eigenvalues and eigenvectors are

$$\hat{\lambda}_1 = 2.680, \hat{\lambda}_2 = .892, \hat{\lambda}_3 = .322, \hat{\lambda}_4 = .106;$$

$$\begin{aligned} \mathbf{e}_1 &= (-.482,\ -.505,\ -.536,\ -.475)', \\ \mathbf{e}_2 &= (-.519,\ -.475,\ .387,\ .596)', \\ \mathbf{e}_3 &= (.624,\ -.524,\ -.420,\ .398)', \\ \mathbf{e}_4 &= (-.329,\ .493,\ -.622,\ .511)'. \end{aligned}$$

Since $(\hat{\lambda}_1 + \hat{\lambda}_2)/\sum_{i=1}^{4} \hat{\lambda}_i = 89.3\%$ we consider a 2-factor model. Hence,

$$\begin{aligned} \mathbf{L} &= (\sqrt{\lambda_1}\mathbf{e}_1, \sqrt{\lambda_2}\mathbf{e}_2) \\ &= \begin{bmatrix} -.789 & -.490 \\ -.827 & -.445 \\ -.877 & .365 \\ -.778 & .563 \end{bmatrix}. \end{aligned}$$

We have,

$$\mathbf{L}\mathbf{L}' = \begin{bmatrix} .863 & .873 & .513 & .338 \\ .873 & .886 & .561 & .391 \\ .513 & .561 & .902 & .887 \\ .338 & .391 & .887 & .921 \end{bmatrix}.$$

Hence,

$$\psi_1 = .137,\ \psi_2 = .114,\ \psi_3 = .098,\ \psi_4 = .078.$$

Also,

$$\mathbf{R} - (\mathbf{L}'\mathbf{L} - \mathbf{\Psi}) = \begin{bmatrix} 0 & -.123 & -.062 & -.062 \\ -.123 & 0 & .039 & -.041 \\ -.063 & .039 & 0 & -.087 \\ -.062 & -.041 & -.087 & 0 \end{bmatrix}$$

which is taken to be approximately a null matrix.

9.3.2 *Principal factor solution*

This is a modification of the principal component method. We describe the procedure in terms of the correlation matrix **R** although the procedure is equally applicable to **S**, when it is > 0. At first step preliminary estimates $\tilde{h}_i^2$ of communalities $h_i^2 (i = 1, \ldots, p)$ are made. The following methods are available.

(1) The simplest estimate of h_i^2 is the highest observed absolute value of correlation of X_i with any of the remaining $p-1$ variables. This method is generally useful for large values of p.

Table 9.3.1

Variable	Factor Loadings F_1	Factor Loadings F_2	Communalities
1	-.789	-.490	.863
2	-.827	-.449	.886
3	-.877	.365	.902
4	-.777	.562	.922
$V_j = \sum_{i=1}^4 l_{ij}^2$	2.679	.891	3.572
%age of total variance	67.01	22.30	89.31

(2) Another simple method is

$$\tilde{h}_i^2 = \frac{r_{ij} r_{ik}}{r_{jk}}, \tag{9.3.9}$$

where r_{ij}, r_{ik} are the two highest correlations with X_i.

(3) A third method is

$$\tilde{h}_i^2 = \sum_{j(\neq i)=1}^{p} \frac{r_{ij}}{p-1}. \tag{9.3.10}$$

(4) The most popular estimate is the square of the multiple correlation (SMC) coefficient of X_i with the remaining $(p-1)$ variables. The SMC coefficient provides a measure of the shared variance in that it measures the proportion of the observed total variance in X_i that is explained by its regression on the remaining $(p-1)$ variables. Thus, an estimate of the communality is

$$\tilde{h}_i^2 = \begin{cases} 1 - \frac{1}{r^{ii}} & \text{if } \mathbf{R} \text{ is used,} \\ s_{ii} - \frac{1}{s^{ii}} & \text{if } \mathbf{S} \text{ is used,} \end{cases} \tag{9.3.11}$$

where $\mathbf{R}^{-1} = ((r^{ij})), \mathbf{S}^{-1} = ((s^{ij}))$. For correlation matrix, $0 \leq \tilde{h}_i^2 \leq 1$. Certain methods for estimating communality can, however, may result in estimates which are greater than one, for example, when $r_{ij}r_{ik} > r_{jk}$ in (9.3.9). Such results are referred to as the *Heyward cases.* The usual procedure is to set $\tilde{h}_i^2 = .99$ or 1 in these cases.

We compute the reduced correlation matrix

$$\mathbf{R}^* = \begin{bmatrix} \tilde{h}_1^2 & r_{12} & \dots & r_{1p} \\ r_{21} & \tilde{h}_2^2 & \dots & r_{2p} \\ . & . & \dots & . \\ r_{p1} & r_{p2} & \dots & \tilde{h}_p^2 \end{bmatrix}, \tag{9.3.12}$$

ensuring that $\tilde{h}_i^2 \leq 1$ and $\tilde{\psi}_i = 1 - \tilde{h}_i^2 \geq 0$. We then factorize

$$\mathbf{R}^* = \mathbf{L}^*\mathbf{L}^{*'}$$

where

$$\mathbf{L}^* = (\sqrt{\lambda_1^*}\hat{\mathbf{e}}_1^*, \sqrt{\lambda_2^*}\hat{\mathbf{e}}_2^*, \dots, \sqrt{\lambda_m^*}\hat{\mathbf{e}}_m^*) \tag{9.3.13}$$

and

$$(\lambda_i^*, \hat{\mathbf{e}}_i^*), (i = 1, \dots, p), \lambda_1^* \geq \dots \geq \lambda_p^*$$

are the (eigenvalues, eigenvector) pairs of $\mathbf{L}^*$ and $m(< p)$ is the number of factors.

We recalculate the communality

$$\tilde{\tilde{h}}_i^2 = \sum_{j=1}^{p} l_{ij}^{*2},$$

and the new reduced correlation matrix $\mathbf{R}_1^*$ obtained by replacing $\tilde{h}_i^2$ by $\tilde{\tilde{h}}_i^2$ and keeping all the other elements in $\mathbf{R}^*$ unchanged. We factor $\mathbf{R}_1^*$ to obtain the new loading matrix $\mathbf{L}^{**}$. The iteration follows until at the kth stage of iteration all the elements of the residual correlation matrix

$$\mathbf{R}_{(k)}^* = \mathbf{R}^* - \mathbf{R}_{k-1}^*$$

are nearly zeros.

Often, the communalities placed on the diagonal of $\mathbf{R}^*$ will be less than one. Since $\mathbf{R}^*$ will no longer be positive semidefinite, some of the eigenvalues will be negative and their corresponding eigenvectors imaginary. These eigenvalues and eigenvectors are discarded. Ideally, one should consider the number of factors as the rank of the reduced correlation matrix $\mathbf{R}^*$.

Again, this rank is not always determined from $\mathbf{R}^*$ and some subjective judgement is necessary.

Again, since the sum of the eigenvalues of $\mathbf{R}^*$ is the total communality $V = \sum_{i=1}^{p} \tilde{h}_i^2$, the sum of positive eigenvalues of $\mathbf{R}^*$ will exceed V. A rule of thumb is to take the number of factors as m for which the sum of the first m positive eigenvalues is approximately equal to V.

The factor solution obtained by principal factor method is not scale-invariant.

If in the principal factor method, the communalities are replaced by unity or by numbers close to unity and if $p \approx m$, then the solution will be very close to that obtained by the principal component method. In practice, both the methods produce comparable factor loadings if n is large and m small.

EXAMPLE 9.3.2: Consider the matrix $\mathbf{R}$ of example 9.3.1. Here

$$\mathbf{R}^{-1} = \begin{bmatrix} 2.625 & -2.189 & .998 & -1.082 \\ -2.189 & 3.511 & -2.331 & 1.511 \\ .998 & -2.331 & 4.489 & -3.174 \\ -1.082 & 1.511 & -3.174 & 3.443 \end{bmatrix}.$$

Calculations of initial estimates of h_i^2 are given below.

Method	$\tilde{h}_1^2$	$\tilde{h}_2^2$	$\tilde{h}_3^2$	$\tilde{h}_4^2$
First	.75	.75	.80	.80
Second	.84 4	1.000	1.371	.711
Third	.533	.567	.617	.517
Fourth	.619	.715	.777	.710

We consider the estimates $\tilde{h}_i^2$ given by the fourth method, that is, $\tilde{h}_i^2 = 1 - \frac{1}{r^{ii}}$. Hence,

$$\mathbf{R}^* = \begin{bmatrix} .62 & .75 & .45 & .40 \\ .75 & .72 & .60 & .35 \\ .45 & .60 & .78 & .80 \\ .40 & .35 & .80 & .71 \end{bmatrix}.$$

We shall consider a 2-factor model. Its first two eigenvalues are

$$\lambda_1^* = 2.394,\ \lambda_2^* = .592,\ \lambda_3^* = -.173,\ \lambda_4^* = .017$$

with eigenvectors are

$$\mathbf{e}_1^* = (-.461,\ -.504,\ -.553,\ -.477)',$$
$$\mathbf{e}_2^* = (-.492,\ -.523,\ .389,\ .577)'.$$

Therefore, $\mathbf{L}^* = (\sqrt{\lambda_1^*}\mathbf{e}_1^*,\ \sqrt{\lambda_2^*}\mathbf{e}_2^*)'$ and the first estimate of communalities are

$$h_1^{*2} = .651,\ h_2^{*2} = .771,\ h_3^{*2} = .821,\ h_4^{*2} = .742,$$

given by the diagonal elements of $\mathbf{L}^*\mathbf{L}^{*'}$.

We construct $\mathbf{R}^{**}$ by substituting h_i^{*2}'s at the diagonal and keeping the off-diagonal elements as those of $\mathbf{R}^*$. Its first two eigenvalues and eigenvectors are found out and similarly, the second series of estimates of communalities are seen to be

$$h_1^{**2} = .661, h_2^{**2} = .802, h_3^{**2} = .841, h_4^{**2} = .757.$$

Similarly, $\mathbf{R}^{***}$ is constructed with h_2^{**2} as its diagonal elements and the process is repeated to give

$$\hat{h}_1^2 = .661, \hat{h}_2^2 = .823, \hat{h}_3^2 = .850, \hat{h}_4^2 = .769.$$

We stop at this step. We find that the residual matrix is

$$\mathbf{R} - \mathbf{R}^{****} - \mathbf{\Psi} = \begin{bmatrix} 0 & .013 & -.049 & .045 \\ .013 & 0 & .047 & -.042 \\ -.049 & .047 & 0 & .009 \\ .045 & -.042 & .009 & 0 \end{bmatrix},$$

which is close to the null matrix. The factor loadings and the communalities are given below.

Table 9.3.2

Variable	Factor Loadings F_1	Factor Loadings F_2	Communalities
1	-.717	.383	.661
2	-.797	-.432	.823
3	-.865	.317	.850
4	-.744	.464	.769
$V_j = \sum_{i=1}^{4} l_{ij}^2$	2.451	.649	3.103
%age of total variance	61.28	16.25	77.58

We note that the principal component solution gives higher values of the communalities than the principal factor solution. The loading vectors are similar in both the cases. The residual matrix in the factor solution is closer to null than that from the component solution.

9.3.3 *The maximum likelihood solution*

Assume that $\mathbf{F}$ and ϵ are independently multinormally distributed with parameters as given in (9.2.3). Then $\mathbf{X} \cap N_p(\mu, \mathbf{\Sigma} = \mathbf{LL}' + \mathbf{\Psi})$. We need to find maximum likelihood estimates of $\mathbf{L}$ and $\mathbf{\Psi}$.

The log-likelihood function is given by

$$\begin{aligned}\ln L(\mathbf{L}, \mathbf{\Psi}) &= \ln f_{\mathbf{X}}(\mathbf{x}; \mu, \mathbf{\Sigma}) \\ &= -\tfrac{n}{2}[\ln|\mathbf{\Sigma}| + \text{ tr } (\mathbf{\Sigma}^{-1}\mathbf{S})] \\ &= -\tfrac{n}{2}[\ln(|\mathbf{LL}' + \mathbf{\Psi}|) + \text{ tr}\{(\mathbf{LL}' + \mathbf{\Psi})^{-1}\mathbf{S}\}].\end{aligned} \tag{9.3.14}$$

Maximizing the above expression with respect to the elements of $\mathbf{L}$ and $\mathbf{\Psi}$ leads to the following equations for maximum likelihood estimation.

$$\mathbf{S}\hat{\mathbf{\Psi}}^{-1}\hat{\mathbf{L}} = \hat{\mathbf{L}}(\mathbf{I} + \hat{\mathbf{L}}'\hat{\mathbf{\Psi}}^{-1}\hat{\mathbf{L}})$$

and

$$\hat{\mathbf{\Psi}} = \text{ diag } (\mathbf{S} - \hat{\mathbf{L}}\hat{\mathbf{L}}')$$

where diag(.) denotes the diagonal matrix formed from (.) by replacing its nondiagonal elements with zeros. However, these equations do not have a unique solution as $L(\mathbf{L}, \mathbf{\Psi}) = L(\mathbf{LT}, \mathbf{\Psi})$ where $\mathbf{T}$ is an orthogonal $m \times m$ matrix. To uniquely determine $\mathbf{L}$ and $\mathbf{\Psi}$ we therefore impose the restriction, as noted in (9.2.11a),

$$\hat{\mathbf{L}}'\hat{\mathbf{\Psi}}^{-1}\hat{\mathbf{L}} = (\mathbf{D}) \tag{9.3.15}$$

where $(\mathbf{D})$ is a diagonal matrix with elements equal to the diagonal elements of $\mathbf{L}'\mathbf{\Psi}^{-1}\mathbf{L}$.

The elements in $\mathbf{L}$ can now be solved analytically, that is, numerical methods are not necessary and the iterative procedures are to be used for only the m nonzero elements of $\mathbf{L}'\mathbf{\Psi}^{-1}\mathbf{L}$. For further discussion on the matter the reader may refer to Lawley and Maxwell (1967), Jȯreskog (1977) and Lee (1979).

EXAMPLE 9.3.3: For the correlation matrix of example 9.2.1, the maximum likelihood solution is given in Table 9.3.3.

9.3.4 *Other extraction procedures*

The computer packages like BMDP, SPSS and SAS provide other extraction procedures, like, little jiffy method, minimum residual method, image analysis, Rao's canonical factoring, α-analysis, etc. If the communalities are near unity, all the extraction procedures produce nearly identical results. Again, if the number of variables is large, nearly identical results are produced by all the procedures. It is recommended that principal component method be first used to find initial factors. Maximum likelihood method, which is theoretically very sound may be used as the final method of analysis.

Table 9.3.3

Variable	Factor Loadings F_1	Factor Loadings F_2	Communality
1	.758	-.052	.577
2	.957	-.217	.982
3	.734	.497	.785
4	.544	.777	.880
$V_j = \sum_{i=1}^4 l_{ij}^2$	2.325	.901	3.224
%age of total variance	58.12	22.52	80.60

9.3.5 *Different types of rotation of factors*

We have seen in Section 9.2 that a solution to the factor analysis model is not unique and a different set of factors is obtained through an orthogonal rotation of the factors. Now, factor rotation may be *orthogonal* or *oblique*. The orthogonal rotation preserves the original orientation of the factors, so that if the factors were orthogonal before rotation, they are till orthogonal after rotation. Thus, the new factors are uncorrelated if the old factors were so. With oblique rotation, factor axes can be rotated independently so that they are not necessarily perpendicular to one another after rotation. The new factors may, therefore, be correlated.

Geometrically, the m common factors may be regarded as coordinate axes $F_1, \ldots, F_m$. The point $(l_{i1}, \ldots, l_{im})$, then, represents the coordinate of the variable X_i in the factor-space of $\mathbf{F}$. Thus each variable X_i corresponds to a point $\mathbf{l}_i = (l_{i1}, l_{i2}, \ldots, l_{im})'$ in the factor space $(i = 1, \ldots, p)$. (Note that the coordinate axes $X_1, \ldots, X_p$ are not necessarily perpendicular (they are perpendicular if $\mathbf{\Sigma}$ is diagonal)). The factor analysis therefore converts each variable to a point $\mathbf{l}_i$ in the factor space $(i = 1, \ldots, p)$.

If the variables are grouped in non-overlapping clusters, factor analysis wants to produce the factor coordinate axes which pass close to these clusters of points. This would associate each group of variables with a new factor (axis) and make interpretation more objective.

If we can attain a rotation in which every point (variable) is close to an axis, then each variable has high load on the factor corresponding to the axis and small loadings on the remaining factors. Such a situation is called a *simple structure*. The interpretation is simple in this case. We observe which variables are associated with which factors and the factors are named accordingly.

The number of factors in which each variable has moderate or high loadings is called the *complexity* of the variable. In case of simple structure, each variable has a complexity of one.

An orthogonal rotation therefore corresponds to a rigid rotation of the coordinate axes such that the axes, after rotation, pass as closely to the clusters as possible. An oblique rotation corresponds to a non-rigid rotation of the coordinate axes such that the rotated axes pass as closely as possible to the clusters of points. Now, since the rotation is not orthogonal (that is, is nonrigid) there is greater freedom of choice of the rotated axes.

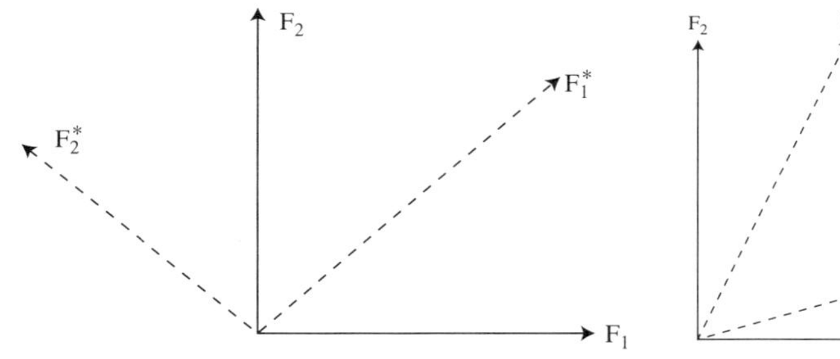

Fig. 9.1: Orthogonal rotation

Fig. 9.2: Oblique rotation

In all the types of rotation, however, it is desirable that the factor loading of the variable X_i in terms of the factor F_j should be either close to zero or very different from zero, i.e. l_{ij} should be approximately zero or very different from zero. A near zero l_{ij} means that X_i is not strongly related to F_j. A large absolute value of l_{ij} means that X_i is determined by F_j to a large extent. If each variable is strongly related to some factors, but are very weakly related to the other factors, then the factors can be easily interpreted (in terms of the variables to which they are strongly related).

As noted before, an orthogonal rotation preserves communalities. However, the variance accounted for by each factor ($\sum_{i=1}^{p} l_{ij}^2$) will change, as will the corresponding proportion ($\lambda_j/$ tr $(\mathbf{S})$). Again, the proportions due to rotated loadings will not necessarily be in the decreasing order.

When there are only two factors ($m = 2$), an orthogonal rotation may be obtained by choosing a transformation matrix

$$\mathbf{T} = \begin{bmatrix} \cos\phi & -\sin\phi \\ \sin\phi & \cos\phi \end{bmatrix}$$

where ϕ is an angle obtained by a visual inspection of graph of loadings $\mathbf{l}_i (i = 1, \ldots, p)$ in the old factor space. The angle ϕ is so chosen that if the axes F_1, F_2 are rotated through ϕ, the new axes F_1^*, F_2^* pass close to these points.

There are three types of orthogonal rotation: (i) *varimax* (ii) *quartimax* (iii) *equimax*.

The *varimax* rotation, proposed by Kaiser (1958) is the most popular one and is often used to rotate the solution obtained by the principal component method. It is assumed that the interpretability of the factor F_j is measured by the variance of the square of its loadings on the variables, i.e. by V_j, the variance among $l_{1j}^2, \ldots, l_{pj}^2$. If $l_{ij}^2/h_i^2 (j = 1, \ldots, p)$ all tend to one or zero, then V_j will tend to be large. Varimax rotation maximizes the average of these variances V_j, i.e. it maximizes

$$V = \frac{1}{p} \sum_{j=1}^{m} [\sum_{i=1}^{p} l_{ij}^2 - (\sum_{i=1}^{p} l_{ij}^2)^2/p] \tag{9.3.16}$$

Computer packages like SPSS, BMDP, SAS use varimax rotation either with raw factor loadings or with normalized loadings.

The *quartimax* method rotates in a fashion such that for a given variable it accomplishes major loadings for a factor. The method has a undesirable tendency to generate a general factor with all or most of the variables having high loadings.

The *equimax* method attempts to achieve simple structure with respect to both the rows and columns of the factor loading matrix.

Instead of the orthogonal rotation matrix $\mathbf{T}$, an oblique rotation uses a general nonsingular transformation matrix $\mathbf{Q}$ to obtain $\mathbf{F}^* = \mathbf{Q}'\mathbf{F}$. Hence,

$$\text{Cov}\,(\mathbf{F}^*) = \mathbf{Q}'\mathbf{I}\mathbf{Q} = \mathbf{Q}'\mathbf{Q} \neq \mathbf{I}.$$

Thus the new factors are correlated. Also, the communalities for $\mathbf{F}^*$ are different from those for $\mathbf{F}$.

Since the rotated axes are not required to be perpendicular, they can easily pass through the major clusters of points in the factor space.

Note 9.3.1: We have seen that for deciding on rotation, the factor loadings should be very close to zero or very different from zero. To assess significance of factor loadings l_{ij} obtained from $\mathbf{R}$, a threshold value of .3 has been suggested. In practice, a critical value of .3 is very low and will yield variables of complexity greater than one. A target value of .5 or .6 should be more acceptable. If l_{ij} exceeds .6, the loading on the jth factor may be considered as high. The .3 value is based on the critical value for significance of an ordinary correlation coefficient r. However, the distribution of the sample loadings is not the same as the distribution of r arising from a bivariate normal distribution. Again, since $h_i^2 = \sum_{j=1}^{m} l_{ij}^2 < 1$, an increase in value of m reduces the average squared loading in a row. Thus, if m is large, critical value should be reduced.

9.4 Factor Scores

We are now required to determine the value of the factors, $\mathbf{F} = (F_1, F_2, \ldots, F_m)'$ for different observations $\mathbf{X}$. For the jth observation $\mathbf{x}_j$ we shall denote the value of $\mathbf{F}$ as $\mathbf{f}_j$, the factor score for this observation $(j = 1, \ldots, n)$.

We assume that we have already obtained the estimated loadings $\hat{l}_{ij}$ and specific variances $\hat{\psi}_i$ by some suitable procedure (principal component, maximum likelihood, etc.) with an useful rotation (eg. verimax) and for subsequent calculations these values will be assumed fixed. There are two methods for calculating the factor scores (a) the weighted least square method (b) regression method.

9.4.1 *Weighted least square method*

Consider the model

$$\mathbf{X} - \mu = \mathbf{LF} + \epsilon,$$

$$\text{Cov }(\epsilon) = \text{ Diag. } (\psi_1, \ldots, \psi_p) = \mathbf{\Psi},$$

taking $\mathbf{F}$ as a vector of unknown constants. It is assumed that estimates, $\hat{\mathbf{L}}, \hat{\mathbf{\Psi}}$ are available. The constants in $\mathbf{F}$ are estimated by minimizing the sum of squares of the weighted residuals

$$\sum_{i=1}^{p} \frac{\epsilon_i^2}{\psi_i} = \epsilon' \mathbf{\Psi}^{-1} \epsilon = (\mathbf{X} - \mu - \mathbf{LF})' \mathbf{\Psi}^{-1} (\mathbf{x} - \mu - \mathbf{LF}). \tag{9.4.1}$$

This is given by

$$\hat{\mathbf{F}} = (\mathbf{L}' \mathbf{\Psi}^{-1} \mathbf{L})^{-1} \mathbf{L}' \mathbf{\Psi}^{-1} (\mathbf{X} - \mu). \tag{9.4.2}$$

An estimate of the factor score for the jth observation $\mathbf{x}_j$ is, therefore,

$$\begin{aligned} \hat{\mathbf{f}}_j &= (\hat{\mathbf{L}}' \hat{\mathbf{\Psi}}^{-1} \hat{\mathbf{L}})^{-1} \hat{\mathbf{L}}' \hat{\mathbf{\Psi}}^{-1} (\mathbf{x}_j - \bar{\mathbf{x}}), \\ &= \hat{\mathbf{\Delta}}^{-1} \hat{\mathbf{L}}^{-1} \hat{\mathbf{\Psi}}^{-1} (\mathbf{x}_j - \bar{\mathbf{x}}), \ j = 1, \ldots, n, \end{aligned} \tag{9.4.3}$$

where, by (9.2.11a), $\hat{\mathbf{\Delta}}$ is a diagonal matrix.

If we consider the standardized variables $\mathbf{Z} = \mathbf{V}^{-1/2}(\mathbf{X} - \bar{\mathbf{x}})$,

$$\hat{\mathbf{f}}_j = (\hat{\mathbf{L}}_z' \hat{\Psi}_z^{-1} \hat{\mathbf{L}}_z)^{-1} \hat{\mathbf{L}}_z' \hat{\mathbf{\Psi}}_z^{-1} \mathbf{z}_j, \tag{9.4.4}$$

where $\mathbf{z}_j = \mathbf{V}^{-1/2}(\mathbf{x}_j - \bar{\mathbf{x}})$ and $\hat{\mathbf{L}}_z, \hat{\mathbf{\Psi}}_z$ are given in Section 9.2.1.

It can be verified that $\frac{1}{n} \sum_{j=1}^n \mathbf{f}_j = \mathbf{0}$ and $\frac{1}{n-1} \sum_{j=1}^n \hat{\mathbf{f}}_j \hat{\mathbf{f}}_j' =$ a diagonal matrix.

If the loadings are estimated by the principal component method, it is customary to estimate the factor scores by using the unweighted least squares method. Then

$$\hat{\mathbf{f}}_j = (\hat{\mathbf{L}}' \hat{\mathbf{L}}')^{-1} \hat{\mathbf{L}}' (\mathbf{x}_j - \bar{\mathbf{x}}) \tag{9.4.5}$$

or for standardized data

$$\hat{\mathbf{f}}_j = (\hat{\mathbf{L}}_z' \hat{\mathbf{L}}_z)^{-1} \hat{\mathbf{L}}_z' \mathbf{z}_j. \tag{9.4.6}$$

Since $\hat{\mathbf{L}} = (\sqrt{\hat{\lambda}_1} \hat{\mathbf{e}}_1, \ldots, \sqrt{\hat{\lambda}_m} \mathbf{e}_m)$,

$$\hat{\mathbf{f}}_j = \begin{bmatrix} \hat{\mathbf{e}}_1' (\mathbf{x}_j - \bar{\mathbf{x}}) / \sqrt{\hat{\lambda}_1} \\ \cdot \\ \cdot \\ \hat{\mathbf{e}}_m' (\mathbf{x}_j - \bar{\mathbf{x}}) / \sqrt{\hat{\lambda}_m} \end{bmatrix}, \tag{9.4.7}$$

if (9.4.5) is used. Similar expression can be obtained for the standardized data. Thus the factor scores are the first m standardized (so that variance of each component is one) principal components.

It can be checked that for (9.4.7),

$$\frac{1}{n}\sum_{j=1}^{n}\hat{\mathbf{f}}_j = \mathbf{0}, \quad \frac{1}{n-1}\sum_{j=1}^{n-1}\hat{\mathbf{f}}_j\hat{\mathbf{f}}_j' = \mathbf{I}.$$

EXAMPLE 9.4.1: For the correlation matrix in Example 9.3.1, the factor score for $\mathbf{z}_j = (.50, \ 1.25, \ -.75, \ 1.00)'$ is $(\hat{\mathbf{L}}_z'\hat{\mathbf{L}}_z)^{-1}\hat{\mathbf{L}}_z'\mathbf{z}_j = (-4.1227, \ -.45866)'$ where $\hat{\mathbf{L}}_z$ is given in Table 9.3.1.

9.4.2 *The regression method*

We assume that the loading matrix $\mathbf{L}$ and specific variance $\mathbf{\Psi}$ are known.

When the common factors $\mathbf{F}$ and specific factors ϵ are independently normally distributed $\mathbf{F} \cap (\mathbf{0}, \mathbf{I}_m), \epsilon \cap (\mathbf{0}, \mathbf{\Psi}),$, the linear combination $\mathbf{X} - \mu = \mathbf{LF} + \epsilon$ and $\mathbf{F}$ has jointly an $N_{p+m}(\mathbf{0}, \mathbf{\Sigma}^*)$ distribution, where

$$\mathbf{\Sigma}^* = \begin{bmatrix} \mathbf{\Sigma} = \mathbf{LL}' + \mathbf{\Psi} & \mathbf{L} \\ \mathbf{L}' & \mathbf{I}_m \end{bmatrix}. \tag{9.4.8}$$

We find that the conditional distribution of $\mathbf{F}$ given $\mathbf{X}$ is multivariate normal with

$$E(\mathbf{F} \mid \mathbf{X}) = \mathbf{L}'\mathbf{\Sigma}^{-1}(\mathbf{X} - \mu) = \mathbf{L}'(\mathbf{LL}' + \mathbf{\Psi})^{-1}(\mathbf{X} - \mu), \tag{9.4.9}$$

$$\text{Cov}(\mathbf{F} \mid \mathbf{X}) = \mathbf{I}_m - \mathbf{L}'\mathbf{\Sigma}^{-1}\mathbf{L} = \mathbf{I}_m - \mathbf{L}'(\mathbf{LL}' + \mathbf{\Psi})^{-1}\mathbf{L}. \tag{9.4.10}$$

Equation (9.4.9) gives the regression of $\mathbf{F}$ on $\mathbf{X}$. For $\mathbf{X} = \mathbf{x}_j$, the factor score $\mathbf{f}_j$ is, therefore, obtained from (9.4.9) by substituting the maximum likelihood estimates of $\mu, \mathbf{L}$, and $\mathbf{\Psi}$. This is

$$\hat{\mathbf{f}}_j^R = \hat{\mathbf{L}}'(\hat{\mathbf{L}}\hat{\mathbf{L}}' + \hat{\mathbf{\Psi}})^{-1}(\mathbf{x}_j - \bar{\mathbf{x}}), j = 1, \ldots, n. \tag{9.4.11}$$

Using the matrix identity

$$\hat{\mathbf{L}}'(\hat{\mathbf{L}}\hat{\mathbf{L}}' + \hat{\mathbf{\Psi}})^{-1} = (\mathbf{I} + \hat{\mathbf{L}}'\hat{\mathbf{\Psi}}^{-1}\hat{\mathbf{L}})^{-1}\hat{\mathbf{L}}'\hat{\mathbf{\Psi}}^{-1},$$

we have

$$\hat{\mathbf{f}}_j^R = (\mathbf{I} + \hat{\mathbf{L}}'\hat{\mathbf{\Psi}}^{-1}\hat{\mathbf{L}})^{-1}\hat{\mathbf{L}}'\hat{\mathbf{\Psi}}^{-1}(\mathbf{x}_j - \bar{\mathbf{x}}). \tag{9.4.12}$$

Now, from (9.4.2), the least square estimate of $\mathbf{f}_j$ is

$$\begin{aligned}\hat{\mathbf{f}}_j^{LS} &= (\hat{\mathbf{L}}'\hat{\mathbf{\Psi}}^{-1}\hat{\mathbf{L}})^{-1}\hat{\mathbf{L}}'\hat{\mathbf{\Psi}}^{-1}(\mathbf{x}_j - \bar{\mathbf{x}}) \\ &= (\hat{\mathbf{L}}'\hat{\mathbf{\Psi}}^{-1}\hat{\mathbf{L}})^{-1}(\mathbf{I} + \hat{\mathbf{L}}'\hat{\mathbf{\Psi}}^{-1}\hat{\mathbf{L}})\hat{\mathbf{f}}_j^R \\ &= [(\hat{\mathbf{L}}'\hat{\mathbf{\Psi}}^{-1}\hat{\mathbf{L}})^{-1} + \mathbf{I}]\hat{\mathbf{f}}_j^R.\end{aligned} \tag{9.4.13}$$

For maximum likelihood estimates $(\hat{\mathbf{L}}'\hat{\mathbf{\Psi}}^{-1}\hat{\mathbf{L}}')^{-1}$ is a diagonal matrix and if the elements of this diagonal matrix are close to zero, the regression and the generalized least square method will give nearly the same factor scores.

Sometimes to reduce the effect of incorrect determination of the number of factors, the sample covariance matrix is used in place of $(\hat{\mathbf{L}}\hat{\mathbf{L}}' + \hat{\mathbf{\Psi}})$ in (9.4.11). Then

$$\hat{\mathbf{f}}_j^R = \hat{\mathbf{L}}'\mathbf{S}^{-1}(\mathbf{x}_j - \bar{\mathbf{x}}), j = 1, \ldots, n. \tag{9.4.14}$$

EXAMPLE 9.4.2: For the correlation matrix in Example 9.3.1 and $\mathbf{z}_j = (.50,\ 1.25,\ -.75,\ 1.00)'$, the factor score is $f_j^R = (-.13818, -1.38736)'$ by formula (9.4.11).

9.5 Determination of the Number of Factors

An important problem in fitting a factor analysis model is the assessment of the the number of factors m to be extracted. We have seen that in the principal component solution m is determined by the number of first m eigenvalues of $\mathbf{S}$ or $\mathbf{R}$ which account for a substantial portion of the total variance, tr $(\mathbf{S})$ or tr $(\mathbf{R})$. In the principal factor method m is such that the sum of the first m positive eigenvalues of $\mathbf{S} - \hat{\mathbf{\Psi}}$ or $\mathbf{R} - \hat{\mathbf{\Psi}}$ is approximately equal to the communality.

Some other criteria for choice of m are: (1) Choose m equal to the number of eigenvalues greater than the average eigenvalue. This is 1 for $\mathbf{R}$ and $\sum_{j=1}^{p} \lambda_j/p$ for $\mathbf{S}$. (2) Use the scree test based on a plot of eigenvalues of $\mathbf{S}$ or $\mathbf{R}$. If there is an elbow, i.e., if the graph drops sharply, followed by a straight line with a much smaller slope, choose m equal to the number of eigenvalues before the straight line begins.

We give below a large sample test for the number of common factors.

We assume that $\mathbf{X}_1, \ldots, \mathbf{X}_n$ is a random sample from a $N_p(\mu, \mathbf{\Sigma})$ population. If the m-factor model holds, $\mathbf{\Sigma} = \mathbf{L}\mathbf{L}' + \mathbf{\Psi}$ where $\mathbf{L}$ is $p \times m$. Testing adequacy of a m-factor model is equivalent to testing

$$H_0 : \mathbf{\Sigma} = \mathbf{L}\mathbf{L}' + \mathbf{\Psi} \tag{9.5.1}$$

against $H_1 : \mathbf{\Sigma}$ is any positive definite matrix.

When $\mathbf{\Sigma}$ does not have any special form, maximum likelihood estimate (m.l.e.) of $\mathbf{\Sigma}$ is $\mathbf{S}_n = (n-1)\mathbf{S}/n$ and the maximum of the likelihood functions is proportional to

$$|\mathbf{S}_n|^{-n/2} e^{-np/2}, \tag{9.5.2}$$

(vide (3.6.14)). Under H_0, the m.l.e. of $\mathbf{\Sigma}$ is $\hat{\mathbf{\Sigma}} = \hat{\mathbf{L}}\hat{\mathbf{L}}' + \hat{\mathbf{\Psi}}$ where $\hat{\mathbf{L}}$ and $\hat{\mathbf{\Psi}}$ are m.l.e.'s of $\mathbf{L}$ and $\mathbf{\Psi}$ respectively. In this case, the maximum of the likelihood function is proportional to

$$|\hat{\mathbf{\Sigma}}|^{-n/2} \exp\{-\frac{1}{2} \text{ tr } [\hat{\mathbf{\Sigma}}^{-1}(\sum_{i=1}^{n}(\mathbf{x}_i - \bar{\mathbf{x}})(\mathbf{x}_i - \bar{\mathbf{x}})')]\}$$

$$= |\hat{\mathbf{\Sigma}}|^{-n/2} \exp\{-\frac{1}{2} n \text{ tr } [\hat{\mathbf{\Sigma}}^{-1}\mathbf{S}_n]\}. \tag{9.5.3}$$

Hence,

$$-2\ln\lambda = -2\ln\frac{\sup_{H_0} L}{\sup_{\text{unrestricted}} L} = -2\ln(\frac{|\hat{\mathbf{\Sigma}}|}{|\mathbf{S}_n|})^{-n/2} + n[\text{ tr } \hat{\mathbf{\Sigma}}^{-1}\mathbf{S}_n - p] \tag{9.5.4}$$

follows a χ^2 distribution with d.f. $t = \frac{(p-m)^2}{2} - \frac{p+m}{2}$ (vide (9.2.12)). It can be seen that $\text{tr } (\hat{\mathbf{\Sigma}}^{-1}\mathbf{S}_n) - p = 0$ provided $\hat{\mathbf{\Sigma}}$ is the m.l.e. of $\mathbf{\Sigma}$. Thus we have

$$-2\ln\lambda = n\ln(\frac{|\hat{\mathbf{\Sigma}}|}{\mathbf{S}_n|}). \tag{9.5.5}$$

Bartlett (1954) has shown that the chi-square approximation to the sampling distribution of $-2\ln\lambda$ can be improved by replacing n in (9.5.5) by $\{n - 1 - (2p + 4m + 5)/6\}$.

Using Bartlett's approximation, we reject H_0 at the level of significance α if

$$\{n - 1 - (2p + 4m + 5)/6\} \ln\frac{|\hat{\mathbf{L}}\hat{\mathbf{L}}' + \hat{\mathbf{\Psi}}|}{|\mathbf{S}_n|} > \chi^2_{[(p-m)^2 - p - m]/2;\alpha} \tag{9.5.6}$$

provided that n and $n-p$ are large. Since the number of degrees of freedom $[(p-m)^2 - (p-m)]/2$ must be positive, we must have

$$m < \frac{1}{2}[2p + 1 - \sqrt{8p + 1}] \tag{9.5.7}$$

in order the test (9.5.6) to apply.

Note 9.5.1: If we work with the sample correlation matrix $\mathbf{R}$, then the test statistic is

$$\frac{|\hat{\mathbf{\Sigma}}|}{|\hat{\mathbf{S}}_n|} = \frac{|\hat{\mathbf{L}}_z\hat{\mathbf{L}}_z' + \hat{\mathbf{\Psi}}_z|}{|\mathbf{R}|} \tag{9.5.8}$$

where $\mathbf{L}_z, \mathbf{\Psi}$ have been defined in Subsection 9.2.1.

Note 9.5.2: If n is large and m is small relative to p, the hypothesis H_0 in (9.5.1) will usually be rejected, indicating that more common factors must be retained. However, even after adding more factors, $\hat{\mathbf{\Sigma}}$ and $\mathbf{S}_n$ will usually be close, so that the test does not give any further insight though the new factors may be significant. Some subjective judgement must be done for the choice of m.

EXAMPLE 9.5.1: Consider the following correlation matrix,

$$\mathbf{R} = \begin{bmatrix} 1.000 & .515 & .578 & .614 & .638 & .605 \\ .578 & 1.000 & .425 & .478 & .485 & .460 \\ .578 & .425 & 1.000 & .985 & .875 & .880 \\ .614 & .478 & .985 & 1.000 & .888 & .895 \\ .638 & .485 & .875 & .888 & 1.000 & .935 \\ .605 & .460 & .880 & .895 & .935 & 1.000 \end{bmatrix}.$$

By fitting a 2-factor model by maximum likelihood method we get $\hat{\mathbf{l}}_z$ and $\hat{\mathbf{\Psi}}_z$. Here

$$|\hat{\mathbf{L}}_z\hat{\mathbf{L}}_z' + \hat{\mathbf{\Psi}}| = \begin{bmatrix} 1.0000 & .3556 & .5884 & .6021 & .6398 & .6363 \\ .3556 & 1.0000 & .4441 & .4564 & .4997 & .4960 \\ .5884 & .4441 & 1.0000 & .9801 & .8741 & .8800 \\ .6021 & .4564 & .9801 & 1.0000 & .8900 & .8950 \\ .6398 & .4997 & .8741 & .8900 & 1.0000 & .9106 \\ .6363 & .4960 & .8800 & .8950 & .9106 & 1.0000 \end{bmatrix}.$$

Using Bartlett's correction, we calculate the test statistic in (9.5.6), taking $n = 100$,

$$[n - 1 - (2p + 4m + 5)/6] \ln \frac{|\hat{\mathbf{L}}_z'\hat{\mathbf{L}}_z + \hat{\mathbf{\Psi}}|}{|\mathbf{R}|}$$

$$= \frac{569}{6} \ln(2.03529) = 5.26.$$

Since $\frac{1}{2}[(p-m)^2 - p - m] = 4$, the 5% critical value $\chi^2_{4;.05} = 9.49$ is not exceeded, we fail to reject H_0. We conclude that the data do not contradict a two-factor model.

9.6 Comparison Between Factor Analysis and Principal Component Analysis

We have seen that factor analysis and principal component analysis are very much related, both having the goal of reducing dimensionality of the data.

However, there are some major differences between the two approaches. In principal component analysis, principal components are linear functions of the variables, while in factor analysis, the variables are expressed as linear combinations of factors. In component analysis, the emphasis is on explaining the total variance $\sum_{i=1}^{p} s_{ii}$, while factor analysis attempts to explain the full covariance matrix $\mathbf{S}$. Principal component analysis does not require any essential assumption, while factor analysis model requires several assumptions(vide (9.2.2), (9.2.3)). If the eigenvalues of $\mathbf{S}$ are distinct, principal components are unique, while different sets of factors can always be obtained through rotation. If more principal components are worked out, the existing principal components do not change, while if we change the number of factors, the estimated factors change.

The main advantage of factor analysis is that it is always possible to rotate the factors until a set of easily interpretable factors is obtained. The principal component analysis does not have this scope. The principal components are linear constructs of the original variables and are suitable for use as inputs in another analysis (for example, we can find multiple regression on the principal components, rather than on the original variables). The factor analysis attempts to locate some inherent factors which constitute the data. If the goal is to find such factors, factor analysis is the ideal tool provided the data fit the model and a set of easily interpretable factors is obtained, while principal component analysis generally serves as an intermediate stage in much larger investigation with data analysis.

9.7 Exercises and Complements

9.1 Consider an one factor model for the correlation matrix

$$\rho = \begin{bmatrix} 1 & .83 & .78 \\ .83 & 1 & .67 \\ .78 & .67 & 1 \end{bmatrix}.$$

Show that there is a unique choice of $\mathbf{L}$ and $\mathbf{\Psi}$ with $\rho = \mathbf{LL}' + \mathbf{\Psi}$ and the solution is admissible.

9.2 Consider the correlation matrix

$$\mathbf{R} = \begin{bmatrix} 1 & .553 & .547 & .410 & .389 \\ .553 & 1 & .610 & .485 & .437 \\ .547 & .610 & 1 & .711 & .665 \\ .410 & .485 & .711 & 1 & .607 \\ .389 & .437 & .665 & .607 & 1 \end{bmatrix}.$$

Show that if $m > 2, t$ given in (9.2.12) is < 0 and the factor model is not well-defined. Estimating the ith communality by $h_i^2 = \max_j \mid r_{ij} \mid$, find principal factor solution with $m = 1$ and $m = 2$.

9.3 The following correlation matrix was obtained from a sample of size 100.

$$\mathbf{R} = \begin{bmatrix} 1.000 & .578 & .525 & .385 & .482 \\ .578 & 1.000 & .612 & .475 & .325 \\ .525 & .612 & 1.000 & .436 & .428 \\ .385 & .475 & .436 & 1.000 & .525 \\ .482 & .325 & .428 & .525 & 1.000 \end{bmatrix}.$$

Fit a 2-factor model by the (i) Principal Component method (ii) Principal Factor method (iii) Maximum Likelihood method. Interpret the results. Find the factor scores for the observation $\mathbf{z}_j = (.0156, .0048, .0657, .0534, .0038).'$ Test for the adequacy of the model.

9.4 Show that one way to do the factor analysis is as follows: Let $\mathbf{X}_1, \ldots, \mathbf{X}_p$ be standardized variables and let $(\lambda_j, \mathbf{e}_j)$ be the eigenvalue, eigenvector pairs of $\rho, (\lambda_1 \geq \lambda_2 \geq \ldots \geq \lambda_p > 0), j = 1, \ldots, p$. The p-principal components are

$$Y_1 = e_{11}X_1 + e_{12}X_2 + \ldots + e_{1p}X_p,$$

$$\cdots$$

$$Y_p = e_{p1}X_1 + e_{p2}X_2 + \ldots + e_{pp}X_p$$

where $\mathbf{e}_i = (e_{i1}, \ldots e_{ip})'$. The transformation from $\mathbf{X}$ to $\mathbf{Y}$ being orthogonal, we can write the inverse relationship as

$$X_1 = e_{11}Y_1 + e_{21}Y_2 + \ldots + e_{p1}Y_p,$$

$$\cdots$$

$$X_p = e_{1p}Y_1 + e_{2p}Y_2 + \ldots + e_{pp}Y_p.$$

For a factor analysis only $m(< p)$ of the principal components are retained, so the last set of equations become

$$X_1 = e_{11}Y_1 + e_{21}Y_2 + \ldots + e_{m1}Y_m + \epsilon_1,$$

$$X_2 = e_{12}Y_1 + e_{22}Y_2 + \ldots + e_{m2}Y_m + \epsilon_2,$$

$$\ldots$$

$$X_p = e_{1p}Y_1 + e_{2p}Y_2 + \ldots + e_{mp}Y_m + \epsilon_p.$$

Table 9.E.1:Ramus Bone Length at Four Ages of 20 Boys

Individual	8yr (x_1)	$8\frac{1}{2}$ yr (x_2)	9 yr (x_3)	$9\frac{1}{2}$ yr (x_4)
1	47.8	48.8	49.0	49.7
2	46.4	47.3	47.7	48.4
3	46.3	46.8	47.8	48.5
4	45.1	45.3	46.1	47.2
5	47.6	48.5	48.9	49.3
6	52.5	53.2	53.3	53.7
7	51.2	53.0	54.3	54.5
8	49.8	50.0	50.3	52.7
9	48.1	50.8	52.3	54.4
10	45.0	47.0	47.3	48.3
11	51.2	51.4	51.6	51.9
12	48.5	49.2	53.0	55.5
13	52.1	52.8	53.7	55.0
14	48.2	48.9	49.3	49.8
15	49.6	50.4	51.2	51.8
16	50.7	51.7	52.7	53.3
17	47.2	47.7	48.4	49.5
18	53.3	54.6	55.1	55.3
19	46.2	47.5	48.1	48.4
20	46.3	47.6	51.3	51.8

(Source: Elston and Grizzle, 1962)

Now, writing $F_j = Y_j/\sqrt{\lambda_j}$ so that F_j has now zero mean and unit variance $(j = 1, \ldots, p)$ we have

$$X_1 = e_{11}\sqrt{\lambda_1}F_1 + e_{21}\sqrt{\lambda_2}F_2 + \ldots + e_{m1}\sqrt{\lambda_m}F_m + \epsilon_1,$$

$$X_2 = e_{12}\sqrt{\lambda_1}F_1 + e_{22}\sqrt{\lambda_2}F_2 + \ldots + e_{m2}\sqrt{\lambda_2}F_m + \epsilon_2,$$

$$\ldots$$

$$X_p = e_{1p}\sqrt{\lambda_1}F_1 + e_{2p}\sqrt{\lambda_2}F_2 + \ldots + e_{mp}\sqrt{\lambda_m}F_m + \epsilon_p.$$

Therefore, the loading $l_{ij} = e_{ij}\sqrt{\lambda_j}$, that is, $\mathbf{l}_{(j)} = \sqrt{\lambda_j}\mathbf{e}_j, i = 1, \ldots, p; j = 1, \ldots, m$.

9.5 Table 9.E.1 gives the data on ramus bone length at four ages of 20 boys.

Table 9.E.2: Ranks of 16 Individuals in Six Tests

Individual	Mathematics	Physics	Literature	Music	Table Tennis	Car Racing
1	1	6	9	8	8	7
2	2	5	11	10	14	11
3	3	3	14	16	16	13
4	4	2	10	9	15	14
5	5	1	12	11	1	1
6	6	4	13	15	2	6
7	7	8	15	14	6	5
8	8	7	16	13	7	16
9	9	9	6	12	9	9
10	10	10	8	7	10	8
11	11	11	7	5	11	10
12	12	13	4	2	13	15
13	13	12	5	6	12	12
14	14	15	1	4	4	3
15	15	14	3	3	5	4
16	16	16	2	1	3	2

(a) Extract two factors by the principal component method and perform a varimax rotation. (b) Also extract two factors by the principal factor method and carry out a varimax rotation. Compare the results of parts (a) and (b).

9.6 Ranks in performance of 16 individuals in tests on Mathematics, Physics, Literature, Music, Table Tennis and Car Racing are given in Table 9.E.2. Rank 1 means the highest performance in the subject.

(a) Using a scree plot, the number of eigenvalues greater than 1, and the percentages , can you find a clear choice of m?
(b) Extract three factors by the principal component method and carry out a varimax rotation.
(c) Extract three factors by the principal principal factor method and perform a varimax rotation.
(d) Compare the results of parts (b) and (c).
(e) Compute factor scores.

Chapter 10

Canonical Correlation

10.1 Introduction

The objective of the canonical correlation analysis is to give a simple description of the structure of correlation between two sets of variables. We often measure two sets of variables on units, for example, a set of student behavior and a set of teacher behavior, a set of aptitude variables and a set of achievement variables, a set of ecological variables and a set of environmental variables. In this chapter we consider a measure of overall correlation between two sets of variables. Hotelling (1935, 1936), who initially developed the technique, worked with the example of relating arithmetic speed and arithmetic power to reading speed and reading power.

Assume that the two sets of variables $\mathbf{X}^{(1)} = (X_1^{(1)}, \ldots, X_p^{(1)})', \mathbf{X}^{(2)} = (X_1^{(2)}, \ldots, X_q^{(2)})'$ have a joint distribution with mean $\mu = (\mu^{(1)'}, \mu^{(2)'})'$ and covariance matrix

$$\mathbf{\Sigma} = \begin{bmatrix} \mathbf{\Sigma}_{11} & \mathbf{\Sigma}_{12} \\ \mathbf{\Sigma}_{21} & \mathbf{\Sigma}_{22} \end{bmatrix}$$

and correlation matrix

$$\rho = \begin{bmatrix} \rho_{11} & \rho_{12} \\ \rho_{21} & \rho_{22} \end{bmatrix}$$

where $\mathbf{\Sigma}_{12}$ and ρ_{12} are of order $p \times q$. We assume that $p \leq q$. The pq elements of $\mathbf{\Sigma}_{12}$ measure the association between the two sets of variables $\mathbf{X}^{(1)}$ and $\mathbf{X}^{(2)}$. When p and q are large, interpreting the elements of $\mathbf{\Sigma}_{12}$ collectively is impossible. The main aim of canonical correlation analysis is to summarize the association between $\mathbf{X}^{(1)}$ and $\mathbf{X}^{(2)}$ in terms of a few suitably chosen covariances or correlations, rather than the pq covariances in $\mathbf{\Sigma}_{12}$. The analysis searches for a pair of linear combinations $\mathbf{a}_1'\mathbf{X}^{(1)}$ and

$\mathbf{b}_1'\mathbf{X}^{(2)}$ with maximum correlation. This correlation is the first canonical correlation and the corresponding pair of linear combinations the first pair of canonical variables. Having found such a pair, the analysis is pursued one step further by searching for a second pair of linear combinations, $\mathbf{a}_2'\mathbf{X}^{(1)}$ and $\mathbf{b}_2'\mathbf{X}^{(2)}$ with maximum correlation among all those linear combinations uncorrelated with the first pair. The correlation found is the second canonical correlation and the corresponding pair of linear combinations the second pair of canonical variables. The argument is repeated until all the possible correlations are exhausted. A more formal definition is given in Section 10.2.

The canonical correlation measures the strength of association between two subsets of variables. The maximization aspect of the technique is an attempt to translate the relationship between the two sets of variables into the relationship between a few pair of canonical variables.

10.2 Canonical Variables and Canonical Correlations

We first consider definitions of population canonical variables and population canonical correlations.

DEFINITION 10.2.1: Consider the linear combinations $U = \mathbf{a}'\mathbf{X}^{(1)}$ and $V = \mathbf{b}'\mathbf{X}^{(2)}$, for some pairs of coefficient vectors $\mathbf{a}, \mathbf{b}$. Then

$$\text{Corr. } (U, V) = \frac{\mathbf{a}'\mathbf{\Sigma}_{12}\mathbf{b}}{\sqrt{\mathbf{a}'\mathbf{\Sigma}_{11}\mathbf{a}}\sqrt{\mathbf{b}'\mathbf{\Sigma}_{22}\mathbf{b}}}. \tag{10.2.1}$$

The first pair of canonical variables is the pair of linear combinations U_1, V_1 having unit variances, which maximize the correlation (10.2.1).

The second pair of canonical variables is the pair of linear combinations $U_2 = \mathbf{a}'\mathbf{X}^{(1)}, V_2 = \mathbf{b}'\mathbf{X}^{(2)}$ having unit variances, which maximize the correlation (10.2.1) among all linear combinations which are uncorrelated with the first pair of canonical variables.

In general, the kth pair of canonical variables are the linear combinations $U_k = \mathbf{a}'\mathbf{X}^{(1)}, V_k = \mathbf{b}'\mathbf{X}^{(2)}$ having unit variances, which maximize the correlation (10.2.1) among all linear combinations which are uncorrelated with the previous $(k-1)$ variable pairs.

The correlation between the kth pair of canonical variables is called the kth canonical correlation $\rho_k (k = 1, 2, \ldots, p)$.

We now consider a few preliminary results.

Assume that $\mathbf{\Sigma}_{11}$ and $\mathbf{\Sigma}_{22}$ are nonsingular and $\mathbf{\Sigma}_{12}$ is a real matrix of order $p \times q$. Without loss of generality suppose that $p \leq q$. Write $\mathbf{\Sigma}_{jj} = \mathbf{A}_j^2$ where $\mathbf{A}_j = \mathbf{\Sigma}_{jj}^{1/2} > 0, j = 1, 2$. Using Singular Value Decomposition Theorem (Theorem A.12.2),

$$\mathbf{F} = \mathbf{A}_1^{-1}\mathbf{\Sigma}_{12}\mathbf{A}_2^{-1} = \mathbf{G}[\mathbf{D}_\rho \ \mathbf{0}]\mathbf{H}' \tag{10.2.2}$$

where

$$\mathbf{G} = (\mathbf{g}_1, \ \mathbf{g}_2 \ \ldots \mathbf{g}_p)$$

has p orthogonal eigenvectors of $\mathbf{FF}'$ as its columns and

$$\mathbf{H} = (\mathbf{h}_1 \ \mathbf{h}_2 \ \ldots \mathbf{h}_q)$$

has q orthogonal eigenvectors of $\mathbf{F}'\mathbf{F}$ as its columns. Here, $\mathbf{D}_\rho = \text{Diag}(\rho_1, \rho_2, \ldots, \rho_p)$ and $|\mathbf{FF}' - \rho^2\mathbf{I}| = 0$, where ρ_i's are the singular values of $\mathbf{F}$, i.e. ρ_i^2's are the eigenvalues of $\mathbf{FF}', \rho_1^2 \geq \rho_2^2 \geq \ldots \geq \rho_p^2$. These are also the p largest non-zero eigenvalues of $\mathbf{F}'\mathbf{F}$. Note that

$$\begin{aligned} \mathbf{FF}' &= \mathbf{\Sigma}_{11}^{-1/2}\mathbf{\Sigma}_{12}\mathbf{\Sigma}_{22}^{-1}\mathbf{\Sigma}_{21}\mathbf{\Sigma}_{11}^{-1/2} = \mathbf{M} \text{ (say)} , \\ \mathbf{F}'\mathbf{F} &= \mathbf{\Sigma}_{22}^{-1/2}\mathbf{\Sigma}_{21}\mathbf{\Sigma}_{11}^{-1}\mathbf{\Sigma}_{12}\mathbf{\Sigma}_{22}^{-1/2} = \mathbf{N} \text{ (say)} . \end{aligned} \tag{10.2.3}$$

Theorem 10.2.1: Consider the linear combinations $\mathbf{a}'\mathbf{X}^{(1)}$ and $\mathbf{b}'\mathbf{X}^{(2)}$. Then

$$\max{}_{\mathbf{a},\mathbf{b}} \text{ Corr. } (\mathbf{a}'\mathbf{X}^{(1)}, \mathbf{b}'\mathbf{X}^{(2)}) = \rho_1 \tag{10.2.4}$$

and this is attained when

$$\mathbf{a}' = \mathbf{a}_1' = \mathbf{g}_1'\mathbf{A}_1 = \mathbf{g}_1'\mathbf{\Sigma}_{11}^{-1/2}, \ \mathbf{b}' = \mathbf{b}_1' = \mathbf{h}_1'\mathbf{A}_2 = \mathbf{h}_1'\mathbf{\Sigma}_{22}^{-1/2}, \tag{10.2.5}$$

when ρ_1^2 is the largest eigenvalue of $\mathbf{M}(= \mathbf{FF}')$ and $\mathbf{g}_1$ is the associated normalized eigenvector. Again, ρ_1^2 is also the maximum eigenvalue of $\mathbf{N}(= \mathbf{F}'\mathbf{F})$ and $\mathbf{h}_1$ is its associated normalized eigenvector.

Let us call the linear combinations

$$\mathbf{g}_1'\mathbf{A}_1\mathbf{X}^{(1)} = U_1 \ \text{ and } \ \mathbf{h}_1'\mathbf{A}_2\mathbf{X}^{(2)} = \mathbf{V}_1. \tag{10.2.6}$$

These are the first pair of canonical variables. Also, $V(U_1) = V(V_1) = 1$.

Consider now the linear combinations $\mathbf{a}'\mathbf{X}^{(1)}$ and $\mathbf{b}'\mathbf{X}^{(2)}$ which are uncorrelated with U_1, V_1. Then, among such linear combinations

$$\max{}_{\mathbf{a},\mathbf{b}} \text{ Corr. } (\mathbf{a}'\mathbf{X}^{(1)}, \mathbf{b}'\mathbf{X}^{(2)}) = \rho_2 \tag{10.2.7}$$

and this is attained when

$$\mathbf{a}' = \mathbf{a}_2' = \mathbf{g}_2'\mathbf{A}_1 = \mathbf{g}_2'\mathbf{\Sigma}_{11}^{-1/2}, \ \mathbf{b}' = \mathbf{b}_2' = \mathbf{h}_2'\mathbf{A}_2 = \mathbf{h}_2'\mathbf{\Sigma}_{22}^{-1/2} \tag{10.2.8}$$

where ρ_2^2 is the second largest normalized eigenvalue of $\mathbf{M}$ and $\mathbf{g}_2$ is the associated eigenvector. The quantity ρ_2^2 is also the second largest normalized eigenvalue of $\mathbf{N}$ and $\mathbf{h}_2$ is its associated eigenvector.

The linear combinations

$$\mathbf{g}_2'\mathbf{A}_1\mathbf{X}^{(1)} = U_2 \quad \text{and } \mathbf{h}_2'\mathbf{A}_2\mathbf{X}^{(2)} = V_2 \tag{10.2.9}$$

are the second pair of canonical variables. Also,

$$V(U_2) = V(V_2) = 1,$$

$$\text{Cov } (U_1, U_2) = \text{ Cov } (V_1, V_2) = \text{ Cov } (U_1, V_2) = \text{ Cov } (V_1, U_2) = 0. \tag{10.2.10}$$

Similarly, the kth pair of canonical variables is

$$U_k = \mathbf{g}_k'\mathbf{A}_1\mathbf{X}^{(1)}; \; V_k = \mathbf{h}_k'\mathbf{A}_2\mathbf{X}^{(2)} \tag{10.2.11}$$

with

$$\text{Corr. } (U_k, V_k) = \rho_k, k = 3, \ldots, p.$$

Here ρ_k is the kth largest eigenvalue of $\mathbf{M}$ and also of $\mathbf{N}$. Also, $\mathbf{g}_k, \mathbf{h}_k$ are the associated normalized eigenvectors of $\mathbf{M}$ and $\mathbf{N}$ respectively. Also

$$\begin{aligned} V(U_k) \quad &= V(V_k) = 1, k = 3, \ldots, p, \\ \text{Cov } (U_k, U_l) &= \text{ Cov } (V_k, V_l) = 0, \\ \text{Cov } (U_k, V_l) &= 0, k \neq l = 1, \ldots, p. \end{aligned} \tag{10.2.12}$$

Proof. Let $\mathbf{c} = \mathbf{A}_1\mathbf{a}, \mathbf{d} = \mathbf{A}_2\mathbf{b}$. Then

$$\begin{aligned} \text{Corr. } (\mathbf{a}'\mathbf{X}^{(1)}, \mathbf{b}'\mathbf{X}^{(2)}) &= \frac{\mathbf{a}'\mathbf{\Sigma}_{12}\mathbf{b}}{\sqrt{\mathbf{a}'\mathbf{\Sigma}_{11}\mathbf{a}}\sqrt{\mathbf{b}'\mathbf{\Sigma}_{22}\mathbf{b}}} \\ &= \frac{\mathbf{c}'\mathbf{A}_1^{-1}\mathbf{\Sigma}_{12}\mathbf{A}_2^{-1}\mathbf{d}}{\sqrt{\mathbf{c}'\mathbf{c}}\sqrt{\mathbf{d}'\mathbf{d}}}. \end{aligned} \tag{10.2.13}$$

By Cauchy-Schwarz inequality (Section A.3)

$$\mathbf{c}'\mathbf{A}_1^{-1}\mathbf{\Sigma}_{12}\mathbf{A}_2^{-1}\mathbf{d} \leq \{\mathbf{c}'\mathbf{A}_1^{-1}\mathbf{\Sigma}_{12}\mathbf{\Sigma}_{22}^{-1}\mathbf{\Sigma}_{21}\mathbf{A}_1^{-1}\mathbf{c}\}^{1/2}\{\mathbf{d}'\mathbf{d}\}^{1/2}. \tag{10.2.14}$$

Hence by (10.2.13)

$$\text{Corr. } (\mathbf{a}'\mathbf{X}^{(1)}, \mathbf{b}'\mathbf{X}^{(2)}) \leq \left[\max{}_{\mathbf{a}}\left\{\frac{\mathbf{c}'\mathbf{A}_1^{-1}\mathbf{\Sigma}_{12}\mathbf{\Sigma}_{22}^{-1}\mathbf{\Sigma}_{21}\mathbf{A}_1^{-1}\mathbf{c}}{\mathbf{c}'\mathbf{c}}\right\}\right]^{1/2}.$$

Now, since $\mathbf{A}_1^{-1}\mathbf{\Sigma}_{12}\mathbf{\Sigma}_{22}^{-1}\mathbf{\Sigma}_{21}\mathbf{A}_1^{-1}$ is a $p \times p$ symmetric matrix

$$\max{}_{\mathbf{c}} \left\{ \frac{\mathbf{c}'\mathbf{A}_1^{-1}\mathbf{\Sigma}_{12}\mathbf{\Sigma}_{22}^{-1}\mathbf{\Sigma}_{21}\mathbf{A}_1^{-1}\mathbf{c}}{\mathbf{c}'\mathbf{c}} \right\} = \lambda_1 \tag{10.2.15}$$

where λ_1 is the largest eigenvalue of $\mathbf{A}_1^{-1}\mathbf{\Sigma}_{12}\mathbf{\Sigma}_{22}^{-1}\mathbf{\Sigma}_{21}\mathbf{A}_1^{-1} = \mathbf{M}$. Clearly, $\rho_1^2 = \lambda_1$.

The equality in (10.2.15) is attained if $\mathbf{c} = \mathbf{g}_1$, the eigenvector of $\mathbf{M}$ corresponding to λ_1 (vide (A.8(14)). In this case

$$\mathbf{a} = \mathbf{A}_1^{-1}\mathbf{g}_1. \tag{10.2.16}$$

Also, equality in (10.2.14) is satisfied if

$$\mathbf{d} \propto \mathbf{A}_2^{-1}\mathbf{\Sigma}_{21}\mathbf{A}_1^{-1}\mathbf{g}_1$$

i.e. if

$$\mathbf{b} = \mathbf{A}_2^{-1}\mathbf{d} \propto \mathbf{\Sigma}_{22}^{-1}\mathbf{\Sigma}_{21}\mathbf{A}_1^{-1}\mathbf{g}_1. \tag{10.2.17}$$

Now,

$$(\mathbf{FF}' - \lambda_1\mathbf{I})\mathbf{g}_1 = \mathbf{0}$$

i.e.,

$$(\mathbf{\Sigma}_{11}^{-1/2}\mathbf{\Sigma}_{12}\mathbf{\Sigma}_{22}^{-1}\mathbf{\Sigma}_{21}\mathbf{\Sigma}_{11}^{-1/2} - \lambda_1\mathbf{I})\mathbf{g}_1 = \mathbf{0}.$$

Pre-multiplying both sides by $\mathbf{\Sigma}_{22}^{-1/2}\mathbf{\Sigma}_{21}\mathbf{\Sigma}_{11}^{-1/2}$,

$$(\mathbf{\Sigma}_{22}^{-1/2}\mathbf{\Sigma}_{21}\mathbf{\Sigma}_{11}^{-1}\mathbf{\Sigma}_{12}\mathbf{\Sigma}_{22}^{-1/2} - \lambda_1\mathbf{I})\mathbf{\Sigma}_{22}^{-1/2}\mathbf{\Sigma}_{21}\mathbf{\Sigma}_{11}^{-1/2}\mathbf{g}_1 = \mathbf{0}. \tag{10.2.18}$$

Therefore, λ_1 is an eigenvalue (the largest eigenvalue) of $\mathbf{\Sigma}_{22}^{-1/2}\mathbf{\Sigma}_{21}\mathbf{\Sigma}_{11}^{-1}\mathbf{\Sigma}_{12}\mathbf{\Sigma}_{22}^{-1/2} = \mathbf{N}$ and its associated eigenvector $\mathbf{h}_1$ is proportional to $\mathbf{\Sigma}_{22}^{-1/2}\mathbf{\Sigma}_{21}\mathbf{\Sigma}_{11}^{-1/2}\mathbf{g}_1 = \mathbf{\Sigma}_{22}^{-1/2}\mathbf{\Sigma}_{21}\mathbf{a}_1$, where the sign is selected to make the correlation (10.2.13) positive and the constant of proportionality is so chosen that $\mathbf{h}'\mathbf{h} = 1$. Hence, by (10.2.17), $\mathbf{b}_1 = \mathbf{\Sigma}_{22}^{-1/2}\mathbf{h}_1$.

Therefore,

$$U_1 = \mathbf{g}_1'\mathbf{\Sigma}_{11}^{-1/2}\mathbf{X}^{(1)}, V_1 = \mathbf{h}_1'\mathbf{\Sigma}_{22}^{-1/2}\mathbf{X}^{(2)}.$$

Also,

$$V(U_1) = \mathbf{g}_1'\mathbf{\Sigma}_{11}^{-1/2}\mathbf{\Sigma}_{11}\mathbf{\Sigma}_{11}^{-1/2}\mathbf{g}_1 = 1, V(V_1) = 1.$$

By (10.2.13) and (10.2.15),

$$\begin{aligned}\max_{\mathbf{a},\mathbf{b}}[\text{ Corr. } (\mathbf{a}'\mathbf{X}^{(1)}, \mathbf{b}'\mathbf{X}^{(2)})] &= \text{ Corr. } (U_1, V_1)\\ &= \sqrt{\lambda_1} = \rho_1.\end{aligned}$$

Also,

$$\text{Cov } (U_1, V_1) = \rho_1.$$

We note that U_1 and any other linear combination $\mathbf{a}'\mathbf{X}^{(1)} = \mathbf{c}'\mathbf{\Sigma}_{11}^{-1/2}\mathbf{X}^{(1)}$ are uncorrelated if

$$0 = \text{Cov}\ (U_1, \mathbf{a}'\mathbf{X}^{(1)}) = \mathbf{g}_1'\mathbf{\Sigma}_{11}^{-1}\mathbf{\Sigma}_{11}\mathbf{\Sigma}_{11}^{-1/2}\mathbf{c} = \mathbf{g}_1'\mathbf{c}$$

i.e., if

$$\mathbf{c}_{\perp}\mathbf{g}_1. \tag{10.2.19}$$

Similarly V_1 and any other linear combination $\mathbf{b}'\mathbf{X}^{(2)} = \mathbf{d}'\mathbf{\Sigma}_{22}^{-1/2}\mathbf{X}^{(2)}$ are uncorrelated if $\mathbf{d} \perp \mathbf{h}_1$.

Also U_1 and any linear combination $\mathbf{b}'\mathbf{X}^{(2)} = \mathbf{d}'\mathbf{\Sigma}_{22}^{-1}\mathbf{X}^{(2)}$ are uncorrelated if

$$\begin{aligned} 0 = \text{Cov}\ (U_1, \mathbf{b}'\mathbf{X}^{(2)}) &= \mathbf{g}_1'\mathbf{\Sigma}_{11}^{-1/2}\mathbf{\Sigma}_{12}\mathbf{\Sigma}_{22}^{-1/2}\mathbf{d} \\ &= \mathbf{g}_1'\mathbf{G}(\mathbf{D}_\rho\ \mathbf{0})\mathbf{H}'\mathbf{d} = \rho_1\mathbf{h}_1'\mathbf{d}, \end{aligned} \tag{10.2.20}$$

i.e., if $\mathbf{d} \perp \mathbf{h}_1$. Similarly, Cov $(\mathbf{a}'\mathbf{X}^{(1)}, V_1) = 0$ if $\mathbf{c} \perp \mathbf{g}_1$. Proceeding as in (10.2.13), (10.2.14), Corr. $(\mathbf{a}'\mathbf{X}^{(1)}, \mathbf{b}'\mathbf{X}^{(2)})$ attains its maximum if

$$\mathbf{c}'\mathbf{\Sigma}_{11}^{-1/2}\mathbf{\Sigma}_{12}\mathbf{\Sigma}_{22}^{-1/2}\mathbf{d} = \{\mathbf{c}'\mathbf{\Sigma}_{11}^{-1/2}\mathbf{\Sigma}_{12}\mathbf{\Sigma}_{22}^{-1}\mathbf{\Sigma}_{21}\mathbf{\Sigma}_{11}^{-1/2}\mathbf{c}\}^{1/2}\{\mathbf{d}'\mathbf{d}\}^{1/2} \tag{10.2.21}$$

where $\mathbf{c} \perp \mathbf{g}_1, \mathbf{d} \perp \mathbf{h}_1$ and $\mathbf{c} = \mathbf{A}_1\mathbf{a}$ and $\mathbf{d} = \mathbf{A}_2\mathbf{b}$. The first factor on the right side of (10.2.21) attains its maximum value $\sqrt{\lambda_2}$, where λ_2 is the second largest eigenvalue of $\mathbf{M}$, when $\mathbf{c} = \mathbf{g}_2$, the associated eigenvector and hence $\mathbf{a} = \mathbf{\Sigma}_{11}^{-1/2}\mathbf{g}_2$.

Again, the equality in (10.2.21) is attained if

$$\mathbf{d} \propto \mathbf{\Sigma}_{22}^{-1/2}\mathbf{\Sigma}_{21}\mathbf{\Sigma}_{11}^{-1/2}\mathbf{g}_2$$

i.e.,

$$\mathbf{b} \propto \mathbf{\Sigma}_{22}^{-1}\mathbf{\Sigma}_{21}\mathbf{\Sigma}_{11}^{-1/2}\mathbf{g}_2, \tag{10.2.22}$$

taking without loss of generality the constant of proportionality to be unity. Proceeding exactly as in (10.2.18) we see that λ_2 is also the second largest eigenvalue of $\mathbf{N}$ and its associated eigenvector $\mathbf{h}_2$ is proportional to $\mathbf{\Sigma}_{22}^{-1/2}\mathbf{\Sigma}_{21}\mathbf{\Sigma}_{11}^{-1}\mathbf{g}_2$, where the sign is selected to make the correlation positive. Hence, after normalization so that $V(\mathbf{b}_2'\mathbf{X}^{(2)}) = 1$,

$$\mathbf{b} = \mathbf{\Sigma}_{22}^{-1/2}\mathbf{h}_2. \tag{10.2.23}$$

Therefore,

$$U_2 = \mathbf{g}_2'\mathbf{\Sigma}_{11}^{-1/2}\mathbf{X}^{(1)},\ V_2 = \mathbf{h}_2'\mathbf{\Sigma}_{22}^{-1/2}\mathbf{X}^{(2)}. \tag{10.2.24}$$

Clearly,

$$V(U_2) = V(V_2) = 1, \quad \text{Corr. } (U_2, V_2) = \sqrt{\lambda_2} = \rho_2.$$

Also,

$$\begin{aligned}\text{Cov } (U_1, V_2) &= \mathbf{g}_1' \boldsymbol{\Sigma}_{11}^{-1/2} \boldsymbol{\Sigma}_{12} \boldsymbol{\Sigma}_{22}^{-1/2} \mathbf{h}_2 \\ &= \mathbf{g}_1' \mathbf{G}[\mathbf{D}_\rho \ \mathbf{0}]\mathbf{H}' \mathbf{h}_2 \\ &= \rho_1 \mathbf{h}_1' \mathbf{h}_2 = 0.\end{aligned} \tag{10.2.25}$$

Proof for the general case follows similarly. □

Note 10.2.1: Note that the choice of U_k, V_k would not be unique if $\rho_k^2 = \rho_{k+1}^2$.

Note 10.2.2: The matrix $\mathbf{N} = \boldsymbol{\Sigma}_{22}^{-1/2} \boldsymbol{\Sigma}_{21} \boldsymbol{\Sigma}_{11}^{-1/2} \boldsymbol{\Sigma}_{12} \boldsymbol{\Sigma}_{22}^{-1/2}$ has p nonzero eigenvalues and the remaining $(q - p)$ eigenvalues are equal to zero. This is because, rank of $\mathbf{N}$ is p, as $\boldsymbol{\Sigma}^{-1/2}$ has rank q and $\boldsymbol{\Sigma}_{21} \boldsymbol{\Sigma}_{11}^{-1} \boldsymbol{\Sigma}_{12} \boldsymbol{\Sigma}_{22}^{-1/2}$ has rank p and $p < q$.

Note 10.2.3: The variables $U_1, \ldots, U_p$ are uncorrelated, but they are not orthogonal ($E(U_i U_j) \neq 0, i \neq j$). Similar result follows for $V_1, \ldots, V_q$.

Corollary 10.2.1.1: If the original variables are standardized, $\mathbf{Z}^{(1)} = (Z_1^{(1)}, \ldots, Z_p^{(1)})', \mathbf{Z}^{(2)} = (Z_1^{(2)}, \ldots, Z_q^{(2)})', p \leq q$, then $\boldsymbol{\Sigma}_{11} = \rho_{11}, \boldsymbol{\Sigma}_{12} = \rho_{12}, \boldsymbol{\Sigma}_{22} = \rho_{22}$. Hence, canonical variables are

$$\begin{aligned}U_k &= \mathbf{g}_k' \rho_{11}^{-1/2} \mathbf{Z}^{(1)}, \\ V_k &= \mathbf{h}_k' \rho_{22}^{-1/2} \mathbf{Z}^{(2)}, k = 1, \ldots, p\end{aligned} \tag{10.2.26}$$

where $\mathbf{g}_k$ and $\mathbf{h}_k$ are eigenvalues of $\rho_{11}^{-1/2} \rho_{12} \rho_{22}^{-1} \rho_{21} \rho_{11}^{-1/2}$ and $\rho_{22}^{-1/2} \rho_{21} \rho_{11}^{-1} \rho_{12} \rho_{22}^{-1/2}$, respectively. Here

$$\text{Corr. } (U_k, V_k) = \rho_k$$

where $\rho_1 \geq \rho_2 \geq \ldots \geq \rho_p$ are the nonzero eigenvalues of $\rho_{11}^{-1/2} \rho_{12} \rho_{22}^{-1} \rho_{21} \rho_{11}^{-1/2}$ (or equivalently of $\rho_{22}^{-1/2} \rho_{21} \rho_{11}^{-1} \rho_{12} \rho_{22}^{-1/2}$).

Note 10.2.4: If $\mathbf{a}_k$ is the coefficient vector for the kth canonical variable U_k,

$$\begin{aligned}U_k &= \mathbf{a}_k' \mathbf{X}^{(1)} \\ &= \mathbf{a}_k' (\mathbf{X}^{(1)} - \mu^{(1)}) + \mathbf{a}_k' \mu^{(1)} \\ &= a_{k1} \sqrt{\sigma_{11}^{(1)}} \{ \tfrac{X_1^{(1)} - \mu_1^{(1)}}{\sqrt{\sigma_{11}^{(1)}}} \} + \ldots + a_{kp} \sqrt{\sigma_{pp}^{(1)}} \{ \tfrac{X_p^{(1)} - \mu_p^{(1)}}{\sqrt{\sigma_{pp}^{(1)}}} \} + \mathbf{a}_k' \mu^{(1)} \\ &= \mathbf{a}_k' \tilde{\mathbf{V}}_{11}^{1/2} \mathbf{Z}^{(1)} + \mathbf{a}_k' \mu^{(1)} \\ &= \mathbf{a}_{kz}' \mathbf{Z}^{(1)} + \mathbf{a}_k' \mu^{(1)}\end{aligned} \tag{10.2.27}$$

where

$$\begin{aligned}
\mathbf{a}_{kz} &= \tilde{\mathbf{V}}_{11}^{1/2}\mathbf{a}_k \\
\tilde{\mathbf{V}}_{11}^{1/2} &= \text{Diag}\,(\sqrt{\sigma_{11}^{(1)}}, \ldots, \sqrt{\sigma_{pp}^{(1)}}), \sigma_{ii}^{(1)} = V(X_i^{(1)}), i = 1, \ldots, p, \\
Z_i^{(1)} &= (X_i^{(1)} - \mu_i^{(1)})/\sqrt{\sigma_{ii}^{(1)}}, \mathbf{Z}^{(1)} = (Z_1^{(1)}, \ldots, Z_p^{(1)})'.
\end{aligned}$$

Therefore, if $\mathbf{a}_k$ is the coefficient vector for the kth canonical variable $U_k = \mathbf{a}_k'\mathbf{X}^{(1)}$, then the coefficient vector for the canonical variable U_{kz} obtained from the standardized variable $\mathbf{Z}^{(1)}$ is $\mathbf{a}_{kz} = \tilde{\mathbf{V}}_{11}^{1/2}\mathbf{a}_k$.

Similarly for the kth canonical variable $V_{kz} = \mathbf{b}_k'\mathbf{X}^{(2)}$, the canonical variable obtained from the standardized vector $\mathbf{Z}^{(2)} = (Z_1^{(2)}, \ldots, Z_q^{(2)})'$ is

$$\begin{aligned}
V_{kz} &= \mathbf{b}_k'\tilde{\mathbf{V}}_{22}^{1/2}\mathbf{Z}^{(2)} + \mathbf{b}_k'\mu^{(2)} \\
&= \mathbf{b}_{kz}'\mathbf{Z}^{(2)} + \mathbf{b}_k'\mu^{(2)}
\end{aligned}$$

where

$$\mathbf{b}_{kz} = \tilde{\mathbf{V}}_{22}^{1/2}\mathbf{b}_k,$$

$$\tilde{\mathbf{V}}_{22} = \text{Diag}\,(\sigma_{11}^{(2)}, \ldots, \sigma_{qq}^{(2)}), \sigma_{jj}^{(2)} = V(X_j^{(2)}), j = 1, \ldots, q,$$

$$Z_j^{(2)} = (X_j^{(2)} - \mu_j^{(2)})/\sqrt{\sigma_{jj}^{(2)}}.$$

Note that correlation coefficient remains unchanged under standardization of variables.

Note 10.2.5: The canonical variables have no physical meaning. If the original variables $\mathbf{X}^{(1)}, \mathbf{X}^{(2)}$ are used, the canonical coefficients $\mathbf{a}$ and $\mathbf{b}$ have units proportional to those of $\mathbf{X}^{(1)}$ and $\mathbf{X}^{(2)}$ sets. If the original variables are standardized to have zero means and unit variances, the canonical coefficients have no units of measurements, and they should be interpreted in terms of the standardized variables.

EXAMPLE 10.2.1: Suppose $\mathbf{Z}^{(1)} = (Z_1^{(1)}, Z_2^{(1)})', \mathbf{Z}^{(2)} = (Z_1^{(2)}, Z_2^{(2)})'$ are two sets of standardized variables and $\mathbf{Z} = (\mathbf{Z}^{(1)'}, \mathbf{Z}^{(2)'})'$. Let

$$\text{Cov}\,(\mathbf{Z}) = \begin{bmatrix} 1.0 & 0.5 & 0.6 & 0.5 \\ 0.5 & 1.0 & 0.4 & 0.3 \\ 0.6 & 0.4 & 1.0 & 0.6 \\ 0.5 & 0.3 & 0.6 & 1.0 \end{bmatrix}.$$

Here

$$\rho_{11} = \begin{bmatrix} 1.0 & 0.5 \\ 0.5 & 1.0 \end{bmatrix}, \ \rho_{11}^{-1/2} = \begin{bmatrix} 1.11536 & -0.29885 \\ -0.29885 & 1.11536 \end{bmatrix}.$$

Also,

$$\rho_{11}^{-1/2}\rho_{12}\rho_{22}^{-1}\rho_{21}\rho_{11}^{-1/2} = \begin{bmatrix} 0.331995 & 0.152089 \\ 0.152089 & 0.072183 \end{bmatrix}.$$

Its eigenvalues are $\rho_1^2 = 0.4021, \rho_2^2 = 0.0021$ and the corresponding eigenvectors are

$$\mathbf{g}_1 = (.908150,\ 0.418644)',\ \mathbf{g}_2 = (-0.418644,\ 0.908150)'.$$

Therefore,

$$\mathbf{a}_1 = \rho_{11}^{-1/2}\mathbf{g}_1 = \begin{bmatrix} 0.887802 \\ 0.195538 \end{bmatrix}.$$

Similarly,

$$\rho_{22}^{-1/2}\rho_{21}\rho_{11}^{-1}\rho_{12}\rho_{22}^{-1/2} = \begin{bmatrix} .277081 & .185415 \\ .185415 & .127083 \end{bmatrix}$$

having eigenvectors $\rho_1^2 = .4021$ and $\rho_2^2 = .0021$ with eigenvectors

$$\mathbf{h}_1 = (.829150,\ .559026),\ \mathbf{h}_2 = (-.559026,\ .829150)'.$$

Therefore,

$$\mathbf{b}_1' = (.762277,\ .335173)'.$$

Hence, the first pair of canonical variables are

$$U_{1z} = .888Z_1^{(1)} + .196Z_2^{(1)},$$

$$V_{1z} = .762Z_1^{(2)} + .335Z_2^{(2)}$$

with first canonical correlation $\rho_1 = \sqrt{.4021} = .634$. This is the largest correlation possible between the linear combinations of variables from the $\mathbf{Z}^{(1)}$ and $\mathbf{Z}^{(2)}$ sets.

The second pair of canonical variables are

$$U_{2z} = -.738Z_1^{(1)} + 1.138Z_2^{1)},$$

$$V_{2z} = -.991Z_1^{(2)} + 1.204Z_2^{(2)},$$

and the second canonical correlation is $\sqrt{.0020} = .045$. This is very small and consequently, the second pair of canonical variables convey very little information about the association between sets.

10.2.1 *Correlation between canonical variables and original variables*

Let the vector of canonical variables be expressed as

$$\mathbf{U} = \begin{bmatrix} U_1 \\ \cdot \\ \cdot \\ U_p \end{bmatrix} = \begin{bmatrix} \mathbf{a}_1' \\ \cdot \\ \cdot \\ \mathbf{a}_p' \end{bmatrix} \mathbf{X}^{(1)} = \mathbf{A}\mathbf{X}^{(1)}, \ \mathbf{V} = \mathbf{B}\mathbf{X}^{(2)} \tag{10.2.28}$$

where the rows of $\mathbf{A}$ are the coefficient vectors of U_k's and similarly for $\mathbf{B}$. Note that

$$\mathbf{A} = \mathbf{G}'\boldsymbol{\Sigma}_{11}^{-1/2} \tag{10.2.29}$$

where $\mathbf{G}$ has been defined in (10.2.2). We are primarily interested in the first p canonical variables in $\mathbf{V}$. Now

$$\text{Cov}\ (\mathbf{U}, \mathbf{X}^{(1)}) = \text{Cov}\ (\mathbf{A}\mathbf{X}^{(1)}, \mathbf{X}^{(1)}) = \mathbf{A}\boldsymbol{\Sigma}_{11}. \tag{10.2.30}$$

Also,

$$\text{Cov}\ (\mathbf{U}, \mathbf{X}^{(1)}) = \text{Cov}\ (\mathbf{U}, \mathbf{A}^{-1}\mathbf{U}) = \mathbf{A}^{-1}. \tag{10.2.31}$$

Again,

$$\text{Corr. }(U_i, X_k^{(1)}) = \frac{\text{Cov}\ (U_i, X_k^{(1)})}{\sqrt{\sigma_{kk}^{(1)}}}.$$

Thus, Corr. $(U_i, X_k^{(1)}) =$ Cov. $(U_i, X_k^{(1)}/\sqrt{\sigma_{kk}^{(1)}})$. Therefore, correlation matrix between $\mathbf{U}$ and $\mathbf{X}^{(1)}$ is

$$\begin{aligned} \rho_{\mathbf{U},\mathbf{X}^{(1)}} &= \text{Corr. }(\mathbf{U}, \mathbf{X}^{(1)}) \\ &= \text{Cov}\ (\mathbf{U}, \tilde{\mathbf{V}}_{11}^{-1/2}\mathbf{X}^{(1)}) \\ &= \text{Cov}\ (\mathbf{A}\mathbf{X}^{(1)}, \tilde{\mathbf{V}}_{11}^{-1/2}\mathbf{X}^{(1)}) \\ &= \mathbf{A}\boldsymbol{\Sigma}_{11}\tilde{\mathbf{V}}_{11}^{-1/2}. \end{aligned} \tag{10.2.32}$$

Similarly,

$$\begin{aligned} \rho_{\mathbf{U},\mathbf{X}^{(2)}} &= \mathbf{A}\boldsymbol{\Sigma}_{12}\tilde{\mathbf{V}}_{22}^{-1/2}, \\ \rho_{\mathbf{V},\mathbf{X}^{(1)}} &= \mathbf{B}\boldsymbol{\Sigma}_{21}\tilde{\mathbf{V}}_{11}^{-1/2}, \\ \rho_{\mathbf{V},\mathbf{X}^{(2)}} &= \mathbf{B}\boldsymbol{\Sigma}_{22}\tilde{\mathbf{V}}_{22}^{-1/2} \end{aligned} \tag{10.2.33}$$

where $\tilde{\mathbf{V}}_{22}$ has been defined before. If the original variables are standardized variables $\mathbf{Z}^{(1)}, \mathbf{Z}^{(2)}$ (each component variable has zero mean and unit standard deviation) and the vector of canonical variables are $\mathbf{U}_z =$

$\mathbf{A}_z\mathbf{Z}^{(1)}, \mathbf{V}_z = \mathbf{B}_z\mathbf{Z}^{(2)}$, where $\mathbf{A}_z, \mathbf{B}_z$ are matrices whose rows contain coefficients in canonical variables from $\mathbf{Z}^{(1)}, \mathbf{Z}^{(2)}$, respectively, then it can be shown that

$$\begin{aligned} \rho_{\mathbf{U}_z,\mathbf{Z}^{(1)}} &= \mathbf{A}_z\rho_{11}, \quad \rho_{\mathbf{U}_z,\mathbf{Z}^{(2)}} = \mathbf{A}_z\rho_{12}, \\ \rho_{\mathbf{V}_z,\mathbf{Z}^{(2)}} &= \mathbf{B}_z\rho_{21}, \quad \rho_{\mathbf{V}_z,\mathbf{Z}^{(2)}} = \mathbf{B}_z\rho_{22} \end{aligned} \tag{10.2.34}$$

where

$$\mathbf{A}_z = \mathbf{A}\tilde{\mathbf{V}}_{11}^{1/2}, \ \mathbf{B}_z = \mathbf{B}\tilde{\mathbf{V}}_{22}^{1/2}$$

(by (10.2.27)). However, $\rho_{\mathbf{U},\mathbf{X}^{(1)}} = \rho_{\mathbf{U},\mathbf{Z}^{(1)}}$. This is, because $\rho_{\mathbf{U},\mathbf{X}^{(1)}} = \mathbf{A}\mathbf{\Sigma}_{11}\bar{\mathbf{V}}_{11}^{-1/2} = (\mathbf{A}\bar{\mathbf{V}}_{11}^{1/2})(\bar{\mathbf{V}}_{11}^{-1/2}\mathbf{\Sigma}_{11}\bar{\mathbf{V}}_{11}^{-1/2}) = \mathbf{A}_z\rho_{11}$. Similarly, $\rho_{\mathbf{U},\mathbf{X}^{(2)}} = \rho_{\mathbf{U},\mathbf{Z}^{(2)}}, \rho_{\mathbf{V},\mathbf{X}^{(1)}} = \rho_{\mathbf{V},\mathbf{Z}^{(1)}}, \rho_{\mathbf{V},\mathbf{X}^{(2)}} = \rho_{\mathbf{V},\mathbf{Z}^{(2)}}$. The canonical correlations remain unaffected by standardization of variables.

Correlation between the original variables and the canonical variables reflect how much the canonical variables summarize the information contained in the original variables. If $\rho_{U_i,X_k^{(1)}}$ is high it means U_i is a good reflection of the variable $X_k^{(1)}$.

Example 10.2.2: For the data in Example 10.2.1, compute the correlations between U_{1z} and $\mathbf{Z}^{(1)}$; V_{1z} and $\mathbf{Z}^{(2)}$; U_{1z} and $\mathbf{Z}^{(2)}$; V_{1z} and $\mathbf{Z}^{(1)}$.

From (10.2.34),

$$\rho_{U_{1z},\mathbf{Z}^{(1)}} = \mathbf{a}'_{1(z)}\rho_{11} = (.887, \ .196)\begin{bmatrix} 1 & .5 \\ .5 & 1 \end{bmatrix} = (.985, \ .640);'$$

$$\rho_{V_{1z},\mathbf{Z}^{(2)}} = \mathbf{b}'_{1z}\rho_{22} = (.763, \ .335)\begin{bmatrix} 1 & .6 \\ .6 & 1 \end{bmatrix} = (.963, \ .793)';$$

$$\rho_{U_{1z},\mathbf{Z}^{(2)}} = \mathbf{a}'_{1(z)}\rho_{12} = (.887, \ .196)\begin{bmatrix} .6 & .5 \\ .4 & .3 \end{bmatrix} = (.611, \ .503)';$$

$$\rho_{V_{1z},\mathbf{Z}^{(1)}} = \mathbf{b}'_{1z}\rho_{21} = (.762, \ .335)\begin{bmatrix} .6 & .4 \\ .5 & .3 \end{bmatrix} = (.625, \ .405)'.$$

Of the two variables in the set $\mathbf{Z}^{(1)}$, the first is more closely associated with the canonical variable U_{1z}. Similarly, of the two variables in the set $\mathbf{Z}^{(2)}$, the first is more closely associated with the canonical variable V_{1z}. This pattern is consistent with the pattern of the coefficients $Z_1^{(1)}, Z_2^{(1)}$ in U_{1z} and $Z_1^{(2)}, Z_2^{(2)}$ in V_{1z}, respectively.

10.2.2 *Relation between canonical correlation and multiple correlation*

When $\mathbf{X}^{(1)}, \mathbf{X}^{(2)}$ each has one component ($p = q = 1$),

$$|\text{Corr. } (X^{(1)}, X^{(2)})| = |\text{Corr.}(aX^{(1)}, bX^{(2)})| \;\forall\; a, b.$$

Hence, canonical variables are $U_1 = X^{(1)}, V = X^{(2)}$ and canonical correlation is

$$|\text{Corr. } (X^{(1)}, X^{(2)})| = \rho_1$$

which is equal to the ordinary product moment correlation between $X^{(1)}, X^{(2)}$.

When $X^{(1)}, X^{(2)}$ each has more than one component, setting $\mathbf{a}' = (0, \ldots, 0, 1, 0, \ldots, 0)$ with 1 in the ith position and $\mathbf{b}' = (0, \ldots, 0, 1, 0, \ldots, 0)$ with 1 in the kth position

$$\begin{aligned} |\text{Corr. } (X_i^{(1)}, X_k^{(2)})| &= |\text{Corr. } (\mathbf{a}'\mathbf{X}^{(1)}, \mathbf{b}'\mathbf{X}^{(2)})| \\ &\leq \max{}_{\mathbf{a},\mathbf{b}} |\text{Corr. } (\mathbf{a}'\mathbf{X}^{(1)}, \mathbf{b}'\mathbf{X}^{(2)})| = \rho_1, \end{aligned}$$

i.e., the absolute value of product-moment correlation between $X_i^{(1)}, X_k^{(2)}$, namely, $|\rho_{X_i^{(1)}, X_k^{(2)}}| \leq$ canonical correlation ρ_1 for $\mathbf{X}^{(1)}, \mathbf{X}^{(2)}$. Thus the absolute value of each entry in $\rho_{12} \leq \rho_1$.

If $p = 1$ and $q > 1$, then we know that the multiple correlation coefficient of $X_1^{(1)}$ on the variables in the set $\mathbf{X}^{(2)}$ is

$$\begin{aligned} \rho_{X_1^{(1)}(\mathbf{X}^{(2)})} &= \max{}_{\mathbf{b}} \text{ Corr. } (X_1^{(1)}, \mathbf{b}'\mathbf{X}^{(2)}) \\ &= \text{(first) canonical correlation between } X_1^{(1)} \text{ and } \mathbf{X}^{(2)} \\ &= \rho_1. \end{aligned}$$

Thus, in this case, multiple correlation is a special case of canonical correlation.

When $p > 1, q > 1, \rho_1$ is larger than each of multiple correlation coefficient of $X_i^{(1)}$ on $\mathbf{X}^{(2)} (i = 1, \ldots, p)$ or multiple correlation coefficient of $X_k^{(2)}$ on $\mathbf{X}^{(1)} (k = 1, \ldots, q)$.

Again, multiple correlation of U_k on $\mathbf{X}^{(2)}$,

$$\begin{aligned} \rho_{U_k(\mathbf{X}^{(2)})} &= \max{}_{\mathbf{b}} \text{ Corr. } (U_k, \mathbf{b}'\mathbf{X}^{(2)}) \\ &= \text{Corr. } (U_k, V_k) = \rho_k, k = 1, \ldots, p. \end{aligned}$$

Similarly,

$$\begin{aligned} \rho_{V_k(\mathbf{X}^{(1)})} &= \max{}_{\mathbf{a}} \text{ Corr. } (\mathbf{a}'\mathbf{X}^{(1)}, V_k) \\ &= \text{Corr. } (U_k, V_k) = \rho_k, k = 1, \ldots, p. \end{aligned}$$

Thus, canonical correlations are also the multiple correlation of U_k on $\mathbf{X}^{(2)}$ or V_k on $\mathbf{X}^{(1)}$.

Therefore, following the property of multiple correlation coefficient ρ_k^2 can be interpreted as the proportion of variance of U_k that is explained by the variables in $\mathbf{X}^{(2)}$. Similarly, ρ_k^2 is also the proportion of variance in V_k that is explained by the variables in $\mathbf{X}^{(1)}$.

10.3 The Sample Canonical Variables and the Sample Canonical Correlations

The results given in the previous section are completely theoretical, as they depend heavily on the parameters which are ordinarily unknown. Here we derive the corresponding sample-based results.

A random sample of n observations on the $(p+q)$ variables $\mathbf{X}^{(1)}$ and $\mathbf{X}^{(2)}$ is selected. The observations can be assembled as follows:

$$
\begin{aligned}
\mathbf{x} &= \begin{bmatrix} \mathbf{x}^{(1)} \\ \mathbf{x}^{(2)} \end{bmatrix} \\
&= \begin{bmatrix}
x_{11}^{(1)} & x_{12}^{(1)} & \cdots & x_{1n}^{(1)} \\
x_{21}^{(1)} & x_{22}^{(1)} & \cdots & x_{2n}^{(1)} \\
\cdot & \cdot & \cdots & \cdot \\
\cdot & \cdot & \cdots & \cdot \\
x_{p1}^{(1)} & x_{p2}^{(1)} & \cdots & x_{pn}^{(1)} \\
x_{11}^{(2)} & x_{12}^{(2)} & \cdots & x_{1n}^{(2)} \\
x_{21}^{(2)} & x_{22}^{(2)} & \cdots & x_{2n}^{(2)} \\
\cdot & \cdot & \cdots & \cdot \\
\cdot & \cdot & \cdots & \cdot \\
x_{q1}^{(2)} & x_{q2}^{(2)} & \cdots & x_{qn}^{(2)}
\end{bmatrix} \\
&= [\mathbf{x}_1 \ \mathbf{x}_2 \ \ldots \mathbf{x}_n]
\end{aligned}
$$

where

$$
\mathbf{x}_j = \begin{bmatrix} \mathbf{x}_j^{(1)} \\ \mathbf{x}_j^{(2)} \end{bmatrix}, j = 1, \ldots n. \tag{10.3.1}
$$

The vector of sample means can be written as

$$
\bar{\mathbf{x}} = \begin{bmatrix} \bar{\mathbf{x}}^{(1)} \\ \bar{\mathbf{x}}^{(2)} \end{bmatrix} \quad \text{where } \bar{\mathbf{x}}^{(1)} = \begin{bmatrix} \bar{x}_1^{(1)} \\ \cdot \\ \cdot \\ \bar{x}_p^{(1)} \end{bmatrix}, \ \bar{\mathbf{x}}^{(2)} = \begin{bmatrix} \bar{\mathbf{x}}_1^{(2)} \\ \cdot \\ \cdot \\ \bar{\mathbf{x}}_q^{(2)} \end{bmatrix},
$$

$$\bar{x}_i^{(1)} = \frac{1}{n}\sum_{j=1}^{n} x_{ij}^{(1)},\ \bar{x}_i^{(2)} = \frac{1}{n}\sum_{j=1}^{n} x_{ij}^{(2)}.$$

Similarly, the sample covariance matrix $\mathbf{S}$ can be written as

$$\mathbf{S} = \begin{bmatrix} \mathbf{S}_{11} & \mathbf{S}_{12} \\ \mathbf{S}_{21} & \mathbf{S}_{22} \end{bmatrix}$$

where

$$\mathbf{S}_{kl} = \frac{1}{n-1}\sum_{j=1}^{n}(\mathbf{x}_j^{(k)} - \bar{\mathbf{x}}^{(k)})(\mathbf{x}_j^{(l)} - \bar{\mathbf{x}}^{(l)})',\ k,l = 1, \tag{10.3.2}$$

$$\mathbf{S}_{11} = \begin{bmatrix} S_{11}^{(1)} & S_{12}^{(1)} & \dots & S_{1p}^{(1)} \\ . & . & \dots & . \\ S_{p1}^{(1)} & S_{p2}^{(1)} & \dots & S_{pp}^{(1)} \end{bmatrix},\ S_{ij}^{(1)} = \frac{1}{n-1}\sum_{k=1}^{n}(x_{ik}^{(1)} - \bar{x}_i^{(1)})(x_{jk}^{(1)} - \bar{x}_j^{(1)}),$$

etc.

DEFINITION 10.3.1: Consider the linear combinations $\hat{U} = \mathbf{t}'\mathbf{x}^{(1)}$ and $\hat{V} = \mathbf{w}'\mathbf{x}^{(2)}$, where, $\mathbf{x}^{(1)} = (x_1^{(1)}, x_2^{(1)}, \dots, x_p^{(1)})$, the observed values of the variable $\mathbf{X}^{(1)}$ for a typical observation and $\mathbf{x}^{(2)}$ has a similar meaning.

The sample correlation between $\hat{U}$ and $\hat{V}$ is

$$r_{\hat{U},\hat{V}} = \frac{\mathbf{t}'\mathbf{S}_{12}\mathbf{w}}{\sqrt{\mathbf{t}'\mathbf{S}_{11}\mathbf{t}}\sqrt{\mathbf{w}'\mathbf{S}_{22}\mathbf{w}}}. \tag{10.3.3}$$

The first pair of sample canonical variables is the pair of linear combinations $\hat{U}_1, \hat{V}_1$ having unit sample variances that maximize (10.3.3).

In general, the kth pair of the sample canonical variables is the pair of linear combinations $\hat{U}_k, \hat{V}_k$ having unit sample variances that maximize (10.3.3) among those linear combinations which are uncorrelated with the previous $k-1$ sample canonical variables $(\hat{U}_i, \hat{V}_i), i = 1, \dots, k-1; k = 1, \dots, p$.

The sample correlation between $\hat{U}_k$ and $\hat{V}_k$ is called the kth sample canonical correlation.

Theorem 10.3.1: For coefficient vectors $\mathbf{t}$ and $\mathbf{w}$ form linear combinations $\mathbf{t}'\mathbf{x}^{(1)}$ and $\mathbf{w}'\mathbf{x}^{(2)}$. Then

$$\max{}_{\mathbf{t},\mathbf{w}} \text{ Sample Corr. } (\mathbf{t}'\mathbf{x}^{(1)}, \mathbf{w}'\mathbf{x}^{(2)}) = r_1 \tag{10.3.4}$$

and this is attained when

$$\mathbf{t} = \mathbf{t}_1 = \hat{\mathbf{g}}_1'\mathbf{S}_{11}^{-1/2},\ \mathbf{w} = \mathbf{w}_1 = \hat{\mathbf{h}}_1'\mathbf{S}_{22}^{-1/2} \tag{10.3.5}$$

where r_1^2 is the maximum eigenvalue of $\mathbf{S}_{11}^{-1/2}\mathbf{S}_{12}\mathbf{S}_{22}^{-1}\mathbf{S}_{21}\mathbf{S}_{11}^{-1/2}$ and $\hat{\mathbf{g}}_1$ is the associated normalized eigenvector. The quantity r_1^2 is also the maximum eigenvalue of $\mathbf{S}_{22}^{-1/2}\mathbf{S}_{21}\mathbf{S}_{11}^{-1}\mathbf{S}_{12}\mathbf{S}_{22}^{-1/2}$ and $\hat{\mathbf{h}}_1$ is its associated normalized eigenvector.

The first pair of sample canonical variables are, therefore,

$$\hat{U}_1 = \hat{\mathbf{g}}_1'\mathbf{S}_{11}^{-1/2}\mathbf{x}^{(1)}, \ \hat{V}_1 = \hat{\mathbf{g}}_2'\mathbf{S}_{22}^{-1/2}\mathbf{x}^{(2)}. \tag{10.3.6}$$

Also, $\text{Var}(\hat{U}_1) = \text{ Var }(\hat{V}_1) = 1$.

Consider now all linear combinations $\mathbf{t}'\mathbf{x}^{(1)}$ and $\mathbf{w}'\mathbf{x}^{(2)}$ which are uncorrelated with $\hat{U}_1, \hat{V}_1$. Then among all such linear combinations

$$\max{}_{\mathbf{t},\mathbf{w}} \text{ Sample Corr. } (\mathbf{t}'\mathbf{x}^{(1)}, \mathbf{w}'\mathbf{x}^{(2)}) = r_2 \tag{10.3.7}$$

and this is attained when

$$\mathbf{t}' = \mathbf{t}_2' = \hat{\mathbf{g}}_2'\mathbf{S}_{11}^{-1/2}, \ \mathbf{w}' = \mathbf{w}_2' = \hat{\mathbf{h}}_2'\mathbf{S}_{22}^{-1/2} \tag{10.3.8}$$

where r_2^2 is the second largest eigenvalue of $\mathbf{S}_{11}^{-1/2}\mathbf{S}_{12}\mathbf{S}_{22}^{-1}\mathbf{S}_{21}\mathbf{S}_{11}^{-1/2}$ and $\hat{\mathbf{g}}_2$ is the associated normalized eigenvector. The quantity r_2^2 is also the second largest eigenvalue of $\mathbf{S}_{22}^{-1/2}\mathbf{S}_{21}\mathbf{S}_{11}^{-1}\mathbf{S}_{12}\mathbf{S}_{22}^{-1/2}$ and $\hat{\mathbf{h}}_2$ is its associated normalized eigenvector.

The linear combinations

$$\hat{U}_2 = \hat{\mathbf{g}}_2'\mathbf{S}_{11}^{-1/2}\mathbf{x}^{(1)} \text{ and } \hat{V}_2 = \hat{h}_2'\mathbf{S}_{22}^{-1/2}\mathbf{x}^{(2)} \tag{10.3.9}$$

are the second pair of sample canonical variables. Also,

$$\begin{aligned} \text{Var }(\hat{U}_2) &= \text{Var}(\hat{V}_2) = 1 \\ \text{Cov }(\hat{U}_1, \hat{U}_2) &= \text{ Cov }(\hat{V}_1, \hat{V}_2) = \text{ Cov }(\hat{U}_i, \hat{V}_j) = 0 (i \neq j = 1, 2). \end{aligned} \tag{10.3.10}$$

Similarly, the kth pair of Canonical variables are

$$\hat{U}_k = \hat{\mathbf{g}}_k'\mathbf{S}_{11}^{-1/2}\mathbf{x}^{(1)}$$

$$\hat{V}_k = \hat{\mathbf{h}}_k'\mathbf{S}_{22}^{-1/2}\mathbf{x}^{(2)}$$

with

$$\text{Sample Corr. } (\hat{U}_k, \hat{V}_k) = r_k, k = 3, \ldots, p.$$

Here r_k^2 is the kth largest eigenvalue of $\mathbf{S}_{11}^{-1/2}\mathbf{S}_{12}\mathbf{S}_{22}^{-1}\mathbf{S}_{21}\mathbf{S}_{11}^{-1/2}$ and also of $\mathbf{S}_{22}^{-1/2}\mathbf{S}_{21}\mathbf{S}_{11}^{-1}\mathbf{S}_{12}\mathbf{S}_{22}^{-1/2}$. Also, $\hat{\mathbf{g}}_k, \hat{\mathbf{h}}_k$ are the associated normalized eigenvectors respectively. Also,

$$\begin{aligned} \text{Var }(\hat{U}_k) &= \text{ Var }(\hat{V}_k) = 1, k = 3, \ldots, p \\ \text{Cov }(\hat{U}_k, \hat{U}_l) &= \text{ Cov }(\hat{V}_k, \hat{V}_l) = 0, k \neq l = 1, \ldots, p \\ \text{Cov }(\hat{U}_k, \hat{V}_l) &= 0. \end{aligned} \tag{10.3.11}$$

The quantities $r_1, \ldots, r_p$ are called the sample canonical correlations.

Proof. Proof follows as in Theorem 10.2.1. Note that we must have $n \geq q+1$ for $\mathbf{S}_{11}^{-1}, \mathbf{S}_{22}^{-1}$ to exist.

Remark 10.3.1: The effect of pq covariances between $\mathbf{x}^{(1)}$ and $\mathbf{x}^{(2)}$ in $\mathbf{S}_{12}$ has been replaced by p canonical correlations. Theses describe the relationship between $\mathbf{x}^{(1)}$ and $\mathbf{x}^{(2)}$ much more succinctly than the pq correlations. In fact we do not need to consider all the p canonical correlations. We can judge the importance of each correlation r_j by its relative size

$$\frac{r_j^2}{\sum_{i=1}^p r_i^2}.$$

If we find $k(< p)$ eigenvalues r_j^2 that account for most of $\sum_{i=1}^p r_i^2$, the remaining canonical correlations can be neglected.

10.3.1 *Sample correlation between original variables and sample canonical variables*

The computation of correlation between original variables and canonical variables is often helpful for interpretation of the canonical variables. Defining

$$\mathbf{T} = \begin{bmatrix} \mathbf{t}_1' \\ \cdot \\ \cdot \\ \mathbf{t}_p' \end{bmatrix} \quad \text{and } \mathbf{W} = \begin{bmatrix} \mathbf{w}_1' \\ \cdot \\ \cdot \\ \mathbf{w}_q' \end{bmatrix}$$

we have the vector of sample canonical variables $\hat{\mathbf{U}} = (\hat{U}_1, \ldots, \hat{U}_p)', \hat{\mathbf{V}} = (\hat{V}_1, \ldots, \hat{V}_q)'$ as

$$\hat{\mathbf{U}} = \mathbf{T}\mathbf{x}^{(1)}, \ \hat{\mathbf{V}} = \mathbf{W}\mathbf{x}^{(2)}.$$

Writing the matrix of correlations between $\hat{\mathbf{U}}$ and $\mathbf{x}^{(1)}$ as $\mathbf{R}_{\hat{\mathbf{U}},\mathbf{x}^{(1)}}$ and other correlation matrices similarly, we have, as in subsection 10.2.1,

$$\begin{aligned} \mathbf{R}_{\hat{\mathbf{U}},\mathbf{x}^{(1)}} &= \mathbf{T}\mathbf{S}_{11}\mathbf{D}_{11}^{-1/2}, \\ \mathbf{R}_{\hat{\mathbf{U}},\mathbf{x}^{(2)}} &= \mathbf{T}\mathbf{S}_{12}\mathbf{D}_{22}^{-1/2}, \\ \mathbf{R}_{\hat{\mathbf{V}},\mathbf{x}^{(1)}} &= \mathbf{W}\mathbf{S}_{21}\mathbf{D}_{11}^{-1/2}, \\ \mathbf{R}_{\hat{\mathbf{V}},\mathbf{x}^{(2)}} &= \mathbf{W}\mathbf{S}_{22}\mathbf{D}_{22}^{-1/2} \end{aligned}$$

where

$$\mathbf{D}_{11}^{-1/2} = \text{ Diag } (1/\sqrt{S_{11}^{(1)}}, \ldots, 1/\sqrt{S_{pp}^{(1)}})$$

and

$$\mathbf{D}_{22}^{-1/2} = \text{Diag}\,(\sqrt{S_{11}^{(2)}}, \ldots, 1/\sqrt{S_{qq}^{(2)}}).$$

If the observations are standardized, the data can be assembled as

$$\mathbf{z} = \begin{bmatrix} \mathbf{z}^{(1)} \\ \mathbf{z}^{(2)} \end{bmatrix} = [\mathbf{z}_1 \ \mathbf{z}_2 \ldots \mathbf{z}_n] \ \text{ and } \mathbf{z}_j = \begin{bmatrix} \mathbf{z}_j^{(1)} \\ \mathbf{z}_j^{(2)} \end{bmatrix},$$

$$\mathbf{z}_j^{(1)} = (z_{1j} \ z_{2j} \ldots z_{pj})', \ \mathbf{z}_j^{(2)} = (z_{1j}^{(2)} \ z_{2j}^{(2)} \ldots z_{qj}^{(2)})'.$$

The vector of sample canonical correlations can be written as

$$\hat{\mathbf{U}}_z = \mathbf{T}_z \mathbf{z}^{(1)}, \ \hat{\mathbf{V}}_2 = \mathbf{W}_z \mathbf{z}^{(2)}$$

and

$$\mathbf{T}_z = \mathbf{T}\mathbf{D}_{11}^{1/2}, \ \mathbf{W}_z = \mathbf{W}\mathbf{D}_{22}^{1/2}.$$

The proof follows as in Note 10.2.4.

Also,

$$\mathbf{R}_{\hat{\mathbf{U}}_z, \mathbf{z}^{(1)}} = \mathbf{T}_z \rho_{11}, \ \mathbf{R}_{\hat{\mathbf{U}}_z, \mathbf{z}^{(2)}} = \mathbf{T}_z \rho_{12},$$

$$\mathbf{R}_{\hat{V}_z, \mathbf{z}^{(1)}} = \mathbf{W}_z \rho_{21}, \ \mathbf{R}_{\hat{V}_z, \mathbf{z}^{(2)}} = \mathbf{W}_z \rho_{22}.$$

The sample correlations remain unaffected by standardization. It can be checked that

$$\begin{aligned} \mathbf{R}_{\hat{\mathbf{U}}, \hat{\mathbf{V}}} &= \mathbf{R}_{\hat{\mathbf{U}}_z, \hat{\mathbf{V}}_z}, \\ \mathbf{R}_{\hat{\mathbf{U}}, \mathbf{x}^{(1)}} &= \mathbf{R}_{\hat{\mathbf{U}}_z, \mathbf{z}^{(1)}}, \\ \mathbf{R}_{\hat{\mathbf{U}}, \mathbf{x}^{(2)}} &= \mathbf{R}_{\hat{\mathbf{U}}_z, \mathbf{z}^{(2)}}, \\ \mathbf{R}_{\hat{\mathbf{V}}, \mathbf{x}^{(1)}} &= \mathbf{R}_{\hat{\mathbf{V}}_z, \mathbf{z}^{(1)}}, \\ \mathbf{R}_{\hat{\mathbf{V}}, \mathbf{x}^{(2)}} &= \mathbf{R}_{\hat{\mathbf{V}}_z, \mathbf{z}^{(2)}}. \end{aligned}$$

Note 10.3.1: We have thus noted that the canonical correlations are invariant to change of scale. For example, if the measurement scale is changed from inches to centimeters, the canonical correlations will not change. However, the corresponding eigenvectors and hence the canonical variates will change.

10.3.2 *Sample covariance in terms of canonical coefficients and canonical correlation*

Let $\mathbf{t}^{(i)}$ and $\mathbf{w}^{(i)}$ denote the ith column of $\mathbf{T}^{-1}$ and $\mathbf{W}^{-1}$ respectively. Now,

$$\hat{\mathbf{U}} = \mathbf{T}\mathbf{x}^{(1)}, \ \hat{\mathbf{V}} = \mathbf{W}\mathbf{x}^{(2)}.$$

Hence,

$$\mathbf{x}^{(1)} = \mathbf{T}^{-1}\hat{\mathbf{U}}, \ \mathbf{x}^{(2)} = \mathbf{W}^{-1}\hat{\mathbf{V}}. \tag{10.3.12}$$

Again,

$$\begin{aligned} &\text{Cov }(\hat{\mathbf{U}}, \hat{\mathbf{V}}) = \mathbf{T}\mathbf{S}_{12}\mathbf{W}', \quad \text{Cov }(\hat{\mathbf{U}}) = \mathbf{T}\mathbf{S}_{11}\mathbf{T}' = \mathbf{I}, \\ &\text{Cov }(\hat{\mathbf{V}}) \quad\ \ = \mathbf{W}\mathbf{S}_{22}\mathbf{W}' = \mathbf{I}. \end{aligned}$$

Therefore,

$$\begin{aligned} \mathbf{S}_{12} &= \mathbf{T}^{-1} \text{ Cov }(\hat{\mathbf{U}}, \hat{\mathbf{V}})(\mathbf{W}^{-1})' \\ &= r_1\mathbf{t}^{(1)}\mathbf{w}^{(1)'} + r_2\mathbf{t}_2\mathbf{w}^{(2)'} + \ldots + r_p\mathbf{t}^{(p)}\mathbf{w}^{(p)'}. \end{aligned} \tag{10.3.13}$$

Similarly,

$$\mathbf{S}_{11} = \mathbf{T}^{-1}(\mathbf{T}^{-1})' = \mathbf{t}^{(1)}\mathbf{t}^{(1)'} + \ldots \mathbf{t}^{(p)}\mathbf{t}^{(p)'}. \tag{10.3.14}$$

$$\mathbf{S}_{22} = \mathbf{W}^{-1}(\mathbf{W}^{-1})' = \mathbf{w}^{(1)}\mathbf{w}(1)' + \ldots + \mathbf{w}^{(q)}\mathbf{w}^{(q)'}. \tag{10.3.15}$$

The expressions (10.3.13), (10.3.14), (10.3.15) give sample covariances in terms of canonical coefficients and canonical correlations. Again,

$$\text{Cov }(\mathbf{x}^{(1)}, \hat{\mathbf{U}}) = \ \text{Cov }(\mathbf{T}^{-1}\hat{\mathbf{U}}, \hat{\mathbf{U}}) = \mathbf{T}^{-1}.$$

Hence, the first r columns of $\mathbf{T}^{-1}$ give $\text{Cov}(\mathbf{x}^{(1)}, \hat{U}_1)$, $\text{Cov }(\mathbf{x}^{(1)}, \hat{U}_2), \ldots,$ $\text{Cov }(\mathbf{x}^{(1)}, \hat{U}_r)$. Similarly, the first r columns of $\mathbf{W}^{-1}$ give $\text{Cov }(\mathbf{x}^{(2)}, \hat{V}_1),$ $\text{Cov }(\mathbf{x}^{(2)}, \hat{V}_2), \ldots,$ $\text{Cov }(\mathbf{x}^{(2)}, \hat{V}_r)$.

10.3.3 *Approximating sample covariances by first r canonical correlations*

Since the first r columns of $\mathbf{T}^{-1}$ give $\text{Cov }(\mathbf{X}^{(1)}, \hat{U}_1), \ldots$ $\text{Cov }(\mathbf{X}^{(1)}, \hat{U}_r)$, if we consider only the first r canonical variables $\hat{U}_1, \ldots, \hat{U}_r$, we can write, following (10.3.12),

$$(\mathbf{t}^{(1)} \ \mathbf{t}^{(2)} \ldots \mathbf{t}^{(r)}) \begin{bmatrix} \hat{U}_1 \\ \hat{U}_2 \\ \cdot \\ \hat{U}_r \end{bmatrix} = \tilde{\mathbf{x}}^{(1)} \tag{10.3.16}$$

where $\tilde{\mathbf{x}}^{(1)}(p \times 1)$ is a vector of variables, close to $\mathbf{x}^{(1)}$. Similarly,

$$\mathbf{w}^{(1)}\hat{V}_1 + \mathbf{w}^2\hat{V}_2 + \ldots + \mathbf{w}^{(r)}\hat{V}_r = \tilde{\mathbf{x}}^{(2)} \tag{10.3.17}$$

where $\tilde{\mathbf{x}}_2(q \times 1)$ is a vector of variables close to $\mathbf{x}^{(2)}$. From (10.3.16), (10.3.17) approximate covariances between $\mathbf{x}^{(1)}, \mathbf{x}^{(2)}$ is the covariance between $\tilde{\mathbf{x}}_1$ and $\tilde{\mathbf{x}}_2$ if we only consider the first r pairs of canonical variables $(\hat{U}_i, \hat{V}_i)(i = 1, \ldots, r)$. From (10.3.13) - (10.3.15), the matrices of error of approximation of covariances are

$$\mathbf{S}_{11} - \mathbf{t}^{(1)}\mathbf{t}^{(1)'} - \ldots - \mathbf{t}^{(r)}\mathbf{t}^{(r)'} = \mathbf{t}^{(r+1)}\mathbf{t}^{(r+1)'} + \ldots \mathbf{t}_p\mathbf{t}_p'. \tag{10.3.18}$$

$$\mathbf{S}_{22} - \mathbf{w}^{(1)}\mathbf{w}^{(1)'} - \ldots - \mathbf{w}^{(r)}\mathbf{w}^{(r)'} = \mathbf{w}^{(r+1)}\mathbf{w}^{(r+1)'} + \ldots + \mathbf{w}_p\mathbf{w}_p'. \tag{10.3.19}$$

$$\mathbf{S}_{12} - r_1\mathbf{t}^{(1)}\mathbf{t}^{(1)'} - \ldots - r_r\mathbf{t}^{(r)}\mathbf{w}^{(r)'} = r_{r+1}\mathbf{t}^{(r+1)}\mathbf{w}^{(r+1)'} + \ldots + r_p\mathbf{t}^{(p)}\mathbf{w}^{(p)'}. \tag{10.3.20}$$

The approximation error matrices indicate how well the first r sample canonical variances reproduce the sample covariance matrices. Patterns of large entries in a row and / or a column indicate a poor fit to the corresponding variable.

Usually, the error in approximating $\mathbf{S}_{12}$ given in (10.3.20) is small compared to errors in approximating $\mathbf{S}_{11}$ and $\mathbf{S}_{22}$ as given in (10.3.18), (10.3.19) respectively. This is because, the residual matrix in the former case depends on the last $p-r$ canonical correlations which are usually close to zero. The residual matrices associated with errors of approximating $\mathbf{S}_{11}, \mathbf{S}_{22}$ depend only on the last $(p-r)$ or $(q-r)$ canonical coefficient vectors whose entries are likely to be larger and hence the elements in these matrices are usually large.

EXAMPLE 10.3.1: Consider Example 10.2.1. We assume that the correlation matrix given there is a sample correlation matrix. The first two pairs of canonical variables are then,

$$\hat{U}_{1z} = .888z_1^{(1)} + .196z_2^{(1)},$$

$$\hat{V}_{1z} = .762z_1^{(2)} + .335z_2^{(2)},$$

$$\hat{U}_{2z} = -.738z_1^{(1)} + 1.138z_2^{(1)},$$

$$\hat{V}_{2z} = -.991z_1^{(2)} + 1.204z_2^{(2)}$$

with $r_1 = .634, r_2 = .045$. Hence

$$\mathbf{T}_z^{-1} = \begin{bmatrix} .888 & .196 \\ -.738 & 1.138 \end{bmatrix}^{-1} = \begin{bmatrix} .985 & -.170 \\ 639 & .769 \end{bmatrix};$$

$$\mathbf{W}_z^{-1} = \begin{bmatrix} .763 & -.991 \\ .335 & 1.204 \end{bmatrix}^{-1} = \begin{bmatrix} .963 & -.268 \\ .793 & .610 \end{bmatrix}.$$

If we only consider the first pair of canonical variables for approximating the variables $\mathbf{z}^{(1)}, \mathbf{z}^{(2)}$, then by formulae (10.3.16) and (10.3.17), the approximating variables are

$$\tilde{\mathbf{z}}_1 = \mathbf{t}_z^{(1)}\hat{U}_{1z} = \begin{bmatrix} .985 \\ .639 \end{bmatrix} (.888z_1^{(1)} + .196Z_2^{(1)}) = \begin{bmatrix} .875z_1^{(1)} + .193z_2^{(1)} \\ .567z_2^{(1)} + .125z_2^{(2)} \end{bmatrix};$$

$$\tilde{\mathbf{z}}_2 = \mathbf{w}_z^{(1)}\hat{V}_{1z} = \begin{bmatrix} ..963 \\ .793 \end{bmatrix} (.762z_1^{(2)} + .335z_2^{(2)}) = \begin{bmatrix} .734z_1^{(2)} + .323z_2^{(2)} \\ .604z_1^{(2)} + .266z_2^{(2)} \end{bmatrix}.$$

The matrices of error of approximation of covariances are, by formulae (10.3.18) - (10.3.20),

$$\begin{aligned} \mathbf{R}_{12} - \text{ sample Cov } (\tilde{\mathbf{z}}_1, \tilde{\mathbf{z}}_2) &= r_2\mathbf{t}_z^{(2)}\mathbf{w}_z^{(2)'} \\ &= .045 \begin{bmatrix} -.170 \\ .769 \end{bmatrix} (-.268,\ .610) \\ &= \begin{bmatrix} .002 & .005 \\ .009 & .021 \end{bmatrix}; \end{aligned}$$

$$\begin{aligned} \mathbf{R}_{11} - \text{ sample Cov } (\tilde{\mathbf{z}}_1) = \mathbf{t}_z^{(2)}\mathbf{t}_z^{(2)'} &= \begin{bmatrix} -.170 \\ .769 \end{bmatrix} (-.170,\ .769) \\ &= \begin{bmatrix} .029 & -.131 \\ -.131 & .591 \end{bmatrix}; \end{aligned}$$

$$\begin{aligned} \mathbf{R}_{22} - \text{ sample Cov } (\tilde{\mathbf{z}}_2) = \mathbf{w}_z^{(2)}\mathbf{w}_z^{(2)'} &= \begin{bmatrix} -.268 \\ .610 \end{bmatrix} (-.268,\ .610) \\ &= \begin{bmatrix} .072 & .163 \\ .163 & .372 \end{bmatrix}. \end{aligned}$$

We see that the first pair of canonical variables effectively summarizes the correlation matrix $\mathbf{R}_{12}$ between $\mathbf{z}^{(1)}$ and $\mathbf{z}^{(2)}$. However, this is not true for individual variables $\hat{U}_{1z}$ and $\hat{V}_{1z}$. The variable $\tilde{\mathbf{z}}^{(1)}$ and $\tilde{\mathbf{z}}^{(2)}$ do not effectively reproduce $\mathbf{R}_{11}$ and $\mathbf{R}_{22}$.

10.3.4 The proportion of total sample variance explained by the canonical variables

When the observations are standardized, all the results in (10.3.12) - (10.3.20) hold with $\hat{\mathbf{U}}$ replaced by $\hat{\mathbf{U}}_z, \hat{\mathbf{V}}$ by $\hat{\mathbf{V}}_z, \mathbf{T}$ by $\mathbf{T}_z, \mathbf{W}$ by $\mathbf{W}_z, \mathbf{S}_{kl}$ by $\mathbf{R}_{kl}, \mathbf{t}^{(i)}$ by $\mathbf{t}_z^{(i)}, \mathbf{w}^{(i)}$ by $\mathbf{w}_z^{(i)}, \mathbf{x}^{(1)}$ by $\mathbf{z}^{(1)}, \mathbf{x}^{(2)}$ by $\mathbf{z}^{(2)}$. Specifically,

$$\text{Sample Cov }(\mathbf{z}^{(1)}, \hat{\mathbf{U}}_z) = \mathbf{T}_z^{-1}$$
$$\text{Sample Cov }(\mathbf{z}^{(2)}, \hat{\mathbf{U}}_z) = \mathbf{W}_z^{-1}.$$

Also the first r columns of $\mathbf{T}_z^{-1}$ give $\text{Cov}\,(\mathbf{z}^{(1)}, \hat{U}_{1z}), \ldots, \text{Cov}\,(\mathbf{z}^{(1)}, \hat{U}_{rz})$. Similarly, the first r columns of $\mathbf{W}_z^{-1}$ give $\text{Cov}\,(\mathbf{z}^{(1)}, \hat{V}_{1z}), \ldots, \text{Cov}\,(\mathbf{z}^{1)}, \hat{V}_{rz})$.

Also, from (10.3.14), the total sample standardized variance of $\mathbf{z}^{(1)}$

$$\text{tr }(\mathbf{R}_{11}) = \text{ tr }[\mathbf{t}_z^{(1)}\mathbf{t}_z^{(1)'} + \ldots + \mathbf{t}_z^{(p)}\mathbf{t}_z^{(p)'}] = p,$$

because tr $(\mathbf{R}_{11}) = p$. Similarly, the total sample standardized variance of $\mathbf{z}^{(2)}$

$$\text{tr }(\mathbf{R}_{22}) = \text{ tr }[\mathbf{w}_z^{(1)}\mathbf{w}_z^{(1)'} + \ldots + \mathbf{w}_z^{(q)}\mathbf{w}_z^{(q)'}] = q.$$

Since $\mathbf{t}_z^{(1)}, \ldots, \mathbf{t}_z^{(r)}$ involve only the sample correlations of the first r canonical variables $\hat{U}_{1z}, \ldots, \hat{U}_{rz}$ with $\mathbf{z}^{(1)}$ the contribution of the first r canonical variables $\hat{U}_{1z}, \ldots, \hat{U}_{rz}$, to the total variance of $\mathbf{z}^{(1)}$ may be defined as

$$\text{tr }[\mathbf{t}_z^{(1)}\mathbf{t}_z^{(1)'} + \ldots + \mathbf{t}_z^{(r)}\mathbf{t}_z^{(r)'}] = \sum_{k=1}^{p}\sum_{i=1}^{r} r^2_{\hat{U}_{iz}, z_k^{(1)}}. \qquad (10.3.21)$$

Similarly, the contribution of the first r canonical variables $\hat{V}_{1z}, \ldots, \hat{V}_{rz}$ to the total variance of $\mathbf{z}^{(2)}$ may be defined as

$$\text{tr }[\mathbf{w}_z^{(1)}\mathbf{w}_z^{(1)'} + \ldots + \mathbf{w}_z^{(r)}\mathbf{w}_z^{(r)'}] = \sum_{k=1}^{p}\sum_{i=1}^{r} r^2_{\hat{V}_{iz}, z_k^{(2)}}. \qquad (10.3.22)$$

Hence, the proportion of the total standardized sample variance of $\mathbf{z}^{(1)}$ explained by $\hat{U}_{1z}, \hat{U}_{2z}, \ldots, \hat{U}_{rz}$ is

$$R^2_{z^{(1)}|\hat{U}_{1z},\ldots,\hat{U}_{rz}} = \frac{1}{p}\sum_{k=1}^{p}\sum_{i=1}^{r} r^2_{\hat{U}_{iz}, z_k^{(i)}}. \qquad (10.3.23)$$

Similarly, the proportion of the total standardized sample variance of $\mathbf{z}^{(2)}$ explained by $\hat{V}_{1z}, \hat{V}_{2z}, \ldots, \hat{V}_{rz}$ is

$$R^2_{z^{(2)}|\hat{V}_{1z},\ldots,\hat{V}_{rz}} = \frac{1}{q}\sum_{k=1}^{p}\sum_{i=1}^{r} r^2_{\hat{V}_{iz}, z_k^{(2)}}. \qquad (10.3.24)$$

EXAMPLE 10.3.2: For the data in example 10.3.1, the proportion of the total standardized sample variance of $\mathbf{z}^{(1)}$ explained by $\hat{U}_{1z}$ is

$$\frac{1}{2}(r^2_{\hat{U}_{1z},z_1^{(1)}} + r^2_{\hat{U}_{1z},z_2^{(1)}}) = \frac{1}{2}[(.985)^2 + (.639)^2] = 0.69.$$

Similarly

$$R^2_{\mathbf{z}^{(2)}|\hat{V}_{1z}} = \frac{1}{2}(r^2_{\hat{V}_{1z},z_1^{(2)}} + r^2_{\hat{V}_{1z},z_2^{(2)}}) = \frac{1}{2}[(.963)^2 + (.793)^2] = 0.78.$$

We may conclude that $\hat{V}_{1z}$ is a better representative of the set $\mathbf{z}^{(2)}$ than $\hat{U}_{1z}$ of its set $\mathbf{z}^{(1)}$.

10.4 Tests of Independence

Suppose we have a sample of size n from a $N_{p+q}(\mu, \mathbf{\Sigma})$ population when

$$\mathbf{\Sigma} = \begin{bmatrix} \mathbf{\Sigma}_{11} & \mathbf{\Sigma}_{12} \\ \mathbf{\Sigma}_{21} & \mathbf{\Sigma}_{22} \end{bmatrix} \tag{10.4.1}$$

where $\mathbf{\Sigma}_{12}$ is of order $p \times q (p \leq q)$. Assume that $n \geq q+1$. We construct a test of independence reflected by the hypothesis $H_0 : \mathbf{\Sigma}_{12} = \mathbf{0}$.

When H_0 holds, $\mathbf{a}'\mathbf{X}^{(1)}$ and $\mathbf{b}'\mathbf{X}^{(2)}$ have covariance $\mathbf{a}'\mathbf{\Sigma}_{12}\mathbf{b} = 0$ for all vectors $\mathbf{a}$ and $\mathbf{b}$. Hence, all canonical correlations must be zero. Conversely, if all canonical correlations are zero, $\mathbf{\Sigma}_{12} = \mathbf{0}$. Therefore, we must have

$$H_0 : \mathbf{\Sigma}_{12} = \mathbf{0} \Leftrightarrow H_0 : \rho_1 = \rho_2 = \ldots \rho_p = 0. \tag{10.4.2}$$

We test H_0 against all the possible alternatives. Note that the hypothesis $H_0(\mathbf{\Sigma}_{12} = \mathbf{0})$ was earlier considered in Subsection 5.8.1.

Let us write

$$(n-1)\mathbf{S} = \mathbf{Q} = \begin{bmatrix} \mathbf{Q}_{11} & \mathbf{Q}_{12} \\ \mathbf{Q}_{21} & \mathbf{Q}_{22} \end{bmatrix}. \tag{10.4.3}$$

It is known that $\mathbf{Q} \cap W_{p+q}(n-1, \mathbf{\Sigma})$. When there is no restriction, the m.l.e. of $\mathbf{\Sigma}$ is $\hat{\mathbf{\Sigma}} = \mathbf{Q}/n$ and the m.l.e. of $\mathbf{\Sigma}_{ij}$ is $\hat{\mathbf{\Sigma}}_{ij} = \mathbf{Q}_{ij}/n (i, j = 1, 2)$. Note that $\mathbf{S}_{ij} = \mathbf{Q}_{ij}/(n-1)$.

Under H_0, the m.l.e's are given by

$$\hat{\hat{\mathbf{\Sigma}}}_{11} = \hat{\mathbf{\Sigma}}_{11}, \ \hat{\hat{\mathbf{\Sigma}}}_{22} = \hat{\mathbf{\Sigma}}_{22}, \ \hat{\hat{\mathbf{\Sigma}}}_{12} = \mathbf{0}. \tag{10.4.4}$$

The likelihood ratio test takes the form

$$\begin{aligned} \Lambda &= \frac{L(\bar{\mathbf{x}}, \hat{\hat{\mathbf{\Sigma}}}_{11}, \hat{\hat{\mathbf{\Sigma}}}_{12}, \hat{\hat{\mathbf{\Sigma}}}_{22})}{L(\bar{\mathbf{x}}, \hat{\mathbf{\Sigma}}_{11}, \hat{\mathbf{\Sigma}}_{12}, \hat{\mathbf{\Sigma}}_{22})} \\ &= \frac{|\hat{\mathbf{\Sigma}}_{11}|^{-n/2} |\hat{\mathbf{\Sigma}}_{22}|^{-n/2}}{|\hat{\mathbf{\Sigma}}|^{-n/2}}. \end{aligned} \tag{10.4.5}$$

Hence,

$$U = \lambda^{2/n} = \frac{|\hat{\mathbf{\Sigma}}|}{|\hat{\mathbf{\Sigma}}_{11}||\hat{\mathbf{\Sigma}}_{22}|} = \frac{|\mathbf{Q}|}{|\mathbf{Q}_{11}||\mathbf{Q}_{22}|}. \tag{10.4.6}$$

Since $\mathbf{Q} > 0$ with probability one, $\mathbf{Q}_{11}, \mathbf{Q}_{22}$ are nonsingular and from (A.5.6) and (A.5.7),

$$|\mathbf{Q}| = |\mathbf{Q}_{11}|.|\mathbf{Q}_{11.2}| = |\mathbf{Q}_{22}|.|\mathbf{Q}_{11.2}|$$

where $\mathbf{Q}_{22.1} = \mathbf{Q}_{22} - \mathbf{Q}_{21}\mathbf{Q}_{11}^{-1}\mathbf{Q}_{12}$ and similar expression for $\mathbf{Q}_{11.2}$. Hence,

$$U = \frac{|\mathbf{Q}_{22.1}|}{|\mathbf{Q}_{22}|} = \frac{|\mathbf{Q}_{11.2}|}{|\mathbf{Q}_{11}|}$$

follows Wilk's $\Lambda(q; n-p-1, p)$ distribution which is the same as $\Lambda(p; n-q-1, q)$ distribution (vide Subsection 5.8.2).

Again,

$$U = \Pi_{i=1}^{p}(1 - r_i^2) \tag{10.4.7}$$

where r_i^2 are the ordered roots of the equation

$$|\mathbf{Q}_{12}\mathbf{Q}_{22}^{-1}\mathbf{Q}_{21}\mathbf{Q}_{11}^{-1} - r^2\mathbf{I}| = 0,$$

that is, are the roots of

$$|\mathbf{S}_{11}^{-1/2}\mathbf{S}_{12}\mathbf{S}_{22}^{-1}\mathbf{S}_{21}\mathbf{S}_{11}^{-1/2} - r^2\mathbf{I}| = 0 \tag{10.4.8}$$

(vide A.8(8)). Hence,

$$-2\ln\Lambda = -n\ln\Pi_{i=1}^{p}(1 - r_i^2), \tag{10.4.9}$$

is a function of the sample canonical correlations r_i's. For large n, the test statistic $-2\ln\Lambda$ is approximately distributed as $\chi^2_{(pq)}$.

Bartlett suggested replacing the multiplicative factor n in the LR statistic (10.4.9) by the factor $n-1-(p+q+1)/2$ to improve the χ^2-approximation to the sampling distribution of $-2\ln\Lambda$. We reject $H_0 : \mathbf{\Sigma}_{12} = \mathbf{0}$(i.e. $\rho_1 = \ldots = \rho_p = 0$) at the significance level α if

$$-(n - 1 - \frac{1}{2}(p + q + 1))\ln\Pi_{i=1}^{p}(1 - r_i^2) > \chi^2_{pq;\alpha} \tag{10.4.10}$$

where $\chi^2_{pq;\alpha}$ is the upper 100α percent point of the $\chi^2_{(pq)}$ distribution.

Alternatively, we can use Rao's (1951) F approximation given in (4.4.10) and (4.4.11). We can also use Roy's maximum root test, Pillai's statistic and Lawley-Hotelling statistic considered in Section 4.5.

If the null hypothesis $H_0(\rho_1 = \rho_2 = \ldots = \rho_p = 0)$ is rejected, we test for the individual canonical correlations. We begin by first assuming $\rho_1 \neq 0$ and the remaining $p-1$ correlations are zero. If this hypothesis is rejected, we assume $\rho_1 \neq 0, \rho_2 \neq 0$ and test for $H_0^{(3)} : \rho_3 = \ldots = \rho_p = 0$. In general, at the $(k+1)$th step we test for the hypothesis

$$H_0^{(k+1)} : \rho_1 \neq 0, \ldots, \rho_k \neq 0, \rho_{k+1} = \ldots = \rho_p = 0$$

against the alternative hypothesis

$$H_1^{(k+1)} : \rho_{k+i} > 0 \text{ for some } i > k+1,$$

given that $\rho_1 \neq 0, \ldots, \rho_k \neq 0$. Bartlett has given the following test. Reject $H_0^{(k+1)}$ at significance level α if

$$-(n-1-\frac{1}{2}(p+q+1)) \ln \Pi_{i=k+1}^{p}(1-r_i^2) > \chi^2_{(p-k)(q-k);\alpha}. \tag{10.4.11}$$

On the other hand, if the sequence of hypotheses $H_0, H_0^{(1)}, H_0^{(2)}, \ldots,$ are tested one at a time until $H_0^{(t)}$ is not rejected for some t, the overall significance level would be different from α and difficult to work out.

EXAMPLE 10.4.1: Assume that the data in Example 10.2.1 are sample correlations obtained from a sample of size 50. Here, $p = 2, q = 2, r_1^2 = .4021, r_2^2 = .0021$. For testing $H_0(\rho_{12} = \mathbf{0})$ (i.e., $\rho_1 = \rho_2 = 0$), the test statistic is

$$-(n-1-\frac{1}{2}(p+q+1))\sum_{i=1}^{2}\ln(1-r_i^2) = \frac{-93}{2}[\ln .5979 + \ln .9979] = 24.0142$$

$$> 9.49 = \chi^2_{4;.05}.$$

Hence H_0 is rejected. We can therefore assume that $\rho_1 \neq 0$ and test for the hypothesis $H_0^1 : \rho_2 = 0$ against the alternative $H_1^1 : \rho_2 \neq 0$. The value of the test statistic

$$\frac{-93}{2} \ln .9979 = .0978 < 3.84 = \chi^2_{1;.05}.$$

Hence, H_0^1 is accepted. We conclude that $\rho_1 \neq 0, \rho_2 = 0$.

10.5 Exercises and Complements

10.1 Prove the results in equation (10.2.33).

10.2 Show that the canonical correlations are invariant under nonsingular linear transformation of the $\mathbf{X}^{(1)}, \mathbf{X}^{(2)}$ variables of the from $\mathbf{CX}^{(1)}$ and $\mathbf{DX}^{(2)}$.

10.3 Show that the nonzero ρ_i^2's are the nonzero solutions λ of the equation

$$|\mathbf{\Sigma}_{12}\mathbf{\Sigma}_{22}^{-1}\mathbf{\Sigma}_{21}\mathbf{\Sigma}_{11}^{-1} - \lambda\mathbf{I}| = 0.$$

10.4 Let $\mathbf{X}^{(1)}(p \times 1)$ and $\mathbf{X}^{(2)}(q \times 1), p \leq q$ have a correlation matrix of equal correlation structure so that

$$\rho_{11} = \begin{bmatrix} 1 & \rho & \dots & \rho \\ \rho & 1 & \dots & \rho \\ \cdot & \cdot & \dots & \cdot \\ \rho & \rho & \dots & 1 \end{bmatrix}_{(p\times p)}, \quad \rho_{12} = \begin{bmatrix} \rho & \rho & \dots & \rho \\ \rho & \rho & \dots & \rho \\ \cdot & \cdot & \dots & \cdot \\ \rho & \rho & \dots & \rho \end{bmatrix}_{(p\times q)},$$

$$\rho_{22} = \begin{bmatrix} 1 & \rho & \dots & \rho \\ \rho & 1 & \dots & \rho \\ \cdot & \cdot & \dots & \cdot \\ \rho & \rho & \dots & 1 \end{bmatrix}_{(q\times q)}.$$

Determine the canonical variables corresponding to the nonzero canonical correlations.

[*Hints*: It is known that the latest roots of ρ_{11} are $\lambda_1 = 1 + (p-1)\rho, \lambda_2 = \dots = \lambda_p = 1 - \rho$. Now, $\rho_{11}\mathbf{1}/\sqrt{p} = [1 + (p-1)\rho]\mathbf{1}/\sqrt{p}$. Hence, $\mathbf{g}_1 = \mathbf{1}/\sqrt{p}$. Therefore, $U_1 = \rho_{11}^{-1/2}\mathbf{1}/\sqrt{p} = [1 + (p-1)\rho]^{-1/2}\mathbf{1}/\sqrt{p}$. Similarly, $V_1 = [1+(q-1)\rho]^{-1/2}\mathbf{1}/\sqrt{q}$. Any set of $(p-1)$ mutually orthogonal vectors of unit length, each of which is orthogonal to $\mathbf{1}/\sqrt{p}$ are eigenvectors corresponding to latent roots $\lambda_2, \dots, \lambda_p$ of ρ_{11}. For example, the columns of Helmert matrix of order $p \times p$ can be used for this purpose.]

10.5 Prove the results in Subsection 10.3.1.

10.6 In order to monitor the discharges of the waste-water treatment plants samples of effluent were divided and sent to two different laboratories for testing. One half of each sample was sent to the Government laboratory and the other half in a private commercial laboratory. Measurements of biochemical oxygen demand (BOD)(x) and suspended solids (SS)(y) were obtained for each half-sample from the two laboratories. The data are displayed in Table 10.E.1.

Table 10.E.1: Effluence Data

x_1	y_1	x_2	y_2
26	47	29	18
16	23	28	16
28	44	37	25
18	54	25	39
21	40	18	35
34	85	54	64
18	26	42	35
61	34	44	64
33	54	44	58
43	37	23	10
30	14	29	22

Assuming that the two sets of variables are (x_1, y_1) and (x_2, y_2), determine the sample canonical variates and their correlations. Interpret these quantities. Are the first canonical variates good summary measures of their respective sets of variables? Explain. Test for the significance of the canonical relations with $\alpha = 0.05$.

Chapter 11

Classification and Discrimination

11.1 Introduction

The problem of classification is as follows. Given that an object is known to come from one of g distinct groups $G_i (i = 1, \ldots, g)$ in a population $\mathcal{P}$, we wish to assign the object on the basis of the observation of a p-vector $\mathbf{x}$ associated with it. The assignment rule should be optimal in some sense, for example, having minimum total probability of misclassification, minimum expected cost of misclassification, etc. Clearly, observations should be available on the units sampled from the different groups to derive an optimal classification rule. Some examples of the application of the problem are:

(1) A student either passes or fails the final examination. Thus he belongs to either of the two groups ($g = 2$). Due to some reasons, a student is unable to appear in the final test. The authority wants to take a decision, whether to promote the student or ask him to repeat the course on the basis of his performance in the class tests and other assignments.

(2) Five body measurements of four species of female sparrows are available ($p = 5, g = 4$). A new measurement $\mathbf{x}$ is obtained. The zoologist wants to classify the bird in one of these groups in a proper way.

Discriminant analysis, initiated by Fisher (1936), on the other hand, describes the separation of groups. Here, $s(\leq p)$ independent linear functions of the variables are often used to describe or elucidate the differences between two or more groups. The linear functions of the variables are so chosen that separate the groups as sharply as possible, giving the directions of the differences among the groups. Thus if we have a six-variate ($p = 6$) observation on units sampled from 3 ($g = 3$) groups, we may want to find two ($s = 2$) linear functions of the variables that will segregate the groups as distinctly as possible. This is the problem of discrimination.

Clearly, the classification rules may be applied for discrimination and the discrimination rules for classification. Thus the problems are interrelated.

Sections 11.2, 11.3 consider the problem of classification with $g = 2$ groups and Section 11.6 for more than two groups. Optimal classification rules for multivariate normal populations, multinomial populations, when their parameters are known or unknown, have been included. Different error rates have been considered. Section 11.5 as also Section 11.6 address logistic discrimination. Fisher's discrimination analysis has been considered in Section 11.4 (for $g = 2$ groups) and in Section 11.7 for more than two groups. In the last section we consider the problem of selection of variables.

We have not considered the cases where probability density functions are unknown and require to be estimated by some nonparametric methods, like, kernel method, nearest neighbor method. Also, some nonparametric methods of classification, like the partitioning methods, distance methods, rank procedures, sequential discrimination have not been considered. For details the reader may refer to Lachenbruch (1975), Breiman,*et al.* (1977), Hora (1980), among many others.

11.2 Classification in Two Groups with Known Distributions and Known Parameters

First we consider the case of two groups with known distributions and known parameters.

Suppose we have a population $\mathcal{P}$ with π_1 proportion of units in Group G_1 and π_2 proportion in Group $G_2, (\pi_1 + \pi_2 = 1)$. The quantities π_1, π_2 are the prior probabilities that a unit selected at random from $\mathcal{P}$ belongs to G_1, G_2 respectively. Let $f_i(\mathbf{x})$ be the probability or probability density of an observation $\mathbf{x}$ if the observation comes from group $G_i (i = 1, 2)$.

Consider the following classification rule for classifying an object selected from $\mathcal{P}$ with observation $\mathbf{x}$ into one of the groups G_1, G_2. Assign to G_i if $\mathbf{x} \in \mathcal{R}_i (i = 1, 2)$ where R_1, R_2 are mutually exclusive subspaces of the sample space, $R, R_1 \cup R_2 = R$.

We can make one of the two errors: assign $\mathbf{x}$ to G_1, when it actually belongs to G_2 or assign $\mathbf{x}$ to G_2 when it actually belongs to G_1.

The probability of error of the first type is

$$P(1|2) = \int_{R_1} f_2(\mathbf{x})d\mathbf{x}. \tag{11.2.1}$$

The probability of error of the second type is

$$P(2|1) = \int_{R_2} f_1(\mathbf{x})d\mathbf{x}. \tag{11.2.2}$$

Now

$$P[\text{ observation is incorrectly specified in } G_2]$$

$$= P[\text{ observation comes from } G_1]P[\text{ it is classified as } G_2|G_1]$$

$$= \pi_1 P(2|1);$$

$$P[\text{ observation is incorrectly specified in } G_1]$$

$$= P[\text{ observation comes from } G_2]P[\text{ it is classified as } G_1|G_2]$$

$$= \pi_2 P(1|2).$$

Hence the total probability of misclassification (TPM) for the classification rule $\mathbf{R} = (R_1, R_2)'$ with respect to the probability distribution $\mathbf{f} = (f_1, f_2)'$ is

$$P(\mathbf{R}, \mathbf{f}) = \pi_1 P(2|1) + \pi_2 P(1|2). \tag{11.2.3}$$

In subsections 11.2.1 through 11.2.5 we shall consider different rules of classification. Our problem is to classify an object in one of the two groups.

11.2.1 *Minimizing the total probability of misclassification (TPM)*

We shall consider the classification rule for which the total probability of misclassification $P(\mathbf{R}, \mathbf{f})$ is minimum for a given $\mathbf{f}$. For this, we consider the following lemma.

Lemma 11.2.1: The integral $\int_{W_1} g(\mathbf{x})d\mathbf{x}$ is minimum with respect to choice of W_1 if $W_1 = W_{01} = \{\mathbf{x} : g(\mathbf{x}) < 0\}$. Here W_1, W_2 are two mutually exclusive subsets of the sample space R.

Proof. Let $W_{02} = \{\mathbf{x} : g(\mathbf{x}) \geq 0\}$. For general W_1 we can write

$$W_1 = (W_1 \cap W_{01}) \cup (W_1 \cap W_{02})$$

so that

$$g(\mathbf{x}) < 0 \text{ in } W_{01} - W_1 \cap W_{01} (\subset W_{01})$$

and

$$g(\mathbf{x}) \geq 0 \text{ in } W_1 - W_1 \cap W_{01} (\subset W_{02}).$$

Hence,

$$\begin{aligned} \int_{W_1} g(\mathbf{x}) d\mathbf{x} &= \int_{W_1 \cap W_{01}} g(\mathbf{x}) d\mathbf{x} + \int_{W_1 - W_1 \cap W_{01}} g(\mathbf{x}) d\mathbf{x} \\ &\geq \int_{W_1 \cap W_{01}} g(\mathbf{x}) d\mathbf{x} \\ &= \int_{W_{01}} g(\mathbf{x}) d\mathbf{x} - \int_{W_{01} - W_1 \cap W_{01}} g(\mathbf{x}) d\mathbf{x} \\ &\geq \int_{W_{01}} g(\mathbf{x}) d\mathbf{x}. \end{aligned}$$

Hence the proof. □

Note 11.2.1: Let $B = \{\mathbf{x} : g(\mathbf{x}) = 0\}$ and $A \subset B$. Then the integral or summation of $g(\mathbf{x})$ over $W_{01} \cup A$ is the same as that over W_{01}. Hence, the choice of W_{01} is not unique. There are many such choices for the boundary points of W_{01}.

We shall now consider a classification rule which minimizes the total probability of misclassification (TPM) $P(\mathbf{R}, \mathbf{f})$, given in (11.2.3).

From (11.2.1) - (11.2.3),

$$\begin{aligned} P(\mathbf{R}, \mathbf{f}) &= \pi_1 (1 - \int_{R_1} f_1(\mathbf{x}) d\mathbf{x}) + \pi_2 \int_{R_1} f_2(\mathbf{x}) d\mathbf{x} \\ &= \pi_1 + \int_{R_1} \{\pi_2 f_2(\mathbf{x}) - \pi_1 f_1(\mathbf{x})\} d\mathbf{x}. \end{aligned} \tag{11.2.4}$$

By Lemma 11.2.1 this is minimized if we choose $R_1 = R_{01}$ where $R_{01} = \{\mathbf{x} : \pi_2 f_2(\mathbf{x}) - \pi_1 f_1(\mathbf{x}) < 0\}$.

The classification rule which minimizes the TPM is therefore: assign $\mathbf{x}$ to G_1 if

$$\frac{f_1(\mathbf{x})}{f_2(\mathbf{x})} > \frac{\pi_2}{\pi_1}, \tag{11.2.5}$$

and to G_2, otherwise. As noted above, the assignment on the boundary where $f_1(\mathbf{x})/f_2(\mathbf{x}) = \pi_2/\pi_1$ can be arbitrary. We have, here, arbitrarily assigned $\mathbf{x}$ to G_2 when $f_1(\mathbf{x}) = f_2(\mathbf{x})$.

Alternatively, for each group G_i, we obtain the *classification function*

$$d_i(\mathbf{x}) = \pi_i f_i(\mathbf{x}), \; i = 1, 2. \tag{11.2.6}$$

If for a given $\mathbf{x}$, the score $d_1(\mathbf{x}) > d_2(\mathbf{x})$ we assign $\mathbf{x}$ to G_1; if $d_1(\mathbf{x}) \leq d_2(\mathbf{x})$ we assign $\mathbf{x}$ to the group G_2.

Some authors call (11.2.6) as a discriminating function. We shall term this function as classification function, reserving the term *discriminating function* for Fisher's discriminants (vide Section 11.4, 11.7) and similar functions (vide Subsection 11.3.2, Section 11.6, etc.).

The rule (11.2.5) is due to Welch (1939).

EXAMPLE 11.2.1: Suppose that $f_i(\mathbf{x})$ is the density function of a $N_p(\mu_i, \mathbf{\Sigma})$-population where $\mu_i, \mathbf{\Sigma}$ are known (i=1,2). Here

$$f_i(\mathbf{x}) = (2\pi)^{-p/2}|\mathbf{\Sigma}|^{-1/2}\exp[-\frac{1}{2}(\mathbf{x}-\mu_i)'\mathbf{\Sigma}^{-1}(\mathbf{x}-\mu_i)].$$

Also,

$$\begin{aligned} f_1(\mathbf{x})/f_2(\mathbf{x}) &= \exp[-\tfrac{1}{2}(\mathbf{x}-\mu_1)'\mathbf{\Sigma}^{-1}(\mathbf{x}-\mu_1) + \tfrac{1}{2}(\mathbf{x}-\mu_2)'\mathbf{\Sigma}^{-1}(\mathbf{x}-\mu_2)] \\ &= \exp[(\mu_1-\mu_2)'\mathbf{\Sigma}^{-1}\mathbf{x} - \tfrac{1}{2}(\mu_1-\mu_2)'\mathbf{\Sigma}^{-1}(\mu_1+\mu_2)] \\ &= \exp[\alpha'\{\mathbf{x} - \tfrac{1}{2}(\mu_1+\mu_2)] \end{aligned} \tag{11.2.7}$$

where

$$\alpha = \mathbf{\Sigma}^{-1}(\mu_1-\mu_2). \tag{11.2.8}$$

Taking logarithm of both sides of (11.2.7), the classification rule is, therefore, as follows. Assign $\mathbf{x}$ to G_1 if

$$\begin{aligned} \ln[f_1(\mathbf{x})/f_2(\mathbf{x})] = D(\mathbf{x}) &= \alpha'[\mathbf{x} - \tfrac{1}{2}(\mu_1+\mu_2)] \\ &> \ln(\pi_2/\pi_1) \end{aligned} \tag{11.2.9}$$

and to G_2 otherwise. We shall see in Section 11.4 that the vector α has a special role in discriminating between two groups.

We shall now find the probabilities of misclassification, (11.2.1) and (11.2.2). Now

$$P(2|1) = Pr[D(\mathbf{x}) \leq \ln(\pi_2/\pi_1)|\mathbf{x} \in G_1] \tag{11.2.10}$$

To find this we shall derive $E[D(\mathbf{x})|\mathbf{x} \in G_i]$ and $V[D(\mathbf{x})|\mathbf{x} \in G_i]$. Now,

$$\begin{aligned} E[D(\mathbf{x})|\mathbf{x} \in G_i] &= \alpha'[E(\mathbf{x}|\mathbf{x} \in G_i) - \tfrac{1}{2}(\mu_1+\mu_2)] \\ &= \alpha'[\mu_i - \tfrac{1}{2}(\mu_1+\mu_2)] \\ &= D(\mu_i) \\ &= \tfrac{1}{2}(-1)^{i+1}\Delta^2 \end{aligned} \tag{11.2.11}$$

where

$$\Delta^2 = (\mu_1 - \mu_2)'\mathbf{\Sigma}^{-1}(\mu_1 - \mu_2) = \alpha'(\mu_1 - \mu_2) \tag{11.2.12}$$

is the square of Mahalanobis distance between μ_1 and μ_2 (vide equation (2.8.12)). Also,

$$V[D(\mathbf{x})|\mathbf{x} \in G_i] = V[\alpha'\mathbf{x}] = \alpha'\mathbf{\Sigma}\alpha = \Delta^2. \tag{11.2.13}$$

Since $\mathbf{x}$ is normally distributed, $D(\mathbf{x})$ has a normal distribution, and from (11.2.11), (11.2.12),

$$\begin{aligned} P(2|1) &= P[D(\mathbf{x}) \leq \ln(\pi_2/\pi_1)|\mathbf{x} \in G_1] \\ &= Pr[\tfrac{D(\mathbf{x})-\Delta^2/2}{\Delta} \leq \tfrac{\ln(\pi_2/\pi_1)-\Delta^2/2}{\Delta}|\mathbf{x} \in G_1] \\ &= Pr[\tau \leq \{\ln(\pi_2/\pi_1)\} - \Delta^2/2\}/\Delta\} \\ &= \Phi(\{\ln(\pi_2/\pi_1) - \Delta^2/2\}/\Delta), \end{aligned} \tag{11.2.14}$$

where τ is the standard normal variable with distribution function Φ and Δ is the positive square root of Δ^2.

Similarly,

$$\begin{aligned} P(1|2) &= \Phi(\{\ln(\pi_1/\pi_2) - \Delta^2/2)\}/\Delta) \\ &= \Phi(-\{\ln(\pi_2/\pi_1) + \Delta^2/2\}/\Delta). \end{aligned} \tag{11.2.15}$$

If $\pi_1 = \pi_2 = 1/2$, then $P(2|1) = \Phi(-\Delta/2) = P(1|2)$. In this case we assign $\mathbf{x}$ to G_1 if

$$\alpha'\mathbf{x} > \frac{1}{2}\alpha'(\mu_1 + \mu_2),$$

that is, if

$$(\mu_1 - \mu_2)'\mathbf{\Sigma}^{-1}[\mathbf{x} - \frac{1}{2}(\mu_1 + \mu_2)] > 0. \tag{11.2.16}$$

Fisher (1936) obtained the formula (11.2.16) though his approach is distribution-free. His approach is discussed in Section 11.4.

EXAMPLE 11.2.2: Suppose that $\mathbf{x}$ follows multivariate normal distribution as in Example 11.2.1, but $\mathbf{\Sigma} \neq \mathbf{\Sigma}_2$. Now,

$$\begin{aligned} Q(\mathbf{x}) &= \ln[f_1(\mathbf{x})/f_2(\mathbf{x})] \\ &= \tfrac{1}{2}\ln(|\mathbf{\Sigma}_2|/|\mathbf{\Sigma}_1|) - \tfrac{1}{2}(\mathbf{x} - \mu_1)'\mathbf{\Sigma}_1^{-1}(\mathbf{x} - \mu_1) + \\ &\quad \tfrac{1}{2}(\mathbf{x} - \mu_2)'\mathbf{\Sigma}_2^{-1}(\mathbf{x} - \mu_2) \\ &= K - \tfrac{1}{2}[\mathbf{x}'(\mathbf{\Sigma}_1^{-1} - \mathbf{\Sigma}_2^{-1})\mathbf{x} - 2\mathbf{x}'(\mathbf{\Sigma}_1^{-1}\mu_1 - \mathbf{\Sigma}_2^{-1}\mu_2)], \end{aligned} \tag{11.2.17}$$

where

$$K = (1/2)\ln(|\mathbf{\Sigma}_2|/|\mathbf{\Sigma}_1|) - \frac{1}{2}(\mu_1'\mathbf{\Sigma}^{-1}\mu_1 - \mu_2'\mathbf{\Sigma}_2^{-1}\mu_2).$$

The optimum rule is therefore as follows. Assign $\mathbf{x}$ to G_1 if $Q(\mathbf{x}) > \ln(\pi_2/\pi_1)$ and to G_2 otherwise. In contrast to $D(\mathbf{x}), Q(\mathbf{x})$ is a quadratic function.

Example 11.2.3: *Independent Bernoulli Variables*: Suppose that $\mathbf{x} = (x_1, x_2)'$ is a pair of independent Bernoulli variables taking the values 1 or 0. Let Pr.$[x_j = 1] = p_{ij}$, Pr $[x_j = 0] = 1 - p_{ij}, j = 1, 2$, if x_j comes from group $G_i (i = 1, 2)$. Then

$$\begin{aligned}\frac{f_1(\mathbf{x})}{f_2(\mathbf{x})} &= \frac{p_{11}^{x_1}(1-p_{11})^{1-x_1}p_{12}^{x_2}(1-p_{12})^{1-x_2}}{p_{21}^{x_1}(1-p_{21})^{1-x_1}p_{22}^{x_2}(1-p_{22})^{1-x_2}}\\ &= \frac{(1-p_{11})(1-p_{12})}{(1-p_{21})(1-p_{22})}\{\frac{p_{11}}{1-p_{11}}\frac{1-p_{21}}{p_{21}}\}^{x_1}\{\frac{p_{12}}{1-p_{12}}\frac{1-p_{22}}{p_{22}}\}^{x_2}.\end{aligned}$$

Therefore,

$$\ln[\frac{f_1(\mathbf{x})}{f_2(\mathbf{x})}] = \beta_0 + \beta_1 x_1 + \beta_2 x_2$$

where

$$\beta_0 = \ln[\frac{(1-p_{11})(1-p_{12})}{(1-p_{21})(-p_{22})}],$$

etc. The rule (11.2.5) reduces to as follows. Assign $\mathbf{x}$ to G_1 if $\beta_0 + \beta_1 x_1 + \beta_2 x_2 > \ln(\pi_2/\pi_1)$ and to G_2 otherwise.

The above rule generalized to the case of a p-dimensional random variable becomes as follows. Assign $\mathbf{x}$ to G_1 if $\beta_0 + \beta_1 x_1 + \ldots + \beta_p x_p > \ln(\pi_2/\pi_1)$ where β's are functions of $2p$ parameters $p_{ij}, i = 1, 2; j = 1, \ldots, p$.

11.2.2 *The likelihood ratio method*

When π_1 is unknown, an intuitive rule is to assign $\mathbf{x}$ to the group G_i for which the likelihood $f_i(\mathbf{x})$ is maximum. Thus, we assign $\mathbf{x}$ to G_1 if $f_1(\mathbf{x})/f_2(\mathbf{x}) > 1$. This is a special case of the rule (11.2.5), namely, when $\pi_1 = \pi_2$.

11.2.3 *Minimizing the expected cost of misclassification (ECM)*

Let $C(1|2)$ be the cost of misclassifying a unit of G_2 as in G_1 and $C(2|1)$ the cost of misclassifying a unit of G_1 as in G_2. Then the expected cost of misclassification (ECM) is

$$C_T = C(2|1)P(2|1)\pi_1 + C(1|2)P(1|2)\pi_2. \tag{11.2.18}$$

A reasonable classification rule should have an ECM as small as possible or nearly so. It follows from Lemma 11.2.1 that C_T is minimum if we take R_1 as $R_1 = \{\mathbf{x} : C(1|2)\pi_2 f_2(\mathbf{x}) < C(2|1)\pi_1 f_1(\mathbf{x})\}$. The rule then becomes as follows. Assign $\mathbf{x}$ to G_1 if

$$\frac{f_1(\mathbf{x})}{f_2(\mathbf{x})} > \frac{\pi_2 C(1|2)}{\pi_1 C(2|1)} \tag{11.2.19}$$

and to G_2 otherwise (Exercise 11.1). The rule reduces to the rule (11.2.5) if the costs are equal.

11.2.4 *Maximizing the posterior probability*

We could also allocate an observation $\mathbf{x}_0$ to the group with the largest posterior probability $P[G_i|\mathbf{x}_0]$.

Suppose that $\mathbf{x}$ is a discrete random vector. Then by Bayes theorem, the posterior probability that $\mathbf{x} \in G_i$ given that $\mathbf{x} = \mathbf{x}_0$, is

$$\begin{aligned} q_i(\mathbf{x}_0) &= P[\mathbf{x} \in G_i|\mathbf{x} = \mathbf{x}_0] \\ &= \frac{P[\mathbf{x}\in G_i]P[\mathbf{x}=\mathbf{x}_0|\mathbf{x}\in G_i]}{\sum_{j=1}^{2} P[\mathbf{x}\in G_j]P[\mathbf{x}=\mathbf{x}_0|\mathbf{x}\in G_j]} \\ &= \frac{\pi_i f_i(\mathbf{x}_0)}{\pi_1 f_1(\mathbf{x}_0)+\pi_2 f_2(\mathbf{x}_0)}. \end{aligned} \tag{11.2.20}$$

If $\mathbf{x}$ is continuous, then we have the approximation

$$P[\mathbf{x}_0 \le \mathbf{x} \le \mathbf{x}_0 + d\mathbf{x}_0|\mathbf{x} \in G_i] = \int_{\mathbf{x}_0}^{\mathbf{x}_0+d\mathbf{x}_0} f_i(\mathbf{x})d\mathbf{x} \approx f_i(\mathbf{x}_0)d\mathbf{x}_0.$$

Using this approximation and letting $d\mathbf{x}_0 \to 0$, we see that (11.2.20) still holds so that (11.2.20) is the posterior probability of $G_i, \mathbf{x}$ continuous or discrete.

Therefore, the allocation rule is: assign $\mathbf{x}$ to G_1 if

$$q_1(\mathbf{x}) > q_2(\mathbf{x}),$$

that is, if

$$\frac{f_1(\mathbf{x})}{f_2(\mathbf{x})} > \frac{\pi_2}{\pi_1}.$$

This is the same as the rule (11.2.5) of minimizing the total probability of misclassification.

11.2.5 *Minimax classification*

A rule that minimizes the total probability of misclassification may not work well for individual classes. We can use a *minimax rule* that allocates $\mathbf{x}$ so as to minimize the maximum individual probability of misclassification, i.e., which minimizes the greater of $P(1|2)$ and $P(2|1)$. Now, for $0 \leq \alpha \leq 1$,

$$\max\ \{P(1|2), P(2|1)\} \geq (1-\alpha)P(2|1) + \alpha P(1|2). \tag{11.2.21}$$

By Lemma 11.2.1, the right side of (11.2.21) is minimized when

$$R_1 = R_{01(\alpha)} = \{\mathbf{x} : f_1(\mathbf{x})/f_2(\mathbf{x}) > \alpha/(1-\alpha) = c \ \text{ say }\} \tag{11.2.22}$$

and $R_{02(\alpha)}$ is defined similarly (Exercise 11.2). Clearly, for $\mathbf{R}_{0(\alpha)} = (R_{01(\alpha)}R_{02(\alpha)})', P(2|1), P(1|2)$ will depend on α. Therefore, for given α,

$$(1-\alpha)P(2|1) + \alpha P(1|2) \geq (1-\alpha)P_\alpha(2|1) + \alpha P_\alpha(1|2). \tag{11.2.23}$$

We choose $\alpha \in [0, 1]$ such that the right side of (11.2.23) is minimum. This is achieved by choosing c (and hence $\alpha = \alpha_0$ say) such that the misclassification probabilities for $\mathbf{R}_{0(\alpha)}$ are equal, namely, $P_0(1|2) = P_0(2|1)$. Then,

$$\begin{aligned}(1-\alpha)P_\alpha(2|1) + \alpha P_\alpha(1|2) &\geq (1-\alpha_0)P_0(2|1) + \alpha_i P_0(1|2) \\ &= P_0(2|1).\end{aligned} \tag{11.2.24}$$

By (11.2.21), therefore,

$$\max\ \{P(1|2), P(2|1)\} \geq P_0(2|1) = \ \max\ (P_0(2|1), P_0(1|2)\}.$$

The minimax rule therefore is to choose c and α such that $P_\alpha(2|1) = P_\alpha(1|2)$ under the minimum TPM rule and then follow the rule $\mathbf{R}_{0(\alpha)}$ for this α. Note that the prior probabilities π_1, π_2 are not involved in this rule.

EXAMPLE 11.2.4: The two distributions are normal with common covariance matrix $\mathbf{\Sigma}$. Then the minimax rule is: allocate $\mathbf{x}$ to G_1 if $D(\mathbf{x}) > \ln c$ where $D(\mathbf{x})$ is given by (11.2.9) and c is so chosen that

$$\Phi\{\frac{\ln c - \Delta^2/2}{\Delta}\} = \Phi\{\frac{-\ln c - \Delta^2/2}{\Delta}\}.$$

The above equation has solution $\ln c = 0$ or $c = 1$. Therefore, the minimax rule is same as the likelihood ratio method.

11.3 Classification in Two Groups with Known Distributions but Unknown Parameters

11.3.1 *General methods*

In practice, the parameters of the probability functions will remain unknown and have to be estimated.

We shall use the notation $f_i(\mathbf{x}|\theta_i)$ instead of $f_i(\mathbf{x})$, where θ_i is the vector of parameters for group G_i.

Let $\mathbf{x}_{i1}, \ldots, \mathbf{x}_{in_i}$ be the sample from group $G_i, (i = 1, 2)$. In case $\theta = (\theta_1', \theta_2')'$ is unknown we can replace θ by $\hat{\theta} = (\hat{\theta}_1(\mathbf{z})', \hat{\theta}(\mathbf{z})')'$ where $\hat{\theta}_i(\mathbf{z}$ is an estimate of θ_i based on the pooled data $\mathbf{z} = (\mathbf{x}_{11}', \ldots, \mathbf{x}_{1n_1}', \mathbf{x}_{21}', \ldots, \mathbf{x}_{2n_2}')', i = 1, 2$.

We have so far seen in Section 11.2 that when θ is known all the classification rule takes the form: assign $\mathbf{x}$ to G_1 if $f_1(\mathbf{x}|\theta_1)/f_2(\mathbf{x}|\theta_2) > c$, where c is a suitably chosen constant and to G_2 otherwise. When θ is unknown, the optimum rule $\mathbf{R}_0 = (R_{01}, R_{02})'$ is, therefore, estimated by $\hat{\mathbf{R}}_0 = (\hat{R}_{01}, \hat{R}_{02})'$ where $\hat{R}_{01} = \{\mathbf{x} : f_1(\mathbf{x}|\hat{\theta}_1)/f_2(\mathbf{x}|\hat{\theta}_2) > c\}$. We would like $\hat{R}_{01}$ to be close to R_{01} and this would be the case for sufficiently large samples. Lachenbruch and Goldstein (1979) suggested that as a rule of thumb n_i should exceed three times the number of parameters in θ_i.

Instead of estimating $f_i(\mathbf{x}|\theta_i)$ by $f_i(\mathbf{x}|\hat{\theta}_i)$ where $\hat{\theta}_i = \hat{\theta}_i(\mathbf{z})$ is an estimate based on the pooled sample data $\mathbf{z}$ defined before, one may make use of an alternative approach, called *predictive approach*, which uses

$$h_i(\mathbf{x}|\mathbf{z}) = \int_\Theta f_i(\mathbf{x}|\theta_i)g_i(\theta_i|\mathbf{z})d\theta_i. \tag{11.3.1}$$

Here g_i can be regarded as either some weighting function based on the sample data $\mathbf{z}$ or a full Bayesian posterior density function for θ_i based on the prior $g_{2i}(\theta_i)$ and the likelihood $g_{1i}(\mathbf{z}|\theta_i)$, i.e.

$$g_i(\theta_i|\mathbf{z}) \propto g_{2i}(\theta_i)g_{1i}(\mathbf{z}|\theta_i). \tag{11.3.2}$$

In the Bayesian inference, the expression (11.3.1) is the predictive density for a future observation $\mathbf{x}$ from group G_i assessed on the basis of the past data $\mathbf{z}$.

It may be noted that once the parameters are estimated, there is no guarantee that the resulting rule will retain the desired optimum properties (like minimum TPM, ECM). However, it is reasonable to expect that it would perform well if the sample size is large.

11.3.2 *Normal populations*

A Linear Classification Function: Consider the case of Example 11.2.1. Here, $f_i(\mathbf{x}|\theta_i)$ is the density function of a $N_p(\mu_i, \mathbf{\Sigma}_i)$ population with $\mathbf{\Sigma}_1 = \mathbf{\Sigma}_2 = \mathbf{\Sigma}$. Let $\mathbf{x}_{i1}, \ldots, \mathbf{x}_{in_i}$ be a random sample of size n_i from the group $G_i (i = 1, 2)$. We can estimate μ_i by $\bar{\mathbf{x}}_i = \sum_{j=1}^{n_i} x_{ij}/n_i$ and $\mathbf{\Sigma}$ by the pooled estimate

$$\mathbf{S}_p = \frac{(n_1 - 1)\mathbf{S}_1 + (n_2 - 1)\mathbf{S}_2}{n_1 + n_2 - 2} = \frac{\mathbf{Q}}{n_1 + n_2 - 2}, \tag{11.3.3}$$

where

$$\mathbf{S}_i = \frac{1}{n_i - 1} \sum_{j=1}^{n_i} (\mathbf{x}_{ij} - \bar{\mathbf{x}}_i)(\mathbf{x}_{ij} - \bar{\mathbf{x}}_i)'.$$

We have thus used the sample, also called the *training sample* to estimate the parameters of the population. Referring to the minimization of the total probability of error rule (in (11.2.9)), we have the allocation rule as follows. Whenever a new observation $\mathbf{x}$ occurs we assign $\mathbf{x}$ to G_1 if

$$D_s(\mathbf{x}) > \ln(\pi_1/\pi_2) = \ln(c) \quad \text{(say)} \tag{11.3.4}$$

where

$$\begin{aligned} D_s(\mathbf{x}) &= \mathbf{a}'[\mathbf{x} - \tfrac{1}{2}(\bar{\mathbf{x}}_1 + \bar{\mathbf{x}}_2)] \\ &= (\bar{\mathbf{x}}_1 - \bar{\mathbf{x}}_2)'\mathbf{S}_p^{-1}[\mathbf{x} - \tfrac{1}{2}(\bar{\mathbf{x}}_1 + \bar{\mathbf{x}}_2)], \end{aligned} \tag{11.3.5}$$

and

$$\mathbf{a} = \mathbf{S}_p^{-1}(\bar{\mathbf{x}}_1 - \bar{\mathbf{x}}_2). \tag{11.3.6}$$

We shall call $D_s(\mathbf{x})$ the *linear classification function* (LCF), the sample estimate of $D(\mathbf{x})$ in (11.2.9).

We assume that $\pi_1 = \pi_2 = 1/2$. The estimated rule for two normal populations therefore amounts to comparing the score

$$y(\mathbf{x}) = \mathbf{a}'\mathbf{x} = (\bar{\mathbf{x}}_1 - \bar{\mathbf{x}}_2)'\mathbf{S}_p^{-1}\mathbf{x} \tag{11.3.7}$$

with the number

$$\begin{aligned} \hat{m} &= \tfrac{1}{2}\mathbf{a}'(\bar{\mathbf{x}}_1 + \bar{\mathbf{x}}_2) \\ &= \tfrac{1}{2}(\bar{y}_1 + \bar{y}_2) \\ &= \tfrac{1}{2}(\bar{\mathbf{x}}_1 - \bar{\mathbf{x}}_2)'\mathbf{S}_p^{-1}(\bar{\mathbf{x}}_1 + \bar{\mathbf{x}}_2) \end{aligned} \tag{11.3.8}$$

where

$$\bar{y}_i = \mathbf{a}'\bar{\mathbf{x}}_i = (\bar{\mathbf{x}}_1 - \bar{\mathbf{x}}_2)'\mathbf{S}_p^{-1}\bar{\mathbf{x}}_i, \ (i = 1, 2). \tag{11.3.9}$$

If $y(\mathbf{x}) > \hat{m}$, the observation $\mathbf{x}$ is assigned to G_1, otherwise it is assigned to G_2. The score $y(\mathbf{x})$ may be called the *linear classification score* of $\mathbf{x}$.

We note that we have transformed each observation $\mathbf{x}_{ij}$ to a scalar variable $y_{ij} = \mathbf{a}'\mathbf{x}_{ij}(i = 1,2; j = 1,\ldots,n_i)$. We calculate $\bar{y}_i = \mathbf{a}'\bar{\mathbf{x}}_i(i = 1,2)$. Whenever a new observation $\mathbf{x}_0$ occurs, we calculate its transformed value $y_0 = \mathbf{a}'\mathbf{x}_0$ and assign it to G_1 or G_2 according as it falls to the right or left of the midpoint $\hat{m}$ between the two univariate means $\bar{y}_1$ and $\bar{y}_2$.

In this context the transformation vector $\mathbf{a}$ has a special interpretation and this will be discussed in Section 11.4.

Alternatively, we can consider

$$\begin{aligned}\ln[\pi_i f_i(\mathbf{x}|\hat{\theta}_i)] &= \ln\pi_i - \tfrac{p}{2}\ln(2\pi) - \tfrac{1}{2}(\mathbf{x}-\bar{\mathbf{x}}_i)'\mathbf{S}_p^{-1}(\mathbf{x}-\bar{\mathbf{x}}_i)\\ &= \ln\pi_i - \tfrac{p}{2}\ln(2\pi) - \tfrac{1}{2}\mathbf{x}'\mathbf{S}_p^{-1}\mathbf{x} + \bar{\mathbf{x}}_i'\mathbf{S}_p^{-1}\mathbf{x} - \tfrac{1}{2}\bar{\mathbf{x}}_i'\mathbf{S}_p^{-1}\bar{\mathbf{x}}_i.\end{aligned} \tag{11.3.10}$$

Subtracting the common part $-\frac{p}{2}\ln(2\pi) - (1/2)\mathbf{x}'\mathbf{S}_p^{-1}\mathbf{x}$ (which occurs in all the other similar terms) we obtain the linear classification function for the ith group,

$$d_{is}(\mathbf{x}) = \ln\pi_i + \bar{\mathbf{x}}_i'\mathbf{S}_p^{-1}(\mathbf{x} - \frac{1}{2}\bar{\mathbf{x}}_i). \tag{11.3.11}$$

We assign $\mathbf{x}$ to the group G_i with the larger value of $d_{is}(\mathbf{x})$.

Instead of $\mathbf{S}_p$ one can also use $\mathbf{Q}/(n_1+n_2)$, the maximum likelihood estimate of $\boldsymbol{\Sigma}$. Some suggestions are to use a ridge-type estimator $\mathbf{S}_p + k\mathbf{I}_p$, instead of $\mathbf{S}_p$ as the former can lead to a smaller e_{act} (vide subsection 11.3.3)(see for example, Di Pillo, 1999; Campbell, 1980).

The method was first proposed by Fisher (1936), who used it to classify two species of *iris* on the basis of four measurement: sepal length, sepal width, petal length and petal width (Exercise 11.8).

It has been observed that the LCF (11.3.5) remains fairly robust to the discrete distributions of various types and mild departure from the assumption of equality of covariance matrices (e.g., Lachenbruch, 1975; Moore, 1973, Krzanowski, 1977).

Quadratic Classification Function: When $\boldsymbol{\Sigma} \neq \boldsymbol{\Sigma}_2$, a quadratic classification function is appropriate. Replacing μ_i by $\bar{\mathbf{x}}_i$ and $\boldsymbol{\Sigma}_i$ by $\mathbf{S}_i$ in (11.2.17) we get the sample estimate of $Q(\mathbf{x})$ as

$$\begin{aligned}Q_s(\mathbf{x}) &= \tfrac{1}{2}\ln(\tfrac{|\mathbf{S}_1|}{|\mathbf{S}_2|}) - \tfrac{1}{2}(\mathbf{x}-\bar{\mathbf{x}}_1)'\mathbf{S}_1^{-1}(\mathbf{x}-\bar{\mathbf{x}}_1)\\ &\quad + \tfrac{1}{2}(\mathbf{x}-\bar{\mathbf{x}}_2)'\mathbf{S}_2^{-1}(\mathbf{x}-\bar{\mathbf{x}}_2)\\ &= -\tfrac{1}{2}\mathbf{x}'(\mathbf{S}_1^{-1} - \mathbf{S}_2^{-1})\mathbf{x} + (\bar{\mathbf{x}}_1'\mathbf{S}_1^{-1} - \bar{\mathbf{x}}_2'\mathbf{S}_2^{-1})\mathbf{x} + C\end{aligned} \tag{11.3.12}$$

where

$$C = \frac{1}{2}\ln(\frac{|\mathbf{S}_2|}{|\mathbf{S}_1|}) - \frac{1}{2}(\bar{\mathbf{x}}_1'\mathbf{S}_1^{-1}\bar{\mathbf{x}}_1 - \bar{\mathbf{x}}_2'\mathbf{S}_2^{-1}\bar{\mathbf{x}}_2). \tag{11.3.13}$$

We shall call this as the *quadratic classification function* (QCF). The classification rule is then to assign $\mathbf{x}$ to G_1 if $Q_s(\mathbf{x}) > \ln(\pi_1/\pi_2)$ and to G_2 otherwise. Note that the quadratic functions can also arise from non-normal distributions.

EXAMPLE 11.3.1: The first six columns of the following table gives the results of analysis of samples of coal from two coal mines, G_1, G_2. The values of the elements are given as $x_1, \ldots, x_5$. Obtain the estimated minimum TPM classification rule, assuming equal prior probabilities. If a new observation has value $\mathbf{x}_0 = (6.3, 13.0, 0.50, 4.24, 8.27)'$, classify the observation.

We assume that the populations are 5-variate normal with equal covariance matrices. The sample mean vectors are

$$\bar{\mathbf{x}}_1 = \begin{bmatrix} 2.70 \\ 35.20 \\ 0.20 \\ 6.12 \\ 9.15 \end{bmatrix}, \quad \bar{\mathbf{x}}_2 = \begin{bmatrix} 1.40 \\ 12.20 \\ 0.02 \\ 5.22 \\ 2.25 \end{bmatrix}.$$

The pooled covariance matrix is

$$\mathbf{S}_p = \begin{bmatrix} 23.804 & & & & \\ 4.170 & 864.774 & & & \\ 0.535 & 2.428 & 0.416 & & \\ 5.882 & 31.148 & 1.172 & 11.348 & \\ 15.209 & 28.476 & 0.180 & 10.051 & 80.279 \end{bmatrix}.$$

Also

$$\mathbf{S}_p^{-1} = \begin{bmatrix} .0522 & & & & \\ .0008 & .0013 & & & \\ -.0065 & .0034 & 3.5290 & & \\ -.0220 & -.0043 & .0433 & .0142 & \\ -.0074 & -.0001 & .0433 & -.0142 & .0156 \end{bmatrix}.$$

The sample linear classification function (11.3.5) is

$$D_s(\mathbf{x}) = (\bar{\mathbf{x}}_1 - \bar{\mathbf{x}}_2)'\mathbf{S}_p^{-1}\mathbf{x} = (.0143, .0271, .6363, -.1487, .0908)'\mathbf{x}.$$

Also

$$\hat{m} = \frac{1}{2}(\bar{\mathbf{x}}_1 - \bar{\mathbf{x}}_2)'\mathbf{S}_p^{-1}(\bar{\mathbf{x}}_1 + \bar{\mathbf{x}}_2) = .4164.$$

For the new observation $\mathbf{x}_0$, $D_s(\mathbf{x}_0) = y(\mathbf{x}_0) = 0.8812$. Since it is more than $\hat{m}$, it is classified as a member of G_1.

Table 11.3.1: Coal Mines Data

Group	x_1	x_2	x_3	x_4	x_5	$y = \mathbf{a}'\mathbf{x}$	Fitted Group
(1)	(2)	(3)	(4)	(5)	(6)	(7)	(8)
G_1	3.2	49.0	0.25	7.01	12.25	1.603	1
	2.8	51.0	0.09	7.15	12.27	1.531	1
	2.7	36.2	0.40	7.15	11.40	1.247	1
	3.2	44.2	0.09	7.21	13.02	1.439	1
	3.6	46.1	0.08	6.26	10.45	1.370	1
	3.8	42.1	0.09	7.23	12.42	1.306	1
	2.8	35.2	0.02	6.12	9.15	0.928	1
G_2	5.0	48.2	0.08	7.10	6.15	0.932	1
	3.7	31.8	0.12	6.00	4.65	0.521	1
	1.4	12.2	0.02	5.56	3.25	-0.168	2
	8.5	17.5	0.08	6.35	4.60	0.120	2
	4.5	36.2	0.70	0.45	5.05	0.545	1
	4.8	35.5	0.60	5.75	2.25	0.762	1
	3.8	42.0	0.20	5.65	2.95	0.748	1
	4.0	36.2	0.08	6.20	2.28	0.375	2
	7.5	32.2	0.40	8.10	12.86	1.198	1
	4.5	46.2	0.09	7.60	5.80	0.744	1
	3.5	31.2	0.06	5.22	10.85	1.143	1

11.3.3 *Evaluating classification functions: Error rates*

In assessing different classification rules we should consider its *error rates* or misclassification probabilities. When the parent population is completely known, misclassification probabilities can be calculated with relative ease, as shown in Example 11.2.1. Since, as is often the case, the parent population is rarely completely known, we have to calculate the error rates associated with the sample classification functions. We assume that there are enough data so that a part of it can be considered as *training sample*, which is used for estimation of parameters and development of the classification function. The other part, called the *validation sample* may be used to evaluate its performance. The estimation of error rates is discussed below.

(1) *The Optimum Error Rate*: For the classification rule $\mathbf{R}_0$, the optimum

error rate (OER) for the group G_i is

$$\begin{aligned} e_{i,opt} &= P(j|i) \\ &= \int_{R_{0j}} f_i(\mathbf{x}|\theta_i), \end{aligned} \tag{11.3.14}$$

and the optimum error rate is

$$e_{opt} = \pi_1 e_{1,opt} + \pi_2 e_{2,opt} = P(\mathbf{R}_0, \mathbf{f}). \tag{11.3.15}$$

Note that in Example 11.2.1 we have calculated the optimum error rate for two equal covariance normal populations.

(2) *The Actual Error Rate*: For the classification rule $\mathbf{R}_0$, the actual error rate for the group G_i is

$$e_{i,act} = \int_{\hat{R}_{0j}} f_i(\mathbf{x}|\theta_i)d\mathbf{x}, \tag{11.3.16}$$

and the actual error rate (AER) is

$$e_{act} = \pi_1 e_{1,act} + \pi_2 e_{2,act} = P(\hat{\mathbf{R}}_0, \mathbf{f}). \tag{11.3.17}$$

The AER indicates how the sample classification function will perform in future samples. However, like OER, it cannot be generally calculated since the parameters in f_i are generally unknown $(i = 1, 2)$. If a large number of observations are classified using the rule $\hat{\mathbf{R}}_0$, then on an average $100e_{act}\%$ of observations will be misclassified. By definition $e_{opt} \leq e_{act}$.

As will be seen below, *apparent error rate* is an approximate estimate of AER in large sample.

(3) *The Expected Actual Error Rate*: These are $E[e_{i,act}]$ and

$$E[e_{act}] = \pi_1 E[e_{1,act}] + \pi_2 E[e_{2act}]. \tag{11.3.18}$$

(4) *The Plug-In Error Rates*: These are

$$\hat{e}_{i,act} = \int_{\hat{R}_{0j}} f_i(\mathbf{x}|\hat{\theta}_j)d\mathbf{x} \tag{11.3.19}$$

and

$$\hat{e}_{act} = \pi_1 \hat{e}_{i,act} + \pi_2 \hat{e}_{2,act} = P(\hat{\mathbf{R}}_0, \hat{\mathbf{f}}) \tag{11.3.20}$$

obtained by using the estimates of unknown parameters in f_1 and f_2.

(5) *The Apparent Error Rates (APER)*: The classification procedure that is obtained from the data set is applied to each member of the set to reclassify the same. This method is commonly known as 're-substitution'. The apparent error rates are then given by

$$e_{i,app} = \frac{m_i}{n_i} \quad \text{and } e_{app} = \pi_1 e_{1,app} + \pi_2 e_{2,app} \tag{11.3.21}$$

where m_i of the n_i observations from group G_i are misclassified by the classification rule. The APER does not depend on the form of the parent population and can easily be calculated from the *confusion matrix*, which shows actual versus predicted group membership.

Confusion Matrix

	Estimated Membership	
Actual Membership	G_1	G_2
G_1	$n_1 - m_1$	m_1
G_2	m_2	$n_2 - m_2$

If π_1, π_2 are unknown and the $n = n_1 + n_2$ observations are a random sample from the combined group, then initial estimates of π_i is $n_i/(n_i + n_2)$ and the estimated apparent error rate is

$$\begin{aligned} \hat{e}_{app} &= \hat{\pi}_1 e_{i,app} + \hat{\pi}_2 e_{2,app} \\ &= \frac{n_1}{n_1+n_2}\frac{m_1}{n_1} + \frac{n_2}{n_1+n_2}\frac{m_2}{n_2} \\ &= \frac{m_1+m_2}{n_1+n_2}. \end{aligned} \tag{11.3.22}$$

This estimate is used for example in SPSS and BMDO7M Program.

The APER is an estimate of the probability that the classification function based on the present sample will misclassify a future observation. This probability is the actual error rate (AER). Unless the sample size is large, the APER tends to underestimate the AER, because the data set used to compute the classification function is also used to evaluate it (in calculating APER). For large sample the APER has a small amount of bias for estimating the AER and can be used with little concern.

One way to avoid the bias is to split the sample into two parts, a training sample and a validation sample, defined before. We submit each observation vector in the validation sample to the classification function obtained from the training sample. Since these observations are not used in calculating the classification function, the resulting error rate is unbiased. To increase the efficiency, we could reverse the role of training sample and validation sample and evaluate on the basis of the training sample. The two estimates of error could then be averaged to get a better estimate of AER.

Another method of getting a better estimate of AER is the 'hold-out' method given in (6) below.

(6) *The Cross-Validation Error Rate* proposed by Lachenbruch and Mickey (1968): The technique consists in determining the allocation rule using the

sample data minus one observation and then using this allocation rule to determine the group of the left-out observation. Repeating the process for each of the n_i observations from the group G_i, one gets the cross-validation error rate

$$e_{i,c} = \frac{a_i}{n_i} \tag{11.3.23}$$

the proportion from group G_i, that are misclassified. The overall cross-validation error rate is

$$e_c = \pi_1 e_{1,c} + \pi_2 e_{2,c} \tag{11.3.24}$$

which can be estimated as in (11.3.22) by

$$\hat{e}_c = \frac{a_1 + a_2}{n_1 + n_2}. \tag{11.3.25}$$

The estimator $\hat{e}_c$ is a good estimator of the expected value of the actual error rate, $E(e_{act})$. Since the jackknife method is based on leaving one observation at each step, the above estimates are sometimes (wrongly) called as Jackknife estimates. The method is also called the *hold-out* procedure.

(7) *The Bootstrap Estimate* proposed by Efron (1979, 1981). This method corrects the bias in the apparent error rates discussed in (5) above.

A new sample of size n_i is drawn with replacement from the original sample selected for the group $G_i (i = 1, 2)$. The original sample and the new sample for the ith group is used to determine the classification rule and this rule is used to classify these observations. Let m_i^* and m_i^{**} be the number of observations misclassified in these two samples respectively. Let $d_i = (m_i^* - m_I^{**})/n_i$. We find an estimate of $E(d_i)$ by $\bar{d}_i = \sum_{j=1}^K d_{ij}/K$, where d_{ij} is the value of d_i on the jth repetition of the experiment of drawing with replacement a sample of size n_i from the original sample of n_i observations from the group $G_i, j = 1, \ldots, K$ (a large quntity, say, 100) , $i = 1, 2$. The bootstrap-corrected apparent error rates are the bootstrap error rates, given by

$$e_{i,boot} = \frac{m_i}{n_i} + \bar{d}_i, \tag{11.3.26}$$

$$e_{boot} = \pi_1 e_{1,boot} + e_{2,boot}. \tag{11.3.27}$$

It has been found that the estimate $\bar{d}_i$ of the bias of $e_{i,app}$ has good efficiency for the case of groups which are close together (McLachlan, 1980).

We now consider a property of the plug-in error rate, $\hat{e}_{act}$.

The plug-in-estimate $\hat{e}_{act}$ relies heavily on the correct estimation of f_i through $\hat{\theta}_i$. It has also poor small sample properties.

Let $r_i(\mathbf{x}) = \pi_i f_i(\mathbf{x}|\theta_i)$. Then

$$\begin{aligned} R_{01} &= \{\mathbf{x} : f_1(\mathbf{x}|\theta_1)/f_2(\mathbf{x}|\theta_2) > \pi_2/\pi_1\} \\ &= \{\mathbf{x} : r_1(\mathbf{x}) > r_2(\mathbf{x})\} \\ &= \{\mathbf{x} : r_1(\mathbf{x}) = \max{}_{i=1,2} r_i(\mathbf{x})\}. \end{aligned}$$

Let $\hat{r}_i(\mathbf{x}) = \pi_i f(\mathbf{x}|\hat{\theta}_i)$ be a positive unbiased estimate of $r_i(\mathbf{x})$, i.e., $E_{\hat{\theta}}(\hat{r}_i(\mathbf{x})) = r_i(\mathbf{x})$ for almost all $\mathbf{x}$. Also we denote $r(\mathbf{x}) = \max{}_{i=1,2}\{r_i(\mathbf{x})\}$. We have the following lemma.

Lemma 11.3.1: We have

$$E(\hat{e}_{act}) \leq e_{opt} < E(e_{act}). \tag{11.3.28}$$

Let $\hat{r}(\mathbf{x}) = \max{}_{i=1,2}\{\hat{r}_1(\mathbf{x}), \hat{r}_2(\mathbf{x})\}$. Then $R_{01} = \{\mathbf{x} : \hat{r}(\mathbf{x}) = \hat{r}_1(\mathbf{x})\}$, $R_{02} = R - R_{01} = \{\mathbf{x} : \hat{r}(\mathbf{x}) = \hat{r}_2(\mathbf{x})\}$.

Since the expected value of the maximum of several functions is not less than the maximum of the expected values of the functions,

$$\begin{aligned} E_{\hat{\theta}}[\hat{r}(\mathbf{x})] &\geq \max{}_{i=1,2} E_{\hat{\theta}}[\hat{r}_i(\mathbf{x})] \\ &= \max{}_{i=1,2} r_i(\mathbf{x}) \\ &= r(\mathbf{x}). \end{aligned} \tag{11.3.29}$$

Now,

$$\begin{aligned} E_{\hat{\theta}}(\hat{e}_{act}) &= E_{\hat{\theta}}[\pi_1 \hat{e}_{1,act} + \pi_2 \hat{e}_{2,act}] \\ &= E_{\hat{\theta}}[\pi_1 \textstyle\int_{\hat{R}_{02}} f_1(\mathbf{x}|\hat{\theta}_1) d\mathbf{x} + \pi_2 \int_{\hat{R}_{01}} f_2(\mathbf{x}|\hat{\theta}_2) d\mathbf{x}]. \end{aligned}$$

Using

$$\int_{\hat{R}_{02}} f_1(\mathbf{x}|\hat{\theta}_1) d\mathbf{x} = 1 - \int_{\hat{R}_{01}} f_1(\mathbf{x}|\hat{\theta}_1) d\mathbf{x},$$

etc.,

$$\begin{aligned} 1 - E_{\hat{\theta}}[\hat{e}_{act}] &= E_{\hat{\theta}}[\textstyle\int_{\hat{R}_{01}} \hat{r}_1(\mathbf{x}) d\mathbf{x} + \int_{\hat{R}_{02}} \hat{r}_2(\mathbf{x}) d\mathbf{x}] \\ &= E_{\hat{\theta}}[\textstyle\int_{\hat{R}_{01}} \hat{r}(\mathbf{x}) d\mathbf{x} + \int_{\hat{R}_{02}} \hat{r}(\mathbf{x}) d\mathbf{x}] \\ &= E_{\hat{\theta}} \textstyle\int_R \hat{r}(\mathbf{x}) d\mathbf{x} \\ &= \textstyle\int_R E_{\hat{\theta}}[\hat{r}(\mathbf{x})] d\mathbf{x} \text{ (by Fubini's Theorem)} \\ &\geq \textstyle\int_R r(\mathbf{x}) d\mathbf{x} \ \ \text{(by (11.3.29))} \\ &= \textstyle\int_{R_{01}} \pi_1 f_1(\mathbf{x}) d\mathbf{x} + \int_{R_{02}} \pi_2 f_2(\mathbf{x}) d\mathbf{x} \\ &= \pi_1 - \textstyle\int_{R_{02}} \pi_1 f_1(\mathbf{x}) d\mathbf{x} + \pi_2 - \int_{R_{01}} \pi_2 f_2(\mathbf{x}) d\mathbf{x} \\ &= \pi_1 + \pi_2 - e_{opt} \\ &= 1 - e_{opt}. \end{aligned}$$

Hence, $e_{opt} \geq E_{\hat{\theta}}[\hat{e}_{act}]$. Also, $e_{opt} < e_{act}$ (with probability one, since in general, $\hat{\theta} \neq \theta$). Therefore, $e_{opt} < E_{\hat{\theta}}[e_{act}] = E[e_{act}]$.

Corollary 11.3.1.1: The above lemma till holds if $\hat{r}_i(\mathbf{x}) = f_i(\mathbf{x}|\hat{\theta}_i)\hat{\pi}_i$ and $E_{\hat{\theta}}[\hat{r}_i(\mathbf{x})] \geq r_i(\mathbf{x})$.

With a suitable estimation procedure, $f_i(\mathbf{x}|\hat{\theta})$ will be an approximately unbiased estimator of $f_i(\mathbf{x}|\theta)$ and consequently, the Lemma 11.3.1 will hold asymptotically. If $\hat{\pi}_i$ is approximately unbiased, then also the Corollary 11.3.1.1 will hold asymptotically. Therefore, $\hat{e}_{act}$ is expected to be an underestimate of e_{act} and hence of $E(e_{act})$. It has been found that the bias can be very serious. Therefore, $\hat{e}_{act}$ is unsatisfactory as an estimate of e_{act}.

The estimators $e_{i,app}$ and e_{app}, though easy to calculate, also tend to be underestimates of AER. In general, they should be calculated only when n_1, n_2 are reasonably large. It has been recommended that n_i should exceed twice the number of parameters in θ_i. For smaller samples, $e_{i,c}$ and $e_{i,boot}$ are preferred, as they are both almost unbiased estimates of $e_{i,act}$.

EXAMPLE 11.3.2: Consider two equal covariance matrix normal populations with unknown parameters and find its error rates.

For the above case we have already found in (11.3.4) the estimated minimum TPM classification rule $\hat{\mathbf{R}}_0$. Writing as in (11.2.11) and (11.2.13),

$$\begin{aligned} E[D_s(\mathbf{x})|\mathbf{x} \in G_1] &= D_s(\mu_1) \\ V[D_s(\mathbf{x})|\mathbf{x} \in G_1] &= V[\mathbf{a}'\mathbf{x}|\mathbf{x} \in G_1] \\ &= \mathbf{a}'\mathbf{\Sigma}\mathbf{a} = \sigma^2 \quad \text{(say)}, \end{aligned}$$

[Note that when calculating the conditional expectation and variance of $D_s(\mathbf{x})$ we are only calculating expectation and variance over $\mathbf{x}$, treating the other components of $D_s(\mathbf{x})$ as constant] we have, as in (11.2.14), estimated value of

$$\begin{aligned} P(2|1) &= P[D_s(\mathbf{x}) \leq \ln(\pi_2/\pi_1)|\mathbf{x} \in G_1] \\ &= P[\tfrac{D_s(\mathbf{x}) - D_s(\mu_1)}{\sigma} \leq \tfrac{\ln c - D_s(\mu_1)}{\sigma}] \\ &= \Phi(\tfrac{\ln c - D_s(\mu_1)}{\sigma}) \end{aligned}$$

where $c = \ln(\pi_2/\pi_1)$. Similarly, estimated value of

$$P(1|2) = \Phi(\frac{-\ln c + D_s(\mu_2)}{\sigma}).$$

Therefore,

$$e_{act} = \pi_1 \Phi(\frac{\ln c - D_s(\mu_1)}{\sigma}) + \pi_2 \Phi(\frac{-\ln c + D_s(\mu_2)}{\sigma}). \tag{11.3.30}$$

Replacing μ_i by $\bar{\mathbf{x}}_i$ and $\mathbf{\Sigma}$ by $\mathbf{S}_p$, we get

$$\hat{e}_{act} = \pi_1\Phi\{(\ln c - D^2/2)/D\} + \pi_2\Phi\{(-\ln c - D^2/2)/D\} \qquad (11.3.31)$$

where D is the positive square root of

$$D^2 = \mathbf{a}'\mathbf{S}_p\mathbf{a} = (\bar{\mathbf{x}}_1 - \bar{\mathbf{x}}_2)'\mathbf{S}_p^{-1}(\bar{\mathbf{x}}_1 - \bar{\mathbf{x}}_2), \qquad (11.3.32)$$

the Mahalanobis D-squared measure of distance between two samples (Exercise 11.4). However, as already noted $\hat{e}_{act}$ is a poor estimator of e_{act}. The estimator $\hat{e}_{act}$ can be improved upon to get a better estimator. For example, using

$$\tilde{D}^2 = \frac{n_1 + n_2 - p - 3}{n_1 + n_2 - 2}D^2 \qquad (11.3.33)$$

in (11.3.31) gives a better estimator of e_{act} than e_c. Some other references in this area are McLachlan (1976) and Campbell (1978).

EXAMPLE 11.3.3: The rows of the following matrices give the sample observations from two groups. Calculate the apparent error rates and Lachenbruch's cross-classified error rates.

$$\mathbf{X}_1 = \begin{bmatrix} 14 & 28 \\ 25 & 37 \\ 16 & 26 \end{bmatrix}, \quad \mathbf{X}_2 = \begin{bmatrix} 18 & 10 \\ 16 & 22 \\ 14 & 11 \end{bmatrix}.$$

Here

$$\bar{\mathbf{x}}_1 = \begin{bmatrix} 18.33 \\ 30.33 \end{bmatrix}, \quad \bar{\mathbf{x}}_2 = \begin{bmatrix} 16.60 \\ 14.33 \end{bmatrix},$$

$$\mathbf{S}_1 = \begin{bmatrix} 34.33 & 32.33 \\ 32.33 & 34.33 \end{bmatrix}, \quad \mathbf{S}_2 = \begin{bmatrix} 4.0 & -1.0 \\ -1.0 & 44.33 \end{bmatrix},$$

$$\mathbf{S}_p = (\frac{1}{2})[\mathbf{S}_1 + \mathbf{S}_2] = \begin{bmatrix} 19.17 & 15.67 \\ 15.67 & 39.33 \end{bmatrix}, \quad \mathbf{S}_p^{-1} = \begin{bmatrix} 0.07736 & -0.03082 \\ -0.03082 & 0.03777 \end{bmatrix}.$$

By (11.3.8),

$$\hat{m} = (\bar{\mathbf{x}}_1 - \bar{\mathbf{x}}_2)'\mathbf{S}_p^{-1}(\bar{\mathbf{x}}_1 - \bar{\mathbf{x}}_2) = 7.77468.$$

Writing $y_{ij} = \mathbf{a}'\mathbf{x}_{ij}$ (vide formula (11.3.9)) where $\mathbf{x}_{ij}$ denotes the jth observation in group $G_i (j = 1, \ldots, 3; i = 1, 2)$ we find that

$$y_{11} = 10.501, \; y_{12} = 11.842, \; y_{13} = 8.812,$$

$$y_{21} = -0.317, \; y_{22} = 6.686, \; y_{23} = 1.466.$$

Hence, the first three observations are in G_1 and the last three in G_2. The discriminant function (11.3.7) has correctly classified each observation. The apparent error rate is zero.

We now calculate error rate by Lachenbruch's hold-out procedure. We first hold out the observation $x_{11} = (14, 28)'$. The mean and variance of the remaining observations in X_1 are

$$\bar{\mathbf{x}}_{1h(1)} = \begin{bmatrix} 20.5 \\ 31.5 \end{bmatrix}, \ \mathbf{S}_{1h(1)} = \begin{bmatrix} 40.5 & 49.5 \\ 49.5 & 60.5 \end{bmatrix},$$

$$\mathbf{S}_{p(11)} = \frac{1}{3}[\mathbf{S}_{1h(1)} + 2\mathbf{S}_2] = \begin{bmatrix} 16.5 & 15.8 \\ 15.8 & 49.72 \end{bmatrix},$$

$$\mathbf{S}_{p(11)}^{-1} = \begin{bmatrix} .08711 & -.02768 \\ -.02768 & .02891 \end{bmatrix}.$$

By (11.3.8), $\hat{m}_{(11)} = (\bar{\mathbf{x}}_{1h(1)} - \bar{\mathbf{x}}_2)'\mathbf{S}_{p(11)}^{-1}(\bar{\mathbf{x}}_{1h(1)} - \bar{\mathbf{x}}_2) = 6.998$. For the held-out observation $\mathbf{x}_{11}, y_{(11)} = 9.244$. Hence x_{11} is assigned to G_1, as $y_{(11)} > \hat{m}_{(11)}$. Similar calculation leaving out $\mathbf{x}_{12}$ placed $\mathbf{x}_{12}$ in G_1.

We now illustrate the minimum-distance method of assigning an $\mathbf{x}$. Leaving out $\mathbf{x}_{13} = (10, 26)'$ we have

$$\bar{\mathbf{x}}_{1h(3)} = \begin{bmatrix} 19.5 \\ 32.5 \end{bmatrix}, \ \mathbf{S}_{1h(3)} = \begin{bmatrix} 60 & 49.5 \\ 49.5 & 40.5 \end{bmatrix},$$

$$\mathbf{S}_{p(13)} = \begin{bmatrix} 22.67 & 15.83 \\ 15.83 & 43.05 \end{bmatrix}, \ \mathbf{S}_{p(13)}^{-1} = \begin{bmatrix} .05935 & -.02182 \\ -.02183 & .03125 \end{bmatrix}.$$

The distance between the left out observation $\mathbf{x}_{13}$ and $\bar{\mathbf{x}}_{1h(3)}$ is

$$D_{13}(1) = (\bar{\mathbf{x}}_{1h(3)} - \mathbf{x}_{13})'\mathbf{S}_{(h3)}^{-1}(\bar{\mathbf{x}}_{1h(3)} - \mathbf{x}_{13}) = 1.055.$$

Similarly, distance between $\mathbf{x}_{13}$ and $\bar{\mathbf{x}}_2$ is $D_{13}(2) = 4.256$. Since $\mathbf{x}_{13}$ is nearer to $\bar{\mathbf{x}}_{1h(3)}$ than to $\bar{\mathbf{x}}_2$, it is is placed in G_1.

Similar calculation with $\mathbf{x}_{21}, \ldots \mathbf{x}_{23}$ left out respectively, gave the following results.

$$D_{21}(1) = 28.8641, \ D_{21}(2) = 6.257, \ D_{22}(1) = 4.744,$$

$$D_{22}(2) = 14.741, \ D_{23}(1) = 8.465, \ D_{23}(2) = 0.608.$$

Hence, $\mathbf{x}_{21}$ and $\mathbf{x}_{23}$ are placed in G_2, while $\mathbf{x}_{22}$ is placed in G_1. Therefore, the cross-validation error rate is $1/6$.

11.3.4 Some examples

EXAMPLE 11.3.4: *Independent Binary Variables with Parameters Unknown*: Consider Example 11.2.3 where $\mathbf{x}' = (x_1', x_2')$ is a vector of independent Bernoulli variables, each taking the values 1 and 0 with $P[x_j = 1] = p_{ij}$ when $\mathbf{x} \in G_i (i = 1, 2; j = 1, 2)$. Then $\mathbf{x}'(\in G_i)$ takes the four values, $1 = (1,1), 2 = (1,0), 3 = (0,1), 4 = (0,0)$ with respective probabilities

$$\theta_{(i)1} = p_{i1}p_{i2},\ \theta_{(i)2} = p_{i1}(1 - p_{i2}),$$

$$\theta_{(i)3} = (1 - p_{i1})p_{i2},\ \theta_{i(4)} = (1 - p_{i1})(1 - p_{i2}). \tag{11.3.34}$$

Therefore, when $\mathbf{x} \in G_i$, its probability mass function is

$$f_i(\mathbf{x}) = p_{i1}^{x_1}(1 - p_{i1})^{1-x_1} p_{i2}^{x_2}(1 - p_{i2})^{1-x_2},\ x_1, x_2 = 0, 1. \tag{11.3.35}$$

If a value $\mathbf{x} = j$ is observed, then by rule (11.2.5) it is assigned to group G_1 if

$$\frac{f_1(\mathbf{x})}{f_2(\mathbf{x})} = \frac{\theta_{(1)j}}{\theta_{(2)j}} > \frac{\pi_2}{\pi_1}. \tag{11.3.36}$$

The parameters $\theta_{(i)k} (i = 1, 2; k = 1, \ldots, 4)$ have to be estimated from the sample data.

Suppose we have a random sample of size n from the population $\mathcal{P}$ and of the n_i units that fall in $G_i, n_i(k)$ units falls in cell k (corresponding $\mathbf{x}$ values are k each). Then

$$P[\mathbf{x} \in G_i,\ \text{and } \mathbf{x} \text{ fall in cell } k]$$

$$= P[\mathbf{x} \in G_i] P[\mathbf{x} \text{ in cell } k \| \mathbf{x} \in G_i]$$

$$= \pi_i \theta_{(i)k}.$$

Hence, given the data $\{n_{ik}\}$, the likelihood function of the parameters $\{\pi_i, p_{ij}; i = 1, 2; j = 1, 2\}$ is

$$\begin{aligned} L(\{p_{ij}, \pi_i\}) &= \Pi_{i=1}^2 \Pi_{k=1}^4 \{\pi_i \theta_{(i)k}\}^{n_{(i)k}} \\ &= \Pi_{i=1}^2 \pi_i^{n_i} \Pi_{k=1}^4 \theta_{(i)k}^{n_{(i)k}} \end{aligned} \tag{11.3.37}$$

where $\theta_{(i)k}$ are to be replaced by functions of p_{i1} and p_{i2} given in (11.3.34). Maximizing (11.3.37) with respect to p_{ij} and π_i give the m.l.e.'s, $\hat{p}_{ij} = n_{ij}/n_i$, where n_{ij} is the number of observations from group G_i with $x_j = 1, j = 1, 2$ and $\hat{\pi}_i = n_i/n$. These can be plugged into (11.3.36).

Instead of using (11.3.36) we can also use the linear classification function (LCF) of (11.3.5) (arising in case of normal populations) since the LCF is robust to discrete data and mild inequality of dispersion matrices (Exercise 11.3).

The problem can also be solved using the method of *logistic classification* described in Section 11.5.

The above method readily extends to the case of p independent binary variables. The number of parameters to be estimated are now: $2p$ probabilities (including the 2 π_i's) for the multinomial model and $2p + \frac{p(p+1)}{2}$ for the elements of μ_1, μ_2 and $\mathbf{\Sigma}$ for the LCF method.

EXAMPLE 11.3.5: *Correlated Discrete Variables, Parameters Unknown*: Let $\mathbf{x} = (x_1, x_2)'$ be a pair of binary variables as above, but x_1, x_2 are not independent. We cannot therefore express $\theta_{(i)k}$ in terms of p_{ij}'s $(i, j = 1, 2)$. For instance, $\theta_{(i)2} \neq p_{i1}(1-p_{i2})$ as x_1, x_2 are not independent. However, the discriminant rule (11.3.36) still holds. Here m.l.e. of $\theta_{(i)j}$ is $\hat{\theta}_{(i)j} = n_{(i)j}/n$ and of π_i is $\hat{\pi}_i = n_i/n$. The rule therefore reduces to: Assign $\mathbf{x}$ to G_1 if

$$\frac{n_{(1)j}}{n_{(2)j}} > 1. \tag{11.3.38}$$

The extension to p-dimensional vector of binary variables is straightforward with $(2^p - 1)$ cell probabilities $\theta_{(i)k}$ to be estimated for each group.

Suppose now $\mathbf{x} = (x_, x_2)'$ where x_1, x_2 are discrete random variables taking s_1, s_2 values respectively. The multinomial model (11.3.37) still holds. However, there are now $s_1 \times s_2$ cell probabilities to be estimated.

The method is generally unsatisfactory as very large sample is needed to obtain satisfactory estimates of a large number of parameters. To overcome the problem of sparseness various authors suggested methods like imposing additional constraints on the multinomial model, neglecting various 'high order' interactions, using Bayesian models, etc. (vide, e.g., Ott and Kronmal (1976), Lachenbruch and Goldstein (1979)).

EXAMPLE 11.3.6: *Multivariate Discrete-Continuous Distribution*: Krzanowski (1977) developed the following model. Suppose $\mathbf{x} = (\mathbf{x}_1', \mathbf{x}_2')'$ where $\mathbf{x}_2$ is a p_2- dimensional vector of binary random variables generating $2^{p_2} = S$ states and $\mathbf{x}_1$ is a p_1-dimensional multivariate normal vector with mean $\mu^{(s)}$ depending on the state s of $\mathbf{x}_2 (s = 1, \ldots, S)$ and a constant dispersion matrix. Then

$$f(\mathbf{x}) = h(\mathbf{x}_2) g(\mathbf{x}_1 | \mathbf{x}_2).$$

The classification rule (11.2.5) is: assign $\mathbf{x}$ to G_1 if

$$\frac{f_1(\mathbf{x})}{f_2(\mathbf{x})} = \frac{h_1(\mathbf{x}_2)g_1(\mathbf{x}_1|\mathbf{x}_2)}{h_2(\mathbf{x}_2)g_2(\mathbf{x}_1|\mathbf{x}_2)} > \frac{\pi_2}{\pi_1}$$

or

$$\frac{g_1(\mathbf{x}_1|\mathbf{x}_2)}{g_2(\mathbf{x}_1|\mathbf{x}_2)} > \frac{h_2(\mathbf{x}_2)\pi_2}{h_1(\mathbf{x}_2)\pi_1}. \qquad (11.3.39)$$

We assume that if $\mathbf{x} \in G_i, \mathbf{x}_1|\mathbf{x}_2 \cap N_{p_1}(\mu_i^{(s)}, \boldsymbol{\Sigma}), i = 1, 2$. Since g is the density function of a multivariate normal distribution the logarithm of (11.3.39) reduces to, vide (11.2.9),

$$(\mu_1^{(s)} - \mu_2^{(s)})'\boldsymbol{\Sigma}^{-1}\{\mathbf{x}_1 - \frac{1}{2}(\mu_1^{(s)} + \mu_2^{(s)})\} > \ln\{\frac{h_2(\mathbf{x}_2^{(s)})\pi_2}{h_1(\mathbf{x}_2^{(s)})\pi_1}\}, \qquad (11.3.40)$$

where $\mathbf{x}_2$ belongs to the set s and it is denoted as $\mathbf{x}_2^{(s)}$. Thus for each s, there would be a separate classification rule (11.3.40). A very large data set would be required to estimate all the parameters in $\mu_i^{(s)}, h_i(\mathbf{x}_2^{(s)})$ and $\boldsymbol{\Sigma}$. Krzanowski (1975) suggested using a log-linear model to model the probability $h_i(\mathbf{x}_2)$ and a similar expression to model $\mu_i^{(s)}, i = 1, 2$.

Another approach to the problem of mixed discrete and continuous variables which does not require distributional assumption is the *logistic regression* discussed in Section 11.5.

We have not considered the cases where probability density functions are unknown and require to be estimated by some nonparametric method like kernel method, nearest neighbor method A good review of the nonparametric density estimation including kernel estimators has been given by Silverman (1980), who noted that the classification analysis provided the initial motivation for the development of density estimation. Also, some nonparametric methods of classification, like the partitioning methods, distance methods, rank procedures, sequential discrimination have not been considered. For details the reader may refer to Lachenbruch (1975), Breiman,*et al.* (1977), Hora (1980), among many others.

In the next section we shall consider Fisher's (1936) method of discriminating between two groups by a discriminating function and its relation with the linear classification function (11.2.16).

11.4 Fisher's Discriminating Function for Separating Two Groups

Fisher (1936) considered the problem of discriminating between two groups by considering a linear function of the variables which will distinguish the two groups as sharply as possible. His method can also be used to classify a unit selected from this population into one of the two groups. Fisher obtained the same result as in (11.2.16) (or(11.3.8) as its sample counterpart), but without assuming normal distribution.

Consider a linear function

$$Y = \alpha'\mathbf{X} \tag{11.4.1}$$

where α is a $p \times 1$ vector to be suitably chosen. We have

$$\begin{aligned} E(Y|\mathbf{X} \in G_i) &= \alpha'\mu_i = \mu_{iY}, \quad i = 1, 2, \\ V(Y|\mathbf{X} \in G_i) &= \alpha'\mathbf{\Sigma}\alpha = \sigma_Y^2, \quad i = 1, 2, \end{aligned} \tag{11.4.2}$$

where we assume $\mathbf{X}$ has a constant covariance matrix. The overall mean of $Y = \bar{\mu}_Y = \sum_{i=1}^{2} \mu_{iY}/2$. We wish to find α so that the ratio of the squared distance between the means $\mu_{iY} (i = 1, 2)$ to the variance of Y is maximum. This will ensure that the variable Y takes, on an average, widely separated values for $\mathbf{X}$ taking values in the two groups G_1, G_2. Specifically, we want to maximize

$$\Delta_2^2 = \frac{(\mu_{1Y} - \mu_{2Y})^2}{\sigma_Y^2} = \frac{[\alpha'(\mu_1 - \mu_2)]^2}{\alpha'\mathbf{\Sigma}\alpha} = \frac{(\alpha'\mathbf{d})^2}{\alpha'\mathbf{\Sigma}\alpha} \tag{11.4.3}$$

where $\mathbf{d} = \mu_1 - \mu_2$, over α. By Theorem A.14.1, the maximum value is attained when

$$\alpha \propto \mathbf{\Sigma}^{-1}\mathbf{d} = \mathbf{\Sigma}^{-1}(\mu_1 - \mu_2). \tag{11.4.4}$$

In this case the value of Δ_2^2 is

$$\Delta^2 = (\mu_1 - \mu_2)'\mathbf{\Sigma}^{-1}(\mu_1 - \mu_2), \tag{11.4.5}$$

the squared Mahalanobis distance between two group means. The function

$$\alpha'\mathbf{X} = (\mu_1 - \mu_2)'\mathbf{\Sigma}^{-1}\mathbf{X} \tag{11.4.6}$$

is the Fisher's discriminant function between two groups. Note that the same function was obtained in (11.2.16) under normality assumptions.

However, the parameters in (11.4.6) will ordinarily remain unknown and have to be estimated. Let $\mathbf{x}_{ij} (j = 1, \ldots, n_i)$ be a sample of size n_i from group $G_i, i = 1, 2$. Let $y_{ij} = \mathbf{a}'\mathbf{x}_{ij}$ be the corresponding value of the linear function $y = \mathbf{a}'\mathbf{x}$. The sample means are $\bar{y}_i = \mathbf{a}'\bar{\mathbf{x}}_i$ where $\bar{\mathbf{x}}_i = \sum_{j=1}^{n_i} \mathbf{x}_{ij}/n_i (i = 1, 2)$. The pooled sample variance of y, is

$$s_y^2 = \frac{\sum_{j=1}^{n_1}(y_{1j} - \bar{y}_1)^2 + \sum_{j=1}^{n_2}(y_{2j} - \bar{y}_2)^2}{n_1 + n_2 - 2} = \mathbf{a}'\mathbf{S}_p\mathbf{a}$$

where $\mathbf{S}_p$ is the pooled variance of $\mathbf{x}$ and is given in (11.3.3). Under the assumptions of the model $\bar{y}_i$ and s_y^2 are unbiased estimates of μ_{iY} and s_Y^2 respectively $(i = 1, 2)$. As in (11.4.3) we want to maximize

$$D_2^2 = \frac{(\bar{y}_1 - \bar{y}_2)^2}{s_y^2} = \frac{[\mathbf{a}'(\bar{\mathbf{x}}_1 - \bar{\mathbf{x}}_2)]^2}{\mathbf{a}'\mathbf{S}_p\mathbf{a}} = \frac{(\mathbf{a}'\hat{\mathbf{d}})^2}{\mathbf{a}'\mathbf{S}_p\mathbf{a}} \tag{11.4.7}$$

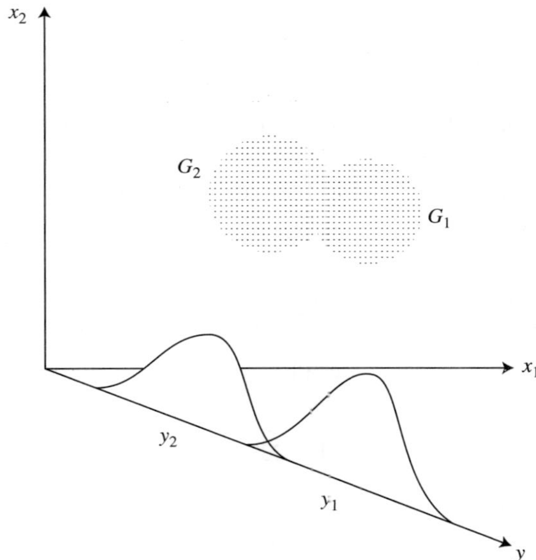

Fig. 11.1: Two-group discriminant analysis

over choice of $\mathbf{a}$ where $\hat{\mathbf{d}} = (\bar{\mathbf{x}}_1 - \bar{\mathbf{x}}_2)$. By Theorem A.14.1, the maximum is attained when

$$\mathbf{a} \propto \mathbf{S}_p^{-1}\hat{\mathbf{d}} = \mathbf{S}_p^{-1}(\bar{\mathbf{x}}_1 - \bar{\mathbf{x}}_2). \tag{11.4.8}$$

In this case the value of D_2^2 is

$$D^2 = (\bar{\mathbf{x}}_1 - \bar{\mathbf{x}}_2)'\mathbf{S}_p^{-1}(\bar{\mathbf{x}}_1 - \bar{\mathbf{x}}_2), \tag{11.4.9}$$

the squared statistical distance between $\bar{\mathbf{x}}_1$ and $\bar{\mathbf{x}}_2$. The function

$$y = \mathbf{a}'\mathbf{x} = (\bar{\mathbf{x}}_1 - \bar{\mathbf{x}}_2)'\mathbf{S}_p^{-1}\mathbf{x} \tag{11.4.10}$$

is Fisher's sample discriminant function for two groups. The same function was obtained in (11.3.8) under normality assumptions.

From (11.4.8) it is seen that the maximizing vector $\mathbf{a}$ is not unique. In fact for any scalar $c(\neq 0)$, the vector $c\mathbf{a} = c\mathbf{S}_p^{-1}(\bar{\mathbf{x}}_1 - \bar{\mathbf{x}}_2)$ will serve as discriminating coefficients.

However, relative values of the coefficients $a_1, \ldots, a_p$ in $\mathbf{a}$ are unique. Hence the direction of $\mathbf{a}$ is unique.

We note that the standardized distance between $\bar{\mathbf{x}}_1$ and $\bar{\mathbf{x}}_2$ is

$$(\bar{\mathbf{x}}_1 - \bar{\mathbf{x}}_2)'\mathbf{S}_p^{-1}(\bar{\mathbf{x}}_1 - \bar{\mathbf{x}}_2) = \frac{(\bar{y}_1 - \bar{y}_2)^2}{s_y^2},$$

the standardized distance between $\bar{y}_1$ and $\bar{y}_2$, for the choice of the discriminant function $y = \mathbf{a}'\mathbf{x}$ as in (11.4.11). Hence the direction given by $\mathbf{a} = \mathbf{S}_p^{-1}(\bar{\mathbf{x}}_1 - \bar{\mathbf{x}}_2)$ is effectively parallel to the distance between $\bar{\mathbf{x}}_1$ and $\bar{\mathbf{x}}_2$. Any direction $\mathbf{a}_1$, other than the one given by $\mathbf{a}$ above would yield a smaller distance between $\mathbf{a}_1'\bar{\mathbf{x}}_1$ and $\mathbf{a}_1'\bar{\mathbf{x}}_2$.

Figure 11.1 represents discriminant analysis between two groups of samples taken from two bivariate normal populations $N_2(\mu_i, \mathbf{\Sigma}), i = 1, 2$. The observations $\mathbf{x}_{ij} = (x_{ij1}, x_{ij2})(j = 1, \ldots, n_i; i = 1, 2)$ are represented by points within the circular regions denoting groups G_1, G_2. Each observation is projected on the line $y = \mathbf{a}'\mathbf{x}$. The distribution of the linear combination $y_{ij} = a_1 x_{ij1} + a_2 x_{ij2}$ is univariate normal. The distributions of y_1 and y_2 are also shown. The vector $\mathbf{a}$ is so chosen that the two distributions are separated as much as possible.

It is emphasized that Fisher's discriminant function (11.4.8) does not depend on the normality assumption.

11.4.1 *Standardized discriminant functions*

The relative importance of the variables $x_1, \ldots, x_p$ (representing the observations on the variables $X_1, \ldots, X_p, \mathbf{X} = (X_1, \ldots, X_p)$) to the separation of two groups G_1, G_2 can be assessed by comparing the coefficients

$a_r(r = 1, \ldots, p)$ in the discriminant function

$$y = \mathbf{a}'\mathbf{x} = a_1x_1 + a_2x_2 + \ldots + a_px_p. \tag{11.4.11}$$

However, such comparisons are useful only when the x's are commensurate, that is, measured on the same scale and have comparable variances.

If the x's are not commensurate, we should first standardize the variables. That is, instead of $\mathbf{x}_{ij} = (x_{ij1}, \ldots, x_{ijp})'$ we should consider $\mathbf{z}_{ij} = (z_{ij1}, \ldots, z_{ijp})'$ where

$$z_{ijk} = \frac{x_{ijk} - \bar{x}_{ik}}{s_k}, \tag{11.4.12}$$

where $\bar{\mathbf{x}}_i = (\bar{x}_{i1}, \ldots, \bar{x}_{ip})', s_k = \frac{1}{n-2}\sum_{i=1}^{2}\sum_{j=1}^{n_i}(x_{ijk} - \bar{x}_{ik})^2$. The corresponding discriminant function should therefore have modified coefficient $a_r^*(r = 1, \ldots, p)$, that is,

$$y_{ij} = a_1^*z_{ij1} + a_2^*z_{ij2} + \ldots + a_p^*z_{ijp}(j = 1, \ldots, n_i; i = 1, 2). \tag{11.4.13}$$

Clearly,

$$a_r^* = a_rs_r, r = 1, \ldots, p.$$

In vector form, this becomes

$$\mathbf{a}^* = \sqrt{Diag\mathbf{S}_p}\mathbf{a} \tag{11.4.14}$$

where Diag $\mathbf{S}_p =$ Diag $(s_1, \ldots, s_p)$.

Scaling of the coefficient vector

We have discussed above the standardization of variables. The vector $\mathbf{a}$ is also frequently scaled or normalized to ease the interpretation of its coefficients.

We assume that all the x variables are commensurate or have been standardized. Two commonly used scaling of $\mathbf{a}$ are as follows.

(i) Choose the coefficient vector such that it has unit length. This is done by choosing $\tilde{\mathbf{a}} = \mathbf{a}/\sqrt{\mathbf{a}'\mathbf{a}}$. In this case all the coefficients $\in [-1, 1]$.

(ii) Choose the coefficient vector such that its first element is 1. This is done by the choice $\tilde{\mathbf{a}} = \mathbf{a}/a_1$. Here the relative importance of the variables $X_2, \ldots, X_p$ vis-a-vis X_1as discriminators can be readily visualized from $\tilde{\mathbf{a}}$.

Interpretation of discriminating functions has been considered in Subsection 11.7.4.

11.4.2 *Test of significance*

For tests of significance we assume that the variable $\mathbf{X}_i$ (for the group G_i) follows $N_p(\mu_i, \mathbf{\Sigma})$ distribution $(i = 1, 2)$.

We have seen that the discriminant function $\alpha'\mathbf{X}$ with $\alpha = \mathbf{\Sigma}^{-1/2}(\mu_1 - \mu_2)$ transforms the squared standardized distance $\Delta^2 = (\mu_1 - \mu_2)'\mathbf{\Sigma}^{-1/2}(\mu_1 - \mu_2)$ into the squared standardized distance between the transformed means $(\mu_{Y1} - \mu_{Y2})^2/\sigma_Y^2$. Testing significance of discriminating function is equivalent to testing $(\mu_{Y1} - \mu_{Y2})^2 = 0$ or $\Delta^2 = 0$.

The statistic to be used is $(\bar{y}_1 - \bar{y}_2)^2/s_Y^2$ or $D^2 = (\bar{\mathbf{x}}_1 - \bar{\mathbf{x}}_2)'\mathbf{S}_p^{-1}(\bar{\mathbf{x}}_1 - \bar{\mathbf{x}}_2)$. It is known from (5.7.1) that under $H_0 : \Delta^2 = 0$, the statistic $\frac{n_1 n_2}{n_1+n_2} D^2 \cap T^2(p, n_1 + n_2 - 2)$. Hence, H_0 is rejected if

$$D_0^2 = \frac{n_1 n_2}{n_1 + n_2} D^2 > \frac{p(n_1 + n_2 - 2)}{n_1 + n_2 - p - 1} F_{p, n_1+n_2-p-1;\alpha}. \tag{11.4.15}$$

In this case the discriminant function coefficient α is significantly different from $\mathbf{0}$.

It may be noted that significant separation does not necessarily imply good classification. The efficacy of a classification procedure can be evaluated independently of any test of separation, by value of an estimate, $\hat{e}_{act}$ of the actual error rate (e.g., e_c, e_{app}, or e_{boot}). However, if the separation is not significant, the search for a useful classification rule may not be fruitful.

11.4.3 *Using Fisher's discriminant function for classification*

Fisher's discriminating functions can also be used for classifying an object selected from the population. The rule is to find the discriminant score $Y = \alpha'\mathbf{X}$ given by (11.4.6) and compare it with the mean score

$$\begin{aligned} \bar{\mu}_Y &= \tfrac{1}{2}(\mu_1 + \mu_2) \\ &= \tfrac{1}{2}(\mu_1 - \mu_2)'\mathbf{\Sigma}^{-1}(\mu_1 + \mu_2). \end{aligned} \tag{11.4.16}$$

If the score Y exceeds $\bar{\mu}_Y, \mathbf{X}$ is allotted to G_1; otherwise, it is allotted to G_2.

In case, the populations parameters are estimated, one calculates the sample discriminant score $y = \mathbf{a}'\mathbf{x}$ given in (11.4.10) and compares it with the mean sample score

$$\begin{aligned} \bar{y} &= \tfrac{1}{2}(\bar{y}_1 + \bar{y}_2) \\ &= \tfrac{1}{2}(\bar{\mathbf{x}}_1 - \bar{\mathbf{x}}_2)'\mathbf{S}_p^{-1}(\bar{\mathbf{x}}_1 + \bar{\mathbf{x}}_2). \end{aligned} \tag{11.4.17}$$

If the sample score $y(\mathbf{x})$ exceeds the mean score $\bar{y}, \mathbf{x}$ is assigned to G_1, otherwise it is assigned to G_2.

EXAMPLE 11.4.1: For the data of Table 11.3.1 the discriminant function is

$$y = .0143x_1 + 0.0271x_2 + 0.6363x_3 - 0.1487x_4 + 0.0908x_5.$$

The y-column in the table shows the discriminant scores for each observation and the next column its fitted group. The *mean* discriminant score $\hat{m}(= (\bar{y}_1 + \bar{y}_2)/2)$ given by (11.4.7)(or (11.3.8)) is 0.4164. Any observation with score exceeding 0.4164 is placed in G_1, otherwise in G_2. Note that equal prior probabilities have been assumed.

It is seen that the discriminant function correctly classifies each observation in group G_1, but fails to do so for 8 observations out of 11 in G_2. The apparent error rate is $8/18 = 0.444$ which is very high.

$$D^2 = [1.30,\ 23.00,\ -0.18,\ 0.90,\ 6.90]\mathbf{S}_p^{-1}\begin{bmatrix} 1.30 \\ 23.00 \\ -0.18 \\ 0.90 \\ 6.90 \end{bmatrix} = 1.2493\,.$$

Hence, $D_0^2 = (77/18)1.2493 = 5.3442 < (80/12)3.11 = 20.733$. Hence $H_0(\Delta^2 = 0)$ is accepted at 5% level. The discriminant function has not efficiently segregated the groups.

11.5 Logistic Classification: $g = 2$

The classification rules developed so far depend on the ratio $f_1(\mathbf{x})/f_2(\mathbf{x})$. We can therefore model the ratio $f_1(\mathbf{x})/f_2(\mathbf{x})$ without specifying it. The *logistic model* assumes that

$$\ln[\frac{f_1(\mathbf{x})}{f_2(\mathbf{x})}] = \delta + \beta'\mathbf{x} \tag{11.5.1}$$

where $\delta, \beta = (\beta_1, \ldots, \beta_p)'$ are parameters to be estimated. The classification rule (11.2.5) becomes: assign $\mathbf{x}$ to G_1 if

$$\delta + \beta'\mathbf{x} > \ln[\pi_2/\pi_1], \tag{11.5.2}$$

i.e., if

$$\delta_0 + \beta'\mathbf{x} > 0 \tag{11.5.3}$$

where

$$\delta_0 = \delta - \ln(\frac{\pi_2}{\pi_1}). \tag{11.5.4}$$

The equation (11.5.3) has a nice relationship with the ratio of posterior probabilities.

The posterior probability that $\mathbf{x}$ comes from G_i is

$$\begin{aligned} q_i(\mathbf{x}) &= P[\mathbf{x} \in G_i | \mathbf{x}] \\ &= \frac{\pi_i f_i(\mathbf{x})}{\pi_1 f_1(\mathbf{x}) + \pi_2 f_2(\mathbf{x})}. \end{aligned} \tag{11.5.5}$$

Using (11.5.1),

$$f_1(\mathbf{x}) = f_2(\mathbf{x}) e^{\delta + \beta' \mathbf{x}}. \tag{11.5.6}$$

Hence

$$\pi_1 f_1(\mathbf{x}) + \pi_2 f_2(\mathbf{x}) = f_2(\mathbf{x})[\pi_1 e^{\delta + \beta' \mathbf{x}} + \pi_2].$$

Therefore,

$$\begin{aligned} q_1(\mathbf{x}) &= \frac{\pi_1 e^{\delta + \beta' \mathbf{x}}}{\pi_1 e^{\delta + \beta' \mathbf{x}} + \pi_2} \\ &= \frac{\exp[\delta + \beta' \mathbf{x} + \ln(\pi_1/\pi_2)]}{\exp[\delta + \beta' \mathbf{x} + \ln(\pi_2/\pi_1)] + 1} \end{aligned} \tag{11.5.7}$$

and

$$q_2(\mathbf{x}) = 1 - q_1(\mathbf{x}).$$

Hence,

$$\begin{aligned} \ln[\frac{q_1(\mathbf{x})}{q_2(\mathbf{x})}] &= \ln\{\frac{\pi_1 f_1(\mathbf{x})}{\pi_2 f_2(\mathbf{x})}\} \\ &= \delta + \beta' \mathbf{x} + \ln(\pi_2/\pi_1) = \delta_0 + \beta' \mathbf{x}. \end{aligned} \tag{11.5.8}$$

This model for the *posterior odds* was first suggested by Cox (1966) and Day and Kerridge (1967) as a basis for discrimination.

An advantage of the logistic approach is that we only need to estimate the $p + 1$ parameters α_0 and β from the sample data without having to specify $f(\mathbf{x})$. The previous models require not only the specification of $f_i(\mathbf{x}|\theta_i)$, but also the estimation of many unknown parameters in $\theta_i (i = 1, 2)$. Furthermore, the family of distributions satisfying (11.5.1) is quite wide. In particular, in includes the multivariate normal distributions with equal dispersion matrices. For this case,

$$\delta + \beta' \mathbf{x} = -\frac{1}{2} \alpha' (\mu_1 + \mu_2) + \alpha' \mathbf{x} \tag{11.5.9}$$

where $\alpha = \mathbf{\Sigma}^{-1}(\mu_1 - \mu_2)$ (vide (11.2.6)). It also includes the multivariate independent binary variables (Example 11.2.3). The logistic model is particularly useful in handling diagnostic data.

EXAMPLE 11.5.1: For the case of two normal populations $N_p(\mu_i, \mathbf{\Sigma}_i)(i = 1, 2)$ we see from (11.2.16) that

$$\begin{aligned}\ln[q_1(\mathbf{x})/q_2(\mathbf{x})] &= \ln(\pi_1/\pi_2) + c_0 + \mathbf{x}'(\mathbf{\Sigma}_1^{-1}\mu_1 - \mathbf{\Sigma}_2^{-1}\mu_2) + \\ &\quad \tfrac{1}{2}\mathbf{x}'(\mathbf{\Sigma}_1^{-1} - \mathbf{\Sigma}_2^{-1})\mathbf{x} \\ &= \gamma_0 + \gamma_1'\mathbf{x} + \mathbf{x}'\mathbf{\Gamma}\mathbf{x},\end{aligned}$$

where $c_0 = (1/2)\ln(|\mathbf{\Sigma}_1|/|\mathbf{\Sigma}_2|)$ and $\mathbf{\Gamma} = ((\gamma_{rs}))$ is symmetric. The above function is linear in the coefficients γ_0, γ and $\gamma_{rs}(r \le s)$ so that it can be written in the form $\beta_0 + \beta'\mathbf{y}$ with $\mathbf{y} = (x_1, \ldots, x_p; x_1^2, x_1x_2, x_2^2, \ldots)'$ and $1 + p + p(p+1)/2$ parameters, β's. However, in this case, there are too many parameters to be estimated.

11.5.1 *Sampling designs*

There are three common sampling designs for estimation of parameters δ, β.

(i) *Mixture Sampling* in which a sample of $n = n_1 + n_2$ units is selected randomly from the population $\mathcal{P}$ so that n_i are random variables (Day and Kerridge, 1967).

(ii) *Separate Sampling* in which for $i = 1, 2$ a sample of fixed size n_i is selected from G_i (Anderson, 1972; Prentice and Pyke, 1979).

(iii) *Conditional Sampling* in which for $j = 1, \ldots, m, n(\mathbf{x}_j)$ members are selected at random from all members of $\mathcal{P}$ with value $\mathbf{x}_j$ and among these, $n_i(\mathbf{x}_j)$ members fall in group $G_i, \sum_{i=1}^2 n_i(\mathbf{x}_j) = n(\mathbf{x}_j), n_i = \sum_{j=1}^m n_i(\mathbf{x}_j), n = \sum_i n_i$.

In all the three cases we use the notation used in (iii). We assume that n observations take m different values $\mathbf{x}_j$ with frequency $n(\mathbf{x}_j)$. Among these, $n_i(\mathbf{x}_j)$ observations fall in G_i. For mixture sampling, n is fixed and $n_i(\mathbf{x}_j), n(\mathbf{x}_j), n_i$ are all random variables. For separate sampling, $n_i = \sum_{j=1}^m n_i(\mathbf{x}_j)(i = 1, 2)$ are fixed and $n_i(\mathbf{x}_j)$ are random. For conditional sampling $n(\mathbf{x}_j)(j = 1, \ldots, m)$ are fixed and $n_i(\mathbf{x}_j)$ are random. Generally, $n(\mathbf{x}_j) = 1$ so that each observation is unique.

We now calculate the likelihood functions under these sampling schemes.

(a) *Conditional Sampling*: Here $n(\mathbf{x}_j) = n_1(\mathbf{x}_j) + n_2(\mathbf{x}_j)$ is fixed and

$n_1(\mathbf{x}_j), n_2(\mathbf{x}_j)$ are random ($j = 1, \ldots, m$). The likelihood function is

$$\begin{aligned} L_C &= \Pi_{j=1}^{m} \Pi_{i=1}^{2} \{P[\mathbf{x} \in G_i | \mathbf{x} = \mathbf{x}_j]\}^{n_i(\mathbf{x}_j)} \\ &= \Pi_{i=1}^{2} \Pi_{j=1}^{m} \{q_i(\mathbf{x}_j)\}^{n_i(\mathbf{x}_j)} \end{aligned} \tag{11.5.10}$$

(vide (11.5.5)). From (11.5.7) we see that $\log L_C$ is a linear function of δ_0 and β from which m.l.e. of δ_0 and β can be found out.

(b) *Mixture Sampling*: Here n is fixed and $n_i(\mathbf{x}_j)$ are random ($i = 1, 2; j = 1, \ldots, m$). The likelihood function is

$$\begin{aligned} L_M &= \Pi_{i=1}^{2} \Pi_{j=1}^{m} \{\pi_i f_i(\mathbf{x}_j)\}^{n_i(\mathbf{x}_j)} \\ &= \Pi_{i=1}^{2} \Pi_{j=1}^{m} \{\tfrac{\pi_i f_i(\mathbf{x}_j)}{\pi_1 f_1(\mathbf{x}_j) + \pi_2 f_2(\mathbf{x}_j)} [\pi_1 f_1(\mathbf{x}_j) + \pi_2 f_2(\mathbf{x}_j)]\}^{n_i(\mathbf{x}_j)} \\ &= \Pi_{i=1}^{2} \Pi_{j=1}^{m} \{q_i(\mathbf{x}_j)\}^{n_i(\mathbf{x}_j)} \Pi_{j=1}^{m} \{\pi_1 f_1(\mathbf{x}_j) + \pi_2 f_2(\mathbf{x}_j)\}^{n(\mathbf{x}_j)} \\ &= L_C L \quad \text{(say)}. \end{aligned} \tag{11.5.11}$$

Since $f_i(\mathbf{x})$ is unspecified no assumption can be made about the form of $f(\mathbf{x}) = \pi_1 f_1(\mathbf{x}) + \pi_2 f_2(\mathbf{x})$, the density function of $f(\mathbf{x})$. We can therefore assume that $f(\mathbf{x})$ contains no useful information about δ_0, β. Therefore, as in conditional sampling, m.l.e.'s are found by maximizing L_C.

(c) *Separate Sampling*: Here n_i is fixed and $n_i(\mathbf{x}_j)$ are random ($i = 1, 2; j = 1, \ldots, m$). The likelihood function is

$$\begin{aligned} L_S &= \Pi_{i=1}^{2} \Pi_{j=1}^{m} \{f_i(\mathbf{x}_j)\}^{n_i(\mathbf{x}_j)} \\ &= \Pi_{i=1}^{2} \Pi_{j=1}^{m} \{\tfrac{\pi_i f_i(\mathbf{x}_j)}{\pi_i}\}^{n_i(\mathbf{x}_j)} = L_M . \Pi_{i=1}^{2} \pi_i^{n_i}. \end{aligned} \tag{11.5.12}$$

If π_1, π_2 are known, this method is equivalent to (11.5.11).

If π_1, π_2 are unknown we proceed as follows (vide Anderson and Blair, 1982). We assume that $\mathbf{x}$ is discrete so that the values of f may be taken as multinomil probabilities. From (11.5.1), $f_1(\mathbf{x}) = f_2(\mathbf{x}) \exp(\delta + \beta'\mathbf{x})$. Hence from (11.5.12),

$$\begin{aligned} L_S &= \Pi_{j=1}^{m} \{f_1(\mathbf{x}_j)\}^{n_1(\mathbf{x}_j)} \{f_2(\mathbf{x}_j)\}^{n_2(\mathbf{x}_j)} \\ &= \Pi_{j=1}^{m} \{f_2(\mathbf{x}_j)\}^{n_1(\mathbf{x}_j) + n_2(\mathbf{x}_j)} \{\exp(\delta + \beta'\mathbf{x}_j)\}^{n_1(\mathbf{x}_j)} \\ &= \Pi_{j=1}^{m} p_{\mathbf{x}_j}^{n(\mathbf{x}_j)} \{\exp[n_1(\mathbf{x}_j)(\delta + \beta'\mathbf{x}_j)]\} \end{aligned} \tag{11.5.13}$$

where $f_2(\mathbf{x}_j) = p_{\mathbf{x}_j}$. The problem is to maximize L_S with respect to $p_{\mathbf{x}_j}$ subject to the constraints that f_2 and f_1 are probability functions, namely,

$$\sum_{j=1}^{m} p_{\mathbf{x}_j} = 1, \tag{11.5.14}$$

$$\sum_{j=1}^{m} p_{\mathbf{x}_j} \exp(\delta + \beta'\mathbf{x}_j) = 1. \tag{11.5.15}$$

Using Lagrange's multipliers, the maximum likelihood estimate of $p_{\mathbf{x}}$ is

$$\hat{p}_{\mathbf{x}} = \frac{n(\mathbf{x})}{n_1 \exp(\delta + \beta'\mathbf{x}) + n_2}. \tag{11.5.16}$$

Substituting $\hat{p}_{\mathbf{x}}$ in L_S we get the maximum value of L_S (for given values of δ, β) as

$$L'_S = L'_C n_1^{-n_1} n_2^{-n_2} \Pi_{j=1}^{m} [n(\mathbf{x}_j)]^{n(\mathbf{x}_j)}$$

where

$$L'_C = \Pi_{i=1}^{2} \Pi_{j=1}^{m} \{\tilde{q}_i(\mathbf{x}_j)\}^{n_i(\mathbf{x}_j)}, \tag{11.5.17}$$

$$\begin{aligned} \tilde{q}_1(\mathbf{x}) &= \tfrac{n_1 \exp(\delta+\beta'\mathbf{x})}{n_1 \exp(\delta+\beta'\mathbf{x})+n_2} \\ &= \tfrac{\exp[\delta+\ln(n_1/n_2)+\beta'\mathbf{x}]}{\exp[\delta+\ln(n_1/n_2)+\beta'\mathbf{x}]+1} \end{aligned} \tag{11.5.18}$$

and

$$\tilde{q}_2(\mathbf{x}) = 1 - \tilde{q}_1(\mathbf{x}).$$

We note that $\tilde{q}_1(\mathbf{x})$ is the same as $q_1(\mathbf{x})$ of (11.5.7) except that $\delta_0 = \delta + \ln(\pi_1/\pi_2)$ is replaced by $\delta + \ln(n_1/n_2)$. Hence with a correct interpretation of δ_0, maimizing L'_C is equivalent to maximizing L_C. We have therefore reduced the problem of maximizing L_S subject to (11.5.14) and (11.5.15) to the problem of maximizing L_C. Prentice and Pyke (1979) suggested that the restriction to discrete variables can be dropped as the above estimates will still have satisfactory properties for continuous variables.

11.6 Classification in More than Two Groups

Suppose we have a population $\mathcal{P}$ consisting of $g(\geq 2)$ mutually exclusive groups and let π_i be the proportion of units in $\mathcal{P}$ in group $G_i(i = 1, \ldots, g; \sum_i \pi_i = 1)$. The theory developed so far for $g = 2$ extends to more than two groups. We define $f_i(\mathbf{x})$ to be the probability or probability density function of $\mathbf{x}$ in $G_i(i = 1, \ldots, g)$. We wish to find a suitable partition $\{R_1, R_2, \ldots, R_g\}$ of the sample space R such that we assign a member to G_i if its value $\mathbf{x} \in R_i$.

The probability of assigning a member to G_j when it actually comes from G_i is

$$P(j|i) = \int_{R_j} f_i(\mathbf{x}) d\mathbf{x}.$$

The probability of misclassifying a member of G_i is, therefore,

$$P(i) = \sum_{j(\neq i)=1}^{g} P(j|i) = 1 - P(i|i).$$

The Total probability of misclassification for the classification rule $\mathbf{R} = (R_1, \ldots, R_g)'$ is

$$P(\mathbf{R}, \mathbf{f}) = \sum_{i=1}^{g} \pi_i P(i) = 1 - \sum_{i=1}^{g} \pi_i P(i|i) \tag{11.6.1}$$

where $\mathbf{f} = (f_1, \ldots, f_g)'$. We shall now consider different rules of misclassification.

11.6.1 *Minimum TPM rule*

When $g = 2$, we see from (11.2.4) that the rule which minimizes $P(\mathbf{R}, \mathbf{f})$ is to assign $\mathbf{x}$ to the group with the larger value of $\pi_i f_i(\mathbf{x})$, i.e., with larger posterior probability $q_i(\mathbf{x})$ since

$$q_i(\mathbf{x}) = \frac{\pi_i f_i(\mathbf{x})}{\sum_{j=1}^{g} \pi_j f_j(\mathbf{x})} = \frac{\pi_i f_i(\mathbf{x})}{f(\mathbf{x})} \quad \text{(say)} \tag{11.6.2}$$

(vide (11.2.20)). We will see that the above result generalizes readily to the case of more than two groups.

Theorem 11.6.1: For the rule which assigns $\mathbf{x}$ to G_i if $\mathbf{x} \in R_i$, the total probability of misclassification (TPM) $P(\mathbf{R}, \mathbf{f})$ is minimized when

$$R_i = R_{0i} = \{\mathbf{x} : q_i(\mathbf{x}) \geq q_j(\mathbf{x}), j = 1, \ldots, g\}.$$

Proof. Define the indicator function $\chi_A(\mathbf{x})$ which takes the value 1 if $\mathbf{x} \in A$ and 0 otherwise. For any partition $\{R_1, \ldots, R_g\}$,

$$\begin{aligned} \textstyle\sum_{i=1}^{g} \pi_i P(i|i) &= \textstyle\sum_{i=1}^{g} \int_{R_i} \pi_i f_i(\mathbf{x}) d\mathbf{x} \\ &= \textstyle\sum_{i=}^{g} \int_{R} \chi_{R_i}(\mathbf{x}) q_i(\mathbf{x}) f(\mathbf{x}) d\mathbf{x} \\ &= \textstyle\int_{R} \sum_{i=1}^{g} \chi_{R_i}(\mathbf{x}) q_i(\mathbf{x}) f(\mathbf{x}) d\mathbf{x} \\ &= \textstyle\int_{R} g(\mathbf{x}) d\mathbf{x} \quad \text{(say)}. \end{aligned} \tag{11.6.3}$$

Let $g_0(\mathbf{x})$ be the corresponding function for the partition $\{R_{01}, \ldots, R_{0g}\}$. Suppose $\mathbf{x} \in R_j$, then $\mathbf{x} \in R_{0m}$ for some m as $R = \cup_j R_{0j}$. Therefore $q_m(\mathbf{x}) \geq q_i(\mathbf{x})\ \forall\ i$ and for this $\mathbf{x}$,

$$\begin{aligned} g_0(\mathbf{x}) &= \textstyle\sum_{i=1}^{g} \chi_{R_{0i}}(\mathbf{x}) q_i(\mathbf{x}) f(\mathbf{x}) \\ &= q_m(\mathbf{x}) f(\mathbf{x}) \\ &= \textstyle\sum_{i=1}^{g} \chi_{R_i}(\mathbf{x}) q_m(\mathbf{x}) f(\mathbf{x}) \\ &\geq \textstyle\sum_{i=1}^{g} \chi_{R_i}(\mathbf{x}) q_i(\mathbf{x}) f(\mathbf{x}) = g(\mathbf{x}). \end{aligned}$$

Thus $g(\mathbf{x})$ is maximized for all $\mathbf{x}$. Hence,

$$P(\mathbf{R}, \mathbf{f}) = 1 - \sum_{i=1}^{g} \pi_i P(i|i) = 1 - \int_R g(\mathbf{x}) d\mathbf{x}$$

is minimized when $R_i = R_{0i}(i = 1, \ldots, g)$. □

Therefore, the optimal classification rule which minimizes the TPM is: Assign $\mathbf{x}$ to G_i if $\pi_i f_i(\mathbf{x}) \geq \pi_j f_j(\mathbf{x})(j = 1, \ldots, g)$, that is, if $\max_j \pi_j f_j(\mathbf{x}) = \pi_i f_i(\mathbf{x})$, with the assignment on the boundary of R_{0i} being arbitrary.

EXAMPLE 11.6.1: *Normal Populations: Common Covariance Matrix*: Suppose that the variable $\mathbf{X}$ when in the ith group follows $N_p(\mu_i, \mathbf{\Sigma})$ distribution. Then

$$\begin{aligned} \ln[\pi_i f_i(\mathbf{x})] &= -\tfrac{1}{2} \ln |\mathbf{\Sigma}| - \tfrac{1}{2}(\mathbf{x} - \mu_i)' \mathbf{\Sigma}^{-1}(\mathbf{x} - \mu_i) + \ln \pi_i \\ &= -\tfrac{1}{2} \ln |\mathbf{\Sigma}| - \tfrac{1}{2}\mathbf{x}' \mathbf{\Sigma}^{-1}\mathbf{x} + \mu_i' \mathbf{\Sigma}^{-1}\mathbf{x} - \tfrac{1}{2}\mu_i' \mathbf{\Sigma}^{-1}\mu_i + \ln \pi_i. \end{aligned} \tag{11.6.4}$$

The first two terms are the same for all groups for a given $\mathbf{x}$ and hence, they can be ignored for the allocation purposes. We therefore define the *linear classification score* for the ith group as

$$d_i(\mathbf{x}) = \mu_i' \mathbf{\Sigma}^{-1}\mathbf{x} - \frac{1}{2}\mu_i' \mathbf{\Sigma}^{-1}\mu_i + \ln \pi_i, i = 1, \ldots, g. \tag{11.6.5}$$

The classification rule is therefore to assign $\mathbf{x}$ to G_k if $d_k(\mathbf{x})$ = largest of $d_1(\mathbf{x}), \ldots, d_g(\mathbf{x})$.

In practice, the parameters μ_i and $\mathbf{\Sigma}_i$ are generally unknown. However a training set of correctly classified observations is generally available for estimation of parameters. Let $\mathbf{x}_{ij}(j = 1, \ldots, n_i)$ be a sample of size n_i drawn from the group $G_i(= 1, \ldots, g), n = \sum_i n_i$. Unbiased estimator are: $\hat{\mu}_i = \bar{\mathbf{x}}_i$,

$$\hat{\mathbf{\Sigma}} = \mathbf{S}_p = \frac{1}{n-g}\mathbf{S}_i, (n_i - 1)\mathbf{S}_i = \sum_{j=1}^{n_i}(\mathbf{x}_{ij} - \bar{\mathbf{x}}_i)(\mathbf{x}_{ij} - \bar{\mathbf{x}}_i)'. \tag{11.6.6}$$

An estimate $d_{is}(\mathbf{x})$ of the linear classification score is $d_i(\mathbf{x})$ is

$$d_{is}(\mathbf{x}) = \bar{\mathbf{x}}_i' \mathbf{S}_p^{-1}\mathbf{x} - \frac{1}{2}\bar{\mathbf{x}}_i' \mathbf{S}_p^{-1}\bar{\mathbf{x}}_i + \ln \pi_i, \; i = 1, \ldots, g. \tag{11.6.7}$$

Estimated minimum TPM rule for equal covariance normal populations is therefore as follows: assign $\mathbf{x}$ to G_k if $d_{ks}(\mathbf{x})$ is the largest of $d_{1s}(\mathbf{x}), \ldots, d_{gs}(\mathbf{x})$.

If only the term $(-1/2)\ln|\mathbf{\Sigma}|$ in (11.6.4) is neglected the resulting function with parameters estimated is $(-1/2)D_i^2(\mathbf{x}) + \ln \pi_i$ where

$$D_i^2(\mathbf{x}) = (\mathbf{x} - \bar{\mathbf{x}}_i)' \mathbf{S}_p^{-1}(\mathbf{x} - \bar{\mathbf{x}}_i), \tag{11.6.8}$$

the squared statistical distance between $\mathbf{x}$ and $\bar{\mathbf{x}}_i$. The estimated minimum TPM rule becomes: assign $\mathbf{x}$ to the group G_k for which

$$-\frac{1}{2}D_k^2(\mathbf{x}) + \ln \pi_k \tag{11.6.9}$$

is largest. If $\pi_i = 1/g \ \forall \ i$, this means allot $\mathbf{x}$ to the 'closest' group, i.e., to the group G_k for which $D_k^2(\mathbf{x})$ is the smallest of $D_1^2(\mathbf{x}), \ldots, D_g^2(\mathbf{x})$.

EXAMPLE 11.6.2: *Normal Populations with Unequal Covariances*: Suppose that $\mathbf{X}$ when it comes from group G_i has the probability density of a $N_p(\mu_i, \mathbf{\Sigma}_i)$ distribution,

$$f_i(\mathbf{x}) = \frac{1}{(2\pi)^{p/2}|\mathbf{\Sigma}_i|^{1/2}} \exp[-\frac{1}{2}(\mathbf{x} - \mu_i)'\mathbf{\Sigma}_i^{-1}(\mathbf{x} - \mu_i)], \ i = 1, \ldots, g. \tag{11.6.10}$$

The minimum TPM rule then becomes: allot $\mathbf{x}$ to G_k if

$$\begin{aligned} \ln[\pi_k f_k(\mathbf{x})] &= \ln \pi_k - \frac{p}{2}\ln(2\pi) - \frac{1}{2}\ln(|\mathbf{\Sigma}_k|) - \frac{1}{2}(\mathbf{x} - \mu_k)'\mathbf{\Sigma}_k^{-1}(\mathbf{x} - \mu_k) \\ &= \max{}_i \ln[\pi_i f_i(\mathbf{x})]. \end{aligned} \tag{11.6.11}$$

The constant $\frac{p}{2}\ln(2\pi)$ in (11.6.11) can be ignored since it is the same for all the g groups. We therefore define the *quadratic classification score* for the ith group to be

$$d_i^Q(\mathbf{x}) = -\frac{1}{2}\ln|\mathbf{\Sigma}_i| - \frac{1}{2}(\mathbf{x} - \mu_i)'\mathbf{\Sigma}_i^{-1}(\mathbf{x} - \mu_i) + \ln \pi_i, \ i = 1, \ldots, g. \tag{11.6.12}$$

The minimum TPM rule of classification then becomes the following: assign $\mathbf{x}$ to G_k if the quadratic score $d_k^Q(\mathbf{x}) = \max{}_i[d_i^Q(\mathbf{x})]$.

In practice the parameters in (11.6.12) will remain unknown. Replacing μ_i by $\bar{\mathbf{x}}_i$ and $\mathbf{\Sigma}_i$ by $\mathbf{S}_i$ we obtain the estimated quadratic classification score for the ith group,

$$\hat{d}_{is}^Q(\mathbf{x}) = -\frac{1}{2}\ln|\mathbf{S}_i| - \frac{1}{2}(\mathbf{x} - \bar{\mathbf{x}}_i)'\mathbf{S}_i^{-1}(\mathbf{x} - \bar{\mathbf{x}}_i) + \ln \pi_i, i = 1, \ldots, g. \tag{11.6.13}$$

The allocation rule becomes: assign $\mathbf{x}$ to G_k if

$$d_{ks}^Q(\mathbf{x}) = \max{}_{i=1,\ldots,g} d_i^Q(\mathbf{x}).$$

11.6.2 *Minimum ECM rule*

We now consider the minimum expected cost of misclassification (ECM) method. Let $C(k|i)$ be the cost of allotting an item to G_k when it actually belongs to $G_i (i \neq k)$. We have $C(i|i) = 0 (i = 1, \ldots, g)$.

The conditional expected cost of misclassifying an object of G_1 is

$$\begin{aligned} ECM(1) &= C(2|1)P(2|1) + C(3|1)P(3|1) + \ldots + C(g|1)P(g|1) \\ &= \textstyle\sum_{k=2}^{g} C(k|1)P(k|1). \end{aligned}$$

This conditional expected cost occurs with prior probability π_1. Similarly we can calculate $ECM(2), \ldots, ECM(g)$. The expected cost of misclassification is

$$\begin{aligned} ECM &= \textstyle\sum_{i=1}^{g} \pi_i ECM(i) \\ &= \textstyle\sum_{i=1}^{g} \pi_i \sum_{k(\neq i)=1}^{g} C(k|i)P(k|i). \end{aligned} \tag{11.6.14}$$

We want to choose mutually exclusive regions $R_1, \ldots, R_g$ such that (11.6.14) is minimum. This is given by the following theorem.

Theorem 11.6.2: The classification regions that minimize the ECM given in (11.6.14) are defined by allocating $\mathbf{x}$ to that group $G_k (k = 1, \ldots, g)$ for which

$$\sum_{i(\neq k)=1}^{g} \pi_i f_i(\mathbf{x}) C(k|i) \tag{11.6.15}$$

is smallest. If a tie occurs $\mathbf{x}$ can be assigned to any of the tied groups.

Proof. Follows as in Theorem 11.6.1 (Exercise 11.5).

If the misclassification costs are equal, the minimum ECM rule reduces to minimum TPM rule given in Theorem 11.6.1.

We have already seen that the minimum TPM rule is equivalent to the maximum posterior probability rule.

11.6.3 *Logistic classification*

The logistic method extends readily to more than two groups. Assume that

$$\ln[\frac{f_i(\mathbf{x})}{f_g(\mathbf{x})}] = \delta_i + \beta_i' \mathbf{x}, \ (i = 1, \ldots, g-1). \tag{11.6.16}$$

This implies that

$$
\begin{aligned}
q_i(\mathbf{x}) &= \frac{\pi_i f_i(\mathbf{x})}{f(\mathbf{x})} \\
&= \frac{\pi_i \exp[\delta_i + \beta_i' \mathbf{x}]}{\sum_{j=1}^{g-1} \pi_j \exp[\delta_j + \beta_j' \mathbf{x}] + \pi_j} \\
&= \frac{\exp]\delta_i + \beta_i' \mathbf{x} + \ln(\pi_i/\pi_g)]}{\sum_{j=1}^{g-1} \exp[\delta_j + \beta_j' \mathbf{x} + \ln(\pi_j/\pi_g)] + 1} \\
&= \frac{e^{z_i}}{1 + \sum_{j=1}^{g-1} e^{z_j}} = \frac{e^{z_i}}{\sum_{j=1}^{g} e^{z_j}}
\end{aligned} \tag{11.6.17}
$$

where

$$
\begin{aligned}
z_i &= z_i(\mathbf{x}) = \delta_i + \beta_i' \mathbf{x} + \ln(\pi_i/\pi_g) \ (i = 1, \ldots, g-1) \\
&= \beta_{0i} + \beta_i' \mathbf{x}, \\
z_g &= 0
\end{aligned} \tag{11.6.18}
$$

where $\delta_i + \ln(\pi_i/\pi_g) = \beta_{0i}$. We assign $\mathbf{x}$ to G_i if

$$\ln[q_i(\mathbf{x})/q_j(\mathbf{x})] \geq 0 \ \forall \ j = 1, \ldots, g,$$

i.e., if

$$z_i - z_j \geq 0$$

or

$$(\beta_{0i} - \beta_{0j}) + (\beta_i - \beta_j)' \mathbf{x} \geq 0 \ \forall \ j = 1, \ldots, g. \tag{11.6.19}$$

For conditional and mixture sampling we can find estimates of β_{0i} and β_i by maximizing the conditional likelihood function

$$L_C = \Pi_{i=1}^{g} \Pi_{j=1}^{m} \{q_i(\mathbf{x})\}^{n_i(\mathbf{x}_j)}.$$

This likelihood can also be used for separate sampling, except that it leads to the estimates of $\delta_i + \ln(n_i/n_g)$ instead of that of $\beta_{0i} = \delta_i + \ln(\pi_i/\pi_g)$.

EXAMPLE 11.6.3. Table 11.6.1 give the data on depth and flow of three streams at different locations. Here

$$\bar{\mathbf{x}}_1 = \begin{bmatrix} 0.4380 \\ 2.0770 \end{bmatrix}, \ \bar{\mathbf{x}}_2 = \begin{bmatrix} 0.8620 \\ 0.2829 \end{bmatrix}, \ \bar{\mathbf{x}}_3 = \begin{bmatrix} 0.6667 \\ 0.1954 \end{bmatrix}, \ \bar{\mathbf{x}} = \begin{bmatrix} 0.6552 \\ 0.8744 \end{bmatrix}$$

$$\mathbf{S}_p = \begin{bmatrix} 0.0123 & 0.1490 \\ 0.1490 & 2.1244 \end{bmatrix}, \ \mathbf{S}_p^{-1} = \begin{bmatrix} 537.731 & -37.715 \\ -37.715 & 3.116 \end{bmatrix}.$$

From (11.6.7), the classification function for the first group is

$$
\begin{aligned}
d_{1s}(\mathbf{x}) &= \bar{\mathbf{x}}_1' \mathbf{S}_p^{-1} \mathbf{x} - \tfrac{1}{2} \bar{\mathbf{x}}_1' \mathbf{S}_p^{-1} \bar{\mathbf{x}}_1 \\
&= (0.4380, \ 2.0770) \mathbf{S}_p^{-1} \mathbf{x} - \tfrac{1}{2} (.4380, \ 2.0770) \mathbf{S}_p^{-1} \begin{bmatrix} .4380 \\ 2.0770 \end{bmatrix} \\
&= -23.97 + 157.03 x_1 - 10.04 x_2.
\end{aligned}
$$

Similarly,

$$d_{2s}(\mathbf{x}) = -190.51 + 452.39x_1 - 31.60x_2,$$

$$d_{3s}(\mathbf{x}) = -114.52 + 350.76x_1 - 24.52x_2.$$

Suppose we want to classify the observation $\mathbf{x}_0 = (0.840,\ 0.172)'$. We find that

$$d_{1s}(\mathbf{x}_0) = 106.208,\ d_{2s}(\mathbf{x}_0) = 184.062,\ d_{3s}(\mathbf{x}_0) = 175.90.$$

Hence $\mathbf{x}_0$ is assigned to the group G_2 as $d_{2s}(\mathbf{x}_0)$ has the largest value. The observation $\mathbf{x}_0$ is actually the sixth observation in G_2 and its actual class is G_2.

As shown in (11.6.8),

$$\tilde{d}_{is}(\mathbf{x}_0) = -\frac{1}{2}\mathbf{x}_0'\mathbf{S}_p^{-1}\mathbf{x}_0 + d_{is}(\mathbf{x}_0) = (-\frac{1}{2})D_i^2(\mathbf{x}_0)$$

where $D_i^2(\mathbf{x}_0) = (\mathbf{x}_0 - \bar{\mathbf{x}}_i)'\mathbf{S}_p^{-1}(\mathbf{x}_0 - \bar{\mathbf{x}}_i)$ is the squared statistical distance between $\mathbf{x}_0$ and $\bar{\mathbf{x}}_i$. Clearly, one should allot $\mathbf{x}_0$ to the group for which $D_i^2(\mathbf{x}_o)$ is the smallest. Here, $D_1^2(\mathbf{x}_0) = 156.196, D_2^2(\mathbf{x}_0) = 0.488, D_3^2(\mathbf{x}_0) = 16.812$. Hence, $\mathbf{x}_0$ should be assigned to G_2.

Table 11.6.1: Stream Data

Depth 1 x_{11}	Flow 1 x_{12}	Depth 2 x_{21}	Flow 2 x_{22}	Depth 3 x_{31}	Flow 3 x_{32}
(1)	(2)	(3)	(4)	(5)	(6)
034	0.636	0.96	0.820	0.71	0.352
0.29	0.319	0.92	0.500	0.72	0.320
0.28	0.734	0.90	0.433	0.64	0.219
0.42	1.327	0.85	0.215	0.64	0.179
0.29	0.487	0.84	0.120	0.67	0.160
0.41	0.924	0.84	0.172	0.61	0.113
0.76	7.350	0.82	0.106	0.56	0.043
0.73	5.890	0.80	0.094	0.73	0.095
0.46	1.979	0.83	0.129	0.72	0.278
0.40	1.124	0.86	0.240		

11.7 Fisher's Method of Discrimination among $g \geq 2$ Populations

Suppose there are g groups (populations), each group being represented by a p-vector variable $\mathbf{X}_i (i = 1, \ldots, g)$. The characteristics measured by the variables are the same for each group. [Thus, in Fisher's experiment (Exercise 11.8), the variables are sepal length, sepal width, petal length, petal width for each of the three groups.] We want to find a smaller number of variables ($< p$) which will distinguish the groups as sharply as possible. Thus we want to find a few variables which will properly discriminate among the groups. This is Fisher's (1936) discriminant analysis.

We assume that

$$\begin{aligned} E(\mathbf{X}_i) &= \mu_i, \\ \text{Cov}\ (\mathbf{X}_i) &= \mathbf{\Sigma},\ i = 1, \ldots, g \end{aligned} \tag{11.7.1}$$

where $\mathbf{\Sigma}$ is a $p \times p$ matrix with full rank. We define

$$\begin{aligned} &\text{the overall group mean vector} = \bar{\mu} = \tfrac{1}{g} \textstyle\sum_{i=1}^{g} \mu_i, \\ &\text{between group SSP matrix} \quad = \mathbf{B}_\mu = \textstyle\sum_{i=1}^{g} (\mu_i - \bar{\mu})(\mu_i - \bar{\mu})'. \end{aligned} \tag{11.7.2}$$

Consider a linear transformation

$$Y_i = \alpha' \mathbf{X}_i,\ i = 1, \ldots, g \tag{11.7.3}$$

where α is a suitably chosen $p \times 1$ vector. Then

$$\begin{aligned} E(Y_i) &= \mu_{iY} = \alpha' \mu_i, i = 1, \ldots, g \\ V(Y_i) &= \alpha' \mathbf{\Sigma} \alpha = \sigma_Y^2 \text{ say }. \end{aligned} \tag{11.7.4}$$

The overall mean of $Y_1, Y_2, \ldots, Y_g$ is

$$\bar{\mu}_Y = \frac{1}{g} \sum_{i=1}^{g} \mu_{iY} = \frac{1}{g} \sum_{i=1}^{g} \alpha' \mu_i = \alpha' \bar{\mu}. \tag{11.7.5}$$

We want to find α such the the ratio of the sum of squares of the deviations of group means μ_{iY}'s from the overall mean $\bar{\mu}_Y$ to the variance of Y is maximum. This will ensure that the variables $Y_i (i = 1, \ldots, g)$ take on an average as distinct values as are as possible.

Specifically, we want to maximize

$$\begin{aligned} \lambda = \frac{\sum_{i=1}^{g} (\mu_{iY} - \bar{\mu}_Y)^2}{\sigma_Y^2} &= \frac{\sum_{i=1}^{g} (\alpha' \mu_i - \alpha' \bar{\mu})^2}{\alpha' \mathbf{\Sigma} \alpha} \\ &= \frac{\alpha' [\sum_{i=1}^{g} (\mu_i - \bar{\mu})(\mu_i - \bar{\mu})'] \alpha}{\alpha' \mathbf{\Sigma} \alpha} = \frac{\alpha' \mathbf{B}_\mu \alpha}{\alpha' \mathbf{\Sigma} \alpha}. \end{aligned} \tag{11.7.6}$$

The result is given by the following theorem.

Theorem 11.7.1: Let $\lambda_1 \geq \lambda_2 \geq \ldots \geq \lambda_s > 0$ denote the $s \leq \min\ (g - 1, p)$ nonzero eigenvalues of $\mathbf{\Sigma}^{-1} \mathbf{B}_\mu$ and $\eta_1, \eta_2, \ldots, \eta_s$ the corresponding eigenvectors scaled such that $\eta_i' \mathbf{\Sigma} \eta_i = 1\ \forall\ i$. Then the vector of coefficients that maximize the ratio (11.7.6) is given by $\alpha_1 = \eta_1$. The linear combination $\alpha_1' \mathbf{X}$ is called the first population discriminant function. The vector $\alpha_2 = \eta_2$

maximizes the ratio (11.7.6) subject to the condition Cov $(\alpha'\mathbf{X}, \alpha_2'\mathbf{X}) = 0$. The linear combination $\alpha_2'\mathbf{X}$ is called the second population discriminant. Continuing the vector $\alpha_k = \eta_k$ maximizes the ratio (11.7.6) subject to the conditions Cov $(\alpha_k'\mathbf{X}, \alpha_i'\mathbf{X}) = 0, i < k$ and $\alpha_k'\mathbf{X}$ is called the kth population discriminant. Also, $V(\alpha_i'\mathbf{X}) = 1,\ i = 1, \ldots, g$.

Proof. By Spectral Decomposition theorem (Theorem A.12.1) we can write

$$\mathbf{\Sigma} = \mathbf{P}'\mathbf{A}\mathbf{P}$$

where $\mathbf{A} =$ Diag $(\gamma_1, \ldots, \gamma_p), \gamma_i(> 0)$ being the eigenvalue of $\mathbf{\Sigma}$. Let $\mathbf{A}^{1/2} =$ Diag $(\sqrt{\gamma_1}, \ldots, \sqrt{\gamma_p})$. Then, by (A.12.4), the symmetric square root matrix $\mathbf{\Sigma}^{1/2} = \mathbf{P}\mathbf{A}^{1/2}\mathbf{P}$ and its inverse is $\mathbf{\Sigma}^{-1/2} = \mathbf{P}\mathbf{A}^{-1/2}\mathbf{P}$ (vide A.12.6).

Let

$$\mathbf{u} = \mathbf{\Sigma}^{1/2}\alpha. \tag{11.7.7}$$

Then

$$\begin{aligned} \alpha'\mathbf{\Sigma}\alpha &= (\alpha'\mathbf{\Sigma}^{1/2})'(\mathbf{\Sigma}^{1/2}\alpha) = \mathbf{u}'\mathbf{u}. \\ \alpha'\mathbf{B}_\mu\alpha &= (\alpha'\mathbf{\Sigma}^{1/2})(\mathbf{\Sigma}^{-1/2}\mathbf{B}_\mu\mathbf{\Sigma}^{-1/2})(\mathbf{\Sigma}^{1/2}\alpha) = \mathbf{u}'(\mathbf{\Sigma}^{-1/2}\mathbf{B}_\mu\mathbf{\Sigma}^{-1/2})\mathbf{u}. \end{aligned} \tag{11.7.8}$$

Therefore, the problem reduces to maximizing

$$\frac{\mathbf{u}'(\mathbf{\Sigma}^{-1/2}\mathbf{B}_\mu\mathbf{\Sigma}^{-1/2})\mathbf{u}}{\mathbf{u}'\mathbf{u}}$$

over $\mathbf{u}$. By the result A.8(14), the maximum of this ratio is λ_1, the largest eigenvalue of $\mathbf{\Sigma}^{-1/2}\mathbf{B}_\mu\mathbf{\Sigma}^{-1/2}$ and this maximum occurs when $\mathbf{u} = \eta_1^*$, the normalized eigenvector associated with λ_1 (We will show below, $\mathbf{\Sigma}^{-1}\mathbf{B}_\mu$ and $\mathbf{\Sigma}^{-1/2}\mathbf{B}_\mu\mathbf{\Sigma}^{-1/2}$ have the same eigenvalues).

Now,

$$\eta_1^* = \mathbf{u} = \mathbf{\Sigma}^{1/2}\alpha_1, \quad \text{or } \alpha_1 = \mathbf{\Sigma}^{-1/2}\eta_1^*. \tag{11.7.9}$$

Also,

$$\begin{aligned} V(\alpha_1'\mathbf{X}) &= \eta_1^{*'}\mathbf{\Sigma}^{-1/2}\mathbf{\Sigma}\mathbf{\Sigma}^{-1/2}\eta_1^* \\ &= \eta_1^{*'}\eta_1^* = 1. \end{aligned} \tag{11.7.10}$$

By A.8(15), $\mathbf{u} \perp \eta_1^*$ minimizes the ratio (11.7.6) when $\mathbf{u} = \eta_2^*$, the normalized eigenvector of $\mathbf{\Sigma}^{-1/2}\mathbf{B}_\mu\mathbf{\Sigma}^{-1/2}$ corresponding to λ_2. For this choice $\alpha_2 = \mathbf{\Sigma}^{-1/2}\eta_2^*$. Also,

$$\begin{aligned} \text{Cov }(\alpha_1'\mathbf{X}, \alpha_2'\mathbf{X}) &= \alpha_1'\mathbf{\Sigma}\alpha_2 = \eta_1^{*'}\mathbf{\Sigma}^{-1/2}\mathbf{\Sigma}\mathbf{\Sigma}^{-1/2}\eta_2^* \\ &= \eta_1^{*'}\eta_2^* = 0, \end{aligned} \tag{11.7.11}$$

since $\eta_2^* \perp \eta_1^*$. Also, $V(\alpha_2'\mathbf{X}) = 1$.

Continuing in this way we see that $\eta_1^{*'}\mathbf{X}, \eta_2^{*'}\mathbf{X}, \ldots, \eta_k^{*'}\mathbf{X}$, ... are the first, second, .., kth,.. discriminant functions of the populations (groups).

Now, if λ and η^* are the eigenvalue, eigenvector pair of $\mathbf{\Sigma}^{-1/2}\mathbf{B}_\mu\mathbf{\Sigma}^{-1/2}$, then

$$\mathbf{\Sigma}^{-1/2}\mathbf{B}_\mu\mathbf{\Sigma}^{-1/2}\eta^* = \lambda\eta^*.$$

Pre-multiplying both sides by $\mathbf{\Sigma}^{-1/2}$,

$$\mathbf{\Sigma}^{-1}\mathbf{B}_\mu\mathbf{\Sigma}^{-1/2}\eta^* = \lambda\mathbf{\Sigma}^{-1/2}\eta^*$$

or

$$\mathbf{\Sigma}^{-1}\mathbf{B}_\mu(\mathbf{\Sigma}^{-1/2}\eta^*) = \lambda(\mathbf{\Sigma}^{-1/2}\eta^*).$$

Therefore, $\mathbf{\Sigma}^{-1}\mathbf{B}_\mu$ has the same eigenvalues as $\mathbf{\Sigma}^{-1/2}\mathbf{B}_\mu\mathbf{\Sigma}^{-1/2}$ but its eigenvector is proportional to $\mathbf{\Sigma}^{-1/2}\eta^*$. Therefore, the eigenvectors of $\mathbf{\Sigma}^{-1}\mathbf{B}_\mu$ are given by $\eta = \mathbf{\Sigma}^{-1/2}\eta^* = \alpha$, as given in the theorem. □

Note 11.7.1: The ith discriminant variable

$$\begin{aligned} Y_i &= \alpha_i'\mathbf{X} \\ &= \eta_i'\mathbf{X} \\ &= \eta^{*'}\mathbf{\Sigma}^{-1/2}\mathbf{X}, i = 1, \ldots, s. \end{aligned} \tag{11.7.12}$$

They show the dimensions or directions of differences among $\mu_{1Y}, \ldots, \mu_{gY}$.

Note 11.7.2: Note that the number of nonzero eigenvalues of $\mathbf{\Sigma}^{-1}\mathbf{B}_\mu$ is $s \leq$ min $(p, g-1)$. This can be seen as follows. Since $\mathbf{\Sigma}^{-1}\mathbf{B}_\mu$ is a $p \times p$ matrix, $s \leq p$. Further, the g vectors $\mu_1 - \bar{\mu}, \ldots, \mu_g - \bar{\mu}$ satisfy $\sum_{i=1}^g(\mu_i - \bar{\mu}) = \mathbf{0}$. Hence the dimension of the space spanned by $\{\mu_1 - \bar{\mu}, \ldots, \mu_g - \bar{\mu}\}$ is $q \leq g-1$. Now, consider a vector $\mathbf{w}$ perpendicular to each of $\mu_1 - \bar{\mu}, \ldots, \mu_g - \bar{\mu}$, i.e. $(\mu_i - \bar{\mu})'\mathbf{w} = 0, \ \forall\ i = 1, \ldots, g$. Then

$$\mathbf{B}_\mu\mathbf{w} = \sum_{i=1}^g(\mu_i - \bar{\mu})(\mu_i - \bar{\mu})'\mathbf{w} = \mathbf{0}$$

or

$$\mathbf{\Sigma}^{-1}\mathbf{B}_\mu\mathbf{w} = \mathbf{0}. \tag{11.7.13}$$

There are $p - q$ such orthogonal eigenvectors $\mathbf{w}$ satisfying (11.7.12). Hence the maximum number of nonzero eigenvalues of $\mathbf{\Sigma}^{-1}\mathbf{B}_\mu$ is $q \leq g-1$. Therefore, $s \leq$ min $(p, g-1)$.

In general $\mu_i (i = 1, \ldots, g)$ and $\mathbf{\Sigma}$ will remain unknown and have to be estimated from the training sample. Let $\mathbf{x}_{ij} (j = 1, \ldots, n_i)$ be a sample of size n_i from the group G_i. Define $\bar{\mathbf{x}}_i = \sum_{j=1}^{n_i} \mathbf{x}_{ij}/n_i$, the sample mean for the ith group; $\bar{\mathbf{x}} = \sum_{i=1}^{g} n_i \bar{\mathbf{x}}_i/n, (n = \sum_i n_i)$, the overall sample mean;

$$\mathbf{S}_i = \frac{1}{n_i - 1} \sum_{j=1}^{n_i} (\mathbf{x}_{ij} - \bar{\mathbf{x}}_i)(\mathbf{x}_{ij} - \bar{\mathbf{x}}_i)',$$

the sample covariance for the group i;

$$\mathbf{S}_p = \frac{1}{n - g} \sum_{i=1}^{g} (n_i - 1)\mathbf{S}_i,$$

the pooled sample covariance. Also, define

$$\mathbf{B} = \sum_{i=1}^{g} n_i (\bar{\mathbf{x}}_i - \bar{\mathbf{x}})(\bar{\mathbf{x}}_i - \bar{\mathbf{x}})',$$

the sample between-group SSP matrix;

$$\begin{aligned} \mathbf{W} &= \textstyle\sum_{i=1}^{g} (n_i - 1)\mathbf{S}_i \\ &= \textstyle\sum_{i=1}^{g} \sum_{j=1}^{n_i} (\mathbf{x}_{ij} - \bar{\mathbf{x}}_i)(\mathbf{x}_{ij} - \bar{\mathbf{x}}_i)', \end{aligned}$$

the sample within-group SSP matrix. It is known that $\bar{\mathbf{x}}_i, \bar{\mathbf{x}}, \mathbf{S}_p = (n - g)^{-1}\mathbf{W}$ are unbiased estimates of $\mu_i, \bar{\mu}, \mathbf{\Sigma}$ respectively.

Replacing the parameters in (11.7.6) by their estimates, we have to find vector $\mathbf{a}$ such that

$$\frac{\mathbf{a}'\mathbf{B}\mathbf{a}}{\mathbf{a}'\mathbf{S}_p\mathbf{a}}$$

or equivalently,

$$\frac{\mathbf{a}'\mathbf{B}\mathbf{a}}{\mathbf{a}'\mathbf{W}\mathbf{a}} \tag{11.7.14}$$

is maximum. The result is given in the following theorem.

Theorem 11.7.2: Let $\hat{\lambda}_1, \hat{\lambda}_2, \ldots, \hat{\lambda}_s > 0$ be the $s \leq \min(p, g - 1)$ nonzero eigenvalues of $\mathbf{W}^{-1}\mathbf{B}$ and $\mathbf{e}_1, \ldots, \mathbf{e}_s$ be the corresponding eigenvectors (scaled so that $\mathbf{e}'\mathbf{S}_p\mathbf{e} = 1$). Then the vector of coefficients that maximize the ratio (11.7.14) is given by $\mathbf{a}_1 = \mathbf{e}_1 = \mathbf{W}^{-1/2}\mathbf{e}_1^*$, where $\mathbf{e}_1^*, \ldots, \mathbf{e}_s^*$ are the corresponding eigenvectors of $\mathbf{W}^{-1/2}\mathbf{B}\mathbf{W}^{-1/2}$. The linear combination $y_1 = \mathbf{a}_1'\mathbf{x} = \mathbf{e}_1'\mathbf{x} = \mathbf{e}_1^{*'}\mathbf{W}^{-1/2}\mathbf{x}$ is called the first sample discriminant. The choice $\mathbf{a}_2 = \mathbf{e}_2$ produces the second discriminant, $y_2 = \mathbf{a}_2'\mathbf{x} = \mathbf{e}_2^{*'}\mathbf{W}^{-1/2}\mathbf{x}$. In this way, we obtain the kth sample discriminant $\mathbf{a}_k'\mathbf{x} = \mathbf{e}_k'\mathbf{x} = \mathbf{e}_k^{*'}\mathbf{W}^{-1/2}\mathbf{x}$.

Proof. Follows as in Theorem 11.7.1.

Note 11.7.3: We have normalized the eigenvectors $\mathbf{e}_1, \ldots, \mathbf{e}_s$ such that $\mathbf{e}_i'\mathbf{S}_p\mathbf{e}_i = 1$ or $\mathbf{e}_i^{*'}\mathbf{W}^{-1/2}\mathbf{S}_p\mathbf{W}^{-1/2}\mathbf{e}_i^* = 1$ or $\mathbf{e}_i^{*'}\mathbf{e}_i^* = n - g, i = 1, \ldots, s$. Thus the sample discriminant functions are

$$y_i = \mathbf{e}_i^{*'}\mathbf{W}^{-1/2}\mathbf{x},$$

with its estimated variance

$$\hat{\text{Var}}\ (y_i) = V(\mathbf{e}_i^{*'}\mathbf{W}^{-1/2}\mathbf{x}) = \mathbf{e}_i^{*'}\mathbf{W}^{-1/2}\mathbf{S}_p\mathbf{W}^{-1/2}\mathbf{e}_i^* = 1, i = 1, \ldots, s.$$

Note 11.7.4: As seen in Subsection 11.4.1, the variables $x_1, \ldots, x_p$ should be commensurate for a meaningful interpretation of discriminating functions $y_1, \ldots, y_s$. If the variables are not commensurate they should be standardized. The value of the mth standardized function corresponding to observation $\mathbf{x}_{ij}$ is

$$\begin{aligned} y_{mij} &= a_{m1}^* z_{ij1} + a_{m2}^* z_{ij2} + \ldots + a_{mp}^* z_{ijp} \\ &(m = 1, \ldots, s; i = 1, \ldots, g; j = 1, \ldots, n_i) \end{aligned}$$

where $a_{im}^*(i = 1, \ldots, p)$ are the modified coefficients corresponding to the discriminant function in the original variable $y_{mij} = a_{m1}x_{ij1} + \ldots + a_p x_{ijp}$ and the standardized variables z_{ijk} have been defined in Subsection 11.4.1. Clearly,

$$a_{mr}^* = a_{mr}s_r \ (r = 1, \ldots, p)$$

where s_r is the within-group standard deviation obtained from the diagonal elements of $\mathbf{S}_p = \mathbf{W}/(n - g)$. □

The following theorem assesses the relative importance of different discriminating functions.

Theorem 11.7.3: The separatory measure

$$\begin{aligned} \Delta_g^2 &= \textstyle\sum_{i=1}^g (\mu_i - \bar{\mu})'\mathbf{\Sigma}^{-1}(\mu_i - \bar{\mu}) \\ &= \lambda_1 + \ldots + \lambda_p \\ &= \lambda_1 + \ldots \lambda_s \end{aligned} \tag{11.7.15}$$

where $\lambda_1, \ldots, \lambda_s$ are the nonzero eigenvalues of $\mathbf{\Sigma}^{-1}\mathbf{B}_\mu$.

Proof. Let $\eta_i^{*'}$ be the row vectors of $\mathbf{P}$ where $\eta_i^*(i = 1, \ldots, p)$ are the eigenvectors of $\mathbf{\Sigma}^{-1/2}\mathbf{B}_\mu\mathbf{\Sigma}^{-1/2}$. Consider

$$\mathbf{Y} = \begin{bmatrix} Y_1 \\ \cdot \\ \cdot \\ Y_s \\ \cdot \\ \cdot \\ Y_p \end{bmatrix} = \begin{bmatrix} \eta_1^{*'}\mathbf{\Sigma}^{-1/2}\mathbf{X} \\ \cdot \\ \cdot \\ \eta_s^{*'}\mathbf{\Sigma}^{-1/2}\mathbf{X} \\ \cdot \\ \cdot \\ \eta_p^{*'}\mathbf{\Sigma}^{-1/2}\mathbf{X} \end{bmatrix} = \mathbf{P}\mathbf{\Sigma}^{-1/2}\mathbf{X}. \tag{11.7.16}$$

Here, $Y_1, \ldots, Y_s$ are the first s discriminating functions. Consider

$$E(\mathbf{Y}|\mathbf{X} \in G_i) = \mu_{iY} = \begin{bmatrix} E(Y_1|G_i) \\ \cdot \\ \cdot \\ E(Y_s|G_i) \\ \cdot \\ \cdot \\ E(Y_p|G_i) \end{bmatrix} = \begin{bmatrix} \mu_{iY_1} \\ \cdot \\ \cdot \\ \mu_{iY_s} \\ \cdot \\ \cdot \\ \mu_{iY_p} \end{bmatrix} = \mathbf{P}\mathbf{\Sigma}^{-1/2}\mu_i. \quad (11.7.17)$$

Also,

$$\bar{\mu}_Y = (\bar{\mu}_{Y_1}, \ldots, \bar{\mu}_{Y_p})' = \frac{1}{g}\sum_{i=1}^{g} \mu_{iY} = \mathbf{P}\mathbf{\Sigma}^{-1/2}\bar{\mu}.$$

Now,

$$\begin{aligned}(\mu_{iY} - \bar{\mu}_Y)'(\mu_{iY} - \bar{\mu}_Y) &= (\mu_i - \bar{\mu})'\mathbf{\Sigma}^{-1}\mathbf{P}\mathbf{P}'\mathbf{\Sigma}^{-1/2}(\mu_i - \bar{\mu}) \\ &= (\mu_i - \bar{\mu})'\mathbf{\Sigma}^{-1}(\mu_i - \bar{\mu}).\end{aligned} \quad (11.7.18)$$

Hence,

$$\begin{aligned}\Delta_g^2 &= \textstyle\sum_{i=1}^{g}(\mu_{iY} - \bar{\mu}_Y)'(\mu_{iY} - \bar{\mu}_Y) \\ &= \Delta_{g1}^2 + \Delta_{g2}^2 + \ldots + \Delta_{gg}^2,\end{aligned}$$

where

$$\begin{aligned}\Delta_{gi}^2 &= (\mu_{Yi} - \bar{\mu}_Y)'(\mu_{Yi} - \bar{\mu}_Y) \\ &= (\mu_{iY_1} - \bar{\mu}_{Y_1})^2 + \ldots + (\mu_{iY_p} - \bar{\mu}_{Y_p})^2.\end{aligned}$$

Hence,

$$\begin{aligned}\Delta_g^2 &= \textstyle\sum_{i=1}^{g}\sum_{j=1}^{p}(\mu_{iY_j} - \bar{\mu}_{Y_j})^2 \\ &= \textstyle\sum_{j=1}^{p}\sum_{i=1}^{g}(\mu_{iY_j} - \bar{\mu}_{Y_j})^2 \\ &= \textstyle\sum_{j=1}^{p}\sum_{i=1}^{g}[\eta_j^{*'}\mathbf{\Sigma}^{-1/2}(\mu_i - \bar{\mu})]^2 \\ &= \textstyle\sum_{j=1}^{p}\eta_j^{*'}\mathbf{\Sigma}^{-1/2}[\sum_{i=1}^{g}(\mu_i - \bar{\mu})(\mu_i - \bar{\mu})']\mathbf{\Sigma}^{-1/2}\eta_j^* \\ &= \textstyle\sum_{j=1}^{p}\eta_j^{*'}\mathbf{\Sigma}^{-1/2}\mathbf{B}_\mu\mathbf{\Sigma}^{-1/2}\eta_j^* \\ &= \textstyle\sum_{j=1}^{p}\lambda_j\end{aligned} \quad (11.7.19)$$

because λ_j is the eigenvalue corresponding to η_j^*.

If $\lambda_{s+1} = \ldots = \lambda_p = 0, \Delta_g^2 = \lambda_1 + \ldots + \lambda_s$. If only the first r discriminants are used, $\Delta_g^2 \approx \sum_{i=1}^{r}\lambda_i$. □

Note 11.7.5: The relative importance of different discriminating functions becomes apparent from the breakdown of Δ_g^2 in terms of the eigenvalues $\lambda_1, \ldots, \lambda_s$. The first discriminant $Y_1 = \eta_1^{*'}\mathbf{\Sigma}^{-1/2}\mathbf{X}$ makes the largest single contribution, λ_1 to the separative measure Δ_g^2. In general, the ith discriminant, $Y_i = \eta_i^{*'} \mathbf{\Sigma}^{-1/2}\mathbf{X}$, contributes λ_i to Δ_g^2. The relative performance of each individual discriminant function Y_i can be assessed by considering the eigenvalue as a proportion of the total,

$$\frac{\lambda_i}{\sum_{j=1}^{s} \lambda_j}. \tag{11.7.20}$$

If the last $s - r$ eigenvalues are such that $\sum_{i=1}^{s-r} \lambda_{r+i}$ is small compared to $\sum_{j=1}^{s} \lambda_j$, then the last discriminants $Y_{r+1}, \ldots, Y_s$ can be neglected without appreciably decreasing the amount of separation. Generally, two or three discriminant functions will suffice to describe the group differences.

However, the parameters in Δ_g^2 will ordinarily remain unknown and have to be estimated. We have therefore, the corresponding result from the sample. The statistic

$$\begin{aligned} \delta_g^2 &= \textstyle\sum_{i=1}^{g} n_i(\bar{\mathbf{x}}_i - \bar{\mathbf{x}})\mathbf{S}_p^{-1}(\bar{\mathbf{x}}_i - \bar{\mathbf{x}}) = \hat{\lambda}_1 + \ldots \hat{\lambda}_g \\ &= \hat{\lambda}_1 + \ldots + \hat{\lambda}_s \end{aligned} \tag{11.7.21}$$

where $\hat{\lambda}_i (i = 1, \ldots, p)$ are the eigenvalues of $\mathbf{W}^{-1}\mathbf{B}$. The relative importance of a sample discriminating function is $y_i = \mathbf{e}_i^{*'} \mathbf{W}^{-1/2}\mathbf{x}$ is measured by $\hat{\lambda}_i / \sum_{j=1}^{s} \hat{\lambda}_j$. If the sum of the last few eigenvalues $\hat{\lambda}_r$'s is small compared to $\sum_{j=1}^{s} \hat{\lambda}_j$, the corresponding discriminant functions can be neglected without appreciably decreasing the amount of separation.

Note 11.7.6: The software 'SPSS' gives under 'discriminant analysis' Fisher's discriminant functions (11.7.12), while 'Minitab' provides 'Classification functions' (11.6.4) for each group.

11.7.0.1 *Scatter Plots*

The discriminant functions reduce the dimension of the problem. The number of large eigenvalues of $\mathbf{W}^{-1}\mathbf{B}$ indicate the dimension of the space spanned by the mean vectors $\bar{\mathbf{x}}_i (i = 1, \ldots, g)$. In many data sets, the first two eigenvalues $\hat{\lambda}_1, \hat{\lambda}_2$ account for most of the $\sum_{j=1}^{s} \hat{\lambda}_j$-value and consequently the sample mean vectors can be plotted in a two-dimensional plane.

The discriminant scores

$$y_{1ij} = \mathbf{a}_1' \mathbf{x}_{ij},\ y_{2ij} = \mathbf{a}_2' \mathbf{x}_{ij}, i = 1, \ldots, g; j = 1, \ldots, n_i$$

can be plotted as a scatter diagram in $y_1 - y_2$ plane, along with the mean scores

$$\bar{y}_{1i} = \mathbf{a}_1' \bar{\mathbf{x}}_1, \bar{y}_{i2} = \mathbf{a}_2' \bar{\mathbf{x}}_i, i = 1, 2.$$

We note that the eigenvalues of $\mathbf{W}^{-1}\mathbf{B}$ indicate the dimensionality of $\bar{\mathbf{x}}_1, \ldots \bar{\mathbf{x}}_g$ and not of $\mathbf{x}_{ij}, i = 1, \ldots, g; j = 1, \ldots, n_i$. The dimensionality of the individual observations $\mathbf{x}_{ij}$ is p, although the actual dimension may be less, that is, because the observations are correlated the points may have more scatter in some dimensions and less in some. The scores (y_{1ij}, y_{2ij}) of the points $\mathbf{x}_{ij}$ represent the projection of $\mathbf{x}_{ij}$ on the two-dimensional discriminant space (assuming $s = 2$).

We noted that the discriminant functions are uncorrelated. However, they are not orthogonal. Thus the angle between $\mathbf{a}_1$ and $\mathbf{a}_2$ is not 90^0. However, the ordinary practice is to plot the discriminant functions on a rectangular co-ordinate system. It is expected that this would not give rise to serious distortion.

EXAMPLE 11.7.1: Consider the stream data of Table 11.6.1. Here, $g = 3, n_1 = n_2 = 10, n_3 = 9$.

$$\mathbf{B} = \begin{bmatrix} .901 & -3.905 \\ -3.905 & 22.110 \end{bmatrix}, \quad \mathbf{W} = \begin{bmatrix} .320 & 3.874 \\ 3.874 & 55.235 \end{bmatrix}.$$

Eigenvectors of $\mathbf{W}$ are:

$$\mathbf{w}_1' = (.070025,\ .997545)', \mathbf{w}_2 = (.997545,\ -.070025)'$$

with corresponding eigenvalues $\omega_1 = 55.5069, \omega_2 = 0.0481$. Hence,

$$\mathbf{W}^{-1/2} = \mathbf{M}' \text{ Diag } (\omega_1^{-1/2},\ \omega_2^{-1/2})\mathbf{M}$$
$$= \begin{bmatrix} 4.54009 & -0.30924 \\ -0.30924 & 0.15593 \end{bmatrix}$$

where $\mathbf{M} = [\mathbf{w}_1,\ \mathbf{w}_2]$. Also,

$$\mathbf{W}^{-1/2}\mathbf{B}\mathbf{W}^{-1/2} = \begin{bmatrix} 31.6533 & -5.4695 \\ -5.4695 & 1.0004 \end{bmatrix}.$$

Its eigenvalues are

$$\hat{\lambda}_1 = 32.6,\ \hat{\lambda}_2 = 0.0537$$

with the corresponding eigenvectors

$$\mathbf{e}_1^* = [.985349,\ -.170552],\ \mathbf{e}_2^* = [.170552,\ .985349].$$

Hence, eigenvectors of $\mathbf{W}^{-1}\mathbf{B}$ are

$$\mathbf{e}_1 = \mathbf{W}^{-1/2}\mathbf{e}_1^* = [4.52632,\ 0.33135]$$
$$\mathbf{e}_2 = \mathbf{W}^{-1/2}\mathbf{e}_2^* = [.46975,\ .10090].$$

The discriminant functions are

$$y_1 = 4.52632x_1 + 0.33135x_2,$$
$$y_2 = .46975x_1 + .10090x_2.$$

We note that $\lambda_2 / \sum_i \lambda_i = .001531$ is very small and hence λ_2 can be neglected. Therefore, the second discriminant function can be neglected.

11.7.1 *Fisher's discriminant procedure for classification*

The above discriminant procedure can be used for classifying a unit selected from the population into one of the g groups.

Let $Y_k = \mathbf{a}_k'\mathbf{X}$ be the kth discriminant, $k \leq s$. Then if $\mathbf{X} \in G_i, E(\mathbf{Y}|\mathbf{X} \in G_i) = \mu_{iY}$ as noted in (11.7.17).

Also, as seen in Theorem 11.7.1, $V(Y_k) = 1$, Cov $(Y_k, Y_l) = 0 (k \neq l = 1, \ldots, s)$ whatever be the group membership of $\mathbf{X}$.

Therefore, whenever $\mathbf{X} \in G_i$, a measure of distance between $\mathbf{y}$ and μ_{iY} is

$$\begin{aligned}(\mathbf{y} - \mu_{iY})'(\mathbf{y} - \mu_{iY}) &= \textstyle\sum_{j=1}^{s}(y_j - \mu_{iY_j})^2 \\ &= \textstyle\sum_{j=1}^{s}(\alpha_j'\mathbf{x} - \alpha_j'\mu_i)^2 \\ &= \textstyle\sum_{j=1}^{s}[\alpha_j'(\mathbf{x} - \mu_i)]^2.\end{aligned} \tag{11.7.22}$$

An appropriate rule will then be to allot $\mathbf{x}$ to the group G_k if

$$\sum_{j=1}^{s}[\alpha_j'(\mathbf{x} - \mu_k)]^2 \leq \sum_{j=1}^{s}[\alpha_j'(\mathbf{x} - \mu_i)]^2, i = 1, \ldots, g. \tag{11.7.23}$$

Replacing the parameters by their estimates, the rule is to allot $\mathbf{x}$ to G_k if

$$\sum_{j=1}^{s}[\mathbf{e}_j'(\mathbf{x} - \bar{\mathbf{x}}_k)]^2 \leq \sum_{j=1}^{s}[\mathbf{e}_j'(\mathbf{x} - \bar{\mathbf{x}}_i)]^2, \; i = 1, \ldots, g. \tag{11.7.24}$$

where $\mathbf{e}_j (j = 1, \ldots, s)$ are the nonzero eigenvalues of $\mathbf{W}^{-1}\mathbf{B}$.

11.7.2 *Relation between normal theory method and Fisher's discrimination method*

In case of equal covariance normal populations we have seen in equation (11.6.9) that the minimum TPM rule is to allot $\mathbf{x}$ to G_k for which

$$\begin{aligned}\tilde{d}_k(\mathbf{x}) &= -\tfrac{1}{2}D_k^2(\mathbf{x}) + \ln \pi_k \\ &= -\tfrac{1}{2}(\mathbf{x} - \mu_k)'\mathbf{\Sigma}^{-1}(\mathbf{x} - \mu_k) + \ln \pi_k\end{aligned}$$

is the largest among $d_1(\mathbf{x}), \ldots, d_g(\mathbf{x})$. We shall show that the rule (11.6.9) and (11.7.23) are equivalent. For this we first consider a theorem.

Theorem 11.7.4: Let $y_j = \alpha_j'\mathbf{x}$ where $\alpha_j = \mathbf{\Sigma}^{-1/2}\eta_j^*$, and η_j^* is the eigenvector of $\mathbf{\Sigma}^{-1/2}\mathbf{B}_\mu\mathbf{\Sigma}^{-1/2}$. Then

$$\begin{aligned}\textstyle\sum_{j=1}^{p}(y_j - \mu_{iY_j})^2 &= \textstyle\sum_{j=1}^{p}[\alpha_j'(\mathbf{x} - \mu_i)]^2 \\ &= (\mathbf{x} - \mu_i)'\mathbf{\Sigma}^{-1}(\mathbf{x} - \mu) \\ &= -2d_i(\mathbf{x}) + \mathbf{x}'\mathbf{\Sigma}^{-1}\mathbf{x} + 2\ln \pi_i.\end{aligned} \tag{11.7.25}$$

If $\lambda_1 \geq \ldots \geq \lambda_s > 0 = \lambda_{s+1} = \ldots = \lambda_p, \sum_{j=s+1}^{p}(y_j - \mu_{iY_j})^2$ is constant for all groups $i = 1, \ldots, g$. Therefore, only the first s discriminants y_j or $\sum_{j=1}^{s}(y_j - \mu_{iY_j})^2$ contribute to the classification rule that uses the sum of squares on the left side of (11.7.25).

Proof. Let $\mathbf{E}^* = [\eta_1^*, \eta_2^*, \ldots, \eta_g^*]$. Then $\mathbf{E}^*\mathbf{E}^{*'} = \mathbf{I}_p$. Now

$$(\mathbf{x} - \mu_i)'\mathbf{\Sigma}^{-1}(\mathbf{x} - \mu_i)$$

$$= (\mathbf{x} - \mu_i)'\mathbf{\Sigma}^{-1/2}\mathbf{E}^*\mathbf{E}^{*'}\mathbf{\Sigma}^{-1/2}(\mathbf{x} - \mu_i). \tag{11.7.26}$$

Again,

$$\mathbf{E}^{*'}\mathbf{\Sigma}^{-1/2}(\mathbf{x} - \mu_i) = \begin{bmatrix} \alpha_1'(\mathbf{x} - \mu_i) \\ \cdot \\ \cdot \\ \alpha_p'(\mathbf{x} - \mu_i) \end{bmatrix} = \begin{bmatrix} y_1 - \mu_{iY_1} \\ \cdot \\ \cdot \\ y_p - \mu_{iY_p} \end{bmatrix}. \tag{11.7.27}$$

Hence, from (11.7.26) and (11.7.27)

$$\begin{aligned}(\mathbf{x} - \mu_i)'\mathbf{\Sigma}^{-1}(\mathbf{x} - \mu_i) &= \textstyle\sum_{j=1}^{p}[\alpha_j'(\mathbf{x} - \mu_i)]^2 \\ &= -2d_i(\mathbf{x}) + \mathbf{x}'\mathbf{\Sigma}^{-1}\mathbf{x} + 2\ln \pi_i\end{aligned}$$

(by (11.7.25)). Again, each $\alpha_j = \mathbf{\Sigma}^{-1}\eta_j^*, j > s$ is an eigenvector of $\mathbf{\Sigma}^{-1}\mathbf{B}_\mu$ with eigenvalue zero. Therefore,

$$\mathbf{\Sigma}^{-1}\mathbf{B}_\mu\alpha_j = \mathbf{\Sigma}^{-1}\sum_{i=1}^{g}(\mu_i - \bar{\mu})(\mu_i - \bar{\mu})'\alpha_j = 0, j > s.$$

Therefore, α_j is orthogonal to each $\mu_i - \bar{\mu}$ and hence to $(\mu_k - \bar{\mu}) - (\mu_i - \bar{\mu}) = (\mu_i - \mu_k)$ for every $i, k = 1, \ldots, g$.

Now,

$$\alpha_j'(\mu_k - \mu_i) = \mu_{kY_j} - \mu_{iY_j} = 0.$$

Hence,

$$y_j - \mu_{kY_j} = y_j - \mu_{iY_j}, j > s, i, k = 1, \ldots, g.$$

Therefore $\sum_{j=s+1}^{p}(y_j - \mu_{iY_j})^2$ is constant for all the groups $i = 1, \ldots, g$. □

Hence, when the prior probabilities $\pi_1 = \ldots = \pi_g = 1/g$, maximizing $d_i(\mathbf{x})$ in the minimum TPM rule is equivalent to minimizing $\sum_{j=1}^{s}[\alpha_j'(\mathbf{x}-\mu_i)]^2$ as in (11.7.23). Similar result holds for their sample counterparts. Therefore, the two rules are equivalent in this case.

If further, only, $r < s$ discriminants are used for classification, there is a loss of amount $\sum_{j=r+1}^{s}[\alpha_j'(\mathbf{x} - \mu_i)]^2$ in the sum of squares to be compared for each group G_i (and for the sample the loss is of amount $\sum_{j=r+1}^{s}[\mathbf{a}_j'(\mathbf{x}-\bar{\mathbf{x}}_i)]^2$ for each group).

EXAMPLE 11.7.2: Consider the data of Example 11.7.1. Suppose we want to classify the observation $\mathbf{x}_0 = (0.840,\ 0.172)'$. For this observation $y_1(\mathbf{x}_0) = 3.85910, y_2(\mathbf{x}_0) = 0.411944$. Denoting by $\bar{y}_{ij}$ the mean score of the discriminant function y_j for the ith group $(i = 1, 2; j = 1, 2)$, we have

$$\bar{y}_{11} = 2.671,\ \bar{y}_{12} = 0.415,$$

$$\bar{y}_{21} = 3.9954,\ \bar{y}_{22} = 0.4335,$$

$$\bar{y}_{31} = 3.0823,\ \bar{y}_{32} = 0.3329.$$

We have

$$(y_1(\mathbf{x}_0) - \bar{y}_{11})^2 + (y_2(\mathbf{x}_0) - \bar{y}_{12})^2 = 10.1640,$$

$$(y_1(\mathbf{x}_0) - \bar{y}_{21})^2 + (y_2(\mathbf{x}_0) - \bar{y}_{22})^2 = .01862,$$

$$(y_1(\mathbf{x}_0) - \bar{y}_{31})^2 + (y_2(x_0) - \bar{y}_{32})^2 = 0.6096.$$

Hence, the observation $\mathbf{x}_0$ is assigned to the group G_2. The same observation was also classified under G_2 by the classification rule (11.6.7) in example 11.6.4.

11.7.3 *Tests of significance*

We noticed in Theorem 11.7.1 that the discriminating ratio $\mathbf{a}'\mathbf{B}\mathbf{a}/\mathbf{a}'\mathbf{W}\mathbf{a}$ is maximized by $\hat{\lambda}_1$, the largest eigenvalue of $\mathbf{W}^{-1}\mathbf{B}$ and the remaining eigenvalues $\hat{\lambda}_2, \ldots, \hat{\lambda}_s$ correspond to other discriminant functions. In testing for the equality of mean vectors $\mu_1, \ldots, \mu_g$ in sampling from g multinormal populations with a common covariance matrix in Subsection 5.11.1, we considered Wilk's statistic

$$\begin{aligned}\Lambda(p; m_E, m_h)_1 &= \frac{|\mathbf{W}|}{|\mathbf{W}+|\mathbf{B}|} \\ &= \Pi_{i=1}^{s}\frac{1}{1+\hat{\lambda}_i}\end{aligned} \tag{11.7.28}$$

where $m_E = n - g, m_H = g - 1, n = \sum_{i=1}^{g} n_i$. (vide equation (4.4.3)). Therefore, all the nonzero eigenvalues of $\mathbf{W}^{-1}\mathbf{B}$ occur in Λ_1 in (11.7.28). It is seen that Λ_1 is small if one or more eigenvalues $\hat{\lambda}$'s are large. Thus Wilk'a Λ tests for the significance of eigenvalues λ's and hence for the discriminant functions.

The s eigenvalues represent s dimensions of separation of mean vectors $\bar{\mathbf{x}}_1, \ldots, \bar{\mathbf{x}}_g$. We are interested in finding which, if any, of these dimensions are significant.

We can use Bartlett's χ^2 approximation of Λ_1 (vide (4.4.9))

$$\begin{aligned} V_1 &= [(n-g) - \tfrac{1}{2}(p-g+2)] \ln \Lambda_1 \\ &= -[n-1-\tfrac{1}{2}(p+g)] \textstyle\sum_{i=1}^{s} \ln(1+\hat{\lambda}_i) \end{aligned} \tag{11.7.29}$$

which approximately follows $\chi^2_{p(g-1)}$ distribution. The test statistic Λ_1 and its approximation (11.7.29) test the significance of all of $\lambda_1, \ldots, \lambda_s$. If this test leads to rejection of $H_0(\lambda_1 = \ldots = \lambda_s = 0)$ we conclude that at least one of the λ's is significantly different from zero. Since λ_1 is the largest we are sure of its significance along with that of the first discriminant function $Y = \alpha'\mathbf{X}$.

To test the significance of $\lambda_2, \ldots, \lambda_s$ we delete $\hat{\lambda}_1$ from Λ_1 and the associated χ^2 approximation to get

$$\begin{aligned} \Lambda_2 &= \Pi_{i=2}^{s} \tfrac{1}{1+\hat{\lambda}_i}, \\ V_2 &= [n-1-\tfrac{1}{2}(p+g)] \textstyle\sum_{i=2}^{s} \ln(1+\hat{\lambda}_i) \end{aligned} \tag{11.7.30}$$

which is approximately χ^2 with $(p-1)(g-2)$ d.f. If this test leads to rejection of H_0 we conclude that at least λ_2 is significant along with the associated discriminating function $Y_2 = \alpha'\mathbf{X}$. We can continue in this fashion, testing each λ in turn until a test fails to reject H_0. The test statistic for the mth step is

$$\Lambda_m = \Pi_{i=m}^{s} \frac{1}{1+\hat{\lambda}_i} \tag{11.7.31}$$

which is distributed as $\Lambda(p-m+1, n-g-m+1, g-m)$. The statistic

$$\begin{aligned} V_m &= -[n-1-\tfrac{1}{2}(p+g)] \ln \Lambda_m \\ &= [n-1-\tfrac{1}{2}(p+g)] \textstyle\sum_{i=m}^{s} \ln(1+\hat{\lambda}_i) \end{aligned} \tag{11.7.32}$$

has an approximate χ^2 distribution with $(p-m+1)(g-m)$ d.f.

If $\hat{\lambda}_i / \sum_j \hat{\lambda}_j$ is small, the associated discriminant functions may not be of interest, even if it is significant.

Following (4.4.10) and (4.4.11) we can use an approximate F-test for each Λ_i.

As in the case of $g = 2$ groups we note that significant discrimination does not necessarily imply that the classification is good. Effectiveness of classification procedures are evaluated by non-parametric measures, like APER,

e_c, e_{boot}, independently of tests of separation. Classification is ordinarily not attempted unless the groups are well separated as evidenced by the above-mentioned tests of significance of λ's. If no significant differences among the mean vectors are found, constructing classification rules may not be worthwhile.

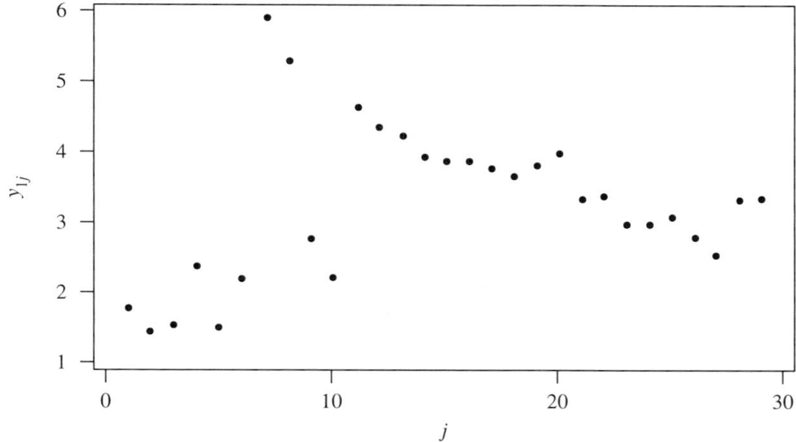

Fig. 11.2: Graph of the first discriminant score

EXAMPLE 11.7.3: Consider again the case of Example 11.7.1. Here $\Lambda_1 = \sum_{i=1}^{2} \frac{1}{1+\lambda_i} = 0.9788, \Lambda_2 = \frac{1}{1+\lambda_2} = 0.9490$. From (11.7.25),

$$\begin{aligned} V_1 &= \tfrac{51}{2}[\ln(1+\lambda_1) + \ln(1+\lambda_2)] = 90.95 \\ &> 9.49 = \chi^4_{4;.05}. \end{aligned}$$

Hence, the first discriminant function effectively separates the groups. Again

$$V_2 = \frac{51}{2}[\ln(1+\lambda_2)] = 1.3391 \ < 3.84 = \chi^2_{1;.05}.$$

Therefore, the second discriminant function is insignificant and can be ignored. Figure 11.2 shows the (first) discriminant score for each observation.

11.7.4 *Contribution of variables in separation of groups*

Contribution of variables in the separation of groups can be assessed in three ways: (i) by standardization of their coefficients in the discriminant functions, (ii) by the partial F statistic of the variables (iii) by the correlation of each variable with the discriminating function. These are also the tools by which discriminant functions are interpreted. We consider them below.

Standardization of coefficients

We have already discussed in Subsections 11.4.1 and Note 11.7.3 the standardization of discriminant coefficients. The standardized coefficients are so adjusted that they apply to the standardized variables $z_{ijk}, i = 1, \ldots, g; j = 1, \ldots, n_i; k = 1, \ldots, p$. The standardized variables are scale free and the standardized coefficients $a_1^*, \ldots, a_p^*$, therefore, correctly reflect the joint contribution of the variables to the discriminant function y as it maximally separates the groups. The discriminant function coefficient vector is an eigenvalue of $\mathbf{W}^{-1}\mathbf{B}$ and as such it takes into account the correlation coefficients among the variables. The coefficients indicate the influence of each variable in the presence of others.

As the standardization is carried out for each of s discriminant functions, pattern of coefficients of the variables $x_1, \ldots, x_p$ will vary and each discriminant function will have a different interpretation. We may rank the variables in order of their contribution to separating the groups, by ranking the average of the absolute values of the coefficients.

Clearly, the contribution of a variable may change if new variables are added or some existing variables are removed. Again, if n/p is too small, the coefficients may be very unstable over different samples.

Partial F-values For any variable x_k we may calculate a partial F-value giving a measure of separation provided by x_k in the presence of the other variables. After completing the partial F-value for each of the variables, the variables can be ranked. If the partial F-value is insignificant, the contribution of x_k to the separation is trivial and as such the variable may be removed from consideration. The partial F also throws light on the choice of variables, considered in Section 11.8.

For two groups, the partial F is given by

$$F = (m_E - p + 1)\frac{T_p^2 - T_{p-1}^2}{m_E + T_{p-1}^2} \tag{11.7.33}$$

where T_p^2 is the two-sample Hotelling's T^2 for $H_0(\mu_1 = \ldots = \mu_p)$ with all p variables, T_{p-1}^2 is the T^2-statistic with all variables, except x_k and $m_E = n_1 + n_2 - 2$. The F-statistic in (11.7.33) is distributed as F_{1,m_E-p+1}.

For the case of more than two groups, the partial Λ for x_k adjusted for the other $p-1$ variables is given by

$$\Lambda(x_k|x_1, \ldots, x_{k-1}, x_{k+1}, \ldots, x_p) = \Lambda_{(k)} = \frac{\Lambda_p}{\Lambda_{p-1}}, \tag{11.7.34}$$

where Λ_p is Wilk's $\Lambda(= \frac{|\mathbf{W}|}{|\mathbf{B}+\mathbf{W}|})$ given by (11.7.28) for all the p variables and Λ_{p-1} involves all variables except x_k. The corresponding partial F is given by

$$F = \frac{1-\Lambda}{\Lambda}.\frac{m_E - p + 1}{m_H}, \tag{11.7.35}$$

where Λ is defined in (11.7.34), $m_E = n - g$ and $m_H = g - 1$. The partial Λ-statistic in (11.7.34) is distributed as $\Lambda(1, m_E - p + 1, m_H)$ and partial F in (11.7.35) as F_{m_H, m_E-p+1}.

Unlike the standardized discrimination function coefficients, the partial F value of a variable gives a measure of the overall contribution of the variables to the separation of groups, taking into account all the s discriminants. However, the partial F values will often rank the variables in the same order as the standardized coefficients in the first discriminant function, specially if $\hat{\lambda}_1 / \sum_{j=1}^{s} \hat{\lambda}_j$ is very large,

A partial index of association for x_k can be defined as

$$R_k^2 = 1 - \Lambda_{(k)}, \ k = 1, \ldots, p \tag{11.7.36}$$

where $\Lambda_{(k)}$ is the partial Λ in (11.7.35) for x_k. This partial R_k^2 is a measure of association between the grouping variable and x_k after adjusting for the other variables.

Correlation between discriminant functions and variables

It is commonly believed that the correlation between a discriminant function and a variable is the best measure of importance of the variable in the discriminant function. It is claimed that this correlation is a better measure of contribution of the variable in the discriminating function in separating the groups in the presence of the other variables than the standardized coefficient of the variable when the variables are commensurate. The correlations are often referred to as 'loadings' or 'structure coefficients' and are provided in many softwares. But the works of Rencher (e.g. 1996) and others have shown that that the correlations in fact show the contributions of each individual variables in a univariate way and does not reflect the contribution of this variable in the presence of other variables. Thus the correlations are misleading if used for the interpretation of discriminant functions.

11.8 Selection of Variables

In the classification analysis, often a large number of variables is measured so as not to miss any variable that might help in predicting the group-membership. However, this approach is likely to include some superfluous variables that do not contribute to allocation. For simplicity and parsimony we should wish to remove the redundant variables. A reduction in the number of variables will often lead to the increase in robustness of linear and quadratic discriminant functions and an improved error rate. Moreover, the cost of measurement is generally reduced if p is reduced. One should therefore choose a subset properly.

First consider the case $g = 2$. Let $\mathbf{x} = (\mathbf{x}^{(1)'}, \mathbf{x}^{(2)'})'$ where $\mathbf{x}^{(1)}$ is of order $k \times 1$. We may wish to test whether the subvector of variables $\mathbf{x}^{(1)}$ will discriminate the groups as good as $\mathbf{x}$. We assume the samples in the two groups arise from multivariate normal populations with equal covariance matrix.

Let $\delta = (\mu_1 - \mu_2), \delta^{(1)} = (\mu^{(1)} - \mu^{(2)})$. Then Mahalanobis squared distances for p and k variables, respectively are $\Delta_p^2 = \delta'\mathbf{\Sigma}^{-1}\delta$ and $\Delta_k^2 = \delta^{(1)'}\mathbf{\Sigma}_{11}^{-1}\delta^{(1)}$. A test for $H_0(\Delta_p^2 = \Delta_k^2)$ is given by

$$F = \frac{n_1 + n_2 - p - 1}{p - k}\left(\frac{D_p^2 - D_k^2}{c + D_k^2}\right) \tag{11.8.1}$$

where

$$D_p^2 = (\bar{\mathbf{x}}_1 - \bar{\mathbf{x}}_2)'\mathbf{S}_p^{-1}(\bar{\mathbf{x}}_1 - \bar{\mathbf{x}}_2),$$

D_k^2 has a similar expression and $c = (n_1 + n_2)(n_1 + n_2 - 2)/(n_1 n_2)$. When H_0 is true, $F \cap F_{(p-k, n_1+n_2-p-1)}$.

Now, by the result in (3.2.20),

$$(\mu_1-\mu_2)'\mathbf{\Sigma}^{-1}(\mu_1-\mu_2) = (\mu_1^{(1)}-\mu_2^{(1)})'\mathbf{\Sigma}_{11}^{-1}(\mu_1^{(1)}-\mu_2^{(1)})+\mathbf{C}'\mathbf{\Sigma}_{22.1}^{-1}\mathbf{C} \quad (11.8.2)$$

where

$$\mathbf{C} = \{\mu_1^{(2)} - \mu_2^{(2)} - \mathbf{\Sigma}_{21}\mathbf{\Sigma}_{11}^{-1}(\mu_1^{(1)} - \mu_2^{(1)})\}. \quad (11.8.3)$$

(Exercise 11.6). Hence H_0 is true if and only if $\mathbf{C} = \mathbf{0}$, that is,

$$\mu_1^{(2)} - \mathbf{\Sigma}_{21}\mathbf{\Sigma}_{11}^{-1}\mu_1^{(2)} = \mu_2^{(2)} - \mathbf{\Sigma}_{21}\mathbf{\Sigma}_{11}^{-1}\mu_2^{(1)}. \quad (11.8.4)$$

Therefore, testing H_0 is equivalent to testing that the conditional distribution of $\mathbf{x}^{(2)}$ given $\mathbf{x}^{(1)}$ is the same for both the groups. This amounts to testing that $\mathbf{x}^{(1)}$ is sufficient for discrimination, as $\mathbf{x}^{(2)}$ provides no further information about the same.

We shall consider the procedures that delete or add variables one at a time. For these procedures the partial F-statistic (11.8.1) for testing $H_0(\Delta_k^2 = \Delta_{k+1}^2)$ for $g = 2$ or the partial F statistic (11.7.35) based on Wilk's Λ for $g > 2$ is used.

In the *forward selection* method the variable entered at each step is the one that maximizes the partial F-statistic. The variables are brought in one at a time and are not tested for removal once they are in the subset. We thus obtain the maximal additional separation of groups above and beyond the separation already attained by the other variables. Since at each step we are dealing with the maximum of several correlated F-values, the choice of significance level for testing becomes a problem. Costanza and Afifi (1979) suggested using a significance level of $0.10 \leq \alpha \leq 0.25$ (preferably the larger values) for testing $\Delta_k^2 = \Delta_{k+1}^2$ in the forward selection procedure. In the *backward selection* procedure the variable that contributes least is deleted at each step, as indicated by the partial F-value.

Stepwise selection is a combination of the forward and backward approaches. Here the $F-$ statistic (11.8.1) for testing $\Delta_k^2 = \Delta_{k+1}^2$ for $g = 2$ or the partial $F-$ statistic (11.7.30) based on Wilk's Λ for $g > 2$ is used to serve as an F-to-enter or F-to-remove. At any given stage the variable with the largest partial F is added to the current subset if its partial Fvalue is larger than F_{IN}, a specified threshold. After a variable has been entered, all the variables in the subset are reexamined and the one with the smallest F to remove is deleted if its F-value is less than F_{OUT}, a second threshold. At each step we do not try to to eliminate the variable that was just entered or add the variable that was just eliminated. This can be ensured by choosing $F_{IN} \geq F_{OUT}$.

Any of the above procedures is commonly known as *stepwise discriminant analysis.* We are in effect doing stepwise multivariate analysis of variance (MANOVA). No discriminant function is being calculated at any step. After the subset-selection is completed, we can calculate the discriminant functions for the selected variables. The variables can also be used in a classification analysis.

11.9 Exercises and Complements

11.1 Prove the relation (11.2.19).

11.2 Prove the relation (11.2.22).

11.3 For the binary data of Example 11.3.4, find the linear discriminant function (LDF) (11.3.5).

11.4 Prove the relation (11.3.31).

11.5 Prove the Theorem 11.6.2.

11.6 Prove the relation (11.8.2).

11.7 The rows of the following matrices give observations in two groups.

$$\mathbf{X}_1 = \begin{bmatrix} 12 & 10 \\ 16 & 10 \\ 14 & 12 \end{bmatrix}, \begin{bmatrix} 24 & 32 \\ 20 & 16 \\ 18 & 27 \end{bmatrix}.$$

Calculate the linear discriminant function (11.3.5). On the basis of this rule reclassify all the observations and obtain apparent error rate. Also, calculate cross-validation error rate. A new observation $\mathbf{x}_0 = (17, 24)'$ has been obtained. Classify this observation.

11.8 The table 11.E.2 give measurements on three species of iris. Find optimal classification functions. Also, perform discriminant analysis. An iris has measurements, $(6.1, 2.8, 4.7, 1.2)'$. Classify the observation.

11.9 The table 11.E.1 show the marks in Mathematics (x) and Statistics (y) of two batches of students in an examination. Find the discriminant function (11.4.10). With the help of this reclassify all the observations. Hence, find the apparent error rate. A new observation $\mathbf{x}_0 = (52, 70)'$

has been obtained. Classify the same. Also, test the significance of the discriminant function.

11.10 The table E.11.3 gives measurements of blood glucose level on three occasions for 52 workers: (i) Fasting (x_1, x_2, x_3), (ii) One hour after sugar intake (y_1, y_2, y_3). Perform discriminant analysis.

Table 11.E.1: Score Data

x_1	y_1	x_2	y_2
(1)	(2)	(3)	(4)
12	34	56	54
45	56	67	66
44	35	49	72
52	63	89	97
8	32	58	76
39	48	53	32
71	84	56	87
38	57	78	98
38	51	64	78
47	62	58	40

Table 11.E.2: Data on Irises

Group	Sepal length x_1	Sepal width x_2	Petal length x_3	Petal width x_4
(1)	(2)	(3)	(4)	(5)
Iris setosa	5.1	3.5	1.4	0.2
	4.9	3.0	1.4	0.2
	4.7	3.2	1.3	0.2
	4.6	3.1	1.5	0.2
	5.0	3.6	1.4	0.2
	5.4	3.9	1.7	0.4
	4.6	3.4	1.4	0.3
	5.0	3.4	1.5	0.2
	4.4	2.9	1.4	0.2
	4.9	3.1	1.5	0.1
	5.4	3.7	1.5	0.2

(Continued)

(Table 11.E.2 continued)

Group	Sepal length x_1	Sepal width x_2	Petal length x_3	Petal width x_4
(1)	(2)	(3)	(4)	(5)
	4.8	3.4	1.6	0.2
	4.8	3.0	1.4	0.1
	4.3	3.0	1.1	0.1
	5.8	4.0	1.2	0.2
	5.7	4.4	1.5	0.4
Iris versicolor	7.0	3.2	4.7	1.4
	6.4	3.2	4.5	1.5
	6.9	3.1	4.9	1.5
	5.5	2.3	4.0	1.3
	6.5	2.8	4.6	1.5
	5.7	2.8	4.5	1.3
	6.3	3.3	4.7	1.6
	4.9	2.4	3.3	1.0
	6.6	2.9	4.6	1.3
	5.2	2.7	3.9	1.4
	5.0	2.0	3.5	1.0
	5.9	3.0	4.2	1.5
	6.0	2.2	4.0	1.0
	6.1	2.9	4.7	1.4
	5.6	2.9	3.6	1.3
Iris virginica	6.3	3.3	6.0	2.5
	5.8	2.7	5.1	1.9
	7.1	3.0	5.9	2.1
	6.3	2.9	5.6	1.8
	6.5	3.0	5.8	2.2
	7.6	3.0	6.6	2.1
	4.9	2.5	4.5	1.7
	7.3	2.9	6.3	1.8
	7.2	3.6	6.1	2.5
	6.5	3.2	5.1	2.0
	6.4	2.7	5.3	1.9
	6.8	3.0	5.5	2.1

(Table abridged from Anderson (1939))

Table 11.E.3: Blood-Glucose Data

x_1	x_2	x_3	y_1	y_2	y_3
(1)	(2)	(3)	(4)	(5)	(6)
60	69	62	97	69	98
56	53	84	103	78	107
80	69	76	66	99	130
55	80	90	80	85	114
62	75	68	116	130	91
74	64	66	77	102	130
73	70	64	115	110	109
68	67	75	76	85	119
69	82	74	72	133	127
60	67	71	130	134	121
70	74	78	150	158	100
66	74	78	150	131	142
83	70	74	99	98	105
48	77	75	113	124	97
66	93	97	136	112	122
74	70	76	109	88	105
60	74	76	109	88	105
60	74	71	72	90	71
63	75	66	130	101	90
6	80	86	130	117	144
77	67	74	83	92	107
70	67	100	150	142	146
73	76	81	119	120	119
78	90	77	122	133	149
73	68	80	102	90	122
72	83	68	104	69	96
5	60	70	119	94	89
52	70	76	92	94	100
68	66	90	119	85	109
78	73	75	164	98	138
103	77	77	160	117	121
77	68	74	144	71	153
60	77	68	77	82	89

(Continued)

(Table 11.E.3 continued)

x_1	x_2	x_3	y_1	y_2	y_3
(1)	(2)	(3)	(4)	(5)	(6)
70	70	72	114	93	122
75	65	71	77	70	109
91	74	93	118	115	150
66	75	73	170	147	121
75	82	76	153	132	115
74	71	68	143	105	100
76	70	64	114	113	129
74	90	86	73	106	116
74	77	80	116	81	77
67	71	69	63	87	70
78	75	80	105	132	80
64	66	71	83	94	133
67	71	68	63	87	70
78	75	80	105	132	80
64	66	71	83	94	133
71	80	76	81	87	86
63	75	73	120	89	59
90	103	74	107	109	101
60	76	61	99	111	98

(Source: O'Sullivan and Mahan, 1966)

Table 11.E.4: Flea Beetles Data

x_1	x_2	x_3	x_4	y_1	y_2	y_3	y_4
(1)	(2)	(3)	(4)	(5)	(6)	(7)	(8)
189	245	137	163	181	305	184	209
192	260	132	217	158	237	133	188
217	276	141	192	184	300	166	231
221	299	142	213	171	273	162	213
171	239	128	158	181	297	163	224
192	262	147	173	181	308	160	223
213	278	136	201	177	301	166	221
192	255	128	185	198	308	141	197

(Continued)

(Table 11.E.4 continued)

x_1	x_2	x_3	x_4	y_1	y_2	y_3	y_4
(1)	(2)	(3)	(4)	(5)	(6)	(7)	(8)
170	244	128	192	180	286	146	214
201	276	146	186	177	299	171	192
195	242	128	192	176	317	166	213
205	263	147	192	192	312	166	209
180	252	121	167	176	285	141	200
192	283	138	183	169	287	162	214
200	294	138	188	164	265	147	192
192	277	150	177	181	308	157	204
200	287	136	173	192	276	154	209
181	255	146	183	175	271	140	19
				197	303	170	205

(Source: A.A.Lubischew, 1962)

11.11 The Table 11.E.4 gives four measurements on two species of flea beetles, namely Haltica oleracea (Group G_1) and *Haltica carduorum* (Group G_2). The measurements are: $x_1(y_1)$ = the distance of the transverse groove from the posterior border of the prothorax (in microns); $x_2(y_2)$ = the length of the elytra (in .01 mm); $x_3(y_3)$ = the length of the second antennal joint (in microns); $x_4(y_4)$ = the length of the third antennal joint (in microns). Perform discriminant analysis.

Appendix A: Matrix Theory

We review in this chapter some of the results in the theory of matrices and determinants. Most of these results appear without proof. For details, the reader may refer to, Gantmacher (1959), Hans (1973), Searle (1982), among many others.

A.1 Types of Matrices

A rectangular array of real or complex numbers is called a matrix. It will be assumed throughout that the numbers are real. Let

$$\mathbf{A} = ((a_{ij})) = \begin{bmatrix} a_{11} & a_{12} & \dots & a_{1n} \\ \cdot & \cdot & \dots & \cdot \\ \cdot & \cdot & \dots & \cdot \\ a_{p1} & a_{p2} & \dots & a_{pn} \end{bmatrix}$$

denote a $p \times n$ matrix. If $p = n, \mathbf{A}$ is a square matrix. If $n = 1, \mathbf{A}$ is a column vector ($p \times 1$), if $p = 1, \mathbf{A}$ is a row vector ($1 \times n$). If $a_{ij} = 0\ \forall i, j, \mathbf{A}$ is a zero (null) matrix. If $n = p$ and $a_{ij} = 1(0)$ if $i = j$ (otherwise), $\mathbf{A}$ is an identity matrix, often denoted as $\mathbf{I}_n$.

A square matrix whose off-diagonal elements $a_{ij} = 0\ \forall\ i \neq j$ is called a diagonal matrix,

$$\text{Diag}\ (a_{11}, \dots, a_{pp}).$$

The elements just to the right (left) of those on the diagonals of a square matrix are called the elements of a super (sub) diagonal, while those along the diagonal are called the elements of principal diagonal.

The transpose of $\mathbf{A}$ is the matrix obtained by interchanging the rows and columns of $\mathbf{A}$ and is denoted by $\mathbf{A}'$. A square matrix $\mathbf{A}$ is symmetric if

$\mathbf{A} = \mathbf{A}'$. An orthogonal matrix $\mathbf{A}$ is a $p \times p$ matrix for which

$$\mathbf{A}\mathbf{A}' = \mathbf{A}'\mathbf{A} = \mathbf{I}_p.$$

EXAMPLE A.1.1: The following matrix called *Helmert matrix* is an orthogonal matrix.

$$\mathbf{A} = \begin{bmatrix} \frac{1}{\sqrt{p}} & \frac{1}{\sqrt{1.2}} & \frac{1}{\sqrt{2.3}} & \cdots & \frac{1}{\sqrt{(p-1)p}} \\ \frac{1}{\sqrt{p}} & -\frac{1}{\sqrt{1.2}} & -\frac{1}{\sqrt{2.3}} & \cdots & --\frac{1}{\sqrt{(p-1)p}} \\ \frac{1}{\sqrt{p}} & 0 & -\frac{2}{\sqrt{2.3}} & \cdots & -\frac{1}{\sqrt{(p-1)p}} \\ . & . & . & \cdots & . \\ \frac{1}{\sqrt{p}} & 0 & 0 & \cdots & -\frac{(p-1)}{\sqrt{(p-1)p}} \end{bmatrix}.$$

This matrix has also the property that all the columns except the first sum to zero.

A $p \times p$ symmetric matrix whose off-diagonal elements are common and whose diagonal elements are also common is said to a matrix of the *intra-class correlation pattern.*

$$\mathbf{A} = \begin{bmatrix} a & b & \ldots b \\ b & a & \ldots b \\ . & . & \ldots. \\ b & b & \ldots a \end{bmatrix}, \; b \geq -a/(p-1). \tag{A.1.1}$$

A *Jacobi* matrix is a square matrix in which the elements not on the principal, super or sub-diagonals are zeros.

A symmetric matrix in which all the elements except those on the super, principal and sub-diagonals are zeros is called a *tri-diagonal* matrix. This is given by

$$\begin{bmatrix} a & b & 0 & 0 & \ldots 0 & 0 \\ b & a & b & 0 & \ldots 0 & 0 \\ 0 & b & a & b & \ldots 0 & 0 \\ . & . & . & . & \ldots. & . \\ 0 & 0 & 0 & 0 & \ldots a & b \\ 0 & 0 & 0 & 0 & \ldots b & a \end{bmatrix}. \tag{A.1.2}$$

A *circular* matrix is one which has the structure

$$\begin{bmatrix} a_1 & a_2 & \ldots a_p \\ a_p & a_1 & \ldots a_{p-1} \\ . & . & \ldots. \\ a_2 & a_3 & \ldots a_1 \end{bmatrix}. \tag{A.1.3}$$

A.2 Matrix Algebra

The following rules hold.

(1) Let $\mathbf{A} = ((a_{ij})), \mathbf{B} = ((b_{ij}))$, both of order $p \times n$. Then $\mathbf{A} \pm \mathbf{B} = \mathbf{C}$, where $\mathbf{C} = ((c_{ij})), c_{ij} = a_{ij} \pm b_{ij}$.

(2) Let $\mathbf{D} : p \times n, \mathbf{E} : n \times q$. Then the product $\mathbf{F}((f_{ij})) = \mathbf{DE}$ is defined and

$$f_{ij} = \sum_{\alpha=1}^{n} d_{i\alpha} e_{\alpha j}, i = 1, \ldots, p; \quad j = 1, \ldots, q.$$

Note that even for square matrices it is not generally true that $\mathbf{AB} = \mathbf{BA}$.

(3) The following properties hold.

(i) $c\mathbf{A} = ((ca_{ij}))$ where c is a scalar.
(ii) $(\mathbf{A} + \mathbf{B})' = \mathbf{A}' + \mathbf{B}'$.
(iii) $(\mathbf{AB})' = \mathbf{B}'\mathbf{A}'$ if $\mathbf{A}, \mathbf{B}$ are conformable for multiplication.
(iv) $(\mathbf{A}')' = \mathbf{A}$.
(v) $\mathbf{A}(\mathbf{B} + \mathbf{C}) = \mathbf{AB} + \mathbf{AC}$.
(vi) $\mathbf{AI} = \mathbf{A}$ for all $\mathbf{A}$.

A.3 Vectors, Vector Spaces

As noted in Section A.1, a row vector is a $1 \times n$ matrix and a column vector is a $n \times 1$ matrix. Unless otherwise stated we shall denote by $\mathbf{a}$ the column vector

$$\mathbf{a} = \begin{bmatrix} a_1 \\ a_2 \\ \cdot \\ \cdot \\ a_n \end{bmatrix} = [a_1, a_2, \ldots, a_n]'.$$

Geometrically, a vector $\mathbf{a}$ is a point in the n-dimensional Euclidean space R^n with co-ordinates $x_1 = a_1, \ldots, x_n = a_n$.

Vector addition: The sum of two vectors $\mathbf{a} = (a_1, \ldots, a_n)'$ and $\mathbf{b} = (b_1, \ldots, b_n)'$ is

$$\mathbf{c} = (c_1, \ldots, c_n)' = \mathbf{a} + \mathbf{b},$$

where $c_i = a_i + b_i$.

Scalar multiplication: Let c be an arbitrary scalar and $\mathbf{a} = (a_1, \ldots, a_n)'$. Then the product $c\mathbf{a} = (ca_1, \ldots, ca_n)'$.

Inner product of two vectors: The *inner* or *dot* product of two vectors $\mathbf{a} = (a_1, \ldots, a_n)'$ and $\mathbf{b} = (b_1, \ldots, b_n)'$ is given by $\mathbf{a}'\mathbf{b} = \mathbf{b}'\mathbf{a} = a_1b_1 + a_2b_2 + \ldots a_nb_n$.

Outer or tensor product of two vectors: The *outer* or *tensor* product of two vectors $\mathbf{a}$ and $\mathbf{b}$ is a $n \times n$ matrix $\mathbf{ab}'$.

Vectors have both length and direction.

DEFINITION A.3.1: *Length of a vector*: The length of the column vector $\mathbf{a}$ is defined as

$$L_{\mathbf{a}} = ||\mathbf{a}|| = \sqrt{\mathbf{a}'\mathbf{a}} = \sqrt{\sum_{i=1}^{n} a_i^2}.$$

Clearly, the length of the vector $\mathbf{a}/L_{\mathbf{a}} = \mathbf{a}/\sqrt{\sum_{k=1}^{n} a_k^2}$ is unity.

The length of the vector $c\mathbf{a}$, where c is a scalar is

$$L_{c\mathbf{a}} = |c|L_{\mathbf{a}}.$$

Multiplication of a vector $\mathbf{a}$ by a scalar c does not change its direction if $c > 0$ and creates a vector with direction opposite to that of $\mathbf{a}$ if $c < 0$. The vector $\mathbf{a}/L_{\mathbf{a}}$ has unit length and the same direction as $\mathbf{a}$.

DEFINITION A.3.2: *Distance between two vectors*: Distance between two vectors $\mathbf{a}$ and $\mathbf{b}$ is

$$||\mathbf{a} - \mathbf{b}|| = \{\sum_{i=1}^{n} (a_i - b_i)^2\}.$$

DEFINITION A.3.3: *Angle between two vectors*: The angle θ between two $n \times 1$ vectors $\mathbf{a}$ and $\mathbf{b}$ is given by

$$\cos\theta = \frac{\mathbf{a}'\mathbf{b}}{||\mathbf{a}||||\mathbf{b}||}.$$

Hence, two vectors $\mathbf{a}$ and $\mathbf{b}$ are perpendicular (orthogonal) to each other, if $\mathbf{a}'\mathbf{b} = 0$.

DEFINITION A.3.4: *Line passing through two vectors*: A straight line passing through two vectors $\mathbf{a}, \mathbf{b}$ is represented by

$$\mathbf{x} = \lambda\mathbf{a} + (1 - \lambda)\mathbf{b},$$

where $\lambda \in (0, 1)$.

The equation of a line from origin to $\mathbf{a}$ is $\mathbf{x} = \lambda\mathbf{a}$.

The direction cosine vector of a line from $\mathbf{0}$ to $\mathbf{a}$ is given by $(\cos\theta_1, \ldots, \cos\theta_n)$ where $\cos\theta_i = a_i/||\mathbf{a}||$.

DEFINITION A.3.5: *Projection of a vector*: The projection of a vector $\mathbf{a}(n \times 1)$ on a vector $\mathbf{b}(n \times 1)$ is given by

$$\frac{(\mathbf{a}'\mathbf{b})}{\mathbf{b}'\mathbf{b}}\mathbf{b} = \frac{(\mathbf{a}'\mathbf{b})}{L^2_{\mathbf{b}}}.\mathbf{b}.$$

The length of the projection vector is

$$L_{\mathbf{a}}|\cos\theta|$$

where θ is the angle between $\mathbf{a}$ and $\mathbf{b}$.

If $\mathbf{b}$ has unit length so that $L_{\mathbf{b}} = 1$, projection of $\mathbf{a}$ on $\mathbf{b}$ is $(\mathbf{a}'\mathbf{b})\mathbf{b}$ (Figure A.1).

Cauchy Schwarz Inequality: For two vectors $\mathbf{a}, \mathbf{b}$, the following inequality holds:

$$(\mathbf{a}'\mathbf{b})^2 \leq (\mathbf{a}'\mathbf{a})(\mathbf{b}'\mathbf{b}),$$

the equality is satisfied, *iff* either $\mathbf{a}$ is proportional to $\mathbf{b}$ or $\mathbf{a} = \mathbf{0}$.

(For an extension of the inequality see Section A.14.)

DEFINITION A.3.6: The space of all real $n \times 1$ vectors, with vector addition and scalar multiplication defined above, is called a *vector space*. Thus R^n can be considered as a vector space over the set of real numbers R.

DEFINITION A.3.7: If W is a subset of R^n such that for all $\mathbf{x}, \mathbf{y} \in W$ and scalar $c \in R$,

$$c(\mathbf{x} + \mathbf{y}) \in W,$$

then W is called a *vector subspace* of R^n.

Two simple examples of vector subspaces of R^n are $\{\mathbf{0}\}$ and R^n itself.

DEFINITION A.3.8: The vector $\mathbf{b} = c_1\mathbf{a}_1 + c_2\mathbf{a}_2 + \ldots + c_k\mathbf{a}_k$ is a linear combination of vectors $\mathbf{a}_1, \mathbf{a}_2, \ldots, \mathbf{a}_k$. The set of all linear combinations of $\mathbf{a}_1, \ldots, \mathbf{a}_k$ is the linear space spanned by $\mathbf{a}_1, \ldots, \mathbf{a}_k$.

The linear space spanned by $\mathbf{a}_1, \ldots, \mathbf{a}_k$ will be denoted as $L(\mathbf{a}_1, \ldots, \mathbf{a}_k)$.

DEFINITION A.3.9: A set of vectors $\mathbf{a}_1 \ldots, \mathbf{a}_k$ is said to be *linearly dependent* if there exist numbers $c_1, \ldots, c_k$ not all zeros such that

$$c_1\mathbf{a}_1 + \ldots c_k\mathbf{a}_k = \mathbf{0}.$$

Otherwise, the set of vectors $\mathbf{a}_1, \ldots, \mathbf{a}_k$ is said to be linearly dependent.

EXAMPLE A.3.1: If one of the vectors, say $\mathbf{a}_1$ is $\mathbf{0}$, the set is a dependent set.

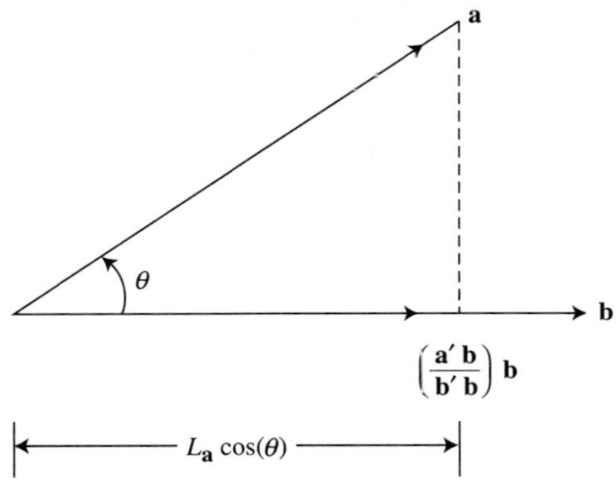

Fig. A.1: The projection of **a** on **b**.

EXAMPLE A.3.2: The set of vectors

$$\mathbf{e}_1 = [1, 0, 0, 0]', \quad \mathbf{e}_2 = [0, 1, 0, 0]',$$

$$\mathbf{e}_3 = [0, 0, 1, 0]', \quad \mathbf{e}_4 = [0, 0, 0, 1]'$$

are linearly independent.

DEFINITION A.3.10: Let W be a subspace of R^n. Then a basis for W is the maximal set of linearly independent vectors, such that each member of W can be expressed as a linear combination of the members of the basis for W. The following properties hold for a basis of W.

(1) Every basis of W contains the same number of members. This number is called the dimension of W and is often denoted as dim(W). The dimension of R^n is n.
(2) Every vector $\mathbf{x} \in W$ can be expressed as a unique linear combination of the elements (vectors) of a given basis for W.

DEFINITION A.3.11: A basis $\{\mathbf{x}_1, \ldots, \mathbf{x}_m\}$ of a m-dimensional subspace W of R^n is called *orthonormal* , if all elements $\mathbf{x}_i (i = 1, \ldots, m)$ have norm unity and are orthogonal to one other, i.e. $\mathbf{x}_i'\mathbf{x}_j = 1(0)$ if $i = (\neq)0$.

If $\mathbf{A}(n \times n)$ is an orthogonal matrix, then the columns of $\mathbf{A}$ form an orthonormal basis of R^n.

The following theorem gives a method of constructing an orthonormal basis for a vector subspace.

Theorem A.3.1: Gram-Schmidt Procedure. Given linearly independent vectors $\mathbf{a}_1, \ldots, \mathbf{a}_k$, we can construct mutually orthogonal vectors $\mathbf{u}_1, \ldots, \mathbf{u}_k$ which span the same linear space as $\mathbf{a}_1, \ldots, \mathbf{a}_k$, i.e. $L(\mathbf{a}_1, \ldots, \mathbf{a}_k) = L(\mathbf{u}_1, \ldots, \mathbf{u}_k)$. This is obtained by setting

$$\begin{aligned}
\mathbf{u}_1 &= \mathbf{a}_1, \\
\mathbf{u}_2 &= \mathbf{a}_2 - \tfrac{(\mathbf{a}_2'\mathbf{u}_1)}{\mathbf{u}_1'\mathbf{u}_1}\mathbf{u}_1, \\
. &= \ldots \\
\mathbf{u}_k &= \mathbf{a}_k - \tfrac{(\mathbf{a}_k'\mathbf{u}_1)}{\mathbf{u}_1'\mathbf{u}_1}\mathbf{u}_1 - \ldots - \tfrac{(\mathbf{a}_k'\mathbf{u}_{k-1})}{\mathbf{u}_{k-1}'\mathbf{u}_{k-1}}\mathbf{u}_{k-1}.
\end{aligned}$$

The vectors $\mathbf{u}_1, \ldots, \mathbf{u}_k$ may be normalized by setting $\mathbf{z}_j = \mathbf{u}_j/\sqrt{\mathbf{u}_j'\mathbf{u}_j}$, $(j = 1, \ldots, k)$. The vectors $\mathbf{z}_1, \ldots, \mathbf{z}_k$ form an orthonormal basis of $L(\mathbf{a}_1, \ldots, \mathbf{a}_k)$. Here, the projection of $\mathbf{a}_k$ on $\mathbf{z}_j$ is $(\mathbf{a}_k'\mathbf{z}_j)\mathbf{z}_j$ and $\sum_{j=1}^{k-1}(\mathbf{a}_k'\mathbf{z}_j)\mathbf{z}_j$ is the projection of $\mathbf{a}_k$ on the linear span $L(\mathbf{a}_1, \ldots, \mathbf{a}_{k-1})$.

A.4 Partitioned Matrices

A matrix $\mathbf{A}$ of the form

$$\mathbf{A} = \begin{bmatrix} \mathbf{A}_{11} & \mathbf{A}_{12} & \ldots & \mathbf{A}_{1k} \\ \ldots & \ldots & \ldots & \ldots \\ \mathbf{A}_{k1} & \mathbf{A}_{k2} & \ldots & \mathbf{A}_{kk} \end{bmatrix}$$

where $\mathbf{A}_{ij}$'s are submatrices of $\mathbf{A}$ is called a *partitioned matrix.*

If

$$\mathbf{A} = \begin{bmatrix} \mathbf{A}_{11} & \mathbf{A}_{12} \\ \mathbf{A}_{21} & \mathbf{A}_{22} \end{bmatrix}, \quad \mathbf{B} = \begin{bmatrix} \mathbf{B}_{11} & \mathbf{B}_{12} \\ \mathbf{B}_{21} & \mathbf{B}_{22} \end{bmatrix} \qquad (A.4.1)$$

where $\mathbf{A}_{11}$ is $k \times l$, $\mathbf{A}_{12}$ is $k \times (q-l)$, $\mathbf{A}_{21}$ is $(p-k) \times l$, $\mathbf{A}_{22}$ is $(p-k) \times (q-l)$ and $\mathbf{B}_{ij}$'s are of similar order,

$$\mathbf{A} + \mathbf{B} = \begin{bmatrix} \mathbf{A}_{11} + \mathbf{B}_{11} & \mathbf{A}_{12} + \mathbf{B}_{12} \\ \mathbf{A}_{21} + \mathbf{B}_{21} & \mathbf{A}_{22} + \mathbf{B}_{22} \end{bmatrix}.$$

Also, if $\mathbf{C}$ is $q \times r$ matrix partitioned as

$$\mathbf{C} = \begin{bmatrix} \mathbf{C}_{11} & \mathbf{C}_{12} \\ \mathbf{C}_{21} & \mathbf{C}_{22} \end{bmatrix},$$

where $\mathbf{C}_{11}$ is $l \times m, \mathbf{C}_{12}$ is $l \times (r-m), \mathbf{C}_{21}$ is $(q-l) \times m, \mathbf{C}_{22}$ is $(q-l) \times (r-m)$, then

$$\mathbf{AC} = \begin{bmatrix} \mathbf{A}_{11} & \mathbf{A}_{12} \\ \mathbf{A}_{21} & \mathbf{A}_{22} \end{bmatrix} \begin{bmatrix} \mathbf{C}_{11} & \mathbf{C}_{12} \\ \mathbf{C}_{21} & \mathbf{C}_{22} \end{bmatrix}$$

$$= \begin{bmatrix} \mathbf{A}_{11}\mathbf{C}_{11} + \mathbf{A}_{12}\mathbf{C}_{21} & \mathbf{A}_{11}\mathbf{C}_{12} + \mathbf{A}_{12}\mathbf{C}_{22} \\ \mathbf{A}_{21}\mathbf{C}_{11} + \mathbf{A}_{22}\mathbf{C}_{21} & \mathbf{A}_{21}\mathbf{C}_{12} + \mathbf{A}_{22}\mathbf{C}_{22} \end{bmatrix}.$$

For partitioned $\mathbf{A}$, as in (A.4.1),

$$\mathbf{A}' = \begin{bmatrix} \mathbf{A}'_{11} & \mathbf{A}'_{21} \\ \mathbf{A}'_{12} & \mathbf{A}'_{22} \end{bmatrix}.$$

If $\mathbf{A}$ is a symmetric matrix, $a_{ij} = a_{ji}$ so that

$$\mathbf{A}' = \mathbf{A}.$$

A.5 Determinants

A square matrix $\mathbf{A} = ((a_{ij}))$ of order m, has associated with it a scalar real-valued function of its elements, called its *determinant* denoted by $|\mathbf{A}|$,

$$|\mathbf{A}| = \sum \pm a_{1i_1} a_{2i_2} \dots a_{mi_m},$$

where the summation is taken over all the $m!$ permutations $(i_1, i_2, \dots, i_m)$ of the integers $(1, \dots, m)$. The plus (minus) sign applies if the numbers of subscripts rearranged in permutation is even or zero (otherwise).

Elementary properties

(1) $|\mathbf{A}| = 0$ if all the elements of a row (column) are zero, or if all the elements of a row (column) are proportional to the elements of another row (column).

(2) $|\mathbf{A}| = |\mathbf{A}'|$.

(3) The value of $|\mathbf{A}|$ remains unchanged if a linear combination of any number of rows (columns) is added to any other row (column).

(4) If two rows (columns)are interchanged, the determinant is multiplied by (-1).

(5) If all the elements of a row (column) are multiplied by a constant factor, the determinant is multiplied by the same factor.

(6) If every element in any row (column) is the sum of two quantities, the determinant maybe expanded as the sum of two determinants of the same order. Thus, if

$$\mathbf{B} = \begin{vmatrix} a_{11} + b_{11} & \dots a_{1n} + b_{1n} \\ a_{21} & \dots a_{2n} \\ . & \dots . \\ a_{n1} & \dots a_{nn} \end{vmatrix},$$

then

$$|\mathbf{B}| = \begin{vmatrix} a_{11} & \dots a_{1n} \\ a_{21} & \dots a_{2n} \\ . & \dots . \\ . & \dots . \\ a_{n1} & \dots a_{nn} \end{vmatrix} + \begin{vmatrix} b_{11} & \dots b_{1n} \\ a_{21} & \dots a_{2n} \\ . & \dots . \\ . & \dots . \\ a_{n1} & \dots a_{nn} \end{vmatrix}.$$

Minors and Cofactors

The minor of a_{ij} of $\mathbf{A}$ is the value of the determinant obtained from $\mathbf{A}$ by deleting the ith row and the jth column.

The cofactor of a_{ij}, denoted as $\mathbf{A}_{ij} = (-1)^{i+j}$. minor of a_{ij}.

A principal minor of $\mathbf{A}$ is a minor obtained by deleting from $\mathbf{A}$ the same row and the same column.

Laplace Expansion of a determinant:

$$\begin{aligned} |\mathbf{A}| &= \textstyle\sum_{j=1}^{p} a_{ij}\mathbf{A}_{ij} \quad \text{for any fixed } i = 1, \dots, p \\ &= \textstyle\sum_{i=1}^{p} a_{ij}\mathbf{A}_{ij} \quad \text{for any fixed } j = 1, \dots, p. \end{aligned} \tag{A.5.1}$$

However,

$$\sum_{k=1}^{p} a_{ik}\mathbf{A}_{jk} = 0, \quad i \neq j.$$

Some special properties

(1) $|c\mathbf{A}| = c^p|\mathbf{A}|$.

(2) Let $\mathbf{A}_1, \ldots, \mathbf{A}_k$ be each of order $p \times p$. Then

$$|\mathbf{A}_1 . \ldots . \mathbf{A}_k| = |\mathbf{A}_1| . \ldots |\mathbf{A}_k|. \tag{A.5.2}$$

Determinant of some patterned matrices

(3) If $\mathbf{A}$ is orthogonal, then

$$|\mathbf{A}| = \pm 1.$$

(4) If $\mathbf{A}$ is triangular or diagonal, $|\mathbf{A}| = \Pi_{i=1}^{p} a_{ii}$.

(5) If $\mathbf{A}(p \times p)$ has a intraclass correlation structure (A.1.1),

$$|\mathbf{A}| = (a - b)^{p-1}[a + b(p-1)]. \tag{A.5.3}$$

More generally, if $\mathbf{G}(pk \times pk)$ has the structure of the *matrix intraclass correlation matrix*

$$\mathbf{G} = \begin{bmatrix} \mathbf{A} & \mathbf{B} & \ldots \mathbf{B} \\ \mathbf{B} & \mathbf{A} & \ldots \mathbf{B} \\ \ldots & \ldots & \ldots\ldots \\ \mathbf{B} & \mathbf{B} & \ldots \mathbf{A} \end{bmatrix}, \tag{A.5.4}$$

where $\mathbf{A}, \mathbf{B}$ are each of order $k \times k$, then

$$|\mathbf{G}| = |\mathbf{A} - \mathbf{B}|^{p-1}[|\mathbf{A} + (p-1)\mathbf{B}|]. \tag{A.5.5}$$

Determinant of a partitioned matrix

The following results are easy to verify.

(6)

$$\begin{vmatrix} \mathbf{A} & \mathbf{0} \\ \mathbf{0}' & \mathbf{B} \end{vmatrix} = |\mathbf{A}|.|\mathbf{B}|.$$

(7) If $|\mathbf{A}| \neq 0$,

$$\begin{vmatrix} \mathbf{A} & \mathbf{C} \\ \mathbf{0}' & \mathbf{B} \end{vmatrix} = |\mathbf{A}|.|\mathbf{B}|.$$

(8) Suppose $\mathbf{A}$ is a partitioned matrix as in (A.4.1). Then

(i) if $|\mathbf{A}_{22}| \neq 0$,

$$|\mathbf{A}| = |\mathbf{A}_{22}|.|\mathbf{A}_{11} - \mathbf{A}_{12}\mathbf{A}_{22}^{-1}\mathbf{A}_{21}| = |\mathbf{A}_{22}|.|\mathbf{A}_{11.2}|; \tag{A.5.6}$$

(ii) if $|\mathbf{A}_{11}| \neq 0$,

$$|\mathbf{A}| = |\mathbf{A}_{11}|.|\mathbf{A}_{22} - \mathbf{A}_{21}\mathbf{A}_{11}^{-1}\mathbf{A}_{12}| = |\mathbf{A}_{11}|.|\mathbf{A}_{22.1}|. \tag{A.5.7}$$

To prove (A.5.6) and (A.5.7), simplify $\mathbf{BAB}'$ and then take its determinant where

$$\mathbf{B} = \begin{bmatrix} \mathbf{I} & -\mathbf{A}_{12}\mathbf{A}_{22}^{-1} \\ \mathbf{0} & \mathbf{I} \end{bmatrix}.$$

(9) Taking in (A.4.1), $\mathbf{A}_{11} = \mathbf{A}, \mathbf{A}_{12} = \mathbf{x}', \mathbf{A}_{22} = c$, (by (A.5.7))

$$\begin{vmatrix} \mathbf{A} & \mathbf{x} \\ \mathbf{x}' & c \end{vmatrix} = |\mathbf{A}|(c - \mathbf{x}'\mathbf{A}^{-1}\mathbf{x}). \qquad (A.5.8)$$

(10) For $\mathbf{B}(p \times n)$ and $\mathbf{C}(n \times p)$, and non-singular $\mathbf{A}(p \times p)$,

$$|\mathbf{A} + \mathbf{BC}| = |\mathbf{A}||\mathbf{I}_p + \mathbf{A}^{-1}\mathbf{BC}| = |\mathbf{A}|.|\mathbf{I}_n + \mathbf{CA}^{-1}\mathbf{B}|. \qquad (A.5.9)$$

To prove this we have,

$$\begin{vmatrix} \mathbf{A} & -\mathbf{B} \\ \mathbf{C} & \mathbf{I}_n \end{vmatrix} = |\mathbf{A}| \begin{vmatrix} \mathbf{I}_p & -\mathbf{A}^{-1}\mathbf{B} \\ \mathbf{C} & \mathbf{I}_n \end{vmatrix}$$

and then use (A.5.6) and (A.5.7).

As a special case of (A.5.9) we see that for non-singular $\mathbf{A}$,

$$|\mathbf{A} + \mathbf{bb}'| = |\mathbf{A}|(1 + \mathbf{b}'\mathbf{A}^{-1}\mathbf{b}). \qquad (A.5.10)$$

Also, for $\mathbf{B}(p \times n)$ and $\mathbf{C}(n \times p)$,

$$|\mathbf{I}_p + \mathbf{BC}| = |\mathbf{I}_n + \mathbf{CB}|. \qquad (A.5.11)$$

A.6 Matrix Inversion

For a square matrix $\mathbf{A} = ((a_{ij}))$ of order p, the transpose of the matrix $((\mathbf{A}_{ij}))$ is called the *adjugate* of $\mathbf{A}$. This will be denoted by adj. $\mathbf{A}$.

If $|\mathbf{A}| \neq 0, \mathbf{A}$ has associated with it an unique $(p \times p)$ matrix, denoted by $\mathbf{A}^{-1} = ((a^{ij}))$ such that

$$\mathbf{A}^{-1}\mathbf{A} = \mathbf{AA}^{-1} = \mathbf{I}. \qquad (A.6.1)$$

$\mathbf{A}^{-1}$ is called the inverse of $\mathbf{A}$. Here

$$a^{ij} = \mathbf{A}_{ij}/|\mathbf{A}|.$$

If $\mathbf{A}$ is singular, $\mathbf{A}^{-1}$ does not exist.

Properties of Matrix Inversion

(1) $(\mathbf{A}')^{-1} = (\mathbf{A}^{-1})'$.

(2) Let $\mathbf{A}, \mathbf{B}$ be both non-singular and of order p. Then

$$(\mathbf{AB})^{-1} = \mathbf{B}^{-1}\mathbf{A}^{-1}. \qquad (A.6.2)$$

(3) If $\mathbf{A}$ is non-singular,

$$|\mathbf{A}^{-1}| = |\mathbf{A}|^{-1}. \qquad (A.6.3)$$

(4) If $\mathbf{D} =$ Diag $(\lambda_1, \ldots, \lambda_p)$, then

$$\mathbf{D}_\lambda^{-1} = \text{ Diag } (\lambda_1^{-1}, \ldots, \lambda_p^{-1}).$$

(5) If $\mathbf{A}$ is an orthogonal matrix, $\mathbf{A}^{-1} = \mathbf{A}'$.

(6) *Binomial Inversion Theorem* : Assume $\mathbf{A} : p \times p, \mathbf{U} : p \times q, \mathbf{B} : q \times q, \mathbf{V} : q \times p$. Then, if $\mathbf{A}, \mathbf{B}$ are non-singular,

$$(\mathbf{A} + \mathbf{UBV})^{-1} = \mathbf{A}^{-1} - \mathbf{A}^{-1}\mathbf{UB}(\mathbf{B} + \mathbf{BVA}^{-1}\mathbf{UB})^{-1}\mathbf{BVA}^{-1}. \quad (A.6.4)$$

In particular, let $\mathbf{B} = \mathbf{I}$ and $q = 1$. Then $\mathbf{U}$ is a column vector $\mathbf{u}$ and $\mathbf{V}$ is a row vector $\mathbf{v}'$. The theorem implies that

$$(\mathbf{A} + \mathbf{uv}')^{-1} = \mathbf{A}^{-1} - \frac{\mathbf{A}^{-1}\mathbf{uv}'\mathbf{A}^{-1}}{1 + \mathbf{v}'\mathbf{A}^{-1}\mathbf{u}}. \qquad (A.6.5)$$

If further, $\mathbf{A} = \mathbf{I}$,

$$(\mathbf{I} + \mathbf{uv}')^{-1} = \mathbf{I} - \frac{\mathbf{uv}'}{1 + \mathbf{v}'\mathbf{u}}. \qquad (A.6.6)$$

Consider the intraclass correlation matrix defined in (A.1.1). The matrix $\mathbf{A}$ may be written as

$$\mathbf{A} = (a - b)\mathbf{I} + b\mathbf{11}',$$

where $\mathbf{1}$ is a column vector of ones. Hence, if $\mathbf{A}$ is of order p, applying (A.6.6),

$$\mathbf{A}^{-1} = \frac{\mathbf{I}}{a - b} - \frac{b\mathbf{11}'}{(a - b)[a + (p - 1)b]}, \qquad (A.6.7)$$

which is again a matrix with intraclass correlation structure.

Inverse of a partitioned matrix

Assume $\mathbf{A}$ is an arbitrary $(p + n) \times (p + n)$ matrix with $|\mathbf{A}| \neq 0$. Partition $\mathbf{A}$ as

$$\mathbf{A} = \begin{bmatrix} \mathbf{A}_{11} & \mathbf{A}_{12} \\ \mathbf{A}_{21} & \mathbf{A}_{22} \end{bmatrix} \qquad (A.6.8)$$

where $\mathbf{A}_{11}$ is $p \times p$ and $\mathbf{A}_{22}$ is $n \times n$. Let

$$\mathbf{B} = \mathbf{A}^{-1} = \begin{bmatrix} \mathbf{B}_{11} & \mathbf{B}_{12} \\ \mathbf{B}_{21} & \mathbf{B}_{22} \end{bmatrix}$$

where $\mathbf{B}_{11}$ is $p \times p, \mathbf{B}_{22}$ is $n \times n$. It follows that

$$\mathbf{B}_{11} = (\mathbf{A}_{11} - \mathbf{A}_{12}\mathbf{A}_{22}^{-1}\mathbf{A}_{21})^{-1} = \mathbf{A}_{11.2}^{-1}; \tag{A.6.9}$$

$$\mathbf{B}_{22} = (\mathbf{A}_{22} - \mathbf{A}_{21}\mathbf{A}_{11}^{-1}\mathbf{A}_{12})^{-1} = \mathbf{A}_{22.1}^{-1}; \tag{A.6.10}$$

$$\mathbf{B}_{12} = -\mathbf{A}_{11}^{-1}\mathbf{A}_{12}\mathbf{A}_{22.1}^{-1}; \tag{A.6.11}$$

$$\mathbf{B}_{21} = -\mathbf{A}_{22}^{-1}\mathbf{A}_{21}\mathbf{A}_{11.2}^{-1}. \tag{A.6.12}$$

It can be easily checked that if $\mathbf{A}_{12} = \mathbf{A}_{21} = \mathbf{0}$, then

$$\mathbf{A}^{-1} = \begin{bmatrix} \mathbf{A}_{11}^{-1} & \mathbf{0} \\ \mathbf{0} & \mathbf{A}_{22}^{-1} \end{bmatrix}.$$

Generalized inverses

Let $\mathbf{A}$ be of order p and assume $|\mathbf{A}| = 0$. Hence, $\mathbf{A}^{-1}$ does not exist. However, it is possible to define a generalization of the usual inverse, $\mathbf{A}^{-}$, as a matrix that exists even when $\mathbf{A}^{-1}$ does not, that equals $\mathbf{A}^{-1}$ when $\mathbf{A}^{-1}$ exists and that satisfies the property

$$\mathbf{A}\mathbf{A}^{-}\mathbf{A} = \mathbf{A}. \tag{A.6.13}$$

The matrix $\mathbf{A}^{-}$ is called a generalized inverse (g-inverse) of $\mathbf{A}$. This matrix is not necessarily unique, although it is possible, by imposing additional constraints, to define it uniquely.

Moore (1935) and Penrose (1955) defined a particular generalized inverse (often called a pseudo-inverse) as a matrix $\mathbf{A}^{-}$ satisfying

$$(a)\mathbf{A}\mathbf{A}^{-}\mathbf{A} = \mathbf{A}, \quad (b)\mathbf{A}^{-}\mathbf{A}\mathbf{A}^{-} = \mathbf{A}^{-},$$

$$(c)(\mathbf{A}\mathbf{A}^{-})' = \mathbf{A}\mathbf{A}^{-}, \quad (d)(\mathbf{A}^{-}\mathbf{A})' = \mathbf{A}^{-}\mathbf{A}.$$

Such a matrix not only exists, but it is unique. For details, the reader may refer to Rao (1962), Ben-Israel and Greville (1980), among many others. We give below some methods of finding a pseudo-inverse.

(a) Let $\mathbf{A}$ be a $p \times n$ matrix ($p \leq n$). Using orthogonal factorization technique (vide A.12), decompose $\mathbf{A}$ as

$$\mathbf{A} = \mathbf{B}(\mathbf{D}_\lambda \;\; \mathbf{0})\mathbf{C}'.$$

Replace non-zero elements of $\mathbf{D}_\lambda$ by their reciprocals and call the resulting matrix as $\mathbf{D}^*_\lambda$. Then

$$\mathbf{A}^- = \mathbf{C}(\mathbf{D}^*_\lambda \;\; \mathbf{0})\mathbf{B}'.$$

EXAMPLE A.6.1: Let

$$\mathbf{A} = \begin{bmatrix} 4 & 2 \\ 2 & 1 \end{bmatrix}.$$

Latent roots of $\mathbf{A}$ are $\lambda_1 = 5,\ \lambda_2 = 0$. The matrix with columns as the latent vectors of $\mathbf{A}$ is

$$\mathbf{B} = \frac{1}{\sqrt{5}} \begin{bmatrix} 2 & -1 \\ 1 & 2 \end{bmatrix}.$$

Here $\mathbf{C} = \mathbf{B}$. Hence,

$$\mathbf{A}^- = \mathbf{B}\mathbf{D}^*_\lambda\mathbf{B}' = \frac{1}{25}\mathbf{A}.$$

(b) If $r(\mathbf{A}) = r$, we rearrange the rows and columns of $\mathbf{A}(p \times n)$ and partition $\mathbf{A}$ so that $\mathbf{A}_{11}$ is an $r \times r$ non-singular matrix. Then it can be verified that

$$\mathbf{A}^- = \begin{bmatrix} \mathbf{A}_{11}^{-1} & \mathbf{0} \\ \mathbf{0} & \mathbf{0} \end{bmatrix}$$

is a g-inverse.

(c) If $\mathbf{A}(p \times p)$ is symmetric of rank r, then by spectral decomposition theorem (Theorem A.12.1), we have

$$\mathbf{A} = \mathbf{\Gamma}\mathbf{\Lambda}\mathbf{\Gamma}',$$

where $\mathbf{\Gamma}_{p\times r} = (\gamma_1, \ldots, \gamma_r), \mathbf{\Lambda} =$ Diag $(\lambda_1 \ldots, \lambda_r), \gamma_i$ being the eigenvector corresponding to the non-zero eigenvalue $\lambda_i, (i = 1, \ldots, r)$. It can be easily checked that

$$\mathbf{A}^- = \mathbf{\Gamma}\mathbf{\Lambda}^{-1}\mathbf{\Gamma}'$$

is a g-inverse.

A.7 Rank

The rank of a matrix $\mathbf{A}$ often denoted as $r(\mathbf{A})$ is the largest order of the submatrices with non-vanishing determinants.

The row (column) rank of a matrix is the maximum number of linearly independent rows (columns), considered as vectors. The row rank and the column rank of a matrix are equal.

We now consider its properties.

(1) If $\mathbf{A} : p \times n$,

$$r(\mathbf{A}) \leq \min\ (p, n).$$

(2) $r(\mathbf{A}) = r(\mathbf{A}')$.

(3) If $\mathbf{D} =$ Diag $(\lambda_1, \ldots, \lambda_p), r(\mathbf{D}) =$ number of non-zero λ's.

(4) Let $\mathbf{A}, \mathbf{B}$ be two matrices conformable for multiplication. then

$$r(\mathbf{AB}) \leq \min\ (r(\mathbf{A}), r(\mathbf{B})).$$

(5) If both $\mathbf{A}$ and $\mathbf{B}$ are $p \times n$, then

$$r(\mathbf{A} + \mathbf{B}) \leq r(\mathbf{A}) + r(\mathbf{B}).$$

(6) $r(\mathbf{AA}') = r(\mathbf{A}'\mathbf{A}) = r(\mathbf{A})$.

(7) If $\mathbf{P} : m \times m, \mathbf{Q} : n \times n, |\mathbf{P}| \neq 0, |\mathbf{Q}| \neq 0$, then for $\mathbf{A} : m \times n$,

$$r(\mathbf{PAQ}) = r(\mathbf{A}).$$

(8) If $n = p$, then $r(\mathbf{A}) = p$ if and only if $\mathbf{A}$ is non-singular.

(9) If $\mathbf{A} : p \times n, \mathbf{B} : n \times r$ such that $\mathbf{AB} = \mathbf{0}$, then

$$r(\mathbf{B}) \leq n - r(\mathbf{A}).$$

(10) If $\mathbf{A} : n \times m$ is of rank m and $\mathbf{B} : m \times p$ is of rank p, then $\mathbf{AB}$ has rank p.

(11) The rank of a symmetric matrix is equal to the number of non-zero eigenvalues. (Eigenvalues have been considered in subsection A.8).

(12) If $\mathbf{A}$ is symmetric and rank $\mathbf{A}$ =1, then

$$|\mathbf{I} + \mathbf{A}| = 1 + tr\mathbf{A}.,$$

where $tr(.)$ denotes the trace of a matrix (.) The concept of trace has been discussed in Subsection A.9.

(13) Let $\mathbf{X}' = (\mathbf{x}_1, \ldots, \mathbf{x}_n)$ be a $p \times n$ matrix of random variables and let $\mathbf{A}$ be a symmetric matrix of rank r. If the joint distribution of elements of $\mathbf{X}$ is absolutely continuous with respect to np-dimensional Lebesgue measure, then the following results hold: rank $(\mathbf{X}'\mathbf{AX}) =$ min (p, r) with probability one and the nonzero eigenvalues of $\mathbf{X}'\mathbf{AX}$ are distinct.

Defining $\mathcal{M}(\mathbf{A})$ as the vector sub-space in R^n spanned by the columns of $\mathbf{A}(n \times p)$, we have $r(\mathbf{A}) =$ dim $\mathcal{M}(\mathbf{A})$. We may choose linearly independent columns of $\mathbf{A}$ as a basis for $\mathcal{M}(\mathbf{A})$. We note that for any vector, $\mathbf{x} = (x_1, \ldots, x_p)', \mathbf{Ax} = x_1\mathbf{a}_{(1)}, + \ldots + x_p\mathbf{a}_{(p)}$, a linear combination of the columns $\mathbf{a}_{(1)}, \ldots \mathbf{a}_{(p)}$ of $\mathbf{A}$ and hence $\mathbf{x} \in \mathcal{M}(\mathbf{A})$.

DEFINITION A.7.1: The *null space* of $\mathbf{A}(n \times p)$ is given by

$$\mathcal{N}(\mathbf{A}) = \{\mathbf{x} \in R^p : \mathbf{A}\mathbf{x} = \mathbf{0}\}.$$

Thus, $\mathcal{N}(\mathbf{A})$ is a subspace of R^p of dimension u, say. Let $e_1, \ldots, e_p$, each of order $p \times 1$, be a basis for R^p, of which $e_1, \ldots, e_u$ form a basis for $\mathcal{N}(\mathbf{A})$. Then, each of vectors $\mathbf{A}\mathbf{e}_{u+1}, \ldots, \mathbf{A}\mathbf{e}_p$ lie in $\mathcal{M}(\mathbf{A})$ and are linearly independent. Thus, $\mathbf{A}\mathbf{e}_{u+1}, \ldots, \mathbf{A}\mathbf{u}_p$ form a maximal set of independent vectors in $\mathcal{M}(\mathbf{A})$ and hence is a basis for $\mathcal{M}(\mathbf{A})$. Thus,

$$\dim \mathcal{N}(\mathbf{A}) + \dim \mathcal{M}(\mathbf{A}) = p.$$

A.8 Eigenvalues and Eigenvectors

Let $\mathbf{A}$ be a square matrix of order n and $\mathbf{x}$ a $n \times 1$ vector. The transformed vector $\mathbf{A}\mathbf{x}$ may be proportional to the original vector $\mathbf{x}$, i.e.

$$\mathbf{A}\mathbf{x} = \lambda\mathbf{x}$$

or

$$(\mathbf{A} - \lambda\mathbf{I})\mathbf{x} = \mathbf{0}, \tag{A.8.1}$$

where λ is a scalar. The linear system of equations (A.8.1) has a solution $\mathbf{x} \neq \mathbf{0}$ *iff* (vide Section A.16)

$$q(\lambda) = |\mathbf{A} - \lambda\mathbf{I}| = 0,$$

or

$$\lambda^n + b_{n-1}\lambda^{n-1} + \ldots + b_1\lambda + b_0 = 0 \tag{A.8.2}$$

or

$$\Pi_{j=1}^n(\lambda - \lambda_j) = 0.$$

The equation (A.8.2) is called the *characteristic equation* of $\mathbf{A}$. For each of its root λ_j, the equation (A.8.1) has a solution $\mathbf{x}_j \neq \mathbf{0}$. The λ_j's are called the *eigenvalues* (or *latent roots* or *characteristic values*) of $\mathbf{A}$, the vector $\mathbf{x}_j$'s are called the *eigenvectors* (or the *characteristic vectors*) of $\mathbf{A}$. Some of the λ_i's will be equal, if $q(\lambda)$ has multiple roots. An eigenvalue $\mathbf{x}$ with real entries is called *standardized* or *normalized* if $\mathbf{x}'\mathbf{x} = 1$. We have

$$\mathbf{A}\,(\mathbf{x}_1 \ldots \mathbf{x}_n) = (\lambda_1\mathbf{x}_1 \ldots \lambda_n\mathbf{x}_n)$$

or

$$\mathbf{A}\mathbf{X} = \mathbf{X}\mathbf{D} \tag{A.8.3}$$

where

$$\mathbf{X} = (\mathbf{x}_1, \dots \mathbf{x}_n), \quad \mathbf{D} = \text{ Diag } (\lambda_1, \dots, \lambda_n).$$

If $\mathbf{x}_j$'s are linearly independent,

$$\mathbf{X}^{-1}\mathbf{A}\mathbf{X} = \mathbf{D}. \qquad (A.8.4)$$

The following results hold:

(1) $|\mathbf{A}| = \Pi_{j=1}^n \lambda_j$. Therefore, $r(\mathbf{A}) < n$ *iff* at least one latent root of $\mathbf{A}$ is zero. The rank of a symmetric matrix is the number of non-zero latent roots it possesses.

(2) In general, latent roots are complex numbers (occurring in complex conjugate pairs when the roots are not real). However, if $\mathbf{A}$ is a real and symmetric matrix, all the eigenvalues and therefore, all the eigenvectors are real.

(3) If $\mathbf{A}$ has the eigenvalue λ and eigenvector $\mathbf{x}$, then $R(\mathbf{A})$, where $R(.)$ is a rational function of (.), has eigenvalue $R(\lambda)$ and the corresponding eigenvector $\mathbf{x}$. In particular, the matrix $\mathbf{A}^m$ has the eigenvalue λ^m with the same multiplicity.

(4) If $\mathbf{A}$ has the eigenvalue λ and the eigenvector $\mathbf{x}$, then $\mathbf{TAT}^{-1}$ has the eigenvalue λ and the eigenvector $\mathbf{Tx}$ where $\mathbf{T}$ is a non-singular $n \times n$ matrix.

(5) Let α be a real number ($\alpha \in \mathcal{R}$). If $(\lambda_i, \mathbf{x}_i)$ is an eigenvalue, eigenvector pair of $\mathbf{A}$, then $(\lambda_i + \alpha, \mathbf{x}_i)$ is an eigenvalue, eigenvector pair of $\mathbf{A} + \alpha\mathbf{I}$.

(6) The matrices $\mathbf{A}$ and $\mathbf{A}'$ have the same latent roots.

(7) If $\mathbf{A} > 0, \mathbf{B} > 0$ are $n \times n$ matrices, then all the non-zero eigenvalues of $\mathbf{B}^{-1}\mathbf{A}$ are positive.

(8) $\mathbf{AB}$ and $\mathbf{BA}$ have the same eigenvalues while their eigenvectors are $\mathbf{x}_i$ and $\mathbf{Bx}_i$ (or $\mathbf{A}^{-1}\mathbf{x}_i$).

(9) If $\mathbf{A}$ and $\mathbf{B}$ are $n \times n$ and $\mathbf{A}$ is non-singular, then the latent roots of $\mathbf{AB}$ and $\mathbf{BA}$ are the same.

(10) If $\mathbf{A}$ is $m \times n$, the non-zero latent roots of $\mathbf{AA}'$ and $\mathbf{A}'\mathbf{A}$ are the same.

(11) $tr(\mathbf{A}) = \sum_{i=1}^m \lambda_i; tr(\mathbf{A}^m) = \sum_{i=1}^n \lambda_i^m$. (Here, $tr(\mathbf{A})$ means trace of $\mathbf{A}$, defined in Section A.9).
If $\mathbf{A}$ is symmetric, for $m = 2, tr(\mathbf{A}^2) = tr(\mathbf{AA}') = \sum \lambda_i^2$. If, therefore, all $\lambda_i = 0$, all $a_{ik} = 0$, i.e., $\mathbf{A} = \mathbf{0}$.

(12) *Theorem of Cayley*: If $f(\lambda) = 0$ is the characteristic equation of $\mathbf{A}$, then also $f(\mathbf{A}) = \mathbf{0}$, i.e., every matrix satisfies its characteristic equation.

(13) If $\mathbf{x}$ and $\mathbf{y}$ are eigenvectors for λ_i and $\alpha \in \mathcal{R}$, then $\mathbf{x}+\mathbf{y}$ and $\alpha\mathbf{x}$ are also eigenvectors for λ_i. Thus, the set of all eigenvectors for λ_i form a sub-space which is called the *eigenspace* of $\mathbf{A}$ for λ_i.

Let λ_a denote a particular eigenvalue of $\mathbf{A}(n \times n)$, with eigenspace H of dimension r. If k denotes the multiplicity of λ_a in $q(\lambda)$, then $1 \leq r \leq k$.

If $\mathbf{A}$ is symmetric, then $r = k$.

If $r = 1$, then the eigenspace for λ_a has dimension 1 and the standardized eigenvector for λ_a is unique (up to sign).

Extremum properties of the eigenvalues

A symmetric matrix $\mathbf{A}$ has n real eigenvalues λ_i and real eigenvectors $\mathbf{x}_i (i = 1, \ldots, n)$, which are characterized by the following properties:

(14) The largest eigenvalue λ_1 is the maximum of the quadratic form $\mathbf{x}'\mathbf{A}\mathbf{x}$ for all vectors $\mathbf{x}$ of length 1, i.e.

$$\lambda_1 = \max_{\mathbf{x}} \frac{\mathbf{x}'\mathbf{A}\mathbf{x}}{\mathbf{x}'\mathbf{x}}.$$

The vector $\mathbf{x}_1$ which gives this maximum value is the eigenvector corresponding to λ_1:

$$\lambda_1 = \frac{\mathbf{x}_1'\mathbf{A}\mathbf{x}_1}{\mathbf{x}_1'\mathbf{x}_1} \quad \text{and} \quad \mathbf{A}\mathbf{x}_1 = \lambda_1 \mathbf{x}_1.$$

(15) The eigenvalue λ_i is the maximum value of the expression $\mathbf{x}'\mathbf{A}\mathbf{x}$ for all $\mathbf{x}$ of length unity, for which

$$\mathbf{x}'\mathbf{x}_1 = \mathbf{x}'\mathbf{x}_2 = \ldots = \mathbf{x}'\mathbf{x}_{i-1} = 0.$$

The vector $\mathbf{x}_i$ for which the form $\mathbf{x}'\mathbf{A}\mathbf{x}$ assumes this maximum value is the eigenvector corresponding to λ_i.

(16) The smallest eigenvalue λ_n is the minimum value of the quadratic form $\mathbf{x}\mathbf{A}\mathbf{x}$ for all vectors $\mathbf{x}$ of length unity, i.e.,

$$\lambda_n = \min_{\mathbf{x}} \frac{\mathbf{x}'\mathbf{A}\mathbf{x}}{\mathbf{x}'\mathbf{x}}.$$

A related result is given Theorem A.14.2.

Latent vectors associated with distinct latent roots of a symmetric matrix are orthogonal. If a latent root has multiplicity m, it is always possible to find m mutually orthogonal latent vectors of unit length that are also orthogonal to the latent vectors corresponding to the distinct latent roots.

If $\mathbf{A}$ is orthogonal, all its latent roots have unit modulus (i.e., all its latent roots lie on the unit circle in the complex plane). Also, the reciprocals of its latent roots are latent roots.

If $\mathbf{A}$ is a $p \times p$ matrix of intraclass correlation structure, as given in (A.1.1), its latent roots are given by

$$\lambda_1 = a + (p-1)b, \ \lambda_2 = \ldots = \lambda_p = a - b.$$

The latent vectors of the matrix $\mathbf{A}$ of intraclass correlation structure does not depend on the elements of $\mathbf{A}$. Any p mutually orthogonal vectors of unit length, the first of which have components that are all identical, will do. For example, the columns of the Helmert matrix in example A.1.1 are a set of latent vectors for $\mathbf{A}$.

Kantorovich inequality: Let $\mathbf{A}$ be an $n \times n$ real symmetric positive definite matrix (defined in Section A.11) and $\mathbf{x}$ be an $n \times 1$ real vector satisfying $\mathbf{x}'\mathbf{x} = 1$. Then

$$\underset{\sim}{x}'\mathbf{A}\mathbf{x}\mathbf{x}'\mathbf{A}^{-1} \leq \frac{(\lambda_1 + \lambda_n)^2}{4\lambda_1\lambda_n}$$

whenre $\lambda_1 \geq \ldots \geq \lambda_n$ are the eigenvalues of $\mathbf{A}$. This inequality has been generalized by Bloomfield and Watson (1975), Knot (1975) Khatri and Rao (1975), among others.

Theorem A.8.1: Let $\mathbf{A}$ be a $p \times p$ symmetric matrix of rank $r(\leq p)$ and let $\mathbf{x}$ be any vector belonging to $\mathcal{M}(\mathbf{A})$, the column-space of $\mathbf{A}$. Then, there exists an orthogonal transformation such that

$$\mathbf{x}'\mathbf{A}^-\mathbf{x} = \sum_{i=1}^{r} \frac{z_i^2}{\lambda_i},$$

where $\lambda_1, \ldots, \lambda_r$ are the non-zero eigenvalues of $\mathbf{A}$.

Proof. Since $\mathbf{x} \in \mathcal{M}(\mathbf{A})$, we can write $\mathbf{x} = \mathbf{A}\mathbf{y}$ for some $\mathbf{y}$. Hence,

$$\mathbf{x}'\mathbf{A}^-\mathbf{x} = \mathbf{y}'\mathbf{A}\mathbf{A}^-\mathbf{A}\mathbf{y} = \mathbf{y}'\mathbf{A}\mathbf{y}.$$

Now, by spectral decomposition theorem, (Theorem A.12.1),

$$\mathbf{A} = \sum_{i=1}^{r} \lambda_i \gamma_i \gamma_i',$$

where γ_i is the eigenvector corresponding to the non-zero eigenvalue λ_i of $\mathbf{A}$, $(i = 1, \ldots, r)$. Since $\mathbf{x} \in \mathcal{M}(\mathbf{A})$, we can write $\mathbf{x} = \mathbf{\Gamma z}$ for some $\mathbf{z}$. Hence,

$$\mathbf{x}'\mathbf{A}^-\mathbf{x} = \mathbf{z}'\mathbf{\Gamma}'\mathbf{A}^-\mathbf{\Gamma z} = \mathbf{z}'\mathbf{\Lambda}^-\mathbf{z} = \sum_{i=1}^{r} z_i^2/\lambda_i$$

where $\mathbf{\Gamma} = (\gamma_1, \ldots, \gamma_r)$ and $\mathbf{\Lambda} =$ Diag $(\lambda_1, \ldots, \lambda_r)$.

A.9 Trace of a Matrix

The sum of the diagonal elements of a square matrix $\mathbf{A}$ is called its trace. Thus, if $\mathbf{A}(p \times p) = ((a_{ij}))$

$$tr(\mathbf{A}) = \sum_{j=1}^{p} c_{jj}. \tag{A.9.1}$$

Properties :

(1) If $\mathbf{A} : p \times n, \mathbf{B} : n \times p$, then $tr(\mathbf{AB}) = tr(\mathbf{BA})$.

(2) The trace of a square matrix is equal to the sum of its latent roots.

(3) If $\mathbf{A} : p \times n, \mathbf{B} : p \times n$, then
$$tr(\mathbf{A} + \mathbf{B}) = tr(\mathbf{A}) + tr(\mathbf{B}).$$

(4) If α is a scaler and $\mathbf{A} : p \times p$, then
$$tr(\alpha\mathbf{A}) = \alpha tr(\mathbf{A}).$$

(5) Let $\mathbf{S}$ be a non-singular $p \times p$ matrix. Then
$$tr(\mathbf{S}^{-1}\mathbf{AS}) = tr(\mathbf{A}).$$
The transformation $\mathbf{S}^{-1}\mathbf{AS}$ is often called a *similarity transformation.*

(6) If $\mathbf{A} : p \times p$ be of rank 1, then its only non-zero latent root is $tr(\mathbf{A})$.

(7) Let $\mathbf{A}$ be a $k \times k$ symmetric matrix and $\mathbf{x}$ be a $k \times 1$ vector. Then
$$\mathbf{x}'\mathbf{Ax} = tr(\mathbf{Axx}').$$
The result follows by the property (1) above.

A.10 Direct Product

Let $\mathbf{A}$ be of order $p \times p$ and $\mathbf{B}$ of order $n \times n$. Then the $pn \times pn$ matrix $\mathbf{M} = \mathbf{A} \otimes \mathbf{B}$ is called the direct product of $\mathbf{A}$ and $\mathbf{B}$ and is given by

$$\mathbf{M} = \mathbf{A} \otimes \mathbf{B} = \begin{bmatrix} a_{11}\mathbf{B} & a_{12}\mathbf{B} & \ldots & a_{1p}\mathbf{B} \\ \ldots & \ldots & \ldots & \ldots \\ a_{p1}\mathbf{B} & a_{p2}\mathbf{B} & \ldots & a_{pp}B \end{bmatrix}. \tag{A.10.1}$$

Properties :

(1) $\alpha(\mathbf{A} \otimes \mathbf{B}) = (\alpha \mathbf{A}) \otimes \mathbf{B} = \mathbf{A} \otimes (\alpha \mathbf{B})$ for all scalar α. Hence this can be written as $\alpha \mathbf{A} \otimes \mathbf{B}$.
(2) $\mathbf{A} \otimes (\mathbf{B} \otimes \mathbf{C}) = (\mathbf{A} \otimes \mathbf{B}) \otimes \mathbf{C}$.
(3) $(\mathbf{A} + \mathbf{B}) \otimes \mathbf{C} = \mathbf{A} \otimes \mathbf{C} + \mathbf{B} \otimes \mathbf{C}$.
(4) $\mathbf{A} \otimes (\mathbf{B} + \mathbf{C}) = \mathbf{A} \otimes \mathbf{B} + \mathbf{A} \otimes \mathbf{C}$.
(5) $(\mathbf{A} \otimes \mathbf{B})' = \mathbf{A}' \otimes \mathbf{B}'$.
(6) Let $\mathbf{A} : p \times p$ and $\mathbf{B} : n \times n.$, $(\mathbf{A} \otimes \mathbf{B})^{-1} = \mathbf{A}^{-1} \otimes \mathbf{B}^{-1}$ for non-singular $\mathbf{A}$ and $\mathbf{B}$.
(7) Let $\mathbf{A} : p \times p$ and $\mathbf{B} : n \times n$. Then $|\mathbf{A} \otimes \mathbf{B}| = |\mathbf{A}|^n \times |\mathbf{B}|^p$.
(8) If $\mathbf{A}$ and $\mathbf{C}$ are of order $p \times p$ and $\mathbf{B}$ and $\mathbf{D}$ are of order $n \times n$, then

$$(\mathbf{A} \otimes \mathbf{B})(\mathbf{C} \otimes \mathbf{D}) = (\mathbf{AC}) \otimes (\mathbf{BD}).$$

(9) suppose the latent roots of $\mathbf{A} : p \times p$ are $\lambda_1, \ldots, \lambda_p$ and those of $\mathbf{B} : n \times n$ are $\mu_1, \ldots, \mu_n$. Then

$$|\mathbf{A} \otimes \mathbf{B}| = (\lambda_1 \ldots \lambda_p)(\mu_1 \ldots \mu_n).$$

(10) $tr(\mathbf{A} \otimes \mathbf{B}) = (tr\mathbf{A})(tr\mathbf{B})$.

Definition A.10.1: If $\mathbf{X} = (\mathbf{x}_{(1)}, \ldots, \mathbf{x}_{(p)})$ is an $(n \times p)$ matrix, let $vec(\mathbf{X})$ denote the $np \times 1$- vector obtained by vectorizing $\mathbf{X}$, i.e.,

$$vec(\mathbf{X}) = \mathbf{X}^V = \begin{bmatrix} \mathbf{x}_{(1)} \\ . \\ . \\ . \\ \mathbf{x}_{(p)} \end{bmatrix}.$$

It follows that

$$(\mathbf{AXB})^V = (\mathbf{B}' \otimes \mathbf{A})\mathbf{X}^V.$$

A.11 Quadratic Forms and Definitions

Let $\mathbf{x} = (x_1, \ldots, x_p)'$ denote a $p \times 1$ vector and $\mathbf{A} = ((a_{ij}))$ denote a $p \times p$ symmetric matrix. Then

$$Q = \mathbf{x}'\mathbf{A}\mathbf{x} = \sum_{i=1}^{p} \sum_{j=1}^{p} a_{ij} x_i x_j \qquad (A.11.1)$$

is called a *quadratic form* in the variables $x_1, \ldots, x_p$.

If $Q = \mathbf{x}'\mathbf{A}\mathbf{x} > 0 (< 0)$ for all vectors $\mathbf{x}(\neq \mathbf{0}), Q$ is called a *positive (negative) definite quadratic form* and the matrix $\mathbf{A}$ is called a *positive (negative) definite* matrix. It is then written as $\mathbf{A} > (<) 0$. If $Q \geq 0 \ \forall \ \mathbf{x}(\neq \mathbf{0}), Q$ is called a *positive semidefinite (p.s.d.) quadratic form* and $\mathbf{A}$ is called a *p.s.d. matrix* ($\mathbf{A} \geq 0$). Similarly, if $Q \leq 0 \ \forall \ \mathbf{x}(\neq \mathbf{0}), Q$ is called a *negative semidefinite (n.s.d.) quadratic form* and $\mathbf{A}$ a *n.s.d. matrix* ($\mathbf{A} \leq 0$). None of these conditions hold, Q is called an *indefinite quadratic form* and $\mathbf{A}$ an *indefinite matrix* .

Consider all submatrices of a symmetric matrix $\mathbf{A}(p \times p)$ whose principal diagonals are parts of the principal diagonal of $\mathbf{A}$. The determinants of these *principal submatrices* are called *principal minors* of $\mathbf{A}$. A necessary and sufficient for $\mathbf{A}$ to be positive definite ($\mathbf{A} > 0$) is that $|\mathbf{A}| > 0$ and all principal minors of $\mathbf{A}$ are positive.

Properties of positive definite matrices

(1) If $\mathbf{A} \geq 0$ and $|\mathbf{A}| \neq 0$, then $\mathbf{A} > 0$.

(2) If

$$\mathbf{A} > 0, \quad \mathbf{A}^{-1} > 0. \tag{A.11.2}$$

(3) Let $\mathbf{A} : n \times n, \mathbf{B} : p \times n (p \leq n), \mathbf{A} > 0$ and $\text{rank}(\mathbf{B}) = r$. Then

$$\mathbf{BAB}' > 0 \ \text{ if } r = p \ \text{ and } \mathbf{BAB}' \geq 0 \ \text{ if } r < p. \tag{A.11.3}$$

(4) If $\mathbf{A} > 0, \mathbf{B} > 0$,

$$|\mathbf{A} + \mathbf{B}| \leq |\mathbf{A} + \mathbf{B}| \tag{A.11.4}$$

with equality *iff* $\mathbf{A}$ is proportional to $\mathbf{B}$.

(5) If $\mathbf{A} > 0, \mathbf{B} > 0$ and $\mathbf{A} - \mathbf{B} > 0$, then

$$|\mathbf{A}| > |\mathbf{B}|. \tag{A.11.5}$$

(6) If $\mathbf{A} \geq 0$, there exists at least one matrix $\mathbf{B}$ such $\mathbf{A} = \mathbf{BB}'$. Conversely, if for some $\mathbf{B}, \mathbf{A} = \mathbf{BB}'$, then $\mathbf{A} \geq 0$.

(7) *Hadamard's Inequality (1893):* Let $\mathbf{A} = ((a_{ij}))$ be $p \times p$ and assume $\mathbf{A} > 0$. Then

$$|\mathbf{A}| \leq \Pi_{i=1}^{p} a_{ii}. \tag{A.11.6}$$

(8) If $\mathbf{A} > 0$, there exists a non-singular matrix $\mathbf{M}$ such that

$$\mathbf{MAM}' = \mathbf{I}. \tag{A.11.7}$$

(9) If $\mathbf{A} > 0$, all its latent rots are positive and conversely.

(10) If $\mathbf{A} > 0, \mathbf{B} > 0$ and $\mathbf{A} - \mathbf{B} > 0$, then

$$\mathbf{B}^{-1} - \mathbf{A}^{-1} > 0. \qquad (A.11.8)$$

(11) Let $\lambda_i(\mathbf{A}), \lambda_i(\mathbf{B})$ denote the ith latent roots of $\mathbf{A}$ and $\mathbf{B}$, respectively, where $\mathbf{A} = \mathbf{A}', \mathbf{B} = \mathbf{B}'$. Then, if $\mathbf{A} - \mathbf{B} \geq 0, \lambda_i(\mathbf{A}) \geq \lambda_i(\mathbf{B}) \ \forall \ i$.

(12) *Cholesky decomposition*: If $\mathbf{A}$ is p.d., there exists a unique upper triangular matrix

$$\mathbf{U} = \begin{bmatrix} u_{11} & u_{12} & \dots & u_{1n} \\ 0 & u_{22} & \dots & u_{2n} \\ . & . & \dots & . \\ 0 & 0 & \dots & u_{nn} \end{bmatrix}$$

with positive diagonal elements ($u_{ii} > 0$) such that $\mathbf{A} = \mathbf{U}'\mathbf{U}$. Note that $|\mathbf{U}| = u_{11}.u_{12}.\dots u_{nn}$.

Similarly, there exists a lower triangular matrix $\mathbf{L}$ with positive diagonal elements such that $\mathbf{A} = \mathbf{L}'\mathbf{L}$.

(13) Let $\mathbf{X}' = (\mathbf{X}_1, \dots, \mathbf{X}_n)$ where $\mathbf{X}_i$ are n independent p-dimensional random vectors and let $\mathbf{A}$ be a positive semi-definite $n \times n$ matrix of rank $r \geq p$. Suppose that for each $\mathbf{X}_i$, all $\mathbf{a}(\neq 0)$ and b, Prob. $(\mathbf{a}'\mathbf{X}_i = b) = 0$. Then Prob. $[\mathbf{X}'\mathbf{A}\mathbf{X} > \mathbf{0}] = 1$, i.e., $\mathbf{X}'\mathbf{A}\mathbf{X}$ is positive definite with probability one.

A.12 Factorization of Matrices

Factorization of symmetric matrices

Theorem A.12.1: Spectral Decomposition (or Jordanian Decomposition) If $\mathbf{A} = \mathbf{A}'$, there exists an orthogonal matrix $\mathbf{\Gamma}$ such that

$$\mathbf{A} = \mathbf{\Gamma}\mathbf{\Lambda}\mathbf{\Gamma}' \qquad (A.12.1)$$

where $\mathbf{\Lambda} =$ Diag $(\lambda_1, \dots, \lambda_p)$ and λ's are latent roots of $\mathbf{A}$ (some of which may be zeros and some of which may be identical). Here $\mathbf{\Gamma} = (\gamma_1 \dots \gamma_p)$ where $\gamma_i(p \times 1)$ is the normalized eigenvector corresponding to $\lambda_i, \gamma_i'\gamma_j = 1(0)$ if $i = (\neq)j$. Hence,

$$\mathbf{A} = \sum_{i=1}^{p} \lambda_i \gamma_i \gamma_i'. \qquad (A.12.2)$$

The matrix $\gamma_i\gamma_i'$ is called the ith spectral decomposition of $\mathbf{A}$ and the decomposition (A.12.2) the *Spectral decomposition* or (*Jordoanian decomposition*) of $\mathbf{A}$.

The following results hold:

(a) If $\mathbf{A}$ is a non-singular symmetric matrix, then for any integer k, positive or negative,

$$\mathbf{A}^k = \sum_{i=1}^{p} \lambda_i^k \gamma_i \gamma_i' = \mathbf{\Gamma}\mathbf{\Lambda}^k\mathbf{\Gamma}' \tag{A.12.3}$$

where $\mathbf{\Lambda}^k = \text{ Diag }(\lambda_1^k, \ldots, \lambda_p^k)$.

(b) If all the eigenvalues of (a symmetric matrix) $\mathbf{A}$ are positive, we can define the rational powers

$$\mathbf{A}^{r/s} = \mathbf{\Gamma}\mathbf{\Lambda}^{r/s}\mathbf{\Gamma}' \tag{A.12.4}$$

for any integer $s(>0)$ and r, where $\mathbf{\Lambda}^{r/s} = \text{ Diag }(\lambda_1^{r/s}, \ldots, \lambda_p^{r/s})$.

(c) If $\mathbf{A}$ is a symmetric singular matrix (at least one eigenvalue of $\mathbf{A}$ is zero), then (A.12.3) and (A.12.4) hold if the exponents (k or r/s) are restricted to be non-negative.

The result (A.12.3) follows easily, since

$$\mathbf{A}^2 = \mathbf{A}.\mathbf{A} = (\mathbf{\Gamma}\mathbf{\Lambda}\mathbf{\Gamma}')(\mathbf{\Gamma}\mathbf{\Lambda}\mathbf{\Gamma}') = \mathbf{\Gamma}\mathbf{\Lambda}^2\mathbf{\Gamma},$$

and the proof follows by induction. Thus,

$$\mathbf{A}^{-2} = \mathbf{\Gamma}\mathbf{\Lambda}^{-2}\mathbf{\Gamma}. \tag{A.12.5}$$

If $\lambda_i > 0 \;\forall\; i$,

$$\mathbf{A}^{-1/2} = \mathbf{\Gamma}\mathbf{\Lambda}^{-1/2}\mathbf{\Gamma}. \tag{A.12.6}$$

If $\lambda_i \geq 0 \;\forall\; i$,

$$\mathbf{A}^{1/2} = \mathbf{\Gamma}\mathbf{\Lambda}^{1/2}\mathbf{\Gamma}. \tag{A.12.7}$$

The decomposition (A.12.7) is called the *symmetric square root decomposition* of $\mathbf{A}$.

Orthogonal Factorization

Theorem A.12.2: Singular Value Decomposition Theorem For $\mathbf{A} : p \times n, p \leq n$, there exist orthogonal matrices $\mathbf{B} : p \times p, \mathbf{C} : n \times n$ such that

$$\mathbf{A} = \mathbf{B}(\mathbf{D}_\lambda \mathbf{0})\mathbf{C}', \tag{A.12.8}$$

where $\mathbf{D}_\lambda = \text{ Diag }(\lambda_1, \ldots, \lambda_p)$ and $|\mathbf{A}\mathbf{A}' - \lambda^2\mathbf{I}| = 0$.

The constants λ's are called the *singular values* of $\mathbf{A}$. Here $\mathbf{B}$ has p orthogonal eigenvectors of $\mathbf{AA}'$ as its columns and $\mathbf{C}$ has n orthogonal eigenvectors of $\mathbf{A}'\mathbf{A}$ as its columns.

The singular value decomposition can also be expressed as a matrix expansion that depends on the rank of $\mathbf{A}$. Specifically, there exist positive constants $\lambda_1, \ldots, \lambda_r$, r orthogonal $p \times 1$ normalized vectors $\mathbf{u}_1, \ldots, \mathbf{u}_r$ and r orthogonal $n \times 1$ normalized vectors $\mathbf{v}_1, \ldots, \mathbf{v}_r$ such that

$$\mathbf{A} = \sum_{i=1}^{r} \lambda_i \mathbf{u}_i \mathbf{v}_i' = \mathbf{U}_r \mathbf{\Lambda}_r \mathbf{V} r' \qquad (A.12.9)$$

where $\mathbf{U}_r = [\mathbf{u}_1 \ldots \mathbf{u}_r], \mathbf{V}_r = [\mathbf{v}_1 \ldots \mathbf{v}_r]$ and $\mathbf{\Lambda}_r =$ Diag $(\lambda_1 \ldots \lambda_r)$. Here $\mathbf{AA}'$ has (eigenvalue, eigenvector) pair $(\lambda_i^2, \mathbf{u}_i)$, so

$$\mathbf{AA}'\mathbf{u}_i = \lambda_i^2 \mathbf{u}_i$$

with $\lambda_1^2, \ldots, \lambda_r^2 > 0 = \lambda_{r+1}^2 = \lambda_{r+2}^2 = \ldots$. Similarly, $(\lambda_i^2, \mathbf{v}_i)(i = 1, \ldots, r)$ are the (eigenvalue, eigenvector) pairs of $\mathbf{A}'\mathbf{A}$.

Note that if $n = p$ and $\mathbf{A} = \mathbf{A}'$, taking $\mathbf{\Gamma} = \mathbf{B} = \mathbf{C}$, the result (A.12.8) reduces to the result (A.12.2).

EXAMPLE A.12.1: Let

$$\mathbf{A} = \begin{bmatrix} 2 & -1 & 1 \\ -1 & 2 & 1 \end{bmatrix}; \quad \text{hence} \quad \mathbf{AA}' = \begin{bmatrix} 1 & -3 \\ -3 & 6 \end{bmatrix}.$$

Here $\lambda_1^2 = 9, \ \mathbf{u}_1 = [\frac{1}{\sqrt{2}} \ -\frac{1}{\sqrt{2}}]'; \ \lambda_2^2 = 3, \ \mathbf{u}_2 = [\frac{1}{\sqrt{2}}, \ \frac{1}{\sqrt{2}}]'$.

Again,

$$\mathbf{A}'\mathbf{A} = \begin{bmatrix} 5 & -4 & 1 \\ -4 & 5 & 1 \\ 1 & 1 & 2 \end{bmatrix},$$

$$\begin{aligned} \lambda_1^2 &= 9, \quad \mathbf{v}_1 = [\tfrac{1}{\sqrt{2}}, \ -\tfrac{1}{\sqrt{2}}, \ 0]'; \\ \lambda_2^2 &= 3, \quad \mathbf{v}_2 = [\tfrac{1}{6}, \ \tfrac{1}{\sqrt{6}}, \ \tfrac{2}{\sqrt{6}}]'; \\ \lambda_3^2 &= 0, \quad \mathbf{v}_3 = [\tfrac{1}{\sqrt{3}}, \ \tfrac{1}{\sqrt{3}}, \ -\tfrac{1}{\sqrt{3}}]'. \end{aligned}$$

Singular value decomposition of $\mathbf{A}$ is

$$\mathbf{A} = 3 \begin{bmatrix} \frac{1}{\sqrt{2}} \\ -\frac{1}{\sqrt{2}} \end{bmatrix} [\frac{1}{\sqrt{2}}, -\frac{1}{\sqrt{2}}, 0] + \sqrt{3} \begin{bmatrix} \frac{1}{\sqrt{2}} \\ \frac{1}{\sqrt{2}} \end{bmatrix} [\frac{1}{\sqrt{6}}, \frac{1}{\sqrt{6}} \frac{2}{\sqrt{6}}].$$

Simultaneous Diagonalization of Matrices

We have the following results.

(a) Let $\mathbf{A} > 0$ and $\mathbf{B} \geq 0$. Then there exists a matrix $\mathbf{V}$, $|\mathbf{V}| \neq 0$, such that

$$\mathbf{VAV}' = \mathbf{I} \text{ and } \mathbf{VBV}' = \text{ Diag } (\lambda_1, \ldots, \lambda_p), \tag{A.12.10}$$

where the λ's are the roots of $|\mathbf{B} - \lambda\mathbf{A}| = 0$.

(b) If $\mathbf{A} = \mathbf{A}', \mathbf{B} = \mathbf{B}'$ and $\mathbf{AB} = \mathbf{BA}$, there exists an orthogonal matrix $\mathbf{P}$ such that

$$\begin{aligned} \mathbf{P}'\mathbf{AP} &= \text{ Diag } (\lambda_1, \ldots, \lambda_p), \\ \mathbf{P}'\mathbf{BP} &= \text{ Diag } (\mu_1, \ldots, \mu_p), \end{aligned} \tag{A.12.11}$$

where λ, μ's are the latent roots of $\mathbf{A}, \mathbf{B}$ respectively.

A.13 Idempotent Matrices

A square matrix $\mathbf{A}$ is said to be *idempotent* if $\mathbf{A}^2 = \mathbf{A}$. If $\mathbf{y} = \mathbf{Ax}$ and $\mathbf{A}$ is idempotent, then $\mathbf{y}$ is a projection of $\mathbf{x}$ on the space spanned by the columns of $\mathbf{A}$. A symmetric idempotent matrix is called a *projection matrix.*

Properties

(1) If $\mathbf{A}$ is idempotent its latent roots are either zeros or ones.
(2) If $\mathbf{A}$ is idempotent and $\mathbf{A} \geq 0$, then $r(\mathbf{A}) = tr(\mathbf{A})$.
(3) If $\mathbf{A}$ is idempotent, $(\mathbf{I} - \mathbf{A})$ is also idempotent.
(4) If $\mathbf{A}$ is idempotent and $\mathbf{A} = \mathbf{A}$, then $\mathbf{A} \geq 0$.
(5) If $\mathbf{A}$ is idempotent and $\mathbf{A} \geq 0$, there exists an orthogonal matrix $\mathbf{B}$ such that

$$\mathbf{BAB}' = \begin{pmatrix} \mathbf{I}_r & \mathbf{0} \\ \mathbf{0} & \mathbf{0} \end{pmatrix},$$

where $r(\mathbf{A}) = r$.
(6) If $\mathbf{A}$ is a projection matrix of rank r, then it can be expressed in the form

$$\mathbf{A} = \sum_{i=1}^{r} \gamma_i \gamma_i'$$

where $\gamma_1, \ldots, \gamma_r$ form an orthonormal set.
(7) If $\mathbf{X}(n \times p)$ is of rank $p, \mathbf{P} = \mathbf{X}(\mathbf{X}'\mathbf{X})^{-1}\mathbf{X}'$ is a projection matrix, and $\mathbf{PX} = \mathbf{X}$ and $tr\mathbf{P} = tr[(\mathbf{X}'\mathbf{X})^{-1}\mathbf{X}'\mathbf{X}] = tr[\mathbf{I}_p] = p$.

A.14 Some Matrix Inequalities

In Section A.3 we noted the Cauchy-Schwarz inequality. Here we consider two more inequalities.

Extended Cauchy-Schwarz inequality: Let $\mathbf{a}$ and $\mathbf{b}$ be two $p \times 1$ vectors and $\mathbf{C}$ be any p.d. matrix. Then

$$(\mathbf{a}'\mathbf{b})^2 \leq (\mathbf{a}'\mathbf{C}\mathbf{a})(\mathbf{b}'\mathbf{C}^{-1}\mathbf{b}) \tag{A.14.1}$$

with equality *iff* $\mathbf{a} = k\mathbf{C}^{-1}\mathbf{b}$ (or $\mathbf{b} = k\mathbf{C}\mathbf{a}$) for some constant k.

For $C = \mathbf{I}$, the equality (A.14.1) reduces to the Cauchy-Schwarz inequality.

Proof. Define $\mathbf{C}^{-1/2} = \sum_{i=1}^{p} \frac{1}{\sqrt{\lambda_i}} \mathbf{e}_i \mathbf{e}_i'$, where $\mathbf{e}_i$ is the normalized eigenvector corresponding to the eigenvalue λ_i of $\mathbf{C}(i = 1, \ldots, p)$. Now

$$\mathbf{a}'\mathbf{b} = \mathbf{a}'\mathbf{I}\mathbf{b} = \mathbf{a}'\mathbf{C}^{1/2}\underset{\sim}{C}^{-1/2}\mathbf{b} = (\mathbf{C}^{1/2}\mathbf{a})'(\mathbf{C}^{-1/2}\mathbf{b}).$$

Now apply Cauchy-Schwarz inequality to the vectors $(\mathbf{C}^{1/2}\mathbf{a})$ and $(\mathbf{C}^{-1/2}\mathbf{b})$.

The above inequality gives the following result.

Theorem A.14.1: Let $\mathbf{C}(p \times p)$ be any p.d. matrix and $\mathbf{b}(p \times 1)$ a given vector. Then for any arbitrary non-zero vector $\mathbf{x}$,

$$\max_{\mathbf{x}} \frac{(\mathbf{x}'\mathbf{b})^2}{\mathbf{x}'\mathbf{C}\mathbf{x}} = \mathbf{b}'\mathbf{C}^{-1}\mathbf{b},$$

the maximum value is attained when $\mathbf{x} = k\mathbf{C}^{-1}\mathbf{b}$ for some constant $k \neq 0$.

Theorem A.14.2: Let $\mathbf{A}$ and $\mathbf{B}$ be two symmetric matrices and suppose that $\mathbf{B} > 0$. Then the maximum (minimum) value of $\mathbf{x}'\mathbf{A}\mathbf{x}$, subject to the condition

$$\mathbf{x}'\mathbf{B}\mathbf{x} = 1 \tag{A.14.2}$$

is attained when $\mathbf{x}$ is the eigenvector of $\mathbf{B}^{-1}\mathbf{A}$ corresponding to the largest (smallest) eigenvalue of $\mathbf{B}^{-1}\mathbf{A}$. Moreover,

$$\lambda_1 = \max_{\mathbf{x} \neq \mathbf{0}} \frac{\mathbf{x}'\mathbf{A}\mathbf{x}}{\mathbf{x}'\mathbf{B}\mathbf{x}}.$$

$$\lambda_p = \min_{\mathbf{x} \neq \mathbf{0}} \frac{\underset{\sim}{x}'\mathbf{A}\mathbf{x}}{\mathbf{x}'\mathbf{B}\mathbf{x}},$$

where λ_1, λ_p are, respectively, the largest and the smallest eigenvalues of $\mathbf{B}^{-1}\mathbf{A}$.

Proof. Let $\mathbf{B}^{1/2}$ be the symmetric square root of $\mathbf{B}$ and $\mathbf{y} = \mathbf{B}^{1/2}\mathbf{x}$. We are to find

$$\max_{\mathbf{y}} \mathbf{y}'\mathbf{B}^{-1/2}\mathbf{A}\mathbf{B}^{-1/2}\mathbf{y} \text{ subject to the condition } \mathbf{y}'\mathbf{y} = 1. \tag{A.14.3}$$

Let a spectral decomposition of the matrix $\mathbf{B}^{-1/2}\mathbf{A}\mathbf{B}^{-1/2}$ be $\mathbf{\Gamma\Lambda\Gamma}'$, i.e.,

$$\mathbf{B}^{-1/2}\mathbf{A}\mathbf{B}^{-1/2} = \mathbf{\Gamma\Lambda\Gamma}',$$

where $\Lambda = \text{Diag}\,(\lambda_1, \ldots, \lambda_p), \mathbf{\Gamma} = (\gamma_1, \ldots, \gamma_p), \lambda_1 \geq \ldots \geq \lambda_p$ being the eigenvalues and $\gamma_1, \ldots, \gamma_p$ the corresponding eigenvectors of $\mathbf{B}^{-1/2}\mathbf{A}\mathbf{B}^{-1/2}$. Again, $\mathbf{B}^{-1/2}\mathbf{A}\mathbf{B}^{-1/2}$ and $\mathbf{B}^{-1}\mathbf{A}$ have the same eigenvalues (by A.8(8)).

We make the transformation $\mathbf{z} = \mathbf{\Gamma}'\mathbf{y}$. Then, from (A.14.3), the problem reduces to finding the maximum value,

$$\max{}_{\mathbf{z}} \mathbf{z}'\mathbf{\Lambda}\mathbf{z}$$

subject to the condition $\mathbf{z}'\mathbf{z} = 1$. By (A.8(14)), the maximum is attained when $\mathbf{z} = \mathbf{z}_1 = (1, 0, \ldots, 0)'$, the eigenvector corresponding to the eigenvalue λ_1 of $\mathbf{\Lambda}$ and the maximum value is λ_1. Now

$$\mathbf{z} = \mathbf{z}_1 \Rightarrow \mathbf{y} = \gamma_1 \Rightarrow \mathbf{x} = \mathbf{B}^{-1/2}\gamma_1.$$

The largest eigenvector of $\mathbf{B}^{-1/2}\mathbf{A}\mathbf{B}^{-1/2}$ satisfies

$$(\mathbf{B}^{-1/2}\mathbf{A}\mathbf{B}^{-1/2})\gamma_1 = \lambda_1\gamma_1,$$

i.e.,

$$(\mathbf{B}^{-1}\mathbf{A})(\mathbf{B}^{-1/2}\gamma_1) = \lambda_1(\mathbf{B}^{-1/2}\gamma).$$

Hence, $\mathbf{B}^{-1/2}\gamma_1$ is the eigenvalue of $\mathbf{B}^{-1}\mathbf{A}$ corresponding to the largest eigenvalue λ_1 of $\mathbf{B}^{-1}\mathbf{A}$.

The proof can similarly be extended for finding the minimum value of the quadratic form.

Theorem A.14.3: Let $\mathbf{E}, \mathbf{F}$ be positive definite matrices. Then

$$\text{Sup}_{\mathbf{x},\mathbf{y}} \left\{ \frac{(\mathbf{x}'\mathbf{D}\mathbf{y})^2}{(\mathbf{x}'\mathbf{E}\mathbf{x})(\mathbf{y}'\mathbf{F}\mathbf{y})} \right\} = \theta_{\max} \tag{A.14.4}$$

where $\theta_{\max}$ is the largest eigenvalue of $\mathbf{E}^{-1}\mathbf{D}\mathbf{F}^{-1}\mathbf{D}'$ and of $\mathbf{F}^{-1}\mathbf{D}'\mathbf{E}^{-1}\mathbf{D}$. The supremum occurs when $\mathbf{x}$ is an eigenvector of $\mathbf{E}^{-1}\mathbf{D}\mathbf{F}^{-1}\mathbf{D}'$ corresponding to $\theta_{\sup}$ and $\mathbf{y}$ is an eigenvector of $\mathbf{F}^{-1}\mathbf{D}'\mathbf{E}^{-1}\mathbf{D}$ corresponding to $\theta_{\max}$.

Proof. Let $\mathbf{z} = \mathbf{D}\mathbf{y}$. Then, assuming non-zero vectors $\mathbf{x}, \mathbf{y}$, we have

$$\text{Sup}_{\mathbf{x},\mathbf{y}} \left\{ \frac{(\mathbf{x}'\mathbf{D}\mathbf{y})^2}{(\mathbf{x}'\mathbf{E}\mathbf{x})(\mathbf{y}'\mathbf{F}\mathbf{y})} \right\} = \text{Sup}_{\mathbf{y}} \left\{ \frac{1}{\mathbf{y}'\mathbf{F}\mathbf{y}} \sup{}_{\mathbf{x}} \left[\frac{(\mathbf{x}'\mathbf{z})^2}{\mathbf{x}'\mathbf{E}\mathbf{x}} \right] \right\} \tag{A.14.5}$$

$$= \sup{}_{\mathbf{y}} \left\{ \frac{\mathbf{z}'\mathbf{E}^{-1}\mathbf{z}}{\mathbf{y}'\mathbf{F}\mathbf{y}} \right\} \quad \text{(by Theorem A.14.1)}$$

$$= \text{Sup}_{\mathbf{y}} \left\{ \frac{\mathbf{y}'\mathbf{D}'\mathbf{E}^{-1}\mathbf{D}\mathbf{y}}{\mathbf{y}'\mathbf{F}\mathbf{y}} \right\} = \theta_{\max} , \qquad (A.14.6)$$

where, by Theorem A.14.2, $\theta_{\max}$ is the maximum eigenvalue of $\mathbf{F}^{-1}\mathbf{D}'\mathbf{E}^{-1}\mathbf{D}$ and therefore of $\mathbf{E}^{-1}\mathbf{D}\mathbf{F}^{-1}\mathbf{D}'$ (by property (8) of Section A.8).

The supremum of [ ] with respect to $\mathbf{x}$ in (A.14.5) ia attained when $\mathbf{x} = k.\mathbf{E}^{-1}\mathbf{z}$ for some constant $k(\neq 0)$.

The supremum in (A.14.6) is attained when $\mathbf{y}$ is an eigenvector of $\mathbf{F}^{-1}\mathbf{D}'\mathbf{E}^{-1}\mathbf{D}$ corresponding to $\theta_{\max}$ (by Theorem A.14.2). This means $\mathbf{F}^{-1}\mathbf{D}'\mathbf{E}^{-1}\mathbf{D}\mathbf{y} = \theta_{\max}\,\mathbf{y}$.

Premultiplying both sides by $\mathbf{E}^{-1}\mathbf{D}$,

$$\mathbf{E}^{-1}\mathbf{D}\mathbf{F}^{-1}\mathbf{D}'\mathbf{E}^{-1}\mathbf{D}\mathbf{y} = \theta_{\max}\,\mathbf{E}^{-1}\mathbf{D}\mathbf{y},$$

i.e.,

$$\mathbf{E}^{-1}\mathbf{D}\mathbf{F}^{-1}\mathbf{D}'\mathbf{x} = \theta_{\max}\,\mathbf{x},$$

i.e., $\mathbf{x}$ is an eigenvector of $\mathbf{E}^{-1}\mathbf{D}\mathbf{F}^{-1}\mathbf{D}'$ corresponding to $\theta_{\max}$. □

A.15 Matrix Differentiation

The derivative of a scalar function $f(\mathbf{X})$ with respect to $\mathbf{X}(n \times p) = ((x_{ij}))$ is defined as

$$\frac{\partial f(\mathbf{X})}{\partial \mathbf{X}} = ((\frac{\partial f(\mathbf{X})}{\partial x_{ij}})).$$

The derivative of a matrix $\mathbf{X} = ((x_{ij}))$ with respect to an element y is defined as

$$\frac{\partial \mathbf{X}}{\partial y} = ((\frac{\partial x_{ij}}{\partial y})).$$

We have the following results.

(1) $\frac{\partial \mathbf{a}'\mathbf{x}}{\partial \mathbf{x}} = \mathbf{a}$.
(2) $\frac{\partial \mathbf{x}'\mathbf{x}}{\partial \mathbf{x}} = 2\mathbf{x}$, $\frac{\mathbf{x}'\mathbf{A}\mathbf{x}}{\partial \mathbf{x}} = 2\mathbf{A}\mathbf{x}$, $\frac{\partial \mathbf{x}'\mathbf{A}\mathbf{y}}{\partial \mathbf{x}} = \mathbf{A}\mathbf{y}$.
(3) $\frac{\partial |\mathbf{X}|}{\partial \mathbf{X}} = |\mathbf{X}|.\mathbf{X}^{-1}$ for $\mathbf{X} : p \times p, |\mathbf{X}| \neq 0$.
(4) $\frac{\partial |\mathbf{X}|^a}{\partial \mathbf{X}} = a|\mathbf{X}|^{a-1}.\frac{\partial |\mathbf{X}|}{\partial \mathbf{X}} = a|\mathbf{X}|^a\mathbf{X}^{-1}$.
(5) $\frac{\partial}{\partial \mathbf{X}}(f(\mathbf{X})g(\mathbf{X})) = f(\mathbf{X})\frac{\partial g(\mathbf{X})}{\partial \mathbf{X}} + g(\mathbf{X})\frac{\partial f(\mathbf{X})}{\partial \mathbf{X}}$ for scalar functions $f(.)$ and $g(.)$.

(6) If $\mathbf{X} : p \times p, |\mathbf{X}| \neq 0, \mathbf{a} : p \times 1$, then

$$\frac{\partial}{\partial \mathbf{X}}(\mathbf{a}'\mathbf{X}^{-1}\mathbf{a}) = -\mathbf{X}^{-1}\mathbf{a}\mathbf{a}'\mathbf{X}^{-1}.$$

(7) $\frac{\partial}{\partial \mathbf{X}} tr(\mathbf{A}'\mathbf{X}) = \mathbf{A}$.

(8) $\frac{\partial}{\partial \mathbf{X}} tr(\mathbf{A}\mathbf{X}') = \mathbf{A}$.

(9)

$$\frac{\partial tr(\mathbf{XY})}{\partial \mathbf{X}} = \begin{cases} \mathbf{Y}' & \text{if all elements of } \mathbf{X}_{(n\times p)} \text{ are distinct,} \\ \mathbf{Y} + \mathbf{Y}' - \text{ Diag } (\mathbf{Y}) & \text{if } \mathbf{X}(n \times n) \text{ is symmetric.} \end{cases}$$

(10) If $\mathbf{X}$ is a matrix with distinct elements,

$$\begin{aligned} \tfrac{\partial tr(\mathbf{AXB})}{\partial \mathbf{X}} &= \mathbf{A}'\mathbf{B}', \\ \tfrac{\partial tr(\mathbf{AX}'\mathbf{B})}{\partial \mathbf{X}} &= \mathbf{BA}, \\ \tfrac{\partial tr(\mathbf{X}'\mathbf{AXB})}{\partial \mathbf{X}} &= 2\mathbf{AXB} \ (\mathbf{A}, \mathbf{B} \text{ symmetric }), \\ \tfrac{\partial tr(\mathbf{XBX}'\mathbf{A})}{\partial \mathbf{X}}) &= 2\mathbf{AXB} \ (\mathbf{A}, \mathbf{B} \text{ symmetric}). \end{aligned}$$

(11) If $\mathbf{I}_{ij}$ be a $p \times p$ matrix with 1 in the (i, j)th place and 0 elsewhere, then for $\mathbf{X}(p \times p)$ and $|\mathbf{X}| \neq 0$,

$$\frac{\partial \mathbf{X}^{-1}}{\partial x_{ij}} = -\mathbf{X}^{-1}\mathbf{I}_{ij}\mathbf{X}^{-1},$$

if all elements of $\mathbf{X}$ are distinct. If $\mathbf{X}$ is symmetric,

$$\frac{\partial \mathbf{X}^{-1}}{\partial x_{ij}} = \begin{cases} -\mathbf{X}^{-1}\mathbf{I}_{ii}\mathbf{X}^{-1}, & i = j, \\ -\mathbf{X}^{-1}(\mathbf{I}_{ij} + \mathbf{I}_{ji})\mathbf{X}^{-1}, & i \neq j. \end{cases}$$

(12) If all elements of $\mathbf{X}(p \times p)$ are distinct, $\frac{\partial |\mathbf{X}|}{\partial x_{ij}} = \mathbf{X}_{ij}$.

If $\mathbf{X}$ is symmetric,

$$\frac{\partial |\mathbf{X}|}{\partial x_{ij}} = \begin{cases} 2\mathbf{X}_{ii}, \ i = j, \\ 2\mathbf{X}_{ij} \ , i \neq j. \end{cases}$$

where $\mathbf{X}_{ij}$ is the cofactor of x_{ij} in $\mathbf{X}$.

Second derivative of a scalar function of a vector with respect to a vector:

Let $f(\mathbf{x})$ be a scalar function of a $p \times 1$ vector $\mathbf{x}$. The derivative of $f(\mathbf{x})$ with respect to $\mathbf{x}$ is a $p \times 1$ vector

$$\frac{\partial f(\mathbf{x})}{\partial \mathbf{x}} = \left(\frac{\partial f(\mathbf{x})}{\partial x_i}, i = 1, \ldots, p \right).$$

The second derivative of $f(\mathbf{x})$ with respect to $\mathbf{x}$ is obtained by differentiating each element of $\frac{\partial f(\mathbf{x})}{\partial \mathbf{x}}$ with respect to the row vector $\mathbf{x}'$. Thus,

$$\frac{\partial^2 f(\mathbf{x})}{\partial \mathbf{x}' \partial \mathbf{x}} = \frac{\partial}{\partial \mathbf{x}'}(\frac{\partial f(\mathbf{x})}{\partial \mathbf{x}}) = \left[\frac{\partial}{\partial x_1}(\frac{\partial f(x)}{\partial \mathbf{x}}), \ldots, \frac{\partial}{\partial x_p}(\frac{\partial f(x)}{\partial \mathbf{x}})\right].$$

It can be easily checked that if $f(\mathbf{x}) = \mathbf{x}'\mathbf{x}$, then

$$\frac{\partial^2 f(\mathbf{x})}{\partial \mathbf{x}' \partial \mathbf{x}} = 2\mathbf{I}_p.$$

If $\mathbf{x}$ is $p \times 1$ and $\mathbf{A}$ is $p \times p$, then

$$\frac{\partial^2 (\mathbf{x}'\mathbf{A}\mathbf{x})}{\partial \mathbf{x}' \partial \mathbf{x}} = \mathbf{A}.$$

The symmetric matrix

$$\frac{\partial^2 f(\mathbf{x})}{\partial \mathbf{x} \partial \mathbf{x}'} = \mathbf{H} = ((h_{ij})) \text{ (say)}$$

is defined as the *Hessian matrix* of the scalar function $f(\mathbf{x})$.

If $\left.\frac{\partial f(\mathbf{x})}{\partial \mathbf{x}}\right]_{\mathbf{x}_0} = 0$, then $\mathbf{x}_0$ is a stationary point of $f(\mathbf{x})$.

If in addition, $H > (<) 0$ for all $\mathbf{x}$, then $\mathbf{x}_0$ corresponds to a global minimum (maximum) of $f(\mathbf{x})$. If $\mathbf{H} > (<) 0$, at least in a neighborhood of a point $\mathbf{x}_0$, then $\mathbf{x}_0$ corresponds to local minimum (maximum).

Functions $f(\mathbf{x})$ for which $\mathbf{H} > (<) 0$ are called convex (concave) functions.

In general, $f(\mathbf{x})$ is convex (concave) if for any two points $\mathbf{x}_1$ and $\mathbf{x}_2$ and for any $\lambda \in (0,1)$, $f[\lambda \mathbf{x}_1 + (1-\lambda)\mathbf{x}_2] \leq (\geq) \lambda f(\mathbf{x}_1) + (1-\lambda) f(\mathbf{x}_2)$.

A.16 System of Equations

Consider a set of m linear equations in n unknowns $x_1, \ldots, x_n$:

$$\mathbf{A}\mathbf{x} = \mathbf{b} \tag{A.16.1}$$

where $\mathbf{A} = ((a_{ij}))_{m \times n}, \mathbf{x} = (x_1, \ldots, x_n)', \mathbf{b} = (b_1, \ldots, b_m)'$.

The equations are consistent (i.e., at least one set of values of $\mathbf{x}$ exists that satisfy all the equations) *iff* $r = r'$, where

$$r = \text{ rank } (\mathbf{A}), \; r' = \text{ rank } (\mathbf{B}) \text{ and } \mathbf{B} = (\mathbf{A}, \mathbf{b}).$$

If rank $(r) = r = r' < n$, there exists an infinite number of solutions of the equations (A.16.1). If $n = r = r'$, the equations have a unique solution, $\mathbf{x} = \mathbf{A}^{-1}\mathbf{b}$. (The general solution of equations (A.16.1) are given below).

Consider the set of m homogeneous equations

$$\mathbf{Ax} = \mathbf{0}. \tag{A.16.2}$$

Here $r = r'$ always and hence the equations are always consistent. When $r = n$, the only solution of (A.16.2) is $\mathbf{x} = \mathbf{0}$. When $r < n$ we reduce the equations (A.16.2) to the equations

$$\sum_{t=1}^{r} a_{it}x_t = -\sum_{t=r+1}^{n} a_{it}x_t (i = 1, \ldots, r) \tag{A.16.3}$$

such that the determinant of the coefficients on the left-hand side are not zeros. In this case, all solutions of (A.16.2) are obtained by giving arbitrary values to $x_{r+1}, \ldots, x_n$ and then solving the equations (A.16.3) for $x_1, \ldots, x_r$.

Let $\mathbf{X}_K$ be the solution obtained by putting $x_{r+1} = 0, \ldots, x_{r+K-1} = 0, x_{r+K} = 1, \ldots, x_n = 0 (K = 1, \ldots, n-r)$. The general solution of (A.16.2) is given by

$$\mathbf{X} = \sum_{K=1}^{n-r} x_{r+K}\mathbf{X}_K \tag{A.16.4}$$

where $x_{r+1}, \ldots, x_n$ are given arbitrary values. Solutions $\mathbf{X}_1, \ldots, \mathbf{X}_{n-r}$ are all linearly independent (i.e., no member of it can be expressed as a sum of multiples of the others). Such a set of solutions is called a fundamental set of solutions. The rank of the matrix $\mathbf{X}(n \times (n-r))$ whose columns form a fundamental set is $(n-r)$.

Theorem A.16.1: A particular solution of the consistent equation

$$\mathbf{Ax} = \mathbf{b} \tag{A.16.5}$$

is

$$\mathbf{x} = \mathbf{A}^{-}\mathbf{b}. \tag{A.16.6}$$

where $\mathbf{A}^{-}$ is a g-inverse of $\mathbf{A}$.

Proof.

$$(A.16.5) \Rightarrow \mathbf{AA}^{-}\mathbf{Ax} = \mathbf{AA}^{-}\mathbf{b},$$

i.e.,

$$\mathbf{Ax} = \mathbf{AA}^{-}\mathbf{b}.$$

Hence, $\mathbf{x} = \mathbf{A}^-\mathbf{b}$. It can be shown that a general solution of a consistent equation (A.16.5) is

$$\mathbf{x} = \mathbf{A}^-\mathbf{b} + (\mathbf{I} - \mathbf{A}\mathbf{A}^-)\mathbf{z},$$

where $\mathbf{z}$ is an arbitrary vector. For $\mathbf{b} = \mathbf{0}$, a general solution is $\mathbf{x} = (\mathbf{I} - \mathbf{A}\mathbf{A}^-)\mathbf{z}$.

A.17 Decomposition of a Vector

A.17.1 *Orthogonal Decomposition of a Vector*:

1. Let $\mathcal{M}$ be a vector subspace of $\mathcal{R}^n$ (the n-dimensional Euclidean space). Then every $n \times 1$ vector $\mathbf{y}$ can be decomposed uniquely as

$$\mathbf{y} = \mathbf{u} + \mathbf{v}$$

where $\mathbf{u} \in \mathcal{M}$ and $\mathbf{v} \in \mathcal{M}^\perp$.

Proof. Suppose there are two decompositions $\mathbf{y} = \mathbf{u}_i + \mathbf{v}_i$ $(i = 1, 2)$. Then $\mathbf{u}_1 - \mathbf{u}_2 + \mathbf{v}_1 - \mathbf{v}_2 = \mathbf{0}$. Again, $\mathbf{u}_1 - \mathbf{u}_2 \in \mathcal{M}, \mathbf{v}_1 - \mathbf{v}_2 \in \mathcal{M}^\perp$. Thus, $(\mathbf{u}_1 - \mathbf{u}_2) + (\mathbf{v}_1 - \mathbf{v}_2)$ is an orthogonal decomposition of $\mathbf{0}$. This is possible *iff* $\mathbf{u}_1 = \mathbf{u}_2, \mathbf{v}_1 = \mathbf{v}_2$.

2. The vector $\mathbf{v}$ has the property

$$\inf{}_{\mathbf{x} \in \mathcal{M}} ||\mathbf{y} - \mathbf{x}|| = ||\mathbf{v}||$$

where $||\delta|| = \sqrt{\sum \delta_i^2}$ denotes the length of δ. The minimum is attained at $\mathbf{x} = \mathbf{u}$.

3. The vector $\mathbf{u}$ can be written as $\mathbf{u} = \mathbf{P}\mathbf{y}$ and $\mathbf{P}$ is unique. Here $\mathbf{P}$ is the orthogonal projection matrix which transforms any $\mathbf{y}$ in $\mathcal{R}^n$ into its orthogonal projection onto $\mathcal{M}$. We can write $\mathbf{P} = \mathbf{P}_{\mathcal{M}}$.

Proof. If there exist two such matrices $\mathbf{P}_1, \mathbf{P}_2$ then since the decomposition of $\mathbf{y}$ is unique, $(\mathbf{P}_1 - \mathbf{P}_2)\mathbf{y} = \mathbf{0}\ \forall\ \mathbf{y}$ or $\mathbf{P}_1 - \mathbf{P}_2 = \mathbf{0}$ and $\mathbf{P}$ is unique. The existence of $\mathbf{P}$ has been proved in A.17.1.8.

4. The matrix $(\mathbf{I}_n - \mathbf{P})$ represents the orthogonal projection onto $^\perp$.

Proof. We have $\mathbf{y} = \mathbf{P}\mathbf{y} + (\mathbf{I}_n - \mathbf{P})\mathbf{y}$. Comparing with the unique representation $\mathbf{y} = \mathbf{u} + \mathbf{v}, \mathbf{v} = (\mathbf{I}_n - \mathbf{P})\mathbf{y}$. The matrix $(\mathbf{I}_n - \mathbf{P})$ projects $\mathbf{y}$ onto $\mathcal{M}^\perp$.

5. $\mathbf{P}$ and $(\mathbf{I}_n - \mathbf{P})$ are symmetric and idempotent.

Proof. For any two $n \times 1$ vectors $\mathbf{y}$ and $\mathbf{z}, \mathbf{P}\mathbf{y} \in \mathcal{M}$ and $(\mathbf{I}_n - \mathbf{P})\mathbf{y} \in \mathcal{M}^\perp$. Hence, $\mathbf{y}'\mathbf{P}'(\mathbf{I}_n - \mathbf{P})\mathbf{z} = \mathbf{0}\ \forall\ \mathbf{y}$ or $\mathbf{P}'(\mathbf{I}_n - \mathbf{P}) = \mathbf{0}$, i.e., $\mathbf{P}' = \mathbf{P}'\mathbf{P}$, so that $\mathbf{P}$ is symmetric and $\mathbf{P} = \mathbf{P}^2$. Also, $(\mathbf{I}_n - \mathbf{P})^2 = \mathbf{I}_n - 2\mathbf{P} + \mathbf{P}^2 = \mathbf{I}_n - \mathbf{P}$.

6. $\mathcal{R}(\mathbf{P}) = \mathcal{M}$ and dim $(\mathcal{M}) = tr(\mathbf{P})$, where $\mathcal{R}(\mathbf{P})$ denotes the vector subspace in $\mathcal{R}^n$ spanned by columns of $\mathcal{P}$, also called the range-space of $\mathbf{P}$.

Proof. We have $\mathbf{Py} = \mathbf{u} \in \mathcal{M}$. Hence, $\mathcal{R}(\mathbf{P}) \subset \mathcal{M}$. Again, if $\mathbf{x} \in \mathcal{M}$, then the unique orthogonal decomposition of $\mathbf{x}$ is $\mathbf{x} = \mathbf{x} + \mathbf{0}$ so that $\mathbf{x} = \mathbf{Px} \in \mathcal{R}(\mathbf{P})$. Hence, $\mathcal{M} \subset \mathcal{R}(\mathbf{P})$. Hence the two spaces are identical. Also, dim $(\mathcal{M}) =$ rank $\mathcal{M} =$ rank $(\mathbf{P}) = tr(\mathbf{P})$.

7. If $\mathbf{P}$ is a symmetric idempotent matrix, then $\mathbf{P}$ represents an orthogonal projection onto $\mathcal{R}(\mathbf{P})$.

Proof. Let $\mathbf{y} = \mathbf{Py} + (\mathbf{I}_n - \mathbf{P})\mathbf{y}$. Then $(\mathbf{Py})'(\mathbf{I} - \mathbf{P})\mathbf{y} = \mathbf{y}'(\mathbf{P} - \mathbf{P}^2)\mathbf{y} = 0$ so that this decomposition gives the orthogonal components of $\mathbf{y}$. The result follows by A.17(6).

8. If $\mathcal{M} = \mathcal{R}(\mathbf{X})$ and $\mathbf{X}$ is of full rank, then $\mathbf{P} = \mathbf{X}(\mathbf{X}'\mathbf{X})^{-1}\mathbf{X}'$.

Proof. We have $\mathbf{PX} = \mathbf{X}$. Hence,

$$\mathbf{X}'(\mathbf{y} - \mathbf{Py}) = \mathbf{X}'(\mathbf{I} - \mathbf{Py}) = \mathbf{0}. \qquad (A.17.1)$$

Let θ be the projection of $\mathbf{y}$ onto $\mathcal{M}, \theta = \mathbf{Py} = \mathbf{X}\beta$, say. If β satisfies $\theta = \mathbf{X}\beta$, then (A.17.1) means

$$\mathbf{X}'\mathbf{y} = \mathbf{X}'\mathbf{X}\beta. \qquad (A.17.2)$$

Conversely, if β satisfies (A.17.2), then

$$(\mathbf{X}\beta)'(\mathbf{y} - \mathbf{X}\beta) = \mathbf{0}.$$

Hence, $\mathbf{y} = \mathbf{X}\beta + (\mathbf{y} - \mathbf{X}\beta)$ represents the orthogonal decomposition of $\mathbf{y}$. Therefore, from (A.17.2), $\beta = (\mathbf{X}'\mathbf{X})^{-1}\mathbf{X}'\mathbf{y}$. Hence, $\theta = \mathbf{X}(\mathbf{X}'\mathbf{X})^{-1}\mathbf{X}'\mathbf{y}$. Therefore, $\mathbf{P} = (\mathbf{X}'\mathbf{X})^{-1}\mathbf{X}'$.

If $\mathbf{X}$ is not of full rank, then $\mathbf{P} = (\mathbf{X}'\mathbf{X})^{-}\mathbf{X}'$, where $\mathbf{X}^{-}$ is the generalized inverse of $\mathbf{X}$.

A.17.2 *Orthogonal Complements*

1. For any matrix $\mathbf{A}$, the null space of $\mathbf{A}$, denoted by $\mathcal{N}(\mathbf{A})$ is the orthogonal complement of the range-space of $\mathbf{A}'$, i.e., $\mathcal{N}(\mathbf{A}) = \{\mathcal{R}[\mathbf{A}]\}^{\perp}$.

This is, because,

$$\mathcal{N}(\mathbf{A}) = \{\mathbf{x} : \mathbf{Ax} = \mathbf{0}\},$$

each $\mathbf{x} \in \mathcal{N}(\mathbf{A})$ is orthogonal to all the rows of $\mathbf{A}$, i.e. all the columns of $\mathbf{A}'$.

2. If the columns of $\mathbf{A}'$ are linearly independent, then the orthogonal projector onto $\mathcal{N}(\mathbf{A}), \mathbf{P}_{\mathcal{N}(\mathbf{A})}$ is $\mathbf{I} - \mathbf{A}'(\mathbf{A}\mathbf{A}')^{-1}\mathbf{A}$.

Proof. Let $\mathcal{N}(\mathbf{A}) = \Omega$. Then, $\mathcal{R}(\mathbf{A}) = \Omega^{\perp}$. Then, $\mathbf{P}_{\Omega^{\perp}} = \mathbf{A}'(\mathbf{A}\mathbf{A}')^{-1}\mathbf{A}$ (by A.17.1.8). Now, $\mathbf{P}_{\Omega} = \mathbf{I} - \mathbf{P}_{\Omega^{\perp}} = \mathbf{I} - \mathbf{A}'(\mathbf{A}\mathbf{A}')^{-1}\mathbf{A}$ by (A.17.1.4).

3. Let $\mathbf{B}$ be any $p \times q$ matrix of rank q. Let $\mathcal{V} = \mathcal{R}(\mathbf{B})$. Then there exists a $(p-q) \times p$ matrix $\mathbf{C}$ satisfying $\mathbf{CB} = \mathbf{0}$ such that $\mathcal{V} = \mathcal{N}(\mathbf{C}) = \{\mathcal{R}[\mathbf{X}]\}^{\perp}$.

Proof. Since $\mathcal{V} = \mathcal{R}(\mathbf{B}), \mathbf{P}_{\mathcal{V}} = \mathbf{B}(\mathbf{B}'\mathbf{B})^{-1}\mathbf{B}'$ and is of rank q. Therefore, $\mathbf{I}_p - \mathbf{P}_{\mathcal{V}}$ is a $p \times p$ matrix of rank $(p-q)$. Let $\mathbf{C}$ be any matrix consisting of $(p-q)$ linearly independent rows of $(\mathbf{I}_p - \mathbf{P}_{\mathcal{V}})$. Then $(\mathbf{I}_p - \mathbf{P}_{\mathcal{V}})\mathbf{P}_{\mathcal{V}} = \mathbf{0} \Rightarrow \mathbf{CB} = \mathbf{0}$. Also, $\mathcal{N}[\mathbf{C}] = \{\mathcal{R}[\mathbf{C}']\}^{\perp} = \{\mathcal{R}(\mathbf{I}_p - \mathbf{P}_{\mathcal{V}})\}^{\perp} = \mathcal{N}(\mathbf{I}_p - \mathbf{P}_{\mathcal{V}}) = \mathcal{R}(\mathbf{B}) = \mathcal{V}$.

4. $(\Omega_1 \cap \Omega_2)^{\perp} = \Omega_1^{\perp} \cup \Omega_2^{\perp}$.

Proof. Let $\mathbf{C}_i$ be such that $\mathcal{N}(\mathbf{C}_i) = \Omega_i (i = 1, 2)$. Now

$$\Omega_1 = \{\mathbf{x} : \mathbf{C}_1\mathbf{x} = \mathbf{0}\},$$

$$\Omega_2 = \{\mathbf{x} : \mathbf{C}_2\mathbf{x} = \mathbf{0}.$$

Hence,

$$\Omega_1 \cap \Omega_2 = \{\mathbf{x} : \mathbf{C}_i\mathbf{x} = \mathbf{0}, i = 1, 2\}.$$

Therefore,

$$\begin{aligned}(\Omega_1 \cap \Omega_2)^{\perp} &= \left\{\mathcal{N}\begin{bmatrix}\mathbf{C}_1 \\ \mathbf{C}_2\end{bmatrix}\right\}^{\perp} \\ &= \mathcal{R}[\mathbf{C}_1', \mathbf{C}_2'] \\ &= \mathcal{R}[\mathbf{C}_1'] \cup \mathcal{R}[\mathbf{C}_2'] \\ &= \Omega_1^{\perp} \cup \Omega_2^{\perp}.\end{aligned}$$

A.17.3 *Projection and Subspaces*

1. If $\omega \subset \Omega$, then $\mathbf{P}_{\Omega}\mathbf{P}_{\omega} = \mathbf{P}_{\omega}\mathbf{P}_{\Omega} = \mathbf{P}_{\omega}$.

Proof. We have $\mathcal{R}(\mathbf{P}_{\omega}) = \omega$, i.e., $\mathbf{P}_{\omega}$ is a projection matrix whose range-space is ω. Again, $\omega \subset \Omega$. Therefore, $\mathbf{P}_{\Omega}(\mathbf{P}_{\omega})$ is the projection of $\mathbf{P}_{\omega}$ on Ω which must be $\mathbf{P}_{\omega}$ itself. Thus $\mathbf{P}_{\Omega}(\mathbf{P}_{\omega}) = \mathbf{P}_{\omega}$. Again, $\mathbf{P}_{\Omega}\mathbf{P}_{\omega} = \mathbf{P}_{\omega}\mathbf{P}_{\Omega}$ by symmetry.

2. Let $\omega \subset \Omega$ Then $\mathbf{P}_{\Omega} - \mathbf{P}_{\omega} = \mathbf{P}_{\omega^{\perp} \cap \Omega}$.

Proof. For any $n \times 1$ vector $\mathbf{y}$ we have

$$\mathbf{P}_{\Omega}\mathbf{y} = \mathbf{P}_{\omega}\mathbf{y} + (\mathbf{P}_{\Omega} - \mathbf{P}_{\omega})\mathbf{y}.$$

Now,

$$\mathbf{P}_\omega(\mathbf{P}_\Omega - \mathbf{P}_\omega) = \mathbf{P}_\omega - \mathbf{P}_\omega = \mathbf{0}.$$

Therefore, $\mathbf{P}_\Omega - \mathbf{P}_\omega$ is the projection matrix which projects $\mathbf{y}$ on the space which is orthogonal to ω and which (the space) at the same time is a subspace of Ω. Thus, $\mathbf{P}_\Omega - \mathbf{P}_\omega = \mathbf{P}_{\omega^\perp \cap \Omega}$.

3. Let $\mathbf{A}$ be any square matrix and $\omega = \mathcal{N}(\mathbf{A}) \cap \Omega$. Then

$$\omega^\perp \cap \Omega = \mathcal{R}(\mathbf{P}_\Omega \mathbf{A}').$$

Proof.

$$\begin{aligned}\omega^\perp \cap \Omega &= \{\mathcal{N}(\mathbf{A}) \cap \Omega\}^\perp \cap \Omega \\ &= [\Omega^\perp + \{\mathcal{N}(\mathbf{A})\}^\perp] \cap \Omega \quad \text{(by A.17.2.4)} \\ &= [\Omega^\perp + \mathcal{R}(\mathbf{A}')\} \cap \Omega \quad \text{(by A.17.2.1)}.\end{aligned}$$

Now, if $\mathbf{x} \in \omega^\perp \cap \Omega$ then

$$\begin{aligned}\mathbf{x} = \mathbf{P}_\Omega \mathbf{x} &= \mathbf{P}_\Omega[(\mathbf{I} - \mathbf{P}_\Omega)\alpha + \mathbf{A}'\beta] \\ &= \mathbf{P}_\Omega \mathbf{A}'\beta \in \mathcal{R}(\mathbf{P}_\Omega \mathbf{A}').\end{aligned}$$

Hence,

$$\omega^\perp \cap \Omega \subseteq \mathcal{R}(\mathbf{P}_\Omega \mathbf{A}'). \tag{A.17.3}$$

Conversely, if $\mathbf{x} \in \mathcal{R}[\mathbf{P}_\Omega \mathbf{A}']$, then $\mathbf{x} \in \mathcal{R}[\mathbf{P}_\Omega] = \Omega$. Again, if $\mathbf{z} \in \omega$, then $\mathbf{Az} = \mathbf{0}$, because $\mathbf{z} \in \mathcal{N}(\mathbf{A})$. Also, $\mathbf{x}'\mathbf{z} = \beta'\mathbf{A}\mathbf{P}_\Omega\mathbf{z} = \beta'\mathbf{Az}$ (because $\mathbf{z} \in \omega) = \mathbf{0}$, that is, $\mathbf{x} \in \omega^\perp$. Hence, if $\mathbf{x} \in \mathcal{R}[\mathbf{P}_\Omega \mathbf{A}']$, then $\mathbf{x} \in \Omega \cap \omega^\perp$. Thus,

$$\mathcal{R}[\mathbf{P}_\Omega \mathbf{A}'] \subset \Omega \cap w^\perp. \tag{A.17.4}$$

Combining (A.17.3) and (A.17.4) the result follows.

4. If $\mathbf{A}$ is $q \times n$ matrix of rank q, then rank $(\mathbf{P}_\Omega \mathbf{A}') = q$ *iff* $\mathcal{R}(\mathbf{A}') \cap \Omega^\perp = \mathbf{0}$.

Proof. Rank $(\mathbf{P}_\Omega \mathbf{A}') \leq$ rank $\mathbf{A} = q$. Suppose rank $(\mathbf{P}_\Omega \mathbf{A}') < q$. Let $\mathbf{A} = [\mathbf{a}_1, \mathbf{a}_2, \ldots, \mathbf{a}_q]'$. Then $\mathbf{A}' = [\mathbf{a}_1, \mathbf{a}_2, \ldots, \mathbf{a}_q]$. If rank $(\mathbf{P}_\Omega \mathbf{A}') < q$, then for some constants $c_1, c_2, \ldots, c_q$,

$$\sum_{i=1}^{q} \mathbf{P}_\Omega c_i \mathbf{a}_i = \mathbf{0}, \text{ i.e., } \mathbf{P}_\Omega \sum_{i=1}^{q} c_i \mathbf{a}_i = \mathbf{0}$$

i.e.

$$\sum_{i=1}^{q} c_i \mathbf{a}_i \in \Omega^\perp.$$

also, $\sum_{i=1}^{q} c_i \mathbf{a}_i \subset \mathcal{R}(\mathbf{A}')$. Therefore, $\Omega^\perp \cap \mathcal{R}(\mathbf{A}') \neq \mathbf{0}$, which is a contradiction. Hence the proof.

Exercises and Complements

A.1 Show that the Vandermonde determinant defined by

$$\begin{bmatrix} 1 & 1 & \dots & 1 \\ a_1 & a_2 & \dots & a_n \\ \dots & \dots & \dots & \dots \\ a_1^{n-1} & a_2^{n-1} & \dots & a_n^{n-1} \end{bmatrix}$$

has the value $\Pi_{i<j}(a_i - a_j)$.

A.2 If $\mathbf{A}, \mathbf{B}, \mathbf{C}$ are three matrices such that $\mathbf{AB}, \mathbf{BC}$ and $\mathbf{ABC}$ are defined. Show that

$$\text{rank }(\mathbf{AB}) + \text{ rank }(\mathbf{BC}) \leq \text{ rank }(\mathbf{B}) + \text{ rank }(\mathbf{ABC}).$$

A.3 Let

$$\mathbf{A} = 8\begin{bmatrix} 1 & 1/4 & 1/2 \\ 1/4 & 1 & 1/3 \\ 1/2 & 1/3 & 1 \end{bmatrix}.$$

Find the latent roots and all sets of normalized vectors of $\mathbf{A}$. Is $\mathbf{A}$ singular? What are the latent roots of $\mathbf{A}^2$?.

A.4 (a) Show that if $\mathbf{A}^2 + 3\mathbf{A} + \mathbf{I} = \mathbf{0}$, then $\mathbf{A}^{-1} = -\mathbf{A} - 3\mathbf{I}$.

(b) Let $\mathbf{A}$ and $\mathbf{B}$ be $n \times n$ matrices such that $\mathbf{AB} = \mathbf{0}$. Show that if $\mathbf{B} \neq 0$, then $\mathbf{A}$ is singular.

A.5 (a) Determine the nullspace of the following matrix

$$\begin{bmatrix} 1 & 1 & -1 & 2 \\ 2 & 2 & -3 & 1 \\ -1 & -1 & 0 & -5 \end{bmatrix}.$$

(b) Let S be the set of all ordered pairs of real numbers. Define scalar multiplication and addition on S by

$$\begin{aligned} \alpha(x_1, x_2) &= (\alpha x_1, \alpha x_2) \\ (x_1, x_2 + (y_1, y_2) &= (x_1 + y_1, 0). \end{aligned}$$

Is S a vector space? Justify.

A.6 Let $\mathbf{Axb}$ be a system of linear equations, where $\mathbf{A}$ is a $n \times n$ matrix.

(a) Suppose $\mathbf{x} = \mathbf{0}$ is the only solution of the system when $\mathbf{b} = \mathbf{0}$. How many solutions, if any, does the system have when $\mathbf{b} = (1, 2, \dots, n)$? Justify.

(b) Suppose $\mathbf{x} = \mathbf{0}$ is not the only solution of the system when $\mathbf{b} = \mathbf{0}$. Does the system have precisely one solution when $\mathbf{b}' = (1, 2, \ldots, n)$? Justify.

A.7 Let $\mathbf{A}$ and $\mathbf{B}$ be symmetric real positive definite matrices of order n and let $\mathbf{A} - \mathbf{B}$ be positive. Show that $\mathbf{B}^{-1} - \mathbf{A}^{-1}$ is positive definite.

A.8 Show that the following matrix identities are true:

(a) $(\mathbf{D} + \mathbf{A}\mathbf{B}^{-1}) = \mathbf{D}^{-1} - \mathbf{D}^{-1}\mathbf{A}(\mathbf{I} + \mathbf{B}\mathbf{D}^{-1}\mathbf{A})^{-1}\mathbf{B}\mathbf{D}^{-1}$;
(b) $(\mathbf{D} + \mathbf{E}\mathbf{F}\mathbf{E}')^{-1} = \mathbf{D}^{-1} - \mathbf{D}^{-1}\mathbf{E}(\mathbf{E}'\mathbf{D}^{-1}\mathbf{E} + \mathbf{F}^{-1})\mathbf{E}'\mathbf{D}^{-1}$;
(c) $(\mathbf{D} + \mathbf{B})^{-1} = \mathbf{D}^{-1} - \mathbf{D}^{-1}(\mathbf{D}^{-1} + \mathbf{B}^{-1})^{-1}\mathbf{D}^{-1}$;
(d) $(\mathbf{D} + \mathbf{B})^{-1}\mathbf{B} = \mathbf{I} - (\mathbf{D} + \mathbf{B})^{-1}\mathbf{D}$,

where it is assumed that all matrices are of appropriate (conformable) dimensions and that all the stated inverses exist.

A.9 Let $\mathbf{x}, \mathbf{a}, \mathbf{b}$ be $k \times 1$ vectors and $\mathbf{A}, \mathbf{B}$ be $k \times k$ symmetric matrices such that $(\mathbf{A} + \mathbf{B})^{-1}$ exists. Show that

$$(\mathbf{x} - \mathbf{a})'\mathbf{A}(\mathbf{x} - \mathbf{a}) + (\mathbf{x} - \mathbf{b})'\mathbf{B}(\mathbf{x} - \mathbf{b}) = (\mathbf{x} - \mathbf{c})'(\mathbf{A} + \mathbf{B})(\mathbf{x} - \mathbf{c}) +$$

$$(\mathbf{a} - \mathbf{b})'\mathbf{A}(\mathbf{A} + \mathbf{B})^{-1}\mathbf{B}(\mathbf{a} - \mathbf{b})$$

where $\mathbf{c} = (\mathbf{A} + \mathbf{B})^{-1}(\mathbf{A}\mathbf{a} + \mathbf{B}\mathbf{b})$.

A.10 Let $(\lambda_i, \mathbf{e}_i)(i = 1, \ldots, m)$ be the eigenvalue-eigenvector pairs of a matrix $\mathbf{A}(m \times m)$. Show that

$$\sum_{i=1}^{m} \mathbf{e}_i\mathbf{e}_i' = \mathbf{I}_m.$$

A.11 Let $\lambda_1(\mathbf{C}) \geq \lambda_2(\mathbf{C}) \geq \ldots \lambda_n(\mathbf{C})$ be the eigenvalues of a $n \times n$ symmetric matrix $\mathbf{C}$. Show that for $n \times n$ symmetric matrices $\mathbf{A}$ and $\mathbf{B}$

$$\lambda_i(\mathbf{A}) + \lambda_n(\mathbf{B}) \leq \lambda_i(\mathbf{A} + \mathbf{B}) \leq \lambda_i(\mathbf{A}) + \lambda_i(\mathbf{B}),$$

$$\lambda_n(\mathbf{B}_i)\lambda_i(\mathbf{A}^2) \leq \lambda_i(\mathbf{A}\mathbf{B}\mathbf{A}) \leq \lambda_1(\mathbf{B})\lambda_i(\mathbf{A}^2), i = 1, \ldots, n.$$

(Wang *et al.*, 1994)

A.12 Let $\mathbf{V}$ be a $n \times n$ positive definite symmetric matrix with distinct characteristic roots $\lambda_1, \ldots, \lambda_r$ having multiplicities $m_1, \ldots, m_r$, respectively, $\sum_{i=1}^{r} m_i = n$. Then show that $\mathbf{V}$ can be expresses $\sum_{i=1}^{r} \lambda_i\mathbf{A}_i$

where $\mathbf{A}_i$ are mutually orthogonal symmetric idempotent matrices satisfying $\mathbf{VA}_i = \lambda_i \mathbf{A}_i, i = 1, \ldots, r$. Show also that the matrix $\mathbf{V}^{-1/2}$ defined by

$$\mathbf{V}^{-1/2} = \sum_{i=1}^{r} \lambda_i^{-1/2} \mathbf{A}_i$$

satisfies

$$\mathbf{V}^{-1/2}\mathbf{V}\mathbf{V}^{-1/2} = \mathbf{I}_n.$$

(Fuller and Battese, 1973)

A.13 Let $\lambda_1, \lambda_2, \ldots, \lambda_n$ be the eigenvalues of a symmetric matrix $\mathbf{A}$ such that $|\lambda_1| > |\lambda_2| > \ldots |\lambda_n|$. Show that for large k,

$$\lim_{k\to\infty}(\frac{\mathbf{A}^k}{\lambda_1^k}) = \mathbf{X}_1\mathbf{X}_1', \quad \lim_{k\to\infty}(\frac{\mathbf{A} - \lambda_1\mathbf{X}_1\mathbf{X}_1'}{\lambda_2})^k = \mathbf{X_2}\mathbf{X}_2',$$

$$\lim_{k\to\infty}(\frac{\mathbf{A} - \lambda_1\mathbf{X}_1\mathbf{X}_1' - \lambda_2\mathbf{X}_2\mathbf{X}_2' - \ldots - \lambda_{n-1}\mathbf{X}_{n-1}\mathbf{X}_{n-1}'}{\lambda_n})^k = \mathbf{X}_n\mathbf{X}_n';$$

$$\lambda_1 \approx (\frac{\mathbf{A}^k}{\mathbf{X}_1\mathbf{X}_1'})^{1/k},$$

$$\lambda_2 \approx [\frac{(\mathbf{A} - \lambda_1\mathbf{X}_1\mathbf{X}_1')^k}{\mathbf{X}_2\mathbf{X}_2'}]^{1/k},$$

$$\ldots\ldots\ldots,$$

$$\lambda_n \approx [\frac{(\mathbf{A} - \lambda_1\mathbf{X}_1\mathbf{X}_1' - \lambda_2\mathbf{X}_2\mathbf{X}_2' - \ldots - \lambda_{n-1}\mathbf{X}_{n-1}\mathbf{X}_{n-1}')^k}{\mathbf{X}_n\mathbf{X}_n'}]^{1/k},$$

where $\mathbf{X}_i$ is the characteristic vector corresponding to $\lambda_i (i = 1, \ldots, n)$.

A.14 If rank$(\mathbf{A}) = r < n$, for a real symmetric matrix $\mathbf{A}(n \times n)$ with distinct non-zero characteristic roots $\lambda_i (i = 1, \ldots, r)$ and $\mathbf{X}_i$ is the characteristic vector corresponding to λ_i, then show that

$$\mathbf{A}^* = \sum_{i=1} r\lambda_i \mathbf{X}_i\mathbf{X}_i'$$

is a g-inverse of $\mathbf{A}$.

Appendix B: Statistical Tables

The List of Tables:

The following statistical tables have been annexed.

B.1 Standard Normal Probabilities
B.2 Percentage Points of t-Distributions
B.3 χ^2-Critical Points
B.4 F-Distribution Critical Points ($\alpha = .05, \alpha = .01$)
B.5 Upper Tail Percentage Points for $\sqrt{b_1}$
B.6 Normalizing Transformation of $\sqrt{b_1}$
B.7 Simulated Percentiles for b_2
B.8 Anderson-Darling Test for Normality
B.9 D'Agostino's Test for Normality
B.10 Wilks' Likelihood Ratio Test
B.11 Roy's Maximum Root Statistic
B.12 Lawley-Hotelling Trace Statistic

Notes on the use of Tables

B.5 UPPER TAIL PERCENTAGE POINTS FOR $\sqrt{b_1}$

The table reproduced from Mulholland (1977) by permission of Biometrika Trustees gives the upper tail percentage points of the distribution of sample coefficient of skewness

$$\sqrt{b_1} = \frac{\sqrt{n}\sum_{i=1}^{n}(x_i - \bar{x})^3}{\{\sum_{i=1}^{n}(x_i - \bar{x})^2\}^{3/2}}$$

for a sample of size n from normal distribution.

B.6 NORMALIZING TRANSFORMATION OF $\sqrt{b_1}$

The table reproduced from D'Agostino and Pearson (1973) by permission of the Biometrika Trustees gives the values of δ and $1/\lambda$ such that the transformed

$$Z(\sqrt{b_1}) = \delta \sinh^{-1}(\frac{\sqrt{b_1}}{\lambda})$$

is approximately distributed as $N(0,1)$.

B.7 Simulated Percentage Points of b_2

The table reproduced from D'Agostino and Tietjan (1971) by permission of the Biometrika Trustees gives the percentile points of the sample coefficient of kurtosis,

$$b_2 = \frac{n\sum_{i=1}^{n}(x_i - \bar{x})^4}{\{\sum_{i=1}^{n}(x_i - \bar{x})^2\}^2}$$

for a random sample of size n from a normal distribution.

B.8 Anderson-Darling Test for Normality

The table taken from Pettit (1977) by the permission of the Biometrika Trustees gives the percentage points of the statistic

$$A_n^2 = -\frac{1}{n}\{\sum_{i=1}^{n}(2i-1)[\log z_i + \log(1 - z_{n+1-i})]\} - n,$$

where $z_i = \Phi([x_{(i)} - \bar{x}])/s)$, $s^2 = \sum_i (x_{(i)} - \bar{x})^2/(n-1)$, Φ is the distribution function of the standard normal distribution and $x_{(1)} \leq x_{(2)} \leq \ldots \leq x_{(n)}$ is an ordered random sample of size n from a normal distribution. Given $p = P[A_n^2 \leq a_n], a_n$ is calculated from the expression

$$a_n = a_\infty(1 + c_1 n^{-1} + c_2 n^{-2}).$$

The hypothesis of normality is rejected if A_n^2 is too large.

B.9 D'Agostino's Test for Normality

The table taken from D'Agostino (1971, 1972) with the permission of the Biometrika Trustees gives the percentage points of the statistic

$$Y = \frac{\sqrt{n}[D - (2\sqrt{\pi})^{-1}]}{0.02998598},$$

where

$$D = \frac{\sum_{i=1}^{n}[i - \frac{1}{2}(n+1)]x_{(i)}}{n^{1/2}[\sum_{i=1}^{n}(x_{(i)} - \bar{x})^2]^{1/2}},$$

and $x_{(1)} \leq x_{(2)} \leq \ldots \leq x_{(n)}$ is an ordered random sample of size n from a normal distribution. As the range of Y is small, D should be calculated to five decimal places. The hypothesis of normality is rejected if $|Y|$ is too large.

B10 Wilks' Likelihood Ratio Test

Let $\mathbf{H}$ and $\mathbf{E}$ have independent Wishart distributions $W_d(m_H, \mathbf{\Sigma})$ and $W_d(m_E, \mathbf{\Sigma})$ respectively. The statistic $U = |\mathbf{E}|/(|\mathbf{E} + \mathbf{H}|)$ has Wilks' distribution $U_{d;m_H,m_E}$ (Section 4.4) and

$$-(\frac{f}{C_\alpha}) \ln U \sim \chi^2_{(dm_H)},$$

where

$$f = m_E - \frac{1}{2}(d - m_H + 1) = m_E + m_H - \frac{1}{2}(m_H + d + 1).$$

Given α and $\chi^2_{dm_H;\alpha}$ where $P[\chi^2_{(dm_H)} > \chi^2_{dm_H;\alpha}] = \alpha$, the upper α percentile values for $-f \ln U$ is $C_\alpha \chi^2_{dm_H;\alpha}$. Values of C_α and $\chi^2_{dm_H;\alpha}$ are given in these tables for the parameters $d, m_H, (d \leq m_H, m_E)$ and

$$M = m_E - d + 1 = (m_E + m_H) - m_H - d + 1$$

for $d = 3$ and $d = 4$ only. For large $m_E, C_\alpha \approx 1$.

For $d > m_H, U_{d;m_H,m_E} = U_{m_H;d,m_E+m_H-d}$. The values of f and M remain unchanged. For other values of the parameters we can use F-approximations (4.4.10), (4.4.11) due to Rao (1951). When $s = \min(d, m_H)$ is 1 or 2, we can use the approximations (4.4.13) or (4.4.14).

The values in the tables have been taken from Schatzoff (1966), Lee (1972), Pillai and Gupta (1969) and Davis (1979) by the permission of the Biometrika Trustees.

B11 Roy's Maximum Root Statistic

Let $\theta_{\max}$ be the maximum root of the equation $|\mathbf{H} - \theta(\mathbf{E}+\mathbf{H})| = 0$ where $\mathbf{E}$ and $\mathbf{H}$ have independent Wishart distributions $W_d(m_E, \mathbf{\Sigma})$ and $W_d(m_H, \mathbf{\Sigma})$ respectively. Let $s = \min(d, m_H), \nu_1 = (|d - m_H| - 1)/2$ and $\nu_2 = (m_E - d - 1)/2$. Upper percentage points for $\theta_{\max}$ are given in these tables from Pillai (1960) for $s = 2(1)7$.

Entries for $s = 6(1)10$ and selected values of ν_2 are given in Pearson and Hartley (1972, Table 48) (see Also Pillai (1964, 1965)). Entries for $s = 11, 12$ are available in Pillai (1970) and for $s = 14, 16, 18, 20$ in Pillai (1967).

Percentage points for the odd values are obtained by interpolation. See also Pearson and Hartley (1972, Table 49).

B12 Lawley-Hotelling Trace Statistics

Let $T_g^2 = m_E \text{ tr } [\mathbf{HE}^{-1}]$, where $\mathbf{H}$ and $\mathbf{E}$ have independent Wishart distributions $W_d(m_H, \mathbf{\Sigma})$ and $W_d(m_E, \mathbf{\Sigma})$, respectively. The table, reproduced from Davis (1970 a,b) by permission of the Biometrika Trustees gives the upper percentage points of

$$\frac{T_g^2}{m_H} = (\frac{m_E}{m_h}) \text{ tr } [\mathbf{HE}^{-1}] = m_E U^{(s)}$$

for values of d, m_H and m_E with $d = 2, 3, 4 \leq m_H, m_E$. If $m_H < d$, we make make the transformation $d \to m_H, m_H \to d, m_E \to m_E + m_H - d$. McKeon (1974) has shown that the distribution of $U^{(s)}$ can be approximated by $kF_{(a,b)}$ where

$$a = dm_H, \;\; b = 4 + (a+2)/(B-1), \;\; k = a(b-2)/b(m_E - d - 1).$$

$$B = \frac{(m_E + m_H - d - 1)(m_E - 1)}{(m_E - d - 3)(m_E - d)}.$$

When $m_E \to \infty, T_g^2 \sim \chi^2_{(dm_H)}$.

Table B.1: Standard Normal Probabilities

x	.00	.01	.02	.03	.04	.05	.06	.07	.08	.09
.0	.5000	.5040	.5080	.5120	.5160	.5199	.5239	.5279	.5319	.5359
.1	.5398	.5438	.5478	.5517	.5557	.5596	.5636	.5675	.5714	.5753
.2	.5793	.5832	.5871	.5910	.5948	.5987	.6026	.6064	.6103	.6141
.3	.6179	.6217	.6255	.6293	.6331	.6368	.6406	.6443	.6480	.6517
.4	.6554	.6591	.6628	.6664	.6700	.6736	.6772	.6808	.6844	.6879
.5	.6915	.6950	.6985	.7019	.7054	.7088	.7123	.7157	.7190	.7224
.6	.7257	.7291	.7324	.7357	.7389	.7422	.7454	.7486	.7517	.7549
.7	.7580	.7611	.7642	.7673	.7704	.7734	.7764	.7794	.7823	.7852
.8	.7881	.7910	.7939	.7967	.7995	.8023	.8051	.8078	.8106	.8133
.9	.8159	.8186	.8212	.8238	.8264	.8289	.8315	.8340	.8365	.8389
1.0	.8413	.8438	.8461	.8485	.8508	.8531	.8554	.8577	.8599	.8621
1.1	.8643	.8665	.8686	.8708	.8729	.8749	.8770	.8790	.8810	.8830
1.2	.8849	.8869	.8888	.8907	.8925	.8944	.8962	.8980	.8997	.9015
1.3	.9032	.9049	.9066	.9082	.9099	.9115	.9131	.9147	.9162	.9177
1.4	.9192	.9207	.9222	.9236	.9251	.9265	.9279	.9292	.9306	.9319
1.5	.9332	.9345	.9357	.9370	.9382	.9394	.9406	.9418	.9429	.9441
1.6	.9452	.9463	.9474	.9484	.9495	.9505	.9515	.9525	.9535	.9545
1.7	.9554	.9564	.9573	.9582	.9591	.9599	.9608	.9616	.9625	.9633
1.8	.9641	.9649	.9656	.9664	.9671	.9678	.9686	.9693	.9699	.9706
1.9	.9713	.9719	.9726	.9732	.9738	.9744	.9750	.9756	.9761	.9767
2.0	.9772	.9778	.9783	.9788	.9793	.9798	.9803	.9808	.9812	.9817
2.1	.9821	.9826	.9830	.9834	.9838	.9842	.9846	.9850	.9854	.9857
2.2	.9861	.9864	.9868	.9871	.9875	.9878	.9881	.9884	.9887	.9890
2.3	.9893	.9896	.9898	.9901	.9904	.9906	.9909	.9911	.9913	.9916
2.4	.9918	.9920	.9922	.9925	.9927	.9929	.9931	.9932	.9934	.9936
2.5	.9938	.9940	.9941	.9943	.9945	.9946	.9948	.9949	.9951	.9952
2.6	.9953	.9955	.9956	.9957	.9959	.9960	.9961	.9962	.9963	.9964
2.7	.9965	.9966	.9967	.9968	.9969	.9970	.9971	.9972	.9973	.9974
2.8	.9974	.9975	.9976	.9977	.9977	.9978	.9979	.9979	.9980	.9981
2.9	.9981	.9982	.9982	.9983	.9984	.9984	.9985	.9985	.9986	.9986
3.0	.9987	.9987	.9987	.9988	.9988	.9989	.9989	.9989	.9990	.9990
3.1	.9990	.9991	.9991	.9991	.9992	.9992	.9992	.9992	.9993	.9993
3.2	.9993	.9993	.9994	.9994	.9994	.9994	.9994	.9995	.9995	.9995
3.3	.9995	.9995	.9995	.9996	.9996	.9996	.9996	.9996	.9996	.9997
3.4	.9997	.9997	.9997	.9997	.9997	.9997	.9997	.9997	.9997	.9998

Table B.2: Percentage Points of the t-Distributions

	α					
d.f.	.25	.10	.05	.025	.01	.005
1	1.000	3.078	6.314	12.706	31.821	63.657
2	.816	1.886	2.920	4.303	6.965	9.925
3	.765	1.638	2.353	3.182	4.541	5.841
4	.741	1.533	2.132	2.776	3.747	4.604
5	.727	1.476	2.015	2.571	3.365	4.032
6	.718	1.440	1.943	2.447	3.143	3.707
7	.711	1.415	1.895	2.365	2.998	3.499
8	.706	1.397	1.860	2.306	2.896	3.355
9	.703	1.383	1.833	2.262	2.821	3.250
10	.700	1.372	1.812	2.228	2.764	3.169
11	.697	1.363	1.796	2.201	2.718	3.106
12	.695	1.356	1.782	2.179	2.681	3.055
13	.694	1.350	1.771	2.160	2.650	3.012
14	.692	1.345	1.761	2.145	2.624	2.977
15	.691	1.341	1.753	2.131	2.602	2.947
16	.690	1.337	1.746	2.120	2.583	2.921
17	.689	1.333	1.740	2.110	2.567	2.898
18	.688	1.330	1.734	2.101	2.552	2.878
19	.688	1.328	1.729	2.093	2.539	2.861
20	.687	1.325	1.725	2.086	2.528	2.845
21	.686	1.323	1.721	2.080	2.518	2.831
22	.686	1.321	1.717	2.074	2.508	2.819
23	.685	1.319	1.714	2.069	2.500	2.807
24	.685	1.318	1.711	2.064	2.492	2.797
25	.684	1.316	1.708	2.060	2.485	2.787
26	.684	1.315	1.706	2.056	2.479	2.779
27	.684	1.314	1.703	2.052	2.473	2.771
28	.683	1.313	1.701	2.048	2.467	2.763
29	.683	1.311	1.699	2.045	2.462	2.756
30	.683	1.310	1.697	2.042	2.457	2.750
40	.681	1.303	1.684	2.021	2.423	2.704
60	.679	1.296	1.671	2.000	2.390	2.660
120	.677	1.289	1.658	1.980	2.358	2.617
∞	.674	1.282	1.645	1.960	2.326	2.576

Table B.3: χ^2 Critical Points

ν	$1-\alpha$ 0.90	0.95	0.975	0.99	0.995	0.999
1	2.71	3.84	5.02	6.63	7.88	10.83
2	4.61	5.99	7.38	9.21	10.60	13.81
3	6.25	7.81	9.35	11.34	12.84	16.27
4	7.78	9.49	11.14	13.28	14.86	18.47
5	9.24	11.07	12.83	15.09	16.75	20.52
6	10.64	12.59	14.45	16.81	18.55	22.46
7	12.02	14.07	16.01	18.48	20.28	24.32
8	13.36	15.51	17.53	20.09	21.95	26.12
9	14.68	16.92	19.02	21.67	23.59	27.88
10	15.99	18.31	20.48	23.21	25.19	29.59
11	17.28	19.68	21.92	24.73	26.76	31.26
12	18.55	21.03	23.34	26.22	28.30	32.91
13	19.81	22.36	24.74	27.69	29.82	34.53
14	21.06	23.6	26.12	29.14	31.32	36.12
15	22.31	25.00	27.49	30.58	32.80	37.70
16	23.54	26.30	28.85	32.00	34.27	39.25
17	24.77	27.59	30.19	33.41	35.72	40.79
18	25.99	28.87	31.53	34.81	37.16	42.31
19	27.20	30.14	32.85	36.19	38.58	43.82
20	28.41	31.41	34.17	35.57	40.00	45.31
21	29.62	32.67	35.48	38.93	41.40	46.80
22	30.81	33.92	36.78	40.29	42.80	48.27
23	32.01	35.17	38.08	41.64	44.18	49.73
24	33.20	36.42	39.36	42.98	45.56	51.18
25	34.38	37.65	40.65	44.31	46.93	52.62
26	35.56	38.89	41.92	45.64	48.29	54.05
27	36.74	40.11	43.19	46.96	49.64	55.48
28	37.92	41.34	44.46	48.28	50.99	56.89
29	39.09	42.56	45.72	49.59	52.34	58.30
30	40.26	43.77	46.98	50.89	53.67	59.70
40	51.81	55.76	59.34	63.69	66.77	73.40
50	63.17	67.50	71.42	76.15	79.49	86.66
60	74.40	79.08	83.30	88.38	91.95	99.61
70	85.53	90.53	95.02	100.4	104.2	112.3
80	96.58	101.9	106.6	112.3	116.3	124.8
90	107.6	113.1	118.1	124.1	128.3	137.2
100	118.5	124.3	129.6	135.8	140.2	149.4

For $\nu > 100$, $\sqrt{2\chi^2_\nu} \sim N(\sqrt{2\nu - 1}, 1)$

Abridged from C.M.Thompson (1941): Tables of percentage points of the χ^2 distribution, Biometrika, **32**, 187 - 191 and published here with the kind permission of the Biometrika Trustees.

Table B.4: F-distribution Critical Points

$1-\alpha$	ν_2	ν_1 = 1	2	3	4	5	6	7	8	9	10	12	15	20	30	60	120	∞
0.90		39.9	49.5	53.6	55.8	57.2	58.2	58.9	59.4	59.9	60.2	60.7	61.2	61.7	62.3	62.8	63.1	63.3
0.95		161	200	216	225	230	234	237	239	241	242	244	246	248	250	252	253	254
0.975	1	648	800	864	900	922	937	948	957	963	969	977	985	993	1,000	1,010	1,010	1,020
0.99		4,050	5,000	5,400	5,620	5,760	5,860	5,930	5,980	6,020	6.060	6,110	6,160	6,210	6,260	6,310	6,340	6,370
0.90		8.53	9.00	9.16	9.24	9.29	9.33	9.35	9.37	9.38	9.39	9.41	9.42	9.44	9.46	9.47	9.48	9.49
0.95		18.5	19.0	19.2	19.2	19.3	19.3	19.4	19.4	19.4	19.4	19.4	19.4	19.5	19.5	19.5	19.5	19.5
0.975	2	38.5	39.0	39.2	39.2	39.3	39.3	39.4	39.4	39.4	39.4	39.4	39.4	39.4	39.5	39.5	39.5	39.5
0.99		98.5	99.0	99.2	99.2	99.3	99.3	99.4	99.4	99.4	99.4	99.4	99.4	99.4	99.5	99.5	99.5	99.5
0.90		5.54	5.46	5.39	5.34	5.31	5.28	5.27	5.25	5.24	5.23	5.22	5.20	5.18	5.17	5.15	5.14	5.13
0.95		10.1	9.55	9.28	9.12	9.01	8.94	8.89	8.85	8.81	8.79	8.74	8.70	8.66	8.62	8.57	8.55	8.53
0.975	3	17.4	16.0	15.4	15.1	14.9	14.7	14.6	14.5	14.5	14.4	14.3	14.3	14.2	14.1	14.0	13.9	13.9
0.99		34.1	30.8	29.5	28.7	28.2	27.9	27.7	27.5	27.3	27.2	27.1	26.9	26.7	26.5	26.3	26.2	26.1
0.90		4.54	4.32	4.19	4.11	4.05	4.01	3.98	3.95	3.93	3.92	3.90	3.87	3.84	3.82	3.79	3.78	3.76
0.95		7.71	6.94	6.59	6.39	6.26	6.16	6.09	6.04	6.00	5.96	5.91	5.86	5.80	5.75	5.69	5.66	5.63
0.975	4	12.2	10.6	9.98	9.60	9.36	9.20	9.07	8.98	8.90	8.84	8.75	8.66	8.56	8.46	8.36	8.31	8.26
0.99		21.2	18.0	16.7	16.0	15.5	15.2	15.0	14.8	14.7	14.5	14.4	14.2	14.0	13.8	13.7	13.6	13.5
0.90		4.06	3.78	3.62	3.52	3.45	3.40	3.37	3.34	3.32	3.30	3.27	3.24	3.21	3.17	3.14	3.12	3.11
0.95		6.61	5.79	5.41	5.19	5.05	4.95	4.88	1.82	4.77	4.74	4.68	4.62	4.56	4.50	4.43	4.40	4.37
0.975	5	10.0	8.43	7.76	7.39	7.15	6.98	6.85	6.76	6.68	6.62	6.52	6.43	6.33	6.23	6.12	6.07	6.02
0.99		16.3	13.3	12.1	11.4	11.0	10.7	10.5	10.3	10.2	10.1	9.89	9.72	9.55	9.38	9.20	9.11	9.02
0.90		3.78	3.46	3.29	3.18	3.11	3.05	3.01	2.98	2.96	2.94	2.90	2.87	2.84	2.80	2.76	2.74	2.72
0.95		5.99	5.14	4.76	4.53	4.39	4.28	4.21	4.15	4.10	4.06	4.00	3.94	3.87	3.81	3.74	3.70	3.67
0.975	6	8.81	7.26	6.60	6.23	5.99	5.82	5.70	5.60	5.52	5.46	5.37	5.27	5.17	5.07	4.96	4.90	4.85
0.99		13.7	10.9	9.78	9.15	8.75	8.47	8.26	8.10	7.98	7.87	7.72	7.56	7.40	7.23	7.06	6.97	6.88
0.90		3.59	3.26	3.07	2.96	2.88	2.83	2.78	2.75	2.72	2.70	2.67	2.63	2.59	2.56	2.51	2.49	2.47
0.95		5.59	4.74	4.35	4.12	3.97	3.87	3.79	3.73	3.68	3.64	3.57	3.51	3.44	3.38	3.30	3.27	3.23
0.975	7	8.07	6.54	5.89	5.52	5.29	5.12	4.99	4.90	4.82	4.76	4.67	4.57	4.47	4.36	4.25	4.20	4.14
0.99		12.2	9.55	8.45	7.85	7.46	7.19	6.99	6.84	6.72	6.62	6.47	6.31	6.16	5.99	5.82	5.74	5.65
0.90		3.46	3.11	2.92	2.81	2.73	2.67	2.62	2.59	2.56	2.54	2.50	2.46	2.42	2.38	2.34	2.31	2.29
0.95		5.32	4.46	4.07	3.84	3.69	3.58	3.50	3.44	3.39	3.35	3.28	3.22	3.15	3.08	3.01	2.97	2.93
0.975	8	7.57	6.06	5.42	5.05	4.82	4.65	4.53	4.43	4.36	4.30	4.20	4.10	4.00	3.89	3.78	3.73	3.67
0.99		11.3	8.65	7.59	7.01	6.63	6.37	6.18	6.03	5.91	5.81	5.67	5.52	5.36	5.20	5.03	4.95	4.86

Table B.4 F-Distribution Critical Points

		ν_1																
$1-\alpha$	ν_2	1	2	3	4	5	6	7	8	9	10	12	15	20	30	60	120	∞
0.90		3.36	3.01	2.81	2.69	2.61	2.55	2.51	2.47	2.44	2.42	2.38	2.34	2.30	2.25	2.21	2.18	2.16
0.95		5.12	4.26	3.86	3.63	3.48	3.37	3.29	3.23	3.18	3.14	3.07	3.01	2.94	2.86	2.79	2.75	2.71
0.975	9	7.21	5.71	5.08	4.72	4.48	4.32	4.20	4.10	4.03	3.96	3.87	3.77	3.67	3.56	3.45	3.39	3.33
0.99		10.6	8.02	6.99	6.42	6.06	5.80	5.61	5.47	5.35	5.26	5.11	4.96	4.81	4.65	4.48	4.40	4.31
0.90		3.29	2.92	2.73	2.61	2.52	2.46	2.41	2.38	2.35	2.32	2.28	2.24	2.20	2.15	2.11	2.08	2.06
0.95		4.96	4.10	3.71	3.48	3.33	3.22	3.14	3.07	3.02	2.98	2.91	2.84	2.77	2.70	2.62	2.58	2.54
0.975	10	6.94	5.46	4.83	4.47	4.24	4.07	3.95	3.85	3.78	3.72	3.62	3.52	3.42	3.31	3.20	3.14	3.08
0.99		10.0	7.56	6.55	5.99	5.64	5.39	5.20	5.06	4.94	4.85	4.71	4.56	4.41	4.25	4.08	4.00	3.91
0.90		3.18	2.81	2.61	2.48	2.39	2.33	2.28	2.24	2.21	2.19	2.15	2.10	2.06	2.01	1.96	1.93	1.90
0.95		4.75	3.89	3.49	3.26	3.11	3.00	2.91	2.85	2.80	2.75	2.69	2.62	2.54	2.47	2.38	2.34	2.30
0.975	12	6.55	5.10	4.47	4.12	3.89	3.73	3.61	3.51	3.44	3.37	3.28	3.18	3.07	2.96	2.85	2.79	2.72
0.99		9.33	6.93	5.95	5.41	5.06	4.82	4.64	4.50	4.39	4.30	4.16	4.01	3.86	3.70	3.54	3.45	3.36
0.90		3.07	2.70	2.49	2.36	2.27	2.21	2.16	2.12	2.09	2.06	2.02	1.97	1.92	1.87	1.82	1.79	1.76
0.95		4.54	3.68	3.29	3.06	2.90	2.79	2.71	2.64	2.59	2.54	2.48	2.40	2.33	2.25	2.16	2.11	2.07
0.975	15	6.20	4.77	4.15	3.80	3.58	3.41	3.29	3.20	3.12	3.06	2.96	2.86	2.76	2.64	2.52	2.46	2.40
0.99		8.68	6.36	5.42	4.89	4.56	4.32	4.14	4.00	3.89	3.80	3.67	3.52	3.37	3.21	3.05	2.96	2.87
0.90		2.97	2.59	2.38	2.25	2.16	2.09	2.04	2.00	1.96	1.94	1.89	1.84	1.79	1.74	1.68	1.64	1.61
0.95		4.35	3.49	3.10	2.87	2.71	2.60	2.51	2.45	2.39	2.35	2.28	2.20	2.12	2.04	1.95	1.90	1.84
0.975	20	5.87	4.46	3.86	3.51	3.29	3.13	3.01	2.91	2.84	2.77	2.68	2.57	2.46	2.35	2.22	2.16	2.09
0.99		8.10	5.85	4.94	4.43	4.10	3.87	3.70	3.56	3.46	3.37	3.23	3.09	2.94	2.78	2.61	2.52	2.42
0.90		2.88	2.49	2.28	2.14	2.05	1.98	1.93	1.88	1.85	1.82	1.77	1.72	1.67	1.61	1.54	1.50	1.46
0.95		4.17	3.32	2.92	2.69	2.53	2.42	2.33	2.27	2.21	2.16	2.09	2.01	1.93	1.84	1.74	1.68	1.62
0.975	30	5.57	4.18	3.59	3.25	3.03	2.87	2.75	2.65	2.57	2.51	2.41	2.31	2.20	2.07	1.94	1.87	1.79
0.99		7.56	5.39	4.51	4.02	3.70	3.47	3.30	3.17	3.07	2.98	2.84	2.70	2.55	2.39	2.21	2.11	2.01
0.90		2.79	2.39	2.18	2.04	1.95	1.87	1.82	1.77	1.74	1.71	1.66	1.60	1.54	1.48	1.40	1.35	1.29
0.95		4.00	3.15	2.76	2.53	2.37	2.25	2.17	2.10	2.04	1.99	1.92	1.84	1.75	1.65	1.53	1.47	1.39
0.975	60	5.29	3.93	3.34	3.01	2.79	2.63	2.51	2.41	2.33	2.27	2.17	2.06	1.94	1.82	1.67	1.58	1.48
0.99		7.08	4.98	4.13	3.65	3.34	3.12	2.95	2.82	2.72	2.63	2.50	2.35	2.20	2.03	1.84	1.73	1.60
0.90		2.75	2.35	2.13	1.99	1.90	1.82	1.77	1.72	1.68	1.65	1.60	1.54	1.48	1.41	1.32	1.26	1.19
0.95		3.92	3.07	2.68	2.45	2.29	2.18	2.09	2.02	1.96	1.91	1.83	1.75	1.66	1.55	1.43	1.35	1.25
0.975	120	5.15	3.80	3.23	2.89	2.67	2.52	2.39	2.30	2.22	2.16	2.05	1.94	1.82	1.69	1.53	1.43	1.31
0.99		6.85	4.79	3.95	3.48	3.17	2.96	2.79	2.66	2.56	2.47	2.34	2.19	2.03	1.86	1.66	1.53	1.38
0.90		2.71	2.30	2.08	1.94	1.85	1.77	1.72	1.67	1.63	1.60	1.55	1.49	1.42	1.34	1.24	1.17	1.00
0.95		3.84	3.00	2.60	2.37	2.21	2.10	2.01	1.94	1.88	1.83	1.75	1.67	1.57	1.46	1.32	1.22	1.00
0.975	∞	5.02	3.69	3.12	2.79	2.57	2.41	2.29	2.19	2.11	2.05	1.94	1.83	1.71	1.57	1.39	1.27	1.00
0.99		6.63	4.61	3.78	3.32	3.02	2.80	2.64	2.51	2.41	2.32	2.18	2.04	1.88	1.70	1.47	1.32	1.00

Abridged from M. Merrington, C. M. Thompson and E. S. Pearson (1943): Tables of percentage points of the inverted Beta (F) distribution, Biometrika, **33**, 74–88 and published here with the kind permission of the Biometrika Trustees.

Table B.5: Upper Tail Percentage Points for $\sqrt{b_1}$

	Upper Percentiles					
n	10	5	2.5	1	.5	.1
4	.831	.987	1.070	1.120	1.137	1.151
5	.821	1.049	1.207	1.337	1.396	1.464
6	.795	1.042	1.239	1.429	1.531	1.671
7	.782	1.018	1.230	1.457	1.589	1.797
8	.765	.998	1.208	1.452	1.605	1.866
9	.746	.977	1.184	1.433	1.598	1.898
10	.728	.954	1.159	1.407	1.578	1.906
11	.710	.931	1.134	1.381	1.553	1.899
12	.693	.910	1.109	1.353	1.526	1.882
13	.677	.890	1.085	1.325	1.497	1.859
14	.662	.870	1.061	1.298	1.468	1.832
15	.648	.851	1.039	1.272	1.440	1.803
16	.635	.834	1.018	1.247	1.412	1.773
17	.622	.817	.997	1.222	1.385	1.744
18	.610	.801	.978	1.199	1.359	1.714
19	.599	.786	.960	1.176	1.334	1.685
20	.588	.772	.942	1.155	1.310	1.657
21	.578	.758	.925	1.134	1.287	1.628
22	.568	.746	.909	1.114	1.265	1.602
23	.559	.733	.894	1.096	1.243	1.575
24	.550	.722	.880	1.078	1.223	1.550
25	.542	.710	.866	1.060	1.203	1.526

Table B.6: Normalizing Transformation of $\sqrt{b_1}$

n	δ	1/λ
8	5·563	0·3030
9	4·260	0·4080
10	3·734	0·4794
11	3·447	0·5339
12	3·270	0·5781
13	3·151	0·6153
14	3·069	0·6473
15	3·010	0·6753
16	2·968	0·7001
17	2·937	0·7224
18	2·915	0·7426
19	2·900	0·7610
20	2·890	0·7779
21	2·884	0·7934
22	2·882	0·8078
23	2·882	0·8211
24	2·884	0·8336
25	2·889	0·8452
26	2·895	0·8561
27	2·902	0·8664
28	2·910	0·8760
29	2·920	0·8851
30	2·930	0·8938
31	2·941	0·9020
32	2·952	0·9097
33	2·964	0·9171
34	2·977	0·9241
35	2·990	0·9308
36	3·003	0·9372
37	3·016	0·9433
38	3·030	0·9492
39	3·044	0·9548
40	3·058	0·9601
41	3·073	0·9653
42	3·087	0·9702
43	3·102	0·9750
44	3·117	0·9795
45	3·131	0·9840
46	3·146	0·9882
47	3·161	0·9923
48	3·176	0·9963
49	3·192	1·0001
50	3·207	1·0038
52	3·237	1·0108
54	3·268	1·0174
56	3·298	1·0235
58	3·329	1.0293
60	3·359	1·0348

n	δ	1/λ
62	3·389	1·0400
64	3·420	1·0449
66	3·450	1·0495
68	3·480	1·0540
70	3·510	1·0581
72	3·540	1·0621
74	3·569	1·0659
76	3·599	1·0695
78	3·628	1·0730
80	3·657	1·0763
82	3·686	1·0795
84	3·715	1·0825
86	3·744	1·0854
88	3·772	1·0882
90	3·801	1·0909
92	3·829	1·0934
94	3·857	1·0959
96	3·885	1·0983
98	3·913	1·1006
100	3·940	1·1028
105	4·009	1·1080
110	4·076	1·1128
115	4·142	1·1172
120	4·207	1·1212
125	4·272	1·1250
130	4·336	1·1285
135	4·398	1·1318
140	4·460	1·1348
145	4·521	1·1377
150	4·582	1·1403
155	4·641	1·1428
160	4·700	1·1452
165	4·758	1·1474
170	4·816	1·1496
175	4·873	1·1516
180	4·929	1·1535
185	4·985	1·1553
190	5·040	1·1570
195	5·094	1·1586
200	5·148	1·1602
205	5·202	1·1616
210	5·255	1·1631
215	5·307	1·1644
220	5·359	1·1657
225	5·410	1·1669
230	5·461	1·1681
235	5·511	1·1693
240	5·561	1·1704
245	5·611	1·1714
250	5·660	1·1724

n	δ	1/λ
260	5·757	1·1744
270	5·853	1·1761
280	5·946	1·1779
290	6·039	1·1793
300	6·130	1·1808
310	6·220	1·1821
320	6·308	1·1834
330	6·396	1·1846
340	6·482	1·1858
350	6·567	1·1868
360	6·651	1·1879
370	6·733	1·1888
380	6·815	1·1897
390	6·896	1·1906
400	6·976	1·1914
410	7·056	1·1922
420	7·134	1·1929
430	7·211	1·1937
440	7·288	1·1943
450	7·363	1·1950
460	7·438	1·1956
470	7·513	1·1962
480	7·586	1·1968
490	7·659	1·1974
500	7·731	1·1979
520	7·873	1·1989
540	8·013	1·1998
560	8·151	1·2007
580	8·286	1·2015
600	8·419	1·2023
620	8·550	1·2030
640	8·679	1·2036
660	8·806	1·2043
680	8·931	1·2049
700	9·054	1·2054
720	9·176	1·2060
740	9·297	1·2065
760	9·415	1·2069
780	9·533	1·2073
800	9·649	1·2078
820	9·763	1·2082
840	9·876	1·2086
860	9·988	1·2089
880	10·098	1·2093
900	10·208	1·2096
920	10·316	1·2100
940	10·423	1·2103
960	10·529	1·2106
980	10·634	1·2109
1000	10·738	1·2111

Table B.7: Simulation Percentiles for b_2

Sample size	Percentiles											
	1	2	2·5	5	10	20	80	90	95	97·5	98	99
7	1·25	1·30	1·34	1·41	1·53	1·70	2·78	3·20	3·55	3·85	3·93	4·23
8	1·31	1·37	1·40	1·46	1·58	1·75	2·84	3·31	3·70	4·09	4·20	4·53
9	1·35	1·42	1·45	1·53	1·63	1·80	2·98	3·43	3·86	4·28	4·41	4·82
10	1·39	1·45	1·49	1·56	1·68	1·85	3·01	3·53	3·95	4·40	4·55	5·00
12	1·46	1·52	1·56	1·64	1·76	1·93	3·06	3·55	4·05	4·56	4·73	5·20
15	1·55	1·61	1·64	1·72	1·84	2·01	3·13	3·62	4·13	4·66	4·85	5·30
20	1·65	1·71	1·74	1·82	1·95	2·13	3·21	3·68	4·17	4·68	4·87	5·36
25	1·72	1·79	1·83	1·91	2·03	2·20	3·23	3·68	4·16	4·65	4·82	5·30
30	1·79	1·86	1·90	1·98	2·10	2·26	3·25	3·68	4·11	4·59	4·75	5·21
35	1·84	1·91	1·95	2·03	2·14	2·31	3·27	3·68	4·10	4·53	4·68	5·13
40	1·89	1·96	1·98	2·07	2·19	2·34	3·28	3·67	4·06	4·46	4·61	5·04
45	1·93	2·00	2·03	2·11	2·22	2·37	3·28	3·65	4·00	4·39	4·52	4·94
50	1·95	2·03	2·06	2·15	2·25	2·41	3·28	3·62	3·99	4·33	4·45	4·88

Table B.8: Anderson-Darling Test for Normality

p	c_1	c_2	a_∞
·05	−·512	2·10	·1674
·10	−·552	1·25	·1938
·15	−·608	1·07	·2147
·20	−·643	·93	·2333
·25	−·707	1·03	·2509
·30	−·735	1·02	·2681
·35	−·772	1·04	·2853
·40	−·770	·90	·3030
·45	−·778	·80	·3213
·50	−·779	·67	·3405
·55	−·803	·70	·3612
·60	−·818	·58	·3836
·65	−·818	·42	·4085
·70	−·801	·12	·4367
·75	−·800	−·09	·4695
·80	−·756	−·39	·5091
·85	−·749	−·59	·5597
·90	−·750	−·80	·6305
·95	−·795	−·89	·7514
·975	−·881	−·94	·8728
·99	−1·013	−·93	1·0348
·995	−1·063	−1·34	1·1578

Table B.9: D'Agostino's Test for Normality

	Percentiles of Y									
n	0·5	1·0	2·5	5	10	90	95	97·5	99	99·5
10	−4·66	−4·06	−3·25	−2·62	−1·99	0·149	0·235	0·299	0·356	0·385
12	−4·63	−4·02	−3·20	−2·58	−1·94	0·237	0·329	0·381	0·440	0·479
14	−4·57	−3·97	−3·16	−2·53	−1·90	0·308	0·399	0·460	0·515	0·555
16	−4·52	−3·92	−3·12	−2·50	−1·87	0·367	0·459	0·526	0·587	0·613
18	−4·47	−3·87	−3·08	−2·47	−1·85	0·417	0·515	0·574	0·636	0·667
20	−4·41	−3·83	−3·04	−2·44	−1·82	0·460	0·565	0·628	0·690	0·720
22	−4·36	−3·78	−3·01	−2·41	−1·81	0·497	0·609	0·677	0·744	0·775
24	−4·32	−3·75	−2·98	−2·39	−1·79	0·530	0·648	0·720	0·783	0·822
26	−4·27	−3·71	−2·96	−2·37	−1·77	0·559	0·682	0·760	0·827	0·867
28	−4·23	−3·68	−2·93	−2·35	−1·76	0·586	0·714	0·797	0·868	0·910
30	−4·19	−3·64	−2·91	−2·33	−1·75	0·610	0·743	0·830	0·906	0·941
32	−4·16	−3·61	−2·88	−2·32	−1·73	0·631	0·770	0·862	0·942	0·983
34	−4·12	−3·59	−2·86	−2·30	−1·72	0·651	0·794	0·891	0·975	1·02
36	−4·09	−3·56	−2·85	−2·29	−1·71	0·669	0·816	0·917	1·00	1·05
38	−4·06	−3·54	−2·83	−2·28	−1·70	0·686	0·837	0·941	1·03	1·08
40	−4·03	−3·51	−2·81	−2·26	−1·70	0·702	0·857	0·964	1·06	1·11
42	−4·00	−3·49	−2·80	−2·25	−1·69	0·716	0·875	0·986	1·09	1·14
44	−3·98	−3·47	−2·78	−2·24	−1·68	0·730	0·892	1·01	1·11	1·17
46	−3·95	−3·45	−2·77	−2·23	−1·67	0·742	0·908	1·02	1·13	1·19
48	−3·93	−3·43	−2·75	−2·22	−1·67	0·754	0·923	1·04	1·15	1·22
50	−3·91	−3·41	−2·74	−2·21	−1·66	0·765	0·937	1·06	1·18	1·24
60	−3·81	−3·34	−2·68	−2·17	−1·64	0·812	0·997	1·13	1·26	1·34
70	−3·73	−3·27	−2·64	−2·14	−1·61	0·849	1·05	1·19	1·33	1·42
80	−3·67	−3·22	−2·60	−2·11	−1·59	0·878	1·08	1·24	1·39	1·48
90	−3·61	−3·17	−2·57	−2·09	−1·58	0·902	1·12	1·28	1·44	1·54
100	−3·57	−3·14	−2·54	−2·07	−1·57	0·923	1·14	1·31	1·48	1·59
150	−3·409	−3·009	−2·452	−2·004	−1·520	0·990	1·233	1·423	1·623	1·746
200	−3·302	−2·922	−2·391	−1·960	−1·491	1·032	1·290	1·496	1·715	1·853
250	−3·227	−2·861	−2·348	−1·926	−1·471	1·060	1·328	1·545	1·779	1·927
300	−3·172	−2·816	−2·316	−1·906	−1·456	1·080	1·357	1·528	1·826	1·983
350	−3·129	−2·781	−2·291	−1·888	−1·444	1·096	1·379	1·610	1·863	2·026
400	−3·094	−2·753	−2·270	−1·873	−1·434	1·108	1·396	1·633	1·893	2·061
450	−3·064	−2·729	−2·253	−1·861	−1·426	1·119	1·411	1·652	1·918	2·090
500	−3·040	−2·709	−2·239	−1·850	−1·419	1·127	1·423	1·668	1·938	2·114
550	−3·019	−2·691	−2·226	−1·841	−1·413	1·135	1·434	1·682	1·957	2·136
600	−3·000	−2·676	−2·215	−1·833	−1·408	1·141	1·443	1·694	1·972	2·154
650	−2·984	−2·663	−2·206	−1·826	−1·403	1·147	1·451	1·704	1·986	2·171
700	−2·969	−2·651	−2·197	−1·820	−1·399	1·152	1·458	1·714	1·999	2·185
750	−2·956	−2·640	−2·189	−1·814	−1·395	1·157	1·465	1·722	2·010	2·199
800	−2·944	−2·630	−2·182	−1·809	−1·392	1·161	1·471	1·730	2·020	2·211
850	−2·933	−2·621	−2·176	−1·804	−1·389	1·165	1·476	1·737	2·029	2·221
900	−2·923	−2·613	−2·170	−1·800	−1·386	1·168	1·481	1·743	2·037	2·231
950	−2·914	−2·605	−2·164	−1·796	−1·383	1·171	1·485	1·749	2·045	2·241
1000	−2·906	−2·599	−2·159	−1·792	−1·381	1·174	1·489	1·754	2·052	2·249

Table B.10: Wilk's Likelihood Ratio Test

$d = 3$

$m_H = 3$

$M \backslash \alpha$	0·100	0·050	0·025	0·010	0·005
1	1·322	1·359	1·394	1·437	1·468
2	1·127	1·140	1·153	1·168	1·179
3	1·071	1·077	1·084	1·092	1·098
4	1·045	1·049	1·053	1·058	1·062
5	1·032	1·035	1·037	1·041	1·043
6	1·023	1·026	1·028	1·030	1·032
7	1·018	1·020	1·021	1·023	1·025
8	1·014	1·016	1·017	1·018	1·019
9	1·012	1·013	1·014	1·015	1·016
10	1·010	1·011	1·011	1·012	1·013
12	1·007	1·008	1·008	1·009	1·009
14	1·005	1·006	1·006	1·007	1·007
16	1·004	1·005	1·005	1·005	1·006
18	1·003	1·004	1·004	1·004	1·005
20	1·003	1·003	1·003	1·004	1·004
24	1·002	1·002	1·002	1·002	1·003
30	1·001	1·001	1·001	1·002	1·002
40	1·001	1·001	1·001	1·001	1·001
60	1·000	1·000	1·000	1·000	1·000
120	1·000	1·000	1·000	1·000	1·000
∞	1·000	1·000	1·000	1·000	1·000
$\chi^2_{dm_H}$	14·6837	16·9190	19·0228	21·6660	23·5894

$m_H = 4$

$M \backslash \alpha$	0·100	0·050	0·025	0·010	0·005
1	1·379	1·422	1·463	1·514	1·550
2	1·159	1·174	1·188	1·207	1·220
3	1·091	1·099	1·107	1·116	1·123
4	1·060	1·065	1·070	1·076	1·080
5	1·043	1·046	1·050	1·054	1·057
6	1·032	1·035	1·037	1·040	1·042
7	1·025	1·027	1·029	1·031	1·033
8	1·020	1·022	1·023	1·025	1·026
9	1·017	1·018	1·019	1·021	1·022
10	1·014	1·015	1·016	1·017	1·018
12	1·010	1·011	1·012	1·012	1·013
14	1·008	1·008	1·009	1·010	1·010
16	1·006	1·007	1·007	1·007	1·008
18	1·005	1·005	1·006	1·006	1·006
20	1·004	1·004	1·005	1·005	1·005
24	1·003	1·003	1·003	1·004	1·004
30	1·002	1·002	1·002	1·002	1·002
40	1·001	1·001	1·001	1·001	1·001
60	1·000	1·001	1·001	1·001	1·001
120	1·000	1·000	1·000	1·000	1·000
∞	1·000	1·000	1·000	1·000	1·000
$\chi^2_{dm_H}$	18·5494	21·0261	23·3367	26·2170	28·2995

$m_H = 5$

$M \backslash \alpha$	0·100	0·050	0·025	0·010	0·005
1	1·433	1·481	1·527	1·584	1·025
2	1·191	1·208	1·224	1·245	1·260
3	1·113	1·122	1·131	1·142	1·150
4	1·076	1·082	1·087	1·094	1·099
5	1·055	1·059	1·063	1·068	1·071
6	1·042	1·045	1·048	1·051	1·054
7	1·033	1·035	1·038	1·040	1·042
8	1·027	1·029	1·030	1·033	1·034
9	1·022	1·024	1·025	1·027	1·028
10	1·019	1·020	1·021	1·023	1·024
12	1·014	1·015	1·015	1·017	1·017
14	1·010	1·011	1·012	1·013	1·013
16	1·008	1·009	1·009	1·010	1·011
18	1·007	1·007	1·008	1·008	1·009
20	1 006	1·006	1·006	1·007	1·007
24	1·004	1·004	1·005	1·005	1·005
30	1·003	1·003	1·003	1·003	1·003
40	1·002	1·002	1·002	1·002	1·002
60	1·001	1·001	1·001	1·001	1·001
120	1·000	1·000	1·000	1·000	1·000
∞	1·000	1·000	1·000	1·000	1·000
$\chi^2_{dm_H}$	22·3071	24·9958	27·4884	30·5779	32·8013

$m_H = 6$

$M \backslash \alpha$	0·100	0·050	0·025	0·010	0·005
1	1·482	1·535	1·586	1·649	1·694
2	1·222	1·241	1·259	1·282	1·298
3	1·135	1·145	1·155	1·167	1·176
4	1·092	1·099	1·105	1·113	1·119
5	1·068	1·072	1·077	1·082	1·086
6	1·052	1·056	1·059	1·063	1·066
7	1·041	1·044	1·047	1·050	1·052
8	1·034	1·036	1·038	1·041	1·042
9	1·028	1·030	1·032	1·034	1·035
10	1·024	1·025	1·027	1·028	1·030
12	1·018	1·019	1·020	1·021	1·022
14	1·014	1·014	1·015	1·016	1·017
16	1·011	1·012	1·012	1·013	1·014
18	1·009	1·009	1·010	1·011	1·011
20	1·007	1·008	1·008	1·009	1·009
24	1·005	1·006	1·006	1·006	1·007
30	1·004	1·004	1·004	1·004	1·004
40	1·002	1·002	1·002	1·002	1·003
60	1·001	1·001	1·001	1·001	1·001
120	1·000	1·000	1·000	1·000	1·000
∞	1·000	1·000	1·000	1·000	1·000
$\chi^2_{dm_H}$	25·9894	28·8693	31·5264	34·8053	37·1564

Table B.10: Wilk's Likelihood Ratio Test

$d = 3$

$M \backslash \alpha$	$m_H = 7$ 0·100	0·050	0·025	0·010	0·005	$m_H = 8$ 0·100	0·050	0·025	0·010	0·005
1	1·529	1·585	1·640	1·708	1·758	1·572	1·632	1·690	1·763	1·816
2	1·251	1·272	1·292	1·317	1·335	1·280	1·302	1·324	1·350	1·370
3	1·156	1·168	1·178	1·192	1·202	1·177	1·190	1·201	1·216	1·227
4	1·109	1·116	1·123	1·132	1·138	1·125	1·133	1·141	1·150	1·157
5	1·081	1·086	1·091	1·097	1·102	1·094	1·100	1·105	1·112	1·117
6	1·063	1·067	1·070	1·075	1·078	1·073	1·078	1·082	1·087	1·091
7	1·050	1·053	1·056	1·060	1·062	1·059	1·063	1·066	1·070	1·073
8	1·041	1·044	1·046	1·049	1·051	1·049	1·052	1·054	1·058	1·060
9	1·034	1·037	1·038	1·041	1·043	1·041	1·043	1·046	1·048	1·050
10	1·029	1·031	1·033	1·035	1·036	1·035	1·037	1·039	1·041	1·043
12	1·022	1·023	1·024	1·026	1·027	1·026	1·028	1·029	1·031	1·032
14	1·017	1·018	1·019	1·020	1·021	1·021	1·022	1·023	1·024	1·025
16	1·014	1·014	1·015	1·016	1·017	1·017	1·018	1·018	1·019	1·020
18	1·011	1·012	1·012	1·013	1·014	1·014	1·014	1·015	1·016	1·017
20	1·009	1·010	1·010	1·011	1·011	1·011	1·012	1·013	1·013	1·014
24	1·007	1·007	1·008	1·008	1·008	1·008	1·009	1·009	1·010	1·010
30	1·005	1·005	1·005	1·005	1·006	1·006	1·006	1·006	1·007	1·007
40	1·003	1·003	1·003	1·003	1·003	1·003	1·004	1·004	1·004	1·004
60	1·001	1·001	1·001	1·001	1·002	1·002	1·002	1·002	1·002	1·002
120	1·000	1·000	1·000	1·000	1·000	1·000	1·000	1·000	1·000	1·001
∞	1·000	1·000	1·000	1·000	1·000	1·000	1·000	1·000	1·000	1·000
$\chi^2_{dm_H}$	29·6151	32·6706	35·4789	38·9322	41·4011	33·1963	36·4151	39·3641	42·9798	45·5585

$M \backslash \alpha$	$m_H = 9$ 0·100	0·050	0·025	0·010	0·005	$m_H = 10$ 0·100	0·050	0·025	0·010	0·005
1	1·612	1·676	1·737	1·814	1·871	1·650	1·716	1·781	1·862	1·921
2	1·307	1·331	1·354	1·382	1·403	1·333	1·359	1·383	1·413	1·435
3	1·198	1·211	1·224	1·240	1·251	1·218	1·232	1·245	1·262	1·274
4	1·141	1·150	1·158	1·169	1·176	1·157	1·167	1·175	1·187	1·195
5	1·107	1·113	1·119	1·127	1·132	1·120	1·127	1·133	1·141	1·147
6	1·084	1·089	1·094	1·099	1·103	1·095	1·101	1·106	1·112	1·116
7	1·068	1·072	1·076	1·080	1·084	1·078	1·082	1·086	1·091	1·094
8	1·057	1·060	1·063	1·067	1·069	1·065	1·068	1·072	1·075	1·078
9	1·048	1·051	1·053	1·056	1·058	1·055	1·058	1·061	1·064	1·066
10	1·041	1·043	1·045	1·048	1·050	1·047	1·050	1·052	1·055	1·057
12	1·031	1·033	1·034	1·036	1·038	1·036	1·038	1·040	1·042	1·043
14	1·025	1·026	1·027	1·028	1·030	1·029	1·030	1·031	1·033	1·034
16	1·020	1·021	1·022	1·023	1·024	1·023	1·024	1·025	1·027	1·028
18	1·016	1·017	1·018	1·019	1·020	1·019	1·020	1·021	1·022	1·023
20	1·014	1·014	1·015	1·016	1·016	1·016	1·017	1·018	1·019	1·019
24	1·010	1·011	1·011	1·012	1·012	1·012	1·012	1·013	1·014	1·014
30	1·007	1·007	1·008	1·008	1·008	1·008	1·009	1·009	1·009	1·010
40	1·004	1·004	1·004	1·005	1·005	1·005	1·005	1·005	1·006	1·006
60	1·002	1·002	1·002	1·002	1·002	1·002	1·002	1·003	1·003	1·003
120	1·001	1·001	1·001	1·001	1·001	1·001	1·001	1·001	1·001	1·001
∞	1·000	1·000	1·000	1·000	1·000	1·000	1·000	1·000	1·000	1·000
$\chi^2_{dm_H}$	36·7412	40·1133	43·1945	46·9629	49·6449	40·2560	43·7730	46·9792	50·8922	53·6720

Table B.10: Wilk's Likelihood Ratio Test

$d = 3$

$M \backslash \alpha$	$m_H = 11$ 0·100	0·050	0·025	0·010	0·005	$m_H = 12$ 0·100	0·050	0·025	0·010	0·005
1	1·685	1·754	1·821	1·907	1·969	1·718	1·791	1·860	1·949	2·013
2	1·358	1·385	1·410	1·442	1·466	1·382	1·410	1·437	1·470	1·495
3	1·237	1·252	1·266	1·284	1·297	1·256	1·272	1·287	1·306	1·319
4	1·173	1·183	1·192	1·204	1·213	1·188	1·199	1·209	1·221	1·230
5	1·133	1·140	1·147	1·156	1·162	1·146	1·154	1·161	1·170	1·176
6	1·106	1·112	1·117	1·124	1·128	1·117	1·123	1·129	1·136	1·141
7	1·087	1·092	1·096	1·101	1·105	1·097	1·101	1·106	1·111	1·115
8	1·073	1·077	1·080	1·084	1·087	1·081	1·085	1·089	1·093	1·097
9	1·062	1·065	1·068	1·072	1·074	1·069	1·073	1·076	1·080	1·082
10	1·054	1·056	1·059	1·062	1·064	1·060	1·063	1·066	1·069	1·071
12	1·041	1·043	1·045	1·047	1·049	1·046	1·048	1·050	1·053	1·054
14	1·033	1·034	1·036	1·037	1·039	1·037	1·039	1·040	1·042	1·043
16	1·027	1·028	1·029	1·030	1·031	1·030	1·032	1·033	1·034	1·035
18	1·022	1·023	1·024	1·025	1·026	1·025	1·026	1·027	1·029	1·029
20	1·019	1·020	1·020	1·021	1·022	1·021	1·022	1·023	1·024	1·025
24	1·014	1·014	1·015	1·016	1·016	1·016	1·017	1·017	1·018	1·019
30	1·009	1·010	1·010	1·011	1·011	1·011	1·011	1·012	1·012	1·013
40	1·006	1·006	1·006	1·007	1·007	1·007	1·007	1·007	1·008	1·008
60	1·003	1·003	1·003	1·003	1·003	1·003	1·003	1·004	1·004	1·004
120	1·001	1·001	1·001	1·001	1·001	1·001	1·001	1·001	1·001	1·001
∞	1·000	1·000	1·000	1·000	1·000	1·000	1·000	1·000	1·000	1·000
$\chi^2_{dm_H}$	43·745	47·400	50·725	54·776	57·648	47·2122	50·9985	54·4373	58·6192	61·5812

$M \backslash \alpha$	$m_H = 13$ 0·100	0·050	0·025	0·010	0·005	$m_H = 14$ 0·100	0·050	0·025	0·010	0·005
1	1·750	1·824	1·896	1·988	2·055	1·780	1·857	1·931	2·026	2·095
2	1·405	1·434	1·462	1·497	1·522	1·427	1·458	1·486	1·523	1·549
3	1·274	1·291	1·306	1·326	1·340	1·292	1·309	1·326	1·346	1·361
4	1·203	1·214	1·225	1·238	1·247	1·217	1·229	1·240	1·254	1·264
5	1·158	1·167	1·174	1·184	1·191	1·171	1·179	1·188	1·198	1·205
6	1·128	1·134	1·140	1·148	1·153	1·138	1·145	1·152	1·159	1·165
7	1·106	1·111	1·116	1·122	1·126	1·115	1·121	1·126	1·132	1·136
8	1·089	1·094	1·098	1·102	1·106	1·097	1·102	1·106	1·111	1·115
9	1·076	1·080	1·083	1·088	1·090	1·084	1·088	1·091	1·095	1·099
10	1·066	1·069	1·072	1·076	1·078	1·073	1·076	1·079	1·082	1·085
12	1·052	1·054	1·056	1·059	1·061	1·057	1·059	1·061	1·064	1·066
14	1·041	1·043	1·045	1·047	1·048	1·046	1·048	1·049	1·052	1·053
16	1·034	1·035	1·037	1·038	1·040	1·037	1·039	1·041	1·042	1·044
18	1·028	1·029	1·031	1·032	1·033	1·031	1·033	1·034	1·035	1·036
20	1·024	1·025	1·026	1·027	1·028	1·027	1·028	1·029	1·030	1·031
24	1·018	1·019	1·019	1·020	1·021	1·020	1·021	1·022	1·023	1·023
30	1·012	1·013	1·013	1·014	1·014	1·014	1·015	1·015	1·016	1·016
40	1·008	1·008	1·008	1·009	1·009	1·009	1·009	1·009	1·010	1·010
60	1·004	1·004	1·004	1·004	1·004	1·004	1·004	1·005	1·005	1·005
120	1·001	1·001	1·001	1·001	1·001	1·001	1·001	1·001	1·001	1·001
∞	1·000	1·000	1·000	1·000	1·000	1·000	1·000	1·000	1·000	1·000
$\chi^2_{dm_H}$	50·660	54·572	58·120	62·428	65·476	54·0902	58·1240	61·7768	66·2062	69·3360

Table B.10: Wilk's Likelihood Ratio Test

$d = 3$

M \ α	$m_H = 15$ 0·100	0·050	0·050	0·010	0·005	$m_H = 16$ 0·100	0·050	0·025	0·010	0·005
1	1·808	1·887	1·964	2·061	2·133	1·835	1·916	1·995	2·095	2·169
2	1·449	1·480	1·510	1·547	1·575	1·469	1·501	1·532	1·571	1·599
3	1·309	1·327	1·344	1·365	1·381	1·325	1·344	1·362	1·384	1·400
4	1·232	1·244	1·256	1·270	1·280	1·245	1·258	1·271	1·285	1·296
5	1·183	1·192	1·200	1·211	1·218	1·195	1·204	1·213	1·224	1·232
6	1·149	1·156	1·163	1·171	1·177	1·159	1·167	1·174	1·182	1·188
7	1·124	1·130	1·135	1·142	1·147	1·133	1·139	1·145	1·152	1·157
8	1·105	1·110	1·115	1·120	1·124	1·114	1·119	1·123	1·129	1·133
9	1·091	1·095	1·099	1·103	1·107	1·098	1·102	1·106	1·111	1·115
10	1·079	1·083	1·086	1·090	1·093	1·085	1·089	1·092	1·097	1·099
12	1·062	1·065	1·067	1·070	1·072	1·067	1·070	1·073	1·076	1·078
14	1·050	1·052	1·054	1·056	1·058	1·054	1·057	1·059	1·061	1·063
16	1·041	1·043	1·045	1·047	1·048	1·045	1·047	1·049	1·051	1·052
18	1·035	1·036	1·037	1·039	1·040	1·038	1·039	1·041	1·043	1·044
20	1·030	1·031	1·032	1·033	1·034	1·032	1·034	1·035	1·036	1·037
24	1·022	1·023	1·024	1·025	1·026	1·025	1·026	1·026	1·027	1·028
30	1·016	1·016	1·017	1·017	0·018	1·017	1·018	1·018	1·019	1·020
40	1·010	1·010	1·010	1·011	1·011	1·011	1·011	1·011	1·012	1·012
60	1·005	1·005	1·005	1·005	1·005	1·005	1·006	1·006	1·006	1·006
120	1·001	1·001	1·001	1·001	1·002	1·002	1·002	1·002	1·002	1·002
∞	1·000	1·000	1·000	1·000	1·000	1·000	1·000	1·000	1·000	1·000
$\chi^2_{dm_H}$	57·505	61·656	65·410	69·957	73·166	60·9066	65·1708	69·0226	73·6826	76·9688

M \ α	$m_H = 17$ 0·100	0·050	0·025	0·010	0·005	$m_H = 18$ 0·100	0·050	0·025	0·010	0·005
1	1·861	1·944	2·025	2·127	2·203	1·886	1·971	2·053	2·158	2·235
2	1·489	1·522	1·554	1·594	1·623	1·508	1·542	1·575	1·616	1·646
3	1·341	1·361	1·379	1·402	1·419	1·357	1·377	1·396	1·420	1·437
4	1·259	1·273	1·285	1·300	1·312	1·272	1·286	1·299	1·315	1·327
5	1·206	1·216	1·225	1·237	1·245	1·218	1·228	1·238	1·249	1·258
6	1·169	1·177	1·184	1·193	1·200	1·179	1·188	1·195	1·204	1·211
7	1·142	1·149	1·154	1·162	1·167	1·151	1·158	1·164	1·171	1·177
8	1·122	1·127	1·132	1·138	1·142	1·129	1·135	1·140	1·146	1·151
9	1·105	1·110	1·114	1·119	1·123	1·112	1·117	1·121	1·127	1·130
10	1·092	1·096	1·100	1·104	1·107	1·099	1·103	1·107	1·111	1·114
12	1·073	1·076	1·079	1·082	1·084	1·078	1·081	1·084	1·087	1·090
14	1·059	1·061	1·064	1·066	1·068	1·064	1·066	1·068	1·071	1·073
16	1·049	1·051	1·053	1·055	1·056	1·053	1·055	1·057	1·059	1·061
18	1·041	1·043	1·044	1·046	1·047	1·045	1·046	1·048	1·050	1·051
20	1·035	1·037	1·038	1·040	1·041	1·038	1·040	1·041	1·043	1·044
24	1·027	1·028	1·029	1·030	1·031	1·029	1·030	1·031	1·032	1·033
30	1·019	1·020	1·020	1·021	1·022	1·021	1·021	1·022	1·023	1·023
40	1·012	1·012	1·013	1·013	1·013	1·013	1·013	1·014	1·014	1·015
60	1·006	1·006	1·006	1·006	1·007	1·006	1·007	1·007	1·007	1·007
120	1·002	1·002	1·002	1·002	1·002	1·002	1·002	1·002	1·002	1·002
∞	1·000	1·000	1·000	1·000	1·000	1·000	1·000	1·000	1·000	1·000
$\chi^2_{dm_H}$	64·295	68·669	72·616	77·386	80·747	67·6728	72·1532	76·1920	81·0688	84·5019

Table B.10: Wilk's Likelihood Ratio Test

$d = 3$

M \ α	$m_H = 19$: 0·100	0·050	0·025	0·010	0·005	$m_H = 20$: 0·100	0·050	0·025	0·010	0·005
1	1·909	1·996	2·080	2·188	2·267	1·932	2·021	2·106	2·216	2·297
2	1·526	1·561	1·595	1·637	1·668	1·544	1·580	1·614	1·657	1·689
3	1·372	1·393	1·412	1·437	1·454	1·387	1·408	1·428	1·453	1·472
4	1·285	1·300	1·313	1·330	1·341	1·298	1·313	1·327	1·344	1·356
5	1·229	1·240	1·250	1·262	1·271	1·240	1·251	1·261	1·274	1·283
6	1·189	1·198	1·205	1·215	1·222	1·199	1·208	1·216	1·226	1·233
7	1·160	1·167	1·173	1·181	1·186	1·168	1·176	1·182	1·190	1·196
8	1·137	1·143	1·148	1·155	1·159	1·145	1·151	1·157	1·163	1·168
9	1·119	1·124	1·129	1·134	1·138	1·127	1·132	1·136	1·142	1·146
10	1·105	1·109	1·113	1·118	1·121	1·112	1·116	1·120	1·125	1·128
12	1·084	1·087	1·090	1·093	1·096	1·089	1·092	1·095	1·099	1·102
14	1·068	1·071	1·073	1·076	1·078	1·073	1·075	1·078	1·081	1·083
16	1·057	1·059	1·061	1·063	1·065	1·061	1·063	1·065	1·067	1·069
18	1·048	1·050	1·052	1·054	1·055	1·052	1·053	1·055	1·057	1·059
20	1·041	1·043	1·044	1·046	1·047	1·044	1·046	1·048	1·049	1·050
24	1·032	1·033	1·034	1·035	1·036	1·034	1·035	1·036	1·038	1·039
30	1·022	1·023	1·024	1·025	1·025	1·024	1·025	1·026	1·027	1·027
40	1·014	1·015	1·015	1·016	1·016	1·015	1·016	1·016	1·017	1·017
60	1·007	1·007	1·008	1·008	1·008	1·008	1·008	1·008	1·009	1·009
120	1·002	1·002	1·002	1·002	1·002	1·002	1·002	1·002	1·002	1·003
∞	1·000	1·000	1·000	1·000	1·000	1·000	1·000	1·000	1·000	1·000
$\chi^2_{dm_H}$	71·040	75·624	79·752	84·733	88·236	74·3970	79·0819	83·2976	88·3794	91·9517

M \ α	$m_H = 21$: 0·100	0·050	0·025	0·010	0·005	$m_H = 22$: 0·100	0·050	0·025	0·010	0·005
1	1·954	2·044	2·131	2·243	2·325	1·975	2·067	2·156	2·269	2·353
2	1·561	1·598	1·633	1·677	1·709	1·578	1·616	1·651	1·696	1·729
3	1·401	1·423	1·444	1·470	1·488	1·415	1·438	1·459	1·485	1·504
4	1·310	1·325	1·340	1·357	1·370	1·322	1·338	1·353	1·371	1·384
5	1·250	1·262	1·273	1·286	1·295	1·261	1·273	1·284	1·297	1·307
6	1·208	1·217	1·226	1·236	1·243	1·218	1·227	1·236	1·246	1·254
7	1·177	1·184	1·191	1·200	1·205	1·185	1·193	1·200	1·209	1·215
8	1·153	1·159	1·165	1·172	1·176	1·160	1·167	1·173	1·180	1·185
9	1·133	1·139	1·144	1·150	1·154	1·142	1·147	1·151	1·157	1·161
10	1·118	1·122	1·127	1·132	1·135	1·124	1·129	1·133	1·139	1·141
12	1·094	1·098	1·101	1·105	1·108	1·099	1·103	1·106	1·110	1·113
14	1·077	1·080	1·083	1·086	1·088	1·082	1·085	1·087	1·091	1·093
16	1·065	1·067	1·069	1·072	1·074	1·069	1·071	1·073	1·076	1·078
18	1·055	1·057	1·059	1·061	1·063	1·059	1·061	1·063	1·065	1·066
20	1·048	1·049	1·051	1·053	1·054	1·051	1·052	1·054	1·056	1·057
24	1·036	1·038	1·039	1·040	1·041	1·039	1·040	1·041	1·043	1·044
30	1·026	1·027	1·028	1·029	1·029	1·028	1·029	1·030	1·031	1·031
40	1·016	1·017	1·018	1·018	1·019	1·018	1·018	1·019	1·020	1·020
60	1·008	1·009	1·009	1·009	1·009	1·009	1·009	1·010	1·010	1·010
120	1·002	1·002	1·003	1·003	1·003	1·003	1·003	1·003	1·003	1·003
∞	1·000	1·000	1·000	1·000	1·000	1·000	1·000	1·000	1·000	1·000
$\chi^2_{dm_H}$	77·745	82·529	86·830	92·010	95·649	81·0855	85·9649	90·3489	95·6257	99·3304

Table B.10: Wilk's Likelihood Ratio Test

$d = 4$

$M \backslash \alpha$	$m_H = 4$ 0·100	0·050	0·025	0·010	0·005	$m_H = 5$ 0·100	0·050	0·025	0·010	0·005
1	1·405	1·451	1·494	1·550	1·589	1·435	1·483	1·530	1·589	1·632
2	1·178	1·194	1·209	1·229	1·243	1·199	1·216	1·233	1·253	1·269
3	1·105	1·114	1·122	1·132	1·139	1·121	1·130	1·139	1·150	1·158
4	1·071	1·076	1·081	1·088	1·092	1·083	1·089	1·094	1·101	1·106
5	1·051	1·055	1·058	1·063	1·066	1·061	1·065	1·069	1·074	1·077
6	1·039	1·042	1·044	1·048	1·050	1·047	1·050	1·053	1·056	1·059
7	1·031	1·033	1·035	1·037	1·039	1·037	1·040	1·042	1·044	1·046
8	1·025	1·027	1·028	1·030	1·032	1·030	1·032	1·034	1·036	1·038
9	1·020	1·022	1·023	1·025	1·026	1·025	1·027	1·028	1·030	1·031
10	1·017	1·018	1·019	1·021	1·022	1·021	1·023	1·024	1·025	1·026
12	1·013	1·014	1·014	1·015	1·016	1·016	1·017	1·018	1·019	1·020
14	1·010	1·010	1·011	1·012	1·012	1·012	1·013	1·014	1·014	1·015
16	1·008	1·008	1·009	1·009	1·010	1·010	1·010	1·011	1·012	1·012
18	1·006	1·007	1·007	1·008	1·008	1·008	1·008	1·009	1·009	1·010
20	1·005	1·006	1·006	1·006	1·007	1·007	1·007	1·007	1·008	1·008
24	1·004	1·004	1·004	1·004	1·005	1·005	1·005	1·005	1·006	1·006
30	1·002	1·003	1·003	1·003	1·003	1·003	1·003	1·004	1·004	1·004
40	1·001	1·002	1·002	1·002	1·002	1·002	1·002	1·002	1·002	1·002
60	1·001	1·001	1·001	1·001	1·001	1·001	1·001	1·001	1·001	1·001
120	1·000	1·000	1·000	1·000	1·000	1·000	1·000	1·000	1·000	1·000
∞	1·000	1·000	1·000	1·000	1·000	1·000	1·000	1·000	1·000	1·000
$\chi^2_{dm_H}$	23·5418	26·2962	28·8454	31·9999	34·2672	28·4120	31·4104	34·1696	37·5662	39·9968

$M \backslash \alpha$	$m_H = 6$ 0·100	0·050	0·025	0·010	0·005	$m_H = 7$ 0·100	0·050	0·025	0·010	0·005
1	1·466	1·517	1·566	1·628	1·674	1·497	1·550	1·601	1·667	1·715
2	1·222	1·240	1·257	1·279	1·295	1·244	1·263	1·281	1·305	1·322
3	1·138	1·148	1·157	1·168	1·177	1·155	1·165	1·175	1·188	1·197
4	1·096	1·102	1·108	1·115	1·121	1·109	1·116	1·122	1·130	1·136
5	1·071	1·076	1·080	1·085	1·089	1·082	1·087	1·092	1·097	1·101
6	1·055	1·059	1·062	1·066	1·068	1·064	1·068	1·071	1·076	1·079
7	1·044	1·047	1·049	1·052	1·055	1·052	1·055	1·057	1·061	1·063
8	1·036	1·038	1·040	1·043	1·045	1·043	1·045	1·047	1·050	1·052
9	1·030	1·032	1·034	1·036	1·037	1·036	1·038	1·040	1·042	1·044
10	1·026	1·027	1·029	1·030	1·032	1·031	1·032	1·034	1·036	1·037
12	1·019	1·020	1·021	1·023	1·024	1·023	1·024	1·026	1·027	1·028
14	1·015	1·016	1·017	1·018	1·018	1·018	1·019	1·020	1·021	1·022
16	1·012	1·013	1·013	1·014	1·015	1·015	1·015	1·016	1·017	1·017
18	1·010	1·010	1·011	1·012	1·012	1·012	1·013	1·013	1·014	1·014
20	1·008	1·009	1·009	1·010	1·010	1·010	1·011	1·011	1·012	1·012
24	1·006	1·006	1·007	1·007	1·007	1·007	1·008	1·008	1·008	1·009
30	1·004	1·004	1·004	1·005	1·005	1·005	1·005	1·005	1·006	1·006
40	1·002	1·002	1·003	1·003	1·003	1·003	1·003	1·003	1·003	1·004
60	1·001	1·001	1·001	1·001	1·001	1·001	1·001	1·002	1·002	1·002
120	1·000	1·000	1·000	1·000	1·000	1·000	1·000	1·000	1·000	1·000
∞	1·000	1·000	1·000	1·000	1·000	1·000	1·000	1·000	1·000	1·000
$\chi^2_{dm_H}$	33·1963	36·4151	39·3641	42·9798	45·5585	37·9159	41·3372	44·4607	48·2782	50·9933

Table B.10: Wilk's Likelihood Ratio Test

$d = 4$

	$m_H = 8$					$m_H = 9$				
$M \backslash \alpha$	0·100	0·050	0·025	0·010	0·005	0·100	0·050	0·025	0·010	0·005
1	1·528	1·583	1·636	1·704	1·754	1·557	1·614	1·669	1·740	1·792
2	1·266	1·286	1·305	1·330	1·348	1·288	1·309	1·329	1·355	1·373
3	1·172	1·183	1·193	1·207	1·216	1·189	1·201	1·212	1·226	1·236
4	1·123	1·130	1·137	1·146	1·152	1·137	1·144	1·152	1·161	1·167
5	1·093	1·099	1·103	1·109	1·114	1·105	1·110	1·115	1·122	1·127
6	1·074	1·078	1·081	1·086	1·089	1·083	1·088	1·091	1·096	1·100
7	1·060	1·063	1·066	1·070	1·072	1·068	1·071	1·075	1·078	1·081
8	1·050	1·052	1·055	1·058	1·060	1·057	1·060	1·062	1·065	1·068
9	1·042	1·044	1·046	1·048	1·050	1·048	1·050	1·053	1·055	1·057
10	1·036	1·038	1·039	1·041	1·043	1·041	1·043	1·045	1·047	1·049
12	1·027	1·029	1·030	1·031	1·033	1·032	1·033	1·034	1·036	1·037
14	1·021	1·023	1·023	1·025	1·026	1·025	1·026	1·027	1·029	1·029
16	1·017	1·018	1·019	1·020	1·021	1·020	1·021	1·022	1·023	1·024
18	1·014	1·015	1·016	1·016	1·017	1·017	1·018	1·018	1·019	1·020
20	1·012	1·013	1·013	1·014	1·014	1·014	1·015	1·015	1·016	1·017
24	1·009	1·009	1·010	1·010	1·010	1·010	1·011	1·011	1·012	1·012
30	1·006	1·006	1·007	1·007	1·007	1·007	1·007	1·008	1·008	1·008
40	1·003	1·004	1·004	1·004	1·004	1·004	1·004	1·005	1·005	1·005
60	1·002	1·002	1·002	1·002	1·002	1·002	1·002	1·002	1·002	1·002
120	1·000	1·000	1·000	1·001	1·001	1·001	1·001	1·001	1·001	1·001
∞	1·000	1·000	1·000	1·000	1·000	1·000	1·000	1·000	1·000	1·000
$\chi^2_{dm_H}$	42·5847	46·1943	49·4804	53·4858	56·3281	47·2122	50·9985	54·4373	58·6192	61·5812

	$m_H = 10$					$m_H = 11$				
$M \backslash \alpha$	0·100	0·050	0·025	0·010	0·005	0·100	0·050	0·025	0·010	0·005
1	1·585	1·644	1·701	1·774	1·828	—	—	—	—	—
2	1·309	1·331	1·352	1·379	1·398	1·330	1·352	1·374	1·402	1·422
3	1·206	1·218	1·230	1·244	1·255	1·222	1·235	1·247	1·262	1·274
4	1·150	1·159	1·166	1·176	1·183	1·164	1·173	1·181	1·191	1·198
5	1·116	1·122	1·128	1·134	1·139	1·127	1·134	1·140	1·147	1·152
6	1·093	1·097	1·102	1·107	1·111	1·103	1·107	1·112	1·118	1·122
7	1·076	1·080	1·083	1·088	1·090	1·085	1·089	1·092	1·097	1·100
8	1·064	1·067	1·070	1·073	1·076	1·071	1·075	1·077	1·081	1·084
9	1·054	1·057	1·059	1·062	1·064	1·061	1·064	1·066	1·069	1·071
10	1·047	1·049	1·051	1·054	1·055	1·053	1·055	1·057	1·060	1·062
12	1·036	1·038	1·039	1·041	1·042	1·041	1·043	1·044	1·046	1·047
14	1·029	1·030	1·031	1·033	1·034	1·033	1·034	1·035	1·037	1·038
16	1·023	1·024	1·025	1·026	1·027	1·027	1·028	1·029	1·030	1·031
18	1·019	1·020	1·021	1·022	1·023	1·022	1·023	1·024	1·025	1·026
20	1·016	1·017	1·018	1·019	1·019	1·019	1·020	1·020	1·021	1·022
24	1·012	1·013	1·013	1·014	1·014	1·014	1·015	1·015	1·016	1·016
30	1·008	1·009	1·009	1·009	1·010	1·010	1·010	1·010	1·011	1·011
40	1·005	1·005	1·005	1·006	1·006	1·006	1·006	1·006	1·007	1·007
60	1·002	1·003	1·003	1·003	1·003	1·003	1·003	1·003	1·003	1·003
120	1·001	1·001	1·001	1·001	1·001	1·001	1·001	1·001	1·001	1·001
∞	1·000	1·000	1·000	1·000	1·000	1·000	1·000	1·000	1·000	1·000
$\chi^2_{dm_H}$	51·8050	55·7585	59·3417	63·6907	66·7659	56·369	60·481	64·201	68·710	71·893

Table B.10: Wilk's Likelihood Ratio Test

$d = 4$

$M \backslash \alpha$	$m_H = 12$ 0·100	0·050	0·025	0·010	0·005	$m_H = 13$ 0·100	0·050	0·025	0·010	0·005
1	1·638	1·700	1·760	1·838	1·895	—	—	—	—	—
2	1·350	1·373	1·396	1·424	1·446	1·369	1·393	1·417	1·446	1·468
3	1·238	1·252	1·264	1·280	1·292	1·254	1·268	1·281	1·298	1·310
4	1·177	1·186	1·195	1·205	1·213	1·190	1·200	1·209	1·220	1·228
5	1·139	1·145	1·152	1·159	1·165	1·150	1·157	1·163	1·171	1·177
6	1·112	1·118	1·122	1·128	1·132	1·122	1·127	1·132	1·139	1·143
7	1·093	1·097	1·101	1·106	1·109	1·102	1·106	1·110	1·115	1·118
8	1·079	1·082	1·085	1·089	1·092	1·086	1·090	1·093	1·097	1·100
9	1·068	1·070	1·073	1·076	1·079	1·074	1·077	1·080	1·083	1·086
10	1·059	1·061	1·063	1·066	1·068	1·065	1·067	1·070	1·073	1·075
12	1·046	1·047	1·049	1·051	1·053	1·050	1·052	1·054	1·056	1·058
14	1·037	1·038	1·039	1·041	1·042	1·041	1·042	1·044	1·045	1·047
16	1·030	1·031	1·032	1·033	1·034	1·033	1·035	1·036	1·037	1·038
18	1·025	1·026	1·027	1·028	1·029	1·028	1·029	1·030	1·031	1·032
20	1·021	1·022	1·023	1·024	1·024	1·024	1·025	1·026	1·027	1·027
24	1·016	1·017	1·017	1·018	1·018	1·018	1·019	1·019	1·020	1·020
30	1·011	1·011	1·012	1·012	1·013	1·012	1·013	1·013	1·014	1·014
40	1·007	1·007	1·007	1·008	1·008	1·008	1·008	1·008	1·008	1·009
60	1·003	1·003	1·004	1·004	1·004	1·004	1·004	1·004	1·004	1·004
120	1·001	1·001	1·001	1·001	1·001	1·001	1·001	1·001	1·001	1·001
∞	1·000	1·000	1·000	1·000	1·000	1·000	1·000	1·000	1·000	1·000
$\chi^2_{dm_H}$	60·9066	65·1708	69·0226	73·6826	76·9688	65·422	69·832	73·810	78·616	82·001

$M \backslash \alpha$	$m_H = 14$ 0·100	0·050	0·025	0·010	0·005	$m_H = 15$ 0·100	0·050	0·025	0·010	0·005
1	1·686	1·751	1·814	1·896	1·956	—	—	—	—	—
2	1·388	1·413	1·436	1·467	1·489	1·406	1·432	1·456	1·488	1·511
3	1·269	1·284	1·297	1·314	1·327	1·284	1·299	1·313	1·331	1·344
4	1·203	1·213	1·222	1·234	1·242	1·216	1·226	1·236	1·248	1·256
5	1·161	1·168	1·175	1·183	1·189	1·172	1·179	1·187	1·195	1·202
6	1·131	1·137	1·142	1·149	1·154	1·141	1·147	1·153	1·159	1·164
7	1·110	1·115	1·119	1·124	1·128	1·118	1·123	1·128	1·133	1·137
8	1·094	1·097	1·101	1·105	1·109	1·101	1·105	1·109	1·113	1·116
9	1·081	1·084	1·087	1·091	1·093	1·087	1·091	1·094	1·098	1·101
10	1·071	1·073	1·076	1·079	1·081	1·077	1·080	1·082	1·085	1·088
12	1·055	1·058	1·059	1·062	1·064	1·060	1·063	1·065	1·067	1·069
14	1·045	1·046	1·048	1·050	1·051	1·049	1·051	1·052	1·054	1·056
16	1·037	1·038	1·039	1·041	1·042	1·040	1·042	1·043	1·045	1·046
18	1·031	1·032	1·033	1·034	1·035	1·034	1·035	1·036	1·038	1·039
20	1·026	1·027	1·028	1·029	1·030	1·029	1·030	1·031	1·032	1·033
24	1·020	1·021	1·021	1·022	1·023	1·022	1·023	1·023	1·024	1·025
30	1·014	1·014	1·015	1·015	1·016	1·015	1·016	1·016	1·017	1·017
40	1·009	1·009	1·009	1·009	1·010	1·010	1·010	1·010	1·011	1·011
60	1·004	1·004	1·005	1·005	1·005	1·005	1·005	1·005	1·005	1·005
120	1·001	1·001	1·001	1·001	1·001	1·001	1·001	1·001	1·001	1·001
∞	1·000	1·000	1·000	1·000	1·000	1·000	1·000	1·000	1·000	1·000
$\chi^2_{dm_H}$	69·9185	74·4683	78·5671	83·5134	86·9937	74·397	79·082	83·298	88·379	91·952

Table B.10: Wilk's Likelihood Ratio Test

$d = 4$

$M \backslash \alpha$	$m_H = 16$ 0·100	0·050	0·025	0·010	0·005	$m_H = 17$ 0·100	0·050	0·025	0·010	0·005
1	1·731	1·799	1·864	1·949	2·012	—	—	—	—	—
2	1·423	1·450	1·475	1·507	1·531	1·440	1·468	1·494	1·527	1·551
3	1·299	1·314	1·329	1·347	1·360	1·313	1·329	1·344	1·363	1·377
4	1·228	1·239	1·249	1·261	1·270	1·240	1·252	1·262	1·275	1·284
5	1·182	1·190	1·198	1·207	1·213	1·193	1·201	1·209	1·218	1·225
6	1·150	1·157	1·163	1·169	1·174	1·160	1·166	1·172	1·180	1·185
7	1·127	1·132	1·136	1·142	1·146	1·135	1·140	1·145	1·151	1·155
8	1·108	1·113	1·117	1·121	1·125	1·116	1·120	1·124	1·129	1·133
9	1·094	1·098	1·101	1·105	1·108	1·101	1·105	1·108	1·112	1·115
10	1·083	1·086	1·089	1·092	1·094	1·089	1·092	1·095	1·098	1·101
12	1·065	1·068	1·070	1·073	1·074	1·070	1·073	1·075	1·078	1·080
14	1·053	1·055	1·056	1·058	1·060	1·057	1·059	1·061	1·063	1·065
16	1·044	1·045	1·047	1·049	1·050	1·048	1·049	1·051	1·053	1·054
18	1·037	1·038	1·040	1·041	1·042	1·040	1·042	1·043	1·045	1·046
20	1·032	1·033	1·034	1·035	1·036	1·035	1·036	1·037	1·038	1·039
24	1·024	1·025	1·026	1·027	1·027	1·026	1·027	1·028	1·029	1·030
30	1·017	1·018	1·018	1·019	1·019	1·019	1·019	1·020	1·020	1·021
40	1·011	1·011	1·011	1·012	1·012	1·012	1·012	1·012	1·013	1·013
60	1·005	1·005	1·006	1·006	1·006	1·006	1·006	1·006	1·006	1·007
120	1·001	1·002	1·002	1·002	1·002	1·002	1·002	1·002	1·002	1·002
∞	1·000	1·000	1·000	1·000	1·000	1·000	1·000	1·000	1·000	1·000
$\chi^2_{dm_H}$	78·8597	83·6753	88·0040	93·2168	96·8781	83·308	88·250	92·689	98·028	101·776

$M \backslash \alpha$	$m_H = 18$ 0·100	0·050	0·025	0·010	0·005	$m_H = 19$ 0·100	0·050	0·025	0·010	0·005
1	1·773	1·843	1·911	1·999	2·065	—	—	—	—	—
2	1·457	1·485	1·511	1·545	1·570	1·473	1·502	1·529	1·563	1·588
3	1·327	1·343	1·359	1·378	1·392	1·340	1·357	1·373	1·393	1·408
4	1·252	1·264	1·274	1·287	1·297	1·264	1·276	1·287	1·300	1·310
5	1·203	1·212	1·220	1·230	1·237	1·214	1·223	1·231	1·241	1·248
6	1·169	1·176	1·182	1·189	1·195	1·178	1·185	1·191	1·199	1·205
7	1·143	1·149	1·154	1·160	1·164	1·151	1·157	1·162	1·169	1·173
8	1·123	1·128	1·132	1·137	1·141	1·130	1·135	1·140	1·145	1·149
9	1·107	1·111	1·115	1·119	1·122	1·114	1·118	1·122	1·126	1·130
10	1·095	1·098	1·101	1·105	1·108	1·101	1·104	1·107	1·111	1·114
12	1·075	1·078	1·080	1·083	1·085	1·080	1·083	1·086	1·089	1·091
14	1·061	1·063	1·065	1·068	1·069	1·066	1·068	1·070	1·073	1·074
16	1·051	1·053	1·054	1·056	1·058	1·055	1·057	1·059	1·061	1·062
18	1·044	1·045	1·046	1·048	1·049	1·047	1·048	1·050	1·051	1·053
20	1·037	1·039	1·040	1·041	1·042	1·040	1·042	1·043	1·044	1·045
24	1·029	1·030	1·030	1·031	1·032	1·031	1·032	1·033	1·034	1·035
30	1·020	1·021	1·022	1·022	1·023	1·022	1·023	1·023	1·024	1·025
40	1·013	1·013	1·014	1·014	1·014	1·014	1·014	1·015	1·015	1·015
60	1·006	1·007	1·007	1·007	1·007	1·007	1·007	1·007	1·008	1·008
120	1·002	1·002	1·002	1·002	1·002	1·002	1·002	1·002	1·002	1·002
∞	1·000	1·000	1·000	1·000	1·000	1·000	1·000	1·000	1·000	1·000
$\chi^2_{dm_H}$	87·7431	92·8083	97·3531	102·816	106·648	92·166	97·351	101·999	107·583	111·495

Table B.10: Wilk's Likelihood Ratio Test

$d = 4$

$M \backslash \alpha$	$m_H = 20$ 0·100	0·050	0·025	0·010	0·005	$m_H = 21$ 0·100	0·050	0·025	0·010	0·005
1	1·812	1·884	1·954	2·045	2·113	—	—	—	—	—
2	1·488	1·518	1·545	1·580	1·606	1·504	1·533	1·562	1·598	1·624
3	1·353	1·371	1·387	1·408	1·422	1·367	1·384	1·401	1·422	1·437
4	1·275	1·288	1·299	1·313	1·323	1·287	1·299	1·311	1·325	1·335
5	1·224	1·233	1·241	1·252	1·259	1·234	1·243	1·252	1·262	1·270
6	1·187	1·194	1·201	1·208	1·215	1·196	1·203	1·210	1·218	1·224
7	1·159	1·165	1·170	1·177	1·182	1·167	1·173	1·179	1·186	1·190
8	1·138	1·143	1·147	1·153	1·157	1·145	1·150	1·155	1·160	1·164
9	1·121	1·125	1·129	1·133	1·137	1·127	1·132	1·136	1·140	1·144
10	1·107	1·110	1·114	1·118	1·121	1·113	1·116	1·120	1·124	1·127
12	1·086	1·088	1·091	1·094	1·096	1·091	1·094	1·096	1·099	1·102
14	1·070	1·072	1·074	1·077	1·078	1·075	1·077	1·079	1·082	1·084
16	1·059	1·061	1·062	1·064	1·066	1·063	1·065	1·066	1·069	1·070
18	1·050	1·052	1·053	1·055	1·056	1·054	1·055	1·057	1·059	1·060
20	1·043	1·045	1·046	1·047	1·048	1·046	1·048	1·049	1·051	1·052
24	1·033	1·034	1·035	1·036	1·037	1·036	1·037	1·038	1·039	1·040
30	1·024	1·024	1·025	1·026	1·026	1·025	1·026	1·027	1·028	1·028
40	1·015	1·016	1·016	1·016	1·017	1·016	1·017	1·017	1·018	1·018
60	1·008	1·008	1·008	1·008	1·008	1·008	1·008	1·009	1·009	1·009
120	1·002	1·002	1·002	1·002	1·002	1·002	1·002	1·003	1·003	1·003
∞	1·000	1·000	1·000	1·000	1·000	1·000	1·000	1·000	1·000	1·000
$\chi^2_{dm_H}$	96·5782	101·879	106·629	112·329	116·321	100·980	106·395	111·242	117·057	121·126

$M \backslash \alpha$	$m_H = 22$ 0·100	0·050	0·025	0·010	0·005
1	1·848	1·922	1·994	2·088	2·158
2	1·518	1·549	1·577	1·614	1·641
3	1·379	1·397	1·414	1·436	1·451
4	1·298	1·310	1·322	1·337	1·347
5	1·243	1·253	1·262	1·273	1·281
6	1·204	1·212	1·219	1·228	1·234
7	1·175	1·181	1·187	1·194	1·199
8	1·152	1·157	1·162	1·168	1·172
9	1·134	1·138	1·142	1·147	1·151
10	1·119	1·123	1·126	1·130	1·134
12	1·095	1·098	1·101	1·104	1·107
14	1·079	1·081	1·083	1·086	1·088
16	1·066	1·068	1·070	1·072	1·074
18	1·057	1·058	1·060	1·062	1·063
20	1·049	1·051	1·052	1·053	1·055
24	1·038	1·039	1·040	1·041	1·042
30	1·027	1·028	1·029	1·030	1·030
40	1·017	1·018	1·018	1·019	1·019
60	1·009	1·009	1·009	1·010	1·010
120	1·003	1·003	1·003	1·003	1·003
∞	1·000	1·000	1·000	1·000	1·000
$\chi^2_{dm_H}$	105·372	110·898	115·841	121·767	125·913

Table B.11: Roy's Maximum Root Statistic

$\alpha = 0.05$, $s = 2$

ν_2 \ ν_1	0	1	2	3	4	5	7	10	15
2	.7919	.3514	.8839	.9045	.9189	.9295	.9441	.9573	.9693
3	.7017	.7761	.8197	.8487	.8696	.8853	.9075	.9283	.9478
4	.6267	.7090	.7600	.7953	.8213	.8413	.8702	.8980	.9247
5	.5646	.6507	.7063	.7459	.7758	.7992	.8337	.8676	.9011
6	.5130	.6002	.6585	.7010	.7337	.7598	.7988	.8380	.8775
7	.4696	.5564	.6160	.6604	.6952	.7232	.7658	.8095	.8544
8	.4328	.5182	.5782	.6238	.6599	.6893	.7348	.7822	.8318
9	.4011	.4846	.5445	.5906	.6276	.6581	.7058	.7562	.8100
10	.3737	.4550	.5143	.5606	.5980	.6293	.6786	.7316	.7889
15	.2781	.3477	.4015	.4455	.4826	.5145	.5670	.6266	.6955
20	.2211	.2810	.3287	.3688	.4034	.4339	.4855	.5463	.6198
25	.1835	.2355	.2780	.3143	.3463	.3478	.4239	.4835	.5580
30	.1568	.2027	.2408	.2738	.3031	.3296	.3760	.4333	.5071
35	.1369	.1780	.2124	.2425	.2696	.2924	.3377	.3924	.4644
40	.1242	.1585	.1898	.2175	.2425	.2655	.3064	.3585	.4282
48	.1031	.1352	.1626	.1870	.2093	.2299	.2670	.3150	.3807
60	.0836	.1103	.1333	.1540	.1731	.1909	.2233	.2661	.3260
80	.0638	.0846	.1027	.1192	.1346	.1409	.1756	.2114	.2630
120	.0433	.0577	.0704	.0821	.0931	.1035	.1230	.1498	.1896
240	.0220	.0295	.0362	.0424	.0483	.0540	.0647	.0798	.1030
∞	.0000	.0000	.0000	.0000	.0000	.0000	.0000	.0000	.0000

$\alpha = 0.01$, $s = 2$

ν_2 \ ν_1	0	1	2	3	4	5	7	10	15
2	.8826	.9173	.9358	.9475	.9556	.9615	.9695	.9768	.9834
3	.8074	.8575	.8863	.9051	.9185	.9286	.9427	.9557	.9679
4	.7381	.7989	.8357	.8607	.8789	.8929	.9129	.9318	.9499
5	.6770	.7446	.7873	.8171	.8394	.8568	.8820	.9066	.9306
6	.6237	.6954	.7422	.7758	.8013	.8215	.8514	.8810	.9106
7	.5773	.6512	.7008	.7371	.7652	.7877	.8215	.8556	.8903
8	.5369	.6116	.6630	.7013	.7313	.7556	.7927	.8308	.8702
9	.5014	.5762	.6285	.6683	.6997	.7255	.7652	.8067	.8503
10	.4701	.5443	.5971	.6378	.6703	.6972	.7391	.7834	.8309
15	.3573	.4247	.4757	.5168	.5511	.5803	.6279	.6812	.7418
20	.2876	.3473	.3941	.4329	.4661	.4951	.5435	.5998	.6670
25	.2404	.2935	.3360	.3719	.4032	.4309	.4782	.5347	.6045
30	.2065	.2540	.2926	.3258	.3550	.3812	.4265	.4819	.5521
35	.1811	.2239	.2592	.2898	.3171	.3417	.3847	.4383	.5077
40	.1610	.2000	.2325	.2608	.2863	.3094	.3503	.4017	.4697
48	.1372	.1712	.1999	.2251	.2480	.2689	.3064	.3544	.4193
60	.1117	.1403	.1646	.1863	.2061	.2244	.2576	.3008	.3607
80	.0855	.1080	.1273	.1448	.1609	.1759	.2035	.2402	.2925
120	.0582	.0740	.0877	.1002	.1118	.1228	.1433	.1711	.2120
240	.0297	.0380	.0453	.0520	.0583	.0644	.0758	.0917	.1160
∞	.0000	.0000	.0000	.0000	.0000	.0000	.0000	.0000	.0000

Table B.11: Roy's Maximum Root Statistic

$\alpha = 0.05$, $s = 3$

ν_2 \ ν_1	0	1	2	3	4	5	7	10	15
2	.8646	.8986	.9188	.9322	.9417	.9489	.9590	.9684	.9771
3	.7922	.8386	.8676	.8876	.9022	.9134	.9296	.9450	.9596
4	.7266	.7815	.8172	.8426	.8617	.8766	.8983	.9195	.9402
5	.6689	.7292	.7698	.7994	.8221	.8400	.8668	.8934	.9199
6	.6185	.6820	.7261	.7589	.7844	.8040	.8358	.8672	.8992
7	.5745	.6398	.6861	.7212	.7489	.7715	.8060	.8416	.8785
8	.5359	.6019	.6497	.6864	.7158	.7400	.7774	.8167	.8580
9	.5019	.5679	.6165	.6544	.6850	.7104	.7503	.7926	.8380
10	.4718	.5373	.5862	.6249	.6564	.6828	.7246	.7696	.8185
15	.3620	.4219	.4690	.5079	.5407	.5691	.6158	.6687	.7299
20	.2931	.3465	.3898	.4265	.4582	.4862	.5334	.5889	.6559
25	.2461	.2937	.3332	.3671	.3970	.4237	.4697	.5252	.5944
30	.2120	.2548	.2907	.3221	.3500	.3752	.4192	.4734	.5429
35	.1863	.2250	.2579	.2869	.3129	.3366	.3784	.4308	.4993
40	.1660	.2013	.2316	.2584	.2828	.3050	.3447	.3905	.4620
48	.1417	.1726	.1994	.2234	.2452	.2654	.3018	.3486	.4125
60	.1157	.1417	.1644	.1850	.2042	.2217	.2538	.2961	.3550
80	.0888	.1093	.1274	.1440	.1592	.1740	.2008	.2366	.2880
120	.0606	.0750	.0879	.0999	.1111	.1217	.1415	.1687	.2089
240	.0310	.0386	.0455	.0519	.0580	.0639	.0750	.0905	.1143
∞	.0000	.0000	.0000	.0000	.0000	.0000	.0000	.0000	.0000

$\alpha = 0.01$, $s = 3$

ν_2 \ ν_1	0	1	2	3	4	5	7	10	15
2	.9248	.9441	.9554	.9629	.9682	.9721	.9777	.9828	.9876
3	.8680	.8983	.9170	.9298	.9391	.9462	.9564	.9660	.9751
4	.8113	.8505	.8757	.8934	.9066	.9169	.9318	.9462	.9602
5	.7562	.8040	.8344	.8564	.8731	.8862	.9056	.9247	.9437
6	.7096	.7601	.7947	.8201	.8397	.8554	.8789	.9025	.9263
7	.6657	.7195	.7571	.7853	.8074	.8252	.8523	.8799	.9084
8	.6262	.6821	.7220	.7524	.7764	.7960	.8262	.8576	.8903
9	.5906	.6478	.6893	.7214	.7470	.7681	.8010	.8356	.8723
10	.5586	.6164	.6590	.6923	.7192	.7416	.7767	.8141	.8544
15	.4375	.4937	.5374	.5730	.6029	.6285	.6703	.7172	.7708
20	.3586	.4104	.4519	.4867	.5167	.5428	.5866	.6376	.6985
25	.3034	.3506	.3893	.4223	.4511	.4767	.5203	.5726	.6370
30	.2629	.3058	.3416	.3726	.3999	.4245	.4670	.5189	.5846
35	.2319	.2712	.3043	.3332	.3591	.3824	.4233	.4741	.5397
40	.2073	.2434	.2742	.3012	.3256	.3477	.3869	.4361	.5010
48	.1776	.2095	.2369	.2613	.2835	.3038	.3401	.3865	.4491
60	.1456	.1727	.1963	.2175	.2369	.2549	.2874	.3298	.3883
80	.1121	.1338	.1528	.1701	.1861	.2010	.2284	.2648	.3166
120	.0769	.0922	.1059	.1185	.1302	.1413	.1619	.1899	.2309
240	.0395	.0477	.0551	.0619	.0684	.0746	.0863	.1025	.1272
∞	.0000	.0000	.0000	.0000	.0000	.0000	.0000	.0000	.0000

Table B.11: Roy's Maximum Root Statistic

$\alpha = 0.05$, $s = 4$

ν_2 \ ν_1	0	1	2	3	4	5	7	10	15
2	.9045	.9259	.9393	.9486	.9554	.9606	.9680	.9751	.9818
3	.8463	.8773	.8976	.9121	.9229	.9313	.9436	.9555	.9671
4	.7904	.8287	.8548	.8738	.8884	.8998	.9169	.9337	.9504
5	.7388	.7825	.8132	.8360	.8537	.8679	.8892	.9108	.9326
6	.6920	.7396	.7737	.8000	.8199	.8364	.8616	.8875	.9141
7	.6499	.7000	.7367	.7650	.7875	.8060	.8345	.8643	.8954
8	.6120	.6638	.7024	.7325	.7568	.7769	.8083	.8414	.8767
9	.5779	.6307	.6706	.7021	.7277	.7492	.7830	.8192	.8583
10	.5472	.6004	.6412	.6737	.7004	.7229	.7588	.7976	.8401
15	.4307	.4822	.5235	.5578	.5869	.6121	.6538	.7012	.7561
20	.3543	.4017	.4409	.4742	.5031	.5286	.5719	.6228	.6843
25	.3006	.3439	.3802	.4117	.4395	.4644	.5072	.5590	.6235
30	.2609	.3004	.3341	.3636	.3899	.4137	.4552	.5064	.5720
35	.2306	.2667	.2978	.3254	.3502	.3728	.4127	.4626	.5279
40	.2063	.2396	.2685	.2943	.3177	.3391	.3773	.4256	.4899
48	.1770	.2065	.2323	.2555	.2768	.2964	.3317	.3772	.4391
60	.1454	.1704	.1927	.2129	.2315	.2488	.2805	.3219	.3796
80	.1122	.1322	.1501	.1666	.1820	.1964	.2230	.2586	.3094
120	.0770	.0913	.1042	.1162	.1274	.1381	.1581	.1854	.2257
240	.0397	.0473	.0542	.0608	.0670	.0730	.0843	.1002	.1243
∞	.0000	.0000	.0000	.0000	.0000	.0000	.0000	.0000	.0000

$\alpha = 0.01$, $s = 4$

ν_2 \ ν_1	0	1	2	3	4	5	7	10	15
2	.9473	.9593	.9668	.9719	.9757	.9785	.9826	.9865	.9901
3	.9032	.9231	.9361	.9453	.9521	.9574	.9651	.9726	.9797
4	.8567	.8836	.9017	.9149	.9248	.9327	.9443	.9557	.9670
5	.8110	.8436	.8663	.8830	.8959	.9062	.9216	.9371	.9526
6	.7677	.8049	.8312	.8510	.8665	.8790	.8980	.9174	.9371
7	.7275	.7680	.7973	.8197	.8375	.8519	.8742	.8972	.9211
8	.6903	.7333	.7650	.7895	.8092	.8254	.8505	.8769	.9047
9	.6561	.7010	.7345	.7607	.7820	.7996	.8273	.8567	.8882
10	.6247	.6708	.7057	.7334	.7560	.7748	.8047	.8369	.8717
15	.5016	.5490	.5867	.6177	.6439	.6664	.7034	.7452	.7930
20	.4175	.4627	.4997	.5309	.5579	.5815	.6213	.6678	.7234
25	.3570	.3992	.4343	.4645	.4910	.5146	.5550	.6033	.6631
30	.3117	.3502	.3837	.4125	.4380	.4609	.5007	.5494	.6111
35	.2765	.3126	.3435	.3707	.3951	.4171	.4558	.5039	.5661
40	.2483	.2819	.3108	.3365	.3596	.3807	.4181	.4651	.5270
48	.2138	.2438	.2699	.2933	.3145	.3341	.3691	.4139	.4742
60	.1763	.2021	.2249	.2454	.2643	.2818	.3135	.3548	.4118
80	.1367	.1575	.1760	.1930	.2087	.2234	.2505	.2864	.3373
120	.0943	.1092	.1227	.1350	.1469	.1579	.1785	.2065	.2474
240	.0488	.0568	.0642	.0711	.0777	.0839	.0957	.1122	.1371
∞	.0000	.0000	.0000	.0000	.0000	.0000	.0000	.0000	.0000

Table B.11: Roy's Maximum Root Statistic

$\alpha = 0.05$, $s = 5$

ν_2 \ ν_1	0	1	2	3	4	5	7	10	15
2	.9289	.9432	.9527	.9594	.9645	.9684	.9741	.9796	.9850
3	.8815	.9032	.9181	.9289	.9372	.9437	.9534	.9629	.9724
4	.8338	.8617	.8814	.8960	.9074	.9165	.9302	.9439	.9577
5	.7882	.8210	.8447	.8627	.8768	.8883	.9058	.9236	.9419
6	.7456	.7822	.8091	.8299	.8465	.8600	.8809	.9026	.9252
7	.7063	.7457	.7752	.7983	.8169	.8323	.8563	.8815	.9082
8	.6702	.7117	.7432	.7681	.7884	.8054	.8321	.8605	.8911
9	.6372	.6801	.7131	.7395	.7612	.7795	.8085	.8399	.8740
10	.6069	.6507	.6849	.7125	.7354	.7547	.7858	.8197	.8571
15	.4883	.5328	.5690	.5993	.6252	.6477	.6850	.7277	.7773
20	.4072	.4495	.4847	.5150	.5414	.5647	.6043	.6511	.7077
25	.3488	.3881	.4215	.4507	.4764	.4995	.5394	.5877	.6480
30	.3049	.3413	.3726	.4003	.4250	.4474	.4865	.5349	.5967
35	.2708	.3045	.3338	.3599	.3834	.4049	.4428	.4904	.5525
40	.2434	.2746	.3021	.3267	.3490	.3696	.4061	.4525	.5141
48	.2097	.2377	.2625	.2849	.3054	.3244	.3585	.4026	.4624
60	.1732	.1973	.2188	.2385	.2567	.2736	.3045	.3450	.4013
80	.1344	.1539	.1714	.1877	.2028	.2171	.2433	.2785	.3286
120	.0928	.1068	.1196	.1316	.1428	.1535	.1735	.2008	.2409
240	.0481	.0557	.0627	.0693	.0756	.0816	.0931	.1091	.1335
∞	.0000	.0000	.0000	.0000	.0000	.0000	.0000	.0000	.0000

$\alpha = 0.01$, $s = 5$

ν_2 \ ν_1	0	1	2	3	4	5	7	10	15
2	.9610	.9689	.9742	.9779	.9806	.9828	.9859	.9889	.9918
3	.9258	.9396	.9490	.9559	.9611	.9651	.9712	.9771	.9830
4	.8871	.9064	.9200	.9300	.9378	.9440	.9533	.9625	.9718
5	.8478	.8719	.8892	.9023	.9125	.9208	.9334	.9461	.9591
6	.8094	.8376	.8582	.8739	.8865	.8967	.9124	.9285	.9453
7	.7729	.8043	.8278	.8457	.8603	.8723	.8909	.9103	.9308
8	.7385	.7724	.7980	.8181	.8345	.8480	.8693	.8918	.9158
9	.7062	.7422	.7696	.7914	.8093	.8242	.8479	.8732	.9000
10	.6762	.7136	.7426	.7658	.7850	.8011	.8269	.8548	.8853
15	.5544	.5948	.6274	.6546	.6777	.6977	.7306	.7680	.8111
20	.4677	.5074	.5404	.5685	.5928	.6143	.6505	.6930	.7440
25	.4038	.4415	.4735	.5011	.5255	.5473	.5846	.6295	.6581
30	.3549	.3904	.4208	.4475	.4713	.4927	.5301	.5757	.6337
35	.3165	.3498	.3786	.4041	.4270	.4478	.4844	.5299	.5889
40	.2854	.3166	.3438	.3681	.3900	.4101	.4457	.4906	.5497
48	.2469	.2751	.2999	.3222	.3426	.3614	.3950	.4382	.4962
60	.2048	.2293	.2512	.2710	.2893	.3063	.3371	.3772	.4326
80	.1596	.1796	.1977	.2142	.2296	.2441	.2706	.3059	.3559
120	.1108	.1253	.1386	.1509	.1625	.1735	.1940	.2217	.2624
240	.0577	.0656	.0730	.0799	.0865	.0927	.1047	.1212	.1464
∞	.0000	.0000	.0000	.0000	.0000	.0000	.0000	.0000	.0000

Table B.11: Roy's Maximum Root Statistic

$\alpha = 0.05$ ν_1 / ν_2	s = 6								
	0	1	2	3	4	5	7	10	15
2	.9450	.9551	.9620	.9670	.9709	.9740	.9785	.9829	.9873
3	.9058	.9216	.9328	.9411	.9476	.9528	.9606	.9684	.9763
4	.8649	.8858	.9010	.9126	.9217	.9290	.9402	.9517	.9633
5	.8246	.8499	.8686	.8830	.8945	.9039	.9185	.9335	.9491
6	.7861	.8149	.8365	.8535	.8671	.8784	.8960	.9145	.9340
7	.7499	.7814	.8054	.8245	.8401	.8531	.8735	.8952	.9184
8	.7160	.7496	.7757	.7966	.8138	.8282	.8511	.8758	.9026
9	.6845	.7198	.7474	.7698	.7384	.8041	.8292	.8566	.8867
10	.6552	.6917	.7206	.7442	.7640	.7808	.8078	.8377	.8708
15	.5372	.5759	.6077	.6346	.6577	.6779	.7115	.7500	.7951
20	.4535	.4913	.5231	.5506	.5746	.5960	.6324	.6754	.7278
25	.3919	.4276	.4583	.4852	.5091	.5306	.5677	.6128	.6692
30	.3447	.3782	.4074	.4333	.4565	.4775	.5144	.5600	.6184
35	.3076	.3390	.3665	.3912	.4135	.4338	.4699	.5151	.5743
40	.2775	.3069	.3329	.3563	.3777	.3973	.4322	.4767	.5357
48	.2403	.2668	.2905	.3120	.3317	.3500	.3830	.4265	.4833
60	.1995	.2226	.2434	.2624	.2801	.2966	.3267	.3662	.4210
80	.1556	.1745	.1916	.2075	.2224	.2364	.2623	.2969	.3462
120	.1081	.1218	.1344	.1463	.1574	.1681	.1880	.2151	.2551
240	.0563	.0638	.0708	.0775	.0838	.0899	.1014	.1176	.1422
∞	.0000	.0000	.0000	.0000	.0000	.0000	.0000	.0000	.0000

$\alpha = 0.01$ ν_1 / ν_2	s = 6								
	0	1	2	3	4	5	7	10	15
2	.9699	.9755	.9793	.9820	.9842	.9858	.9883	.9907	.9931
3	.9412	.9512	.9583	.9635	.9676	.9708	.9757	.9805	.9854
4	.9086	.9230	.9334	.9413	.9475	.9525	.9600	.9677	.9756
5	.8745	.8930	.9065	.9169	.9252	.9320	.9424	.9531	.9642
6	.8406	.8625	.8789	.8917	.9019	.9104	.9236	.9373	.9517
7	.8075	.8323	.8512	.8661	.8782	.8883	.9040	.9207	.9384
8	.7758	.8031	.8241	.8409	.8546	.8661	.8842	.9037	.9247
9	.7458	.7750	.7978	.8161	.8313	.8441	.8645	.8865	.9106
10	.7173	.7482	.7724	.7922	.8086	.8225	.8449	.8694	.8964
15	.5986	.6334	.6619	.6858	.7063	.7241	.7535	.7872	.8262
20	.5111	.5462	.5757	.6010	.6231	.6426	.6757	.7147	.7616
25	.4450	.4792	.5081	.5335	.5559	.5760	.6106	.6524	.7042
30	.3936	.4261	.4542	.4789	.5011	.5211	.5561	.5990	.6536
35	.3527	.3835	.4103	.4342	.4557	.4754	.5100	.5531	.6090
40	.3194	.3484	.3740	.3969	.4177	.4367	.4706	.5134	.5698
48	.2775	.3040	.3276	.3488	.3683	.3863	.4187	.4602	.5160
60	.2315	.2548	.2757	.2948	.3125	.3289	.3588	.3977	.4515
80	.1814	.2006	.2181	.2342	.2493	.2634	.2894	.3240	.3730
120	.1266	.1407	.1538	.1659	.1774	.1882	.2085	.2360	.2763
240	.0663	.0741	.0841	.0883	.0949	.1012	.1132	.1298	.1550
∞	.0000	.0000	.0000	.0000	.0000	.0000	.0000	.0000	.0000

Table B.11: Roy's Maximum Root Statistic

$\alpha = 0.05$ ν_2 \ ν_1	s = 7								
	0	1	2	3	4	5	7	10	15
2	.9561	.9635	.9687	.9726	.9757	.9781	.9817	.9854	.9890
3	.9232	.9351	.9438	.9504	.9556	.9598	.9662	.9727	.9793
4	.8879	.9040	.9160	.9253	.9327	.9388	.9481	.9577	.9677
5	.8523	.8722	.8872	.8990	.9085	.9163	.9285	.9413	.9548
6	.8176	.8406	.8582	.8722	.8837	.8932	.9082	.9241	.9410
7	.7843	.8099	.8298	.8457	.8589	.8700	.8875	.9063	.9267
8	.7527	.7804	.8022	.8199	.8345	.8469	.8668	.8884	.9120
9	.7229	.7523	.7757	.7948	.8108	.8244	.8463	.8705	.8972
10	.6949	.7257	.7503	.7707	.7878	.8025	.8263	.8527	.8823
15	.5792	.6130	.6411	.6651	.6858	.7039	.7342	.7692	.8103
20	.4944	.5282	.5570	.5820	.6040	.6236	.6570	.6968	.7453
25	.4305	.4631	.4914	.5162	.5384	.5583	.5930	.6351	.6879
30	.3809	.4119	.4390	.4632	.4850	.5048	.5395	.5825	.6378
35	.3415	.3707	.3965	.4198	.4409	.4603	.4944	.5374	.5939
40	.3093	.3369	.3615	.3837	.4040	.4227	.4561	.4986	.5552
48	.2690	.2941	.3167	.3373	.3562	.3738	.4056	.4466	.5024
60	.2244	.2465	.2665	.2850	.3021	.3181	.3474	.3858	.4392
80	.1759	.1942	.2109	.2264	.2401	.2547	.2801	.3141	.3626
120	.1229	.1363	.1487	.1605	.1715	.1820	.2018	.2287	.2684
240	.0644	.0718	.0788	.0854	.0918	.0979	.1095	.1257	.1504
∞	.0000	.0000	.0000	.0000	.0000	.0000	.0000	.0000	.0000

$\alpha = 0.01$ ν_2 \ ν_1	s = 7								
	0	1	2	3	4	5	7	10	15
2	.9761	.9801	.9830	.9851	.9868	.9881	.9901	.9921	.9940
3	.9522	.9597	.9651	.9693	.9725	.9751	.9791	.9831	.9872
4	.9244	.9354	.9436	.9499	.9549	.9590	.9653	.9718	.9785
5	.8947	.9091	.9199	.9284	.9352	.9408	.9496	.9587	.9682
6	.8645	.8819	.8952	.9057	.9143	.9214	.9325	.9443	.9568
7	.8346	.8546	.8701	.8825	.8927	.9013	.9147	.9292	.9447
8	.8055	.8278	.8452	.8594	.8710	.8809	.8965	.9135	.9320
9	.7776	.8017	.8209	.8365	.8495	.8605	.8782	.8976	.9189
10	.7508	.7766	.7971	.8141	.8282	.8403	.8600	.8815	.9056
15	.6363	.6665	.6915	.7126	.7308	.7467	.7732	.8037	.8392
20	.5491	.5803	.6068	.6297	.6497	.6675	.6978	.7337	.7770
25	.4817	.5125	.5391	.5624	.5831	.6016	.6338	.6726	.7210
30	.4286	.4583	.4843	.5073	.5280	.5467	.5795	.6198	.6713
35	.3858	.4142	.4393	.4617	.4820	.5005	.5332	.5740	.6272
40	.3506	.3777	.4017	.4233	.4431	.4612	.4934	.5342	.5881
48	.3060	.3309	.3533	.3736	.3922	.4094	.4405	.4803	.5341
60	.2565	.2787	.2987	.3171	.3341	.3500	.3789	.4167	.4689
80	.2020	.2205	.2375	.2532	.2678	.2816	.3070	.3409	.3889
120	.1418	.1556	.1683	.1803	.1916	.2023	.2223	.2496	.2894
240	.0747	.0824	.0897	.0966	.1031	.1094	.1214	.1380	.1633
∞	.0000	.0000	.0000	.0000	.0000	.0000	.0000	.0000	.0000

Table B.12: Lawley-Hotelling Trace Statistic

	m_E \ m_H	2	3	4	5	6	8	$d = 2$ 10	12	15	20	25	40	60
5%	2	9.8591†	10.659†	11.098†	11.373†	11.562†	11.804†	11.952†	12.052†	12.153†	12.254†	12.316†	12.409†	12.461†
	3	58.428	58.915	59.161	59.308	59.407	59.531	59.606	59.655	59.705	59.755	59.785	59.830	59.855
	4	23.999	23.312	22.918	22.663	22.484	22.250	22.104	22.003	21.901	21.797	21.733	21.636	21.582
	5	15.639	14.864	14.422	14.135	13.934	13.670	13.504	13.391	13.275	13.156	13.083	12.972	12.909
	6	12.175	11.411	10.975	10.691	10.491	10.228	10.063	9.9489	9.8320	9.7118	9.6381	9.5251	9.4610
	7	10.334	9.5937	9.1694	8.8927	8.6975	8.4396	8.2765	8.1639	8.0480	7.9285	7.8549	7.7417	7.6773
	8	9.2069	8.4881	8.0752	7.8054	7.6145	7.3614	7.2008	7.0896	6.9748	6.8560	6.7826	6.6694	6.6048
	10	7.9095	7.2243	6.8294	6.5702	6.3860	6.1405	5.9837	5.8745	5.7612	5.6433	5.5701	5.4564	5.3910
	12	7.1902	6.5284	6.1461	5.8942	5.7147	5.4744	5.3200	5.2122	5.0997	4.9820	4.9085	4.7938	4.7274
	14	6.7350	6.0902	4.7168	5.4703	5.2941	5.0574	4.9048	4.7977	4.6856	4.5678	4.4939	4.3780	4.3105
	16	6.4217	5.7895	5.4230	5.1804	5.0067	4.7727	4.6213	4.5147	4.4028	4.2846	4.2102	4.0930	4.0243
	18	6.1932	5.5708	5.2095	4.9700	4.7982	4.5663	4.4157	4.3094	4.1976	4.0791	4.0042	3.8855	3.8158
	20	6.0192	5.4046	5.0475	4.8105	4.6402	4.4099	4.2600	4.1539	4.0420	3.9231	3.8477	3.7278	3.6569
	25	5.7244	5.1237	4.7741	4.5415	4.3740	4.1465	3.9977	3.8919	3.7798	3.6598	3.5832	3.4605	3.3868
	30	5.5401	4.9487	4.6040	4.3743	4.2086	3.9829	3.8347	3.7291	3.6166	3.4957	3.4181	3.2926	3.2168
	35	5.4140	4.8291	4.4880	4.2604	4.0959	3.8715	3.7237	3.6181	3.5054	3.3836	3.3051	3.1774	3.1000
	40	5.3224	4.7424	4.4039	4.1778	4.0143	3.7908	3.6433	3.5377	3.4247	3.3022	3.2230	3.0933	3.0140
	50	5.1981	4.6249	4.2900	4.0661	3.9039	3.6817	3.5346	3.4289	3.3154	3.1919	3.1115	2.9787	2.8965
	60	5.1178	4.5490	4.2166	3.9941	3.8328	3.6114	3.4646	3.3588	3.2450	3.1206	3.0392	2.9041	2.8196
	70	5.0616	4.4960	4.1653	3.9439	3.7831	3.5624	3.4157	3.3099	3.1957	3.0706	2.9886	2.8516	2.7652
	80	5.0200	4.4569	4.1275	3.9068	3.7465	3.5262	3.3796	3.2737	3.1594	3.0338	2.9512	2.8126	2.7247
	100	4.9628	4.4030	4.0754	3.8557	3.6961	3.4764	3.3300	3.2240	3.1093	2.9829	2.8994	2.7586	2.6683
	200	4.8514	4.2982	3.9742	3.7567	3.5983	3.3798	3.2336	3.1275	3.0120	2.8838	2.7984	2.6520	2.5559
	∞	4.7442	4.1973	3.8769	3.6614	3.5044	3.2870	3.1410	3.0346	2.9182	2.7879	2.7002	2.5470	2.4428
1%	2	2.4673*	2.6666*	2.7758*	2.8444*	2.8914*	2.9517*	2.9886*	3.0135*	3.0387*	3.0641*	3.0796*	3.1025*	3.1154*
	3	2.0840†	2.0898†	2.9923†	2.9938†	2.9948†	2.9961†	2.9968†	2.9973†	2.9978†	2.9983†	2.9985†	2.9991†	2.9993†
	4	74.275	71.026	69.244	68.116	67.337	66.332	65.712	65.290	64.862	64.427	64.163	63.763	63.538
	5	38.295	35.567	34.070	33.121	32.465	31.615	31.088	30.729	30.364	29.993	29.766	29.422	29.228
	6	26.118	23.794	22.517	21.706	21.143	20.413	19.958	19.648	19.332	19.009	18.812	18.511	18.341
	7	20.388	18.326	17.191	16.469	15.967	15.313	14.905	14.626	14.341	14.049	13.870	13.596	13.442
	8	17.152	15.268	14.229	13.567	13.106	12.504	12.127	11.868	11.603	11.331	11.165	10.909	10.764
	10	13.701	12.038	11.120	10.531	10.121	9.5819	9.2431	9.0096	8.7694	8.5215	8.3688	8.1334	7.9990
	12	11.920	10.388	9.5405	8.9961	8.6148	8.1132	7.7962	7.5770	7.3505	7.1159	6.9707	6.7457	6.6166
	14	10.844	9.3990	8.5974	8.0816	7.7196	7.2419	6.9389	6.7287	6.5109	6.2843	6.1435	5.9244	5.7980
	16	10.128	8.7432	7.9743	7.4786	7.1301	6.6691	6.3758	6.1718	5.9598	5.7385	5.6002	5.3847	5.2596
	18	9.6174	8.2781	7.5334	7.0527	6.7142	6.2655	5.9793	5.7797	5.5718	5.3540	5.2178	5.0040	4.8794
	20	9.2360	7.9316	7.2056	6.7365	6.4057	5.9664	5.6855	5.4893	5.2844	5.0692	4.9342	4.7214	4.5970
	25	8.6044	7.3601	6.6663	6.2169	5.8993	5.4760	5.2042	5.0134	4.8135	4.6021	4.4686	4.2577	4.1314
	30	8.2188	7.0127	6.3394	5.9026	5.5933	5.1801	4.9138	4.7264	4.5293	4.3199	4.1870	3.9745	3.8479
	35	7.9592	6.7796	6.1205	5.6923	5.3888	4.9825	4.7200	4.5348	4.3395	4.1312	3.9985	3.7850	3.6581
	40	7.7727	6.6125	5.9638	5.5420	5.2426	4.8413	4.5816	4.3980	4.2038	3.9962	3.8635	3.6488	3.5191
	50	7.5228	6.3891	5.7545	5.3414	5.0477	4.6533	4.3972	4.2156	4.0230	3.8161	3.6830	3.4661	3.3335
	60	7.3630	6.2465	5.6212	5.2138	4.9238	4.5338	4.2800	4.0998	3.9081	3.7014	3.5676	3.3490	3.2139
	70	7.2520	6.1478	5.5290	5.1255	4.8382	4.4512	4.1991	4.0197	3.8285	3.6219	3.4880	3.2675	3.1303
	80	7.1705	6.0753	5.4613	5.0608	4.7754	4.3908	4.1398	3.9610	3.7703	3.5636	3.4294	3.2074	3.0685
	100	7.0588	5.9761	5.3688	4.9723	4.6896	4.3081	4.0587	3.8808	3.6906	3.4839	3.3491	3.1247	2.9831
	200	6.8435	5.7852	5.1910	4.8025	4.5251	4.1497	3.9034	3.7271	3.5377	3.3304	3.1941	2.9640	2.8152
	∞	6.6385	5.6040	5.0226	4.6419	4.3695	4.0000	3.7566	3.5817	3.3928	3.1845	3.0462	2.8082	2.6492

†Multiply entry by 100. *Multiply entry by 10^4.

Table B.12: Lawley-Hotelling Trace Statistic

$d = 3$

	$m_E \backslash m_H$	3	4	5	6	8	10	12	15	20	25	40	60
5%	3	25·930*	26·996*	27·665*	28·125*	28·712*	29·073*	29·316*	29·561*	29·809*	29·959*	30·19*	30·31*
	4	1·1880*	1·1929*	1·1959*	1·1978*	1·2003*	1·2018*	1·2028*	1·2038*	1·2048*	1·2054*	1·2063*	1·2068*
	5	42·474	41·764	41·305	40·983	40·562	40·300	40·120	39·937	39·750	39·635	39·462	39·366
	6	25·456	24·715	24·235	23·899	23·458	23·182	22·092	22·799	22·600	22·479	22·294	22·190
	7	18·752	18·056	17·605	17·288	16·870	16·608	16·427	16·241	16·051	15·934	15·755	15·653
	8	15·308	14·657	14·233	13·934	13·540	13·290	13·118	12·941	12·758	12·646	12·473	12·375
	10	11·893	11·306	10·921	10·649	10·287	10·057	9·8974	9·7320	9·5603	9·4541	9·2897	9·1955
	12	10·229	9·6825	9·3234	9·0680	8·7271	8·5088	8·3566	8·1982	8·0330	7·9301	7·7700	7·6777
	14	9·2550	8·7356	8·3935	8·1495	7·8225	7·6122	7·4649	7·3110	7·1497	7·0488	6·8908	6·7991
	16	8·6180	8·1183	7·7884	7·5526	7·2355	7·0307	6·8868	6·7360	6·5772	6·4774	6·3204	6·2287
	18	8·1701	7·6851	7·3644	7·1347	6·8251	6·6244	6·4830	6·3343	6·1771	6·0780	5·9212	5·8292
	20	7·8384	7·3649	7·0513	6·8263	6·5224	6·3249	6·1853	6·0383	5·8822	5·7834	5·6266	5·5341
	25	7·2943	6·8407	6·5394	6·3227	6·0287	5·8365	5·7001	5·5555	5·4010	5·3025	5·1446	5·0503
	30	6·9654	6·5245	6·2311	6·0196	5·7319	5·5431	5·4085	5·2654	5·1116	5·0129	4·8535	4·7575
	35	6·7453	6·3132	6·0253	5·8175	5·5341	5·3476	5·2143	5·0720	4·9185	4·8195	4·6586	4·5608
	40	6·5877	6·1621	5·8783	5·6732	5·3929	5·2081	5·0757	4·9340	4·7806	4·6813	4·5189	4·4195
	50	6·3773	5·9606	5·6823	5·4809	5·2050	5·0224	4·8911	4·7502	4·5967	4·4968	4·3319	4·2297
	60	6·2433	5·8324	5·5577	5·3587	5·0856	4·9044	4·7739	4·6334	4·4798	4·3793	4·2123	4·1078
	70	6·1504	5·7436	5·4715	5·2742	5·0031	4·8229	4·6929	4·5526	4·3988	4·2979	4·1292	4·0227
	80	6·0823	5·6786	5·4084	5·2122	4·9426	4·7632	4·6336	4·4935	4·3395	4·2381	4·0680	3·9600
	100	5·9891	5·5896	5·3220	5·1276	4·8601	4·6817	4·5525	4·4126	4·2583	4·1563	3·9840	3·8734
	200	5·8099	5·4186	5·1562	4·9653	4·7017	4·5252	4·3970	4·2574	4·1023	3·9988	3·8212	3·7042
	∞	5·6397	5·2565	4·9992	4·8116	4·5519	4·3773	4·2499	4·1104	3·9541	3·8487	3·6642	3·5384
1%	3	6·4845†	6·7500†	6·9169†	7·0313†	7·1778†	7·2675†	7·3281†	7·3891†	7·4511†	7·4883†	—	—
	4	5·9896*	5·9946*	5·9976*	5·9996*	6·0021*	6·0035*	6·0046*	6·0056*	6·0067*	6·0071*	6·008*	6·008*
	5	1·2738*	1·2420*	1·2219*	1·2080*	1·1901*	1·1790*	1·1715*	1·1638*	1·1561*	1·1514*	1·144*	1·141*
	6	59·507	57·032	55·462	54·377	52·973	52·102	51·509	50·906	50·292	49·918	49·349	49·04
	7	37·994	35·993	34·721	33·840	32·695	31·984	31·498	31·002	30·496	30·188	29·718	29·452
	8	28·308	26·599	25·511	24·755	23·771	23·157	22·737	22·308	21·868	21·599	21·188	20·955
	10	19·737	18·355	17·471	16·855	16·050	15·544	15·197	14·840	14·472	14·246	13·899	13·702
	12	15·973	14·765	13·990	13·448	12·737	12·288	11·978	11·659	11·328	11·124	10·809	10·628
	14	13·905	12·803	12·096	11·599	10·945	10·530	10·243	9·9462	9·6377	9·4463	9·1490	8·9780
	16	12·610	11·581	10·918	10·452	9·8359	9·4444	9·1724	8·8900	8·5955	8·4121	8·1260	7·9605
	18	11·729	10·751	10·120	9·6756	9·0870	8·7117	8·4503	8·1782	7·8934	7·7154	7·4365	7·2743
	20	11·091	10·152	9·5452	9·1173	8·5492	8·1861	7·9325	7·6679	7·3901	7·2159	6·9419	6·7818
	25	10·075	9·2005	8·6339	8·2333	7·6992	7·3560	7·1152	6·8627	6·5958	6·4273	6·1598	6·0019
	30	9·4785	8·6441	8·1022	7·7183	7·2050	6·8739	6·6407	6·3953	6·1346	5·9690	5·7042	5·5464
	35	9·0874	8·2798	7·7548	7·3822	6·8829	6·5598	6·3317	6·0909	5·8339	5·6700	5·4063	5·2478
	40	8·8113	8·0233	7·5108	7·1460	6·6564	6·3392	6·1147	5·8771	5·6227	5·4598	5·1962	5·0367
	50	8·4479	7·6861	7·1894	6·8358	6·3599	6·0503	5·8305	5·5970	5·3457	5·1838	4·9196	4·7578
	60	8·2195	7·4745	6·9882	6·6416	6·1744	5·8696	5·6528	5·4218	5·1722	5·0108	4·7455	4·5815
	70	8·0627	7·3295	6·8504	6·5087	6·0474	5·7460	5·5312	5·3019	5·0535	4·8922	4·6258	4·4598
	80	7·9485	7·2239	6·7502	6·4120	5·9551	5·6562	5·4428	5·2147	4·9670	4·8058	4·5383	4·3706
	100	7·7932	7·0805	6·6141	6·2809	5·8300	5·5344	5·3230	5·0965	4·8497	4·6883	4·4190	4·2484
	200	7·4980	6·8083	6·3561	6·0323	5·5930	5·3037	5·0961	4·8725	4·6270	4·4650	4·1906	4·0124
	∞	7·2220	6·5542	6·1156	5·8009	5·3725	5·0892	4·8849	4·6638	4·4190	4·2557	3·9738	3·7843

* Multiply entry by 100. † Multiply entry by 10^4.

Table B.12: Lawley-Hotelling Trace Statistic

$d = 4$

	$m_E \backslash m_H$	4	5	6	8	10	12	15	20	25	40	60
5%	4	49·964*	51·204*	52·054*	53·142*	53·808*	54·258*	54·71*	55·17*	55·46*	—	—
	5	1·9964*	2·0013*	2·0046*	2·0087*	2·0112*	2·0128*	2·0145*	2·0161*	2·0171*	2·019*	—
	6	65·715	64·999	64·497	63·841	63·432	63·151	62·866	62·573	62·396	62·13	—
	7	37·343	36·629	36·129	35·474	35·064	34·782	34·495	34·200	34·019	33·75	—
	8	26·516	25·868	25·413	24·814	24·437	24·178	23·912	23·639	23·471	23·214	23·072
	10	17·875	17·326	16·938	16·424	16·098	15·872	15·640	15·399	15·250	15·021	14·891
	12	14·338	13·848	13·500	13·037	12·741	12·535	12·321	12·099	11·961	11·747	11·624
	14	12·455	12·002	11·680	11·248	10·972	10·778	10·577	10·366	10·234	10·029	9·9103
	16	11·295	10·868	10·563	10·154	9·8904	9·7054	9·5119	9·3085	9·1810	8·9808	8·8644
	18	10·512	10·104	9·8121	9·4190	9·1647	8·9857	8·7978	8·5996	8·4748	8·2778	8·1626
	20	9·9500	9·5560	9·2736	8·8926	8·6453	8·4708	8·2871	8·0926	7·9696	7·7748	7·6601
	25	9·0585	8·6384	8·4223	8·0616	7·8261	7·6590	7·4821	7·2933	7·1730	6·9805	6·8659
	30	8·5377	8·1825	7·9265	7·5784	7·3502	7·1876	7·0147	6·8291	6·7101	6·5181	6·4026
	35	8·1968	7·8517	7·6026	7·2631	7·0397	6·8801	6·7099	6·5262	6·4079	6·2156	6·0989
	40	7·9566	7·6188	7·3746	7·0413	6·8214	6·6640	6·4955	6·3131	6·1952	6·0023	5·8844
	50	7·6404	7·3125	7·0751	6·7501	6·5350	6·3804	6·2143	6·0334	5·9157	5·7214	5·6011
	60	7·4417	7·1202	6·8872	6·5676	6·3555	6·2027	6·0381	5·8581	5·7403	5·5446	5·4222
	70	7·3054	6·9884	6·7584	6·4426	6·2325	6·0809	5·9173	5·7378	5·6200	5·4230	5·2987
	80	7·2061	6·8924	6·6646	6·3515	6·1430	5·9924	5·8294	5·6503	5·5323	5·3343	5·2084
	100	7·0711	6·7619	6·5372	6·2279	6·0215	5·8721	5·7101	5·5313	5·4131	5·2133	5·0849
	200	6·8143	6·5139	6·2952	5·9933	5·7910	5·6439	5·4836	5·3053	5·1863	4·9819	4·8471
	∞	6·5741	6·2821	6·0692	5·7743	5·5758	5·4309	5·2721	5·0940	4·9737	4·7629	4·6190
1%	4	12·401†	12·800†	13·012†	13·283†	13·449†	13·561†	13·67†	13·79†	13·87†	—	—
	5	9·9992*	10·004*	10·008*	10·012*	10·014*	10·016*	10·018*	10·02*	10·02*	—	—
	6	1·9377*	1·9064*	1·8848*	1·8570*	1·8398*	1·8281*	1·8162*	1·8041*	1·7969*	—	—
	7	85·053	82·731	81·125	79·047	77·759	76·882	75·989	75·082	74·522	—	—
	8	51·991	50·178	48·921	47·290	46·276	45·583	44·877	44·156	43·715	43·04	—
	10	29·789	28·478	27·566	26·376	25·632	25·121	24·597	24·060	23·731	23·224	22·95
	12	21·965	20·889	20·138	19·154	18·534	18·108	17·668	17·215	16·936	16·505	16·261
	14	18·142	17·199	16·539	15·670	15·121	14·742	14·349	13·943	13·691	13·301	13·077
	16	15·916	15·059	14·457	13·662	13·157	12·807	12·444	12·066	11·831	11·466	11·255
	18	14·473	13·674	13·112	12·368	11·894	11·564	11·221	10·863	10·639	10·289	10·086
	20	13·466	12·710	12·177	11·470	11·018	10·703	10·374	10·030	9·8138	9·4748	9·2771
	25	11·924	11·237	10·751	10·103	9·6871	9·3951	9·0890	8·7668	8·5618	8·2383	8·0476
	30	11·055	10·409	9·9509	9·3382	8·9430	8·6646	8·3715	8·0602	7·8626	7·5468	7·3586
	35	10·499	9·8801	9·4405	8·8511	8·4695	8·2000	7·9153	7·6115	7·4177	7·1059	6·9186
	40	10·114	9·5138	9·0872	8·5142	8·1424	7·8791	7·6002	7·3015	7·1102	6·8006	6·6132
	50	9·6141	9·0400	8·6308	8·0795	7·7204	7·4652	7·1938	6·9015	6·7131	6·4054	6·2168
	60	9·3053	8·7472	8·3490	7·8114	7·4603	7·2101	6·9434	6·6548	6·4679	6·1606	5·9704
	70	9·0954	8·5485	8·1578	7·6297	7·2840	7·0373	6·7736	6·4875	6·3015	5·9940	5·8022
	80	8·9437	8·4048	8·0197	7·4984	7·1567	6·9124	6·6510	6·3666	6·1812	5·8733	5·6799
	100	8·7388	8·2111	7·8334	7·3214	6·9851	6·7443	6·4858	6·2036	6·0189	5·7100	5·5139
	200	8·3542	7·8476	7·4842	6·9900	6·6639	6·4293	6·1763	5·8979	5·7138	5·4012	5·1976
	∞	8·0000	7·5132	7·1633	6·6857	6·3691	6·1402	5·8920	5·6164	5·4323	5·1133	4·8981

* Multiply entry by 100. † Multiply entry by 10^4.

Bibliography

Anderson, E. (1939): The Irises of the Gasp'e Peninsula. *Bull. of the Amer. Iris Soc.*, **59**, 2 - 5.

Anderson, J.A. (1972): Separate sample logistic discrimination. *Biometrika*, **59**, 19 - 35.

Anderson, J.A. and Blair, V. (1982): Penalized maximum likelihood estimation in logistic regression and discrimination. *Biometrika*, **69**, 123 - 136.

Anderson, T.W.(1958): *Introduction to Mathematical Statistical Analysis.* Wiley, New York; 2nd edition, 1984.

Anderson, T.W. (1963 a): Asymptotic theory for principal component analysis. *Ann. Math. Stat.*, **34**, 122 - 148.

Anderson, T.W. (1963 b): A test for equality of means when covariance matrices are unequal. *Ann. Math. Stat.*, **34**, 671 - 672.

Andrews, D.F. (1971): A note on the selection of data transformations. *Biometrika*, **58**, 249 - 254.

Atkinson, A.C. (1973): Testing transformations to normality. *J. Roy. Stat. Soc.*, **B 35**, 473 - 479.

Bartlett, M.S. (1938): Further aspect of the theory of multiple regression. *Proc. of the Cambridge Philosophical Society*, **34**, 33-40.

Bartlett, M.S. (1951): The effect of standardization on an approximation in factor analysis. *Biometrika* **38**, 337 - 344.

Bartlett, M.S. (1954): A note on multiplying factors for various chi-square approximations. *J. Roy. Stat. Soc.*, **B 16**, 296 - 298.

Beale, E.M.L. (1970): Note on procedures for variables selection in multiple regression. *Technometrics*, **12**, 909 - 914.

Beale, E.M.L., Kendall, M.G.and Mann, D.W. (1967): The discarding of variables in multivariate analysis. *Biometrika*, **54**, 357 - 366.

Ben-Israel, A. and Greville, T.N.E. (1980): *Generalized Inverses: Theory and Applications.* Huntington, New York: Robert E. Krieger Company.

Bennett, B.M. (1951): Note on a solution to the generalized Behrans-Fisher problem. *Ann. Instt. Stat. Math.*, **2**, 87 - 90.

Bergmann, R. (1957): A study of independence and dependence in multivariate normal analysis. *Mimeograph Ser. No 186*, Institute of Statistics, University of North Carolina.

Berk, K.N. (1978): Comparing subset regression procedures. *Technometrics*, **20**, 1-6.

Bickel, P.J. (1964): On some alternative estimators for shifts in the *p*-variate one-sample problem. *Ann. Math. Stat.*, **35**, 1079 - 1090.

Bloomfield, P. and Watson, G.S. (1975): The efficiency of least square, *Biometrika*, **62**, 121-128.

Bock, R.D. (1960): Components of variable analysis as a structural and dimensional analysis for psychological tests. *British Journal of Statistical Psychology*, **13**(2), 151 - 163.

Box, G.E.P. (1949): A general distribution theory for a class of likelihood criteria. *Biometrika*, **36**, 317 - 346.

Box, G.E.P. (1950): Problems in the analysis of growth and the linear curves. *Biometrika*, **7**, 362 - 389.

Box, G.E.P. and Cox, D.R. (1964): An analysis of transformation (with Discussion), *J. Roy. Stat. Soc.*, **B**, **26**, 211 - 252.

Brauer, A.T. (1953): Bounds for characteristic roots of matrices, in L.S.Paige *et al.* (eds.) *Simultaneous Linear Equations and the Determination of Eigenvalues*, pp 101 - 106, US Department of Commerce, Washington.

Breiman, L., Meisel, W., and Purcell, E. (1977): Variable kernel estimates of multivariate densities. *Technometrics*, **19**, 135 - 144.

Brown, P.J. and Zidek, J.V. (1980): Adaptive multivariate ridge regression. *Ann. Stat.*, **2**, 64 - 74.

Buck, S.F.A. (1960): A method of estimation of missing values in multivariate data suitable for use with an electronic computer. *J. Roy. Stat. Soc.*, **B 22**, 302 - 307.

Campbell, N.A. (1978): The influence function as an aid in outlier detection in discriminant analysis. *Appl. Stat.*, **27**, 251 - 258.

Campbell, N.A. (1980): Shrunken estimators in discriminant and canonical variate analysis. *Appl. Stat.*, **29**, 5 - 14.

Carroll, R.J. (1980): A robust method for testing transformations to achieve approximate normality. *J. Roy. Stat. Soc.*, **B 42**, 71 - 78.

Cattell, R.B. (1966): The Screen test for the number of factors. *Multivariate Behavioral Research*, **1**, 140 - 161.

Chatterjee, S. and Price, B. (1977): *Regression Analysis by Example*. New York: Wiley.

Chen, C.F. (1979): Bayesian inference for a normal dispersion matrix and its applications to stochastic multiple regression analysis. *J. Roy. Stat. Soc.*, **B 41**, 235-248.

Choi, S.C. (1977): Test of equality of dependent correlation coefficient, *Biometrika*, **64**, 645 - 647.

Cochran, W. G. (1934): The distribution of quadratic forms in a normal system, with applications to the analysis of variance. *Proc. Camb. Phil. Soc.*, **30**, 178 - 191.

Cortella, R.B. and Jespers, J. (1967): A general plasmode (No. 30 - 10 - 5 - 2) for factor analytic exercises and research. *Multivariate Research Monographs*,

67(3), 1-212.

Costanza, M.C. and Afifi, A.A. (1979): Comparison of stopping rules in forward stepwise discriminant analysis. *J. Amer. Stat. Assoc.*, **74**, 777 - 785.

Cox, D.R. (1966): Some procedures associated with the logistic qualitative response curve. *In* F.N.David (*Ed.*) *Research Papers in Statistics: Festschrift for J. Neyman,* pp. 55 - 71. Wiley, New York.

Craig, A.T. (1943): A note on the independence of certain quadratic forms. *Ann. Math. Stat.* **14**, 195 - 197.

Cramer, H. (1946): *Mathematical Methods of Statistics*, Princeton University Press.

D'Agostino, R.B. (1971): An omnibus test of normality for moderate and large samples. *Biometrika*, **58**, 341 - 348.

D'Agostino, R.B. (1972): Small sample probability plans for the D test of normality. *Biometrika*, **59**, 221 - 222.

D'Agostino, R.B. and Pearson, E.S. (1973): Tests for departure for normality. Empirical results for the distributions of b_2 and $\sqrt{b_1}$. *Biometrika*, **60**, 613 - 622. [Correction: *Biometrika*, **61**, 647].

D'Agostino, R.B. and Tietjen, G.L. (1971): Simulated probabil ity points of b_2 for small samples. *Biometrika*, **58**, 669 - 672.

Davis, A. W. (1970a): Exact distributions of Hotelling's generalized T_0^2. *Biometrika*, **57**, 187-191.

Davis, A.W. (1970b): Further applications of a differential equation for Hotelling's generalized T_0^2. *Ann. Instt. Stat. Math.*, **22**, 77- 87.

Davis, A.W. (1979): On the differential equation for Meijer's $G_{p,p}^{p,0}$ function, and further tables of Wilk's likelihood criterion. *Biometrika*, **66**, 519 - 531.

Davis, A. W. (1980): Further tabulation of Hotelling's generalized T_0^2. *Comm. Stat. Simul. Comput. B*, **9**, 321- 336.

Day N.E. and Kerridge, D.F. (1967): A general maximum likelihood discriminant. *Biometrics*, **23**, 313 - 323.

Dempster, A.P., Leird, N.M., and Rubin, D.B. (1977): Maximum likelihood from incomplete data in the EM Algorithm. *J. Roy. Stat. Soc.* **B 39**, 1 - 38.

Di Pillo, P.J. (1979): Biased discriminant analysis: evaluation of the optimum probability of misclassification. *Commun. Stat. Theor. Methods. A*, **8**, 1447 - 1457.

Eaton, M.L. (1969): Some remarks on Scheffe"s solution to the Behrans-Fisher problem. *J. Amer. Stat. Assoc.*, **64**, 1318 - 1322.

Effron, B. (1979): Bootstrap methods: another look at the jackknife. *Ann. Sttat.*, **7**, 1 - 26.

Effron, B. (1981): Nonparametric standard errors and confidence intervals. *Can. J. Stat.*, **9**, 139 - 172.

Elston, R.C. and Grizzle, J.E. (1962): Estimation of time-response curves and their confidence bands. *Biometrics*, **18**, 148 - 159.

Ferrer, D.E. and Gleuber, R.R. (1967): Multicollinearity in regression analysis (the problem revisited). *Review of Economic Statistics*, **49**, 92-107.

Fisher, R.A. (1936): The use of multiple measurements in taxonomic problems. *Annals of Eugenics*, **7**, 179 - 188.

Fisher, R.A.(1939): The sampling distribution of some statistics obtained from non-linear equations. *Ann. Eugen.*, **9**, 238-249.

Fre'chet, M. (1937): Reche'rches the'oriques modernes Sur la the'orie des probabile's, *Forms*, **1**(III), Paris.

Fuller, W.A. and Battese, G.E. (1973): Transformation for estimation of linear models with nested error structure, *J. Amer. Stat. Assoc.* , **68**, 626-632.

Furnival, G.M. and Wilson, R.W. (1974): Regression by leaps and bounds. *Technometrics*, **16**, 499 - 511.

Galton, F. (1888): Co-relations and their measurement, chiefly from anthropometric data,. *Proc. Royal Soc.*, **45**, 135 - 140.

Gantmacher, F.R. (1959): *The Theory of Matrices, Vols I & II.* Chesla Pub. Co., New York.

Geisser, S. (1966): Bayesian estimation in multivariate analysis. *Ann. Math. Stat.*, **36**, 150 - 159.

Ghosh, M. and Meeden, G. (1986): Empirical Bayes estimation in finite population sampling. *J. Amer. Stat. Assoc.*, **81**, 1058 - 1062.

Giri, N.C. (1977): *Multivariate Statistical Inference* . Academic Press: New York.

Girshick, M.A. (1936): Principal components. *J. Amer. Stat. Assoc*, **31**, 519 - 528.

Girshick, M. A. (1939): On the sampling theory of roots of determinantal equations, *Ann. Math. Stat.*, **1**, 203 - 224.

Girshick, M.A. (1936): Principal component analysis. *J. Amer. Stat. Assoc.* **31**, 519 - 528.

Gleason, T.C. and Staelin, R. (1975): A proposal for handling missing data. *Psychometrika*, **40**, 229 - 252.

Gleser, H.J. (1976): A canonical representation for the noncentral Wishart distribution useful for simulation. *J. Amer. Stat. Assoc.*, **71**, 690-695.

Gnanadesikan, R. (1977): *Methods for Statistical Data Analysis of Multivariate Observations*, (1997), 2nd edn., John Wiley, New York.

Gnanadesikan, R. and Kettenring, J.R. (1972): Robust estimates, residuals and outlier detection with multiresponse data. *Biometrika*, **28**, 81 - 124.

Hadamard, J. (1893): Re'solution d'une Question Relative aux Determinants, *Bu.. Scientific Mathematics*, **2**, 240 - 248.

Haitovsky, Y. (1969): Multicollinearity in regression analysis: Comments. *Review of Economics and Statistics*, **50**, 486 - 489.

Hans, S. (1973): *Matrices and Linear Algebra.* Holt, Rinehart $ Winston, London.

Heilberger, R.M. (1977): Regression with pairwise-present covariance matrix: A dangerous practice. Technical Report 19, The Wharton School, University of Pennsylvania, Department of Statistics.

Hemel, J.B., van der Voet, H., Hindriks, F.R. and van der Slik, W. (1987): Stepwise deletion: A technique for missing data handling in multivariate analysis. *Analytica Chimica Acta*, **193**, 255 - 268.

Heo, T.Y. (1987): The comparison of eigensystem technique for measuring multicollinearity in multivariate normal data. Master's thesis, Brigham Young University, Deptt. of Statistics.

Heywood, H.B. (1931): On finite sequences of real numbers. it Proceedings of the Royal Society, Ser. A, **134**, 486 - 501.

Hinkley, D. (1975): On power transformations to normality. *Biometrika*, **62**, 101 - 111.

Hinkley, D. (1977): On quick choice of power transformation. *Appl. Stat.*, **26**, 67 - 69.

Hoerl, A.E. and Kennard, R.W. (1970): Ridge regression: Applications to nonorthogonal problems. *Technometrics*, **12**, 69 - 82.

Hogg, R.V. and Craig, A.T. (1965): *Introduction to Mathematical Statistics*, Macmillan and Amerind.

Hora, S.C. (1980): Sequential discrimination. *Commun. in Stat. Theory & Mthods*, A, **9**, 905 - 916.

Horn, J.L. (1965): A rationale and test for the number of factors in factor analysis. *Psychometrika*, **30**, 179..

Hotelling, H. (1931): The generalization of Studen's ratio. *Ann. Math. Stat.*, **2**, 360.

Hotelling, H. (1933): Analysis of a complex statistical variable into principal components. *J. of Education Psychology* **26**, 417 - 441.

Hotelling, H. (1935): The most predictable criterion. *Jour. Educational Psychology*, **26**, 139 - 142.

Hotelling, H. (1936): Relations between two sets of variables. *Biometrika*, **28**, 321 - 377.

Hotelling, H. (1951): A generalized T-test and measure of multivariate dispersion. *Proc. Second Berkley Symp. Math. Stat. Prob.*, **1**, 23 - 41.

Hsu, P.L.(1939): On the distributions of the roots of certain determinantal equation. *Ann. Eugen*, **9**, 256-258.

Huber, P.J. (1977): Robust covariances. In *Statistical Decision Theory and Related Topics 2*(eds. Gupta, S.S. and Moore, D.S.), Academic Press, New York, pp. 165 - 191.

Jackson, J.E. (1959): Quality control methods for several related variables. *Technometrics*, **1**, 359 - 377.

James, G.S. (1954): Tests of linear hypotheses in univariate and multivariate analysis when the ratio of the population variances are unknown. *Biometrika*, **41**, 19 - 43.

James, W. and Stein, C. (1961); Estimation with quadratic loss. *Proc. Fourth Berkeley Symp. Math. Stat. Prob.* , **1**, 362-379.

John, J.A. and Draper, N.R. (1980): An alternative family of transformations. *Appl. Stat.*, **29**, 190 - 197.

John, S. (1976): Fitting sampling distribution agreeing in support and moments and tables of critical values of sphericity criterion. *J. Multivar. Analys.*, **6**, 601 - 607.

John, S. (1977): Unbiased and upper critical values of mean trace of multivariate beta for testing difference of two covariance matrices or several mean vectors. *Comm. Stat. Simul. Comput. B*, **6**, 89 - 96.

Johnson, N.L. and Kotz, S. (1972): *Distributions in Statistics*, vol 4. *Continuous Multivariate Distributions* Wiley: Mew York.

Johnson, R.A. and Wichern, D.W. (1998): *Applied Multivariate Statistical Anal-*

ysis, 4th edn., Prentice-Hall, New Jersey.

Jöreskog, K.G. (1977): Factor analysis by least -squares and maximum-likelihood methods. In K.Enslein, A. Ralston, and H.S.Wilf (Eds.), *Statistical Methods for Digital Computers*, Vol. 3, pp 125 - 153, Wiley: New York.

Kaiser, H.F. (1958): The varimax criterion for analytic rotation in factor analysis. *Psychometrika*, **23**, 187 - 200.

Kaiser, H.F. (1970): A second generation Little Jiffy. *Psychometrika*, **35**, 401 - 415.

Kaiser, H.F. and Rice, J. (1974): Little Jiffy, Mark IV. *Educational and Psychological Measurement*, **34**, 111 - 117.

Khatri, C.G. (1966): A note on a MANOVA model applied to problems in growth curves. *Ann. Instt. Stat. Math.*, **18**, 75-86.

Khatri, C.G. and Rao, C.R. (1975): Some extensions of the Kantorovich inequality and statistical applications. *J. Multivariate Analysis*, **11**, 498-505.

Kingman, A. and Graybill, F.A. (1970): A nonlinear characterization of the normal distribution., *Ann. Math. Stat.*, **41**, 1889 - 1895.

Knott, M. (1975): On minimum efficiency of least squares. *Biometrika*, **62**, 129-132.

Kres, H. (1983): (Ed.) *Statistical Tables for Multivariate Analysis*, Springer-Verlag, New York.

Krzanowski, W.J. (1975): Discrimination and classification using both binary and continuous variables. *J. Amer. Stat. Assoc.*, **70**, 782 - 790.

Krzanowski, W.J. (1977): The performance of Fisher's linear discriminant function under non-optimal conditions, *Technometrics*, **19**(2), 191 - 200.

Lachenbruch, P.A. (1975): *Discriminant Analysis*, Hafner Press, New York.

Lachenbruch, P.A. and Goldstein, M. (1979): Discriminant analysis, *Biometrics*, **35**, 69 - 85.

Lachenbruch, P.A. and Mickey, M.R. (1968): Estimation of error rates in discriminant analysis, *Technometrics*, **10** (1), 1-11.

Laha, R.G. (1957): On a characterization of the normal distribution from properties of suitable linear statistics., *Ann. Math. Stat.*, **28**, 126 - 139.

Lawley, D.N. (1938): A generalization of Fisher's Z test. *Biometrika*, **30**, 180 - 187.

Lawley, D.N. (1956): Tests of significance for the latent roots of covariance and correlation matrices. *Biometrika*, **43**, 128 - 136.

Lawley, D.N. (1963): On testing a set of correlation coefficients for equality. *Ann. Math. Stat.* **34**, 149 - 151.

Lawley, D.N. and Maxwell, A.E. (1971): *Factor Analysis as a Statistical Method.* 2nd ed. Butterwarths: London.

Lee, S.Y. (1979): Constrained estimation in covariance structure models. *Biometrika*, **66**, 539 - 545.

Lee, Y.S. (1972): Some results on the distribution of Wilk's likelihood criterion. *Biometrika*, **95**, 649 - 663.

Little, R.J.A. (1992): Regression with missing X's: A review. *J Amer. Stat. Assoc.***87**, 1227 - 1237.

Lubischew, A.A. (1962): On the use of discriminant functions in taxonomy, *Bio-*

metrics, **18**, 455 - 477.

Lukacs, E. (1956): Characterization of populations by properties of suitable statistics. *Proc. Third Berkley Symp.*, **2**, 195 - 214.

Mallows, C.L. (1973): Some comments on C_p, *Technometrics*, **15**, 661 - 675.

Manly, B.F.J. (1994): *Multivariate Statistical Methods, A Primer*, 2nd edn., Chapman and Hall / CRC, Boca Raton, London, New York, Washington, D.C.

Mardia, K.V. (1970): Measures of multivariate skewness and kurtosis with applications. *Biometrika* , **57**, 519-524.

Mardia, K.V. (1974): Applications of some measures of multivariate skewness and kurtosis for testing normality and robustness studies. *Sankhya* **B** **36**, 115 - 128.

Mardia, K.V. (1975): Assessment of multinormality and the robustness of Hotelling's T^2 test. *Appl. Stat.*, **24**, 163 - 171.

Mardia, K.V. (1980): Tests of univariate and multivariate normality. In *Handbook of Statistics I* (P.R. Krishnaiah, ed.), North-Holland, Amsterdam, pp. 279 - 320.

Mardia, K.V., Kent, J.T., and Bibby, J.M. (1979): *Multivariate Analysis.* Academic Press, London.

Mason, R.L., Gunst, R.F. and Webster, J.T. (1975): Regression analysis and the problem of multicollinearity. *Communications in Statistics* **4**, 277 - 292.

McKeon, J.J. (1974): F approximation to the distribution of Hotelling's T_0^2. *Biometrika*, **61**, 381 - 383.

McLachlan, G.J. (1976): The bias of the apparent error rate in discriminant analysis. *Biometrika*, **63**, 239 - 244.

McLachlan, G.J. (1980): The efficiency of Efron's 'boot-strap' approach applied to error rate estimation in discriminant analysis. *J. Stat. Comput. Simul.*, **11**, 273 - 279.

Meyer, E.P. (1975): A measure of average intercorrelation. *Educational and Psychological Measurement* **35**, 67 - 72.

Miller, R.G., Jr. (1981): *Simultaneous Statistical Inference*, 2nd edition, McGraw-Hill, New York.

Mitra, S. K. (1969): Some characteristic and non-characteristic properties of the Wishart distribution, *Sankhya A*, **31**, 19 - 22.

Mood, A.M. (1941): On the joint distribution of the median in samples from a multivariate population. *Ann. Math. Stat.*, **12**, 268 - 278.

Moore, D.H., II (1973): Evaluation of the five discriminant procedures for binary variables. *J. Amer. Stat. Assoc.*, **68**, 399 - 404.

Moore, E.H. (1935): *General Analysis*, American Phil. Soc., Philadelphia.

Muirhead, R.J. (1982): *Aspects of Multivariate Statistical Theory.* Wiley, New York.

Mukhopadhyay, P.(1998): Small area estimation of population in the districts of Hoogly and Murshidabad, Indian Statistical Institute, Kolkata (Unpublished Survey Report)

Mukhopadhyay, P. (2005): *Applied Statistics*, 2nd edition, Books and Allied, Kolkata, India.

Mulholland, H.P. (1977): On the null distribution of $\sqrt{b_1}$ for samples of size at most 25, with tables. *Biometrika*, **64**, 401 - 409.

Myers, R.H. (1990): *Classical and Modern Regression with Applications*, 2nd edn,, Duxbury Press, Boston.

Nagarsenkar, B.N. (1975): Percentage points of Wilk's L_{vc} criteria. *Comm. Stat.* **4**, 620 - 641.

Ogawa, J. (1949): On the independence of linear and quadratic forms of a random sample from a normal population. *Ann. Instt. Math. Stat.* **1**, 83 - 108.

Olkin, I and Pratt, J.W. (1958): Unbiased estimation of certain correlation coefficients. *Ann. Math. Stat.*, **29**, 201 - 211.

Orchard, T. and Woodbury, M.A. (1972): A missing information principle: Theory and Applications. In *Sixth Berkley Symposium on Mathematical Statistics and Probability.* pp 697 - 715, Berkley, CA: University of California Press.

O'Sullivan, J.B. and Mahan, C.M. (1966): Glucose tolerance tests: Variability in pregnant and non-pregnant women. *Amer. J. Clinical Nuitrition* **19**, 345 - 351.

Ott, J. and Kronmal, R.A. (1976): Some classification procedures for binary data using orthogonal functions. *J. Amer. Stat. Assoc*, **71**, 391 - 399.

Patil, G.P. and Boswell, M.T. (1970): A characteristic property of the multivariate normal density function and some of its applications., *Ann. Math. Stat.*, **41**, 1970 - 1977.

Pearson, E.S. and Hartley, H.O. (eds.) (1966): *Biometrika Tables for Statisticians,* Vol. I, Cambridge University Press.

Pearson, E.S. and Hartley, H.O. (eds.) (1972): *Biometrika Tables for Statisticians,* Vol. II, Cambridge University Press.

Penrose, R. (1955): A Generalized Inverse for Matrices, *Proc. Camb. Phil. Soc.*, **51**, 406-413.

Pettitt, A.N. (1977): Testing the normality of several independent samples using the Anderson-Darling statistic. *Appl. Stat.*, **26**, 156 - 161.

Pillai, K.C.S. (1960): *Statistical Tables for Tests of Multivariate Hypotheses.* Statistical Center, University of Phillipines, Manila.

Pillai, K.C.S.(1964): On the distribution of largest of seven roots of a matrix in multivariate analysis. *Biometrika* , **51**, 270 - 275.

Pillai, K.C.S. (1965): On the distribution of the largest characteristic root of a matrix in multivariate analysis. *Biometrika*, **52**, 405 - 414.

Pillai, K.C.S. and Flury, B.N. (1984): Percentage points of the largest characteristic root of the multivariate Beta matrix. *Comm. Stat., T & M*, **13**(18), 2199 - 2237.

Pillai, K.C.S. and Gupta, A.K. (1969): On the exact distributions of Wilk's criterion. *Biometrika*, **56**, 109 - 118.

Pillai, K.C.S. and Young, D. C. (1971): On the exact distribution of Hotelling's generalized T_0^2. *J. Multivar. Analysis*, **1**, 90 - 107.

Potthoff, R.F. and Roy, S.N. (1964): A generalized multivariate analysis of variance model useful especially for growth-curve problems. *Biometrika*, **51**,

313 - 326.

Prentice, R.L. and Pyke, R. (1979): Logistic disease incidence models and case-control studies. *Biometrika*, **66**, 403 - 411.

Rao, C.R. (1951): An asymptotic expansion of the distribution of Wilk's criterion. *Bull. Inst. Int. Stat.*, **33**, 177 - 180.

Rao, C.R. (1962): A note on generalized inverse of a matrix with applications to problems in the mathematical statistics. *J. Roy. Stat. Soc.*, **24**, 152 - 158.

Rao, C.R. (1964): The use and interpretation of principal component analysis in applied research. *Sankhya* **B 26**, 329 - 358.

Rao, C.R. (1966): Covariance adjustment and related problems in multivariate analysis. In P.R.Krishnaiah (Ed.) *Multivariate Analysis*, Vol **II**, 87-103. Academic Press: New York.

Rao, C.R. (1967): Least square theory using an estimated dispersion matrix and its applications to measurement of signals. *Proc. Fifth Berkeley Symp. Math. Stat. Prob.,* **1**, 355-372.

Rao, C.R. (1969): Some characterization of the multivariate normal distribution. In (ed. P.R.Krishnaiah) *Multivariate Analysis*, vol. II, pp.321 - 328, Academic Press, New York.

Rao, C.R. (1972): Recent trends of research work in multivariate analysis, *Biometrics*, **28**, 3 - 22.

Rao, C.R. (1973): Representation of best linear unbiased estimators in the Gauss-Markoff model with a singular dispersion matrix, *J. Multivariate Analysis*, **3**, 276 - 292.

Rencher, A.C. (1995): *Methods of Multivariate Analysis*. New York: Wiley.

Rencher, A.C. (1998): *Multivariate Statistical Inference and Applications*. New York: Wiley.

Rencher, A.C. and Pun, F.C. (1980): Influence of R^2 in best subset regression. *Technometrics* **22**, 49 - 53.

Roy, S.N. (1939): p-Statistics or some generalizations in analysis of variance appropriate to multivariate problems. *Sankhya*, **4**, 381 - 396.

Roy, S.N. (1957): *Some Aspects of Multivariate Analysis*, Wiley, New York.

Rubin, D.B. (1976): Inference and missing data. *Biometrika* **63**, 581 - 592.

Schatzoff, M. (1966): Exact distributions of Wilk's likelihood ratio criterion. *Biometrika*, **53**, 347 - 358.

Scheffe', H. (1943): On solutions of the Behrans-Fisher problem based on the t-distribution. *Ann. Math. Stat.*, **14**, 35 - 44.

Schuurmann, F.J., Krishnaiah, P.R. and Chattopadhyay, A. K. (1975): Exact percentage points of the distribution of a trace of a multivariate beta matrix. *J. Multivar. Analys.*, **3**, 445 - 453.

Schwertman, N.C. and Allen, D.M. (1973): The smoothing of an indefinite matrix with applications to growth curve analysis with missing observations. Tech. Rep. No. 56, U. of Kentucky, Deptt. of Statistics.

Schwertman, N.C. and Allen, D.M. (1979): Smoothing an indefinite variance-covariance matrix. *J. Stat. Computation and Simulation* **9**, 183 - 194.

Searle, S.R. (1982): *Matrix Algebra Useful for Statistics*. John Wiley, N.Y.

Seber, G.A.F. (1977): *Linear Regression Analysis*, Wiley: New York.

Seber, G.A.F. (1984): *Multivariate Observations*, Wiley: New York.

Shapiro, S.S. and Wilks, N.B. (1965): An analysis of variance test for normality. *Biometrika*, **52**, 591 - 611.

Silverman, B.W. (1986): *Density Estimation for Statistics and Data Analysis*, Chapman and Hall, U.K.

Silvey, S.D. (1970): *Statistical Inference.* Penguin, Baltimore.

Siotani, M , Hayakawa, T. and Fujikoshi, Y. (1985): *Modern Multivariate Statistical Analysis*, Ohio: American Sciences.

Sparkes, R.S., Coutsourides, D., and Troskie, L. (1983): The multivariate C_p, *Comm. Stat, Part A, Theory and Methods*, **12**(15), 1775 - 1793.

Spearman, C. (1904): 'General intelligence' objectively determined and measured. *Amer. Jour. of Psychology*, **15**, 201 - 293.

Stein, C. (1955): Inadmissibility of the usual estimates for the mean of a multivariate normal distribution. in *Proc. of the 3rd Berkeley Symp. on Math. Stat. & Prob., Vol. I*, Berkeley, University of California Press, p. 197 - 202.

Thompson, M.L. (1978a): Selection of variables in multiple regression:Part 1. A review and evaluation. *Int. Stat. Rev.*, **46**, 1 - 19.

Thompson, M.L. (1978b): Selection of variables in multiple regression: Part II. Chosen procedures, computations and examples. *Int. Stat. Rev.* **46**, 129 - 146.

Thurstone, L.L. (1947): *Multiple Factor Analysis.* University of Chicago Press, Chicago.

Thurstone, L.L. and Thurstone, T.G. (1941): Factorial studies of intelligence. *Psychometric Mongr*, **2**.

Tiku, M.L. (1967): Tables of the power of the F-test. *J. Amer. Stat. Assoc.*, **62**, 525-539.

Tiku, M.L. (1972): More tables of the power of the F-test. *J. Amer. Stat. Assoc.*, **67**, 709-710.

Tiku, M.L. and Balakrishnan, N. (1985): Testing the quality of variance-covariance matrices the robust way. *Comm. Stat., T & M.*, **14**(12), 3033-3051.

Tukey, J.W. (1957): On the comparative anatomy of transformations. *Ann. Math. Stat.*, **28**, 602 - 632.

Tukey, J.W. (1960): A survey of sampling from contaminated distributions. In *Contributions to Probability & Statistics.* (eds. I. Olkin, *et al.*), Standford University Press, pp. 448 - 485.

Wang, S.G., Chow, S.C. and Tse, S.K. (1994): On ordinary least squares methods for sample surveys, Stat. Probab. Letters, **20**, 173-182.

Welch, B.L. (1939): Note on discriminant functions, *Biometrika*, **31**, 218 - 220.

Wilks, S.S. (1932): Certain generalizations in the analysis of variance. *Biometrika*, **24**, 471 - 494.

Wright, S. (1954): The interpretation of multivariate systems, in O.Kempthorne et al. (eds.), *Statistics and Mathematics in Biology*, pp. 11 - 33, The Iowa State University Press, Ames, Iowa.

Author Index

Numbers in italics indicate the pages on which the complete reference is given

A

Afifi, A.A., 445, *527*

Allen, D.M., 31, *533*

Anderson, E., 448, *525*

Anderson, J.A., 420, *525*

Anderson, T.W., 64, 148, 177, 298, 308, 312, 318, 319, 320, 321, 322, *525*

Andrews, D.F., 90, *525*

Atkinson, A.T., 90, *525*

B

Balakrishnan, N., 149, *534*

Bartlett, M.S., 120, 149, 318, 321, 355, *525*

Battese, G.E., 491, *526*

Beale, E.M.L., 237, *525*

Ben-Israel, A., 465, *525*

Bennet, B.H., 148, 177, *525*

Bergmann, R., 314, *525*

Berk, K.N., 192, *526*

Bibby, J.M., *531*

Bickel, P.J., 91, *526*

Blair, V., 420, *523*

Bloomfield, P., 471, *526*

Bock, R.D., 32, 314, *526*

Boswell, M.T., 53, *532*

Box, G.E.P., 173, 323, *526*

Brauer, A.T., 315, *526*

Breiman, L., 390, 412, *526*

Brown, P.J., 192, *526*

Buck, S.F.A., 32, *526*

C

Campbell, N.A., 400, 408, *526*

Carroll, R.J., 90, *526*

Cattell, R.B., 327, *526*

Chatterjee, S., 14, *526*

Chattopadhyay, A.K., *533*

Chen, C.F., 75, *526*

Choi, S.C., 323, *526*

Chow, S.C., *534*

Cochran, W.G., 64, *527*

Cortella, R.B., 327, *526*

Costanza, M.C., 445, *527*

Coutsourides, D., *534*

Cox, D.R., 419, *526*, *527*

Craig, A.T., 96, 109, *527*, *529*

Cramer, H., 64, 67, *527*

D

D'Agostino, R.B., 82, 83, 494, 495, *527*

Davis, A.W., 120, 123, 495, 496, *527*

Day, N.E., 419, 420, *527*

Dempster, A.P., 32, *527*

Di Pillo, P.J., 400, *527*

Draper, N.R., 90, *529*

E

Eaton, M.L., 148, 177, *527*

Effron, B., 405, *527*

Elston, R.C., 357, *527*

F

Ferrer, D.E., 13, *527*

Fisher, R.A., 111, 394, 400, 412, 413, 429, *527*

Flury, B.N., 122, *532*

Fre'chet, M., 98, *528*

Fujikoshi, Y., 12, *534*

Furnival, G.M., 234, *528*

Fuller, W.A., 491, *526*

G

Galton, F., 331, *528*

Gantmacher, F.R., 453, *528*

Geisser, S., 298, *528*

Ghosh, M., 99, *528*

Giri, N.C., 73, *528*

Girshick, M.A., 298, 312, 318, *528*

Gleason, T.C., 13, 33, *528*

Gleser, H.J., 111, *528*

Gleuber, R.R., 13, *527*

Gnanadesiken, R., 91, 94, 324, *528*

Goldstein, M., 398, 411, *530*

Graybill, F.A., 53, *530*

Greville, T.N.E., 465, *525*

Grizzle, J.E., 359, *527*

Gunst, R.F., 13, *531*

Gupta, A.K., 120, 495, *532*

H

Hadamard, J., 474, *528*

Haitovsky, H.O., 13, *528*

Hans, S., 453, *528*

Hartley, H.O., 82, 122, 496, *532*

Hayakawa, T., 12, *534*

Heilberger, R.M., 31, *528*

Hemel, J.B., 31, *528*

Heo, J.Y., 14, *528*

Heywood, H.B., 333, *528*

Hindricks, F.R., 31, *528*

Hinkley, D., 90, *528*

Hoerl, A.E., 13, *529*

Hogg, R.V., 94, *529*

Hora, S.C., 390, 412, *529*

Horn, J.L., 327, *529*

Hotelling, H., 111, 123, 239, 298, 363, 529

Hsu, P.L., 111, *529*

Huber, P.J., 92, *529*

J

Jackson, J.E., 329, *529*

James, G.S., 179, *529*

James, W., 73, *529*

Jespers, J, 327, *526*

John, J.A., 90, *529*

John, S., 124, *529*

Johnson, N.L., 122, *529*

Johnson, R.A., 83, *529*

Jòreskog, K.G., 347, *530*

K

Kaiser, H.F., 338, 350, *530*

Kendall, M.G., *525*

Kennard, R.W., 13, *529*

Kent, J.T., *531*

Kerridge, D.F., 419, 420, *527*

Kettenring, J.R., 94, *528*

Khatri, C.G., 251, 471, *530*

Kingman, A., 53, *530*

Knott, M., 471, *530*

Koltz, S., 122, *529*

Kres, H., 124, *530*

Krishnaiah, P.R., *533*

Kronmal, R.A., 411 , *532*

Krzanowski, W.J., 86, 400, 411, 412, *530*

L

Lachenbruch, P.A., 390, 398, 400, 404, 411, 412, *530*

Laha, R.G., 53, *530*

Lawley, D.N., 123, 318, 323, 347, *530*

Lee, S.Y., 120, 347, 495, *530*

Leird, N.M., 32, *527*

Little, R.J.A., 33, *530*

Lubischew, A.A., 451, *530*

Lukacs, E., 53, *531*

M

Mahan, C.M., 450, *532*

Mallows, C.L., 234, 241, *531*

Manly, B.F. J., 37, *531*

Mann, D.W., *525*

Mardia, K.V., 59, 80, 86, 99, 105, *531*

Mason, R.L., 13, *531*

Maxwell, A.E., 347, *530*

McKeon, J.J., 123, 496, *531*

McLachlan, G.J., 405, 408, *531*

Meeden, G., 99, *528*

Meisel, W., *526*

Meyer, E.P., 315, *531*

Mickey, M.R., 404, *530*

Miller, R.G., Jr., 256, *531*

Mitra, S.K., 104, 125, *531*

Mood, A.M., 91, *531*

Moore, D.H., II, 400, *531*

Moore, E.H., 463, *531*

Muirhead, R.J., 50, 111, *531*

Mukhopadhyay, P., 40, 279, *531*

Mulholland, H.P., 82, 494, *532*

Myers, R.H., 235, *532*

N

Nagarsenkar, B.N., 323, *532*

O

Ogawa, J., 110, *532*

Olkin, I., 25, *532*

Orchard, T., 32, *532*

O'Sullivan, J.B., 450, *532*

Ott, J., 411, *532*

P

Patil, G.P., 53, *532*

Pearson, E.S., 82, 122. 494, 496, *527*, *532*

Penrose, R., 465, *532*

Pettit, A.N., 83, 123, 494, *532*

Pillai, K.C.S., 120, 122, 124, 495, *532*

Potthoff, R.F., 249 *532*

Pratt, J.W., 25, *532*

Prentice, R.L., 420, 422, *533*

Price, B, 14, *526*

Purcell, E., *526*

Pun, F.C., 13, *533*

Pyke, R., 420, 422, *533*

R

Rao, C.R., 53, 111, 120, 251, 385, 465, 471, 495, *530*, *533*

Rencher, A,C., 13, 444, *533*, *455*

Rice, J., 339, *530*

Roy, S.N., 122, 132, 176, 208, *532*, *533*

Rubin, D.B., 31, 32, *527*, *533*

S

Schatzoff, M., 120, 495, *533*

Scheffe', H., 148, 177, *533*

Schuurmann, F.J., 124, *533*

Schwertman, N.C., 31, *533*

Searle, S.R., 453, *533*

Seber, G.A.F., 1, 192, 195, *533*

Shapiro, S.S., 85, *534*

Silverman, B.W., 412, *534*

Silvey, S.D., 131, *534*

Siotani, M., 12, *534*

Sparkes, R.S., 241, *534*

Spearman, C., 321, *534*

Staelin, R., 13, 33, *528*

Stein, C., 73, *529*, *534*

T

Thompson, M.L., 192, *534*

Thurstone, L.L., 331, *528*

Thurstone, T.G., 331, *528*

Tietjan, G.L., 82, 494, *527*

Tiku, M.L., 131, 149, *534*

Troskie, L., *534*

Tukey, J.W., 92, *534*,

Tse, S.K., 490, *534*

V

van der Slik, W., 31, *528*

van der Voet., H., 31, *528*

W

Wang, S.G., 490, *534*

Webster, J.T., 13, *531*

Welch, B.L., 393, *534*

Wichern, D.W., 83, *529*

Wilk, M.B., 85, *534*

Wilks, S.S., 32, 118, *534*

Wilson, R.W., 234, *528*

Woodbury, M.A., 32, *532*

Y

Young, D.C., 123, *534*

Z

Zidek, J.V., 192, *526*

Subject Index

Agricultural data, 39

Analysis of Covariance, 6, 279 - 292

- multivariate, 285 - 292
    - and a conditional hypothesis, 293 - 5
    - general theory, 285 - 8
    - one-way model, 288 - 290
- univariate, 279 - 285
    - general theory, 279 - 281
    - two-way model, 237 - 9

Analysis of variance, multivariate, *see* Multivariate analysis of variance

Analysis of variance, univariate (ANOVA)

- contrasts, 263
    - orthogonal, 264
    - sum of squares due to a set of, 264
- model, 183
- one-way model, 208, 253 - 6
    - Bonferroni's procedure, 256
    - Scheffe's procedure, 256
- two-way model, 266 - 8, 270 - 4
    - one observation per cell, 266 - 8
    - $r(> 1)$ observations per cell, 270 - 4
    - unbalanced data, 271

ANOVA, *see* Analysis of variance, univariate

Apparent error rate, *see* Error rate(s)

Association, measures of, 238

Beetles data, 450

Behrans-Fisher problem, 148 - 9, 177 - 9

- Eaton's test, 178
- Jame's test, 179
- T^2 test, 148

Bird data, 36 - 7

Box's M-test, 173

Canonical correlation(s), 6, 363 - 386

- approximating sample covariance matrix, 380 - 1
- canonical variates, *see* Canonical variate
- definition of, 364, 376
- invariance under standardization of variables, 373, 379
- and multiple correlation, 374

tests of independence, 384 - 5
tests of significance, 384 - 5

Canonical variates,

correlation
with original variables, 372 - 3, 378 - 9
definition of, 364, 375
explaining part of the total variance, 375, 383
interpretation of, 370
standardized coefficients, 369 - 370

Centering matrix, 11

Central limit theorem (Multivariate), 64 - 8

Characteristic roots, *see* Eigenvalues

Cholesky decomposition, 475

Classical linear regression model

confidence region for β, 227
estimation of β, 196, 197, 228
Gauss-Markov theorem, 227
multiple correlation coefficient, *see* R^2
selection of variables, 233 - 8

all possible subsets, 234
dependence analysis, 237
interdependence analysis, 237
stepwise selection, 236

tests of hypotheses, 229

Classification analysis (allocation), 390 - 412, 422 - 8

assigning a sample unit to a group, 390
assumptions, 391
and discriminant analysis, 417, 437
discriminant functions used in classification, 417, 437
error rates, 402 - 7, *see* also Error rate(s)
likelihood ratio method of classification, 395
linear classification functions (scores), 399, 424
logistic classification, 411, 418 - 422, 425
maximizing posterior probability, 396
minimax classification, 397
minimizing expected cost of misclassification, 395, 425
minimizing total probability of misclassification (TPM), 391, 423
minimizing TPM for samples from Bernoulli distribution, 395, 402
minimizing TPM for samples from normal distribution, 393 - 5, 397, 399 - 401, 420, 424, 425
posterior odds, 419
quadratic classification functions (scores), 400, 425
training sample, 399, 402
validation sample, 402

Cluster analysis, 7

Coal-mines data, 402

Coefficient of determination, 231, (also *see* R^2)

Communality, *see* Factor analysis, communality

Condition number, 13

Confidence interval(s) for linear functions of $\mu, \mu_1 - \mu_2$, etc.

Bonferroni simultaneous confidence intervals, 140, 142, 144, 151, 154
large sample confidence intervals, 142, 143, 149, 150, 151
T^2-simultaneous confidence intervals, 137 - 140, 144, 151, 153

Confidence region

for μ, 135, 153

for $\mu_1 - \mu_2$, 144

Confusion matrix, 404

Constant density ellipsoid, 45

Contrasts, 263 - 6

orthogonal, 264
sum of squares and products due to a set of, 264, 265

Contrast-matrix, 153, 165

Correlation

bias, 25
canonical, *see* Canonical correlation(s)
of two random variables, 12, 24
unbiased estimate, 25

Correlation matrix

with intraclass correlation structure, 24, 56, 161, 313, 454
of a random vector, 25
relation with covariance matrix, 13, 25
sample, 12
as standardized covariance matrix, 13
test comparing two covariance matrices, 162

Covariance of two linear combinations, 19, 23

Covariance matrix

for one random vector, 23
pooled, 119
population, 23
positive definite, 12, 24
relationship to correlation matrix, 12, 25
sample, 11

distribution of, 108
unbiased, 108

test for equality of, 156
for two random vectors, 23

Criteria measure, 5

Cross-validation error rate, 404

Data

missing, 31 - 34

Data matrix ($\mathbf{X}$), 9, 10

normal data matrix, 58
standardized, 13

Diagonal matrix, 453

Dependence methods, 5

Design matrix, 183

Determinant(s), 460

Dirichlet distribution, 78

Discriminant analysis, 6, 413 - 8, 428 - 442

and classification analysis, 417, 437
contribution of variables to separation of groups, 442
(discriminant) function, 131, 147, 413, 429

standardized, 413, 442
coefficients in, 415

selection of variables in discriminant functions

backward selection, 445
forward selection, 445
stepwise selection, 445

tests of significance, 417, 439

Dispersion matrix, *see* Covariance matrix

Distance measures

Euclidean distance, 28
Mahalanobis distance, 30
properties, 29
squared statistical distance, 29

Eigenvalues, 468 - 472

Eigenvectors, 468 - 472

Ellipsoid, 41

Ellipsoid, constant density, 45

EM algorithm, 33

Error rate(s), 402 - 7

- actual error rate, 403
- apparent error rate, 403
- bootstrap estimate of, 405
- confusion matrix, 404
- cross-validation error rate, 404
- expected actual error rate, 403
- hold-out procedure of estimation, 405
- jackknife estimate, 405
- for normal population, 407
- plug-in error rate, 403
- optimum error rate, 403

Estimability of a parameter function, 196

Estimation

- least square estimate of $\mathbf{B}$, 190 - 2, 196 - 8
- maximum likelihood estimate of
  - μ and $\mathbf{\Sigma}$, 70 - 77
  - μ, when $\mu = k\mu_0$, 73 - 75
  - $\mathbf{\Sigma}$ when $\mathbf{\Sigma} = k\mathbf{\Sigma}_0$, 75

Expected values of

- quadratic forms in random variables, 25
- random vectors, 22

Experimental error rate, 265

F test(s)

- partial F test for discriminant functions, 442, 445
- and T^2, 114

Factor analysis, 3, 7, 331 - 361

- communality (definition), 333
- estimation of loadings, 338 - 348
  - principal component method, 339 - 341
  - principal factor method, 343 - 346
  - maximum likelihood method, 347
- factor scores, 337, 351 - 4
  - regression method, 353
  - weighted least square method, 352
- interpretation of factors, 337
- factor rotation
  - oblique rotation, 349 - 351
  - orthogonal rotation, 349
    - equimax rotation, 350
    - quartimax rotation 350
    - varimax rotation, 350
- Heywood case, 335
- loading matrix, 332
- non-uniqueness, 336
- number of factors, 334 - 6
- pattern matrix, 332
- and principal component analysis, 357
- orthogonal factor model, 332
- scale-invariance, 335
- specific variance (definition), 333

Fre'chet's inequality, 98

Gauss-Markov theorem

- multivariate, 195
- univariate, 227

Generalized inverse, 465

Generalized variance, 13 (definition), 107 (distribution)

Hotelling's T^2, *see* T^2-tests

Growth curve models,

- general solution, 249

in randomized block design, 247
single growth curve, 244 - 5
two growth curves, 246
two-dimensional growth curve, 248

Hypothesis tests, *see* Tests of hypotheses

Idempotent matrix, 478

Identity matrix, 453

Imputation, *see* Missing data

Intercorrelation, index of , 13 - 14

condition number, 13
Interdependence methods, 5

Inverse of a matrix, 463

Jackknife, 405

Largest root test, *see* Roy's test

Latent roots, *see* Eigenvalues

Lawley-Hotelling test,

F-approximation, 123
formal definition, 123
table of critical values, 522-4
for testing equality of mean vectors, 171

Least square estimators, 190 - 6

Length of a vector, 456

Likelihood function, 69

Likelihood ratio test

in canonical correlation, 384
for discriminant functions, 417, 439
formal definition, 130
in multivariate regression, 201 - 8
one-sample T^2-test, 110 - 112
testing equality of covariance matrices, 173
testing equality of mean vectors, 169
testing equality of multinormal populations, 175
two-sample T^2 test, 143
in univariate regression, 229

Linear classification rule, 399, 424

Linear transformation of variables, 20 - 21

Mahalanobis transformation, 20
principal component transformation, 21
scaling transformation, 20

Logistic classification, 411, 418-422, 425

Logit analysis, 6

loglinear models, 7

Mahalanobis distance, 30

MANOVA *see* Multivariate analysis of variance

Matrix (Matrices)

addition of, 455
algebra of, 455 - 6
centering matrix, 12
circular matrix, 454
characteristic equation, 468
Cholesky decomposition, 475
correlation matrix, *see* Correlation matrix
covariance matrix,*see* Covariance matrix
data matrix, 9, 10, 13, 58
definition, 453
determinant, 460
diagonal matrix, 453
differentiation, 481 - 3
direct product, 472
eigenvalues, 468 - 472

characteristic equation, 468
and determinant, 469
extremum properties, 470

of rational functions, 468
of positive definite matrix, 469
of symmetric matrix, 469
and trace, 469

eigenvectors, 468 - 472, *see also*

eigenspace, 472

factoring of,

singular value decomposition, 476 - 7
spectral (Jordanian) decomposition, 475 - 6

idempotent matrix, 478
of intraclass correlation structure, 24, 56, 161, 313, 454, 462
inverse, 463 - 6

generalized inverse, 465
of partitioned matrix, 464

Kantorovich inequality, 471
null matrix, 453
orthogonal matrix, 454
partitioned matrix, 459
positive-definite matrix, 474
positive semi-definite matrix, 474
quadratic forms, 473
random matrix, expected value of, 22
rank, 466
spectral decomposition, 465
square root matrix, 475
trace, 472
vector, *see* Vector (s)

Maximum likelihood estimators

of covariance matrix, 70 - 7
likelihood function, 69
of mean vector, 70 - 7
multivariate normal, 70 - 7
in regression, 185 - 190

Mean Vector

population, 22

sample, 11

Misclassification rates, *see* Error rate (s)

Missing data, 31 - 4, 216 - 7

estimation of missing values (imputation), 32 - 4
listwise deletion, 31
missing at random, 31
missing completely at random, 31
pairwise deletion, 31
smoothing procedure, 31
SVD method, 33 - 4

Moment generating function, 44

Multicollinearity, 45

Multidimensional scaling, 7

Multiple correlation, *see also* R^2

Multivariate analysis, 3

Multivariate analysis of covariance, *see*

Analysis of covariance, multivariate

Multivariate analysis of variance (MANOVA), 6, 256 - 278

contrasts

orthogonal, 264
tests for, 264 - 5

one-way fixed effects model, 256 - 9

comparison of MANOVA tests, 262 - 3
multiple comparison, 260 - 2

two-way fixed effects model, 269 - 270, 274 - 8

multiple comparison, 275 - 6
one observation per cell, 269 - 270
$r(> 1)$ observations per cell, 274 - 8

Multivariate central limit theorem, 64 - 8

Multivariate data, 9

Multivariate kurtosis, 80

Multivariate normal distribution, 45 - 58

- central limit theorem, 64 - 8
- conditional distribution, 53 - 6
- constant density ellipsoid, 45
- contour plot,
- estimation of μ and $\mathbf{\Sigma}$, 68 - 77
- independence and zero covariance, 51 - 2
- likelihood function, 69
- linear functions of, 49
- marginal distribution, 53
- matrix normal distribution, 77
- moment generating function of, 44 - 5
- moments of, 45
- properties of, 48 - 58
- and quadratic forms, 60 - 4
- singular, 47
- tests for multivariate normality, 86 - 7
- transformation of normal data matrix, 58 - 60
- transformation to achieve multivariate normality, 88 - 91
  - power transformation, 88

Multivariate normality, tests for, 86 - 7

Multivariate Poisson distribution, 99

Multivariate regression, 6, 183 - 244

- association, measures of, 238 - 9
  - determinant correlation, 239
  - trace correlation, 238
  - vector alienation, 239
- classical linear regression, *see* Classical linear regression
- estimation of $\mathbf{B}$,
  - with centered x's, 224
  - with covariances, 189 - 190
  - least squares, 190 - 6
  - maximum likelihood, 185 - 190
- estimation of $\mathbf{\Sigma}$, 186 - 190
- Gauss- Markov theorem, 194 - 6
- model,
  - assumption, 184
  - in centered form, 224
  - normal model, 183 - 5
- subset selection, 239 - 244
  - all possible subsets, 241 - 3
  - step-wise selection, 241 - 3
- tests of hypotheses,
  - $\mathbf{E}$ matrix, 205
  - general linear hypotheses $H_0 : \mathbf{A}_1\mathbf{B} = \mathbf{D}$ and $H_0 : \mathbf{A}_1\mathbf{B}\mathbf{L}_1 = \mathbf{D}_1$, 203 - 7
  - $\mathbf{H}$ matrix, 205
  - union-intersection test,*see* Roy's test,
  - Wilk's Λ, 203, 206, 207

Multivariate skewness, 80

Multivariate t-distribution, 77 - 8

Normal distribution,

- multivariate normal, *see* Multivariate normal distribution,

Normality, tests for *See* Tests for normality

$\mathbf{0}$ matrix, 453

Orthogonal matrix, 454
Outliers, 325

Paired T^2-test, 150 - 1
Partial F, 442, 445
Pillai's test
  definition, 124
  F-approximation, 124,
  for testing equality of several mean vectors, 171
  in MANOVA, 258
Principal components (PC), 4, 7, 297 - 327
  definition, 298
  discarding components,
    percent of variance, 302
    scree graph, 327
    tests of significance, 321
    trace of $\mathbf{\Sigma}$, 302
  of equiprobablity covariance matrix, 314
  and factor analysis, 357
  geometrical interpretation of, 306, 316 - 8
  large sample properties, 318 - 321
  last few principal components, 325 - 7
  of multivariate normal distribution, 306
  last few principal components to retain, *see* Principal components, discarding components,
  orthogonality of, 301
  outliers, 325
  percent of variance, 362
  plotting of, 326
  population, 298
  population correlation between a variable and PC, 302
  principal component regression, 298, 326
  sample, 307
  sample correlation between a variable and sample PC, 309
  scale invariance, lack of, 303, 310
  scree plot, 327
  special pattern in **S** or **R**, 312 - 5
    equal correlation, 313
    equal variances and covariances,313
  and squared multiple correlation coefficient, 309
  tests of hypotheses, 321 - 4
Profile analysis, 4, 163 - 8, 217 - 220
  case of two profiles, 163 - 8
  case of more than two profiles, 217 - 220
  coincident profiles, 165
  definition, 164
  parallel profiles, 164 - 5
  same average effect, 166 - 7

Quadratic classification rule, 400, 425
Quadratic forms, 473
Q-Q plot, 82 - 6

R^2 (squared multiple correlation), 230 - 3
Random matrix,
  expected value of, 22
Random variables
  correlation (population), 24
  covariance (population), 23
  sample covariance, 11
  sample correlation, 12

linear functions of, 18, 22
multivariate, *see* Random vectors,
univariate,
sample mean, 11
sample variance, 11

Random vectors

covariance matrix of one random vector, 23
covariance matrix of two random vectors, 23
linear combinations of, 22
mean vector (expected value), 22
sample mean vector, 11
sum of two random vectors, covariance matrix of, 23
expected value of, 22

Randomized block design, 154, 247

Rank of a matrix,

Regression, multiple (one y and several x's),

see Classical linear regression model

Regression, multivariate, *see* Multivariate regression

Repeated measurement design, 153

Robust estimation, 91 - 4

bisquare estimator, 92
Hodges-Lehmann estimator, 91
Huber's estimators, 92
m-estimators, 92
trimmed mean, 91
Winsorized mean, 92

Rotation, *see* Factor analysis

Roy's largest root statistic,

formal definition, 122
tables of percentage points, 516 - 521

in testing equality of several mean vectors, 171

Roy's test, *see* Union-intersection test(s),

table of critical values, 516 - 521

Simultaneous confidence intervals

Bonferroni: for $\mathbf{a}'\mu$, 140 - 2, 154
Bonferroni: for $\mathbf{a}'(\mu_1 - \mu_2)$, 144, 151
large sample: for $\mathbf{a}'\mu$, 142 - 3
large sample: for $\mathbf{a}'(\mu_1 - \mu_2)$, 149, 151
T^2: for $\mathbf{a}'\mu$, 137 - 140, 153
T^2 for $\mathbf{a}'(\mu_1 - \mu_2)$, 144, 151
T^2 for $\mathbf{a}'\mathbf{ABb}$, 223

Summary Statistics, 10 - 19

correlation matrix, 12
covariance matrix, 11 - 2
generalized variance, 13
geometrical interpretation, 14 - 5
mean vector, 11
indices of intercorrelation, 13 - 4
(of) linear combinations of variables, 18 - 9
SSP matrix, 12
total variance, 13

T^2-tests

Behrens-Fisher problem,

multivariate, 148 - 9
univariate, 148

and F-distribution, 114
formal definition of T^2, 111
invariance, 111
likelihood ratio test, 134
non-central distribution, 117
one mean vector

$\mathbf{\Sigma}$ known, 131

$\mathbf{\Sigma}$ unknown, 131
and Mahalanobis statistic, 116
paired observation test, 150
power, 131
two mean vectors,
$\mathbf{\Sigma}_1 = \mathbf{\Sigma}_2$, 143, 171
$\mathbf{\Sigma}_1 \neq \mathbf{\Sigma}_2$, 148 - 9
(and) union-intersection test, 132

Tests for normality, 81 - 87
bivariate, 85 - 6
multivariate, 86 - 7
univariate,m 81 - 5

Tests of hypotheses
T^2-tests, *see* T^2-tests
analysis of covariance
multivariate, *see* Analysis of covariance, multivariate
univariate, *see* Analysis of covariance, univariate
canonical correlation, 384 - 5
covariance matrices, equality of,
Box's M-tests, 173 - 5
in discriminant analysis, 417, 439
equality of components of a mean vector, 152
equality of covariance matrices, 173
equality of two mean vectors, 143, 148 - 9, 215, 216
equality of two mean subvectors, 155 - 6
for equicorrelation matrix, 162 - 3
for independence, 158 - 161
for one covariance matrix, 156 - 8
for several covariance matrices, 173 - 5
for several mean vectors,
equal dispersion matrices, 169 - 174
unequal dispersion matrices, 177 - 9
for two covariance matrices, 162 - 3
for several multinormal populations, 175 - 6
for two covariance matrices, 162 - 3
likelihood ratio test, *see* Likelihood ratio test
union-intersection test, *see* Union-intersection test(s)

Training sample, 399, 402

Transformation (linear) of random variables, 20 - 21
Mahalanobis transformation, 20
principal component transformation, 21
scaling transformation, 20

Union-intersection tests,
in regression, 221 - 2
one mean vector, 132
several mean vectors, 176
two mean vectors, 147

Variables, *see also* Random variables
categorical, 6
commensurate, 4
standardized, 13

Variance
generalized, 13
distribution of, 107
population, 22
sample, 11
total, 13

Variance matrix, *see* Covariance matric,

Variance-covariance matrix, *see* Covariance matrix,

Varimax rotation, 350

Vectors

definition, 455
differentiation, 481
distance between two vectors, 456
distance from origin, 456
linearly dependent, 458
normalized, 456
projection of, 457
space, 457

Wilks' Λ test statistic,

in canonical correlation, 384 - 5
in discriminant analysis, 439 - 441
in MANOVA, 258
χ-square approximation, 120
definition, 118
F-approximation, 120
for testing equality of several mean vectors, 169 - 171
partial, 240, 241, 443
properties of, 119 - 120
table of critical values, 120, 506 - 515
transformation to exact F, 119 - 120

Wishart distribution, 101 - 111

definition, 112
generalized quadratic form, 108 - 116
independence, 105
non-central, 111
partitioned matrices, 103 - 6
properties, 103 - 7
and Wilk's Λ, 118

0 matrix, 453